ANNALS OF
THE NEW YORK ACADEMY
OF SCIENCES

Volume 976

EDITORIAL STAFF

Executive Editor
BARBARA M. GOLDMAN

Managing Editor
JUSTINE CULLINAN

Associate Editors
LINDA H. MEHTA
ANGELA FINK
DONNA LYSAK

The New York Academy of Sciences
2 East 63rd Street
New York, New York 10021

THE NEW YORK ACADEMY OF SCIENCES
(Founded in 1817)

BOARD OF GOVERNORS, September 2002 – September 2003

TORSTEN N. WIESEL, *Chairman of the Board*
JOHN T. MORGAN, *Treasurer*

Honorary Life Governors
WILLIAM T. GOLDEN   JOSHUA LEDERBERG

Governors

ELEANOR BAUM   KAREN E. BURKE
LAWRENCE B. BUTTENWIESER   PRAVEEN CHAUDHARI
BRIAN FERGUSON   GERALD FISCHBACH   JOHN H. GIBBONS
MICHAEL GOLDEN   RONALD L. GRAHAM   MARNIE IMHOFF
JACQUELINE LEO   BRUCE McEWEN   PAUL MARKS
RONAY MENSCHEL   JOHN F. NIBLACK   SANDRA PANEM
PETER RINGROSE   JOHN J. ROCHE   LEE G. VANCE
DEBORAH WILEY

HELENE L. KAPLAN, *Counsel* [ex officio]

CELLULAR AND MOLECULAR PHYSIOLOGY OF SODIUM–CALCIUM EXCHANGE

PROCEEDINGS OF THE FOURTH INTERNATIONAL CONFERENCE

ANNALS OF THE NEW YORK ACADEMY OF SCIENCES
Volume 976

CELLULAR AND MOLECULAR PHYSIOLOGY OF SODIUM–CALCIUM EXCHANGE

PROCEEDINGS OF THE FOURTH INTERNATIONAL CONFERENCE

*Edited by Jonathan Lytton, Paul P.M. Schnetkamp,
Larry V. Hryshko, and Mordecai P. Blaustein*

The New York Academy of Sciences
New York, New York
2002

Copyright © 2002 by the New York Academy of Sciences. All rights reserved. Under the provisions of the United States Copyright Act of 1976, individual readers of the Annals are permitted to make fair use of the material in them for teaching or research. Permission is granted to quote from the Annals provided that the customary acknowledgment is made of the source. Material in the Annals may be republished only by permission of the Academy. Address inquiries to the Permissions Department (editorial@nyas.org) at the New York Academy of Sciences.

Copying fees: For each copy of an article made beyond the free copying permitted under Section 107 or 108 of the 1976 Copyright Act, a fee should be paid through the Copyright Clearance Center, Inc., 222 Rosewood Drive, Danvers, MA 01923 (www.copyright.com).

∞ The paper used in this publication meets the minimum requirements of the American National Standard for Information Sciences—Permanence of Paper for Printed Library Materials, ANSI Z39.48-1984.

Library of Congress Cataloging-in-Publication Data

Cellular and molecular physiology of sodium–calcium exchange :
the proceedings of the fourth international conference / edited by
Jonathan Lytton.
   p.; cm. — (Annals of the New York Academy of Sciences; v. 976)
  Result of a conference sponsored by the American Physiological Society
and held Oct. 10–14, 2001 in Banff, Alberta, Canada.
Includes bibliographical references and index.
    ISBN 1-57331-386-6 (cloth: alk. paper) — ISBN 1-57331-387-4 (paper: alk. paper)
    1. Sodium channels—Congresses. 2. Calcium channels—Congresses.
3. Sodium cotransport systems—Congresses.
I. Lytton, Jonathan. II. American Physiological Society. III. Series.
   [DNLM: 1. Ion Transport—Congresses. 2. Sodium-Calcium Exchanger—physiology
—Congresses. 3. Biological Transport,
Active—Congresses. QU 34 C393 2002]
  Q11.N5 vol. 976
  [QP535.N2]
  500 s—dc21
  [571.6'4]

2002014760

GYAT / PCP
Printed in the United States of America
ISBN 1-57331-386-6 (cloth)
ISBN 1-57331-387-4 (paper)
ISSN 0077-8923

ANNALS OF THE NEW YORK ACADEMY OF SCIENCES

Volume 976
November 2002

CELLULAR AND MOLECULAR PHYSIOLOGY OF SODIUM–CALCIUM EXCHANGE

PROCEEDINGS OF THE FOURTH INTERNATIONAL CONFERENCE

Editors
JONATHAN LYTTON, PAUL P.M. SCHNETKAMP, LARRY V. HRYSHKO, AND MORDECAI P. BLAUSTEIN

This volume is the result of a conference entitled **Cellular and Molecular Physiology of Sodium–Calcium Exchange: The IVth International Conference**, sponsored by the American Physiological Society and held on October 10–14, 2001 in Banff, Alberta, Canada.

CONTENTS

Preface. *By* JONATHAN LYTTON xiii

Keynote Address

The Na^+/Ca^{2+} Exchange Molecule: An Overview. *By* KENNETH D. PHILIPSON, DEBORA A. NICOLL, MICHELA OTTOLIA, BEATE D. QUEDNAU, HANNES REUTER, SCOTT JOHN, AND ZHIYONG QIU 1

Part I. Molecular Physiology

Toward a Topological Model of the NCX1 Exchanger. *By* DEBORA A. NICOLL, MICHELA OTTOLIA, AND KENNETH D. PHILIPSON 11

Probing Ion Binding Sites in the Na^+/Ca^{2+} Exchanger. *By* MUNEKAZU SHIGEKAWA, TAKAHIRO IWAMOTO, AKIRA UEHARA, AND SATOMI KITA 19

Ion Channel–like Properties of the Na^+/K^+ Pump. *By* PABLO ARTIGAS AND DAVID C. GADSBY 31

The Na/Ca-K Exchanger Gene Family. *By* ROBERT T. SZERENCSEI, ROBERT J. WINKFEIN, CONAN B. COOPER, CLEMENS PRINSEN, TASHI G. KINJO, KYEONGJIN KANG, AND PAUL P.M. SCHNETKAMP 41

Poster Papers

Toward Comparative Genomics of Calcium Transporters. *By* ALEXANDER KRAEV AND DAVID H. MACLENNAN 53

Sodium–Calcium Exchanger NCX1, NCX2, and NCX3 Transcripts in Developing Rat Brain. *By* S.K. POLUMURI, A. RUKNUDIN, MARGARET M. MCCARTHY, TARA S. PERROT-SINAL, AND D.H. SCHULZE 60

Differential Expression of Na/Ca Exchanger and Na/Ca+K Exchanger Transcripts in Rat Brain. *By* XIAO-FANG LI AND JONATHAN LYTTON ... 64

Sodium–Calcium Exchangers in Olfactory Tissue. *By* D.H. SCHULZE, M. PYRSKI, A. RUKNUDIN, J.W. MARGOLIS, S.K. POLUMURI, AND F.L. MARGOLIS 67

Molecular Identification of the NCX Isoform Expressed in Tracheal Smooth Muscle of Guinea Pig. *By* R. MEJÍA-ELIZONDO, R. ESPINOSA-TANGUMA, AND V.M. SAAVEDRA-ALANIS 73

Localization and Molecular Characterization of the Crayfish NCX. *By* L.M. STINER, Z. ZHANG, AND M.G. WHEATLY 77

Characterization and Functional Activity of a Truncated Na/Ca Exchange Isoform Resulting from a New Splicing Pattern of NCX1. *By* FRANÇOISE VAN EYLEN, ADAMA KAMAGATE, AND ANDRÉ HERCHUELZ 81

Activation of the Cardiac Na^+/Ca^{2+} Exchanger by DEPC. *By* MICHELA OTTOLIA, SABRINA SCHUMANN, DEBORA A. NICOLL, AND KENNETH D. PHILIPSON 85

Cysteine Residues Are Important for Functional Expression of the Na^+/Ca^{2+} Exchanger NCX1 in HEK 293 Cells. *By* XIAOYAN REN, JUDITH KASIR, AND HANNAH RAHAMIMOFF 89

Topological Studies of the Rat Brain K^+-Dependent Na^+/Ca^{2+} Exchanger NCKX2. *By* XINJIANG CAI, KATHY ZHANG, AND JONATHAN LYTTON .. 90

Studies on the Oligomeric State of the Sodium/Calcium+Potassium Exchanger NCKX2. *By* SEUNGHWA SALLY YOO, STEPHEN LEACH, AND JONATHAN LYTTON 94

Conformational Changes of the 120-kDa Na^+/Ca^{2+} Exchanger Protein upon Ligand Binding: A Fourier Transform Infrared Spectroscopy Study. *By* RAMI I. SABA, ERIK GOORMAGHTIGH, JEAN-MARIE RUYSSCHAERT, AND ANDRÉ HERCHUELZ 97

Sodium–Calcium Exchange Crystallization. *By* CALVIN C. HALE, JULIE BOSSUYT, CHANANADA K. HILL, ELMER M. PRICE, DAN H. SCHULZE, JON W. LEDERER, ROBERTO POLJAK, AND BRADFORD C. BRADEN 100

Proteomics Approach to Na^+/Ca^{2+} Exchangers in Prokaryotes. *By* A. RUKNUDIN AND D.H. SCHULZE 103

Temperature Dependence of Cardiac Na^+/Ca^{2+} Exchanger. *By* CHRISTIAN MARSHALL, CHADWICK ELIAS, XIAO-HUA XUE, HOA DINH LE, ALEXANDER OMELCHENKO, LARRY V. HRYSHKO, AND GLEN F. TIBBITS 109

The Ca/Ca Exchange Mode of the Na/Ca Exchanger Investigated by Photolytic Ca^{2+} Concentration Jumps. *By* ANDREAS HAASE, MICHAEL KAPPL, GEORG NAGEL, PHILIP G. WOOD, AND KLAUS HARTUNG 113

Functional Analysis of Polar Residues Important for Activity of Na^+/H^+ Exchangers. *By* LARRY FLIEGEL, CHRISTINE WIEBE, RAKHILYA MURTAZINA, PAVEL DIBROV, BRENDA J. BOOTH, BONNIE L. BULLIS, EMILY DIBATTISTA, AND DYAL N. SINGH 117

Part II. Stoichiometry

Stoichiometry of the Na^+/Ca^{2+} Exchanger: Models and Implications. *By* SATOSHI MATSUOKA 121

Influence of Na/Ca Exchange Stoichiometry on Model Cardiac Action Potentials. *By* DENIS NOBLE 133

Rat Heart NCX1.1 Stoichiometry Measured in a Transfected Cell System. *By* JONATHAN LYTTON AND HUI DONG 137

Sodium–Calcium Exchange Stoichiometry: Is The Noose Tightening? *By* TONG MOOK KANG, MARK STECIUK, AND DONALD WILLIAM HILGEMANN 142

Stoichiometry: A Discussion 152

Poster Papers

Reexamination of the Stoichiometry of Na^+/Ca^{2+} Exchange with Whole-Cell Voltage Clamp of Guinea Pig Ventricular Myocytes. *By* MASAMITSU HINATA, HISAO YAMAMURA, LIBING LI, YASUHIDE WATANABE, TOMOKAZU WATANO, YUJI IMAIZUMI, AND JUNKO KIMURA 154

Simultaneous Measurement of $[Na]_i$, $[Ca]_i$, and I_{NCX} in Intact Cardiac Myocytes. *By* KENNETH S. GINSBURG, CHRISTOPHER R. WEBER, SANDA DESPA, AND DONALD M. BERS 157

Electrophysiological Studies of the Cloned Rat Cardiac NCX1.1 in Transfected HEK Cells: A Focus on the Stoichiometry. *By* HUI DONG, JEREMY DUNN, AND JONATHAN LYTTON 159

Part III. Functional Regulation

Tissue-Specific Modes of Na/Ca Exchanger Regulation. *By* LARRY V. HRYSHKO 166

NCX1 Surface Expression: A Tool to Identify Structural Elements of Functional Importance. *By* HANNAH RAHAMIMOFF, XIAOYAN REN, CHAVA KIMCHI-SARFATY, SURESH AMBUDKAR, AND JUDITH KASIR ... 176

Functional Regulation of Alternatively Spliced Na^+/Ca^{2+} Exchanger (NCX1) Isoforms. *By* D.H. SCHULZE, S.K. POLUMURI, T. GILLE, AND A. RUKNUDIN 187

The Cardiac Sodium–Calcium Exchanger Associates with Caveolin-3. *By* JULIE BOSSUYT, BONNIE E. TAYLOR, MARILYN JAMES-KRACKE, AND CALVIN C. HALE 197

Poster Papers

Kinetic Model of Transport and Regulation of the Cardiac, Brain, and Kidney Isoforms of NCX1. *By* ALEXANDER OMELCHENKO, CHADWICK L. ELIAS, ROBERT SCATLIFF, PLATON CHOPTIANY, MARK HNATOWICH, AND LARRY V. HRYSHKO 205

Phosphorylation of the Na^+/Ca^{2+} Exchangers by PKA. *By* A. RUKNUDIN AND
D. H. SCHULZE 209

Part IV. Cellular Regulation

Cellular Regulation of Sodium–Calcium Exchange. *By* MADALINA
CONDRESCU, KWABENA OPUNI, BASIL M. HANTASH, AND
JOHN P. REEVES 214

Intracellular Ionic and Metabolic Regulation of Squid Nerve Na^+/Ca^{2+}
Exchanger. *By* REINALDO DIPOLO AND LUIS BEAUGÉ 224

Pathways Regulating Na^+/Ca^{2+} Exchanger Expression in the Heart. *By*
DONALD R. MENICK, LIN XU, CHRISTIANA KAPPLER, WENJING JIANG,
PATRICK R. WITHERS, NEAL SHEPHERD, SIMON J. CONWAY, AND
JOACHIM G. MÜLLER 237

Multiple Modes of Regulation of Na^+/H^+ Exchangers. *By* HISAYOSHI
HAYASHI, KATALIN SZÁSZI, AND SERGIO GRINSTEIN 248

Cyclosporin A Regulates Sodium–Calcium Exchanger (NCX1) Gene Expression *in Vitro* and Cardiac Hypertrophy in NCX1 Transgenic Mice. *By*
MARIA C. JORDAN, BEATE D. QUEDNAU, KENNETH P. ROOS, ROBERT
S. ROSS, KENNETH D. PHILIPSON, AND SUSANNE B. NICHOLAS 259

Role of Sodium–Calcium Exchanger (*Ncx1*) in Embryonic Heart Development: A Transgenic Rescue? *By* SIMON J. CONWAY, AGNIESZKA
KRUZYNSKA-FREJTAG, JIAN WANG, RHONDA ROGERS, PAIGE L.
KNEER, HONGMEI CHEN, TONY CREAZZO, DONALD R. MENICK, AND
SRINAGESH V. KOUSHIK 268

Poster Papers

The Gene Promoter of Human Na^+/Ca^{2+} Exchanger Isoform 3 (SLC8A3) Is
Controlled by cAMP and Calcium. *By* NADIA GABELLINI, STEFANIA
BORTOLUZZI, GIAN A. DANIELI, AND ERNESTO CARAFOLI 282

Role of MAP Kinases in the Na^+/Ca^{2+} Exchanger Gene Expression in Feline
Adult Cardiocytes. *By* LIN XU, JOACHIM G. MULLER, PATRICK R.
WITHERS, CHRISTIANA KAPPLER, WENJING JIANG, AND
DONALD R. MENICK 285

Part V. Calcium Homeostasis

Regulation of Phosphatidylinositol-4,5-Bisphosphate Bound to the Bovine
Cardiac Na^+/Ca^{2+} Exchanger. *By* LUIS BEAUGÉ, CARLA ASTEGGIANO,
AND GRACIELA BERBERIÁN 288

Potassium-Dependent Sodium–Calcium Exchange through the Eye of the Fly.
By R. WEBEL, K. HAUG-COLLET, B. PEARSON, R.T. SZERENCSEI,
R.J. WINKFEIN, P.P.M. SCHNETKAMP, AND N.J. COLLEY 300

Na/Ca Exchange in Function, Growth, and Demise of β-Cells. *By* ANDRÉ
HERCHUELZ, OSCAR DIAZ-HORTA, AND FRANÇOISE VAN EYLEN 315

Binding of the Retinal Rod Na^+/Ca^{2+}-K^+ Exchanger to the cGMP-Gated Channel Indicates Local Ca^{2+}-Signaling in Vertebrate Photoreceptors. *By* PAUL J. BAUER 325

Poster Papers

Exercise Training Enhances Coronary Smooth Muscle Cell Sodium–Calcium Exchange Activity in Diabetic Dyslipidemic Yucatan Swine. *By* E.A. MOKELKE, M. WANG, AND M. STUREK 335

Dysregulation of $[Ca^{2+}]_i$ in OK-PTH Cells Expressing a Mesangial Cell Na^+/Ca^{2+} Exchanger Isoform from Dahl/Rapp Salt-Sensitive Rats. *By* T. UNLAP, E. HWANG, G. KOVACS, J. PETI-PETERDI, B. SIROKY, I. WILLIAMS, AND P.D. BELL 338

Enhanced Susceptibility of a Na^+/Ca^{2+} Exchanger Isoform from Mesangial Cells of Salt-Sensitive Dahl/Rapp Rats to Oxidative Stress Inactivation. *By* T. UNLAP, E.H. HWANG, B.J. SIROKY, J. PETI-PETERDI, G. KOVACS, I. WILLIAMS, AND P.D. BELL 342

Reverse Mode Na^+/Ca^{2+} Exchange in the Collagen Activation of Human Platelets. *By* DIANE E. ROBERTS AND RATNA BOSE 345

Na^+/Ca^{2+} Exchange in Activated and Nonactivated Human Platelets. *By* RATNA BOSE, YUN LI, AND DIANE ROBERTS 350

Effect of Glucose on the Expression Level of the Plasma Membrane Ca^{2+}-ATPase and the Na^+/Ca^{2+} Exchanger in Pancreatic Islet Cells. *By* HELENA MARIA XIMENES, ADAMA KAMAGATE, FRANÇOISE VAN EYLEN, AND ANDRÉ HERCHUELZ 354

Part VI. Neuronal Function

Na/Ca Exchanger and PMCA Localization in Neurons and Astrocytes: Functional Implications. *By* M.P. BLAUSTEIN, M. JUHASZOVA, V.A. GOLOVINA, P.J. CHURCH, AND E.F. STANLEY 356

Immunohistochemical Detection of the Sodium–Calcium Exchanger in Rat Hippocampus Cultures Using Subtype-Specific Antibodies. *By* T. THURNEYSEN, D.A. NICOLL, K.D. PHILIPSON, AND H. PORZIG 367

A Study of the Activity of the Plasma Membrane Na/Ca Exchanger in the Cellular Environment. *By* MARISA BRINI, SABRINA MANNI, AND ERNESTO CARAFOLI 376

K+-Dependent Na^+/Ca^{2+} Exchangers in the Brain. *By* JONATHAN LYTTON, XIAO-FANG LI, HUI DONG, AND ALEXANDER KRAEV 382

Brain Distribution of the Na^+/Ca^{2+} Exchanger-Encoding Genes NCX1, NCX2, and NCX3 and Their Related Proteins in the Central Nervous System. *By* ADRIANA CANITANO, MICHELE PAPA, FRANCESCA BOSCIA, PASQUALINA CASTALDO, STEFANIA SELLITTI, MAURIZIO TAGLIALATELA, AND LUCIO ANNUNZIATO 394

Poster Papers

Neurotransmitter-Induced Activation of Sodium-Calcium Exchange Causes Neuronal Excitation. *By* KRISTER S. ERIKSSON, OLGA A. SERGEEVA, DAVID R. STEVENS, AND HELMUT L. HAAS 405

Na$^+$/Ca^{2+} Exchanger in Na$^+$ Efflux–Ca^{2+} Influx Mode of Operation Exerts a Neuroprotective Role in Cellular Models of *in Vitro* Anoxia and *in Vivo* Cerebral Ischemia. *By* A. TORTIGLIONE, G. PIGNATARO, M. MINALE, A. SECONDO, A. SCORZIELLO, G.F. DI RENZO, S. AMOROSO, G. CALIENDO, V. SANTAGADA, AND L. ANNUNZIATO 408

Mitochondria Buffer Sodium-Dependent Ca^{2+} Influx in Cultured Cerebellar Granule Cells. *By* LECH KIEDROWSKI AND ANETA CZYŻ 413

ATP Stimulation of Na$^+$/Ca^{2+} Exchanger in Bovine Brain Membrane Vesicles Is Similar to That of the Heart and Independent of Ionic Strength of Assay or Preparation. *By* GRACIELA BERBERIÁN, CARLA ASTEGGIANO, AND CUONG PHAM 418

Part VII. Na/Ca Exchanger Function in Disease

Is Na/Ca Exchange during Ischemia and Reperfusion Beneficial or Detrimental? *By* ELIZABETH MURPHY, HEATHER R. CROSS, AND CHARLES STEENBERGEN 421

Simulation of Na/Ca Exchange Activity during Ischemia. *By* DENIS NOBLE .. 431

Role of the Na/Ca Exchanger in Arrhythmias in Compensated Hypertrophy. *By* KARIN R. SIPIDO, PAUL G. A. VOLDERS, MARIEKE SCHOENMAKERS, S. H. MARIEKE DE GROOT, FONS VERDONCK, AND MARC A. VOS 438

Enhanced Sodium–Calcium Exchange in the Infarcted Heart: Effects on Sarcoplasmic Reticulum Content and Cellular Contractility. *By* SHELDON E. LITWIN AND DONGFANG ZHANG 446

Na/Ca Exchange in Heart Failure: Contractile Dysfunction and Arrhythmogenesis. *By* STEVEN M. POGWIZD AND DONALD M. BERS 454

Modulation of Contractility in Failing Human Myocytes by Reverse-Mode Na/Ca Exchange. *By* VALENTINO PIACENTINO III, CHRISTOPHER R. WEBER, JOHN P. GAUGHAN, KENNETH B. MARGULIES, DONALD M. BERS, AND STEVEN R. HOUSER 466

Poster Papers

Decreased β-Adrenergic Responsiveness of Na/Ca Exchange Current in Failing Pig Myocytes. *By* SHAO-KUI WEI, STEPHEN U. HANLON, AND MARK C.P. HAIGNEY 472

Ca Influx via the Na/Ca Exchanger Maintains Sarcoplasmic Reticulum Ca Content in Failing Human Myocytes. *By* VALENTINO PIACENTINO III, KENNETH B. MARGULIES, AND STEVEN R. HOUSER 476

Calcium Influx via I$_{NCX}$ Is Favored in Failing Human Ventricular Myocytes. *By* CHRISTOPHER R. WEBER, VALENTINO PIACENTINO III, KENNETH B. MARGULIES, DONALD M. BERS, AND STEVEN R. HOUSER 478

Overexpression of Na^+/Ca^{2+} Exchanger Attenuates Postinfarcted Myocardial Dysfunction. *By* JIANG-YONG MIN, MATTHEW F. SULLIVAN, VICTOR CHU, JU-FENG WANG, IVO AMENDE, JAMES P. MORGAN, KENNETH D. PHILIPSON, AND THOMAS G. HAMPTON 480

Ischemic Tolerance of Homozygous Transgenic Mouse Hearts Overexpressing the Sodium–Calcium Exchanger. *By* MALCOLM M. BERSOHN, CECIL R. CARMACK, AND KENNETH D. PHILIPSON 483

Na/Ca Exchanger Overexpression Induces Endoplasmic Reticulum–Related Apoptosis and Caspase-12 Release. *By* O. DIAZ-HORTA, A. HERCHUELZ, AND F. VAN EYLEN 487

Part VIII. Cardiac Physiology and Pathophysiology

The Molecular Architecture of Calcium Microdomains in Rat Cardiomyocytes. *By* DAVID R. L. SCRIVEN, AGNIESZKA KLIMEK, KELLY L. LEE, AND EDWIN D. W. MOORE 488

Na/Ca Exchange Function in Intact Ventricular Myocytes. *By* DONALD M. BERS AND CHRISTOPHER R. WEBER 500

Pharmacology of Na^+/Ca^{2+} Exchanger. *By* JUNKO KIMURA, YASUHIDE WATANABE, LIBING LI, AND TOMOKAZU WATANO 513

Functional Consequences of Na/Ca Exchanger Overexpression in Cardiac Myocytes. *By* CESARE TERRACCIANO 520

Poster Papers

Na^+/Ca^{2+} Exchanger Is Functional in Both Ca^{2+} Influx and Efflux Modes in Rat Myocytes. *By* XUE-QIAN ZHANG, JIANLIANG SONG, GEORGE M. TADROS, LAWRENCE I. ROTHBLUM, JEREMY DUNN, JONATHAN LYTTON, AND JOSEPH Y. CHEUNG 528

Volatile Anesthetics and Regulation of Cardiac Na^+/Ca^{2+} Exchange in Neonates versus Adults. *By* Y. S. PRAKASH, LARRY W. HUNTER, INANC SECKIN, AND GARY C. SIECK 530

Modulation of the Na^+/Ca^{2+} Exchanger by Isoprenaline, Adenosine, and Endothelin-1 in Guinea Pig Ventricular Myocytes. *By* YIN HUA ZHANG, ANNABEL K. HINDE, ANDREW F. JAMES, AND JULES C. HANCOX 535

Effect of KB-R7943 on Oscillatory Na^+/Ca^{2+} Exchange Current in Guinea Pig Ventricular Myocytes. *By* LIBING LI AND JUNKO KIMURA 539

Inhibition of the *Drosophila* Na^+/Ca^{2+} Exchanger, CALX1.1, by KB-R7943. *By* MICHAEL R. ISAAC, CHADWICK L. ELIAS, HOA D. LE, ALEXANDER OMELCHENKO, MARK HNATOWICH, AND LARRY V. HRYSHKO 543

Index of Contributors 547

Financial assistance was received from:

- **AMERICAN PHYSIOLOGICAL SOCIETY**
- **ALBERTA HERITAGE FOUNDATION FOR MEDICAL RESEARCH**
- **CANADIAN INSTITUTES OF HEALTH RESEARCH**
- **UNIVERSITY OF CALGARY, FACULTY OF MEDICINE**
- **THE CANADIAN SOCIETY OF BIOCHEMISTRY AND MOLECULAR & CELLULAR BIOLOGY**
- **MERCK & CO. INC—USA**
- **INVITROGEN CORP.**
- **AXON INSTRUMENTS**
- **NEW ENGLAND BIOLABS LTD.**
- **CARL ZEISS CANADA LTD.**
- **MANDEL SCIENTIFIC COMPANY LTD.**
- **VWR CANLAB**

The New York Academy of Sciences believes it has a responsibility to provide an open forum for discussion of scientific questions. The positions taken by the participants in the reported conferences are their own and not necessarily those of the Academy. The Academy has no intent to influence legislation by providing such forums.

Preface

JONATHAN LYTTON

Department of Biochemistry and Molecular Biology, University of Calgary, Health Sciences Centre, Calgary T2N 4N1, Canada

Calcium, acting as a second-messenger signaling molecule, is involved in the regulation of a large number of normal cellular functions and physiological processes. In so doing, intracellular Ca^{2+} levels fluctuate in complex temporal and spatial patterns. It also appears that *dys*regulation of these patterns lies at the root of many diseases and pathological conditions, such as heart failure and stroke. The plasma membrane Na^+/Ca^{2+} exchanger plays a prominent role in the control of Ca^{2+} homeostasis and has garnered considerable attention over the more than three decades since its discovery in heart and nerve.

This conference marks the fourth gathering of international scientists from various fields, all of whom share an intense interest in unraveling the mysteries of Na^+/Ca^{2+} exchanger function and regulation. Previous meetings—in Stowe, England in 1987, in Baltimore, Maryland in 1991, and in Woods Hole, Massachusetts in 1995—were all important focal points for the exchange of new information and ideas. It is hoped that the Banff meeting of 2001 achieved the same goal and will help to project the field forward at the start of the 21st century.

As is evident in these proceedings, there have been many changes and new developments in the Na^+/Ca^{2+} exchanger field over the past five years. Foremost among these are studies on the molecular biology of the molecule. Detailed structure–function work has now reached a new level of maturity, and information is pouring in at a rapid rate. Mechanisms that regulate the function of the Na^+/Ca^{2+} exchanger, both at the molecular level and at the cellular level, have also entered a mature stage and are poised to provide important new information.

The revolution in genomic biology over the past few years has contributed to the Na^+/Ca^{2+} exchanger field in revealing a complexity of related genes and alternatively spliced products expressed in various human tissues and also in model organisms such as *D. melanogaster* and *C. elegans*. An exciting key development has been the application of mouse genetics, by way of both transgenic expression and targeted gene deletion approaches, toward unraveling the multitude of physiological roles for the many different Na^+/Ca^{2+} exchanger gene products. The marriage of powerful molecular biology tools with sophisticated physiological analyses has set the stage for major advances in our understanding of the precise role of Na^+/Ca^{2+} exchanger isoforms in various processes.

Physiological measurements of Na^+/Ca^{2+} exchanger function, and the contribution of this function to both normal and pathological states, has been an intense and controversial area of study. Discussions at the Banff conference point toward a consensus viewpoint on the role of the exchanger, especially in cardiac tissue. Another exciting new development since the last meeting has been the discovery of more highly selective inhibitors of the Na^+/Ca^{2+} exchanger. Several studies touch on these new pharmacological agents and point toward the promise they hold as tools for both the experimentalist and the clinician.

Finally, some developments in the field were not new, but new again! Thought to have been settled a score or more years ago, the issue of the stoichiometry of the Na^+/Ca^{2+} exchanger resurfaced as a hot topic of controversy. Discussion gave rise to a novel theory of exchanger operation to help reconcile new data, in support of a 4 sodium : 1 calcium stoichiometry, with previous work demonstrating a 3 sodium : 1 calcium stoichiometry.

As chair of the organizing committee for the Fourth International Conference on Sodium–Calcium Exchange, I would like to thank my colleagues on the committee—Mordecai Blaustein, Reinaldo DiPolo, Don Hilgemann, Larry Hryshko, Akinori Noma, Ken Philipson, Hannah Rahamimoff, John Reeves, and Paul Schnetkamp—for their helpful and essential advice. I would also like to express our sincere gratitude to the American Physiological Society for sponsoring and managing our meeting. Thanks to the tireless efforts of Martin Frank, Executive Director, Linda Allen, Marcella Jackson, and their staff, our meeting ran smoothly and successfully. And lastly, the meeting would not have been successful without the outstanding contributions of all the participants.

The Na$^+$/Ca^{2+} Exchange Molecule

An Overview

KENNETH D. PHILIPSON, DEBORA A. NICOLL, MICHELA OTTOLIA,
BEATE D. QUEDNAU, HANNES REUTER, SCOTT JOHN, AND ZHIYONG QIU

Departments of Physiology and Medicine and the Cardiovascular Research Laboratory, UCLA School of Medicine, Los Angeles, California 90095, USA

> ABSTRACT: An overview of the molecular physiology of the Na$^+$/Ca^{2+} exchanger is presented. This includes information on the variety of exchangers that have been described and their regulatory properties. Molecular insight is most detailed for the cardiac Na$^+$/Ca^{2+} exchanger (NCX1). Parts of the NCS1 molecule involved in regulation and ion transport have been elucidated, and initial information on the topology and structure is available.
>
> KEYWORDS: Na$^+$/Ca^{2+} exchange; NCX family; NCKX family; α repeat regions; CHX exchangers

Appreciation of the importance of Na$^+$/Ca^{2+} exchange in the cellular physiology of many tissues has been growing. There are now over 2,000 articles on this topic. About 62% of these articles involve the Na$^+$/Ca^{2+} exchanger found in the sarcolemmal membrane of cardiomyocytes. In fact, there are at least 6 recent journal editorials that address the role of the exchanger in cardiac pathophysiology.[1–6] These include editorials with provocative titles (e.g., "Sodium-Calcium Exchange: The Phantom Menace").[3] There have been several recent reviews that exhaustively cover almost all aspects of Na$^+$/Ca^{2+} exchange.[7,8] We will present here a brief overview of molecular aspects of Na$^+$/Ca^{2+} exchange. We will attempt to point out some underappreciated and lesser known aspects of the exchanger with emphasis on results from our own laboratory.

The Na$^+$/Ca^{2+} exchanger is a member of a large superfamily of membrane proteins. Members of this superfamily are defined by the presence of sequence motifs known as α repeats (FIG. 1). The α repeats are regions of intramolecular homology within the transmembrane segments (TMSs) of the proteins.[9] The α repeats probably arose from a gene duplication event, and mutational analysis indicates their importance in ion translocation.[10]

We have divided the superfamily into four families[8]: (1) *The NCX family*. This includes the most intensively studied exchanger—the cardiac sarcolemmal exchanger,

---

Address for correspondence: Dr. Kenneth D. Philipson, Cardiovascular Research Laboratory, MRL 3-645, UCLA School of Medicine, Los Angeles, CA 90095-1760. Voice: 310-825-7677.
kphilipson@mednet.ucla.edu

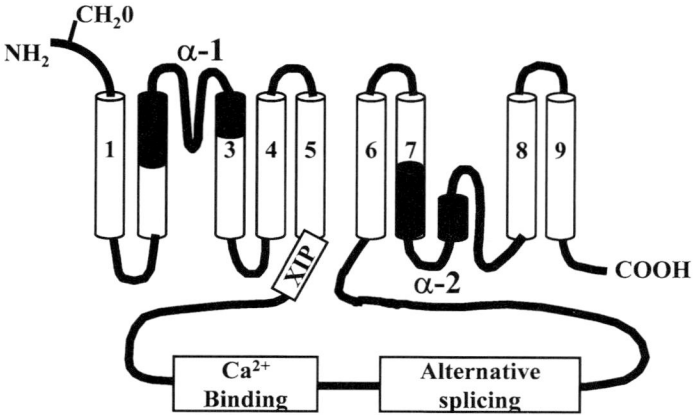

FIGURE 1. Proposed structure of the Na^+/Ca^{2+} exchanger. Shown are 9 transmembrane segments and a large intracellular loop. Almost all aspects of this topology have been experimentally confirmed. The first extracellular segment is glycosylated (CH_2O). Also shown are two proposed reentrant loops, which have some experimental support. The two α repeat regions are shown in black. These are regions of intramolecular homology and have been implicated in ion translocation. The XIP region is involved in an inactivation process. Shown also is a Ca^{2+}-binding site responsible for regulation of the exchanger by intracellular Ca^{2+}. The location where extensive alternative splicing occurs is also denoted.

NCX1. Also, included are two mammalian homologues, NCX2 and NCX3, and homologues from squid, *Drosophila*, and *Arabidopsis*. (2) *The NCKX family*. The prototype is NCKX1, abundantly expressed in the outer segments of rod photoreceptors. These proteins exchange 4 Na^+ for 1 Ca^{2+} plus 1 K^+. (3) *The bacterial family*. Several bacterial genomes encode proteins with α repeats. These proteins may be Ca^{2+}/H^+ exchangers. (4) *The CHX exchangers*. This family includes a Ca^{2+}/H^+ exchanger of yeast vacuoles.

Outside of the α-repeat regions, only very limited sequence similarities exist between the Na^+/Ca^{2+} exchanger and other proteins. (When we refer to the Na^+/Ca^{2+} exchanger, we are referring to the cardiac Na^+/Ca^{2+} exchanger, NCX1, unless otherwise stated.) These are displayed graphically in FIGURE 2. There is no known significance to any of these apparent homologies. Thus, little insight into the exchanger has been gained from studies on other proteins.

Alternative splicing further increases the complexity of Na^+/Ca^{2+} exchangers. NCX1 undergoes extensive alternative splicing within the large intracellular loop (FIG. 3). Six small exons are used in different combinations in different tissues.[11,12] Exons A and B are mutually exclusive, thus maintaining an open reading frame. Exon A is usually present in excitable tissues (e.g., heart and brain), while exon B is almost always present in nonexcitable tissues. The combination ACDEF (NCX1.1) is the dominant combination in heart tissue. The physiological importance of alternative splicing is not clear and is addressed in detail in other chapters in this volume.

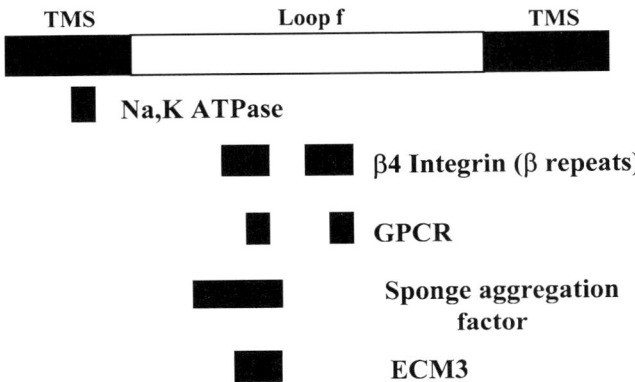

FIGURE 2. Sequence similarities between NCX1 and other proteins. The top bar shows a linear depiction of the arrangement of the Na^+/Ca^{2+} exchanger. There are two domains of transmembrane segments separated by a large intracellular loop (loop f). Shown are regions of similarity to portions of the Na^+,K^+-ATPase, $\beta 4$ integrin, an orphan G protein–coupled receptor, a sponge aggregation factor, and a sea urchin extracellular matrix protein.

What differences exist among the Na^+/Ca^{2+} exchangers? Our laboratory has been involved in describing three quite striking differences. First, we have noted that charge movements associated with ion translocation are different in NCX1 and NCX-SQ1, the squid neuronal exchanger.[13] Charge movements are due to fixed charges on the exchanger protein itself that move through the membrane electrical field during the conformational changes that accompany ion translocation. NCX1 behaves as if there are two fixed negative charges that move in the same direction as translocated ions.[14] Thus, movement of the 3 Na^+ ions through the membrane involves the net movement of one positive charge. The movement of one Ca^{2+} ion occurs with no net movement of charge. In contrast, NCX-SQ1 behaves as if three negative charges on the protein move through the electrical field of the membrane.[13] Charge accompanies Ca^{2+} translocation but not Na^+ translocation. There is a fundamental difference in the protein charge distribution and movement between the mammalian and squid exchangers. (The stoichiometry of the exchanger may not be $3Na^+/1Ca^{2+}$ as used in the argument here. Even if the stoichiometry is not 3/1, however, the fundamental difference between the two exchangers still exists.)

The other two differences between exchangers that we will discuss involve regulatory properties that we will briefly review. The two major intrinsic regulatory properties of the Na^+/Ca^{2+} exchanger are Na^+-dependent inactivation and Ca^{2+} regulation.[15,16] These are both most readily studied using the giant excised patch technique,[17] as demonstrated in FIGURE 4. Na^+-dependent inactivation is a process similar to the inactivations more commonly associated with ion channels. The inactivation is initiated by the binding of Na^+ to transport sites at the intracellular surface

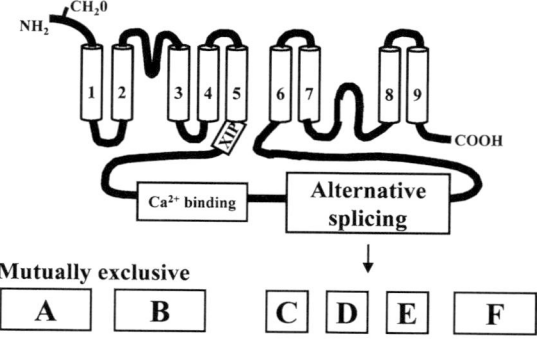

FIGURE 3. Alternative splicing of NCX1. A region of the large intracellular loop of the exchanger is encoded by six exons (A–F) that are used in different combinations in different tissues. Either exon A or B must be present to maintain an open reading frame. The combination ACDEF is dominant in the myocardium, while the combinations AD and BD are typically found in brain and kidney, respectively.

of the exchanger. The bound Na^+ is either transported or induces the exchanger to enter an inactive state. In addition to transporting Ca^{2+}, the exchanger is also regulated by the binding of Ca^{2+} to a distinct high-affinity site on the large intracellular loop. This Ca^{2+} is not transported but does activate NCX1. These two regulatory mechanisms interact; a characteristic feature is that increases in regulatory Ca^{2+} can eliminate the Na^+-dependent inactivation of the cardiac exchanger.

The second major difference that we found was between the Ca^{2+} regulatory properties of NCX1 and a *Drosophila* Na^+/Ca^{2+} exchanger, CALX1. Like the cardiac exchanger NCX1, CALX1 is regulated by Ca^{2+}. However, CALX1 is *inhibited* by regulatory Ca^{2+} rather than stimulated. The responses to Ca^{2+} of the two exchangers are strikingly opposite.[18] NCX1 has no activity in the absence of regulatory Ca^{2+}, whereas CALX1 has no activity in the presence of regulatory Ca^{2+}. Molecular dissection of these different responses has been initiated.[19]

The third distinction between Na^+/Ca^{2+} exchangers that we observed involves splice variants of NCX1. We compared the regulatory properties of splice variants typically found in heart (NCX1.1; exons ACDEF), brain (NCX1.4; exons AD), and kidney (NCX1.3; exons BD). No differences between heart and brain variants were found. However, one basic regulatory property was different in the kidney splice variant. As described above, increases in regulatory Ca^{2+} eliminate Na^+-dependent inactivation in the cardiac and brain splice variants. In contrast, regulatory Ca^{2+} has no effect on inactivation in the kidney splice variant.[20] Again, molecular study of this phenomenon has begun (see a later chapter in this volume).

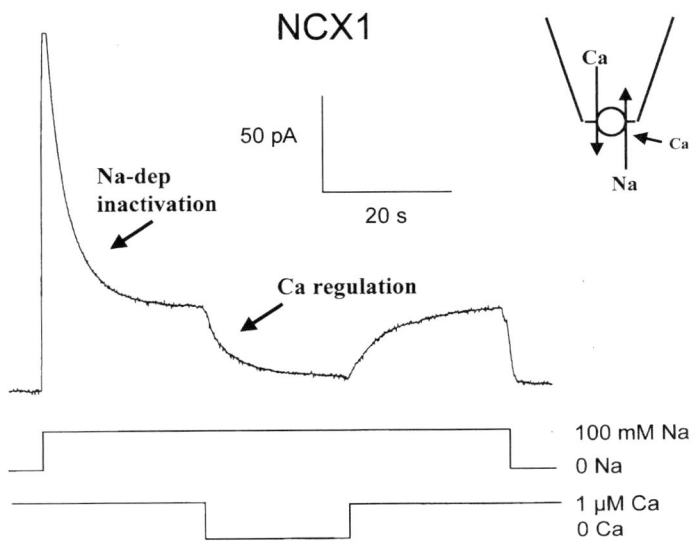

FIGURE 4. Regulation of the Na^+/Ca^{2+} exchanger by Na^+ and Ca^{2+}. Shown are outward Na^+/Ca^{2+} exchange currents generated across a giant patch excised from an oocyte expressing NCX1. Ca^{2+} is present within the pipette at the extracellular surface, and current is initiated by the rapid application of Na^+ into the bath at the intracellular surface. Current peaks and then partially inactivates (Na^+-dependent inactivation). The Na^+/Ca^{2+} exchanger also requires submicromolar levels of nontransported, regulatory Ca^{2+} in the bath for activity (Ca^{2+} regulation). When this Ca^{2+} is removed, current further declines.

Some molecular information on regulation is available. We have identified the binding site for regulatory Ca^{2+} on the large intracellular loop of the exchanger (FIG. 1).[21] This was accomplished through the use of fusion proteins and the $^{45}Ca^{2+}$ overlay technique. The regulatory Ca^{2+} binds within amino acids 375–508 of the exchanger. Furthermore, we identified specific aspartate residues that are likely ligands for the Ca^{2+}. One surprising feature was the large conformational change induced by the binding of Ca^{2+}. This was indicated by a 20-kDa shift in the apparent molecular weight of loop protein induced by the presence of Ca^{2+} as indicated on SDS-PAGE (FIG. 5). Apparently, the binding of Ca^{2+} causes a substantial change in protein folding, and this activates exchange activity. Mutations that effected the $^{45}Ca^{2+}$-binding affinity of loop protein had similar effects on the ability of Ca^{2+} to regulate the exchange activity.[22] Thus, we have confidence that the Ca^{2+}-binding site on the exchanger loop is indeed the Ca^{2+}-regulatory site of the exchanger.

The XIP region of the exchanger (FIG. 1) has been implicated in the Na^+-dependent inactivation process.[23] Nine different mutations of the XIP region all have major effects on inactivation. Furthermore, the XIP region may also be involved in the regulation of the exchanger by PIP_2.[24] PIP_2 stimulates exchange activity by removing inactivation. We have found that PIP_2 can directly bind to XIP peptide, and this interaction may be the basis for the regulatory effects of PIP_2. Nevertheless, a

Ca^{2+}-induced mobility shift

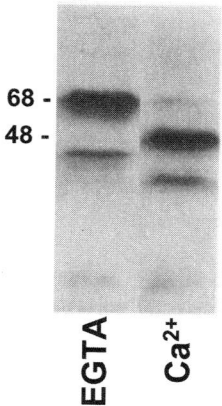

FIGURE 5. Ca^{2+}-induced mobility shift of Na$^+$/Ca^{2+} exchanger loop protein. Shown is the mobility on SDS-PAGE of a fusion protein encompassing amino acids 240–511 of the exchanger in the presence of EGTA (1 mM) or Ca^{2+} (1 mM). Protein was blotted onto nitrocellulose and detected by the ^{45}Ca^{2+} overlay technique.[21]

complete understanding of the relationship between inactivation, the XIP region, and PIP$_2$ is not yet in hand.

We have been able to express Na$^+$/Ca^{2+} exchange activity by coexpressing the two halves of the exchanger in *Xenopus* oocytes or in HEK cells.[25,26] In these experiments, the cDNA coding for the exchanger was cut within the large intracellular loop, and the resulting half exchangers could be expressed individually or together (FIG. 6). No exchange activity was ever seen with half exchanger molecules. This contrasts with results from three other laboratories that observe low exchange activity after the expression of truncated exchangers.[27–29] The reason for this discrepancy is unclear. In some of our experiments, we used half exchanger molecules that were tagged with green fluorescent protein (GFP). We found that both halves of a "split" exchanger were required for proper processing of the protein to the plasma membrane (FIG. 6). Apparently, the two halves assemble within the endoplasmic reticulum, and assembly is required for proper membrane processing. In addition, we found that "split" exchangers retained most of their regulatory properties.

To begin to understand the molecular mechanism of the Na$^+$/Ca^{2+} exchanger, it is necessary to learn about the three-dimensional arrangement of the transmembrane segments within the plasma membrane. We have addressed this using a disulfide cross-linking approach.[30] An exchanger has been made in which all cysteines have been removed by mutagenesis. Two cysteines are then re-introduced into the protein. The two new cysteines are located in two different transmembrane segments. If a disulfide bond forms, this indicates that the two transmembrane segments are in close proximity. We have carried this procedure out with 64 different pairs of cysteines, and our initial results are summarized in FIGURE 7. We specifically find that the α

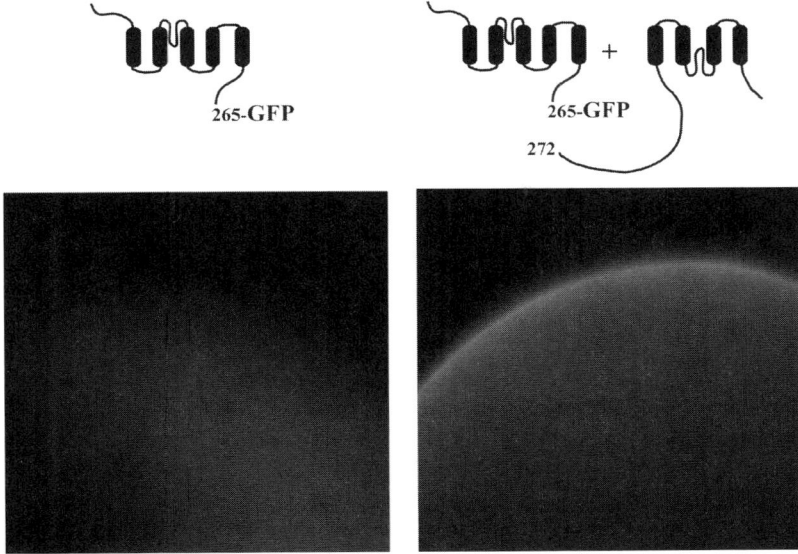

FIGURE 6. Fluorescent images of *Xenopus* oocytes expressing GFP-tagged "split" exchangers. In the *left panel*, a half exchanger that was truncated after amino acid 265 and tagged with GFP (see schematic) was expressed. Only a diffuse staining of the oocyte interior was observed. The *right panel* demonstrates fluorescence from an oocyte in which both halves of the exchanger were expressed. The plasma membrane is strongly labeled.

repeat regions are near one another. That is, transmembrane segments 2, 3, 7, and 8 can be cross-linked together (FIG. 7). These are regions that have previously been implicated as being important in ion translocation. With the arrangement shown in FIGURE 7, the two putative reentrant loops would be facing one another in the center of the protein. Much more detailed work is necessary.

Molecular approaches have also begun to be useful for studying the physiological role of the Na^+/Ca^{2+} exchanger through the use of genetically modified mice. Initial work has been done with transgenic mice overexpressing the cardiac exchanger under the control of the α-myosin heavy chain promoter.[31–33] Although myocytes from transgenic mice appeared to have normal excitation–contraction coupling, contributions of the exchanger to the onset and decay of contraction were enhanced. Under appropriate conditions, the overexpressed Na^+/Ca^{2+} exchanger could generate sufficient reverse activity to trigger Ca^{2+}-induced Ca^{2+} release from the SR and induce a Ca^{2+} transient. Reverse exchange was not as effective as Ca^{2+} current through the Ca^{2+} channel in inducing Ca^{2+} release but was nevertheless effective in this role.[31–33]

The studies just described have all used overexpressing transgenic mice that were heterozygotes. Subsequently, we have generated homozygous exchanger overexpressors. The increase in activity in the myocardium from these mice was 210%, whereas this value was 130% in the myocardium from heterozygotes. Abnormalities can be observed in these mice. First, many of the female homozygous mice develop

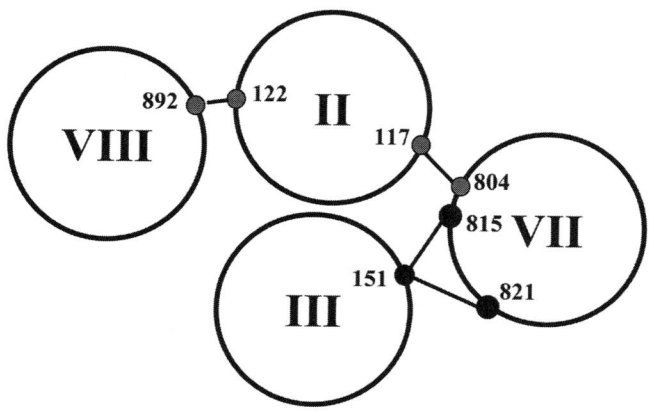

FIGURE 7. Arrangement of transmembrane segments of the Na^+/Ca^{2+} exchanger. Shown is a possible arrangement of transmembrane segments 2, 3, 7, and 8. Shown are sites of disulfide bond formation. For example, a cysteine at position 151 in transmembrane segment 3 can form a disulfide bond with cysteines at either position 815 or 821 in transmembrane segment 7. Sites indicated by *black-filled circles* are modeled to be near the intracellular surface, whereas sites indicated by *grey-filled circles* are modeled to be near the extracellular surface.

hypertrophy and heart failure in the postpartum state. Second, mice homozygous for overexpression of the Na^+/Ca^{2+} exchanger have enhanced I_{Ca} but decreased Ca^{2+} transients, although SR loads are unchanged. That is, there is a decrease in gain. Both of these phenomena are being investigated further.

Lastly, mice have been generated in which NCX1 has been knocked out (Refs. 34–36 and Philipson laboratory, unpublished observation). All studies find that the homozygous knockout mice are embryonic lethal at about day 10.5 p.c., although some differences in the specifics of these mice have been noted. A most interesting observation comes from the study of Koushik *et al.*[35] These authors measured Ca^{2+} transients in heart tubes from wild-type and NCX KO mice at day 9.5 p.c. Surprisingly, apparently normal Ca^{2+} transients could be elicited by electrical stimulation in the KO heart tubes. We have reproduced this observation and are beginning to use this system for excitation–contraction coupling studies.

Overall, both new molecular biological and physiological approaches hold much promise for future research on the Na^+/Ca^{2+} exchanger.

ACKNOWLEDGMENTS

This work has been supported by funds from the National Institutes of Health; the American Heart Association, Western States Affiliate; and by the Laubisch Foundation.

REFERENCES

1. ADACHI-AKAHANE, S. & Y. KURACHI. 2001. New era for translational research in cardiac arrhythmias. Circ. Res. **88:** 1095–1096.
2. BARRY, W.H. 2000. Na^+-Ca^{2+} exchange in failing myocardium: friend or foe? Circ. Res. **87:** 529–531.
3. GOLDHABER, J.I. 1999. Sodium-calcium exchange: the phantom menace. Circ. Res. **85:** 982–984.
4. HOUSER, S.R. 2000. When does spontaneous sarcoplasmic reticulum Ca^{2+} release cause a triggered arrythmia? Cellular versus tissue requirements. Circ. Res. **87:** 725–727.
5. ISENBERG, G. 2001. How can overexpression of Na^+,Ca^{2+}-exchanger compensate the negative inotropic effects of downregulated SERCA? Cardiovasc. Res. **49:** 1–6.
6. POGWIZD, S.M. 2000. Increased Na^+-Ca^{2+} exchanger in the failing heart. Circ. Res. **87:** 641–643.
7. BLAUSTEIN, M.P. & W.J. LEDERER. 1999. Sodium/calcium exchange: its physiological implications. Physiol. Rev. **79:** 763–854.
8. PHILIPSON, K.D. & D.A. NICOLL. 2000. Sodium-calcium exchange: a molecular perspective. Annu. Rev. Physiol. **62:** 111–133.
9. SCHWARZ, E.M. & S. BENZER. 1997. Calx, a Na-Ca exchanger gene of *Drosophila melanogaster*. Proc. Natl. Acad. Sci. USA **94:** 10249–10254.
10. NICOLL, D.A., L.V. HRYSHKO, S. MATSUOKA, et al. 1996. Mutation of amino acid residues in the putative transmembrane segments of the cardiac sarcolemmal Na^+-Ca^{2+} exchanger. J. Biol. Chem. **271:** 13385–13391.
11. KOFUJI, P., W.J. LEDERER & D.H. SCHULZE. 1994. Mutually exclusive and cassette exons underlie alternatively spliced isoforms of the Na/Ca exchanger. J. Biol. Chem. **269:** 5145–5149.
12. QUEDNAU, B.D., D.A. NICOLL & K.D. PHILIPSON. 1997. Tissue specificity and alternative splicing of the Na^+/Ca^{2+} exchanger isoforms NCX1, NCX2, and NCX3 in rat. Am. J. Physiol. **272:** C1250–C1261.
13. HE, Z., Q. TONG, B.D. QUEDNAU, et al. 1998. Cloning, expression, and characterization of the squid Na^+-Ca^{2+} exchanger (NCX-SQ1). J. Gen. Physiol. **111:** 857–873.
14. HILGEMANN, D.W., D.A. NICOLL & K.D. PHILIPSON. 1991. Charge movement during Na^+ translocation by native and cloned cardiac Na^+/Ca^{2+} exchanger. Nature **352:** 715–718.
15. HILGEMANN, D.W., A. COLLINS & S. MATSUOKA. 1992. Steady-state and dynamic properties of cardiac sodium–calcium exchange—secondary modulation by cytoplasmic calcium and ATP. J. Gen. Physiol. **100:** 933–961.
16. HILGEMANN, D.W., S. MATSUOKA, G.A. NAGEL & A. COLLINS. 1992. Steady-state and dynamic properties of cardiac sodium–calcium exchange—sodium-dependent inactivation. J. Gen. Physiol. **100:** 905–932.
17. HILGEMANN, D.W. 1990. Regulation and deregulation of cardiac Na^+–Ca^{2+} Exchange in giant excised sarcolemmal membrane patches. Nature **344:** 242–245.
18. HRYSHKO, L.V., S. MATSUOKA, D.A. NICOLL, et al. 1996. Anomalous regulation of the Drosophila Na^+–Ca^{2+} exchanger by Ca^{2+}. J. Gen. Physiol. **108:** 67–74.
19. DYCK, C., K. MAXWELL, J. BUCHKO, et al. 1998. Structure–function analysis of CALX1.1, a Na^+–Ca^{2+} exchanger from Drosophila. Mutagenesis of ionic regulatory sites. J. Biol. Chem. **273:** 12981–12987.
20. DYCK, C., A. OMELCHENKO, C.L. ELIAS, et al. 1999. Ionic regulatory properties of brain and kidney splice variants of the NCX1 Na^+–Ca^{2+} exchanger. J. Gen. Physiol. **114:** 701–711.
21. LEVITSKY, D.O., D.A. NICOLL & K.D. PHILIPSON. 1994. Identification of the high affinity $Ca(2^+)$-binding domain of the cardiac Na^+–Ca^{2+} exchanger. J. Biol. Chem. **269:** 22847–22852.
22. MATSUOKA, S., D.A. NICOLL, L.V. HRYSHKO, et al. 1995. Regulation of the cardiac Na^+–Ca^{2+} exchanger by Ca^{2+}. Mutational analysis of the Ca^{2+}-binding domain. J. Gen. Physiol. **105:** 403–420.
23. MATSUOKA, S., D.A. NICOLL, Z. HE & K.D. PHILIPSON. 1997. Regulation of cardiac Na^+–Ca^{2+} exchanger by the endogenous XIP region. J. Gen. Physiol. **109:** 273–286.

24. HE, Z., S. FENG, Q. TONG, et al.. 2000. Interaction of PIP_2 with the XIP region of the cardiac Na/Ca exchanger. Am. J. Physiol. **278:** C661–C666.
25. OTTOLIA, M., S. JOHN, Z. QIU & K.D. PHILIPSON. 2001. "Split" Na^+–Ca^{2+} exchangers: implications for function and expression. J. Biol. Chem. **276:** 19603–19609.
26. QIU, Z., J. CHEN, D.A. NICOLL & K.D. PHILIPSON. 2001. A disulfide bond is required for functional assembly of NCX1 from complementary fragments. Biochem. Biophys. Res. Commun. **287:** 825–828.
27. GABELLINI, N., T. IWATA & E. CARAFOLI. 1995. Alternative splicing site modifies the carboxyl-terminal *trans*-membrane domains of the Na^+/Ca^{2+} exchanger J. Biol. Chem. **270:** 6917–6924.
28. LI, X.F. & J. LYTTON. 1999. A circularized sodium–calcium exchanger exon 2 transcript. J. Biol. Chem. **274:** 8153–8160.
29. VAN EYLEN, F., A. KAMAGATE & A. HERCHUELZ. 2001. A new Na/Ca exchanger splicing pattern identified in situ leads to a functionally active 70 kDa NH_2-terminal protein. Cell Calcium **30:** 191–198.
30. QIU, Z., D.A. NICOLL & K.D. PHILIPSON. 2001. Helix packing of functionally important regions of the cardiac Na^+-Ca^{2+} exchanger. J. Biol. Chem. **276:** 194–199.
31. ADACHI-AKAHANE, S., L. LU, Z. LI, et al. 1997. Calcium signaling in transgenic mice overexpressing cardiac Na^+-Ca^{2+} exchanger. J. Gen. Physiol. **109:** 717–729.
32. TERRACCIANO, C.M., A.I. SOUZA, K.D. PHILIPSON & K.T. MACLEOD. 1998. Na^+–Ca^{2+} exchange and sarcoplasmic reticular Ca^{2+} regulation in ventricular myocytes from transgenic mice overexpressing the Na^+–Ca^{2+} exchanger. J. Physiol. **512:** 651–667.
33. YAO, A., Z. SU, A. NONAKA, et al. 1998. Effects of overexpression of the Na^+-Ca^{2+} exchanger on $[Ca^{2+}]_i$ transients in murine ventricular myocytes. Circ. Res. **82:** 657–665.
34. CHO, C.H., S.S. KIM, M.J. JEONG, et al. 2000. The Na^+–Ca^{2+} exchanger is essential for embryonic heart development in mice. Mol. Cells **10:** 712–722.
35. KOUSHIK, S.V., J. WANG, R. ROGERS, et al. 2001. Targeted inactivation of the sodium–calcium exchanger (NCX1) results in the lack of a heartbeat and abnormal myofibrillar organization. FASEB J. **15:** 1209–1211.
36. WAKIMOTO, K., K. KOBAYASHI, O.M. KURO, et al. 2000. Targeted disruption of Na^+/Ca^{2+} exchanger gene leads to cardiomyocyte apoptosis and defects in heartbeat. J. Biol. Chem. **275:** 36991–36998.

Toward a Topological Model of the NCX1 Exchanger

DEBORA A. NICOLL, MICHELA OTTOLIA, AND KENNETH D. PHILIPSON

Cardiovacular Research Laboratories, University of California at Los Angeles, Los Angeles, California 90095-1760, USA

ABSTRACT: The Na^+/Ca^{2+} exchanger (NCX1) catalyzes the counter-transport of sodium and calcium ions. Understanding how this is accomplished requires knowledge of the structure of NCX1 and identifying amino acid residues involved in binding and transport of ions. The amino acid sequence of NCX1 has been known for more than a decade. Based on hydropathy analysis, NCX1 was modeled to contain 12 transmembrane segments. In this model, the α-repeat regions, which are the result of a gene duplication event (see below), are oriented on the extracellular face of NCX1. In the years since NCX1 was sequenced, a considerable amount of effort has gone into testing the initial 12-transmembrane-segment model. Immunologic and protein-processing studies as well as functional analyses of mutants have determined the location of the amino and carboxy termini and several intracellular regions. However, disulfide bond analysis and cysteine mutagenesis coupled with accessibility studies indicate that the structure of NCX1 diverges from a simple membrane protein consisting only of transmembrane α-helical segments. These recent data support a model containing 9 transmembrane α-helices with the α-repeat regions forming nonhelical re-entrant loops. A bacterial protein containing a pair of α-repeat regions but of unknown function has also been shown to have oppositely oriented α-repeats.

KEYWORDS: sodium–calcium exchange; topology; disulfide bond; cysteine mutagenesis; α-repeat regions

THEN AND NOW

When NCX1 was first cloned, a model for how the protein might fold in the membrane was presented.[1] This model was solely based on the hydropathy profile of the amino acid sequence. The model consisted of 12 α-helical transmembrane segments (TMSs) with a very long, intracellular loop between TMSs VI and VII (FIG. 1A). The amino and carboxy termini were in the cytoplasm. Another defining characteristic of this model was that the α repeats map to the extracellular surface at TMSs III–IV and VII–VIII. The α repeats share significant sequence similarity, suggesting that they are the result of a gene duplication event,[2] and appear to be involved in ion bind-

---

Address for correspondence: Debora A. Nicoll, Cardiovacular Research Laboratories, University of California at Los Angeles, 3645 MRL, 675 Young Dr. S., Los Angeles, CA 90095-1760. Voice: 310-825-5137; fax: 310-206-5777.
  dnicoll@mednet.ucla.edu

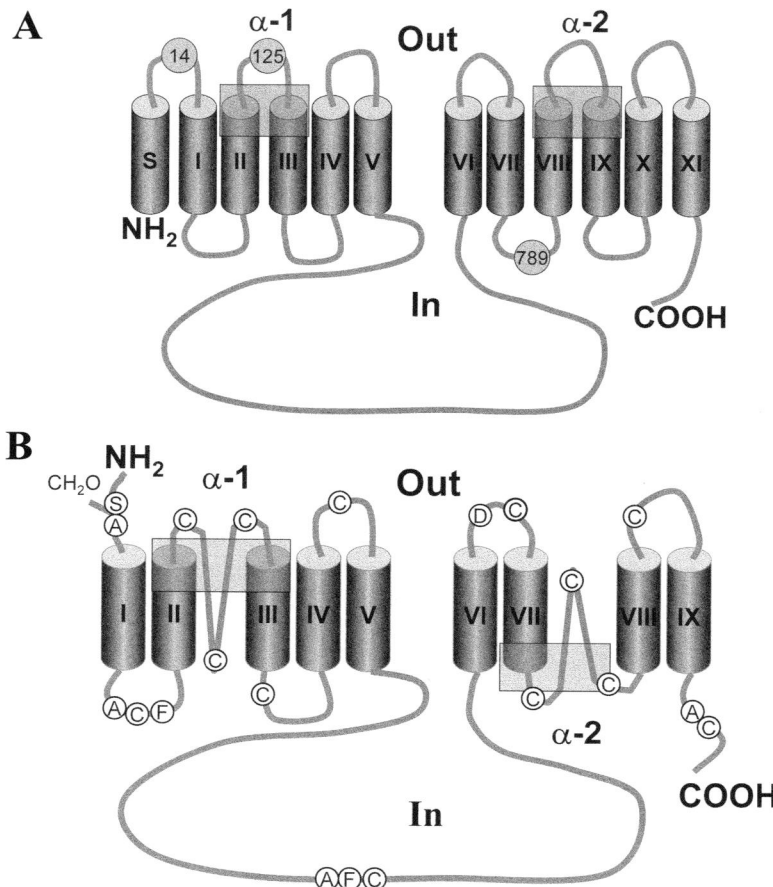

FIGURE 1. Models for NCX1. **(A)** Model based on hydropathy analysis. Locations of specific residues mentioned in the text are circled. **(B)** Model based on experimental data. Regions that have been localized by protein synthesis (S), antibody binding (A), functional activity (F), cysteine mutagenesis, and substitution (C) or disulfide bond analysis (D) are indicated.

ing/transport.[3] Each α repeat is composed of a group of 12 highly conserved residues separated from a group of 9 highly conserved residues by a nonconserved linker of 18–20 residues.

The model has been rigorously tested by many people in several different laboratories. Results from the various experiments indicate that the topological model of NCX1 is quite different from that originally proposed (FIG. 1B). Rather than 12 α-helical TMSs, NCX1 only has 9. The amino terminus contains a cleaved signal sequence and is glycosylated and extracellular. Also, the α repeats are on opposite sides of the membrane and appear to form re-entrant loops that interact in the protein. The data that lead to this new topological model of NCX1 will be reviewed.

THE AMINO TERMINUS OF NCX1

Durkin et al.[4] isolated NCX1 protein from cardiac tissue and performed microsequence analysis of the amino terminus. They found that rather than starting at the initiating methionine predicted from the cDNA sequence, NCX1 actually began about 30 amino acids downstream at a glutamate residue. After comparing the predicted amino acid sequence with a database of cleaved leader sequences, they hypothesized that NCX1 contained a cleaved leader peptide.

Hryshko et al.[5] tested this hypothesis in *in vitro* translation experiments. When the full length or variously truncated cDNAs for NCX1 were translated in the presence of microsomes, the molecular weight of the product was larger than if the cDNAs were translated in the absence of microsomes. Microsomes allow cleavage of signal peptides and core glycosylation of peptides. Therefore, the increased molecular weight of the microsome-translated peptides showed that the exchanger is glycosylated. Following digestion with endoglycosidase H, the NCX1 protein was then smaller than the protein translated in the absence of microsomes. The reduced size was due to cleavage of the leader peptide in the presence of microsomes.

By means of site-directed mutagenesis, it was then possible to determine that NCX1 is glycosylated at an asparagine 9 residues downstream from the site of leader peptide cleavage.[5] These data unequivocally demonstrate that the amino terminus of the exchanger is glycosylated and extracellular.

ANTIBODY BINDING TO NCX1 LOCALIZES TERMINI AND TWO INTRACELLULAR LOOPS

Antibodies have been quite helpful in localizing some portions of NCX1. The approaches to using antibodies have been threefold. First, Porzig et al.[6] generated a panel of monoclonal antibodies against a preparation of purified NCX1. To determine the epitopes of the antibodies, a sublibrary consisting of portions of NCX1 fused to the *E. coli* β-galactosidase protein was generated, and the antibodies absorbed to fusion protein. By sequencing the clones that absorbed antibodies, the epitopes were identified.

The second approach was to generate antibodies to peptides with sequences corresponding to portions of the exchanger.[7] And the third approach was to insert epitope "tags" into the exchanger cDNA and express the chimeric proteins.[7,8]

Once antibodies to known sites in NCX1 were available, the antibodies were used to determine the topology of the epitopes by binding to permeabilized or nonpermeabilized cells expressing NCX1. Cells were permeabilized either by using the detergent Triton X-100 or by homogenizing the cells. One caveat to keep in mind is that, in addition to permeabilizing membranes, Triton X-100 can also sometimes unmask epitopes.

Using these approaches, it has been confirmed that the amino terminus is extracellular, and the loops connecting TMSs I and II and TMSs V and VI are intracellular, as is the carboxy terminus.

FUNCTIONAL EVIDENCE FOR INTRACELLULAR LOCALIZATION OF TWO TMS CONNECTING LOOPS

NCX1 has been shown to undergo two types of regulation, designated I_1 and I_2.[9] Both are most easily seen in the reverse mode of exchange upon ion applications to the intracellular surface. I_1, or Na-dependent inactivation, is seen after activating reverse exchange currents by applying Na$^+$. The NCX1 current slowly inactivates, and the rate and extent of inactivation are dependent on the Na$^+$ concentration. I_2 is a Ca^{2+}-dependent regulation. No exchange currents are seen in the absence of a low level of intracellular, regulatory Ca^{2+}.

Many mutants in the loop connecting TMSs V and VI have been shown to disrupt I_1 and/or I_2.[10–13] These data confirm that the loop is intracellular. Likewise, it has been shown that a mutation of residue asparagine 101, in the loop connecting TMSs I and II, to a cysteine alters the regulatory properties of the exchanger, thereby indicating that this portion of NCX1 is intracellular.[14]

DISULFIDE BOND ANALYSIS ALTERS THE MODEL

The first indication that significant alterations to the topological model of NCX1 were necessary came on the coattails of an experiment designed to examine endogenous disulfide bonds in the exchanger. It had long been known that NCX1 displayed anomalous properties upon SDS-PAGE separation. The NCX1 protein migrates as predicted to 120 kDa but also to 160 kDa. The relative amounts of each band varied from gel to gel. It was determined that the 160-kDa band was dominant under nonreducing conditions, and the 120-kDa band under reducing conditions. These observations are consistent with disulfide bond formation in NCX1.[15]

To identify the residues that formed the disulfide bond, Santacruz-Toloza *et al.* reintroduced pairs of cysteines into a cysteineless version of NCX1.[16] The only pairs of native cysteines that resulted in a shift to the 1600-kDa form of NCX1 under reducing conditions were C14 + C792 or C20 + C792. Thus, the cysteine 792, originally modeled to be in the intracellular loop between TMSs VII and VIII, formed a disulfide bond with either cysteine 14 or 20 in the extracellular amino terminus of NCX1.

Disulfide bonds cannot form across the membrane, so cysteine 792 must be extracellular. The proximity of residue 792 to the α-2 repeat suggests that the α-2 repeat must be at the intracellular surface in the new model.

YrbG HAS α REPEATS ON OPPOSITE SIDES OF THE MEMBRANE

There is an NCX superfamily defined by the existence of a pair of α repeats in each protein. One member of this superfamily is the *E. coli* protein, YrbG, of unknown function. YrbG is clearly the result of a gene duplication event with about 34% identity between the amino and carboxy halves of the protein.

Sääf *et al.*[17] have examined the topology of YrbG in PhoA fusions. PhoA is an *E. coli* alkaline phosphatase that is functional in the periplasm but not in the cytoplasm. By fusing PhoA to truncated YrbG proteins in each of the proposed nonmembranous loops and measuring phosphatase activity, they determined that the α repeats of YrbG are on opposite sides of the membrane.

CYSTEINE SUBSTITUTION MUTAGENESIS AND ACCESSIBILITY STUDIES LOCALIZE NONTRANSMEMBRANE LOOPS

These studies have proven to be the most powerful way to examine NCX1 topology to date. They take advantage of the fact that in several expression systems NCX1 activity is insensitive to most membrane-impermeant, sulfhydryl-modifying reagents. Thus, amino acids can be mutated to cysteine, and, if the mutated NCX1 is still functional, the effects of membrane-impermeant reagents on NCX1 activity can be examined. Reagents that inhibit NCX1 activity can be used to localize the mutated residue to the intra- or extracellular spaces. This technique has been used by both us[18] and Iwamoto et al.[8,19] to identify 4 extracellular and 5 intracellular segments.

These studies confirm that the α repeats are on opposite sides of the membrane and that NCX1 is likely composed of 9 α-helical transmembrane segments.

THE α REPEATS FORM RE-ENTRANT LOOPS

Iwamoto et al.[8] found that one residue in the α-1 repeat, asparagine-125 (FIG. 1A), when substituted with a cysteine, was sensitive to intracellular, but not extracellular, applications of MTSET. Given that other residues tested in the α-1 repeat were sensitive to extracellular reagents, this was surprising. Subsequently, we found that the mutant E120C was sensitive to both extra and intracellular applications of sulfhydryl reagents (FIG. 2).

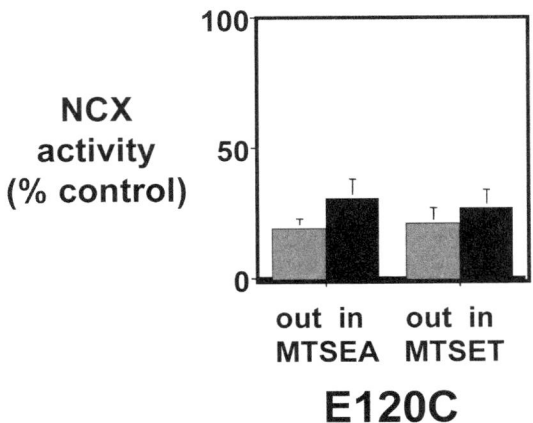

FIGURE 2. Residue E120 is accessible to both intracellular and extracellular application of sulfhydryl-modifying reagents. Residue E120 was mutated to a cysteine, and the mutant exchanger was expressed in *Xenopus* oocytes. Exchanger activities were measured before (control) and after pretreatment with the membrane impermeant sulfhydryl modifying reagent. Activity is expressed as percent of control.

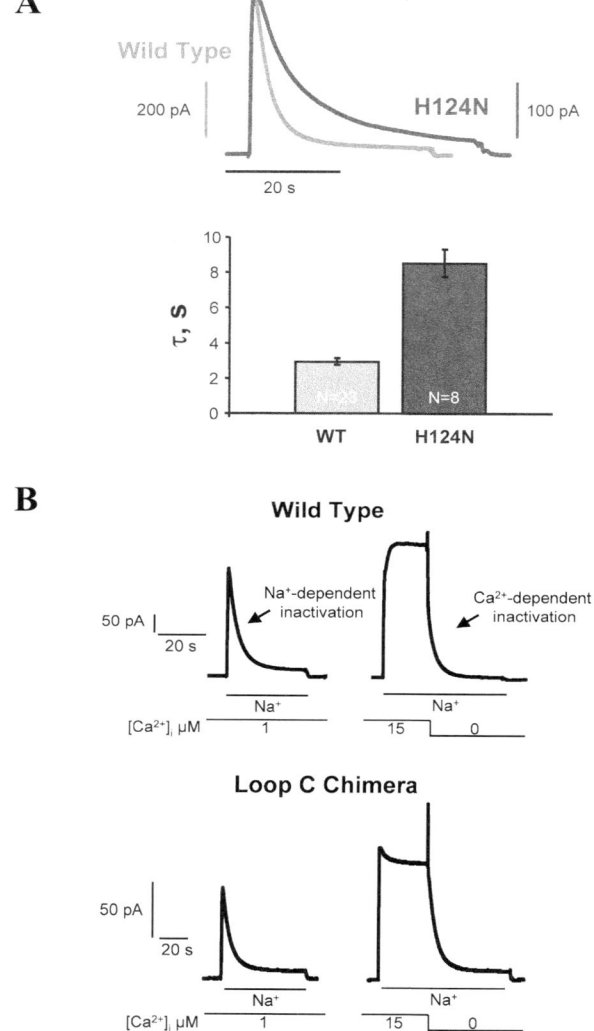

FIGURE 3. Regulatory properties of α-1 repeat mutants. **(A)** Mutant H124N displays slowed I_1 inactivation. **(B)** The dog–squid chimera, consisting of the squid α-1 nonconserved linker region in the dog NCX1 background, displays normal regulatory properties.

In a more detailed study, Iwamoto et al.[19] determined that N125C was accessible from both sides of the membrane, as were mutants D825C and N842C, in the α-2 repeat. They also determined that I847C is accessible from the intracellular surface. Thus, the α repeats apparently form non–α-helical membrane re-entrant loops.

In some respects this is still a surprising conclusion. Each α repeat is composed of 12 highly conserved amino acids separated from another 9 highly conserved residues by an unconserved linker of 18–20 amino acids. Three of the 4 residues that are accessible from both sides of the membrane are in the unconserved linker. It might be expected that re-entrant loops would be functionally important and highly conserved.

To further examine the role of the nonconserved portion of the α-1 repeat, we have been examining mutants in this region and have constructed a chimeric NCX1 protein containing the α-1 linker from the squid NCX. Some mutants in the nonconserved α-1 repeat region have altered I_1 regulatory properties. For example, the mutant H124A displays a significantly slower rate of I_1 (FIG. 3A). On the other hand, the chimera, which has a glutamine at position 124, appears to have normal regulatory properties (FIG. 3B). Other notable differences between NCX1 and squid NCX in this region include a lysine instead of asparagine at position 125 and a glutamine instead of an aspartate at position 130. As noted above, N125 is accessible from both sides of the membrane. Iwamoto et al. have found that mutation of residue D130 results in an altered affinity of the exchanger for extracellular Ca^{2+}, though we have not been able to replicate these data.[19]

HELIX PACKING OF NCX1

Qiu et al.[20] have begun to analyze how the membrane-spanning portions of NCX1 are packed in the membrane. The approach was to take advantage of the gel mobility shift evident in disulfide-bonded forms of the exchanger. Pairs of cysteines were introduced into the cysteineless NCX1 background and examined for mobility shifts in reducing/nonreducing conditions or in the presence of disulfide cross-linking reagents.

Four pairs of interacting residues were detected in functional NCX1 proteins. Residue 122 in TMS II was found to interact with residue 892 in TMS VIII. Residue 117 in TMS II interacts with residue 804 in TMS VII. Also, residue 151 in TMS III interacts with residue 815 or 821 in TMS VII. These data show that TMSs II and III, which are part of α-1, and TMSs VII and VIII, which are part of α-2, are all close to one another. Thus, it is proposed that the α-repeats interact with one another in the folded protein, perhaps to form the ion translocation pathway.

SUMMARY

Experimental data support a model for NCX1 that consists of 9 α-helical transmembrane segments and 2 nonhelical membrane re-entrant loops. The re-entrant loops are formed by amino acids in the α-repeat regions and are on opposite sides of the membrane. The α-repeat regions may interact with one another in the folded protein to mediate ion transport.

ACKNOWLEDGMENTS

This work was supported by grants from the National Institutes of Health, the American Heart Association, and the Laubisch Fund. We appreciated Dr. Beate Quednau's careful reading of this manuscript.

REFERENCES

1. NICOLL, D.A., S. LONGONI & K.D. PHILIPSON. 1990. Molecular cloning and functional expression of the cardiac sarcolemmal Na^+–Ca^{2+} exchanger. Science **250**: 562–565.
2. SCHWARZ, E.M. & S. BENZER. 1997. *Calx*, a Na–Ca exchanger gene of *Drosophila melanogaster*. Proc. Natl. Acad. Sci. USA **94**: 10249–10254.
3. NICOLL, D.A., L.V. HRYSHKO, S. MATSUOKA, *et al*. 1996. Mutation of amino acid residues in the putative transmembrane segments of the cardiac sarcolemmal Na^+–Ca^{2+} exchanger. J. Biol. Chem. **271**: 13385–13391.
4. DURKIN, J.T., D.C. AHRENS, Y.-C.E. PAN & J.P. REEVES. 1991. Purification and amino-terminal sequence of the bovine cardiac sodium–calcium exchanger: evidence for the presence of a signal sequence. Arch. Biochem. Biophys. **290**: 369–375.
5. HRYSHKO, L.V., D.A. NICOLL, J.N. WEISS & K.D. PHILIPSON. 1993. Biosynthesis and initial processing of the cardiac sarcolemmal Na^+–Ca^{2+} exchanger. Biochim. Biophys. Acta **1151**: 35–42.
6. PORZIG, H., Z. LI, D.A. NICOLL & K.D. PHILIPSON. 1993. Mapping of the cardiac sodium-calcium exchanger with monoclonal antibodies. Am. J. Physiol. **265**: C748–756.
7. COOK, O., W. LOW & H. RAHAMIMOFF. 1998. Membrane topology of the rat brain Na^+–Ca^{2+} exchanger. Biochim. Biophys. Acta **1371**: 40–52.
8. IWAMOTO, T., T.Y. NAKAMURA, Y. PAN, *et al*. 1999. Unique topology of the internal repeats in the cardiac Na^+/Ca^{2+} exchanger. FEBS Lett. **446**: 264–268.
9. HIGEMANN, D.W. 1990. Regulation and deregulation of cardiac Na^+–Ca^{2+} exchange in giant excised sarcolemmal membrane patches. Nature **344**: 242–245.
10. MATSUOKA, S., D.A. NICOLL, R.F. REILLY, *et al*. 1993. Initial localization of regulatory regions of the cardiac sarcolemmal Na^+–Ca^{2+} exchanger. Proc. Natl. Acad. Sci. USA **90**: 3870–3874.
11. MATSUOKA, S., D.A. NICOLL, L.V. HRYSHKO, *et al*. 1995. Regulation of the cardiac Na^+–Ca^{2+} exchanger by Ca^{2+}. J. Gen. Physiol. **105**: 403–420.
12. MATSUOKA, S., D.A. NICOLL, Z. HE & K.D. PHILIPSON. 1997. Regulation of the cardiac Na^+–Ca^{2+} exchanger by the endogenous XIP region. J. Gen. Physiol. **109**: 273–286.
13. LI, Z., D.A. NICOLL, A. COLLINS, *et al*. 1991. Identification of a peptide inhibitor of the cardiac sarcolemmal Na^+–Ca^{2+} exchanger. J. Biol. Chem. **266**: 1014–1020.
14. DOERING, A.E., D.A. NICOLL, Y. LU, *et al*. 1998. Topology of a functionally important region of the cardiac Na^+/Ca^{2+} exchanger. J. Biol. Chem. **273**: 778–783.
15. PHILIPSON, K.D., S. LONGONI & R. WARD. 1988. Purification of the cardiac Na^+–Ca^{2+} exchange protein. Biochim. Biophys. Acta **945**: 298–306.
16. SANTACRUZ-TOLOZA, L., M. OTTOLIA, D.A. NICOLL & K.D. PHILIPSON. 2000. Functional analysis of a disulfide bond in the cardiac Na^+–Ca^{2+} exchanger. J. Biol. Chem. **275**: 182–188.
17. SÄÄF, A., L. BAARS & G. VON HEIJNE. 2001. The internal repeats in the Na^+/Ca^{2+} exchanger-related *Escherichia coli* protein YrbG have opposite membrane topologies. J. Biol. Chem. **276**: 18905–18907.
18. NICOLL, D.A., M. OTTOLIA, L. LU, *et al*. 1999. A new topological model of the cardiac sarcolemmal Na^+–Ca^{2+} exchanger. J. Biol. Chem. **274**: 910–917.
19. IWAMOTO, T., A. UEHARA, I. IMANAGA & M. SHIGEKAWA. 2000. The Na^+/Ca^{2+} exchanger NCX1 has oppositely oriented re-entrant loop domains that contain conserved aspartic acids whose mutation alters its apparent Ca^{2+} affinity. J. Biol. Chem. **275**: 38571–38580.
20. QIU, Z., D.A. NICOLL & K.D. PHILIPSON. 2001. Helix packing of functionally important regions of the cardiac Na^+–Ca^{2+} exchanger. J. Biol. Chem. **276**: 194–199.

Probing Ion Binding Sites in the Na$^+$/Ca^{2+} Exchanger

MUNEKAZU SHIGEKAWA,[a] TAKAHIRO IWAMOTO,[a] AKIRA UEHARA,[b] AND SATOMI KITA[a]

[a]*Department of Molecular Physiology, National Cardiovascular Center Research Institute, Suita, Osaka 565-8565, Japan*

[b]*Department of Physiology, School of Medicine, Fukuoka University, Fukuoka 814-0180, Japan*

> ABSTRACT: The membrane domain of the Na$^+$/Ca^{2+} exchanger (NCX) contains two conserved internal repeat sequences, designated the α-1 and α-2 repeats. We have studied the topological disposition of residues in the α repeats and a neighboring region by substituted cysteine accessibility scanning as well as the functional importance of these residues by kinetically evaluating transport activities of cysteine-substituted or other site-directed NCX1 mutants. The results suggest that the α-1 repeat contains a reentrant loop originating from extracellular side of the membrane, while the α-2 repeat and its neighboring region contain a complex reentrant loop structure originating from the cytoplasmic side. We identified several residues in the α-1 repeat loop whose mutation caused significant alterations in the interaction of NCX1 with a transport substrate Ca$^{2+}_o$ and inhibitory ions Ni^{2+} and Co^{2+}. On the other hand, we found residues in the α-2 repeat loop region whose mutation altered the interaction with an activating ion Li$^+$ and an inhibitory drug KB-R7943 in addition to the effect on Ca$^{2+}_o$ and Ni^{2+}. Collectively, our data suggest that these α-repeat regions participate in the formation of ion translocation pathway of the exchanger and that the α-1 repeat loop plays an important role in ion selection.
>
> KEYWORDS: Na$^+$/Ca^{2+} exchanger; ion binding sites; NCX isoforms; α-repeat loops; ion translocation

INTRODUCTION

The mammalian Na$^+$/Ca^{2+} exchanger forms a multigene family comprising three isoforms: NCX1, NCX2, and NCX3.[1] NCX1 is highly expressed in heart, brain, and kidney and at low levels in other tissues, whereas the expression of NCX2 and NCX3 is limited mainly to brain. In cardiomyocytes, NCX1 plays an indispensable role in extrusion of cytoplasmic Ca^{2+} during each cardiac contraction cycle. However, the roles played by NCXs in other cell types still remain to be defined.

Address for correspondence: Dr. Munekaza Shigekawa, Molecular Physiology, National Cardiovascular Center Research Institte, Fujishiro-Dai 5-7-1, Suita 565-8565, Japan. Voice: 81-6-6833-5012; fax: 81-6-6835-5314.
  shigekaw@ri.ncvc.go.jp

Recent topological studies[2,3] suggested that the mature NCX1 protein comprises 9 transmembrane segments (TMs) and a large hydrophilic loop between TMs 5 and 6, with NH_2- and COOH-termini located on the external and internal sides, respectively (see model in FIG. 6A). The large central hydrophilic loop is exposed on the cytoplasm and is involved in the regulation of the exchanger by cytoplasmic factors such as Ca^{2+}, Na^+, and protein kinases.[1,4–8] In the NH_2-terminal and COOH-terminal halves of the membrane domain of NCX1, there are two internal repeat sequences of about 40 amino acids designated the α-1 and α-2 repeats, which presumably comprise TMs and loops.[9] These repeat sequences are phylogenetically conserved in all members of the NCX and related cation exchanger families,[1] suggesting the functional importance of these sequences. The importance of the putative TM segments of α repeats has been suggested by recent reports[10,11] that mutations in these regions cause a large reduction in exchange activity or a change in the current–voltage relationship in NCX1 and that mutation of a residue at the cytoplasmic portion of TM2, which is one of TMs in the α-1 repeat, increases the apparent affinity of NCX1 for intracellular Na^+ (Na^+_i) and also appears to produce Li^+ transport capacity.

We have recently shown that NCX isoforms have distinct differences in their biochemical and pharmacological properties[12] despite the fact that they share ~70% identity in the overall amino acid sequences[1] and that they have very similar apparent affinities for the external transport substrate Ca^{2+} (Ca^{2+}_o) or Na^+ (Na^+_o).[12,13] For example, the inhibitory effect of a divalent cation Ni^{2+} is 10-fold more potent on NCX1 than on NCX2 or NCX3, whereas the effect of an inhibitor KB-R7943 is 3-fold less potent on NCX1 or NCX2 than on NCX3.[12] Furthermore, the extent of stimulation of Na^+/Ca^{2+} exchange by an external monovalent cation Li^+ is 2-fold greater in NCX2 or NCX3 than in NCX1.

In this article, we summarize our recent studies of the structure–function of the α repeats in the Na^+/Ca^{2+} exchanger and discuss the roles of the loop regions of the α repeats in the interaction of NCX1 with transport ions and modifiers.

MATERIALS AND METHODS

Construction of chimeric and mutant exchanger cDNAs, transfection of cDNAs into CCL39 cells, and selection of stable cell clones exhibiting high exchange activity were described elsewhere.[12,14,15] Assays of the rate of Na^+_i-dependent $^{45}Ca^{2+}$ uptake into Na^+-loaded cells and whole-cell exchange currents were performed as described.[2,12,14,15] $^{45}Ca^{2+}$ uptake was assayed after cysteines individually introduced into the NCX1 protein were modified during the last 5–10 min of cell Na^+-loading (to ~85 mM Na^+_i) with following externally applied sulfhydryl probes: 2-trimethylammonioethylmethane thiosulfonate (MTSET), 2-aminoethylmethane thiosulfonate (MTSEA), N-ethylmaleimide (NEM), and metal ions (Zn^{2+} or Hg^{2+}).[2,15] The effect of internal MTSET was assessed during measurement of the whole-cell outward exchange current under conditions as described previously.[2] Modification of single cysteine mutants with 3-(N-maleimidylpropionyl)biocytin (biotin maleimide) was performed before and after permeabilization of CCL39 cells expressing cysteine-less NCX1 with streptolysin O.[15]

RESULTS AND DISCUSSION

Topologies of Internal α Repeats and a Neighboring Region

We studied topologies of the α-1 and α-2 repeats and a region immediately COOH-terminal to the latter by scanning reactivity of cysteine-substituted mutants of NCX1 with sulfhydryl probes applied extracellularly or intracellularly.[2,15] FIGURE 1 shows typical results from these experiments in which we examined whether externally applied various probes cause alteration of the rate of Na^+_i-dependent $^{45}Ca^{2+}$ uptake into CCL39 cells expressing cysteine-substituted mutants. The probes used have different sizes and chemical reactivity. MTSEA is smaller than MTSET, but, like MTSET, is membrane-impermeable under the conditions used.[15] NEM is membrane-permeable, but presumably reacts only with deprotonated cysteine sulfhydryls accessible from aqueous medium. Metal ions such as Zn^{2+} and Hg^+ are much smaller in size compared to MTS reagents or NEM and are membrane-impermeable.

Many substituted cysteines in the α-1 repeat reacted similarly with external MTSET or MTSEA, causing reduction of exchange activity, although MTSEA

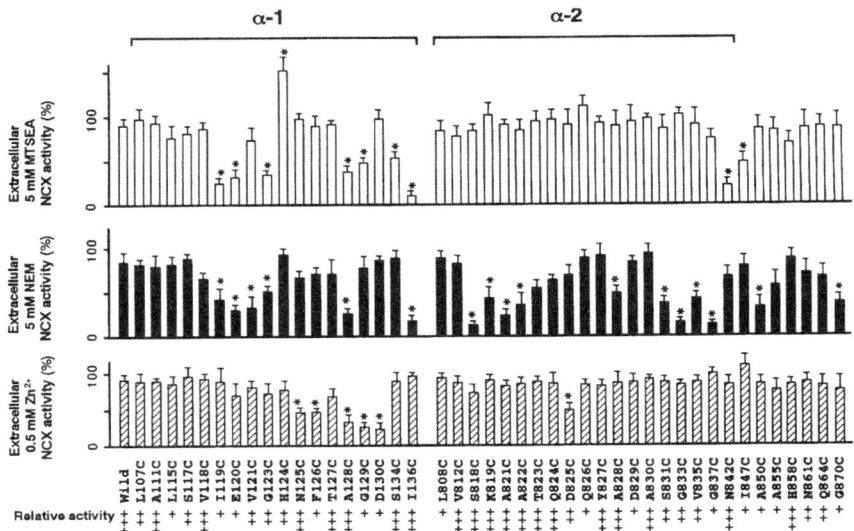

FIGURE 1. Effect of extracellular application of MTSEA, NEM, or Zn^{2+} on the initial rate of Na^+_i-dependent $^{45}Ca^{2+}$ uptake into CCL39 cells expressing cysteine mutants of NCX1. CCL39 cells expressing wild-type NCX1 (*wild*) or cysteine mutants were pretreated with or without either 5 mM MTSEA (*white bars*), 5 mM NEM (*black bars*), or 0.5 mM ZnCl$_2$ (*hatched bars*) during the last 5 (for Zn^{2+} and MTSEA) or 10 min (for NEM) of cell Na$^+$ loading. Data are presented as percentage of the values obtained in the absence of sulfhydryl probes. Uptake activities of transfectant clones are shown below each mutant: +, 0.5 ~ 1 nmol/mg/30 s; ++, 1 ~ 3 nmol/mg/30 s; +++, 3 ~ 8 nmol/mg/30 s. (Modified from Iwamoto et al.[15])

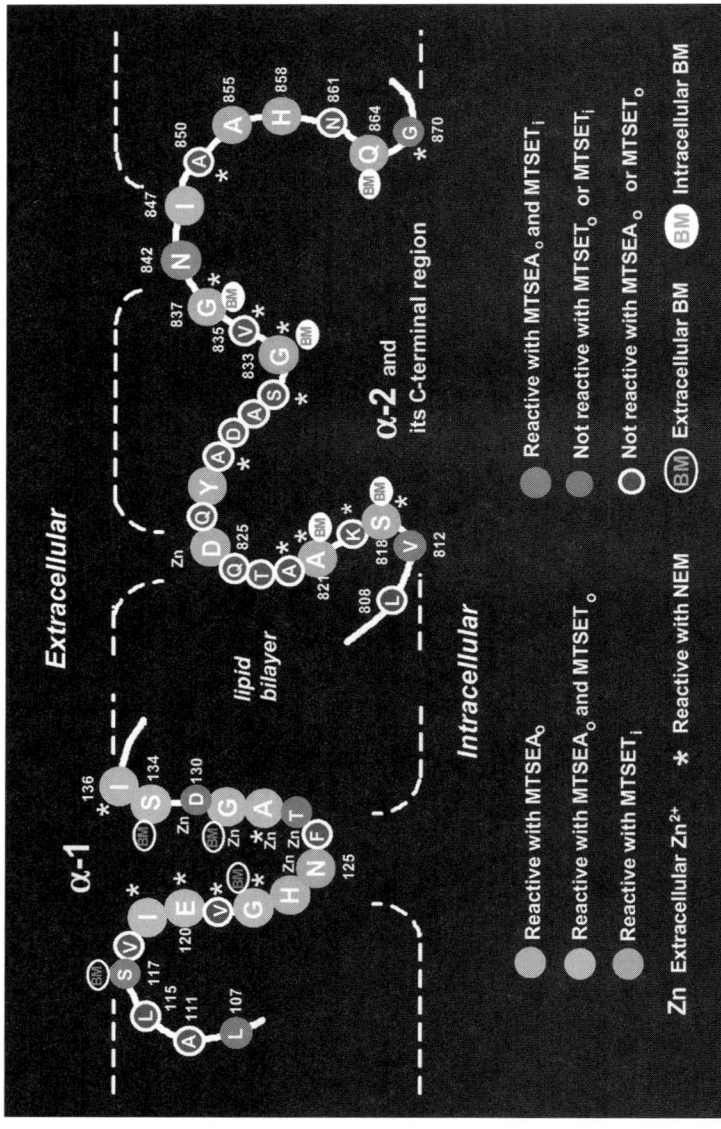

FIGURE 2. See Plate 1 in color section. A summary figure showing accessibility of introduced cysteines in the α-1 and α-2 repeats and the COOH-terminal neighboring region of the latter. Differently colored circles denote differences in the reactivity of substituted cysteines with MTS reagents applied externally or internally. Substituted cysteines labeled with other sulfhydryl probes (Zn^{2+}, NEM, and biotin maleimide [BM]) are also shown in the figure. (Modified from Iwamoto et al.[2,15])

caused an increase in the activity of H124C (FIG. 1; for data with MTSET, see Refs. 2 and 15). These accessible residues therefore are likely to be exposed on the extracellular side. In contrast, substituted cysteines in the α-2 repeat and its COOH-terminal region were mostly inaccessible to these external MTS reagents, although two residues were accessible to MTSEA, but not to MTSET (FIG. 1; see also see Refs. 2 and 15). However, many of the α-2 repeat mutants were inhibited by membrane-permeable NEM, suggesting that these residues are accessible from the intracellular side. Interestingly, one α-2 repeat mutant was inhibited by externally applied metal ions (for data with Hg^+, see Ref. 15).

FIGURE 2 summarizes data from cysteine scanning analyses. The figure contains data from FIGURE 1, those from measurements of the whole-cell outward exchange current in the presence or absence of external or internal MTSET,[2] as well as those from simple chemical labeling experiments in which we examined whether the introduced cysteines into the cysteine-less NCX1 mutant are labeled with membrane-impermeable biotin maleimide applied externally or internally.[15]

In the α-1 repeat, N125C was accessible to internally applied MTSET.[2] On the other hand, this residue and the neighboring residue F126C were accessible to external metal ion, but not to larger external MTS probes or biotin maleimide. D130C was also accessible only to external metal ion. It appears therefore that these residues are located in the narrow, restricted aqueous space accessible only to external small ions. In contrast, T127C was not affected by any of the probes used. The residues labeled light blue in the α-1 repeat (FIG. 2) reacted with external larger MTS probes or biotin maleimide, but not to smaller metal ions, suggesting that they are located on the lining of the larger aqueous space. In contrast, activities of cysteine mutants of residues near the NH_2-terminal end of the α-1 repeat were not affected by these external small and large probes (FIGS. 1 and 2). These accessibility patterns, together with the topological information that the α-1 repeat consists of portions of TM2 and TM3 and a loop connecting these TMs,[2,3] suggest that the α-1 repeat contains a reentrant membrane loop originating from extracellular side. It appears that Asn125, which is accessible from the cytoplasmic side, and some of the neighboring residues might be positioned near the tip of this reentrant structure.

In the α-2 repeat and its immediate COOH-terminal region, none of the cysteine mutants examined were accessible to external MTSET or biotin maleimide, but a few residues were accessible to metal ion (D825C) or MTSEA (N842C and I847C) from the external side. However, 10 cysteine mutants were inhibited by internal MTSET[2] and 11 cysteine mutants were inhibited by NEM. In addition, 5 of 9 mutants were labeled by internal biotin maleimide.[15] Since a total of 16 substituted cysteines were accessible to different probes from the internal side (see FIG. 2), it is likely that the α-2 repeat and its neighboring region contain a complex reentrant loop structure originating from the cytoplasmic side of the membrane. A topology model for NCX1 protein constructed on the basis of these data is shown, together with a model for arrangement of α repeats proposed by Qui et al.[16] (FIGS. 6A and B). The α-2 repeat consists of a portion of putative TM7 and a non-TM loop domain. The arrangement of α repeats, in which two prominent loop structures penetrate the membrane from the opposite sides, reminds us of a somewhat similar structural feature of the water channel aquaporin 1.[17]

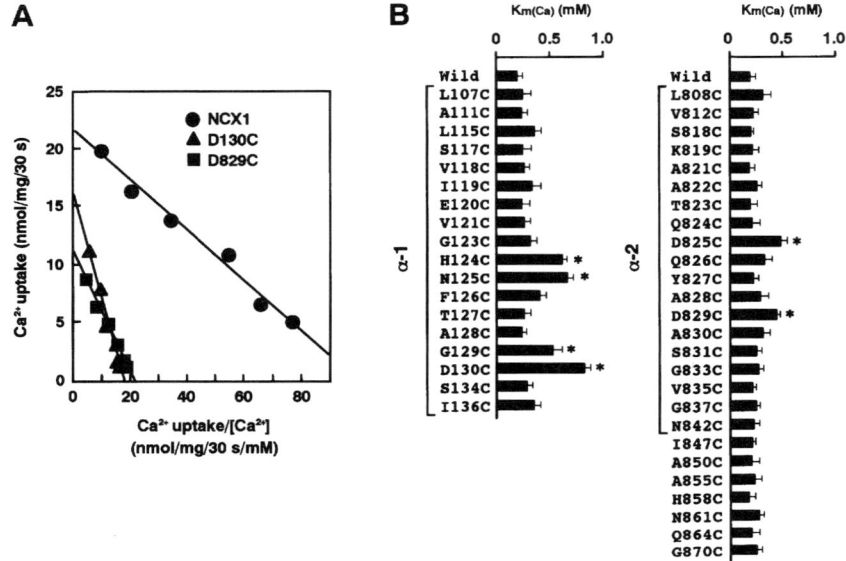

FIGURE 3. The apparent affinity for Ca^{2+}_o of the wild-type NCX1 or cysteine-substituted mutants. Initial rates of Na^+_i-dependent $^{45}Ca^{2+}$ uptake into transfectants were measured as a function of Ca^{2+}_o concentration ranging from 0.067 to 2 mM. $K_{m(Ca)}$ values were determined using an Eadie-Hofstee transformation. (**A**) Eadie-Hofstee plots for the wild-type NCX1, D130C, and D829C. (**B**) $K_{m(Ca)}$ values for cysteine-substituted mutants. Data are mean ± SE. (n = 3). $^*P < 0.05$ *versus* wild-type NCX1. (Data reproduced from Iwamoto *et al.*[15] with permission.)

Interactions of α Repeat Loops with Transport Ions and Modifiers

We examined the functional importance of residues in the α repeat regions by kinetically evaluating alterations in the rate of Na^+_i-dependent Ca^{2+} uptake into Na^+-loaded CCL39 cells expressing cysteine-substituted or other site-directed mutants of NCX1. We searched for residues whose mutation causes change in the affinity for transport substrate Ca^{2+}_o (FIGS. 3A and B). We found that of 44 cysteine-substituted mutants of α repeats, 6 in the loop regions of these repeats exhibited up to 4.6-fold reduction in apparent Ca^{2+}_o affinity of the exchanger, with α-1 repeat mutants exerting greater effects than α-2 repeat mutants. Among these six residues, three aspartic acids (Asps130, 825, and 829) are conserved in the members of the NCX family. Asps130 and 829 were not functionally replaced even with conserved residues Glu or Asn.[15] Asp825, however, could be replaced by Glu or Asn without any effect on Ca^{2+}_o affinity. Asn125 could be replaced by Gly. On the other hand, Ca^{2+}_o affinity decreased further by up to sixfold, when double or triple cysteine substitutions were introduced into these conserved aspartic acids.[15]

Ni^{2+} is a competitive inhibitor with Ca^{2+}_o, while KB-R7943 is a mixed-type inhibitor.[12] As stated in the introduction, Ni^{2+} inhibits NCX1 10-fold more potently than NCX3, while KB-R7943 inhibits NCX1 3-fold less potently compared to

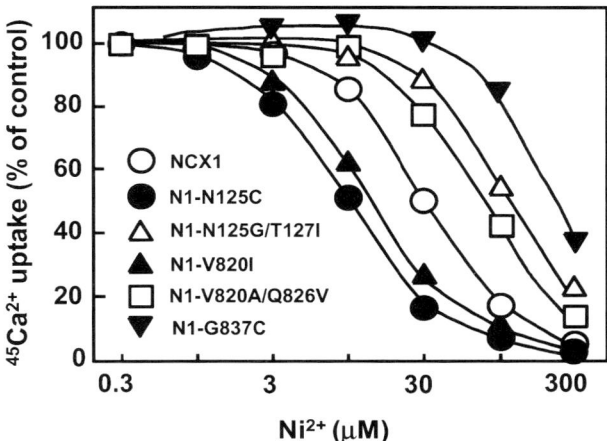

FIGURE 4. Dose–response curves for the effect of Ni^{2+} on the initial rate of Na^+_i-dependent $^{45}Ca^{2+}$ uptake into cells expressing NCX1 mutants. Data are means of 3 independent experiments. (Modified from Iwamoto et al.[14])

NCX3. These NCX isoforms, however, exhibit the same apparent affinity for the transport substrate Ca^{2+}_o or Na^+_o,[12,13] suggesting that the binding sites for substrates and inhibitors are not the same. We employed a chimera strategy between NCX1 and NCX3 to identify the structural domains that confer differential responses to inhibitors. We found that two segments in the exchanger corresponding to α-1 and α-2 repeats are individually responsible for about 4-fold alteration in Ni^{2+} sensitivity, both together accounting for 80% of the difference between the isoforms.[14] In contrast, the segment corresponding to the α-2 repeat fully accounted for the difference in KB-R7943 sensitivity.[18]

Because α repeats are highly conserved, only several residues are different between NCX1 and NCX3. We examined the contribution of each of these residues to Ni^{2+} sensitivity.[14] Swapping of the α-1 repeat between the isoforms was mimicked by simultaneous mutation of Asn125 and Thr127 of NCX1 to the corresponding residues of NCX3 (N125G/T127I), whereas exchange of the α-2 repeat was mimicked by single (Val820) or double mutations of Val820 and Gln826 of NCX1 to the corresponding residues of NCX3 (V820A or V820A/Q826V) (FIG. 4; the result for V820A, not shown). In the case of KB-R7943, the phenotype change between NCX1 and NCX3 occurred when Gln826 in the α-2 repeat loop of NCX1 was exchanged with the corresponding residue of NCX3 (Q826V) (FIG. 5).

Using cysteine-substituted mutants of residues in the α repeats and the surrounding regions, we further searched for amino acid residues that influence Ni^{2+} sensitivity of NCX1. Besides the residues found in the above chimera study, we identified three more residues whose mutation alters Ni^{2+} sensitivity of NCX1 (FIG. 4). D130C and D825C exhibited 8- and 2-fold greater sensitivity to inhibition by Ni^{2+}, respectively, compared to the wild type (data not shown, see Ref. 15), whereas G837C caused a 7-fold reduction in Ni^{2+} sensitivity. Similarly, we searched for residues that

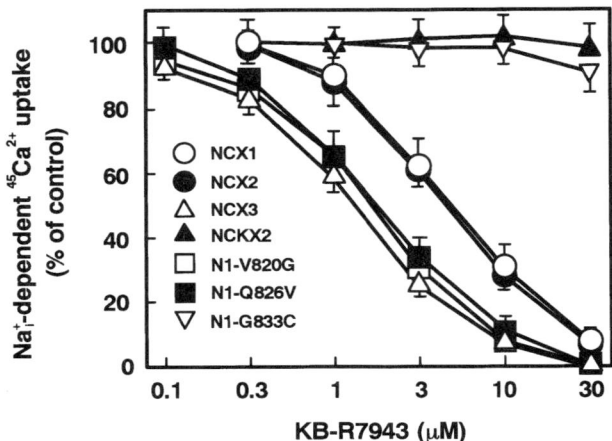

FIGURE 5. Dose–response curves for the effects of KB-R7943 on the initial rate of Na^+_i-dependent $^{45}Ca^{2+}$ uptake into cells expressing wild-type NCX isoforms, NCX1 mutants, and NCKX2. Data are means ± SE of 3 independent experiments. (Data reproduced from Iwamoto et al.[18] with permission.)

influence KB-R7943 sensitivity of NCX1. We found that G833C exhibits a drastically reduced sensitivity to the drug (FIG. 5). When this Gly was mutated to Thr, NCX1 became resistant even to 100 µM KB-R7943, raising an interesting possibility that this residue might form part of the drug-binding site. In contrast, V820G exhibited increased sensitivity to the drug.

Previously, we showed that Ni^{2+} and Co^{2+} exert very similar inhibitory effects on Ca^{2+} transport by NCX isoforms.[12] We found that $^{57}Co^{2+}$ is transported into CCL39 cells expressing NCX3, but not into cells expressing NCX1 (TABLE 1, Kita et al., manuscript in preparation). $^{57}Co^{2+}$ uptake by NCX3 at 3 mM $^{57}Co^{2+}$ at 37°C was inhibited by 70% in the presence of 1 mM Ca^{2+} and by more than 90% in the presence of 10 µM KB-R7943. However, the rate of $^{57}Co^{2+}$ uptake was very slow under these conditions and corresponded to about 1% of the rate of $^{45}Ca^{2+}$ uptake measured at 1 mM $^{45}Ca^{2+}_o$. In order to identify structural domains responsible for the differential response to Co^{2+}, we again used a chimera strategy between NCX1 and NCX3. Because Co^{2+} uptake into cells varied depending on the expression levels of chimeric exchangers, we simultaneously measured the rate of Ca^{2+} uptake at 1 mM Ca^{2+}_o, which is close to V_{max}, and calculated the ratio between the Co^{2+} and Ca^{2+} uptake rates for each chimera. We found that swapping of the α-1 repeat, but not that of the α-2 repeat, between NCX1 and NCX3 permits the occurrence of Co^{2+} uptake in NCX1 (TABLE 1). Simultaneous substitution of Asn125 and Thr127 of the α-1 repeat of NCX1 with the corresponding residues from NCX3 were sufficient for the occurrence of Co^{2+} uptake. These two residues are ones that alter Ni^{2+} sensitivity of NCX1. We also found that the His124 to Cys mutation in the α-1 repeat permits the occurrence of Co^{2+} uptake in NCX1.

TABLE 1. Na^+_i-dependent $^{57}Co^{2+}$ and $^{45}Ca^{2+}$ uptake into CCL39 cells expressing NCX1/NCX3 chimerae and NCX1 variants carrying mutations in the α-1 repeat

	A: $^{57}Co^{2+}$ uptake (nmol/mg/10 min)	B: $^{45}Ca^{2+}$ uptake (nmol/mg/30 s)	Ratio (A/B)
NCX1	0.077 ± 0.079	4.530 ± 0.087	0.017 ± 0.017
N1-3α1[a]	0.660 ± 0.191*	3.697 ± 0.139	0.178 ± 0.051*
N1-3α2[a]	0.060 ± 0.076	1.913 ± 0.095	0.031 ± 0.040
N1-3α1/α2[a]	0.514 ± 0.100*	3.088 ± 0.366	0.166 ± 0.032*
NCX3	0.829 ± 0.084*	3.452 ± 0.044	0.240 ± 0.024*
N1-V118L	0.014 ± 0.003	1.155 ± 0.047	0.012 ± 0.003
N1-N125G	0.006 ± 0.002	0.467 ± 0.109	0.012 ± 0.004
N1-T127I	0.448 ± 0.017*	6.477 ± 0.112	0.069 ± 0.003
N1-N125G/T127I	0.597 ± 0.160*	4.396 ± 0.174	0.136 ± 0.037*
N1-V121C	0.099 ± 0.024	2.629 ± 0.073	0.038 ± 0.009
N1-G123C	0.358 ± 0.035*	3.527 ± 0.081	0.101 ± 0.010
N1-H124C	1.124 ± 0.142*	4.771 ± 0.133	0.236 ± 0.030*
N1-N125C	0.450 ± 0.096*	5.415 ± 0.214	0.083 ± 0.018

[a]N1-3α1, N1-3α2, and N1-3α1/α2 are chimeric exchangers that contain amino acid residues 109–133, 788–829, and 109–133/788–829 of NCX3, respectively, in the background of NCX1. Values are means ± SE.
*$P < 0.05$ vs. NCX1.

Monovalent cations are known to increase the V_{max} of NCX activity.[4] Li^+-dependent activation is 2-fold greater in NCX3 compared to NCX1.[12] Studies with chimeric and site-directed mutant exchangers revealed that such difference in the response is determined by the origin of the α-2 repeat and was fully mimicked by simultaneous substitution of Val820 and Gln826 of NCX1 with the corresponding residues from NCX3.[14]

In our study of the topological disposition of residues by substituted cysteine scanning analysis, we observed that many cysteine mutants in the loops of α repeats react with sulfhydryl reagents with an inhibition of exchange activity (FIGS. 1 and 2). It is thus likely that these susceptible residues are localized at the important positions for exchange activity. FIGURE 6A further shows important residues that were identified in our study of the functions of cysteine-substituted and other site-directed mutants of the α repeats.

In the α-1 repeat loop of NCX1, we identified several residues whose mutation cause up to 5-fold reduction in the apparent affinity for Ca^{2+}_o. Among them, the conserved Asp130 seems to be important for the interaction with Ca^{2+}_o, although it is not essential for exchange activity. When this Asp and Asn125 are mutated to Cys, the intrinsic inhibitory potency of Ni^{2+}, which is an apparent inhibitory constant (K_i) corrected for the competition of this cation with the transport substrate Ca^{2+}_o, increased markedly by 8- and 3-fold, respectively.[14,15] On the other hand, when Asn125 and near-by Thr127 of NCX1 are simultaneously replaced by the corresponding residues (Gly and Ile) from NCX3, the exchanger is now able to transport

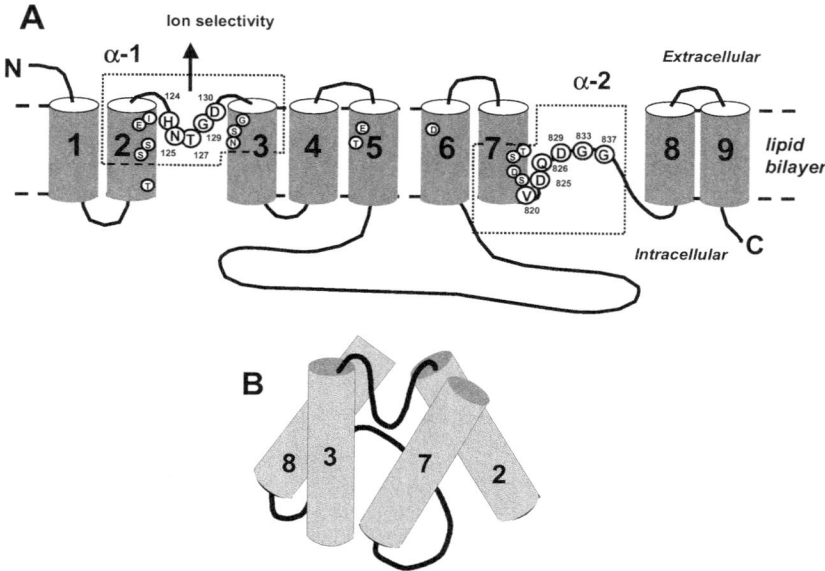

FIGURE 6. Topological models for NCX1 and summary for functional studies with mutants. **(A)** Putative transmembrane helices are indicated by cylinders with Arabic numbers. N and C indicate NH$_2$- and COOH-terminals, respectively. Locations of predicted functionally important residues are shown. **(B)** Model for helix packing of TMs 2, 3, 7, and 8 suggested by cross-linking data. (Modified from Qui et al.[16])

Co^{2+}, indicating that these mutations alter substrate selectivity of the exchanger. Furthermore, these same mutations in NCX1 reduce Ni^{2+} sensitivity about 4-fold without influencing Ca$^{2+}_o$ affinity. Therefore, a portion of the reentrant loop of the α-1 repeat containing these residues seems to play an important role in ion selection, thus forming part of the ion transport pore.

In the loop region of the α-2 repeat of NCX1, mutations of conserved Asp825 and Asp829 cause a small reduction of Ca$^{2+}_o$ affinity with little effect on Ni^{2+} sensitivity.[15] On the other hand, when Val820 of NCX1 is replaced by the corresponding residue Ala from NCX3, Ni^{2+} sensitivity decreases about 4-fold compared to the wild-type NCX1, without influencing the affinity for Ca$^{2+}_o$ or Na^+_o.[14] When the same Val820 to Ala mutation was combined with substitution of Gln826 with Val from NCX3 in the NCX1 background, the V$_{max}$ of Li$^+$-dependent activation increased 2-fold compared to the wild-type NCX1. In addition, Cys substitution of G837 in NCX1 results in a 7-fold reduction in Ni^{2+} sensitivity and at the same time in a large increase in the Li$^+$-dependent activation of the exchanger (Iwamoto et al., unpublished observation). Therefore, the reduction of Ni^{2+} sensitivity and the increase in Li$^+$-dependent activation seem to be somehow related in the α-2 repeat loop, although this is not true for the α-1 repeat loop. Finally, mutations of Val820 and Gln826 of NCX1 individually increased the sensitivity to KB-R7943 by 3-fold, while mutation of Gly833 renders the exchanger practically insensitive to the drug.

It is intriguing to note that mutations at limited residues in the α-2 repeat loop are able to produce complex effects on the interaction of NCX1 with inhibitors and an activator. All these ions obviously act from the external side. We have evidence that KB-R7943 also acts from the external side.[18] Although the location of the actual binding sites for these ions and the drug are currently unknown, available evidence seems to suggest that at least external Ca^{2+}, Ni^{2+}, and Co^{2+} interact with the sites in the α-1 repeat (see above). Therefore, the observed effects of mutations in the α-2 repeat loop on exchange activity seem to be largely indirect, occurring through a conformational change of the α-2 repeat region.

Nicoll et al.[10] reported that exchange activity is lost or greatly reduced when mutation is introduced into acidic or hydroxyl-containing residues in the putative TM segments of the α-1 and α-2 repeats and neighboring TM regions (FIG. 6A). We essentially confirmed these results. As many of these residues in the membrane region are conserved in the NCX and related cation exchanger families, they could play important roles in the interaction of the exchanger with substrate ions and modifiers. However, detailed analysis of the functions of these residues is difficult because of low activities of the mutant exchangers.

CONCLUSION

Ion translocation during Na^+/Ca^{2+} exchange involves reaction steps such as ion binding, ion occlusion/translocation, and ion release at the opposite side of the membrane. At present, we know little about the structural basis for these reactions. In addition, there is little information about the shape and dimension of the overall ion transport pathway in the NCX protein. Our data provide evidence suggesting that the loop of the α-1 repeat plays an important role in ion selection and is involved in the interaction with the transport substrate and inhibitory ions. Some of the residues in TMs of the α-1 and α-2 repeats and neighboring regions are also likely to participate in these ion interactions.

REFERENCES

1. PHILIPSON, K.D. & D.A. NICOLL. 2000. Sodium-calcium exchange: a molecular perspective. Annu. Rev. Physiol. **62:** 111–133.
2. IWAMOTO, T., T.Y. NAKAMURA, Y. PAN, et al. 1999. Unique topology of the internal repeats in the cardiac Na^+/Ca^{2+} exchanger. FEBS Lett. **446:** 264–268.
3. NICOLL, D.A., M. OTTOLIA, L. LU, et al. 1999. A new topological model of the cardiac Na^+-Ca^{2+} exchanger. J. Biol. Chem. **274:** 910–917.
4. SHIGEKAWA, M. & T. IWAMOTO. 2001. Cardiac Na^+–Ca^{2+} exchange: molecular and pharmacological aspects. Circ. Res. **88:** 864–876.
5. KIMURA, J., A. NOMA & H. IRISAWA. 1986. Na–Ca exchange current in mammalian heart cells. Nature **319:** 596–597.
6. DIPOLO, R. & L. BEAUGÉ. 1987. Characterization of the reverse Na/Ca exchange in squid axons and its modulation by Ca_i and ATP: Ca_i-dependent Na_i/Ca_o and Na_i/Na_o exchange modes. J. Gen. Physiol. **90:** 505–525.
7. HILGEMANN, D.W., S. MATSUOKA, G. A. NAGEL & A. COLLINS. 1992. Steady-state and dynamic properties of cardiac sodium-calcium exchange: sodium-dependent inactivation. J. Gen. Physiol. **100:** 905–932.

8. IWAMOTO, T., Y. PAN, S. WAKABAYASHI, et al. 1996. Phosphorylation-dependent regulation of cardiac Na^+/Ca^{2+} exchanger via protein kinase C. J. Biol. Chem. **271:** 13609–13615.
9. SCHWARZ, E.M. & S. BENZER. 1997. Calx, a Na–Ca exchanger gene of *Drosophila melanogaster*. Proc. Natl. Acad. Sci. USA **94:** 10249–10254.
10. NICOLL, D.A., L.V. HRYSHKO, S. MATSUOKA, et al. 1996. Mutation of amino acid residues in the putative transmembrane segments of the cardiac sarcolemmal Na^+–Ca^{2+} exchanger. J. Biol. Chem. **271:** 13385–13391.
11. DOERING, A.E., D.A. NICOLL, Y. LU, et al. 1998. Topology of a functionally important region of the cardiac Na^+/Ca^{2+} exchanger. J. Biol. Chem. **273:** 778–783.
12. IWAMOTO, T. & M. SHIGEKAWA. 1998. Differential inhibition of Na^+/Ca^{2+} exchanger isoforms by divalent cations and isothiourea derivative. Am. J. Physiol. **275:** C423–C430.
13. LINK, B., Z. QIU, Z. HE, et al. 1998. Functional comparison of the three isoforms of the the Na^+/Ca^{2+} exchanger (NCX1, NCX2, NCX3). Am. J. Physiol. **274:** C415–C423.
14. IWAMOTO, T., A. UEHARA, T.Y. NAKAMURA, et al. 1999. Chimeric analysis of Na^+/Ca^{2+} exchangers NCX1 and NCX3 reveals structural domains important for differential sensitivity to external Ni^{2+} or Li^+. J. Biol. Chem. **274:** 23094–23102.
15. IWAMOTO, T., A. UEHARA, I. IMANAGA & M. SHIGEKAWA. 2000. The Na^+/Ca^{2+} exchanger NCX1 has oppositely oriented reentrant loop domains that contain conserved aspartic acids whose mutation alters its apparent Ca^{2+} affinity. J. Biol. Chem. **275:** 38571–38580.
16. QIU, Z., D.A. NICOLL & K.D. PHILIPSON. 2001. Helix packing of functionally important regions of the cardiac Na^+–Ca^{2+} exchanger. J. Biol. Chem. **276:** 194–199.
17. MURATA, K., K. MITSUOKA, T. HIRAI, et al. 2000. Structural determinants of water permeation through aquaporin-1. Nature. **407:** 599–605.
18. IWAMOTO, T., S. KITA, A. UEHARA, et al. 2001. Structural domains influencing sensitivity to isothiourea derivative inhibitor KB-R7943 in cardiac Na^+/Ca^{2+} exchanger. Mol. Pharmacol. **59:** 524–531.

Ion Channel–like Properties of the Na^+/K^+ Pump

PABLO ARTIGAS AND DAVID C. GADSBY

Laboratory of Cardiac/Membrane Physiology, The Rockefeller University, New York, New York 10021, USA

ABSTRACT: Ion pumps and exchangers are considered to be different from ion channels for two principal reasons. Ion pumps move ions *against*, whereas ion channels allow ions to move *with*, the electrochemical potential gradient, and pumps transport ions relatively slowly, ~10^2 s^{-1}, whereas channels conduct ions rapidly, ~10^7 s^{-1}. However, the latter high rate refers only to the open pore, and yet all ion channels contain at least one gate. Not surprisingly, the conformational changes associated with channel gating occur with kinetics similar to those of ion pumping. Indeed, ion pumps may be viewed as ion channels with two gates, one external to, and the other internal to, the ion binding cavity. The simple operational rule for such a pump is that the two gates should never be open simultaneously; otherwise, the pump would become a channel and conduct dissipative fluxes several orders of magnitude larger than, and in the opposite direction to, the active transport fluxes. Analyses of Na^+ ion movements mediated by the Na^+/K^+ pump under various conditions have suggested that in at least one, short-lived, conformation of the pump, an ion-channel-like structure, closed at its intracellular end, connects the extracellular solution with the ion binding sites deep in the protein core. Here we use the marine toxin, palytoxin, to act on Na^+/K^+ pumps in outside-out patches excised from cardiac myocytes and so transform the pumps into nonselective cation channels which we study using macroscopic, and single-channel, recording. We find that gating of the palytoxin-induced channels is regulated by the pump's natural ligands. Thus, external K^+ congeners tend to close, and external Na^+ tends to open, an extracellular gate, whereas ATP acts from the cytoplasmic solution to open an intracellular gate. These gating influences echo the normal ion occlusion and deocclusion reactions that first entrap two extracellular K^+ ions within the interior of the pump (between the two gates) and then release them to the cytoplasmic side in a step accelerated by ATP. These results offer the promise of being able to examine ion occlusion and deocclusion steps at the microscopic level in single Na^+/K^+ pump molecules.

KEYWORDS: Na^+,K^+-ATPase; Na^+/K^+ pump; ion channel; intracellular gate; extracellular gate; gating; conformational change; ion occlusion; deocclusion; palytoxin; ventricular myocyte; outside-out patch; macroscopic current; microscopic current; single-channel current

---

Address for correspondence: David C. Gadsby, Laboratory of Cardiac/Membrane Physiology, The Rockefeller University, 1230 York Avenue, New York, NY 10021-6399. Voice: 212-327-8680; fax: 212-327-7589.
 gadsby@mail.rockefeller.edu

Ann. N.Y. Acad. Sci. 976: 31–40 (2002). © 2002 New York Academy of Sciences.

INTRODUCTION

Ion-motive pumps, ion exchangers, and ion channels are traditionally viewed as being very different molecular entities. This is despite the fact that they are all integral membrane proteins, presumably with predominantly helical membrane domains, and that they all somehow provide an energetically favorable pathway for shepherding ions across the otherwise inhospitable phospholipid bilayer. A major reason for the perception that ion pumps and ion channels are so different is that a channel's ion-selective pore conducts dissipative ion flow, down the electrochemical gradient, at a high rate (e.g., 10^7 to 10^8 s^{-1}), whereas an ATPase pump moves ions against their electrochemical gradient, and far more slowly (e.g. $\sim 10^2$ s^{-1}). But the high flux rates apply only to open ion channels, and all channels contain at least one gate that allows the ion flow to be turned on and off as required. Opening and closing of the gate involve substantial conformational changes that are generally relatively slow ($\sim 10^2$ to 10^3 s^{-1}) and require input of energy; the source varies depending on whether the channel is gated by changes in membrane voltage, by binding and unbinding of an extra- or intracellular ligand, or by direct mechanical distortion. Regardless of the channel type, the conformational changes that accompany its opening and closing need not differ greatly in extent or rate from those associated with the ion movements effected by a pump.

What kind of conformational changes does the Na^+,K^+-ATPase (or Na^+/K^+ exchange pump) undergo during its normal transport cycle, through which it extrudes 3 Na^+ from a cell and recovers 2 K^+ at the expense of a single molecule of ATP (FIG. 1)? Within the cycle the Na^+/K^+ pump adopts two major conformations: E_1, in which binding sites are accessible from the cytoplasmic surface, and E_2, in which access is from the extracellular surface. Both conformations exist in phosphorylated and dephosphorylated forms, and it is the strict coupling of ion binding steps to phosphorylation and dephosphorylation steps that ensures the normal forward cycling that effects Na^+/K^+ exchange. Thus, upon binding of the third cytoplasmic Na^+ ion to the E_1 conformation, the γ-phosphate is transferred from ATP (acting with high apparent affinity, ≤ 1 μM) to form the phosphorylated E_1P (Na_3) state, in which three Na^+ ions are occluded within the protein and are inaccessible from either side of the membrane (symbolized by the parentheses). A spontaneous relaxation to the E_2P state then results in release of the three Na^+ ions to the exterior, reflecting the reduced apparent affinity of the binding sites for external Na^+ ions. Extracellular K^+ ions bind with high affinity to the E_2P conformation, and binding of the second K^+ ion triggers dephosphorylation of the pump and concomitant occlusion of the two K^+ ions. The overall transport cycle is rate-limited by the subsequent conformational transition of the dephosphorylated enzyme, $E_2(K_2) \rightarrow E_1 \cdot K_2$, which deoccludes the K^+ ions, releasing them to the cell interior in exchange for Na^+ ions. That deocclusion step is accelerated by binding of ATP acting with relatively low apparent affinity (~ 100 μM).

The single net positive charge exported in each Na^+/K^+ transport cycle results in a steady-state outward current across the membrane of ~ 20 attoAmp per pump, far too small to observe in individual pumps, but readily measured in cells containing millions of pumps.[1–4] Na^+/K^+ pump current is usually measured as the component of membrane current abolished by sudden and complete inhibition of the pump by a cardiotonic steroid, like ouabain. As evidence suggests that the $3Na^+:2K^+$ transport

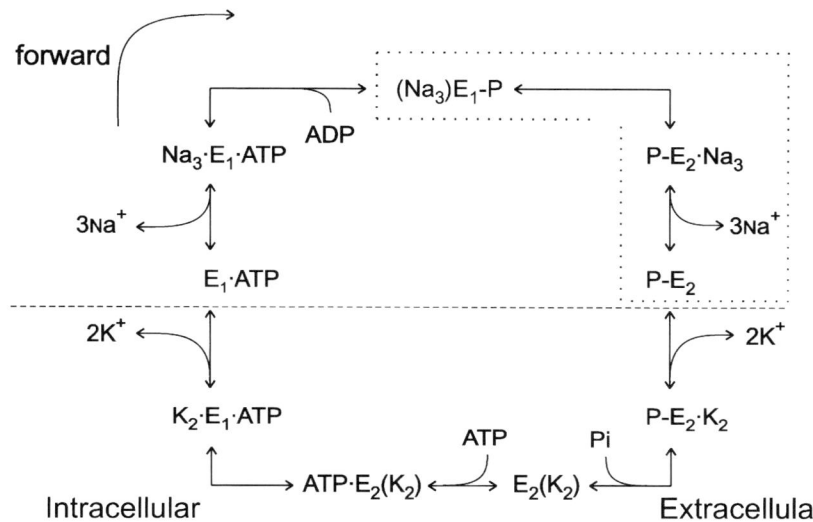

FIGURE 1. Cartoon (taken from Holmgren et al.[21]) showing simplified normal Post-Albers 3 Na^+/2 K^+ transport cycle of the Na^+,K^+-ATPase, which can assume two principal conformations: in E_1 the ion-binding sites face the cytoplasm and show high apparent affinity for Na^+, and in E_2P the ion-binding sites open to the extracellular solution, and apparent affinity is low for Na^+ but high for K^+. Phosphorylation by ATP of the pump in the E_1 form occludes (traps within the protein core) three Na^+ ions, which are released to the external medium after the conformational change to E_2P, whereupon 2 K^+ ions bind, leading to dephosphorylation of the protein and K^+ occlusion. Binding of ATP at a low-affinity site favors the subsequent transition back to the E_1 state, which prompts K^+ release to the cytoplasm, completing the cycle. The *horizontal dashed line* separates Na^+- and K^+-translocation pathways. The states enclosed by the *dotted box* yield voltage-jump-induced charge movements, associated with deocclusion and release to the exterior of the three transported Na^+ ions.

stoichiometry remains fixed over a broad range of conditions,[5–7] the amplitude of Na^+/K^+ pump current provides a direct measure of the turnover rate of the transport cycle. In principle, every step in the cycle is reversible, and with appropriate ion and nucleotide concentrations, the cycle can run backwards,[8] extruding K^+, importing Na^+, making ATP from ADP and P_i, and generating inward pump current.[9–12] In addition, the occluded intermediates E_1P (Na_3) and E_2 (K_2) underlie the well-characterized partial reactions of the Na^+/K^+ pump, electroneutral (i.e., no net charge movement) $3Na^+/3Na^+$ exchange[13–14] (in the absence of K^+ ions) and electroneutral $2K^+/2K^+$ exchange[10,15,16] (in the absence of Na^+ ions), which involve exclusively steps in the upper half, or lower half of FIGURE 1, respectively. Studies of the influence of membrane potential and extracellular [Na^+] on the rates of forward $3Na^+/2K^+$ exchange (size of outward pump current[7,17]), backward $2K^+/3Na^+$ exchange (size of inward pump current[12]), and electroneutral $3Na^+/3Na^+$ exchange (size of pump-mediated ^{22}Na efflux[14]), as well as on the size and kinetics of transient charge movements associated with deocclusion and release of 3 Na^+ ions to the exterior,[18–21] all suggest that in one short-lived conformation of the pump, a narrow channel-like structure connects the extracellular solution with the ion binding sites in the protein core. Moreover, in negotiating that channel at least one of the Na^+ ions traverses more than two-thirds of the membrane's electrical field.[12,14,18–21]

The notion of a channel-like conformation of the Na^+/K^+ pump has also been raised in connection with electrical recordings on purified Na^+/K^+ pumps incorporated into phospholipid bilayers,[22] and on Na^+/K^+ pumps in *Xenopus* oocytes exposed to low pH solutions.[23,24] Acknowledging the idea that pumps and channels may be more closely related than generally assumed, an ion pump can be viewed as a modified ion channel with at least two gates (one external, and the other internal, to the binding sites) that open alternately but never simultaneously.[25] This implies tight coupling between the two gates. The fact that a pump actively transports $\sim 10^2$ ions s^{-1}, whereas if both gates were open simultaneously it would become a channel capable of conducting passive (i.e., in the opposite direction) movement of $\sim 10^7$ ions s^{-1}, argues that the probability of the gates being open simultaneously must be $<10^{-5}$ in a useful pump.

Intriguingly, the marine toxin palytoxin (PTX) has been shown to interact with the Na^+/K^+ pump and thereby generate a relatively nonselective monovalent cation channel.[26] Work with transfected yeast cells, which lack a native Na^+/K^+ pump, confirmed that the target of PTX action is indeed the Na^+/K^+ pump,[27,28] and that phosphorylation of the pump is not required for PTX action.[29] Also, comparison of the results of recent cysteine-scanning accessibility studies[30,31] with inferences based on the crystal structure of the related SERCA Ca^{2+} pump,[32] and with functional consequences of Na^+/K^+ pump mutagenesis,[33,34] suggest that the PTX-induced dissipative flux pathway comprises at least part of the pathway traveled by the normally transported Na^+ and K^+ ions.

Using outside-out patches excised from cardiac myocytes, we have confirmed that PTX elicits dissipative cation flow fast enough ($\sim 10^7$ ions s^{-1}) to be measured in a single Na^+/K^+ pump.[35,36] We find that this ion flow is interrupted by occasional closures of one or the other gate. In particular, external K^+ or Cs^+ tends to close, and external Na^+ to open, the extracellular gate, whereas nucleotides act from the cytoplasmic solution to open the intracellular gate. We conclude that PTX transforms the Na^+/K^+ pump into a channel in which gating seems to reflect the normal ion occlusion and deocclusion reactions that alternately entrap first 3 Na^+, and then 2 K^+, within the interior of the pump, between the two gates.

METHODS

Macroscopic and microscopic membrane currents were recorded at room temperature (22.5–24.5 °C) in outside-out patches excised from guinea-pig ventricular myocytes, which were isolated as described.[37] The 10–30 MΩ pipettes used for both unitary and macroscopic current recording were pulled from thin-walled borosilicate glass (PG52151-4, WPI Sarasota, FL), coated with Sylgard, and used without fire-polishing. The composition of the pipette (cytoplasmic surface) solution was (in mM): 150 Na^+, 130 glutamic acid, 10 HEPES, 10 EGTA, 10 TEACl, 1 $MgCl_2$ without or with 5 mM MgATP as indicated in each figure. The external solution contained (mM): 140 sulfamic acid, 10 HEPES, 10 HCl, 1 $MgCl_2$, 1 $CaCl_2$, 0.5-5 $BaCl_2$, and 160 mM Na^+, K^+, or Cs^+. For all solutions, pH = 7.4 and osmolality = 285–305 mOsm kg^{-1}. An aliquot of 100 μM PTX (*Palythoa toxica*, Calbiochem) stock solution (in water) was defrosted and diluted in the external solution just before each experiment; 0.002% BSA was added to all PTX-containing solutions to minimize binding of PTX to non-glass surfaces.

RESULTS AND DISCUSSION

In the presence of roughly symmetrical 150–160 mM Na$^+$-containing solutions, and with 5 mM MgATP at the cytoplasmic surface (in the pipette), exposure of an outside-out patch of myocyte membrane to 100 nM PTX (a saturating concentration under these conditions) resulted in a large, rapid increase in inward current at the –40 mV holding potential (FIG. 2A). Steady-state current–voltage relationships, obtained by plotting current measured towards the end of 80-ms voltage steps to potentials

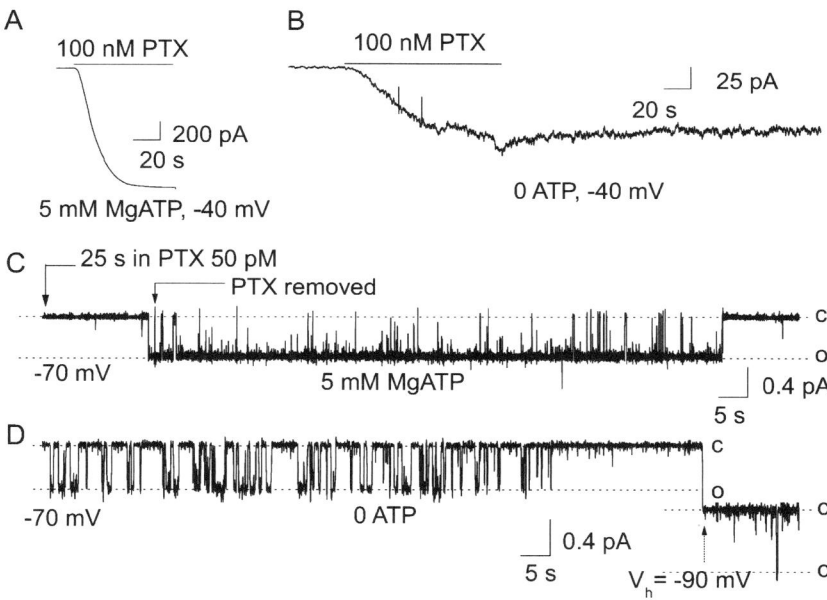

FIGURE 2. (**A** and **B**) Macroscopic currents induced by application of 100 nM PTX to outside-out patches excised from ventricular myocytes, bathed in 150–160 mM Na$^+$ solutions, and held at –40 mV, with 5 mM MgATP (**A**) or with no ATP (**B**) present in the internal solution; note different current scales. (**C** and **D**) Representative PTX-induced single-channel recordings from outside-out myocyte patches, held at –70 mV and bathed in 150–160 mM Na$^+$ solutions. (**C**) A patch with 5 mM MgATP in the pipette was exposed to 20 pM PTX, which was quickly removed once the channel opened. Long open bursts characterized the gating behavior of the channel, which remained active for ~2 hr. (**D**) Trace from a patch without pipette ATP, showing gating behavior shortly after removal of unbound PTX: openings occurred in shorter bursts than seen in the presence of MgATP, and longer closed periods were more frequent.

over the range −100 to +80 mV, with 160 mM Na^+, K^+, or Cs^+ in the external solution showed that this PTX-induced current shift reflected activation of a relatively nonselective cation conductance with a reversal potential near 0 mV. When similarly sized pipettes contained no MgATP, however, the same high concentration of PTX consistently activated a smaller amplitude conductance, and did so somewhat more slowly (FIG. 2B; note expanded current scale).

To examine the microscopic events responsible for these PTX-induced conductance changes, we applied a 2,000-fold lower concentration of PTX to outside-out patches of the usual size (judged from pipette resistance) with 150 mM Na^+ and 5 mM MgATP in the pipette and 160 mM Na^+ in the external solution, waited until a channel opened, and then quickly washed away unbound toxin to prevent formation of other channels (FIG. 2C). Despite withdrawal of toxin from the external solution, the channel in FIGURE 2C continued to open and close for the entire 2-hr period of recording before the seal was lost. A correspondingly slow decline of macroscopic current was seen when PTX was washed out after activating a steady level of conductance under comparable conditions (e.g., as in FIGURE 2A), indicating that dissociation of PTX occurs extremely slowly when Na^+ ions are present on both sides of the membrane and MgATP is available in the cytoplasmic solution. Single-channel current–voltage relationships revealed that the channel conductance at negative potentials was ∼7 pS in these solutions, similar to values previously reported for PTX-induced channels in mammalian myocytes.[35,36] As illustrated by this ∼2-min record (FIG. 2C), the channel spent most of its time in the open state (dashed line labeled "O"), but openings were interrupted by both short (<1 s) and long (several seconds) closed periods. With open probability so high in the presence of Na ions and MgATP, we can use the single-channel current amplitude in FIGURE 2C to estimate that the patch in FIGURE 2A contained almost 5,000 Na^+/K^+ pumps.

The record in FIGURE 2D was obtained from a different outside-out patch, also exposed to 150 mM internal Na^+ and 160 mM external Na^+, but with no MgATP in the pipette. In this case, 100 pM PTX had been applied ∼100 s before the beginning of the record, but had then been removed ∼50 s later once the channel had begun opening and closing. Although current–voltage relationships confirmed that the single-channel conductance (also ∼7 pS) was unaltered, it is evident that the fraction of time spent in the open state was markedly reduced in the absence of MgATP. Presumably, this reduced open probability of PTX-induced channels without nucleotide is a major factor underlying the smaller amplitude of the macroscopic current in 0 ATP compared to that in 5 mM MgATP (FIG. 2A). Although the distinct reopening of the channel 10–15 s after the voltage step to −90 mV confirms that the PTX was still bound to that Na^+/K^+ pump after the prolonged ∼30-s closed period, activity of this channel ceased after a further minute or so, likely because of eventual dissociation of the PTX. The slow but discernible decay of the macroscopic current following withdrawal of PTX in the absence of ATP (FIG. 2B) is similarly indicative of more rapid PTX unbinding in the absence of ATP than in its presence.

Unbinding of PTX is much faster still when extracellular Na^+ ions are replaced by K^+, or its congener Cs^+, ions (FIG. 3). The relatively slow opening of PTX-induced channels (despite the high, 200-nM, PTX concentration) with high [Na^+] on both sides of the membrane but with no ATP, already illustrated in FIGURE 2B, is recapitulated in FIGURE 3A and B under identical conditions. However, in contrast to the fairly slow current decay seen upon washout of PTX in the presence of external

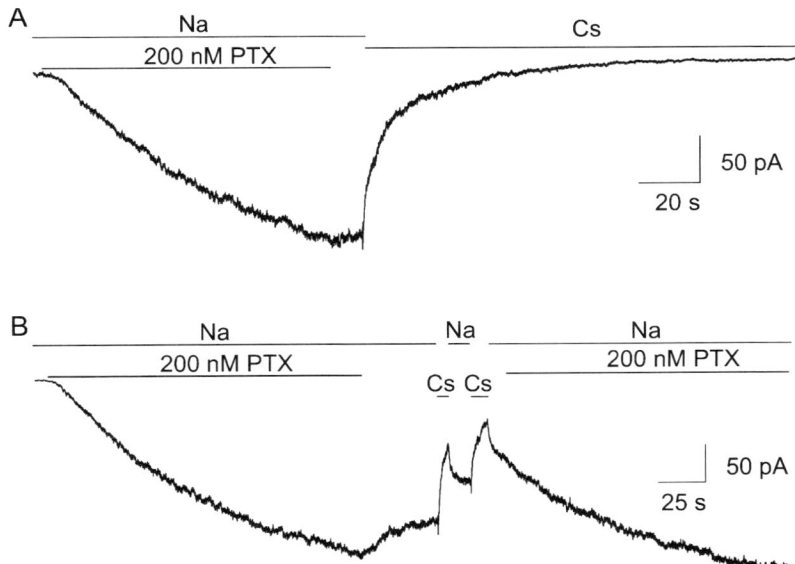

FIGURE 3. (**A**) Fast closing of PTX-induced channels by Cs$^+$ in the absence of internal ATP monitored by decay of macroscopic currents. 200 nM PTX slowly elicited a conductance with characteristics similar to that in FIGURE 2B. Replacing external Na$^+$ with Cs$^+$ caused (after a small transient spike-like increase in current reflecting the slightly higher permeability of Cs$^+$ ions) a double exponential conductance decay. (**B**) Distinction between reversible channel closing and irreversible (due to washout of PTX) unbinding of PTX. Brief exposures to Cs$^+$ induced rapid closing, like the fast component shown in **A**, but upon return to Na$^+$, channels with PTX still bound reopened. Thus, PTX unbinding is faster in external Cs$^+$ than in Na$^+$, but channel closing and PTX unbinding are distinct, though related, phenomena.

Na$^+$ (FIGS. 2B and 3B), a switch to external Cs$^+$ solution caused a rapid decline of the current regardless of whether the Cs was applied soon after the removal of PTX from the bath solution (FIG. 3A), or some time after PTX was withdrawn (FIG. 3B). The full current decay in FIGURE 3A (i.e., closure of all the channels that had been opened by PTX) occurred in two phases, one fast and the other almost an order of magnitude slower. The record in FIGURE 3B demonstrates that these two phases are separable. Thus, a brief exposure to Cs$^+$ ions, long after unbound PTX had been washed away, resulted in a rapid current decline (preceded by a small, very brief, current increase that reflects the slightly higher permeability of the PTX-induced channels to Cs$^+$ ions than to Na$^+$ ions). But, upon switching back to Na$^+$ solution, a large part (almost half) of the current was restored (following the small transient current reduction reflecting the permeability difference just mentioned). A second brief exposure to Cs$^+$, and return to Na$^+$, again caused the same sequence of events. As PTX was absent from the bath, the relatively steady current level during each exposure to Na$^+$ reflected the fraction of Na$^+$/K$^+$ pumps with PTX still bound. This fraction was evidently substantially reduced after each brief sojourn in Cs$^+$ solution, indicating that Cs^+ ions accelerated dissociation of PTX. That the stepwise reduction in current in Na solution reflected unbinding of PTX was confirmed by readmitting PTX in the presence of Na (FIG. 3B, right), when the current gradually increased again just as it had during the initial exposure to PTX.

However, equally important, the restoration of current upon switching back to Na^+ solution from Cs^+ solution must reflect reopening of channels in Na^+ that had been temporarily closed by the preceding exposure to Cs^+. This means that Cs^+ ions can shut some PTX-induced channels, and Na^+ can reopen them, all while PTX remains bound. Taken together with the results in FIGURE 2, which demonstrate that intracellular MgATP enhances opening of PTX-bound channels, the findings presented here suggest that despite the presence of PTX the Na^+/K^+ pump remains capable of undergoing some of its normal reaction steps. These appear to include the occlusion step mediated by extracellular K^+ congeners, and the deocclusion step accelerated by a high cytoplasmic concentration of MgATP (see FIGURE 1, above).

The earliest work with PTX was unable to discern whether the nonselective cation pathway opened by the toxin was closely or distantly related to that traversed by the normally transported ions. The recent demonstration that several cysteines introduced into transmembrane helices 5 and 6 of the Na^+/K^+ pump were accessible to small hydrophilic sulfhydryl reagents,[30,31] some at locations already pinpointed by mutational analysis and corresponding to residues revealed to coordinate the transported Ca^{2+} ions in the related Ca^{2+}-ATPase,[32] implies that the PTX-induced channel includes at least part of the normal transport pathway. The findings briefly summarized here suggest that the PTX-induced channel is also gated by conformational changes that, at least qualitatively, mimic those that accompany occlusion and deocclusion of K^+ ions during normal forward Na^+/K^+ transport by the unmodified pump. If correct, this conclusion implies that PTX may be a powerful tool to examine the molecular mechanisms of ion occlusion and deocclusion in single Na^+/K^+ pump molecules.

ACKNOWLEDGMENTS

We thank C. Basso and M. Holmgren for helpful discussion. This work was supported by NIH Grant HL-36783.

REFERENCES

1. THOMAS, R.C. 1969. Membrane current and intracellular sodium changes in a snail neurone during extrusion of injected sodium. J. Physiol. **201:** 495–514.
2. GADSBY, D.C., J. KIMURA & A. NOMA. 1985. Voltage dependence of Na/K pump current in isolated heart cells. Nature **315:** 63–65.
3. LAFAIRE, A.V. & W. SCHWARZ. 1986. Voltage dependence of the rheogenic Na^+/K^+ ATPase in the membrane of oocytes of *Xenopus laevis*. J. Membrane Biol. **91:** 43–51.
4. BIELEN, F.V., H.G. GLITSCH & F. VERDONCK. 1991. Dependence of Na^+ pump current on external monovalent cations and membrane potential in rabbit cardiac Purkinje cells. J. Physiol. **442:** 169–189.
5. EISNER, D.A. & W.J. LEDERER. 1980. Characterization of the electrogenic sodium pump in cardiac Purkinje fibres. J. Physiol. **303:** 441–474.

6. GADSBY, D.C. 1980. Activation of electrogenic Na^+/K^+ exchange by extracellular K^+ in canine cardiac Purkinje fibers. Proc. Natl. Acad. Sci. USA **77**: 4035–4039.
7. RAKOWSKI, R.F., D.C. GADSBY & P. DE WEER. 1989. Stoichiometry & voltage dependence of the sodium pump in voltage-clamped, internally dialyzed squid giant axon. J. Gen. Physiol. **93**: 903–941.
8. GARRAHAN, P.J. & I.M. GLYNN. 1967. The incorporation of inorganic phosphate into adenosine triphosphate by reversal of the sodium pump. J. Physiol. **192**: 237–256.
9. DE WEER, P. & R.F. RAKOWSKI. 1984. Current generated by backward-running electrogenic Na pump in squid giant axons. Nature **309**: 450–452.
10. BAHINSKI, A., M. NAKAO & D.C. GADSBY. 1988. Potassium translocation by the Na^+/K^+ pump is voltage insensitive. Proc. Natl. Acad. Sci. USA **85**: 3412–3416.
11. EFTHYMIADIS, A. & W. SCHWARZ. 1991 Conditions for a backward-running Na^+/K^+ pump in *Xenopus* oocytes. Biochim Biophys Acta. **1068**: 73–76.
12. DE WEER, P., D.C. GADSBY & R.F. RAKOWSKI. 2001. Voltage dependence of the apparent affinity for external Na^+ of the backward-running sodium pump. J. Gen. Physiol. **117**: 315–328.
13. GARRAHAN, P.J. & I.M. GLYNN. 1967. The behaviour of the sodium pump in red cells in the basence of external potassium. J. Physiol. **192**: 237–256.
14. GADSBY D.C., R.F. RAKOWSKI & P. DE WEER. 1993. Extracellular access to the Na,K pump: pathway similar to ion channel. Science **260**:100–103.
15. GOLDSHLEGER, R., S.J. KARLISH, A. REPHAELI & W.D. STEIN. 1987. The effect of membrane potential on the mammalian sodium-potassium pump reconstituted into phospholipid vesicles. J. Physiol. **387**: 331–355.
16. STÜRMER, W., R. BUHLER, H.-J. APELL & P. LÄUGER. 1991. Charge translocation by the Na,K-pump: II. Ion binding and release at the extracellular face. J. Membrane Biol. **121**: 163–176.
17. NAKAO, M. & D.C. GADSBY. l989. [Na] and [K] dependence of the Na/K pump current-voltage relationship in guinea-pig ventricular myocytes J. Gen. Physiol. **94**: 539–565.
18. HEYSE, S., I. WUDDEL, H.-J. APELL & W. STÜRMER. 1994. Partial reactions of the Na,K-ATPase: determination of rate constants. J. Gen. Physiol. **104**: 197–240.
19. HILGEMANN, D.W. 1994. Channel-like function of the Na,K pump probed at microsecond resolution in giant membrane patches. Science **263**: 1429–1432.
20. WUDDELL, I. & H.-J. APELL. 1995. Electrogenicity of the sodium transport pathway in the Na,K-ATPase probed by charge-pulse experiments. Biophys. J. **69**: 909–921.
21. HOLMGREN M., J. WAGG, F. BEZANILLA, *et al.* 2000. Three distinct and sequential steps in the release of sodium ions by the Na^+/K^+-ATPase. Nature **403**: 898–901.
22. ANNER, B.M. 1985. Interaction of $(Na^+ + K^+)$-ATPase with artificial membranes. II. Expression of partial transport reactions. Biochim. Biophys. Acta **832**: 335–353.
23. EFTHYMIADIS, A., J. RETTINGER & W. SCHWARZ. 1993. Inward-directed current generated by the Na^+,K^+ pump in Na^+- and K^+-free medium. Cell Biol Int. **17**: 1107–1116.
24. WANG, X. & J.-D. HORISBERGER. 1995. A conformation of Na^+-K^+ pump is permeable to proton. Am. J. Physiol. **37**: C590–C595.
25. LÄUGER, P. 1979. A channel mechanism for electrogenic ion pumps. Biochim. Biophys. Acta **552**: 143–61.
26. HABERMANN, E., M. HUDEL & M.E. DAUZENROTH. 1989. Palytoxin promotes potassium outflow from erythrocytes, HeLa and bovine adrenomedullary cells through its interaction with Na^+,K^+-ATPase. Toxicon **27**: 419–430.
27. SCHEINER-BOBIS, G. *et al.* 1994. Palytoxin induces K^+ efflux from yeast cells expressing the mammalian sodium pump. Mol. Pharmacol. **45**: 1132–1136
28. REDONDO, J., B. FIEDLER & G. SCHEINER-BOBIS. 1996. Palytoxin-induced Na^+ influx into yeast cells expressing the mammalian sodium pump is due to the formation of a channel within the enzyme. Mol. Pharmacol. **49**: 49–57.
29. SCHEINER-BOBIS, G. & H. SCHNEIDER. 1997. Palytoxin-induced channel formation within the Na^+/K^+-ATPase does not require a catalytically active enzyme. Eur. J. Biochem. **248**: 717–723.
30. GUENNOUN, S. & J.D. HORISBERGER. 2000. Structure of the 5th transmembrane segment of the Na,K-ATPase alpha subunit: a cysteine-scanning mutagenesis study. FEBS Lett. **482**: 144–148.

31. GUENNOUN, S. & J.D. HORISBERGER. 2002. Cysteine-scanning mutagenesis study of the sixth transmembrane segment of the Na,K-ATPase alpha subunit. FEBS Lett. **513:** 277–281.
32. TOYOSHIMA, C., *et al.* Crystal structure of the calcium pump of sarcoplasmic reticulum at 2.6 Å resolution. 2000. Nature **405:** 647–655.
33. NIELSEN J.M., P.A. PEDERSEN, S.J. KARLISH & P.L. JORGENSEN. 1998. Importance of intramembrane carboxylic acids for occlusion of K^+ ions at equilibrium in renal Na,K-ATPase. Biochemistry **37:** 1961–1968.
34. PEDERSEN P.A., J.M. NIELSEN, J.H. RASMUSSEN & P.L. JORGENSEN. 1998. Contribution to Tl^+, K^+, and Na^+ binding of Asn776, Ser775, Thr774, Thr772, and Tyr771 in cytoplasmic part of fifth transmembrane segment in alpha-subunit of renal Na,K-ATPase. Biochemistry **37:** 17818–17827.
35. MURAMATSU, I. *et al.* 1988. Single ionic channels induced by palytoxin in guinea pig ventricular myocytes. Br. J. Pharmacol. **93:** 811–816.
36. IKEDA, M., K. MITANI & K. ITO. 1988. Palytoxin induces a non-selective cation channel in single ventricular cells of rat. Naunyn-Schmiedeberg's Arch. Pharmacol. **337:** 591–593.
37. GADSBY, D.C. & M. NAKAO. 1989. Steady-state current-voltage relationship of the Na/K pump in guinea-pig ventricular myocytes. J. Gen. Physiol. **94:** 511–537.

The Na/Ca-K Exchanger Gene Family

ROBERT T. SZERENCSEI, ROBERT J. WINKFEIN, CONAN B. COOPER, CLEMENS PRINSEN, TASHI G. KINJO, KYEONGJIN KANG, AND PAUL P.M. SCHNETKAMP

Department of Biophysics & Physiology, Faculty of Medicine, University of Calgary, Calgary, Alberta, Canada

ABSTRACT: Ca^{2+} extrusion driven by both the inward Na^+ gradient as well as the outward K^+ gradient is essential for visual transduction in retinal rod and cone photoreceptors because it removes Ca^{2+} that enters photoreceptors via the cGMP-gated and light-sensitive channels. We have cloned rod and cone Na/Ca-K exchanger (NCKX) cDNAs from several species, and we have cloned NCKX cDNAs from lower organisms that lack vertebrate-type vision. Although *in situ* NCKX physiology has only been documented for vertebrate photoreceptors, it is now clear that NCKX gene products have a much broader distribution pattern. Here, we review some of the structural and functional features that have emerged from our studies on different members of the NCKX gene family.

KEYWORDS: Ca^{2+}; Na/Ca-K exchanger; NCKX proteins; genes; transmembrane-spanning segments; High Five cells; expression systems

INTRODUCTION

Extrusion of Ca^{2+} driven by both the inward sodium gradient and the outward potassium gradient, or Na/Ca-K exchange, was first described in the outer segments isolated from retinal rod photoreceptors;[3,16,19] to date, this remains the only cell for which the *in situ* physiology of the Na/Ca-K exchanger (NCKX) has been documented (reviewed in Refs. 9 and 17). In rod photoreceptors, the NCKX extrudes Ca^{2+} that enters the outer segment in darkness via the cGMP-gated and light-sensitive channels; cGMP-gated channels are closed in bright light, and continued Ca^{2+} extrusion via the NCKX lowers free cytosolic Ca^{2+}. Lowering cytosolic free Ca^{2+} initiates a negative feedback loop by activating guanylyl cyclase, and the increased rate in cGMP synthesis is the main contributor to the process of light adaptation (reviewed in Ref. 30). The NCKX protein was purified from bovine retinal rod outer segments and shown to be a single polypeptide with an apparent molecular weight of 220 kDa.[4] The first rod NCKX1 cDNA was cloned from bovine retina encoding for a protein of 1,199 residues,[14] later revised to 1,216 residues (by the inclusion of an additional repeat of a 17 amino acid motif found in a large cytosolic loop;[28] subsequent cloning of cone NCKX2 cDNAs from chicken and human retinas revealed a distinct

---

Address for correspondence: Paul P.M. Schnetkamp, Department of Physiology & Biophysics, Faculty of Medicine, Health Sciences Centre, University of Calgary, 3330 Hospital Dr. N.W., Calgary, AB T2N 4N1, Canada. Voice: 403-220-6862; fax: 403-283-8731.
pschnetk@ucalgary.ca

and much smaller gene product coding for NCKX proteins of 651 and 661 residues, respectively.[13] NCKX cDNAs were also obtained from lower organisms such as *Caenorhabditis elegans*[26] and *Drosophila*.[7] In this chapter, we review initial studies in our laboratory concerning the structure and function of members of the NCKX gene family. NCKX gene products are identified by sequence similarity that is confined to two sets of putative transmembrane spanning segments (TMs) of NCKX1, which comprise less than 30% of the full-length mammalian rod NCKX1 sequence. We discuss our evidence that the different NCKX gene products indeed all code for proteins carrying out Na/Ca-K exchange transport when expressed in heterologous systems.

METHODS

The various NCKX cDNA clones used in this study have been described elsewhere: chicken and human cone NCKX2 clones;[13] the dolphin rod NCKX1;[6] the *Caenorhabditis elegans* NCKX;[26] and bovine heart NCX1.[1] NCKX and NCX1 cDNAs were expressed in HEK293 cells or in insect cells as described.[6,26]

^{86}Rb and ^{45}Ca uptake in sodium-loaded insect cells expressing the various NCKX proteins was measured with a rapid filtration method using borosilicate glass fiber filters, as described in detail elsewhere.[26]

Rod outer segments were isolated and purified on mixed sucrose-Ficoll400 density gradients.[15]

RESULTS

Primary Structure of the Different NCKX Proteins

The first NCKX cDNAs were cloned from a bovine retinal cDNA library, and the sequence showed little or no sequence similarity with any other protein present in the data base at the time.[14] Subsequent cloning of the rod NCKX1 cDNAs from human,[28] dolphin,[6] rat,[11] and chicken[13] retinas revealed a remarkable range in molecular weights (from 663 residues of chicken rod NCKX1 to almost double this size of 1,216 residues for bovine rod NCKX1), and revealed a remarkable sequence variability for the two large hydrophilic domains of rod NCKX1 (see below). Furthermore, several splice variants were observed for the mammalian rod NCKX1,[11,28] whereas no splice variants were found for the much smaller chicken rod NCKX1.[13] A second NCKX2 cDNA was cloned from rat brain, and from human and chicken retina; NCKX2 transcripts were found in various parts of the brain,[27] and in retinal cone photoreceptors and in retinal ganglion cells.[12,13] Two distinct splice variants were observed for the chicken, human, and rat NCKX2 transcripts, the shorter one lacking 17 amino acids in the large cytosolic loop. A third mammalian NCKX3 was recently described,[8] and NCKX cDNAs have been cloned from lower organisms such as *Caenorhabditis elegans*[26] and *Drosophila*.[7] Hydropathy analysis strongly suggests that all NCKX cDNAs studied to date code for proteins consisting of two large hydrophilic loops and two sets of TMs. The TMs are thought to contain 5 or 6 membrane-spanning alpha helices with short (<25 residues) connecting loops. The

TABLE 1. Properties of NCKX clones

	Length	Splice variants	Protein	Function
Rod NCKX1				
Bovine	1216	2	**	–
Human	1081	2	*	–
Dolphin	1013	ND	****	**
Chicken	663	1	***	**
Cone NCKX2				
Human	661	2	****	****
Chicken	651	2	****	****
C. elegans	596	1	ND	**
Drosophila	856	1	ND	***

NOTE: The number of asterisks indicates increasing amounts of protein expression or function.

first large hydrophilic loop is located at the N-terminus, while the other is located in the cytosol and separates the two TMs. Consistent with this overall topology, removal of the sequences coding for two large hydrophilic loops from the bovine rod NCKX1 cDNA resulted in truncated protein with normal Na/Ca-K exchange function when expressed in cultured insect cells.[25,26] All of the variation in length between the various NCKX clones can be accounted for by differences in size and amino acid sequence of the two large hydrophilic loops; a remarkable sequence variability was observed for the rod NCKX1 isoform within those hydrophilic loops.[6,13] A similar sequence variability is not found in other sodium-coupled transporters and is unexpected in view of the fact that vertebrate rod vision is highly invariant.

Some of the structural parameters of the various NCKX cDNAs obtained in our laboratory are summarized in TABLE 1. Sequence similarity between NCKX cDNAs from *Caenorhabditis elegans* and *Drosophila* and those from retinal rods and cones is limited to the two sets of TMs, with no sequence similarity observed in the two large hydrophilic loops.

Structure of the Rod and Cone NCKX Genes

Rod NCKX1 has been localized to chromosome 15q22 and its genomic organization has been determined.[28] Cone NCKX2 has been localized to chromosome 9p22,[12] while the genomic organization NCKX2 was found to be very similar to that of NCKX1, consisting of 11 and 10 exons, respectively. Exon II contains about half of the coding sequence from the N-terminus through the first set of TMs for both rod NCKX1 and cone NCKX2.[23,28] PCR analysis revealed a number of splice variants for both mammalian NCKX1 and mammalian and avian NCKX2 (see above); mammalian rod NCKX1 splice variants represent various combinations of excisions of exons 3 to 6 located in the N-terminal portion of the large cytosolic loop, whereas the single splice variant observed for cone/brain NCKX2 involves excision of exon 5, located in the middle of the large cytosolic loop. Curiously, only full-length rod NCKX1 transcripts were detected in the chicken retina; the much shorter chicken rod

NCKX1 lacks a large hydrophilic N-terminal hydrophilic loop and lacks the acidic part of the cytosolic loop of mammalian rod NCKX1, but it does contain the better conserved N-terminal part of the large cytosolic loop which is the site of alternate splicing in mammalian rod NCKX1 transcripts.[13]

NCKX and Hereditary Retinal Disease

The rod and cone NCKX genes were analyzed for sequence variations in patients with hereditary retinal disease, and a large number of sequence variations were detected, in particular, for the rod NCKX1 gene. However, pathogenicity could not be established with certainty, although missense mutation observed in two patients caused a near complete loss of transport function when the mutated NCKX cDNA was expressed in insect cells.[23]

Functional Criteria for Na/Ca-K Exchange Transport

The first two rod NCKX1 cDNAs were cloned from bovine and human retinal cDNA libraries, respectively, but neither cDNA resulted in functional protein when expressed in cell lines.[6,26,28] In another study, transfection of bovine rod NCKX1 in HEK293 cells resulted in potassium-independent Na/Ca exchange currents, and it was suggested that K^+ dependence and K^+ transport might require an additional protein.[10] As discussed above, the various members of the NCKX gene family only share sequence identity in the two sets of TMs comprising about 320 residues or only 26% of the total bovine rod NCKX1 sequence. Moreover, it has been suggested that the site coding for K^+ dependence and K^+ transport is located on part of the cytosolic loop of bovine rod NCKX1,[22] a domain not found in cone NCKX2 or in NCKX sequences from lower organisms. Thus, it is important to establish that the various NCKX cDNAs indeed code for Na/Ca exchangers that not only require K^+ but also transport K^+. The standard experimental test for Na/Ca-K exchange (or for Na/Ca exchange) function upon heterologous expression of NCKX (or NCX) cDNAs in cell lines is based on the well-established reversibility of rod Na/Ca-K exchange fluxes (or Na/Ca exchange fluxes), that is, both NCKX and NCX in situ carry out Na_{out}-dependent Ca^{2+} efflux (forward exchange) as well as Na_{in}-dependent Ca^{2+} influx (reverse exchange), dependent on the direction of the transmembrane Na^+ gradient.[15,18] Reverse Na/Ca(-K) exchange is measured in most published work on functional expression of NCKX or NCX cDNAs in cell lines because measurement of forward exchange requires imposing sustained high intracellular Ca^{2+} concentrations which are toxic for cells, but which can be tolerated in organelles such as isolated rod outer segments.[15,21] The top panels of FIGURE 1 compare the K^+ dependence of human cone NCKX2 with that of bovine heart NCX1 expressed in High Five cells. We measured ^{45}Ca uptake associated with reverse Na/Ca-K exchange in transfected cells in three different media: uptake in K^+ medium indicates full Na/Ca(-K) exchange activity; uptake in Na^+ medium represents background uptake (similar to that observed in untransfected cells) because reverse Na/Ca(-K) exchange is fully inhibited by high external Na^+; while uptake in Li^+ medium tests for a specific requirement for K^+ characteristic for reverse Na/Ca-K exchange.[26] A Li^+ medium is used in these experiments in view of the observation that reverse Na/Ca-K exchange in media lacking alkali cations as a major osmotic constituent often

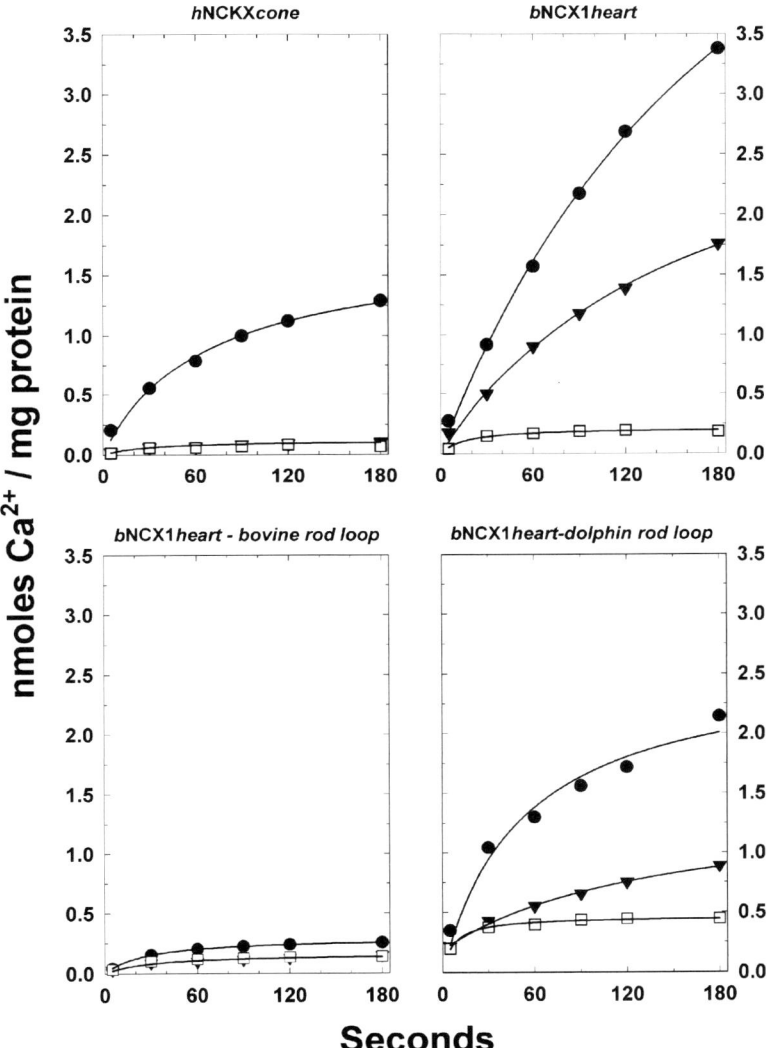

FIGURE 1. K^+ dependence of NCX1, NCKX2, and NCX1-NCKX1 chimeras. Bovine heart NCX1, human cone NCKX2, NCX1 containing the bovine rod NCKX1 cytosolic loop, and NCX1 containing the dolphin rod NCKX1 cytosolic loop were transiently expressed in High Five cells. Reverse Na/Ca(-K) exchange was assayed in sodium-loaded cells in media containing 20 mM Hepes (adjusted to pH 7.4 with arginine), 35 µM $CaCl_2$, 1 µCi ^{45}Ca, 80 mM sucrose, and 150 mM KCl (*circles*), 150 mM LiCl (*inverted triangles*), or 150 mM NaCl (*squares*). Sodium-loading procedure and ^{45}Ca uptake protocol have been described in detail elsewhere.[26] Temperature: 25°C.

shows a significant K^+-independent component of reverse Na/Ca-K exchange.[21,26] Clearly, reverse Na/Ca-K exchange via cone NCKX2 was completely dependent on the presence of K^+, and uptake observed in Li^+ medium was indistinguishable from that observed in Na^+ medium; the latter represents background ^{45}Ca uptake, indistinguishable from that observed in mock-transfected cells. Reverse Na/Ca exchange in cells transfected with bovine heart NCX1 was optimal in K^+ medium, while uptake in Li^+ medium ranged from being very similar to that observed in K^+ medium (when High Five cells were maintained in choline medium) to being about half of that observed in K^+ medium (the worst case scenario illustrated here). The experiments discussed in the next section suggest that some caution should be exercised in interpreting experiments on the K^+ dependence of NCX or NCKX constructs.

Role of the Large Cytosolic Loop of Rod NCKX1 in Functional Expression in Heterologous Systems

Closer examination of NCKX1 cDNAs used in our lab revealed that the bovine rod NCKX1 represented the full-length sequence, the human rod NCKX1 lacked exon 3, while the dolphin rod NCKX1 lacked exons 3 through 6, coding for 114 residues.[29] Interestingly, cDNAs representing the two longer splice variants, bovine rod NCKX1 and human rod NCKX1, did not yield functional expression in heterologous systems, whereas the shortest splice variant represented by the dolphin rod NCKX1 clone yielded strong Na/Ca-K exchange function when expressed in either HEK293, CHO, or insect cell lines.[6,26] This may suggest that the sequences encoded for by exons 3 through 6 somehow prevent functional expression of the rod NCKX1 cDNA in cell lines. Consistent with this, deletion of this part of the sequence from the bovine rod NCKX1 clone restored some function, although not as strong as that observed with dolphin rod NCKX1. Conversely, insertion of the bovine cytosolic loop into the dolphin rod NCKX1 abolished function but not protein expression.[6] Here, we have examined this further by splicing the cytosolic loop of either bovine or dolphin NCKX1 into the bovine heart NCX1 from which the original cytosolic loop was removed. Very little or no function was observed in the bovine heart NCX1-bovine rod NCKX1 chimera, but, in contrast, strong Na/Ca exchange function was observed in the bovine heart NCX1-dolphin rod NCKX1 chimera (FIG. 1, bottom two panels). This experiment demonstrates that sequences within the cytosolic loop of bovine rod NCKX1 can silence transport function when expressed in cell lines, whereas the dolphin NCKX1 cytosolic loop does not. It should be pointed out that the sequence of the rod NCKX1 cytosolic loop is remarkably variable between different mammalian species, and it is possible that domains other than those coded by exons 3 through 6 contribute to the observed difference in functional expression for the different rod NCKX1 cDNAs. In the case of the rat rod NCKX1, the full-length sequence containing exons 3 through 6 showed some transport function, although much lower than that observed with dolphin rod NCKX1.[11]

A surprising feature of the bovine heart NCX1-dolphin rod NCKX1 chimera was that the activity in K^+ medium was always markedly higher than that in Li^+ medium. This may be consistent with observations by Seiler *et al.*, who suggest that K^+ dependence and K^+ transport are located on part of the cytosolic loop of bovine rod NCKX1.[22] Therefore, we examined whether Ca^{2+} uptake via reverse Na/Ca(-K) exchange was associated by K^+ transport as measured with the K^+ congener ^{86}Rb.

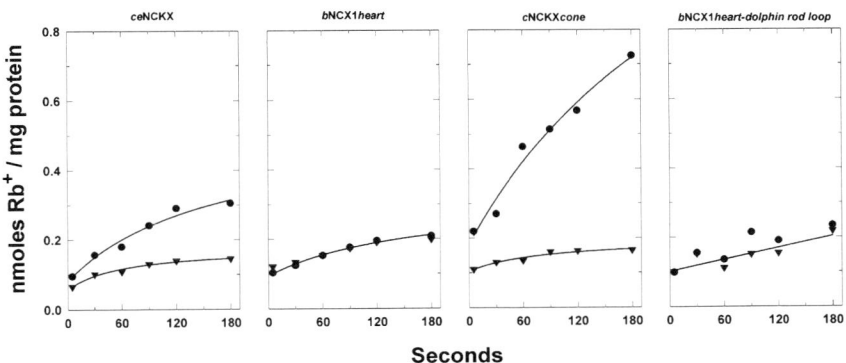

FIGURE 2. Ca^{2+} dependent ^{86}Rb transport via ceNCKX, heart NCX1, cone NCKX2, and NCX1-NCKX1 chimera. Bovine heart NCX1, chicken cone NCKX2, NCKX from *Caenorhabditis elegans*, and NCX1 containing the dolphin rod NCKX1 cytosolic loop were transiently expressed in High Five cells. ^{86}Rb uptake was initiated at time zero by addition of ^{86}Rb, 0.4 mM RbCl, and either 0.4 mM $CaCl_2$ (*solid circles*) or 0.4 mM EDTA (*inverted triangles*) to a suspension of sodium-loaded High Five cells. High Five cells expressed the different NCKX cDNAs as indicated. The suspension medium contained 150 mM choline chloride, 80 mM sucrose, 20 mM Hepes (adjusted to pH 7.4 with arginine), and 0.05 mM EDTA. Temperature: 25°C.

Ca^{2+}-dependent ^{86}Rb transport was clearly observed in cells transfected with the human cone NCKX2, but not in cells transfected with either bovine heart NCX1 or with the bovine heart NCX1-dolphin rod NCKX1 chimera (FIG. 2) despite the fact the ^{45}Ca uptake was at least as high or higher than that observed in cells transfected with the human cone NCKX2.

Heterologous Expression Systems

Structure-function studies on members of the NCKX gene family require cellular expression systems that are amenable to simple and preferably quantitative assay(s) of NCKX function. Furthermore, for biochemical and structural studies an expression system is required that expresses large amounts of protein and can be scaled up. HEK293 cells (transient transfection) and CHO cells (stable transfection) were used for fluorescent calcium imaging, either at the single-cell level or in a suspension of cells.[5,6] This provides for a convenient but qualitative assay of NCKX function; moreover, it proved difficult to generate stable cell lines as protein expression levels rapidly declined within a few passes. We interpret this result to indicate that continued expression of NCKX protein in mammalian cell lines is toxic to the cells, most likely due to the rather low equilibrium free Ca^{2+} concentration that can be reached by NCKX.[20] In contrast, the culture media for insect cells such as High Five cells contain low extracellular Na^+ concentrations that render the NCKX ineffective in lowering cytosolic free Ca^{2+} to very low levels detrimental for proper cell growth; furthermore, High Five cells can be grown in suspension, and billions of cells can be

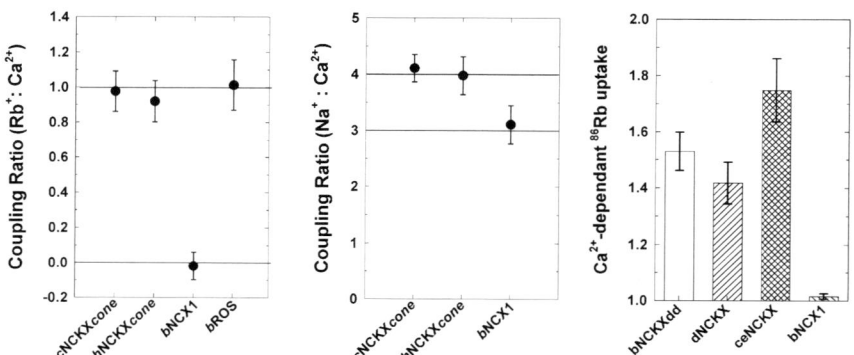

FIGURE 3. Rb:Ca and Na:Ca coupling ratios of the retinal cone NCKX. The Rb:Ca and Na:Ca coupling ratios were measured as described in the text in High Five cells expressing the human or chicken retinal cone NCKX, or expressing bovine heart NCX1. **Left panel:** Average values ± standard deviation are illustrated for 15 experiments with chicken cone NCKX, 13 experiments with human cone NCKX, 9 experiments with bovine heart NCX1, and 8 experiments with bovine retinal rod outer segments. **Middle panel:** Average values ± standard deviation are illustrated for 10 experiments with chicken cone NCKX, 15 experiments with human cone NCKX, and 10 experiments with bovine heart NCX1. **Right-hand panel:** Calcium-dependent ^{86}Rb uptake in High Five cells transformed with different NCKX cDNAs. ^{86}Rb uptake was measured in the presence of 0.4 mM CaCl$_2$ or 0.4 mM EDTA as described in the legend of FIGURE 2. High Five cells were used expressing, *from left to right*, bovine retinal rod NCKX from which the two large hydrophilic loops were removed, dolphin retinal rod NCKX, an NCKX paralog from *C. elegans*, or bovine heart NCX1. Temperature: 25°C. Reprinted (with slight modifications) with permission from *Biochemistry*.[25]

obtained economically. We have used a relatively new expression vector to generate stable High Five cell lines expressing the various NCKX cDNAs. ^{45}Ca influx in sodium-loaded High Five cells presents a simple and quantitative measure for reverse NCKX activity.[13,26] With this system of stably transfected High Five cells, clonal cell lines could be grown for about three months before NCKX expression levels started to decline; in contrast, High Five cells stably transfected with the bovine heart NCX1 did not show any rundown on the timescale of one year. The High Five cells, stably expressing the various NCKX proteins, could be grown in sufficient quantity to permit a determination of the transport stoichiometry of 4 Na$^+$/1 K$^+$ + 1 Ca^{2+} by direct measurements of the Na$^+$/Ca^{2+} and K$^+$/Ca^{2+} coupling ratios for the cone NCKX2 (FIG. 3). Functional expression levels for the other NCKX clones were insufficient to determine directly the transport stoichiometry, but cells expressing the rod NCKX1 or a NCKX cloned from *Caenorhabditis elegans* all showed a Ca^{2+}-dependent ^{86}Rb uptake (used as a K$^+$ congener) associated with reverse Na/Ca-K exchange, whereas cells expressing the bovine heart NCX1 did not show any Ca^{2+}-dependent ^{86}Rb uptake (FIG. 3, right-hand panel). We conclude from these experiments that the different NCKX clones used in our experiments each code for a single polypeptide that is sufficient to carry out Na/Ca-K exchange with a transport stoichi-

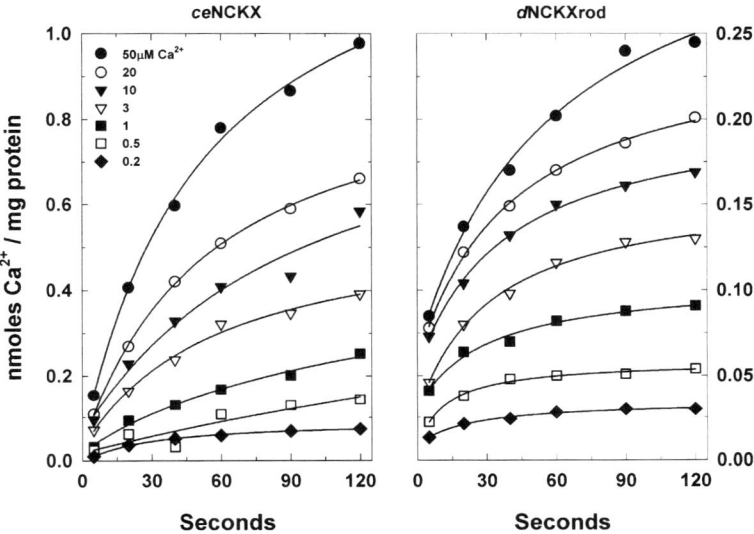

FIGURE 4. Calcium dependence of ^{45}Ca uptake via reverse Na/Ca-K exchange. ^{45}Ca uptake was initiated at time zero by addition of ^{45}Ca and 200 µM CaHEDTA to High Five cells in a medium containing 150 mM KCl, 80 mM sucrose, 20 mM Hepes (adjusted to pH 7.4 with arginine), and various concentrations of HEDTA to yield the following free calcium concentrations (**left-hand and right-hand panels**): 0.2 µM (*filled diamonds*), 0.5 µM (*open squares*), 1 µM (*filled squares*), 3 µM (*open inverted triangles*), 10 µM (*filled inverted triangles*), 20 µM (*open circles*), and 50 µM (*filled circles*). An apparent calcium dissociation constant of 1.5 µM for the CaHEDTA complex was used. **Left-hand panel:** Cells expressing *Caenorhabditis elegans* NCKX; **right-hand panel:** cells expressing dolphin rod NCKX1. In all cases calcium uptake was corrected for uptake observed in cells in a medium containing 150 mM NaCl instead of KCl; this represented background calcium uptake in High Five cells when reverse Na/Ca(-K) exchange was inhibited. Temperature: 25°C.

ometry of 4 Na$^+$/1 K$^+$ + 1 Ca^{2+}. The only sequence elements in common between the rod NCKX1, cone NCKX2, and *C. elegans* NCKX are the two sets of putative TMs, implying that these TMs code for all the residues directly involved in Na/Ca-K exchange transport. Consistent with this, we removed the two large hydrophilic loops from the bovine rod NCKX1 and observed potassium-dependent Na/Ca exchange transport[26] as well as Ca^{2+}-dependent ^{86}Rb uptake (FIGS. 2 and 3).

Functional Characterization of Heterologously Expressed NCKX Clones

We have used three different transport assays to characterize the kinetic parameters of the different NCKX clones expressed in High Five cells, in CHO or in HEK293 cells: (1) Na$_{in}$-dependent Ca^{2+} influx (reverse exchange) was measured in Na$^+$-loaded High Five cells with the use of ^{45}Ca and a rapid filtration technique using borosilicate glass fiber filters;[26] (2) a transport stoichiometry of 4 Na$^+$/1 K$^+$ + 1 Ca^{2+} indicates that NCKX proteins are electrogenic transporters, and we have mea-

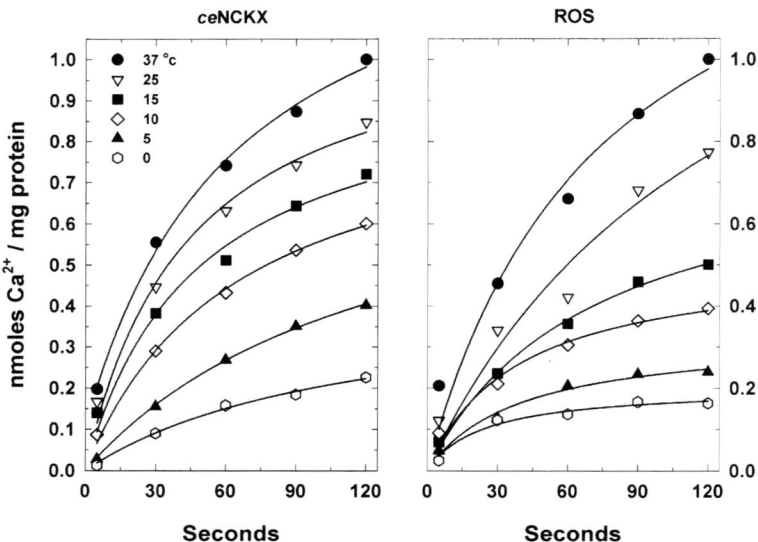

FIGURE 5. Temperature dependence of ^{45}Ca uptake via reverse Na/Ca-K exchange. The temperature dependence of ^{45}Ca uptake via reverse Na/Ca-K exchange observed for NCKX in isolated bovine rod outer segments was compared with that of *Caenorhabditis elegans* NCKX expressed in High Five cells. ^{45}Ca uptake was measured in KCl medium as described in FIGURE 1 legend.

sured and characterized NCKX currents in High Five cells expressing NCKX clones;[24] (3) fluo 3-loaded CHO or HEK293 cells were used to measure NCKX-mediated changes in cytosolic free Ca^{2+}.[5,6] The external Na^+ and K^+ dissociation constants for activating reverse and forward Na/Ca-K exchange, respectively, were very similar for the different NCKX isoforms used, suggesting that the cation binding/transport sites are made up by conserved residues located within the two sets of TMs. One of the *in situ* characteristic features of rod NCKX1 is the high affinity (K_m of ~1 µM) for extracellular Ca^{2+} when no competing cations are present. Similarly, K_m values of 1–3 µM were observed for the rod and cone NCKX clones expressed in High Five cells.[25] Here, we illustrate the external Ca^{2+} dependence of *C. elegans* NCKX expressed in High Five cells, and compare it with the external Ca^{2+} dependence of the dolphin rod NCKX1 (FIG. 4). In both cases a high affinity (low micromolar) toward extracellular Ca^{2+} was observed, characteristic for other members of the NCKX gene family; in contrast, the affinity for extracellular Ca^{2+} for members of the NCX gene family is much lower with dissociation constants of a few hundred micromolar (reviewed in Ref. 2). Another characteristic feature of NCKX is the fairly steep temperature dependence of reverse Na/Ca-K exchange observed for rod NCKX1, both *in situ* and expressed in High Five cells.[26] Here, we show the temperature dependence of *C. elegans* NCKX expressed in High Five cells is very similar compared with that of bovine rod NCKX1 *in situ* (FIG. 5).

ACKNOWLEDGMENTS

This work was supported through an operating grant by the Canadian Institutes for Health Research (to P.P.M.S.). P.P.M.S. is a scientist of the Alberta Heritage Foundation for Medical Research AHFMR. C.P. was recipient of a AHFMR postdoctoral fellowship; T.G.K. and C.B.C. are recipients of AHFMR studentships; and K.K. is recipient of a studentship from the Canadian Foundation Fighting Blindness.

REFERENCES

1. ACETO, J.F., M. CONDRESCU, C. KROUPIS, et al. 1992. Cloning and expression of the bovine cardiac sodium-calcium exchanger. Arch. Biochem. Biophys. **298:** 553–560.
2. BLAUSTEIN, M.P. & W.J. LEDERER. 1999. Sodium/calcium exchange: its physiological implications. Physiol. Rev. **79:** 763–854.
3. CERVETTO, L., L. LAGNADO, R.J. PERRY, et al. 1989. Extrusion of calcium from rod outer segments is driven by both sodium and potassium gradients. Nature **337:** 740–743.
4. COOK, N.J. & U.B. KAUPP. 1988. Solubilization, purification, and reconstitution of the sodium-calcium exchanger from bovine retinal rod outer segments. J. Biol. Chem. **263:** 11382–11388.
5. COOPER, C.B., R.T. SZERENCSEI & P.P.M. SCHNETKAMP. 2000. Spectrofluorometric detection of Na-Ca+K exchange. Methods Enzymol. **315:** 847–864.
6. COOPER, C.B., R.J. WINKFEIN, R.T. SZERENCSEI & P.P.M. SCHNETKAMP. 1999. cDNA-Cloning and functional expression of the dolphin retinal rod Na-Ca+K exchanger NCKX1: comparison with the functionally silent bovine NCKX1. Biochemistry **38:** 6276–6283.
7. HAUG-COLLET, K., B. PEARSON, S. PARK, et al. 1999. Cloning and characterization of a potassium-dependent sodium/calcium exchanger in *Drosophila*. J. Cell Biol. **147:** 659–669.
8. KRAEV, A., B.D. QUEDNAU, S. LEACH, et al. 2001. Molecular cloning of a third member of the potassium-dependent sodium-calcium exchanger gene family, NCKX3. J. Biol. Chem. **276:** 23161–23172.
9. LAGNADO, L. & P.A. MCNAUGHTON. 1990. Electrogenic properties of the Na:Ca exchange. J. Membr. Biol. **113:** 177–191.
10. NAVANGIONE, A., G. RISPOLI, N. GABELLINI & E. CARAFOLI. 1997. Electrophysiological characterization of ionic transport by the retinal exchanger expressed in human embryonic kidney cells. Biophys. J. **73:** 45–51.
11. POON, S., S. LEACH, X.F. LI, et al. 2000. Alternatively spliced isoforms of the rat eye sodium/calcium+potassium exchanger NCKX1. Am. J. Physiol. Cell Physiol. **278:** C651–C660.
12. PRINSEN, C.F.M., C.B. COOPER, R.T. SZERENCSEI, et al. 2002. The retinal rod and cone $Na^+/Ca^{2+}-K^+$ exchangers. *In* Photoreceptors and Calcium. W. Baehr & K. Palczewski, Eds. Landes Bioscience. Georgetown, TX.
13. PRINSEN, C.F.M., R.T. SZERENCSEI & P.P.M. SCHNETKAMP. 2000. Molecular cloning and functional expression the potassium-dependent sodium-calcium exchanger from human and chicken retinal cone photoreceptors. J. Neurosci. **20:** 1424–1434.
14. REILÄNDER, H., A. ACHILLES, U. FRIEDEL, et al. 1992. Primary structure and functional expression of the Na/Ca,K-exchanger from bovine rod photoreceptors. EMBO J. **11:** 1689–1695.
15. SCHNETKAMP, P.P.M. 1986. Sodium-calcium exchange in the outer segments of bovine rod photoreceptors. J. Physiol. **373:** 25–45.
16. SCHNETKAMP, P.P.M. 1989. Na-Ca or Na-Ca-K exchange in the outer segments of vertebrate rod photoreceptors. Prog. Biophys. Mol. Biol. **54:** 1–29.
17. SCHNETKAMP, P.P.M. 1995. Calcium homeostasis in vertebrate retinal rod outer segments. Cell Calcium **18:** 322–330.

18. SCHNETKAMP, P.P.M., D.K. BASU, X.B. LI & R.T. SZERENCSEI. 1991. Regulation of intracellular free Ca^{2+} concentration in the outer segments of bovine retinal rods by Na-Ca-K exchange measured with Fluo-3. II. Thermodynamic competence of transmembrane Na^+ and K^+ gradients and inactivation of Na^+-dependent Ca^{2+} extrusion. J. Biol. Chem. **266:** 22983–22990.
19. SCHNETKAMP, P.P.M., D.K. BASU & R.T. SZERENCSEI. 1989. Na-Ca exchange in the outer segments of bovine rod photoreceptors requires and transports potassium. Am. J. Physiol. (Cell Physiol.) **257:** C153–C157.
20. SCHNETKAMP, P.P.M., D.K. BASU & R.T. SZERENCSEI. 1991. The stoichiometry of Na-Ca^+K exchange in rod outer segments isolated from bovine retinas. Ann. N. Y. Acad. Sci. **639:** 10–21.
21. SCHNETKAMP, P.P.M., X.B. LI, D.K. BASU & R.T. SZERENCSEI. 1991. Regulation of free cytosolic Ca^{2+} concentration in the outer segments of bovine retinal rods by Na-Ca-K exchange measured with Fluo-3. I. Efficiency of transport and interactions between cations. J. Biol. Chem. **266:** 22975–22982.
22. SEILER, E.P., D. GUERINI, F. GUIDI & E. CARAFOLI. 2000. The N-terminal portion of the main cytosolic loop mediates K^+ sensitivity in the retinal rod Na^+/Ca^{2+}-K^+-exchanger. Eur. J. Biochem. **267:** 2461–2472.
23. SHARON, D., H. YAMAMOTO, T.L. MCGEE, et al. 2002. Mutated alleles of the rod and cone Na/Ca+K exchanger genes in patients with retinal diseases. IOVS **43:** 1971–1979.
24. SHENG, J.-Z., C.F.M. PRINSEN, R.B. CLARK, et al. 2000. Na^+-Ca^{2+}-K^+ Currents measured in insect cells transfected with the retinal cone or rod Na^+-Ca^{2+}-K^+ exchanger cDNA. Biophys. J. **79:** 1945–1953.
25. SZERENCSEI, R.T., C.F.M. PRINSEN & P.P.M. SCHNETKAMP. 2001. The stoichiometry of the retinal cone Na/Ca-K exchanger heterologously expressed in insect cells: comparison with the bovine heart Na/Ca exchanger. Biochemistry **40:** 6009–6015.
26. SZERENCSEI, R.T., J.E. TUCKER, C.B. COOPER, et al. 2000. Minimal domain requirement for cation transport by the potassium-dependent Na/Ca-K exchanger: comparison with an NCKX paralog from *Caenorhabditis elegans*. J. Biol. Chem. **275:** 669–676.
27. TSOI, M., K.-H. RHEE, D. BUNGARD, et al. 1998. Molecular cloning of a novel potassium-dependent sodium-calcium exchanger from rat brain. J. Biol. Chem. **273:** 4155–4162.
28. TUCKER, J.E., R.J. WINKFEIN, C.B. COOPER & P.P.M. SCHNETKAMP. 1998. cDNA Cloning of the human retinal rod Na-Ca+K exchanger: comparison with a revised bovine sequence. IOVS **39:** 435–440.
29. TUCKER, J.E., R.J. WINKFEIN, S.K. MURTHY, et al. 1998. Chromosomal localization and genomic organization of the human retinal rod Na-Ca+K exchanger. Hum. Genet. **103:** 411–414.
30. YAU, K.-W. & D.A. BAYLOR. 1989. Cyclic GMP-activated conductance of retinal photoreceptor cells. Annu. Rev. Neurosci. **12:** 289–327.

Toward Comparative Genomics of Calcium Transporters

ALEXANDER KRAEV AND DAVID H. MacLENNAN

Charles H. Best Institute, University of Toronto, Toronto, M5G 1X5 Canada

KEYWORDS: calcium transporters; mammalian genome; *C. elegans*

INTRODUCTION

Studies of calcium transporters have focused historically on mammalian organisms (for recent reviews, see Refs. 1–6) as the ones expected to provide data of immediate relevance to medicine. Many other experimental systems were chosen on the basis of their suitability to specific methods, such as methods of classic genetics or electrophysiology. As a result, fragmentary data have accumulated across taxonomically unrelated phyla on various aspects of calcium handling that are difficult to integrate into a coherent model. Comparison of genome sequences of model organisms now offers a possibility to build a conceptual framework for reinterpretation and integration of these data. In 1996 one of us started to analyze the public data from the *Caenorhabditis elegans* Genome Project[7] for the presence of genes, encoding putative calcium transporter proteins. As a result of this effort, new calcium transporters have been identified in *C. elegans*[8] and in mammalian genomes.[9] The analysis is being extended to other completed genomes. Our presentation is a progress report of this ongoing effort.

EXPERIMENTAL PROCEDURES

Computer Analysis

Draft genome sequencing data were probed with protein signature sequences[10] using BLAST algorithm.[11] The candidate gene transcripts were first completed *in silico* and subjected to further analysis for criteria of orthology.[12] When judged as erroneously predicted or incomplete, transcript sequences were verified and/or completed using RACE.[13] The experimentally confirmed transcript sequences were submitted to public sequence databases.

The databases and conventions used in this analysis are as follows: (1) *C. elegans* genome (www.sanger.ac.uk/Projects/C_elegans); (2) *Drosophilia melanogaster* genome (flybase.bio.indiana.edu); (3) human genome (www.ncbi.nlm.nih.gov/genome/guide/human/ or www.ensembl.org).

Address for correspondence: Alexander Kraev, Samuel Lunenfeld Research Institute, Mt. Sinai Hospital, 600 University Avenue, Toronto, ON, Canada M5G 1X5. Voice: 416-586-8543; fax: 416-586-8588.
 alexander.kraev@utoronto.ca

Please note that plasma membrane calcium channel genes from *Drosophila* and from man have been omitted from this analysis, since specialized reviews have appeared recently in the literature.[2,14-16]

Prediction Validation Procedures

C. elegans N2 propagation, RNA isolation, and RT-PCR amplifications were done essentially as described.[8] Two-step temperature profiles (cycling between 95° and 70°C without an annealing step), and a fast thermocycling machine capable of achieving ramp speed >2°/s, were found to be essential for reproducible amplification of low-abundance transcripts from total RNA and for avoiding sequencing artefacts. Automated sequencing reactions were run on sequencing gels at the local core facility (Institute of Biochemistry, ETH Zurich, Switzerland, and Hospital of Sick Children, Toronto, Canada). Raw sequencing data were edited and assembled into contigs using Sequencher (Gene Codes, MA).

RESULTS AND DISCUSSION

Current gene inventories of the worm, fly, and human calcium transporters are presented in TABLES 1–3. All major transporter families previously defined in mammalian genomes are found in other model organisms as well, but the number of members in each family does not vary between genomes in a consistent manner. The most striking observation is that while the number of sarcoplasmic reticulum (SR) -based transporter genes can be explained as a result of two asymmetrical duplications (three genes for RYR, IP3R, and SERCA in man versus one in both nematode and fly), the size of some plasma membrane-based transporter families, namely, PMCA and NCX, either remains constant or contracts. Unexpectedly, while both man and worm have several genes corresponding to the Na/Ca-exchanger family (as exemplified by the extensively studied "cardiac" Na/Ca-exchanger, NCX1[4]), and the plasma membrane Ca^{2+}-ATPase family, there is only one member in each family in fly. Intriguingly, an exception to this rule is found in the case of the potassium-dependent Na/Ca-exchanger family (NCKX), previously believed to be specific to the retinal photoreceptors of the eye,[17,18] but recently shown to be expressed quite ubiquitously.[9] While the nematode has only two family members, both man and *Drosophila* have four. The TSR (Golgi) calcium pump is a notable newcomer to the family of SR calcium pumps. This transporter was first identified in rat,[19] but was ignored for almost a decade until recently rediscovered.[20,21] Here there is only one family member in both fly and worm, and two in man.

Space constraints prohibit a discussion of the potential of alternative splicing to generate even further diversity in transporter structure and function. Such analysis on the genome scale is still very far from complete. Another intriguing fact that remains to be explained is why several nematode genes[8,22,23] are apparently driven by more complex promoters than are their human counterparts.[24,25] The use of alternative promoters may contribute substantially to protein diversity, as well as to complexity in spatial and temporal distribution.

Recent studies in nonmammalian systems, demonstrating the key role of calcium transporters in the processes of neuronal migration,[26] larval molting,[27] and various

TABLE 1. Inventory of genes, encoding calcium transporting proteins of the nematode *C. elegans*

Gene name	Synonym	Chr	Genomic clone	EST clone	cDNA Accession	Promoters	Reference	Proposed function
mca-1		IV	W09C2	yk49f8	AJ223616	4	Kraev et al.[8]	PM Ca^{2+}-ATPase
mca-2		IV	Y76B12	yk44a6	AJ010708	2	Kraev et al.[8]	PM Ca^{2+}-ATPase
mca-3		IV	Y67D8/R05C11	yk186e4	AJ010646	4	Kraev et al.[8]	PM Ca^{2+}-ATPase
mca-4	sca-1	III	K11D9	yk247b7	AJ012296	1	Zwaal et al.[30]	SR Ca-ATPase
mca-5	CePMR1	I	ZK256, CC4	yk3345	AJ303081	N.D.	Van Baelen et al.[21]	TSR Ca-ATPase
ncx-1		V	Y102G3	yk24h3	X91803	1	Kraev et al.[31]	PM Na/Ca-exchanger
ncx-2		V	C10G8	yk462d8	AJ001181	1	Kraev et al., unpublished	PM Na/Ca-exchanger
ncx-3		IV	R102, ZC168	yk532f1	None	N.D.	Kraev et al., unpublished	PM Na/Ca-exchanger
ncx-4	CeNCKX	I	F35C12	None	AJ005701	1	Szerencsei et al.[32]	PM Na/Ca,K-exchanger
ncx-5		V	Y32F6B	None	None	1	Kraev et al., unpublished	PM Na/Ca,K-exchanger
itr-1		IV	F33D4, C08G9	yk33g8	AJ243179-82	3	Baylis et al.[22]	SR Ca-channel (IP3R)
ryr-1	CeRYR	V	K11C4	yk3h7	D45899 (gene)	1	Maryon et al.[33]	SR Ca-channel (RYR)
egl-19		IV	C48A7, B0496	yk605d11	AF023602	N.D.	Lee et al.[34]	L-type VGCC
unc-2		X	T02C5	yk131b1	U25119 (gene)	N.D.	Schafer and Kenyon[35]	N/P/Q-type VGCC
cca-1		X	C54D2	yk899c11	AF368920	N.D.	Mittman, unpublished	T-type VGCC
(cax-1a)		III	C07A9.4	None	None	N.D.	Kraev et al., unpublished	(Ca/X-exchanger)
(cax-1b)		III	C07A9.11	None	None	N.D.	Kraev et al., unpublished	(Ca/X-exchanger)
(cax-2a)		V	C13D9.7	None	None	N.D.	Kraev et al., unpublished	(Ca/X-exchanger)
(cax-2b)		V	C13D9.8	None	None	N.D.	Kraev et al., unpublished	(Ca/X-exchanger)

ABBREVIATIONS: chr—chromosom; PM—plasma membrane; SR—sarco/endoplasmic reticulum; TSR—transitional (Golgi) reticulum; IP3R—inositol-1,4,5-trisphosphate receptor; RYR—ryanodine receptor; VGCC—voltage-gated calcium channel.

TABLE 2. Inventory of genes encoding calcium transporting proteins of the fruit fly *D. melanogaster*[a]

Gene name	Synonym	Chromosome, band	cDNA accession	Reference	Proposed function	Mutant phenotypes
Ca-P60A	CG3725	2R, 60A10–12	M62892	Magyar and Varadi[36]	SR Ca-ATPase	Affect eye, leg, wing; LLR
Ca-76F	CG7651	3L, 76F3	None	Kraev, unpublished	TSR Ca-ATPase	Unknown
Ca-102B	CG2165	4, 102B5	None	Kraev, unpublished	PM Ca-ATPase	Unknown
Calx	CG5685	3R, 91B1–3	AF009897	Schwartz and Benzer[37]	PM Na/Ca-exchanger	Unknown
NCKX30C	CG18660	2L, 30C7–D1	AF190455	Haug-Collet et al.[38]	PM Na/Ca,K-exchanger	Unknown
NCKXnnX	CG12061	Unknown	None	Kraev, unpublished	PM Na/Ca,K-exchanger	Unknown
NCKX82A	CG1090	3R, 82A5–B1	None	Kraev, unpublished	PM Na/Ca,K-exchanger	Unknown
NCKXnnY	CG2893	Unknown	None	Kraev, unpublished	PM Na/Ca,K-exchanger	Unknown
Rya-r44F	CG10844	2R, 44F3-8	D17389	Takeshima et al.[39]	SR Ca-channel (RYR)	Affect embryonic/larval muscle
Itp-r83A	CG1063	3R, 83A7–B1	AJ238949	Sinha and Hasan[40]	SR Ca-channel (IP3R)	Affect moulting, LLR

[a]Excluding plasma membrane calcium channels.
ABBREVIATIONS: PM—plasma membrane; SR—sarco/endoplasmic reticulum; TSR—transitional (Golgi) reticulum; IP3R—inositol-1,4,5-trisphosphate receptor; RYR—ryanodine receptor; LLR—larval lethal recessive.

TABLE 3. Inventory of genes, encoding calcium-transporting proteins of *Homo sapiens*[a]

Gene name	Synonym	Chromosome	Refseq accession	Proposed function
ATP2B1	pmca1	12	NM_001682	PM Ca-ATPase
ATP2B2	pmca2	3	NM_001683	PM Ca-ATPase
ATP2B3	pmca3	X	NM_021949	PM Ca-ATPase
ATP2B4	pmca4	1	NM_001584	PM Ca-ATPase
ATP2A1	serca1	16	NM_00168	SR Ca-ATPase
ATP2A2	serca2	12	NM_001681	SR Ca-ATPase
ATP2A3	serca3	17	NM_005173	SR Ca-ATPase
ATP2C1	spca1	3	NM_014382	TSR (Golgi) Ca-ATPase
ATP2C2[b]	spca2[b]	16	NM_014861	TSR (Golgi) Ca-ATPase
SLC8A1	ncx1	2	NM_021097	PM Na/Ca-exchanger
SLC8A2	ncx2	19	XM_038969	PM Na/Ca-exchanger
SLC8A3	ncx3	14	NM_033262	PM Na/Ca-exchanger
SLC24A1	nckx1	15	NM_004727	PM Na/Ca,K-exchanger
SLC24A2	nckx2	9	NM_020344	PM Na/Ca,K-exchanger
SLC24A3	nckx3	20	NM_020689	PM Na/Ca,K-exchanger
SLC24A4[c]	nckx4[c]	14	None	PM Na/Ca,K-exchanger
RYR1	ryr1	19	NM_000540	SR Ca-release channel (RYR)
RYR2	ryr2	1	NM_001035	SR Ca-release channel (RYR)
RYR3	ryr3	15	NM_001036	SR Ca-release channel (RYR)
ITPR1	ip3r1	3	NM_002222	SR Ca-release channel (IP3R)
ITPR2	ip3r2	12	NM_002223	SR Ca-release channel (IP3R)
ITPR3	ip3r3	6	NM_002224	SR Ca-release channel (IP3R)

[a]Excluding plasma membrane calcium channels.
[b]Kraev, unpublished observation.
[c]Kraev and Lytton, unpublished observation.
ABBREVIATIONS: PM–plasma membrane; SR–sarco/endoplasmic reticulum; TSR–transitional (Golgi) reticulum; RYR–ryanodine receptor; IP3R–inositol-1,4,5-trisphosphate receptor.

aspects of embryo development,[28,29] indicate the importance of investigating in greater detail the role of these proteins in the early stages of mammalian development. Such studies may help reveal why seemingly small alterations in coding sequences of these genes bring about dramatic disease phenotypes in man.[3]

REFERENCES

1. BERRIDGE, M.J. 1998. Neuronal calcium signaling. Neuron **21**: 13–26.
2. CATTERALL, W.A. 2000. Structure and regulation of voltage-gated Ca^{2+} channels. Annu. Rev. Cell. Dev. Biol. **16**: 521–555.

3. MACLENNAN, D.H. 2000. Ca^{2+} signalling and muscle disease. Eur. J. Biochem. **267:** 5291–5297.
4. PHILIPSON, K.D. & D.A. NICOLL. 2000. Sodium-calcium exchange: a molecular perspective. Annu. Rev. Physiol. **62:** 111–133.
5. CARAFOLI, E., L. SANTELLA, D. BRANCA & M. BRINI. 2001. Generation, control, and processing of cellular calcium signals. Crit. Rev. Biochem. Mol. Biol. **36:** 107–260.
6. BERCHTOLD, M.W., H. BRINKMEIER & M. MUNTENER. 2000. Calcium ion in skeletal muscle: its crucial role for muscle function, plasticity, and disease. Physiol. Rev. **80:** 1215–1265.
7. THE C. ELEGANS CONSORTIUM. 1998. Genome sequence of the nematode C. elegans: a platform for investigating biology. Science **282:** 2012–2018.
8. KRAEV, A., N. KRAEV & E. CARAFOLI. 1999. Identification and functional expression of the plasma membrane calcium ATPase gene family from Caenorhabditis elegans. J Biol. Chem. **274:** 4254–4258.
9. KRAEV, A., B.D. QUEDNAU, S. LEACH, et al. 2001. Molecular cloning of a third member of the potassium-dependent sodium-calcium exchanger gene family, NCKX3. J. Biol. Chem. **276:** 23161–23172.
10. APWEILER, R., T.K. ATTWOOD, A. BAIROCH, et al. 2000. InterPro—an integrated documentation resource for protein families, domains and functional sites. Nucl. Acids Res. **16:** 1145–1150.
11. ALTSCHUL, S.F., W. GISH, W. MILLER, et al. 1990. Basic local alignment search tool. J. Mol. Biol. **215:** 403–410.
12. TATUSOV, R.L., A.R. MUSHEGIAN, P. BORK, et al. 1996. Metabolism and evolution of Haemophilus influenzae deduced from a whole-genome comparison with Escherichia coli. Curr. Biol. **6:** 279–291.
13. FROHMAN, M.A. 1993. Rapid amplification of complementary DNA ends for generation of full-length complementary DNAs: thermal RACE. Methods Enzymol. **218:** 340–356.
14. LITTLETON, J.T. & B. GANETZKY. 2000. Ion channels and synaptic organization: analysis of the Drosophila genome. Neuron **26:** 35–43.
15. RANDALL, A. & C.D. BENHAM. 1999. Recent advances in the molecular understanding of voltage-gated Ca^{2+}-channels. Mol. Cell. Neurosci. **14:** 255–272.
16. ELLIOTT, A.C. 2001. Recent developments in non-excitable cell calcium entry. Cell Calcium **30:** 73–93.
17. REILÄNDER, H., A. ACHILLES, U. FRIEDEL, et al. 1992. Primary structure and functional expression of the Na/Ca,K-exchanger from bovine rod photoreceptors. EMBO J. **11:** 1689–1695.
18. PRINSEN, C.F., R.T. SZERENCSEI & P.P. SCHNETKAMP. 2000. Molecular cloning and functional expression of the potassium-dependent sodium-calcium exchanger from human and chicken retinal cone photoreceptors. J. Neurosci. **20:** 1424–1434.
19. GUNTESKI-HAMBLIN, A.-M., D.M. CLARKE & G.E. SHULL. 1992. Molecular cloning and tissue distribution of alternatively spliced mRNAs encoding possible mammalian homologs of the yeast secretory pathway calcium pump. Biochemistry **31:** 7600–7608.
20. SUDBRAK, R., J. BROWN, C. DOBSON-STONE, et al. 2000. Hailey-Hailey disease is caused by mutations in ATP2C1 encoding a novel Ca(2+) pump. Hum. Mol. Genet. **9:** 1131–1140.
21. VAN BAELEN, K., J. VANOEVELEN, L. MISSIAEN, et al. 2001. The Golgi PMR1 P-type ATPase of Caenorhabditis elegans. Identification of the gene and demonstration of calcium and manganese transport. J. Biol. Chem. **276:** 10683–10691.
22. BAYLIS, H.A., T. FURUICHI, F. YOSHIKAWA, et al. 1999. Inositol 1,4,5-trisphosphate receptors are strongly expressed in the nervous system, pharynx, intestine, gonad and excretory cell of Caenorhabditis elegans and are encoded by a single gene (itr-1). J. Mol. Biol. **294:** 467–476.
23. GOWER, N.J., G.R. TEMPLE, J.E. SCHEIN, et al. 2001. Dissection of the promoter region of the inositol 1,4,5-trisphosphate receptor gene, itr-1, in C. elegans: a molecular basis for cell-specific expression of IP3R isoforms. J. Mol. Biol. **306:** 145–157.
24. HILFIKER, H., M.A. STREHLER-PAGE, T.P. STAUFFER, et al. 1993. Structure of the gene encoding the human plasma membrane calcium pump isoform 1. J. Biol. Chem. **268:** 19717–19725.

25. TAMURA, T., M. HASHIMOTO, J. ARUGA, et al. 2001. Promoter structure and gene expression of the mouse inositol 1,4,5-trisphosphate receptor type 3 gene. Gene **275:** 169–176.
26. TAM, T., E. MATHEWS, T.P. SNUTCH & W.R. SCHAFER. 2000. Voltage-gated calcium channels direct neuronal migration in Caenorhabditis elegans. Dev. Biol. **226:** 104–117.
27. ACHARYA, J.K., K. JALINK, R.W. HARDY, et al. 1997. InsP3 receptor is essential for growth and differentiation but not for vision in Drosophila. Neuron **18:** 881–887.
28. SULLIVAN, K.M.C., K. SCOTT, C.S. ZUKER & G.M. RUBIN. 2000. The ryanodine receptor is essential for larval development in Drosophila melanogaster. Proc. Natl. Acad. Sci. USA **97:** 5942–5947.
29. ZWAAL, R.R., K. VAN BAELEN, J.T. GROENEN, et al. 2001. The sarco-endoplasmic reticulum Ca^{2+} ATPase is required for development and muscle function in C. elegans. J. Biol. Chem. In press.
30. ZWAAL, R.R., K. VAN BAELEN, J.T. GROENEN, et al. 2001. The sarco-endoplasmic reticulum Ca^{2+} ATPase is required for development and muscle function in Caenorhabditis elegans. J. Biol. Chem. **276:** 43557–43563.
31. KRAEV, A., I. CHUMAKOV & E. CARAFOLI. 1996. The organization of the human gene NCX1 encoding the sodium–calcium exchanger. Genomics **37**(1): 105–112.
32. SZERENCSEI, R.T., J.E. TUCKER, C.B. COOPER, et al. 2000. Minimal domain requirement for cation transport by the potassium-dependent Na/Ca-K exchanger. Comparison with an NCKX paralog from Caenorhabditis elegans. J. Biol. Chem. **275:** 669–676.
33. MARYON, E.B., R. CORONADO & P. ANDERSON. 1996. unc-68 encodes a ryanodine receptor involved in regulating C. elegans body-wall muscle contraction. J. Cell Biol. **134:** 885–893.
34. LEE, R.Y., L. LOBEL, M. HENGARTNER, et al. 1997. Mutations in the alpha1 subunit of an L-type voltage-activated Ca^{2+} channel cause myotonia in Caenorhabditis elegans. EMBO J. **16:** 6066–6076.
35. SCHAFER, W.R. & C.J. KENYON. 1995. A calcium-channel homologue required for adaptation to dopamine and serotonin in Caenorhabditis elegans. Nature **375:** 73–78.
36. MAGYAR, A. & A. VARADI. 1990. Molecular cloning and chromosomal localization of a sarco/endoplasmic reticulum-type Ca2(+)-ATPase of Drosophila melanogaster. Biochem. Biophys. Res. Commun. **173:** 872–877.
37. SCHWARZ, E.M. & S. BENZER. 1997. Calx, a Na–Ca exchanger gene of Drosophila melanogaster. Proc. Natl. Acad. Sci. USA **94:** 10249–10254.
38. HAUG COLLET, K., B. PEARSON, R. WEBEL, et al. 1999. Cloning and characterization of a potassium-dependent sodium/calcium exchanger in Drosophila. J. Cell Biol. **147:** 659–670.
39. TAKESHIMA, H., M. NISHI, N. IWABE, et al. 1994. Isolation and characterization of a gene for a ryanodine receptor/calcium release channel in Drosophila melanogaster. FEBS Lett. **337:** 81–87.
40. SINHA, M. & G. HASAN. 1999. Sequencing and exon mapping of the inositol 1,4,5-trisphosphate receptor cDNA from Drosophila embryos suggests the presence of differentially regulated forms of RNA and protein. Gene **233:** 271–276.

Sodium–Calcium Exchanger NCX1, NCX2, and NCX3 Transcripts in Developing Rat Brain

S.K. POLUMURI,[a] A. RUKNUDIN,[a] MARGARET M. McCARTHY,[b] TARA S. PERROT-SINAL,[b] AND D.H. SCHULZE[a]

[a]*Department of Microbiology and Immunology,* [b]*Department of Physiology, School of Medicine, University of Maryland, Baltimore, Maryland 21201, USA*

KEYWORDS: NCX genes; quantitative RT-PCR; brain development

INTRODUCTION

Calcium plays an important role in regulating a variety of neuronal processes, and neurons have low resting levels of free intracellular calcium, like other cells. After physiological events trigger calcium influx into neurons, calcium must be extruded, because sustained increase in the cytosol may lead to cell death.[1] Two systems, plasma membrane Ca^+ ATPase[2] and the sodium–calcium (Na^+/Ca^{2+}) exchanger,[3] can result in efflux of intracellular calcium. The Na^+/Ca^{2+} exchanger, which is particularly active in heart and neurons, uses the Na^+ gradient generated by the Na^+/K^+-ATPase to remove Ca^{2+} from the cytosol.[4] Three Na^+/Ca^{2+} exchanger genes have been cloned in mammals.[5–7] NCX1 is expressed in all tissues, whereas significant amounts of NCX2 and NCX3 mRNA have only been detected in brain and skeletal muscles.[4,8] The fact that there is little functional difference in these three Na^+/Ca^{2+} exchangers[9] suggests that NCX gene expression could be regulated during development. This study was designed to determine the relative amounts of expression of 3 NCX genes in rat brain during development, using a novel method of quantitative end-labeled RT-PCR. We are summarizing the results obtained in 2 regions of the rat brain telencephalon and spinal cord.

METHODS

Total RNA was isolated from the two regions of brain, telencephalon and spinal cord of male and female rats during three stages (neonatal/new born, juvenile/21 days, and adult/2 months) of development. One microgram of total RNA was used to prepare the cDNA according to the manufacturer's (Invitrogen) instructions for analysis.

Address for correspondence: S. K. Polumuri, Department of Microbiology and Immunology, University of Maryland School of Medicine, 655 W. Baltimore Street, Baltimore, MD 21201. Voice: 410-706-5180; fax: 410-706-2129.
polumuri@yahoo.com

To determine the relative amounts of related gene products, we have modified our quantitative end-labeled reverse transcriptase PCR (QERT-PCR) method to study relative amounts of NCX family genes.[10] The primers for the NCX gene family were designed in regions of greatest sequence identity that flank areas in which there are natural differences in the length between the genes. The PCR products for NCX1 to NCX2 differ by 15 bases, and NCX2 and NCX3 differ by 12 bases. PCR was performed using the following oligonucleotides: NCX1-F (rat NCX1, 744–761) tggagcatctttgcctat, NCX1-R (rat NCX1, 1191–1210) cggtaaaatgctcggctctt, NCX2-F (rat NCX2, 610–627) tggagcatctttgcctat, NCX2-R (rat NCX2, 1042–1061) cggtagaaggcacggctctt, NCX3-F (rat NCX3, 1458–1475) tggagcgtctttgcctat, and NCX3-R (rat NCX3, 1878–1897) cggtagaaagcacggctctt. For an internal control, S16 ribosomal protein gene was used. The following oligonucleotides were used for the control: rat S16-F (rat S16, 91–111) cggacgcaagaaaacagccac and S16-R (rat S16, 473–493) ttatcggtaggatttctggta. A mixture of the forward primers were end-labeled with ^{32}P and T_4 kinase and used in the PCR for 30 cycles (94°C for 1 min, 40°C for 1 min, 72°C for 1.5 min). Samples were withdrawn at different PCR cycles and run on a 5% sequencing gel. After electrophoresis, the dried gels were analyzed using the phosphoimager (Amersham Pharmacia Biotech.).

RESULTS AND DISCUSSION

Quantitation of Relative Amounts of Three NCX Gene Transcripts

To evaluate the method, cloned cDNA samples from the three NCX genes (used as templates) were mixed in different ratios, and PCR amplification was performed. When the ratios of the three templates were varied, the band intensities of the PCR products agreed with the relative amounts of starting DNAs, confirming the reliability of the method (data not shown).

The telencephalon (forebrain) differentiates into major components of the brain in the adult, including the cerebral cortex, basal ganglia (striatum and septum), hippocampus, and olfactory bulb. Our analysis showed dramatic changes in the gene products for the three genes (NCX1-3) in the telencephalon during development (see FIG. 1A). As is seen in all brain tissues at early development (neonatal), NCX1 is the predominant transcript. In the juvenile and adult, there is a change in NCX1 and NCX2. The relative proportion of NCX1 decreases with respect to NCX2. The level of NCX3 appears unchanged with respect to the other two transcripts. These results are in agreement with recent reports[4,11] that show a predominance of NCX2 transcripts in the telencephalon. In adult telencepahalon, both NCX1 and NCX2 were equally expressed, whereas the NCX3 was in lower amounts (FIG. 1A and C). This result differs from recent reports in which the adult cerebral cortex, striatum/septum, and hippocampus were shown to have NCX2 levels higher than NCX1 and NCX3.[12] One possible explanation for this difference could be use of the entire telencephalon in this study. NCX2 level was lower than NCX1 in some of the differentiated parts independently (i.e., olfactory bulb results are not shown).

In contrast, in the spinal cord NCX1 levels were predominant throughout development (FIG. 1B and D). These results are in agreement with earlier reports in which in spinal cord/brain stem, cultured neuronal cells, and astrocytes showed the same

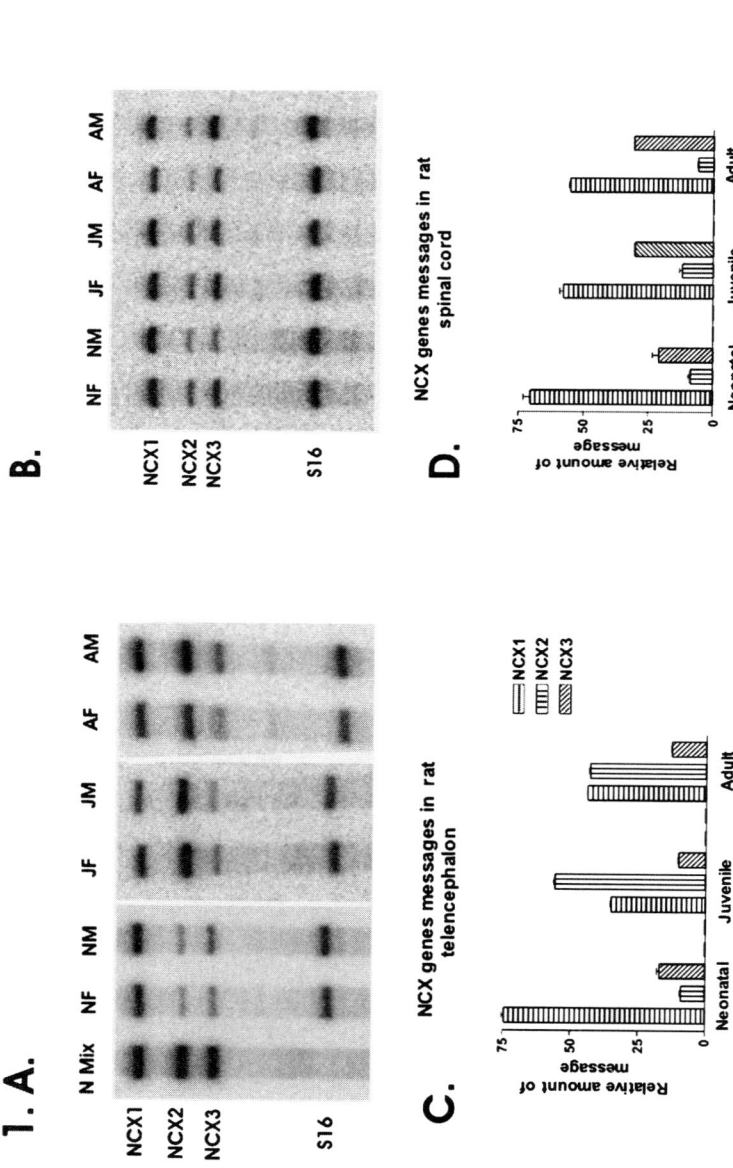

FIGURE 1. Expression levels of NCX1, NCX2 and NCX3 in rat telencephalon and spinal cord. (**A**) and (**B**) are the autoradiographs for telencephalon and spinal cord, respectively. The results of the QERT-PCR are shown, and lanes are marked to show the age (N = neonatal, J = juvenile, A = adult) and sex (F = female, M = male) of brain tissue. Nmix is the control, an equal ratio of NCX1, NCX2, and NCX3 cDNAs. Relative expression levels of NCX1, NCX2, and NCX3 in rat telencephalon and spinal cord are shown in (**C**) and (**D**), respectively (average of male and female samples, $n = 4–8$).

pattern.[11,12] There were no sex-specific differences in NCX1-3 transcript levels during the development of the brain.

CONCLUSIONS

Quantitation of NCX1, 2, and 3 transcripts in the brains of different age groups was determined by using QERT-PCR. NCX2 transcript was increased in juvenile rat telencephalon, whereas in spinal cord NCX1 levels were predominant throughout development. There were no sex-specific differences in NCX1-3 transcript levels during development of the brain. The results demonstrate the developmental and regional differences in the expression among the members of the NCX gene family in the rat brain.

REFERENCES

1. BERRIDGE, M.J. 1998. Neuron **21:** 13–26.
2. CARAFOLI, E. 1994. FASEB J. **8:** 993–1002.
3. PHILIPSON, K.D. & D.A. NICOLL. 1992. Curr. Opin. Cell Biol. **4:** 678–683.
4. LI, L., D. GUERINI & E. CARAFOLI. 2000. J. Biol. Chem. **275:** 20903–20910.
5. NICOLL, D.A., S. LONGONI & K.D. PHILIPSON. 1990. Science **250:** 562–565.
6. LOW, W., J. KASIR & H. RAHAMIMOFF. 1993. FEBS Lett. **316:** 63–67.
7. NICOLL, D.A., B.D. QUEDNAU, Z. QUI, et al. 1996. J. Biol. Chem. **271:** 24914–24921.
8. FRAYSSE, B., T. ROUAUD, M. MILLOUR, et al. 2001. Am. J. Physiol. Cell Physiol. **280:** 146–154.
9. LINK, B., Z. QIU, Z. HE, et al. 1998. Am. J. Physiol. Cell. Physiol. **274:** 415–423.
10. HE, S., A. RUKNUDIN, L.L. BAMBRICK, et al. 1998. J. Neurosci. **18:** 4833–4841.
11. SAKAUE, M., H. NAKAMURA, I. KANEKO, et al. 2000. Brain Res. **881:** 212–216.
12. LI, Y. & R.A. COLVIN. 1997. Mol. Brain Res. **50:** 285–292.

Differential Expression of Na/Ca Exchanger and Na/Ca+K Exchanger Transcripts in Rat Brain

XIAO-FANG LI AND JONATHAN LYTTON

Cardiovascular Research Group, Department of Biochemistry and Molecular Biology, University of Calgary, Calgary, Alberta, Canada T2N 4N1

KEYWORDS: *in situ* hybridization; Northern blotting; developmental expression

INTRODUCTION

Sodium–calcium exchanger–driven Ca^{2+} extrusion across the plasma membrane is an important determinant of intracellular Ca^{2+} control. Detailed structural and functional studies have revealed the existence of two classes of exchanger proteins: sodium–calcium exchanger, NCX, and potassium-dependent sodium–calcium exchanger, NCKX. Three NCX genes (NCX1, NCX2, and NCX3) and three NCKX genes (NCKX1, NCKX2, and NCKX3) have been cloned, and all but NCKX1 are expressed in brain.[1–6] The individual physiological roles for each of these different exchanger gene products in intracellular calcium handling are currently unclear. Recent studies indicate that, like NCKX1 in rod outer segments of eye, NCKX2 and probably also NCKX3 transport K^+ in addition to Na^+ and Ca^{2+}.[7–10] Such an altered transport stoichiometry compared to the NCX family suggests that NCKX family members may be designed to operate in environments with unique Ca^{2+} handling requirements. This led us to explore the differential expression of NCX and NCKX transcripts in brain.

DISTRIBUTION STUDIES

In situ hybridization was used to study the distribution of each transcript in brain using digoxigenin-labeled probes and paraffin-embedded serial coronal sections of the whole rat brain and sagittal sections of mouse brain, as previously described.[5,6] Overall results demonstrated that NCX1, NCX2, NCKX2, and NCKX3 genes were highly expressed in brain; and a moderate level of NCX3 was detected only in me-

Address for correspondence: Jonathan Lytton, Ph.D., Department of Biochemistry & Molecular Biology, University of Calgary Health Sciences Centre, Room 2518, 3330 Hospital Drive NW, Calgary, AB, Canada T2N 4N1. Voice: 403-220-2893; fax: 403-283-4841.
jlytton@ucalgary.ca

dulla oblongata region. The distribution of the four highly expressed isoforms was to a great extent complementary, and the expression pattern of each gene was unique. Only a few regions, including CA1, CA2, CA3, dentate gyrus of hippocampus, tenia tecta, and the piriform cortex, displayed abundant expression of all 4 isoforms. Cells with the highest NCX1 mRNA levels included all layers of cortex, granule cells of olfactory bulb, septum, nucleus of lateral olfactory tract, amygdaloid, thalamus (reticular nucleus, medial-dorsal nucleus and parafasicular nucleus), hippocampus, pontine nucleus, oculomotor nucleus, most medulla oblongata area, inferior olive nucleus, and Purkinje cells and Golgi cells in cerebellum. Labeling for NCX2 was highest in all 6 layers of cortex, hippocampus, and striatum. High levels of NCKX2 were widely distributed in all 6 layers of cortex; striatum; most thalamic nuclei (including medial habenular, ventral, posterior, ventral-posterior, lateral-posterior, ventral-anterior, medial-dorsal, reticular, geniculate, and central-lateral nucleus); subthalamic nucleus; ventral part of zona incerta; hippocampus; medial genicula nucleus; anterior pretectal nucleus; pontine nucleus; most medulla oblongata area; basket, stellate, and Purkinje cells of cerebellum; and cerebellar nuclei. Strong staining for NCKX3 was observed in layer IV of cortex, thalamus (antero-dorsal nucleus, medial-dorsal nucleus, and parafascicular nucleus), hippocampus, inferior olive nucleus, and basket and stellate cells of cerebellum.

DEVELOPMENTAL STUDIES AND QUANTITATIVE ANALYSIS

Northern blotting was used for developmental studies and quantitative analysis. For the developmental experiment, total RNA preparations were isolated from whole brain of various ages of rats. Both NCX1 and NCKX3 transcripts were expressed in brain embryonically (17–19 days dpc). After birth, the level of these transcripts increased slightly and reached a plateau level in the adult. For NCX2 and NCKX2 transcripts, expression was very low until 2 weeks following birth. By the end of week 2, transcript abundance increased dramatically and then leveled off in the adult. Quantitative analysis of exchanger transcript expression was performed on RNA isolated from adult rat brain, using cold sense transcripts as internal standards. The order of abundance was NCX2 > NCX1 > NCKX2 > NCKX3 ≫ NCX3.

SUMMARY AND CONCLUSION

In situ hybridization studies demonstrated that each exchanger gene transcript had a unique distribution pattern in brain, with NCX1 and NCKX2 expressed most widely and NCX2 and NCKX3 relatively localized. Northern blot studies indicated that each exchanger had a distinct developmental pattern. Although NCX2 was relatively localized, quantitative analysis revealed that transcripts of the NCX2 gene were the most abundant, followed in order of abundance by NCX1, NCKX2, and NCKX3. NCX3 was the least abundant isoform. In conclusion, these data clearly demonstrate distinct distribution, expressional level, and developmental patterns for 5 NCX and NCKX genes, which may underlie distinct physiological roles in calcium handling in brain.

REFERENCES

1. NICOLL, D.A., S. LONGONI & K.D. PHILIPSON. 1990. Molecular cloning and functional expression of the cardiac Na,Ca-exchanger. Science **250:** 562–565.
2. LI, Z., S. MATSUOKA, L.V. HRYSHKO, et al. 1994. Cloning of the NCX2 isoform of the plasma membrane Na–Ca exchanger. J. Biol. Chem. **269:** 17434–17439.
3. NICOLL, D.A., B.D. QUEDNAU, Z. QUI, et al. 1996. Cloning of a third mammalian Na^+–Ca^{2+} exchanger, NCX2. J. Biol. Chem. **271:** 24914–24921.
4. REILÄNDER, H., A. ACHILLES, U. FRIEDEL, et al. 1992. Primary structure and functional expression of the Na/Ca,K-exchanger from bovine rod photoreceptors. EMBO J. **11:** 1689–1695.
5. TSOI, M., K.-H. RHEE, D. BUNGARD, et al. 1998. Molecular cloning of a novel potassium-dependent sodium–calcium exchanger from rat brain. J. Biol. Chem. **273:** 4115–4162.
6. KRAEV, A., B.D. QUEDNAU, S. LEACH, et al. 2001. Molecular cloning of a third member of the potassium-dependent sodium–calcium exchanger gene family, NCKX3. J. Biol. Chem. **276:** 23161–23172.
7. CERVETTO, L., L. LAGNADO, R.J. PERRY, et al. 1989. Extrusion of calcium from rod outer segments is driven by both sodium and potassium gradients. Nature **337:** 740–743.
8. SCHNETKAMP, P.P.M., D.K. BASU & R.T. SZERENCSEI. 1989. Na^+–Ca^{2+} exchange in bovine rod outer segments requires and transports K^+. Am. J. Physiol. Cell. Physiol. **257:** C153–C157.
9. DONG, H., P.E. LIGHT, R.J. FRENCH & J. LYTTON. 2001. Electrophysiological characterization and ionic stoichiometry of the rat brain K^+-dependent Na^+/Ca^{2+} exchanger, NCKX2. J. Biol. Chem. **276:** 25919–25928.
10. SZERENCSEI, R.T., C.F. PRINSEN & P.P. SCHNETKAMP. 2001. Stoichiometry of the retinal cone Na/Ca–K exchanger heterologously expressed in insect cells: comparison with the bovine heart Na/Ca exchanger. Biochemistry **40:** 6009–6015.

Sodium–Calcium Exchangers in Olfactory Tissue

D.H. SCHULZE,[a] M. PYRSKI,[b] A. RUKNUDIN,[a] J.W. MARGOLIS,[b]
S.K. POLUMURI,[a] AND F.L. MARGOLIS[b]

[a]*Department of Microbiology-Immunology, University of Maryland School of Medicine, Baltimore, Maryland 21201, USA*

[b]*Department of Anatomy and Neurobiology, University of Maryland School of Medicine, Baltimore, Maryland 21201, USA*

KEYWORDS: olfactory; olfactory bulb; olfactory epithelium; olfactory neuron; OMP; olfactory marker protein; *in situ* hybridization

INTRODUCTION

Role of Ca^{2+} in Olfactory Receptor Neurons (ORNs)

Cytosolic Ca^{2+} is a critical participant in intracellular signaling cascades in virtually all eucaryotic cells. In ORNs, entry of Ca^{2+} in response to odor–receptor interaction is a critical early step in the transduction of chemical to electrical signals.[1–5] In addition, the elevated intracellular level of Ca^{2+} in response to odor stimulation participates in the control of excitation and adaptation of ORNs.[4–9] Ca^{2+} amplifies its effect and increases the transduction current by activating a Ca^{2+}-activated chloride channel.[4,10] In conjunction with calmodulin, Ca^{2+} mediates desensitization of the cAMP-gated channels.[2,9,14] Ca^{2+} also influences the activities of the cAMP-phosphodiesterase[14] and adenylate cyclase as well as other regulatory events in the ORNs.[4,11,12] Thus, regulation of intracellular Ca^{2+} levels is a critical cellular activity. A great deal of effort has been devoted to characterizing the molecules associated with these early events in chemoreception and calcium entry in ORNs of many vertebrate species. Less attention has been paid to the mechanisms by which the ORNs remove or sequester the intracellular Ca^{2+} (Ca_i^{2+}).[3,4,15–21] This is of particular interest, since the cilia of ORNs, the site where the actual chemotransduction takes place, contain little cytoplasm and no discernable endoplasmic reticulum (ER) that could serve as a sink for the entering Ca^{2+}.[1,5,22–25] Furthermore, the cilia emanate from a dendritic knob, whose proximal region is also deficient in ER and mitochondria, although the dendrite itself and the soma contain abundant ER and mitochondria. Thus, the reduction in Ca_i^{2+} most likely depends on the removal of the Ca_i^{2+} by transmembrane transporters in the cilia.

Address for correspondence: Frank L. Margolis Ph.D., Department of Anatomy and Neurobiology, Health Science Facility, University of Maryland School of Medicine, 685 West Baltimore Street, Baltimore, MD 21201. Voice: 410-706-8913; fax: 410-706-2512.
fmargoli@umaryland.edu

Na^+/Ca^{2+} Exchangers in ORN Function

Characterization of the downstream events in the transduction process have lagged behind studies of the initial events of receptor–ligand activation of G proteins and adenylate cyclase, gating of channels, influx of Ca^{2+}, and their regulation. Nevertheless, the mechanisms that regulate adaptation and termination of the response and reset Ca_i^{2+} back to baseline are as important to the overall transduction process as is the initial receptor-mediated Ca^{2+} influx. Reisert and Matthews[16,17] have redirected attention to the role of Na^+/Ca^{2+} exchange in termination of the transduction current. Indeed, manipulation of Na^+/Ca^{2+} exchange efficiency by altering extracellular [Na^+] resulted in effects that are reminiscent of those seen in the olfactory marker protein (OMP)-null mouse.[26,27] Therefore, it was important to address mechanism(s) by which elevated Ca_i^{2+} is returned to baseline and the potential role of OMP as a modulator in this process.

A recent review[4] addresses this question and states that "...the Na^+/Ca^{2+} exchanger is the main mechanism that returns cytoplasmic Ca^{2+} concentration to basal levels after stimulation.... Other mechanisms of Ca^{2+} removal, such as diffusion into the cell body or a Ca^{2+}/ATPase are also likely to be present in the olfactory neuron, but are probably quantitatively less significant." Similar opinions have been expressed by others,[20,21] and functional studies[17,18,28] strongly implicate Na^+/Ca^{2+} exchange as an important component of this process. Recent immunocytochemical evidence for the presence of Na^+/Ca^{2+} exchangers on the cilia and dendritic knobs of ORNs of squid and rats[19,21] supports this view. The antibody used by Lucero *et al.*[19] is directed against the squid Na^+/Ca^{2+} exchanger (NCX-SQ1), whereas that used by Noe *et al.*[21] is directed against the bovine retinal-rod Na^+/Ca^{2+}–K^+ exchanger (NCKX1). This difference in exchanger gene expression in ORNs between the two species could either reflect species variation of exchanger type in ORNs, or else the presence of multiple exchangers in ORNs. The latter seems possible since compartmentalization of Ca^{2+} extrusion mechanisms have been reported in the photoreceptor, another highly specialized sensory neuron.[30] Despite these initial reports, the expression and distribution of Na^+/Ca^{2+} exchangers in ORNs and their roles in maintaining the intracellular levels of Ca^{2+} in ORNs have not been adequately investigated in the vertebrate olfactory system.

RESULTS

Expression of Na^+/Ca^{2+} Exchangers in Rat Olfactory Tissue

There are 3 Na^+/Ca^{2+} exchanger genes (NCX1, 2, 3)[31,32] and 3 Na^+/Ca^{2+}–K^+ exchanger genes (NCKX1, 2, 3).[33–36] Transcription generates multiple mRNA splice variants, as many as 32 for NCX1,[36–38] some of which are expressed in tissue-specific profiles.[31,39–44] The virtual absence of molecular information about which of the NCX and/or NCKX exchanger genes and their mRNA splice variants are expressed by ORNs prompted us to begin to characterize NCX and NCKX gene expression in olfactory tissue. We have demonstrated that mRNAs for all 6 genes are present in rat olfactory mucosa and bulb, 4 of which are illustrated (FIG. 1). In addition we observed the presence of splice variants for NCX1 and NCKX1. Our NCKX1 primers generated amplicons of 351, 275, and 175 bases. Upon separation

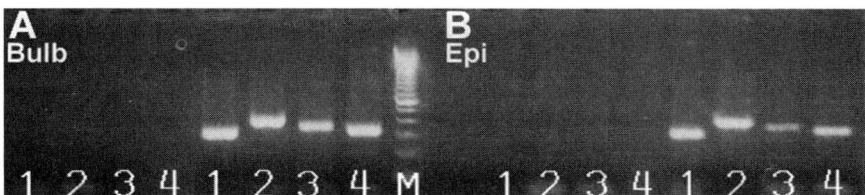

FIGURE 1. Na^+/Ca^{2+} exchanger expression in rat olfactory tissues. PCR products of cDNA generated from olfactory mRNA, in the presence or absence of reverse transcriptase (RT) were electrophoresed in agarose gel and stained with ethidium bromide. **Panel A** (bulb) represents olfactory bulb and **panel B** (epi) represents olfactory epithelium; M = molecular weight markers. Lanes: 1 = NCKX2, 2 = NCX1, 3 = NCX2, 4 = NCX3. In each panel the lanes on the left are from cDNA −RT, and the lanes on the right are +RT. RNA isolated from rat olfactory tissue was treated with DNAse and random primed cDNA prepared. PCR products generated with gene-specific primers demonstrated that mRNAs of all 6 genes are expressed in rat olfactory mucosa and bulb, 4 of which are illustrated. The absence of PCR products when RT was omitted during cDNA synthesis confirmed the absence of contaminating genomic DNA. In each case the amplicon was of the predicted size and was susceptible to appropriate restriction enzyme digestion, confirming its identiy.

and sequencing, these products were found to represent the mRNA splice variants ABCD, BCD, and CD, indicating that at least these 3 mRNA splice variants of NCKX1[36] are expressed in olfactory tissue. For NCX1, the technique of QERT-PCR[41] enabled us to demonstrate the presence of several splice variants of NCX1 containing either the A or B exon. Thus, we have demonstrated that all 6 Na^+/Ca^{2+} exchanger genes are expressed in rat olfactory epithelium and olfactory bulb and mRNAs for at least 2 of them are present as several splice variants.

Localization of Na^+/Ca^{2+} Exchanger mRNAs in Mouse Olfactory Mucosa

One of our ultimate aims is to characterize the regulation of Na^+/Ca^{2+} exchanger activity in OMP-null mice and the profile of cellular localization of these exchangers in olfactory tissue. Therefore, it was essential to characterize the expression of these genes in mice as well as in rats. RT-PCR using gene-specific oligonucleotide primer pairs demonstrated the presence of all 6 Na^+/Ca^{2+} exchanger-gene mRNA transcripts in mouse olfactory tissue mRNA. Thus, we have demonstrated the presence of mRNAs for NCX1, 2, 3 and NCKX1, 2, 3 in tissue from rat and mouse nasal olfactory mucosa and olfactory bulbs.

The cellular complexity of the olfactory epithelium precludes an *a priori* assignment as to which cell types in the tissue express which of the exchanger genes. To address the cell-specificity of exchanger gene expression, we have begun to characterize the mRNA expression pattern by *in situ* hybridization. Analysis of the NCX1 expression pattern in mouse olfactory epithelium shows that NCX1 mRNA is present in virtually all olfactory neurons (FIG. 2). By contrast, there is essentially no detectable expression in sustentacular cells or in other nonneuronal components of the olfactory neuroepithelium. Comparison with the expression pattern of OMP mRNA, a marker of mature ORNs, demonstrates that the expression of NCX1 mRNA in the olfactory epithelium is restricted to ORNs and that its expression is independent of ORN maturity.

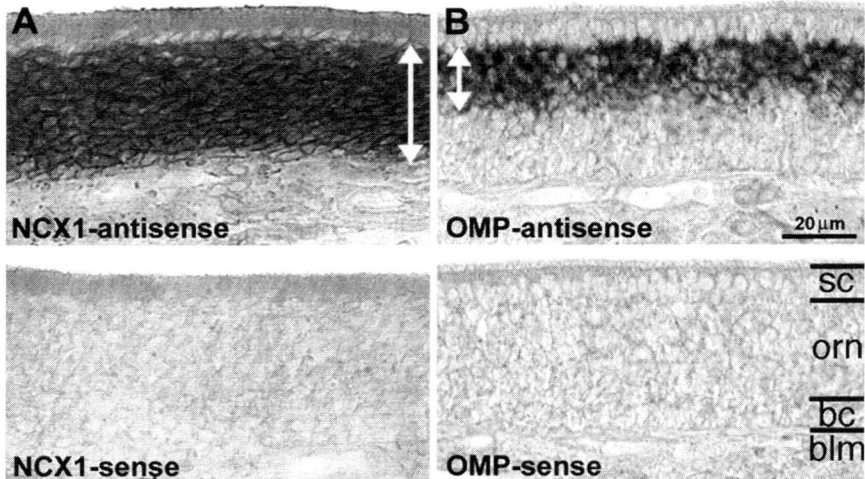

FIGURE 2. *In situ* localization of NCX1 and OMP mRNAs in mouse olfactory epithelium using digoxigenin-labeled probes. *In situ* hybridization for **(A)** NCX1 and **(B)** OMP in 15-μm coronal cryosections of mouse olfactory epithelium using digoxigenin-labeled sense and antisense riboprobes. Riboprobes were generated by *in vitro* transcription using subcloned fragments from the coding region of the appropriate gene. Expression of both mRNAs is exclusively restricted to ORNs. Whereas a signal for NCX1 mRNA **(A)** is present throughout the depth of the neuroepithelium, indicating expression in both mature and immature ORNs (*long arrow, top panel*), OMP mRNA **(B)** is only evident in mature ORNs (*short arrow, top panel*). No hybridization signals were obtained with sense probe controls for either OMP or NCX1. The scale bar is as indicated. Abbreviations: sc, sustentacular cells; orn, olfactory receptor neurons; bc, basal cells; blm, basal lamina.

SUMMARY

We have demonstrated that mRNAs for NCX1,2,3 (also see Polumuri *et al.* and Schulze *et al.* in this volume) and NCKX1,2,3 are all expressed in the olfactory epithelium of adult rats and mice. Furthermore, we demonstrate the presence of NCX1 mRNA in mature and immature ORNs by *in situ* hybridization. Characterization of the cellular expression of other Na^+/Ca^{2+} exchanger mRNAs is in progress. Our molecular data confirm the reports of Na^+/Ca^{2+} exchanger activity in ORNs and provide the basis for further characterization of the expression and function of Na^+/Ca^{2+} exchangers in olfactory chemotransduction.

ACKNOWLEDGMENTS

This work was supported in part by National Institutes of Health Grants DC-03112 (FLM) and HL62521(DHS).

REFERENCES

1. LEINDERS-ZUFALL, T., et al. 1998. Imaging odor-induced calcium transients in single olfactory cilia: specificity of activation and role in transduction. J. Neurosci. **18:** 5630–5639.
2. SCHILD, D. & D. RESTREPO. 1998. Transduction mechanisms in vertebrate olfactory receptor cells. Physiol Rev. **78:** 429–466.
3. TAREILUS, E., J. NOE & H. BREER. 1995. Calcium signals in olfactory neurons. Biochim. Biophys. Acta **1269:** 129–138.
4. MENINI, A. 1999. Calcium signalling and regulation in olfactory neurons. Curr. Opin. Neurobiol. **9:** 419–426.
5. BOEKHOFF, I., C. KRONER & H. BREER. 1996. Calcium controls second-messenger signalling in olfactory cilia. Cell. Signal. **8:** 167–171.
6. TORRE, V.J.F., et al. 1995. Transduction and adaptation in sensory receptor cells. J. Neurosci. **15:** 7757–7768.
7. KURAHASHI, T. & A. MENINI. 1997. Mechanism of odorant adaptation in the olfactory receptor cell. Nature **385:** 725–729.
8. LEINDERS-ZUFALL, T., M. MA & F. ZUFALL. 1999. Impaired odor adaptation in olfactory receptor neurons after inhibition of Ca^{2+}/calmodulin kinase II. J. Neurosci. **19:** RC19.
9. REICH, G., et al. 1999. Calcium regulation of cyclic nucleotide signaling in lobster olfactory receptor neurons. J. Neurochem. **73:** 147–152.
10. LOWE, G. & G.H. GOLD. 1993. Nonlinear amplification by calcium-dependent chloride channels in olfactory receptor cells. Nature **366:** 283–286.
11. ZUFALL, F. & T. LEINDERS-ZUFALL. 2000. The cellular and molecular basis of odor adaptation. Chem. Senses **25:** 473–481.
12. CHEN, S., et al. 2000. Blocking adenylyl cyclase inhibits olfactory generator currents induced by "IP(3)-odors." J. Neurophys. **84:** 575–580.
13. ZUFALL, F., G.M. SHEPHERD & S. FIRESTEIN. 1991. Inhibition of the olfactory cyclic nucleotide gated ion channel by intracellular calcium. Proc. R. Soc. London Ser. B: Biol. Sci. **246:** 225–230.
14. BORISY, F.F., et al. 1992. Calcium/calmodulin-activated phosphodiesterase expressed in olfactory receptor neurons. J. Neurosci. **12:** 915–923.
15. ZUFALL, F., T. LEINDERS-ZUFALL & C.A. GREER. 2000. Amplification of odor-induced $Ca^{(2+)}$ transients by store-operated Ca^{2+} release and its role in olfactory signal transduction. J. Neurophys. **83:** 501–512.
16. REISERT, J. & H.R. MATTHEWS. 2001. Response properties of isolated mouse olfactory receptor cells. J. Physiol. **530:** 113–122.
17. REISERT, J. & H.R. MATTHEWS. 1998. Na^+-dependent Ca^{2+} extrusion governs response recovery in frog olfactory receptor cells. J. Gen. Physiol. **112:** 529–535.
18. DANACEAU, J.P. & M.T. LUCERO 2000. Electrogenic Na^+/Ca^{2+} exchange. A novel amplification step in squid olfactory transduction. J. Gen. Physiol. **115:** 759–768.
19. LUCERO, M.T., W. HUANG & T. DANG. 2000. Immunohistochemical evidence for the Na^+/Ca^{2+} exchanger in squid olfactory neurons. Phil. Trans. R. Soc. London Ser. B: Biol. Sci. **355:** 1215–1218.
20. JUNG, A., et al. 1994. Sodium/calcium exchanger in olfactory receptor neurones of *Xenopus laevis*. Neuroreport **5:** 1741–1744.
21. NOE, J., et al. 1997. Sodium/calcium exchanger in rat olfactory neurons. Neurochem. Int. **30:** 523–531.
22. NAKAMURA, T. & G.H. GOLD. 1987. A cyclic nucleotide-gated conductance in olfactory receptor cilia. Nature **325:** 442–444.
23. BOEKHOFF, I., et al. 1990. Rapid activation of alternative second messenger pathways in olfactory cilia from rats by different odorants. EMBO J. **9:** 2453–2458.
24. SCHILD, D. & J. BISCHOFBERGER. 1991. Ca^{2+} modulates an unspecific cation conductance in olfactory cilia of *Xenopus laevis*. Exp. Brain Res. **84:** 187–194.
25. MATSUZAKI, O., et al. 1999. Localization of the olfactory cyclic nucleotide-gated channel subunit 1 in normal, embryonic and regenerating olfactory epithelium. Neuroscience **94:** 131–140.

26. BUIAKOVA, O.I., et al. 1996. Olfactory marker protein (OMP) gene deletion causes altered physiological activity of olfactory sensory neurons. Proc. Natl. Acad. Sci. USA **93:** 9858–9863.
27. IVIC, L., et al. 2000. Adenoviral vector-mediated rescue of the OMP-null phenotype *in vivo*. Nat. Neuro. **3:** 1113–1120.
28. PIPER, D.R. & M.T. LUCERO. 1999. Calcium signalling in squid olfactory receptor neurons. Biol. Signals Recept. **8:** 329–337.
29. LUCERO, M., W. HUANG & T. DANG. 2000. Immunohistochemical evidence for the Na^+/Ca^{2+} exchanger in squid olfactory neurons. Phil. Trans. R. Soc. London Ser. B: Biol. Sci. **355:** 1215–1218.
30. KRIZAJ, D. & D.R. COPENHAGEN. 1998. Compartmentalization of calcium extrusion mechanisms in the outer and inner segments of photoreceptors. Neuron **21:** 249–256.
31. BLAUSTEIN, M.P. & W.J. LEDERER. 1999. Sodium/calcium exchange: its physiological implications. Physiol. Rev. **79:** 763–854.
32. SHIGEKAWA, M. & T. IWAMOTO. 2001. Cardiac Na^+-Ca^{2+} exchange: molecular and pharmacological aspects. Circ. Res. **88:** 864–876.
33. SCHNETKAMP, P. 1995. How does the retinal rod Na-Ca+K exchanger regulate cytosolic free Ca^{2+}? J. Biol. Chem. **270:** 13231–13239.
34. SCHNETKAMP, P., J.E. TUCKER & R.T. SZERENCSEI. 1995. Ca^{2+} influx into bovine retinal rod outer segments mediated by $Na^+/Ca^{2+}/K^+$ exchange. Am. J. Physiol. **269:** C1153–1159.
35. PRINSEN, C., R.T. SZERENCSEI & P.P. SCHNETKAMP. 2000. Molecular cloning and functional expression of the potassium-dependent sodium–calcium exchanger from human and chicken retinal cone photoreceptors. J. Neurosci. **20:** 1424–1434.
36. POON, S., S. LEACH, X.F. LI, et al. 2000. Alternatively spliced isoforms of the rat eye sodium/calcium$^+$potassium exchanger NCKX1. Am. J. Physiol. Cell. Physiol. **278:** C651–660.
37. SCHULZE, D.H., et al. 1996. Alternative splicing of the Na^+-Ca^{2+} exchanger gene, NCX1. Ann. N.Y. Acad. Sci. **779:** 46–57.
38. KOFUJI, P., W.J. LEDERER & D.H. SCHULZE. 1994. Mutually exclusive and cassette exons underlie alternatively spliced isoforms of the Na/Ca exchanger. J. Biol. Chem. **269:** 5145–5149.
39. LEDERER, W.J., et al. 1996. The molecular biology of the Na^+-Ca^{2+} exchanger and its functional roles in heart, smooth muscle cells, neurons, glia, lymphocytes, and nonexcitable cells. Ann. N.Y. Acad. Sci. **779:** 7–17.
40. RUKNUDIN, A., et al. 2000. Functional differences between cardiac and renal isoforms of the rat Na^+-Ca^{2+} exchanger NCX1 expressed in *Xenopus* oocytes. **529**(Pt. 3): 599–610.
41. HE, S., et al. 1998. Isoform-specific regulation of the Na^+/Ca^{2+} exchanger in rat astrocytes and neurons by PKA. J. Neurosci. **18:** 4833–4841.
42. QUEDNAU, B., D.A. NICOLL & K.D. PHILIPSON. 1997. Tissue specificity and alternative splicing of the Na^+/Ca^{2+} exchanger isoforms NCX1, NCX2, and NCX3 in rat. Am. J. Physiol. **272**(4 Pt. 1): C1250–1261.
43. SAKAUE, M., et al. 2000. Na^+-Ca2$^+$ exchanger isoforms in rat neuronal preparations: different changes in their expression during postnatal development. Brain Res. **881:** 212–216.
44. VAN EYLEN, F., A. BOLLEN & A. HERSCHUELZ. 2001. NCX1 Na/Ca exchanger splice variants in pancreatic islet cells. J. Endocrinol. **168:** 517–526.

Molecular Identification of the NCX Isoform Expressed in Tracheal Smooth Muscle of Guinea Pig

R. MEJÍA-ELIZONDO,[a] R. ESPINOSA-TANGUMA,[b] AND V.M. SAAVEDRA-ALANIS[a]

Departments of Biochemistry[a] and Physiology,[b] School of Medicine, University of San Luis Potosí, San Luis Potosí, SLP, Mexico, 78210

KEYWORDS: NCX1.3; RT-PCR; alternatively spliced mRNA

INTRODUCTION

In airway smooth muscle (ASM), the intracellular calcium concentration $[Ca^{2+}]_i$ is raised by acetylcholine and histamine mainly by release from sarcoplasmic reticulum stores and to a lesser degree by entry through voltage-operated Ca^{2+} channels (VOC) and possibly receptor-operated Ca^{2+} channels (ROC). During relaxation, $[Ca^{2+}]_i$ returns to basal levels mainly by reuptake into the sarcoplasmic reticulum via a Ca^{2+} ATPase and by extrusion across the sarcolemmal membrane via a Ca^{2+} ATPase and a Na^+/Ca^{2+} exchanger (NCX).[1]

Evidence for NCX activity in ASM has been largely indirect, and the results controversial. NCX mRNA was identified in human tracheal muscle using RT-PCR[2] and shown to correspond to the major renal isoform, NCX1.3.

In our laboratory we also have found evidence that suggests the presence of NCX activity in guinea pig tracheal rings. Because guinea pig trachea represents a very useful model of contraction–relaxation in ASM, expression of NCX at the mRNA level was examined.

METHODS

Reverse Transcription and Polymerase Chain Reaction

Guinea pig trachealis smooth muscle was dissected from epithelium and connective tissue under a stereomicroscope. RNA was extracted from 20 mg of ASM tissue, and 3 µg of RNA were used for cDNA synthesis with AMV reverse transcriptase and oligo dT as a primer. Amplification of the transmembrane constant region of NCX

Address for correspondence: Dr. Ricardo Espinosa-Tanguma, Departamento de Fisiología, Facultad de Medicina, UASLP, Carranza 2405, San Luis Potosi, SLP 78210, Mexico. Voice: 52-444-826-2345, ext. 552; fax: 52-444-817-6976.
espinosr@uaslp.mx

was performed using sense 5'-tgtttcgattatgtgatgcactttct-3' and antisense 5'-cccacattgatgaaagcaaagatggt-3' primers with Taq DNA polymerase. Amplification of the alternatively spliced variable region used sense 5'-tcttctgaggcttctgaggat-3' and antisense 5'-accaattgccgacaaacaga-3' primers. A 150-bp *Eco*RI fragment spanning the alternatively spliced region was obtained by PCR with *Pfu* DNA polymerase and ligated into a pGEM® 3Zf(+) vector (Promega). The cDNA insert was sequenced manually using the Sanger's dideoxy NTP method with [α-^{35}S]dATP and detected by autoradiography.

RESULTS

RT-PCR of smooth muscle RNA using primers from the conserved carboxy-terminal transmembrane segments of NCX1 produced a product of 450 bp, which was similar in size to the fragments amplified from heart and brain NCX (FIG. 1). Amplification of the NCX1 intracellular loop resulted in a DNA band of about 460 bp in ASM, distinct from the 570-bp band from heart and the 530-bp band from brain (FIG. 2). This size difference suggested that the isoform expressed in trachea was different from the major isoform expressed in heart (NCX1.1), which includes exons ACDEF or the major isoforms expressed in brain (NCX1.5 and NCX1.7), which contain the exons ADF and BDF, respectively.

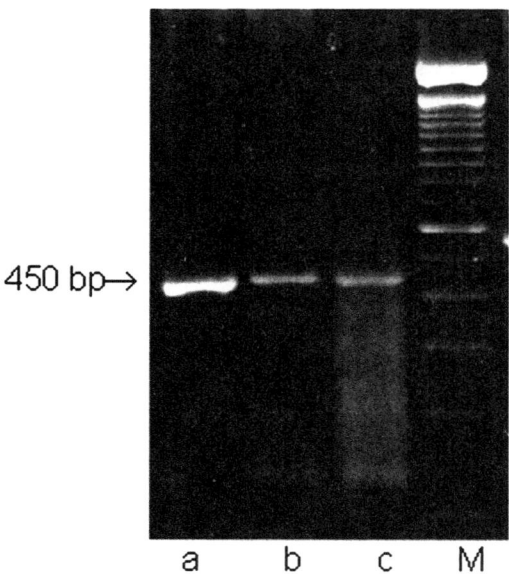

FIGURE 1. RT-PCR for conserved transmembranal NCX region of guinea pig. *Lane a*, heart NCX; *lane b*, trachea smooth muscle NCX; *lane c*, brain NCX. M, 100-bp ladder DNA marker.

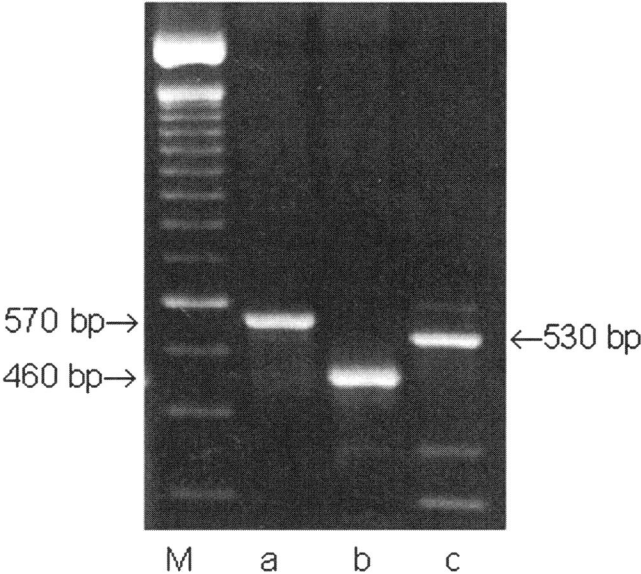

FIGURE 2. RT-PCR for the alternatively spliced NCX region of guinea pig. *Lane a*, heart; *lane b*, trachea smooth muscle; *lane c*, brain. M, 100-bp ladder DNA marker.

We cloned the spliced variant into the pGEM® vector and sequenced it manually, which revealed a sequence corresponding to exons B and D. This result confirms the isoform NCX1.3 reported for human trachea.[2]

DISCUSSION

We have shown that guinea pig tracheal smooth muscle expresses the same isoform, NCX1.3, reported for the human tissue. This alternatively spliced isoform is identical to the major isoform expressed in kidney.[3] In contrast to the human ASM sequence, the guinea pig sequence contains the amino acid phenylalanine at position 20 of exon B, instead of leucine.

Because the mRNA expression level in ASM was qualitatively not much less than that detected in other tissues, like heart and brain, it will be important to investigate whether it is translated at the protein level. Since the alternative isoforms expressed in heart, brain, and kidney seem to be modulated differentially by $[Ca^{2+}]_i$, $[Na^+]_I$, and phosphorylation, it will be interesting to know the role of the isoform reported here in the contractile function of airway smooth muscle.[4,5]

ACKNOWLEDGMENTS

This work was supported by grants from CONACYT to R.E-T. (N-P4094 and 33401-N) and V.M.S-A. (32755-M).

REFERENCES

1. HAKONARSON, K. & M.M. GRUNSTEIN. 1998. Regulation of second messengers associated with airway smooth muscle contraction and relaxation. Am. J. Respir. Crit. Care Med. **158:** S115–122.
2. PITT, A. & A.J. KNOX. 1996. Molecular characterization of the human airway smooth muscle Na^+/Ca^{2+} exchanger. Am. J. Respir. Cell Mol. Biol. **15:** 726–730.
3. TSURUYA, Y., M.M. BERSOHN, Z. LI, D.A. NICOLL & K.D. PHILIPSON. 1994. Molecular cloning and functional expression of the guinea pig cardiac Na^+–Ca^{2+} exchanger. Biochim. Biophys. Acta **1196:** 97–99.
4. QUEDNAU, B.D., A.D. NICOLL & D.K. PHILIPSON. 1997. Tissue specificity and alternative splicing of the Na^+/Ca^{2+} exchanger isoforms NCX1, NCX2, and NCX3 in rat. Am. J. Physiol. **272:** C1250–1261.
5. DYCK, C., A. OMELCHENKO, CH.L. ELIAS, *et al.* 1999. Ionic regulatory properties of brain and kidney splice variants of the NCX1 Na^+–Ca^{2+} exchanger. J. Gen. Physiol. **114:** 701–711.
6. RUKNUDIN, A., S. HE, W.J. LEDERER & D.H. SCHULZE. 2000. Functional differences between cardiac and renal isoforms of the rat Na^+–Ca^{2+} exchanger NCX1 expressed in *Xenopus* oocytes. J. Physiol. **529**(Pt. 3): 599–610.

Localization and Molecular Characterization of the Crayfish NCX

L.M. STINER, Z. ZHANG, AND M.G. WHEATLY

Biological Sciences, Wright State University, Dayton, Ohio 45435, USA

KEYWORDS: NCX; *Procambarus clarkii*; antibody generation; immunohistochemistry; crayfish; calcium homeostasis

INTRODUCTION

The molting cycle of the freshwater crayfish *Procambarus clarkii* has emerged as an ideal model to study the cellular/molecular mechanisms of calcium (Ca) homeostasis.[1] The model is based on the finding that transepithelial Ca flux, which is negligible in intermolt (Ca balance), shows pronounced changes in magnitude and directionality around ecydsis (shedding). In premolt, cuticular Ca is reabsorbed and excreted (negative Ca balance). In the initial 4 days postmolt, crayfish exhibit dramatic net Ca uptake (2–10 mmol/kg/h). In crustaceans, the Ca^{2+} transporting epithelia are the gills (passive diffusional loss or active uptake), antennal gland (kidney, filtration, and reabsorption), hepatopancreas (liver, lumenal Ca storage/dietary uptake), and cuticular hypodermis (demineralization or mineralization). Arthropod molting is coordinated by the steroid ecdysone. Postmolt provides a biological "switch" for activation/upregulation of the Ca^{2+} translocating proteins involved in transepithelial Ca^{2+} influx and associated IC Ca homeostasis, possibly through steroidal regulation of gene expression.

The existing model for crustacean Ca homeostasis is based on *in vitro* studies performed on intermolt animals and extrapolated to other molting stages. The generally accepted model for unidirectional influx (postmolt gill and hepatopancreas/premolt hypodermis/intermolt antennal gland) is that Ca^{2+} enters the apical membrane passively through carrier-mediated facilitated diffusion (Ca^{2+}/nNa^+ [or H^+] exchanger, NCX, that may be electroneutral or electrogenic) or through simple diffusion through verapamil-inhibited Ca^{2+} channels. Active basolateral efflux involves a vanadate-sensitive, high-affinity but low-capacity, calmodulin-dependent Ca^{2+} ATPase (K_m 0.2 µM, J_{max} 1–10 nmol/min/mg) and a low-affinity high-capacity electrogenic NCX whose activity is fueled by the Na^+ pump (K_m 2 µM, J_{max} <10 nmol/mg/min).

Address for correspondence: Dr. L.M. Stiner, Biological Sciences, Wright State University, 3640 Colonel Glenn Hwy., 235A BH, Dayton, OH 45435. Voice: 937-775-2655; fax 937-775-3320.

lstiner788@aol.com

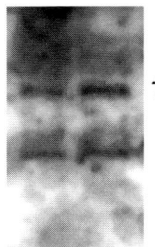

124 kDa

89 kDa

FIGURE 1. Cardiac muscle eluate from the DEAE column showed two bands (MW 124 and 89 kDa) as predicted for NCX.

Intermolt kinetics have suggested that the Ca pump serves a "housekeeping" role in regulating intracellular (IC) [Ca], whereas the NCX is the "workhorse" primarily responsible for basolateral efflux. Subcellular Ca homeostasis involves sequestration in SR/ER (using a Ca pump, SERCA), mitochondria, or vesicles, or binding to Ca binding proteins.

The objective of this study was to confirm crayfish NCX antiserum specificity via immunoblot and immunohistochemistry in hepatopancreas, antennal gland, and cardiac muscle. Polyclonal antibodies were generated as outlined previously.[2,3]

IMMUNOBLOTTING

NCX protein was isolated from cardiac muscle through alkaline extraction and was eluted on a DEAE-sepharose chromatography column and a wheat germ agglutinin affinity column.[4] Samples were then separated by SDS PAGE (9%) and either visualized using Coomassie blue or transferred to nitrocellulose for immunoblotting with primary antibody (1°Ab [1:1000]; 2°Ab was goat anti-rabbit IgG conjugated with alkaline phosphatase). The muscle eluate from the DEAE column showed two bands (MW 124 and 89 kDa, as predicted for NCX) (FIG. 1). However, the sample eluted from the wheat germ column did not show any bands of the expected size. This result suggests that we have partially purified crayfish NCX after the DEAE column but that the conditions of the wheat germ column may require additional modifications to optimize binding and elution.

IMMUNOHISTOCHEMISTRY

Antibody specificity and NCX localization were studied using immunohistochemistry and bright-field light microscopy. The protocol employed permanent colorimetric visualization of paraffin sections. Tissues were fixed in periodate-lysine-paraformaldehyde (PLP), dehydrated, cleared, and embedded in paraffin wax. Paraffin sections (6–8 μm) were cut with a microtome, collected on positively charged slides, and heat dried. The staining protocol was described previously.[2] NBT/BCIP plus suppressor (Pierce, Rockford, IL) purple substrate solution was used for visualization. Sections were counterstained in 1% fast green for contrast (FIG. 2).

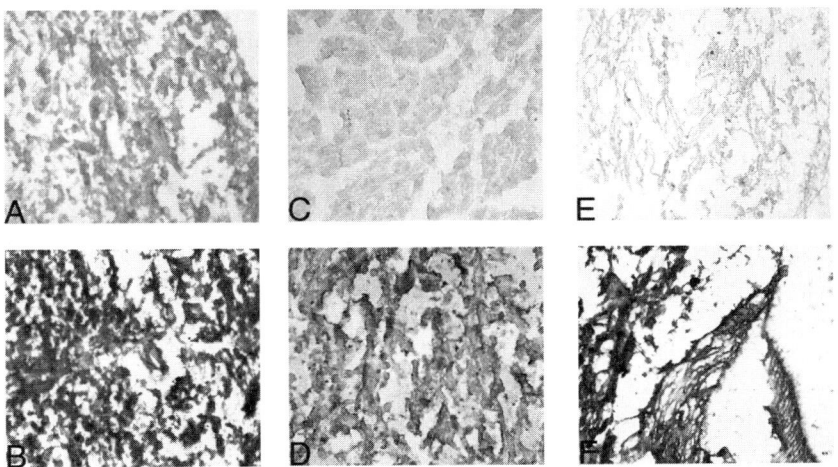

FIGURE 2. NCX was localized in axial muscle, antennal gland, and cardiac muscle compared with negative controls. (**Left**) Axial muscle: (**A**) negative control; (**B**) primary antibody. (**Middle**) Antennal gland: (**C**) negative control; (**D**) primary antibody. (**Right**) Cardiac: (**E**) negative control; (**F**) primary antibody.

DISCUSSION

Immunohistochemical analysis revealed NCX antibody crossreactivity with axial muscle, antennal gland, and cardiac muscle. Although we visualized NCX bands from a partially purified sample, crude homogenate samples revealed only a 70-kDa band corresponding to one of the NCX subunits of antennal gland and hepatopancreas. Moreover, sporadically both bands corresponding to the NCX protein were visualized in axial and cardiac muscles. In light of this new evidence we cannot definitively conclude that binding visualized via immunohistochemical analysis was due to binding of our antibody to the NCX. Nonetheless, there is a possibility that NCX binding sites exposed during immunohistochemical analysis are conducive to binding, whereas binding sites revealed during Western analysis prohibit binding. For this reason we cannot definitely rule out the possibility that the NCX was visualized during the immunostaining procedures.

REFERENCES

1. WHEATLY, M.G. 1999. Calcium homeostasis in crustacea: the evolving role of branchial, renal, digestive and hypodermal epithelia. J. Exp. Zool. **283**: 620–640.
2. WHEATLY, M.G., Z. ZHANG, J.R. WEIL, et al. 2001. Novel subcellular and molecular tools to study Ca^{2+} transport mechanisms during the elusive moulting stages of crustaceans: flow cytometry and polyclonal antibodies. J. Exp. Biol. **204**: 959–966.

3. BOERSMA, W.J.A., J.J. HAAIJMAN & E. CLAASSEN. 1993. Use of synthetic peptide determinants for the production of antibodies. *In* Immunohistochemistry. A.C. Cuello, Ed. vol. II: 1–78. Wiley. New York.
4. DURKIN, J., D. AHRENS, Y. PAN & J. REEVES. 1991. Purification and amino-terminal sequence of the bovine cardiac sodium–calcium exchanger: evidence for the presence of a signal sequence. Arch. Biochem. Biophys. **290:** 369–375.

Characterization and Functional Activity of a Truncated Na/Ca Exchange Isoform Resulting from a New Splicing Pattern of NCX1

FRANÇOISE VAN EYLEN, ADAMA KAMAGATE, AND ANDRÉ HERCHUELZ

Laboratory of Pharmacology, Brussels University School of Medicine, B-1070, Brussels, Belgium

KEYWORDS: Na/Ca exchange isoform; NCX1

INTRODUCTION

The Na/Ca exchanger (NCX) is a ubiquitous transporter that plays an important role in regulating cellular Ca^{2+} balance. NCX1 is composed of 938 amino acids and presents 9 hydrophobic transmembrane segments and a large intracellular loop (550 amino acids) located between transmembrane segments 5 and 6.[1,2] Electrophoresis gels show two protein bands at 160 and 120 kDa under nonreducing conditions. Reducing agents decrease the intensity of the 160-kDa band and lead to the appearance of a 70-kDa protein band.[3] It is currently accepted that the 120 kDa band represents the native protein, the 70-kDa band being a proteolytic fragment or a subunit of the native protein, and the 160 kDa an alternate form of the 120-kDa protein.[3,4]

Studies on the organization of the human NCX1 gene have revealed that it is comprised of at least 14 exons spread out over more than 200 kb of genomic DNA.[5,6] Alternative splicing generates tissue-specific variants of NCX1, arising from a region of the large intracellular loop encoded by 6 small exons: A to F (also referred to as exons 3 to 8).[6] To maintain an open reading frame, all splice variants must include either exon A or B, which are mutually exclusive. The arrangement of different exons could allow for the generation of up to 32 different NCX1 isoforms. Thirteen NCX1 isoforms have been identified so far.[7,8]

Using antibodies directed against different parts of the exchanger, Saba *et al.*[9] recently showed that the 70-kDa polypeptide seen in native sarcolemmal vesicle preparations corresponds to the NH_2-terminal portion of the exchanger. By contrast, Iwata *et al.*,[10] expressing the exchanger as a fusion protein with a poly-His tag at its COOH terminus, showed that the 70-kDa band is the COOH-terminal portion of the protein. Therefore, the nature of the 70-kDa protein remains unclear.

Address for correspondence: Dr. André Herchuelz, Laboratory of Pharmacology, Brussels University School of Medicine, Bât. GE, 808 route de Lennik, B-1070, Brussels, Belgium. Voice: 32-2-555-6201; fax: 32-2-555-6370.
herchu@ulb.ac.be

A

```
                                              Exon X
Nt seq.   ~~~~~~~TCC  AGAATGATGA  AATTGTAGTA  TCAAAATGCA  TGAGAGTGCC
Aa seq.   595~~~~  Q     N  D  E    I  V  V    S  K  C  M    R  V  P

          TGCTCAGATA  ACCATCAGAT  GAATATGAAG  CTTGCTCATC  CCTGGGAGTG
           A  Q  I    T  I  R  *    I  *  S    L  L  I    P  G  S  A

          CAGAAATCAG  CTCCTTTTAC  CTCTTGAAGA  AGTGGATACA  GCATCCTCTA
           E  I  S    S  F  Y    L  L  K  K    W  I  Q     H  P  L

                                                              Exon B
          CTGCTCAATT  GCATAAACCA  GCACTTGTCC  TTGAAGATC   ATTACCATTA
           L  L  N  C    I  N  Q    H  L  S    L  K  I    I  T  I  R

          GAATATTTGA  CCGTGAGGAA  TATGAGAAAG  AGTGCAGTTT  CTCCCTTGTG
           I  F  D    R  E  E    Y  E  K  E    C  S  F    S  L  V

                                                              Exon D
          CTTGAGGAAC  CAAAATGGAT  AAGAAGGGGA  ATGAAAGGTG  GCTTCACAAT
           L  E  E  P    K  W  I    R  R  G    M  K  G  G    F  T  I

          AACAGACGAA  TATGATGACA  AGCAGCCACT  GACCAGCAAA  GAGGAA~~~~
           T  D  E    Y  D  D  K    Q  P  L    T  S  K    E  E  ~~~~
```

B

Exon X sequence

```
Nt seq.  AGTATCAAAATGCATGAGAGTGCCTGCTCAGATAACCATCAGATGAATATGAAGCTTGCT
Aa seq.   S  I  K  M  H  E  S  A  C  S  D  N  H  Q  M  N  M  K  L  A
         CATCCCTGGGAGTGCAGAAATCAGCTCCTTTTACCTCTTGAAGAAGTGGATACAGCATCC
          H  P  W  E  C  R  N  Q  L  L  L  P  L  E  E  V  D  T  A  S
         TCTACTGCTCAATTGCATAAACCAGCACTTGTCCTTG
          S  T  A  Q  L  H  K  P  A  L  V  L
```

FIGURE 1. (A) Nucleotide and amino acid sequences of the cloned human eye NCX1.33 exchanger splice variant at the splice site. Sequences of the newly identified exon X are in bold. Stop codons in exon X are indicated by an *asterisk*. Exons B and D are also shown between ■, ◆, and → ←, respectively. *Wavy dashes* make reference to NCX1 sequences. The amino acid numbering scheme assumes the deduced initiator methionine as the first amino acid. (B) Nucleotide and deduced amino acid sequences of the exon X. (From Van Eylen et al.[11] Reprinted with permission from *Cell Calcium*.)

RESULTS AND DISCUSSION

A new NCX1 splicing pattern was identified during the cloning of NCX1 isoforms from the human eye carried out using the right transcription-polymerase chain reaction (RT-PCR) method. The insertion of a newly identified sequence of 157 bp, which we called X, upstream exons B and D of the NCX1.3 isoform (FIG. 1), generates a stop codon in frame with the NCX1 coding sequence that should lead to a truncated Na/Ca exchanger (which we called NCX1.33), comprising only the N-terminal portion of the exchanger and a shortened intracellular loop. Insulin-secreting BRIN-BD11 cells were stably transfected with NCX1.33.

The transcription of NCX1.33 in the transfected cells was determined by the RT-PCR method. Primers were designed to anneal to conserved sequences flanking the

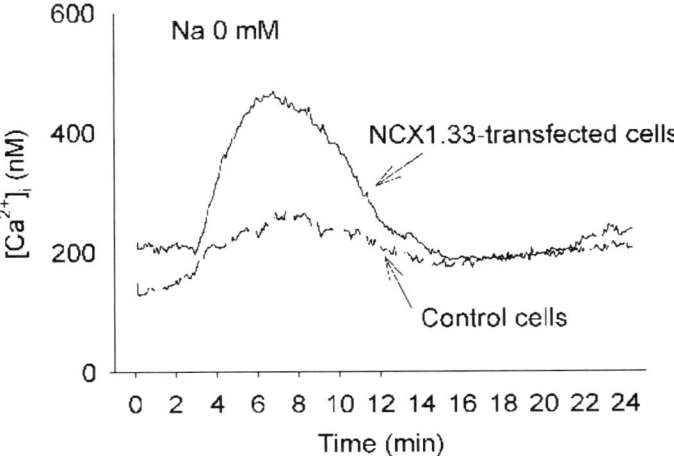

FIGURE 2. Effect of extracellular Na^+ removal on $[Ca^{2+}]_i$ in control and NCX1.33-transfected BRIN-BD11 cells. After loading with fura-2 for 1 hour, cells on coverslips were mounted onto an inverted fluorescence microscope for epifluorescence. $[Ca^{2+}]_i$ was calculated from the ratio of the 340- and 380-nm signals, using an *in vitro* calibration. The period of exposure to Na^+-free medium is indicated by a bar above the curves. The curves shown are the mean values of 62 (control cells) and 85 traces (NCX1.33) of individual cells recorded in two and three separate experiments, respectively. (From Van Eylen *et al.*[11] Reprinted with permission from *Cell Calcium*.)

putative splicing area in the cytoplasmic loop of NCX1. In nontransfected cells, PCR amplification yielded two bands of 379 and 310 bp, indicating the presence of exons BDF and BD characterizing NCX1.7 and NCX1.3 isoforms, respectively. In NCX1.33-transfected cells, PCR amplification yielded three bands. Two of them corresponded to NCX1.3 and NCX1.7, the third upper band of 467 bp representing the NCX1.33 transcript that showed marked overexpression. Overexpression was assessed at the protein level by Western blot analysis and immunofluorescence using a monoclonal antibody directed against the canine cardiac Na/Ca exchange. The truncated exchanger migrated as a 70-kDa band and was appropriately targeted to the plasma membrane. To obtain functional evidence of NCX1.33 overexpression, Na^+_i-dependent $^{45}Ca_o$ uptake and $[Ca^{2+}]_i$ changes induced by extracellular Na^+ removal (reverse Na/Ca exchange) were examined. FIGURE 2 shows the effect of extracellular Na^+ removal on $[Ca^{2+}]_i$. In control cells, Na^+ removal induced a modest increase in $[Ca^{2+}]_i$. By contrast, in NCX1.33-transfected cells, extracellular Na^+ removal induced a more rapid and marked increase in $[Ca^{2+}]_i$. Thus, at the peak, the increase in $[Ca^{2+}]_i$ was 305 ± 24 nM ($n = 85$) and 136 ± 59 nM ($n = 62$), respectively ($P < 0.005$).

CONCLUSIONS

Taken as a whole, the current study indicates the existence of a new splicing pattern of NCX1. This leads to expression of a truncated NCX1 exchanger consisting of the 5 N-terminal set of transmembrane domains and a shortened intracellular loop. This truncated isoform is functionally active and migrates as a 70-kDa protein. Our data confirm the view that the native 70-kDa exchanger protein is the N-terminal rather than the C-terminal portion of the protein.

ACKNOWLEDGMENT

This work was supported by the Belgian Fund for Scientific Research (FRSM 3.4562.00), in which F.V.E . is a Postdoctoral Researcher.

REFERENCES

1. BLAUSTEIN, M.P. & W.J. LEDERER. 1999. Physiol. Rev. **79:** 763–854.
2. PHILIPSON, K.D. & D.A. NICOLL. 2000. Annu. Rev. Physiol. **62:** 111–133.
3. PHILIPSON, K.D., S. LONGONI & R. WARD. 1988. Biochim. Biophys. Acta **945:** 298–306.
4. DURKIN, J.T., D.C. AHRENS, Y.-C. PAN & J.P. REEVES. 1991. Arch. Biochem. Biophys. **290:** 369–375.
5. SCHELLER, T., A. KRAEV, S. SKINNER & E. CARAFOLI. 1998. J. Biol. Chem. **273:** 7643–7649.
6. KRAEV, A., I. CHUMAKOV & E. CARAFOLI. 1996. Genomics **37:** 105-112.
7. QUEDNAU, B.D., D.A. NICOLL & K.D. PHILIPSON. 1997. Am. J. Physiol. **272:** C1250–C1261.
8. VAN EYLEN, F., A. BOLLEN & A. HERCHUELZ. 2001. J. Endocrinol. **168:** 517–526.
9. SABA, R.I., A. BOLLEN & A. HERCHUELZ. 1999. Biochem. J. **338:** 139–145.
10. IWATA, T., C. GALLI, P. DAINESE, *et al.* 1995. Cell Calcium **17:** 263–269.
11. VAN EYLEN, F., A. KAMAGATE & A. HERCHUELZ. 2001. Cell Calcium **30:** 191–198.

Activation of the Cardiac Na^+/Ca^{2+} Exchanger by DEPC

MICHELA OTTOLIA, SABRINA SCHUMANN, DEBORA A. NICOLL, AND KENNETH D. PHILIPSON

Department of Physiology and Cardiovascular Research Laboratories, University of California at Los Angeles, Los Angeles, California 90095-1760, USA

KEYWORDS: Na^+/Ca^{2+} exchanger; diethyl pyrocarbonate; DEPC; histidine

The identification of amino acid residues within the exchanger involved in regulatory properties contributes to the general understanding of how this transporter works. The large intracellular loop, connecting transmembrane segments 5 and 6, has been extensively studied, and several regulatory regions have been found. These parts of the protein are fundamental in controlling Na^+ and Ca^{2+} regulation.[1] It is also known that the loop between the first two transmembrane segments is involved in the regulation of the exchanger[2] and that removal of all exchanger cysteines does not alter the biophysical properties of the transporter.[3] Recently, we observed that application of diethyl pyrocarbonate (DEPC) modifies the regulatory properties of NCX1.1 measured in patches and augments the Na^+-dependent uptake of $^{45}Ca^{2+}$ into *Xenopus* oocytes expressing the exchanger. The effect of DEPC on $^{45}Ca^{2+}$ uptake was more prominent when measurements were performed at acidic pH. For the wild-type exchanger, activity increased by 4 ± 0.8-fold ($n = 8$) at pH 6.0 and 1.4 ± 0.1-fold ($n = 4$) when measured at pH 7.0.

The regulatory properties of the exchanger were studied using the giant patch technique.[4] Currents were evoked by rapid application of 100 mM internal Na^+ in the presence of 8 mM Ca^{2+} in the pipette. Upon cytoplasmic application of 5 mM DEPC, the extent of Na^+-dependent inactivation (I_1) decreased from 80 ± 5% to 30 ± 12% ($n = 5$, $[Ca^{2+}]_i = 1$ μM) and Ca^{2+} regulation from 96 ± 1% to 60 ± 8% ($n = 4$). Since DEPC is known to react specifically with histidine residues,[5] we investigated their role in the regulation of the exchanger by either mutating or deleting them. Most of the histidine-mutated exchangers were active and responded to DEPC in a manner similar to that of the wild-type exchanger (FIG. 1). An exception is the mutant in which histidine at position 165 was replaced with a lysine. The ionic current recorded from oocytes expressing H165K lack both Na^+ and Ca^{2+} regulation,

---

Address for correspondence: Dr. Kenneth D. Philipson, Department of Physiology and Cardiovascular Research Laboratories, University of California at Los Angeles, 3645 MRL, 675 Young Dr. S., Los Angeles, California 90095-1760. Voice: 310-825-7679; fax: 310-206-5777.
kphilipson@mednet.ucla.edu

Ann. N.Y. Acad. Sci. 976: 85–88 (2002). © 2002 New York Academy of Sciences.

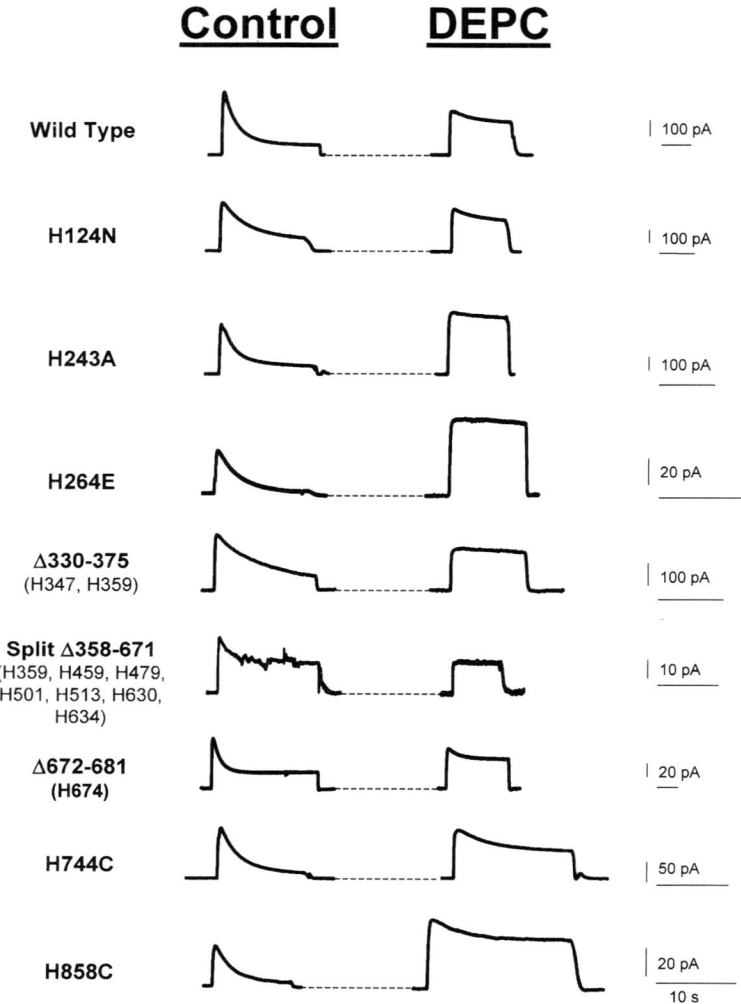

FIGURE 1. Wild-type and histidine-mutant exchangers activated by DEPC. Outward currents recorded from oocytes expressing the indicated construct in control conditions (*right*) and after DEPC application (*left*). $V_H = 0$ mV, $[Ca^{2+}]_i = 1$ μM. Histidines 302, 789, and 933 have not yet been examined.

resembling the phenotype of DEPC-activated exchanger currents and was not modified by DEPC (FIG. 2). These results indicate that histidine 165 is involved in the exchanger regulatory mechanisms and suggest that this histidine could be the residue modified by DEPC. Interestingly, DEPC still stimulated H165K activity in Na^+-dependent $^{45}Ca^{2+}$ uptake measurements. Indeed, a twofold increase in activity

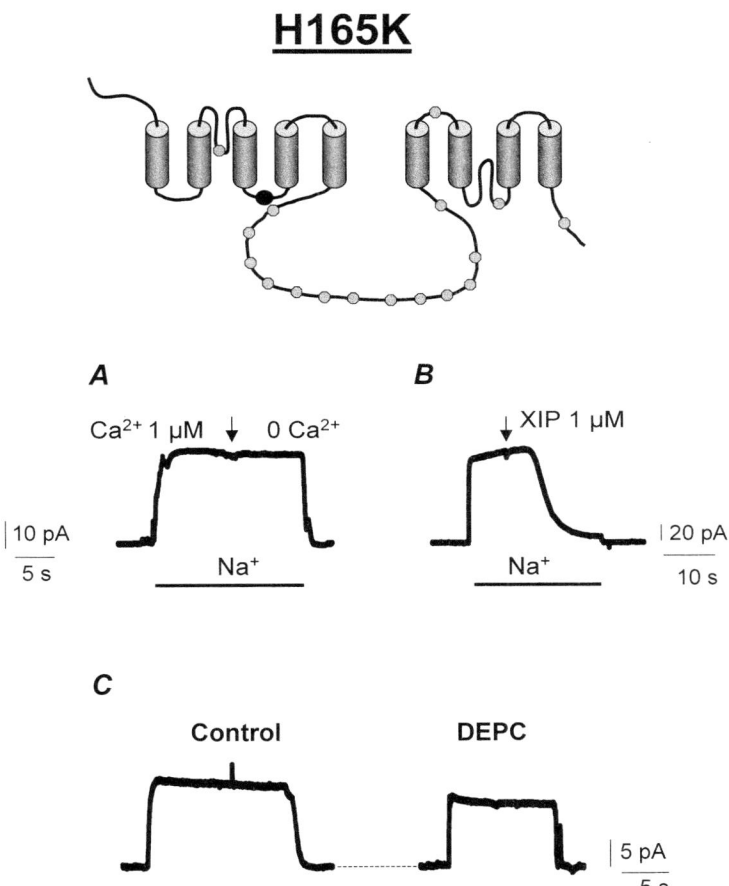

FIGURE 2. Removal of histidine 165 renders the exchanger insensitive to regulatory Na^+ and Ca^{2+}. Outward currents recorded from oocytes expressing H165K exchanger mutant. H165K, which lacks I_1 and I_2 (**A**), is blocked by XIP (**B**) and is not activated by DEPC (**C**).

measured as $^{45}Ca^{2+}$ uptake was observed in oocytes expressing H165K after DEPC treatment ($n = 2$). The reason for the discrepancy between the giant patch and $^{45}Ca^{2+}$ uptake measurement is unknown. It is possible that two different phenomena are under investigation.

Currently, additional mutagenesis studies are ongoing to determine the function of the remaining histidines (H302, 789, and 933) and to determine if the charge at position 165 plays an important role in exchanger regulation.

REFERENCES

1. MATSUOKA, S., D.A. NICOLL, R.F. REILLY, *et al.* 1993. Initial localization of regulatory regions of the cardiac sarcolemmal Na^+-Ca^{2+} exchanger. Proc. Natl. Acad. Sci. USA **90:** 3870–3874.
2. DOERING, A.E., D.A. NICOLL, Y. LU, *et al.* 1998. Topology of a functionally important region of the cardiac Na^+/Ca^{2+} exchanger. J. Biol. Chem. **273:** 778–783.
3. SANTACRUZ-TOLOZA, L., M. OTTOLIA, D.A. NICOLL & K.D. PHILIPSON. 2000. Functional analysis of a disulfide bond in the cardiac Na^+-Ca^{2+} exchanger. J. Biol. Chem. **275:** 182–188.
4. HILGEMANN, D.W. & C.C. LU. 1998. Giant membrane patches: improvements and application. Methods Enzymol. **293:** 267–280.
5. LUNDBLAD, L.U. & C.M. NOYES. 1984. Modification of histidine residues. *In* Chemical Reagents for Protein Modification. Vol. **1:** 105–125. CRC Press. Boca Raton, FL.

Cysteine Residues Are Important for Functional Expression of the Na^+/Ca^{2+} Exchanger NCX1 in HEK 293 Cells

XIAOYAN REN, JUDITH KASIR, AND HANNAH RAHAMIMOFF

Department of Biochemistry, Hebrew University-Hadassah Medical School, Jerusalem 91120, Israel

KEYWORDS: cysteine residues; Na/Ca exchanger; NCX1; HEK 293 cells

Rat isoforms of the NCX1 protein, RBE-1 (NCX1.4), RBE-2 (NCX1.5), and RHE-1 (NCX1.1), have 14 cysteine residues. Based on the current topology model, four of these cysteines, C14, C20, C122, and C780, are modeled to be on, or close to, the extracellular face of the membrane, two are modeled to be intramembranous (C151 and C210), and the remaining cysteines are facing the cytoplasmic side of the membrane. Studies in excitable cells and membrane preparations suggested that Na^+/Ca^{2+} exchange was modulated by sulfhydryl reagents. Yet treatment of Xenopus oocytes expressing the dog sarcolemmal Na^+/Ca^{2+} exchanger with MTS reagents suggested, that the transport activity was not sensitive to sulfhydryl reagents. We recently reexamined the sensitivity of rat isoforms of the Na^+/Ca^{2+} exchanger NCX1 expressed in HEK 293 cells. We showed that sulfhydryl reagents inhibit the transport activity of the protein. The inhibition also persisted when all putative external cysteines were exchanged with Ala or Ser. To determine covalent modification of which of the cysteine residues inhibited the transport activity, each of the 14 cysteines of the rat Na^+/Ca^{2+} exchanger RBE-2 (NCX1.5) alone or in different combinations was mutated. Each mutant was expressed in HEK 293 cells, and the effects of MTSEA on transport activity were determined. Our results suggest that each individual cysteine residue can be exchanged with Ala or Ser without significant loss of transport activity. When, however, sequential exchange starting with Cys14 is done, a gradual loss of transport activity occurs, and when 7 initial cysteine residues are replaced, only traces of transport activity are detected. Surface biotinylation suggests that surface expression correlates with transport activity. Reconstitution of transfected cell proteins did not rescue the transport activity of these mutants.

Address for correspondence: Dr. Xiaoyan Ren, Department of Biochemistry, Hebrew University–Hadassah Medical School, Box 12272, Jerusalem 91120, Israel. Voice: 9772-2-6757510; fax: 972-2-6757061.
 renxiaovan@yahoo.com

Topological Studies of the Rat Brain K⁺-Dependent Na⁺/Ca²⁺ Exchanger NCKX2

XINJIANG CAI, KATHY ZHANG, AND JONATHAN LYTTON

Cardiovascular Research Group, Department of Biochemistry and Molecular Biology, University of Calgary, Calgary, Alberta T2N 4N1, Canada

KEYWORDS: cysteine mutagenesis; epitope tagging; transmembrane orientation

INTRODUCTION

The second member of the K^+-dependent Na^+/Ca^{2+} exchanger family, NCKX2, is believed to play an important role in maintaining Ca^{2+} homeostasis in many brain regions.[1] The current topological model of NCKX2 contains a putative signal peptide (M0), an NH_2-terminal hydrophobic domain containing proposed transmembrane segments (TMS) M1 to M5, and a COOH-terminal hydrophobic domain containing TMSs M6-M11. The hydrophobic domains are separated from one another by an intracellular loop, and the mature protein is predicted to have an extracellular NH_2 terminus and an intracellular COOH terminus (FIG. 1). To date, this topological model of NCKX-type exchangers has not been tested experimentally. Recently, examination of the hydropathy profile for a new member of the NCKX exchanger family, NCKX3, gave rise to a new topological model in which the COOH-terminal hydrophobic domain contains only 5 TMSs, thus placing the COOH terminus of the protein outside the cell.[2] In this study, we used the substituted cysteine accessibility method (SCAM) and epitope-tagging approach to explore which of these two models more accurately reflects the actual topology of NCKX2.

RESULTS AND DISCUSSION

Cysteine Mutants. There are eight native cysteine residues in the NCKX2 molecule (FIG. 1). Three constructs with combined cysteine-to-alanine mutations were made: C1-4 (C14A, C24A, C154A, and C224A), C5-8 (C395A, C614A, C633A, and C666A), and a cys-less exchanger, C1-8. Western blotting showed C1-4 was expressed well in HEK-293 cells, whereas C5-8 and C1-8 expression was too low to be detected. *In vitro* translation demonstrated that the C5-8 construct could be translated, suggesting that mutation of the last four cysteine residues might affect protein

---

Address for correspondence: Jonathan Lytton, Ph.D., Department of Biochemistry & Molecular Biology, University of Calgary Health Sciences Centre, Room 2518, 3330 Hospital Drive NW, Calgary, AB, Canada T2N 4N1. Voice: 403-220-2893; fax: 403-283-4841.
jlytton@ucalgary.ca

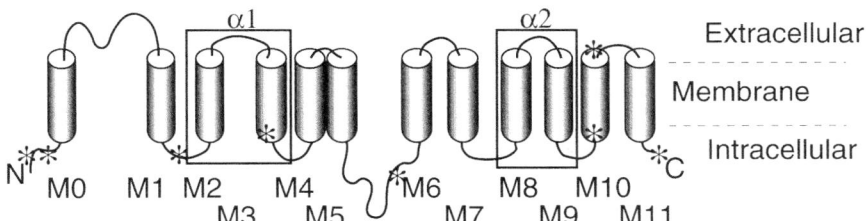

FIGURE 1. Topological model of NCKX2 based on hydropathy analysis. The rat brain NCKX2 contains 8 endogenous cysteine residues, indicated by *asterisks*: Cys16, Cys 24, Cys 154, Cys 224, Cys 395, Cys 614, Cys 633, and Cys 666. The proposed transmembrane segments are indicated M0–M10, and the two internally homologous α-repeats, thought to be critical in forming the ion transport pathway, are boxed and labeled α-1 and α-2.

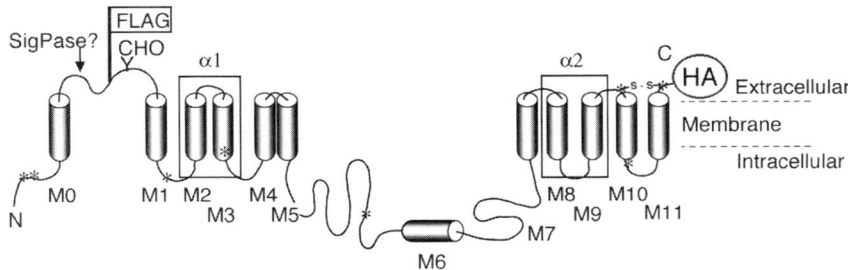

FIGURE 2. A new putative topological model for NCKX2. In this model the C-terminus is extracellular, Cys614 and Cys666 are engaged in a disulfide bond, and putative transmembrane segment 6 is placed inside the cytoplasm to form a part of the large intracellular loop. Sites of the FLAG and HA epitopes are shown, as well as the site of glycosylation (CHO), and the putative signal peptide cleavage site (SigPase?). Other details are similar to those in FIGURE 1.

stability in transfected HEK-293 cells. The single cysteine mutants, C614A and C666A, were also not expressed in HEK-293 cells, whereas a mutant named C1-5&7, in which all cysteines except C614 and C666 were mutated to alanine, was well expressed in HEK-293 cells and was functionally active.

Thus, we speculated that C614 and C666 may form a structurally and functionally important cystine disulfide bond. Disulfide bonds are usually found extracellularly, as the cytoplasm is a reducing environment. This speculation regarding cysteines C614 and C666 would place the COOH terminus of NCKX2 outside the cell, in conflict with the original topological model for NCKX2, the new revised topology model for NCX1 (see Nicoll *et al.*, in this volume), but consistent with that of NCKX3 (see FIG. 3 for the revised model).

Epitope Tagging. The location of the NCKX2 COOH terminus was tested by adding an HA epitope at the COOH terminus of an NCKX2 construct that was also tagged with a FLAG epitope in the NH_2-terminal extracellular loop (FIG. 2). Immu-

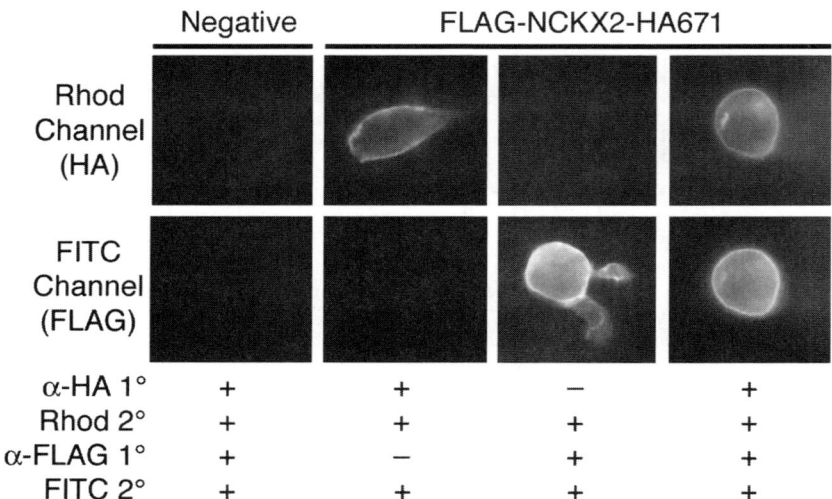

FIGURE 3. Nonpermeabilized HEK-293 cells transfected with vector alone (control) or with FLAG-NCKX2-HA671 were stained using monoclonal anti-FLAG antibody followed by FITC-conjugated anti-mouse antibody, using polyclonal anti-HA antibody followed by rhodamine-conjugated anti-rabbit antibody or double stained using both sets of antibodies, as indicated.

nofluorescent analysis of HEK-293 cells transfected with this construct (FLAG-NCKX2-HA671) showed that both the HA and FLAG tags could be recognized with specific antibodies without permeabilization of the cell membrane (FIG. 3). This finding supports the new putative NCKX-type topology model, based on hydropathy analysis of NCKX3, in which the C-terminus is extracellular.[2]

Cysteine-Selective Labeling. The accessibility of the cysteine residues in NCKX2 was tested by covalently modifying the protein expressed in transfected HEK-293 cells with 3-(N-maleimidylpropionyl) biocytin (MPB). Results demonstrated that cysteine residue(s) in wild-type NKCX2 and C1-4 become accessible for MPB labeling only after β-mercaptoethanol (β-ME) treatment. Interestingly, C1-5&7 could not be labeled even after β-ME treatment. Thus we conclude that C614 and C666 are not accessible for labeling by the hydrophilic MPB even after treatment with β-ME, and so may be concealed within the hydrophobic portion of the membrane.

SUMMARY

In this study, we demonstrated that C614 and C666 were essential for the expression and function of the rat brain NCKX2 in transfected HEK-293 cells. The NCKX2 protein was labeled with a cysteine-selective reagent only after prior reduction with β-mercaptoethanol, but this labeling appeared not to involve either C614 or C666. We also found that the COOH terminus of the NCKX2 exchanger was located extracellularly, supporting a new putative NCKX-type exchanger model. This model is distinct from the revised NCX1 topology, and it remains an intriguing possibility that structural difference between NCX and NCKX-type exchangers may underlie their different ion specificity.

REFERENCES

1. TSOI, M., K.-H. RHEE, D. BUNGARD, et al. 1998. Molecular cloning of a novel potassium-dependent sodium-calcium exchanger from rat brain. J. Biol. Chem. **273:** 4115–4162.
2. KRAEV, A., B.D. QUEDNAU, S. LEACH, et al. 2001. Molecular cloning of a third member of the potassium-dependent sodium-calcium exchanger gene family, NCKX3. J. Biol. Chem. **276:** 23161–23172.

Studies on the Oligomeric State of the Sodium/Calcium+Potassium Exchanger NCKX2

SEUNGHWA SALLY YOO, STEPHEN LEACH, AND JONATHAN LYTTON

Cardiovascular Research Group, Department of Biochemistry and Molecular Biology, University of Calgary, Calgary, Alberta T2N 4N1, Canada

KEYWORDS: NCKX; immunoprecipitation; density gradient centrifugation

INTRODUCTION

Many ion channels and exchangers function as oligomers in a cell. However, the oligomeric structure of the Na^+/Ca^{2+} exchangers, a primary Ca^{2+} extruding system in heart, brain, and kidney, has not been investigated. Inasmuch as exploring intramolecular interactions of ion channels and exchangers can convey significant insight into the structure, function, and physiology of these molecules, the oligomeric state of the sodium/calcium+potassium exchanger NCKX2 was examined by employing coimmunoprecipitation and measuring hydrodynamic properties.

RESULTS AND DISCUSSION

Human embryonic kidney cells (HEK-293) were transiently transfected with FLAG-tagged NCKX2 and/or wild-type NCKX2 constructs, and microsomes were prepared by homogenization and differential centrifugation. Coimmunoprecipitation studies using antibodies against the FLAG epitope and the wild-type exchanger demonstrated an association between coexpressed proteins when solubilized in 1% CHAPS, implying the existence of an oligomeric NCKX2 species (FIG. 1).

To determine the number of subunits that associate in this complex, hydrodynamic studies were carried out. Microsomes from FLAG-tagged NCKX2 transfected HEK-293 cells were solubilized in either 1% CHAPS or 2% Triton X-100, and the proteins were separated on a 5–20% linear sucrose gradient together with three marker proteins, catalase, aldolase, and bovine serum albumin. When solubilized in Triton X-100, NCKX2 migrated at a position consistent with a protein monomer bound to a micelle of detergent (FIG. 2A). However, when solubilized in CHAPS containing buffer, NCKX2 migrated on the gradient in several peaks, consistent with the presence of an oligomeric species (FIG. 2B). Repeated experiments indicated

---

Address for correspondence: Jonathan Lytton, Ph.D., Department of Biochemistry & Molecular Biology, University of Calgary Health Sciences Centre, Room 2518, 3330 Hospital Drive NW, Calgary, AB, Canada T2N 4N1. Voice: 403-220-2893; fax: 403-283-4841.
jlytton@ucalgary.ca

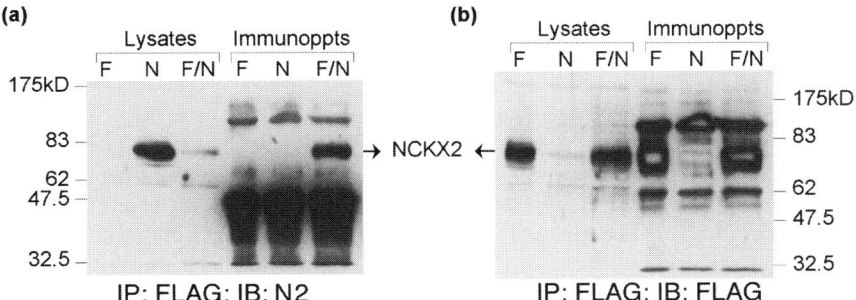

FIGURE 1. Coimmunoprecipitation of FLAG-tagged NCKX2 (F) and wild-type NCKX2 (N) suggests coassociation. Wild-type and FLAG-tagged NCKX2 proteins were expressed alone or together in HEK-293 cells, solubilized in 1% CHAPS buffer, immunoprecipated (IP) with α-FLAG antibody. Samples of the cell lysates and the immunoprecipitates were analyzed by immunoblotting (IB) with **(a)** α-N2 antibody and **(b)** α-FLAG antibody. Note that the epitope for the α-N2 antibody overlaps the FLAG epitope insertion site, and hence N2 recognizes only wild-type and not FLAG-tagged NCKX2.

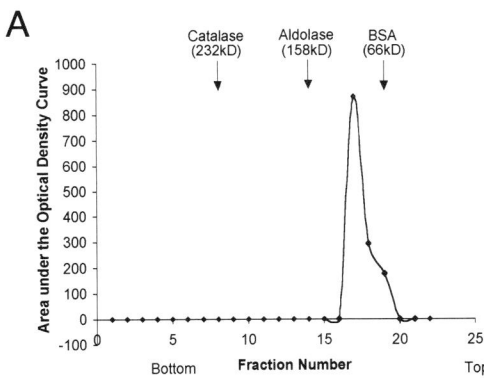

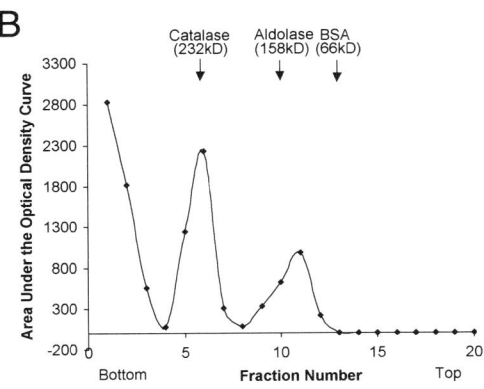

FIGURE 2. Sucrose density gradient centrifugation of detergent-solubilized FLAG-tagged NCKX2. **(A)** FLAG-tagged NCKX2 solubilized in 2% Triton X-100. NCKX2 protein is located at a fraction corresponding to an apparent molecular weight of about 100 kDa. **(B)** FLAG-tagged NCKX2 solubilized in 1% CHAPS. The NCKX2 protein is distributed in several peaks across the gradient corresponding to apparent molecular weights of about 240 and 120 kDa plus material at the bottom of the tube.

variability in the fraction of NCKX2 that migrated in each identifiable peak, suggesting that the oligomeric complex was only partially stable under these conditions.

CONCLUSIONS

Immunoprecipitation and sucrose density centrifugation experiments suggest that NCKX2 may form an oligomeric structure. Further experiments under carefully controlled conditions will be required to define the stability of this complex and its precise composition.

Conformational Changes of the 120-kDa Na$^+$/Ca^{2+} Exchanger Protein upon Ligand Binding

A Fourier Transform Infrared Spectroscopy Study

RAMI I. SABA,[a] ERIK GOORMAGHTIGH,[b] JEAN-MARIE RUYSSCHAERT,[b] AND ANDRÉ HERCHUELZ[a]

[a]*Laboratoire de Pharmacodynamie and* [b]*Laboratoire de Chimie Physique des Macromolecules aux Interfaces, Facultés de Médecine et des Sciences, Brussels, Belgium*

KEYWORDS: Na/Ca exchanger; ligand binding; Fourier transform infrared spectroscopy study

The Na/Ca exchanger is the major cytoplasmic Ca^{2+} extrusion mechanism in cardiomyocytes.[1,2] The exchange of Na$^+$ and Ca^{2+} occurs in separate steps that implicate at least two major conformational states (E_1 and E_2).[3] These conformational changes are poorly characterized. Na$^+$ and Ca^{2+} regulate the exchanger protein at binding sites separate from the ion translocation sites. The Ca^{2+} regulatory site is between amino acids 445-455 in the cytoplasmic loop. The Na$^+$ regulatory site is associated with a highly basic stretch of 20 amino acids (residues 251-271) in the cytoplasmic loop, named XIP (exchange inhibitory peptide). A synthesized XIP peptide exogenously inhibits the Na/Ca exchanger.

The Na/Ca exchanger protein was purified and reconstituted[4–6] and was associated with 160-, 120-, and 70-kDa polypeptides. The 160- and 120-kDa proteins have the same NH$_2$-terminal sequence as the cloned exchanger.[4–6] However, the nature of the 70-kDa fragment remains controversial.

In the first part of our work, we purified, reconstituted, and characterized the active 70-kDa Na/Ca exchanger.[6] We characterized the 70-kDa fragment and demonstrated that it has the same NH$_2$-terminal sequence as does the cloned 120-kDa bovine cardiac exchanger. Moreover, we determined that the 70-kDa protein is composed of ~620 amino acids and hence contains the endogenous XIP and the regulatory Ca^{2+} binding domains.[6]

In the second part, we used Fourier transform infrared attenuated total reflection spectroscopy (FTIR-ATR) to analyze the secondary and tertiary structure of the purified 70-kDa Na/Ca exchanger.[7] FTIR-ATR showed that the protein is composed of 44% α-helices, 25% β-sheets, 16% β-turns, and 15% random structures. Next, we

Address for correspondence: Dr. André Herchuelz, Laboratoire de Pharmacodynamie, Faculté de Médicine et des Sciences, Bât. GE, route de Lennik 808, B-1070 Brussels, Belgium. Voice: 32-2-555-6201; fax: 32-2-555-6370.
herchu@ulb.ac.bc

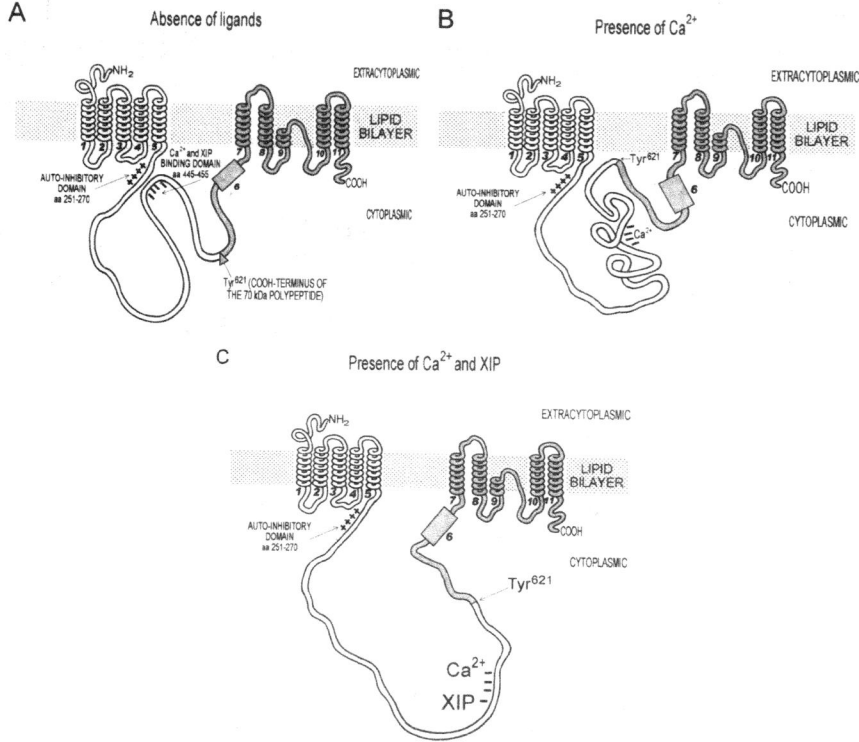

FIGURE 1. Drawing illustrating the conformational changes of the 120-kDa Na^+/Ca^{2+} exchanger induced by its regulators. This illustration is based on the new topology model.[9,10] Transmembrane segments are shown as helices and numbered in bold-italic characters, except segments *6* and *9,* drawn as a box, to denote the possibility that they are not membrane spanning.[9,10] The protein portion, shown in *gray*, is the *318* amino acid polypeptide corresponding to the COOH-terminal portion of the 120-kDa protein that is lacking in the 70-kDa one. The 70-kDa fragment is the NH_2-terminal portion of the 120-kDa protein and ends around Tyr^{621}, as indicated in the panels. (**A**) In the absence of ligands, the negatively charged domain (aas 445-455) binds the endogenous XIP domain (aas 251-270), thus preventing exogenous XIP binding. (**B**) Ca^{2+} binding to its regulatory site dislocates it from the autoinhibitory region and induces an active protein conformation characterized by the folding of some secondary structures into a more compact configuration, making the XIP binding domain accessible to exogenous XIP. (**C**) When XIP is added in the presence of Ca^{2+}, an important conformational change occurs with an effect opposite to that observed upon Ca^{2+} addition, that is, the protein adopts an unfolded conformation, leading to its inhibition. N.B.: The rearrangements of the intracellular loop illustrated in the figure are solely indicative of *changes* in the *tertiary* structure of the protein and must not be taken either as changes in *secondary* structure or as the actual conformational changes that occur upon ligand binding.

evaluated the tertiary structure modifications induced by Ca^{2+} addition to the reconstituted 70-kDa protein by measuring hydrogen/deuterium (H/D) exchange rates. In the absence of ligands, 40% of the protein hydrogen atoms is inaccessible to exchange with deuterium. When Ca^{2+} was added (Ca^{2+}:protein ratio of ~10:1 mole:mole), a major conformational change occurred characterized by the shielding of at least 93 amino acids.[7] The observed conformation is most probably related to a Ca^{2+}-induced activated conformation of the protein.[7]

The last part of our work was to purify and reconstitute the native 120-kDa Na/Ca exchanger and to study its secondary and tertiary structure by FTIR-ATR in the presence of regulatory Ca^{2+} and inhibitory XIP.[8] Comparison of the results obtained for the 120- and the 70-kDa proteins allowed better knowledge of the different regions implicated in the activation/deactivation mechanisms occurring throughout the ion exchange cycles. The 120-kDa protein is composed of 39% α-helices, 20% β-sheets, 25% β-turns, and 16% random coils without modifications upon ligand addition.[8] This is compatible with a topology model having 4–6 transmembrane segments in the COOH-terminal portion of the *protein*.[9,10] The secondary structure of the NH_2-portion of the *cytoplasmic loop* was different from that of the COOH-portion.[8] The tertiary structure modifications induced by Ca^{2+} and/or XIP were evaluated by measuring the H/D exchange rate.[8] In the absence of ligands, 51% of the protein hydrogen atoms is exchanged. In the presence of Ca^{2+}, a conformational change occurred, characterized by the shielding of at least 103 amino acids, comparable to what we observe for the 70-kDa fragment.[7] Ca^{2+} and XIP increased the accessibility to 66%. XIP alone induced no tertiary structure modifications. No changes were observed when XIP was added to the 70 kDa in the presence and in the absence of Ca^{2+}.[8] Accordingly, we proposed a mechanism for the regulation of the exchanger protein by Ca^{2+} and XIP, summarized in FIGURE 1.

In conclusion, our work describes for the first time the conformational changes of the Na/Ca exchanger induced by XIP and Ca^{2+} and suggest that the XIP inhibitory mechanism implicates the COOH-terminal portion of the intracellular loop. Moreover, XIP and Ca^{2+} bind at different sites, where exogenous XIP binding occurs only in the presence of Ca^{2+}.

REFERENCES

1. MAXWELL, K., J. SCOTT, A. OMELCHENKO, *et al.* 1999. Am. J. Physiol. **277:** H2212–H2221.
2. EISNER, D.A. & W.J. LEDERER. 1985. Am. J. Physiol. **248:** C189–C202.
3. KHANANSHVILI, D., E. WEIL-MASLANSKY & D. BAAZOV. 1996. Ann. N.Y. Acad. Sci. **779:** 217–235.
4. PHILIPSON, K.D., S. LONGONI & R. WARD. 1988. Biochim. Biophys. Acta **945:** 298–306.
5. DURKIN, J.T., D.C. AHRENS, Y.C. PAN & J.P. REEVES. 1991. Arch. Biochem. Biophys. **290:** 369–375.
6. SABA, R.I., A. BOLLEN & Z. HERCHUELZ. 1999. Biochem. J. **338:** 139–145.
7. SABA, R.I., J.M. RUYSSCHAERT, A. HERCHUELZ E. GOORMAGHTIGH. 1999. J. Biol. Chem. **274:** 15510–15518.
8. SABA, R.I., E. GOORMAGHTIGH, J.M. RUYSSCHAERT & A. HERCHUELZ. 2001. Biochemistry **40:** 324-332.
9. NICOLL, D.A., M. OTTOLIA, L. LU, *et al.* 1999. J. Biol. Chem. **274:** 910–917.
10. IWAMOTO, T., T.Y. NAKAMURA, Y. PAN, *et al.* 1999. FEBS Lett. **446:** 264–268.

Sodium–Calcium Exchange Crystallization

CALVIN C. HALE,[a] JULIE BOSSUYT,[a] CHANANADA K. HILL,[a] ELMER M. PRICE,[a] DAN H. SCHULZE,[b] JON W. LEDERER,[c] ROBERTO POLJAK,[d] AND BRADFORD C. BRADEN[d,e]

[a]*Dalton Cardiovascular Research Center, Department of Biomedical Sciences, University of Missouri, Columbia, Missouri 65211, USA*

[b]*Department of Microbiology and Immunology, University of Maryland School of Medicine, Baltimore, Maryland 21201, USA*

[c]*Molecular Biology and Biophysics, University of Maryland School of Medicine, Baltimore, Maryland 21201, USA*

[d]*Center for Advance Research in Biotechnology, University of Maryland, Rockville, Maryland 20850, USA*

[e]*Natural Sciences Department, Bowie State University, Bowie, Maryland 20715, USA*

KEYWORDS: crystallization; sodium-calcium exchange; NCX1

As with many mammalian membrane proteins, the cardiac sodium–calcium exchanger (NCX1) has eluded physical analysis, such as crystallization and X-ray diffraction, in part because of low copy number and difficulty in purification. In an attempt to overcome these problems, we expressed a baculovirus vector construct containing NCX1 in *Trichoplusia ni* larvae (cabbage looper caterpillars) rather than in cultured insect cells.[1] To facilitate affinity purification of larva-expressed recombinant NCX1 protein, a polyhistidine tag was engineered on the C-terminus of bovine NCX1 cDNA and subcloned into a baculovirus vector. This construct was used to infect early fourth instar *T. ni* larvae. Three days postinfection, larvae were frozen in liquid N_2 and maintained at −70°C. Pooled larvae (~20 gm) were thawed and homogenized and a subcellular membrane fraction was isolated.[1] Recombinant NCX1 containing a COOH-terminus polyhistidine tag supported Na^+ and Ca^{2+} exchange transport activity in larval membranes, albeit at a lower level than that of wild-type protein without the tag.[2] Full-length 120-kDa recombinant NCX1 was verified on non-reducing immunoblots.[2] Under reducing conditions, a doublet ~70 kDa was observed by Coomassie blue visualization. For affinity purification, larva vesicles containing recombinant, his-tagged NCX1 were solubilized with sodium cholate detergent and applied to a commercially available chelated Ni^{2+} column following

Address for correspondence: Calvin C. Hale, Dalton Cardiovascular Research Center, Department of Biomedical Sciences, University of Missouri, Columbia, MO 65211. Voice: 573-882-3244; fax: 573-884-4232.
chale@missouri.edu

FIGURE 1. NCX1 crystal grown in octylglucoside detergent in an alkaline sodium phosphate buffer. The crystal is an elongated (10 µm) octagon. Diffraction of this crystal indicated areas of organization and disorganization (transmembrane helices and extramembranal segments?), but the crystal quality was too poor to yield data sufficient to determine a cell (coordinates indicating the size of an NCX1 molecule).

the manufacturer's guidelines (HisTrap, Pharmacia Biotech, Uppsala, Sweden). Eluted NCX1 protein was active after reconstitution into soybean phospholipid proteoliposomes.

For initial crystal screens, 1,500 larvae were infected and subjected to the aforementioned membrane and protein purification. Affinity-purified NCX1 was dialyzed and concentrated to a volume of 100 µl and a final protein concentration of 21.5 mg/ml. Initial crystal screens using the vapor-diffusion method yielded small (10 µm) crystals at alkaline pH in sodium cholate or octylglucoside detergents (FIG. 1).[3] Crystals have been produced under detergent-free conditions in the presence of polyethylene glycol and at alkaline pH (FIG. 2). In other screens, microcrystals have been produced that may be suitable for crystal seeding (not shown).

Crystals shown in FIGURES 1 and 2 have been subject to X-ray diffraction analysis. All NCX1 crystals diffracted thus far have been of low quality and have failed to yield structural data. The microcrystals are presently being used in crystal seeding experiments in attempts to grow higher quality protein crystals.

The larva expression system may have widespread application for other low-copy, difficult-to-purify membrane proteins.

FIGURE 2. NCX1 crystal grown under detergent-free conditions. The crystal shown grew in the presence of polyethylene glycol in an alkaline sodium phosphate buffer. The crystal shape is a flattened hexagon. Diffraction of this crystal indicated that it was formed by several smaller, disorganized crystals, all of poor quality.

REFERENCES

1. HALE, C.C., J.A.. ZIMMERSCHIED & E.M. PRICE. 1999. Large-scale expression of recombinant cardiac sodium-calcium in insect larvae. Protein Expression Purific. **15:** 121–126.
2. HALE, C.C., C.K. HILL, E.M. PRICE & J. BOSSUYT. 2002. Expressing and purifying membrane transport proteins in high yield. J. Biochem. Biophys. Meth. **50:** 233–234.
3. GARAVITO, R.M., D. PICOT & P.J. LOLL. 1996. Strategies for crystallizing membrane proteins. J. Bioenerg. Biomemb. **21:** 13–27.

Proteomics Approach to Na^+/Ca^{2+} Exchangers in Prokaryotes

A. RUKNUDIN AND D.H. SCHULZE

Department of Microbiology and Immunology, University of Maryland, Baltimore, Maryland 21201, USA

KEYWORDS: proteomics; Na/Ca exchangers; prokaryotes

INTRODUCTION

Amino acid sequence alignments are widely used in the analysis of a protein's structure, function, and evolutionary relationships. The accurate alignment of related protein sequences is the key to the homology-based comparative methods.[1] Many of the most interesting functional and evolutionary relationships among proteins are so ancient that they can be reliably detected through sequence analysis. The conserved motifs identified by sequence analysis can be mirrored in the tertiary structures of these proteins.[2] Proteins within a superfamily usually share the same protein folds and may possess related functions. These structural and functional similarites are reflected in the alignment conservation patterns. Conserved positions are usually clustered in distinct motifs surrounded by sequence segments of low conservation.[3] However, poorly conserved regions might also arise from the imperfections in multiple alignment algorithms and thus indicate possible alignment errors if observed in closely related proteins.

One of the difficult problems in sequence analysis is the presentation of the results in an easily understandable way. Quantification of conservation by attributing a conservation index to each aligned position makes motif detection possible and easy to present.[4] As the number of sequences to be aligned increases, the complexity of the alignment process becomes computationally expensive, and display of the results can become more difficult to interpret. For visual comparison, a gapped alignment of even 10 sequences written out in one letter code often prevents easy comprehension of the overall architecture of the protein family and can obscure the significance of the results. Therefore, visual aids that condense and summarize the information content are essential for proper and easy understanding of sequence features and their relationships.[5] The Visual Sequence Analysis (VISA) program employs such a visual presentation method. We used VISA software for analysis of the Na^+/Ca^{2+} exchanger proteins and their relatives from various sequenced genomes.

Address for correspondence: Abdul Ruknudin Ph.D., Department of Microbiology and Immunology, School of Medicine University of Maryland, 655 W. Baltimore Street, Baltimore, MD 21201. Voice: 410-706-1379; fax: 410-706-2129.
  aruknudi@umaryland.edu

> Motif I   TxxxxGxSxPExxxS
> Motif II   Ax GTSx PE
> Motif III   GNxxGxN
> x represents any amino acid

FIGURE 1. Protein sequence motifs identified by VISA in the family of Na^+/Ca^{2+} exchanger genes.

VISA displays global similarities within a set of related protein sequences. This program identifies amino acid patterns that are common to many members of a set of sequences and displays the results as a series of histograms. Each dominant peak can be assigned a color and their peaks correspond to conserved sequence motifs.[6]

IDENTIFYING THE PROTEIN MOTIFS

Using VISA, we analyzed the Na^+/Ca^{2+} exchanger and its related gene sequences obtained from the GeneBank. A set of related sequences was searched for, locating common amino acid patterns. The basic pattern was initiated by representing triplets of amino acids separated by two short (variable length up to 15 AA) runs of nonspecific residues.[7,8] The following example illustrates the methodology used.

AA1 ----- up to 15 AA ---- **AA2** ----- up to 15 AA ---- **AA3** ----

Three sequence motifs have been maintained in mammalian Na^+/Ca^{2+} exchangers NCX1, NCX2, and NCX3, in *Drosophila* Na^+/Ca^{2+}, in squid Na^+/Ca^{2+}, in an open reading frame (ORF) in the genome *Methanococcus jannaschii* (Archeae), and in an ORF found in *Escherichia coli*. One or more of these motifs were found in several deduced proteins of other microorganisms, but we did not include them here for the sake of clarity. The observed patterns suggest significant local similarities within the sequence set. We observed the following three motifs.

Having seen the common motifs in the Na^+/Ca^{2+} exchanger and its orthologous gene in *E. coli*, we thought the product of the *E. coli* gene might function like the NCX gene products. So we cloned the *E. coli* gene YrbG, which contained these three motifs (FIG. 1).

CLONING OF THE *E. COLI* OPEN READING FRAME

Primers N-terminal GATCCTCGAGATGCTTTTAGCTACGGCAC and C-terminal GCGACTAGTTTATTCAACGAGTATTGGCG containing XhoI and SpeI restriction sites (underlined), respectively, were used to amplify genomic DNA purified from the *E. coli* strain MJ1655. The polymerase chain reaction (PCR) was per-

FIGURE 2. Derived amino acid sequence of the *E. coli* YrbG gene. The three motifs identified by our search are highlighted in the *E. coli* exchanger sequence.

```
Bacterial exchanger protein sequence

MLLATALLIVGLLLVVYSADRLVFAASILCRTFGIPPLII

GMTVVSIGTSLPEVIVSLAASLHEQRDLAVGTALGSNIIN

ILLILGLAALVRPFTVHSDVLRRELPLMLLVSVVAGSVLY

DGQLSRSDGIFLLFLAVLWLLFIVKLARQAERQGTDSLTR

EQLAELPRDGGLPVAFLWLGIALIIMPVATRMVVDNATVL

ANYFAISELTMGLTAIAIGTSLPELATAIAGVRKGENDIA

VGNIIGANIFNIVIVLGLPALITPGEIDPLAYSRDYSVML

LVSIIFALLCWRRSPQPGRGVGVLLTGGFIVWLAMLYWLS

PILVE
```

formed using 25 cycles of amplification with the primers noted above with a 50°C annealing temperature. The PCR products were directly subcloned into the vector pSD64TF. The clone was sequenced and was found to be identical to the published sequence (NCBI Accession #AAC76228) with the exception of a single silent mutation. The deduced amino acid is presented in FIGURE 2.

ANALYZING THE FUNCTION AND EXPRESSION OF THE BACTERIAL EXCHANGER

Measuring Na^+-Dependent Ca^{2+} Uptake

To investigate the function of the YrbG gene product, cRNA from the cloned *E. coli* cDNA was synthesized using mMessage Machine (Ambion). *Xenopus* oocytes were injected with cRNA using standard procedures. For comparison, rat cardiac NCX1 cRNA was injected into another set of oocytes.[9] After 3–5 days of incubation at 17°C, the oocytes were peeled off the follicle layer, incubated for 30 minutes in ND96 containing 100 μM ouabain to increase the $[Na^+]_i$, and transferred to a solu-

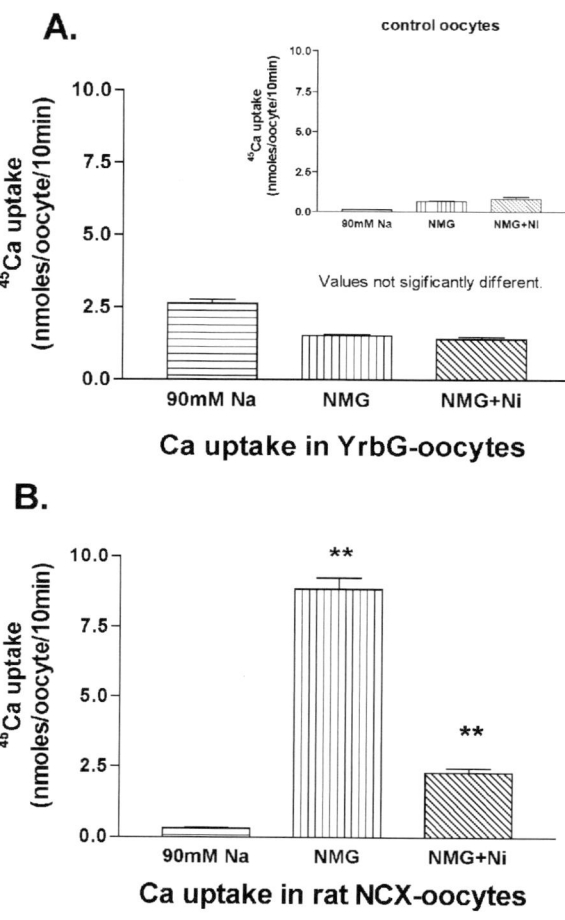

FIGURE 3. Absence of Na^+-dependent Ca^{2+} uptake in YrbG cRNA injected oocytes. Experiments were done at 32°C. *Asterisks* indicate significant differences ($P = <0.001$).

tion containing *N*-methyl glucamine (NMG), replacing Na^+ in ND96 and radioactive $^{45}Ca^{2+}$ isotope (20 µl/ml). After 10 minutes, Ca^{2+} uptake was stopped by adding $LaCl_3$, and the oocytes were washed in EGTA solution two times. The oocytes were solubilized, and the amount of $^{45}Ca^{2+}$ was counted in a liquid scintillation counter (Beckman). The results of the Ca^{2+} influx experiments are presented in FIGURE 3.

The oocytes injected with YrbG cRNA did not show any significant Ca^{2+} influx on changing the extracellular Na^+ to 0 mM. Some Ca^{2+} influx was observed in the presence of 90 mM Na^+ (FIG. 3A). However, under identical conditions, the oocytes injected with rat cardiac NCX1 cRNA showed significant Ca^{2+} influx on changing the extracellular Na^+ to 0 mM, and this Ca^{2+} influx was inhibited by 5 mM Ni^{2+}

FIGURE 4. Identification of bacterial exchanger protein in the *Xenopus* oocyte membrane. The expressed bacterial exchanger protein is clearly identified by the VSV antibody. Control oocytes do not show such VSV-antibody staining.

(FIG. 3B). The uninjected oocytes showed negligible amounts of Ca^{2+} influx under these conditions (FIG. 3A inset).

Identifying the Expressed Bacterial Exchanger in Oocyte Membrane

To see whether the *E. coli* YrbG was expressed and targeted to the oocyte membrane, an epitope tag was attached to the YrbG ORF. This enabled us to detect YrbG protein expression in the absence of any specific antibody to the YrbG protein. To the full-length *E. coli* exchanger in the pSD64TF vector, a fragment from a human ezrin cDNA clone containing a vesicular stomatis virus (VSV) epitope tag (Algrain 1993) was added. Sequence analysis confirmed that the clone contained the full-length *E. coli* exchanger with five amino acids of the VSV epitope (NRLGK) before the stop codon. cRNA made from this clone was injected into the oocytes, and immunocytochemistry studies were done using polyclonal antibodies (MBL Co.) to the VSV epitope. After being stained with the secondary antibody conjugated with fluorescein, the oocytes were observed with a confocal microscope (Zeiss LSM410). Immunostaining of the VSV indicates that the protein product of the YrbG gene has been made successfully in the oocytes and is transported to the oolemma (FIG. 4B), even though we could not detect any Na^+-dependent Ca^{2+} uptake in these oocytes.

CONCLUSION

Using a novel proteomic approach we identified conservative motifs in the Na^+/Ca^{2+} exchanger gene family. The motifs were also found in some ORFs from prokaryotic genomes. One ORF in the *E. coli* genome was cloned and expressed in *Xenopus* oocytes. However, the *E. coli* exchanger did not function like the Na^+/Ca^{2+} exchanger. We are investigating the function of this *E. coli* protein.

REFERENCES

1. MYLVAGANAM, S.E., M. PRABHAKARAN, S.S. TUDOR, *et al.* 2002. Structural proteomics: methods in deriving protein structural information and issues in data management. Biotechniques Suppl. March: 42–46.
2. JONASSEN, I., I. EIDHAMMER, D. CONKLIN & W.R. TAYLOR. 2002. Structure motif discovery and mining the PDB. Bioinformatics **18:** 362–367.
3. COGHLAN, A., D.A. MAC DONAILL & N.H. BUTTIMORE. 2001. Representation of amino acids as five-bit or three-bit patterns for filtering protein databases. Bioinformatics **17:** 676–685.
4. PEI, J. & N.V. GRISHIN. 2001. AL2CO: calculation of positional conservation in a protein sequence alignment. Bioinformatics **17:** 700–712.
5. KUMAR, S., X. CHENG, S. KLIMASAUSKAS, *et al.* 1994. The DNA (cytosine-5) methyltransferases. Nucleic Acids Res. **22:** 1–10.
6. POSFAI, J., Z. SZARAZ & R.J. ROBERTS. 1994. VISA: Visual Sequence Analysis for the comparison of multiple amino acid sequences. Comput. Appl. Biosci. **10:** 537–544.
7. POSFAI, J., A.S. BHAGWAT, G. POSFAI & R.J. ROBERTS. 1989. Predictive motifs derived from cytosine methyltransferases. Nucleic Acids Res. **17:** 2421–2435.
8. SMITH, H.O., T.M. ANNAU & S. CHANDRASEGARAN. 1990. Finding sequence motifs in groups of functionally related proteins. Proc. Natl. Acad. Sci. USA **87:** 826–830.
9. RUKNUDIN, A., S. HE, W.J. LEDERER & D.H. SCHULZE. 2000. Functional differences between cardiac and renal isoforms of the rat Na^+-Ca^{2+} exchanger NCX1 expressed in *Xenopus* oocytes. J. Physiol. **529** (Pt 3): 599–610.

Temperature Dependence of Cardiac Na$^+$/Ca^{2+} Exchanger

CHRISTIAN MARSHALL,[a,*] CHADWICK ELIAS,[b,*] XIAO-HUA XUE,[a,*] HOA DINH LE,[b] ALEXANDER OMELCHENKO,[b] LARRY V. HRYSHKO,[b] AND GLEN F. TIBBITS[a,c,*]

[a]*Cardiac Membrane Research Laboratory, Simon Fraser University, Burnaby, BC, V5A 1S6, Canada*

[b]*Institute of Cardiovascular Sciences, St. Boniface General Hospital Research Centre, The University of Manitoba, Winnipeg, MB, R2H 2A6, Canada*

[c]*Cardiovascular Sciences, BC Research Institute for Children's and Women's Health, Vancouver, BC, V5Z 4H4, Canada*

KEYWORDS: NCX; temperature dependence; Na/Ca exchange

INTRODUCTION

The Na$^+$/Ca^{2+} exchanger (NCX) is an integral membrane protein that plays an important role in the regulation of Ca^{2+} concentration in the cytosol. Active transporters, such as NCX1.1, involved in ion translocation in the mammalian heart, are highly temperature dependent. For example, it has been demonstrated that the Q$_{10}$ (fold change in activity for a 10°C change in temperature) for NCX1.1 is in the range of 2.2–4.0 in mammals.[1,2] Cardiac function in active salmonid species such as rainbow trout (*Oncorhynchus mykiss*) is distinguished by its ability to maintain adequate contractility under hypothermic conditions that are cardioplegic to mammals. The recent cloning of the trout cardiac NCX (NCX-TR1.0) provides us with a molecular model for investigating the temperature dependence of the NCX molecule.[3] The NCX-TR1.0 has a predicted topology similar to that of mammalian NCX1.1 based on hydropathy analysis and sequence identity. At the amino acid level, the NCX-TR1.0 shows ~75% overall identity to NCX1.1. However, sequence identity between these exchangers is significantly higher in regions of the molecule known to be functionally important, such as the α-repeats (~92%), XIP site (~85%), and regulatory Ca^{2+} binding domains (~86%).[3] In a previous study, we characterized in detail the temperature dependencies of NCX1.1 and TR-NCX1.0 wild-type exchangers expressed in oocytes by measuring outward currents using the giant excised patch technique.[4] The peak outward current of NCX1.1 exhibited typical mammalian

---

*These authors contributed equally to this study.
Address for correspondence: Dr. Glen F. Tibbits, Cardiac Membrane Research Laboratory, Simon Fraser University, Burnaby, BC, V5A 1S6, Canada. Voice: 604-291-3658; fax: 604-291-3040.
tibbits@sfu.ca

Ann. N.Y. Acad. Sci. 976: 109–112 (2002). © 2002 New York Academy of Sciences.

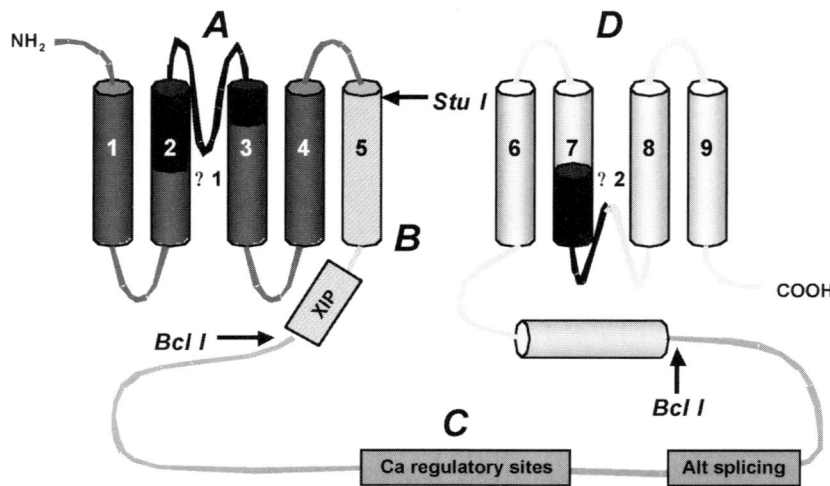

FIGURE 1. NCX topology and strategy for chimera construction. The NCX is predicted to have nine transmembrane segments (TMSs) separated by a large intracellular loop. TMSs are represented by cylinders and are numbered accordingly. Within the intracellular loop are the XIP site, Ca^{2+} regulatory site, and alternative splicing region. The α-regions (α1 and α2) play a crucial role in ion translocation and are shaded in *black*. Note that the *shaded cylinder* within the intracellular loop was once modeled to be in the membrane. Arrows (with corresponding restriction enzyme) denote where wild-type dog NCX1.1 and trout NCX-TR1.0 cDNA were cut for chimera construction. The resulting sections are indicated as letters *A* (*dark gray*), *B* (*hatched*), *C* (*gray*), and *D* (*light gray*). Chimera nomenclature is as follows: DTT (dog AB, trout CD); TDD (trout AB, dog CD); DTD (dog ABD, trout C); and DTTT (dog A, trout BCD).

temperature sensitivities with a Q_{10} value of 2.4, whereas the NCX-TR1.0 peak current was relatively temperature insensitive with a Q_{10} value of 1.2.[4] Furthermore, it was found that the disparities in temperature dependence between these two exchanger isoforms are unlikely due to either differences in inactivation kinetics or NCX regulatory mechanisms.[4] The purpose of this study was to delineate the regions of the NCX molecule responsible for its temperature sensitivity.

METHODS

Our strategy was to examine further the temperature dependence of the NCX involved the construction of chimeric NCX proteins (FIG. 1). Initially, three chimeras were constructed. Wild-type dog and trout exchanger cDNA were cut twice at homologous places into three domains: an N-terminal domain that is comprised of the first five transmembrane segments and the XIP site; the intracellular loop, containing the Ca^{2+} binding domain and alternative splicing site; and a C-terminal domain that includes the final four transmembrane segments. Rearranging these domains produced three initial chimera constructs: DTT, TDD, and DTD. A fourth chimera, named DTTT, was produced to isolate the effect of sequence differences within the

XIP site and the TM5 regions on the differential temperature dependence. Expression of chimeric exchangers in *Xenopus* oocytes was carried out as described previously.[4,5] Outward currents for wild-type and chimeric exchangers were measured in *Xenopus* oocytes over a temperature range of 7–30°C using the giant excised patch technique.[4,5]

RESULTS AND DISCUSSION

Qualitative examination of outward exchange currents for the chimeras DTD, DTT, and TDD over a range of temperatures placed the region responsible for NCX temperature dependence within the N-terminal TM segments and XIP site (data not

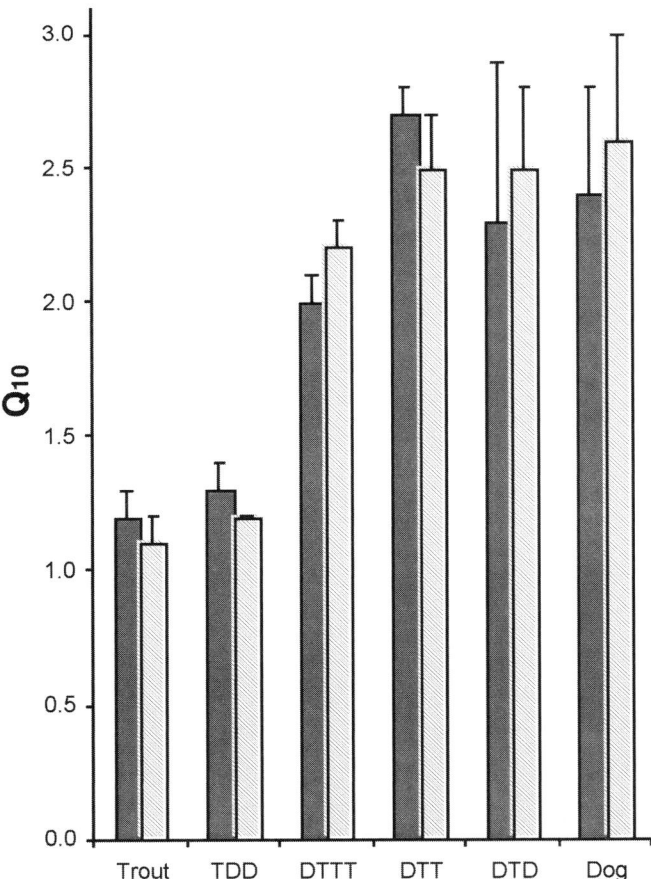

FIGURE 2. Q_{10} values for wild-type and chimeric exchangers. Q_{10} values from peak currents are shown in *solid bars,* and *hatched bars* represent Q_{10} values from steady-state currents.

shown). The fourth chimera, DTTT, revealed that differences within TM5 and the XIP regions have an effect on the temperature dependence of the NCX molecule, but the contribution is relatively small in comparison with the first four TM segments. This supports the earlier finding that deregulation of both wild-type trout and dog exchangers with chymotrypsin treatment had no significant effect on the temperature dependencies of the NCX isoforms.[4] The Q_{10} values (FIG. 2) were calculated for wild-type and chimeric exchangers as indices of temperature dependence. Consistent with qualitative observations, the origin of the N-terminal region determines the temperature-dependent phenotype exhibited by the chimeric exchangers. These data indicate that the TM5 and XIP regions play a relatively minor role in the temperature dependence of the NCX compared to the first four transmembrane segments; however, this region does significantly contribute to the temperature sensitivity of NCX. Further mutational analysis is needed to determine the specific amino acids involved in the temperature dependence of the molecule, but we speculate that a series of amino acid substitutions within the first four TM segments are responsible for the high activity of trout NCX at low temperatures.

REFERENCES

1. ELIAS, C.L., X.H. XUE, C.R. MARSHALL, et al. 2001. Temperature dependence of cloned mammalian and salmonid cardiac Na^+/Ca^{2+} exchanger isoforms. Am. J. Physiol. Cell Physiol. **281:** C993–C1000.
2. HILGEMANN, D.W., A. COLLINS & S. MATSUOKA. 1992. Steady-state and dynamic properties of cardiac sodium-calcium exchange. Secondary modulation by cytoplasmic calcium and ATP. J. Gen. Physiol. **100:** 933–961.
3. KIMURA, J., S. MIYAMAE & A. NOMA. 1987. Identification of sodium-calcium exchange current in single ventricular cells of guinea-pig. J. Physiol. (Lond.) **384:** 199–222.
4. OMELCHENKO, A., C. DYCK, M. HNATOWICH, et al. 1998. Functional differences in ionic regulation between alternatively spliced isoforms of the Na^+-Ca^{2+} exchanger from *Drosophila melanogaster*. J. Gen. Physiol. **111:** 691–702.
5. XUE, X.H., L.V. HRYSHKO, D.A. NICOLL, et al. 1999. Cloning, expression, and characterization of the trout cardiac Na^+-Ca^{2+} exchanger. Am. J. Physiol. **277:** C693–700.

The Ca/Ca Exchange Mode of the Na/Ca Exchanger Investigated by Photolytic Ca^{2+} Concentration Jumps

ANDREAS HAASE, MICHAEL KAPPL,[a] GEORG NAGEL, PHILIP G. WOOD, AND KLAUS HARTUNG

Max-Planck-Institut für Biophysik, Kennedyallee 70, D-60596 Frankfurt, Germany

KEYWORDS: DM-nitrophen; caged Ca^{2+}; antiporter

The Na/Ca exchanger catalyzes the countertransport of Na^+ and Ca^{2+} as well as the self-exchange of either Na^+ or Ca^{2+}, that is, Na/Na and Ca/Ca exchange. These self-exchange modes offer a unique possibility to investigate partial reactions of the exchanger. Previously, using a combination of patch-clamp techniques and photolytic Ca^{2+} concentration jumps, we showed that the exchanger generates a transient current under conditions promoting Ca/Ca exchange when the Ca^{2+} concentration on the cytolosolic side is rapidly increased.[1] Thus, contrary to previous suggestions, charge translocation is involved in Ca^{2+} translocation, although stationary Ca/Ca exchange is electroneutral. A cytosolic Ca^{2+} jump generates inward, not outward, current. Thus, it has been suggested that the movement of overall negative charge is correlated with the outward translocation of Ca^{2+}. The time course of the current signal is fairly rapid: time to peak is less than 0.1 ms, and the duration is <1 ms. Thus, very fast Ca^{2+} concentration jumps and recording techniques are required if this signal is to be analyzed quantitatively. Photoinduced Ca^{2+} release from the chelator DM-nitrophen occurs with at least 38,000 s^{-1} (2.6 μs) and the bandwidth of the patch clamp is about 10 kHz (15 μs).[1,2] To learn more about the mechanism of Ca^{2+} translocation, we investigated the voltage and Ca^{2+} dependence of the transient current. Initial experiments were performed with membrane patches excised from guinea pig cardiac myocytes.[1,3] More detailed investigations were conducted with NCX1 from guinea pig expressed in *Xenopus* oocytes. No significant differences between either system have been observed so far. A major advantage of the oocyte system is that on the average, signal amplitudes are much larger than ones obtained with membranes from cardiac myocytes.

---

[a]Present address: Universität Siegen, Physikalische Chemie, Adolf-Reichwein-Str., D-57076 Siegen, Germany.

Address for correspondence: Dr. K. Hartung, Max-Planck-Institut für Biophysik, D-60596 Frankfurt/Main, Germany. Voice: 49-69-6303-307; fax: 49-69-6303-305.

hartung@mpibp-frankfurt.mpg.de

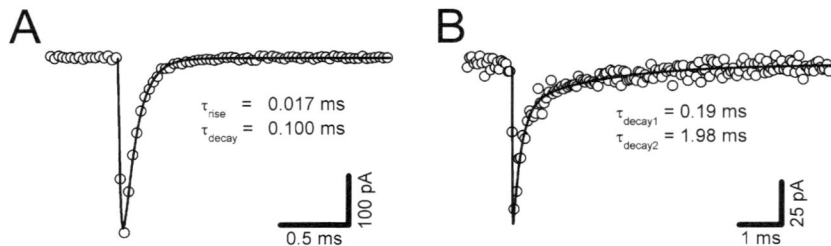

FIGURE 1. Current records obtained after a photolytic Ca^{2+} concentration jump. The patch pipette contained either 5 mM Ca^{2+} (**A**) or 5 mM Ba^{2+} (**B**), promoting either Ca/Ca or Ba/Ca exchange. The Ca^{2+} concentration jump was 400 µM in **A** and 100 µM in **B**.

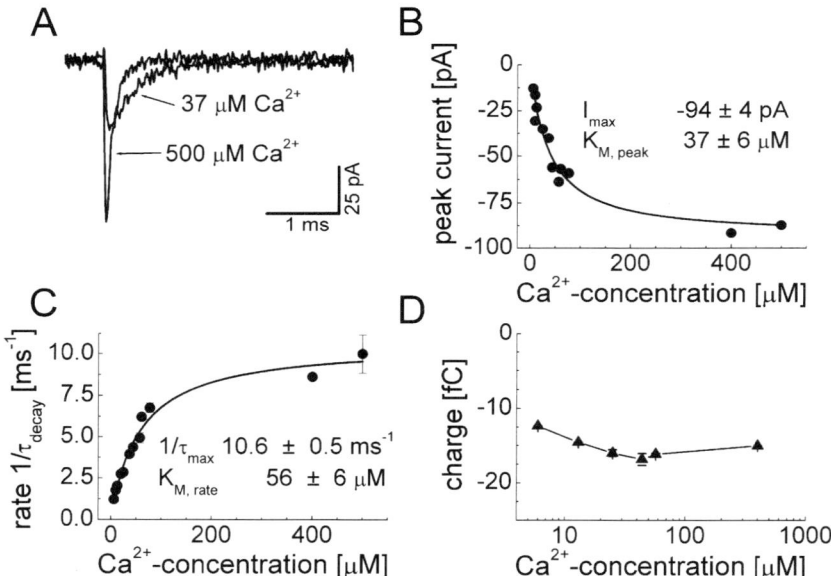

FIGURE 2. Ca^{2+} dependence of the transient current signal. (**A**) Superimposed current signals obtained by different Ca^{2+} concentration jumps. (**B**, **C**, and **D**) The Ca^{2+} dependence of peak current, decay rate constant, and integrated current signal or charge translocated.

FIGURE 1A shows a current signal generated by a large (400 µM) concentration jump at 0 mV. The current rises rapidly, within less than 0.1 ms, to a maximum and then decays with a time constant of 0.1 ms ($1/t = 10{,}000$ s^{-1}) to zero. The rising phase cannot be evaluated quantitatively because it overlaps with an artefact generated by the laser flash.[1] FIGURE 2 shows the Ca^{2+} concentration dependence of the peak current, the decay rate constant, and the integrated current signal, that is, the amount of charge translocated. The Ca^{2+} dependence of the peak current and of the

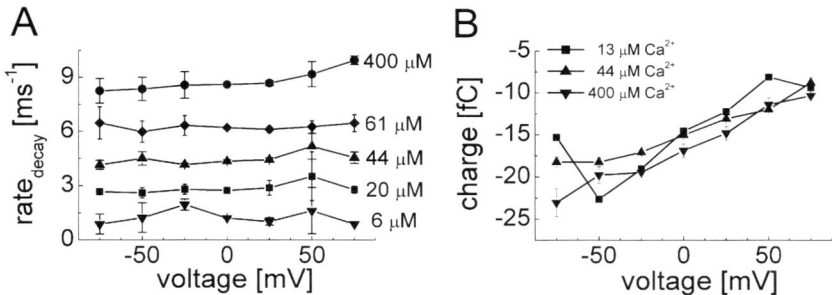

FIGURE 3. Voltage dependence of the transient current. (**A**) The decay rate constant is voltage independent at different Ca^{2+} concentrations. (**B**) More charge is translocated at negative potentials.

rate constant can be described by Michaelis-Menten kinetics with $K_m = 37$ and 56 µM, respectively, whereas the amount of charge translocated is constant over the whole range of Ca^{2+} concentrations. Assuming that the Ca/Ca exchange mode can be described by a linear reaction scheme of forward and backward reactions, the exchanger molecules are in true equilibrium, determined by the Ca^{2+} concentration on both sides, the membrane potential and other parameters. If charge translocation is related to one of these reaction steps, the amount of charge translocated is directly proportional to a change of the equilibrium distribution of states. Vice versa, the constancy of the amount of charge translocated indicates that the equilibrium distribution of states is fairly independent of the cytoplasmic Ca^{2+} concentration. Saturation of the relaxation rate and the peak amplitude indicate that Ca^{2+} binding is not electrogenic. In case of electrogenic binding both parameters should increase linearly with the Ca^{2+} concentration.

Performing Ca^{2+} concentration jumps at different voltages shows that the magnitude of the inward current increases at negative potentials; the relaxation rate constant, however, is constant between ± 75 mV (FIG. 3A). Thus, the amount of charge translocated increases at negative potentials (FIG. 3B). The voltage independence of the relaxation rate constant and the voltage dependence of the amount of charge translocated indicate that the time course is determined by a voltage independent reaction step, whereas the equilibrium is determined by a different voltage-dependent step that is too fast to be observed. The voltage dependence of the charge translocated (FIG. 3B) and the voltage independence of the relaxation rate constant (FIG. 3A) are maintained over a wide range of Ca^{2+} concentrations from 6–400 µM. These results show that the relaxation rate constant (voltage independent) and the charge translocation (voltage dependent) are related to different reaction steps. The equilibrium distribution of states is voltage but not Ca^{2+} dependent, whereas the relaxation rate constant is Ca^{2+} but not voltage dependent.

The voltage and Ca^{2+} dependence of the exchanger in the Ca/Ca exchange mode can be modeled by a five-state reaction scheme (*Scheme 1*) with electroneutral Ca^{2+} binding on both sides, one electroneutral and rate-limiting conformational transition

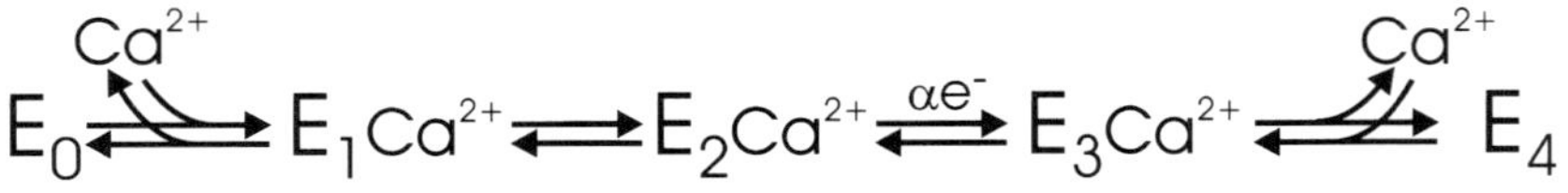

$$E_0 \xrightleftharpoons{Ca^{2+}} E_1Ca^{2+} \rightleftharpoons E_2Ca^{2+} \xrightleftharpoons{\alpha e^-} E_3Ca^{2+} \xrightleftharpoons{Ca^{2+}} E_4$$

SCHEME 1

following intracellular Ca^{2+} binding and a second conformational transition, which is fast equilibrating and electrogenic.[1]

Further information on the Ca^{2+} translocating branch of the reaction cycle may be obtained by the variation of the ion condition on the extracellular side. Several other divalent cations can substitute for Ca^{2+}, but with different efficiencies. Sr^{2+}, for example, catalyzes Na^+ translocation as well as Ca^{2+}. Performing a cytoplasmic Ca^{2+} concentration jump with Sr^{2+} in the pipette instead of Ca^{2+} generates a transient current that is indistinguishable from that with Ca^{2+} in the pipette. Ba^{2+}, on the other hand, catalyzes Na^+ translocation very poorly.[4] Performing a Ca^{2+} concentration jump with Ba^{2+} on the extracellular side generates a large current comparable to that obtained with extracellular Ca^{2+}, but the decay of the current occurs in two phases with time constants 0.19 and 1.98 ms (FIG. 1B). A similar result is obtained with low Ca^{2+} (0.1 mM) on the extracellular side, indicating that a variation of the initial conditions may allow the detection of further details on the mechanism of Ca^{2+} translocation.

REFERENCES

1. KAPPL, M. & K. HARTUNG. 1996. Rapid charge translocation by the Na^+-Ca^{2+} exchanger after a Ca^{2+} concentration jump. Biophys. J. **71:** 2473–2485.
2. ELLIS-DAVIES, G.C.R., J.H. KAPLAN & R.J. BARSOTTI. 1996. Laser photolysis of caged calcium: rates of calcium release by nitrophenyl-EGTA and DM-nitrophen. Biophys. J. **70:** 1006–1016.
3. KAPPL, M., G. NAGEL & K. HARTUNG. 2001. Voltage and Ca^{2+} dependence of pre-steady-state currents of the Na-Ca exchanger generated By Ca^{2+} concentration jumps. Biophys. J. **81:** 2628–2638.
4. TRAC, M., C. DYCK, M. HNATOWICH, et al. 1997. Transport and regulation of the cardiac Na-Ca exchanger, NCX1. Comparison between Ca^{2+} and Ba^{2+}. J. Gen. Physiol. **109:** 361–369.

Functional Analysis of Polar Residues Important for Activity of Na$^+$/H$^+$ Exchangers

LARRY FLIEGEL, CHRISTINE WIEBE, RAKHILYA MURTAZINA,
PAVEL DIBROV, BRENDA J. BOOTH, BONNIE L. BULLIS,
EMILY DiBATTISTA, AND DYAL N. SINGH

Department of Biochemistry, Faculty of Medicine, CIHR Membrane Protein Group, University of Alberta, Edmonton, Alberta, Canada T6G 2H7

KEYWORDS: cation coordination; cation transport; Na$^+$/H$^+$ exchanger; membrane proteins; pH regulation

INTRODUCTION

Na$^+$/H$^+$ exchangers are a family of integral membrane proteins that exchange protons for Na$^+$. They exist in yeast, *Escherichia coli,* and vertebrates. In mammals, seven isoforms of the Na$^+$/H$^+$ exchanger are known. The first isoform to be discovered, the NHE1 isoform, is ubiquitous and transports one intracellular proton out of the cell in exchange for one extracellular Na$^+$. Another isoform of the Na$^+$/H$^+$ exchanger is the yeast protein *Sod2* that catalyzes the reverse process in *Schizosaccharomyces pombe,* the removal of an intracellular Na$^+$ in exchange for an extracellular proton.[1]

Little is known about the amino acids involved in cation coordination and transport by the eukaryotic Na$^+$/H$^+$ exchangers. Examination of the amino acid sequence of the Na$^+$/H$^+$ exchangers revealed conserved polar amino acids in some transmembrane segments. Polar amino acids of membrane proteins have been involved in the cation binding and transport of a number of ion pumps and transporters. We therefore examined the effect of mutation of several amino acids of conserved transmembrane segments of the exchanger. Our specific hypothesis was that specific polar amino acids that are within or are associated with specific transmembrane segments of the membrane domains are responsible for cation binding and translocation of Na$^+$ and H$^+$. To test this hypothesis we used site-specific mutagenesis to examine the role of specific amino acids that could be involved in the transport of cations. The human Na$^+$/H$^+$ exchanger NHE1 isoform was mutated and stably transfected into AP-1 cells, as described earlier.[2,3] Na$^+$/H$^+$ exchanger activity including V_{max} and K_m was analyzed after ammonium chloride-induced prepulse, as we described earlier.[4] For the yeast Na$^+$/H$^+$ exchanger sod2, mutagenesis and activity measurements were as described earlier.[5]

Address for correspondence: Dr. Larry Fliegel, Department of Biochemistry, Faculty of Medicine, CIHR Membrane Protein Group, University of Alberta, 347 Medical Science Building, Edmonton, Alberta, Canada, T6G 2H7.
lfliegel@gpu.srv.ualberta.ca

Ann. N.Y. Acad. Sci. 976: 117–120 (2002). © 2002 New York Academy of Sciences.

In the case of the NHE1 isoform of the human Na^+/H^+ exchanger we added a hemagglutinin tag to the C-terminus of the protein,[3] so that we could easily detect the amount of protein in Western blots and check the localization by immunocytochemistry. The mutations made were D238N, P239A, E262Q, E262D, S263A, N266A, D267N, D267E, S359A, SS387, 388AA, S390A, E391Q, E391D, T392V, S401A, T402V, and S406A. All of the mutant proteins expressed at similar levels, and all were targeted to the plasma membrane.

When we examined the activity of the control and mutant proteins, only mutation of Glu262, Asp267, and E391 affected Na^+/H^+ exchanger activity. The E262Q and D267N mutants showed no activity equivalent to the mock transfected AP-1 cells. The E391Q mutant showed greatly reduced activity compared to that of the controls. The conservative substitutions of E262D, D267E, and E391D all restored activity, although the E262D mutant had slightly depressed activity relative to the wild type. We found that the E262D mutant also had an increase in the K_m for Li^+ (but not for Na^+). This result suggested that a change had occurred in the relative affinity for the two cations, possibly indicating an alteration in the coordination site of the protein.

Other conserved amino acids were tested for their contribution to Na^+/H^+ exchanger activity. Ser359 of transmembrane segment IX was mutated to Ala; however, this had no effect on activity. Similarly, mutation of Pro239 and Asn266 did not affect activity. It is surprising that mutation of Pro239 did not affect activity, because prolines usually function as helix breakers within the membrane,[6] and we hypothesized that such a large change in conformation could disrupt the protein. However, this residue proved to be unimportant to the function of the protein. Asp238 together with Pro239 comprise a well-conserved motif in almost all the mammalian Na^+/H^+ exchangers and in the yeast Na^+/H^+ exchanger sod2. Thus, it was surprising that this residue was not important to the function of the mammalian Na^+/H^+ exchanger.

To determine if polar amino acids are important in different Na^+/H^+ exchangers similar experiments were done with the yeast Na^+/H^+ exchanger sod2. All eight His residues were mutated to Arg, and the conserved amino acids Asp145, Asp241, and Asp266,267 were mutated to Asn residues. We characterized the ability of the sod2 protein to allow growth of *S. pombe* in LiCl-containing medium. In addition, we examined the ability to expel sodium in acid medium and the ability to carry out sodium-dependent proton influx at pH 6.1, as described earlier.[5] Of the His residues, only His367 was essential for activity. In addition, mutation of amino acids Asp145, Asp241, and Asp266,267 all impaired, at least partially, sod2 activity.

FIGURE 1 illustrates the relative location of the mutated amino acids in human NHE1 and sod2. The topology of NHE1 is based on recent analysis by cysteine scanning mutagenesis.[7] That of sod2 is theoretical and is based on hydrophobicity analysis.[1] Whereas a complete comparison still awaits further analysis of the protein structure, it is of interest that in both cases amino acid D267 is important in function. In addition, surrounding amino acids of this region are also significant in function. It was of particular interest that conservation of the side chains in NHE1 conserved activity. We hypothesized earlier that Na^+/H^+ exchangers may act by coordination of substrate cations through a crown ether-like cluster of polar amino acids.[8] Our current results support the hypothesis that the oxygen in the side chains of E262, D267, and E391 could serve this role for the mammalian Na^+/H^+ exchanger. Although it is impossible at this stage to determine if the amino acids are influencing transport through effects on structure, experiments using limited trypsinolysis (not shown)

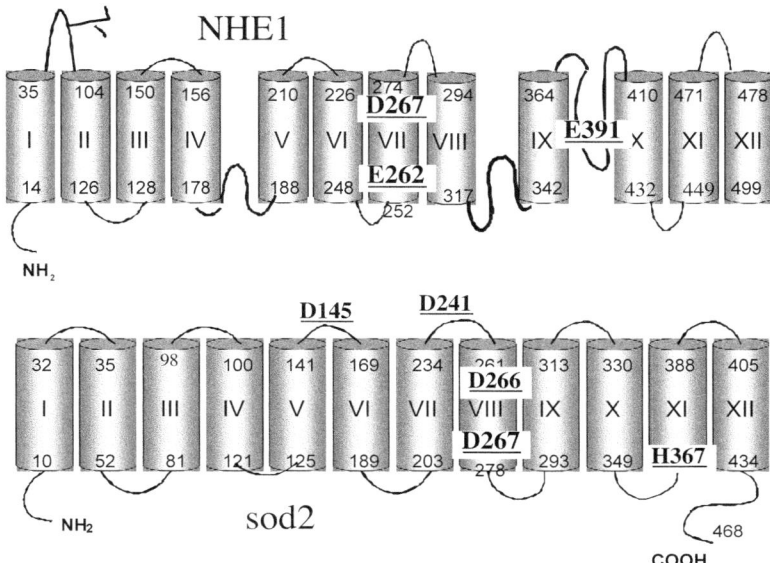

FIGURE 1. Topological models of the transmembrane region of the NHE1 isoform of the human Na^+/H^+ exchanger and the yeast Na^+/H^+ exchanger sod2. Residues that are important in activity are underlined.

suggest that this may not be the case. Future experiments are necessary to determine the exact conformation of these parts of the protein and their role in cation binding and transport.

SUMMARY

Na^+/H^+ exchangers are a family of ubiquitous membrane proteins. In mammals the NHE1 isoform of the protein is widely distributed through all tissues and regulates cytosolic pH by removing an intracellular H^+ in exchange for an extracellular Na^+. In fission yeast, excess levels of intracellular Na^+ are detrimental to these cells, and the Na^+/H^+ exchanger, sod2, plays a major role in the regulation of internal sodium concentration. We examined the functional role of conserved, polar, amino acid residues occurring in membrane-associated segments of the Na^+/H^+ exchanger proteins. For the mammalian Na^+/H^+ exchanger, mutant proteins of transmembrane segments VI and VII and the membrane-associated segment from amino acids 387 to 406 were assessed by characterization of intracellular pH changes in stably transfected cells that lacked an endogenous Na^+/H^+ exchanger. All of the mutant proteins were expressed and were targeted properly to the plasma membrane. Mutation of amino acid residues Glu262, Asp267, and E391 affected the Na^+/H^+ exchanger. Conservative substitutions with alternative acidic residues restored Na^+/H^+ exchanger activity. The Glu262Asp mutant had a decreased affinity for Li^+. For the yeast

Na$^+$/H$^+$ exchanger, similar experiments examined the effect of mutagenesis of several conserved polar amino acids. Of all Histidine residues, only His 367 was significant for activity. The Asp266,267 pair were mutated simultaneously, and sod2 function was found to be significantly impaired. Results also indicated that residues Asp145 and Asp241 are important for proper function of sod2. The results support the hypothesis that side chain oxygen atoms in a few, critically placed amino acids are important in various kinds of Na$^+$/H$^+$ exchanger activity.

REFERENCES

1. WIEBE, C.A., E.R. DIBATTISTA & L. FLIEGEL. 2001. Functional role of amino acid residues in Na$^+$/H$^+$ exchangers. Biochem. J. **357:** 1–10.
2. MURTAZINA, B.R., B.J. BOOTH, B.L. BULLIS, *et al.* 2001. Functional analysis of polar amino acid residues in membrane associated regions of the NHE1 isoform of the Na$^+$/H$^+$exchanger. Eur. J. Biochem. **268:** 1–13.
3. WANG, H., D. SINGH & L. FLIEGEL. 1998. Functional role of cysteine residues in the Na$^+$/H$^+$exchanger. Arch. Biochem. Biophys. **358:** 116–124.
4. SILVA, N.L.C.L., H. WANG, C.V. HARRIS, *et al.* 1997. Characterization of the Na$^+$/H$^+$ exchanger in human choriocarcinoma (BeWo) cells. Eur. J. Physiol. **433:** 792–802.
5. DIBROV, P., P.G. YOUNG & L. FLIEGEL. 1998. Functional analysis of amino acid residues essential for activity in the Na$^+$/H$^+$ exchanger of fission yeast. Biochemistry **36:** 8282–8288.
6. VON HEIJNE, G. 1991. Proline kinks in transmembrane alpha-helices. J. Mol. Biol. **218:** 499–503.
7. WAKABAYASHI, S., T. PANG, X. SU & M. SHIGEKAWA. 2000. A novel topology model of the human Na$^+$/H$^+$ exchanger isoform 1. J. Biol. Chem. **275:** 7942–7949.
8. DIBROV, P. & L. FLIEGEL. 1998. Comparative molecular analysis of Na$^+$/H$^+$ exchangers: a unified model for Na$^+$/H$^+$ antiport? FEBS Lett. **424:** 1–5.

Stoichiometry of the Na^+/Ca^{2+} Exchanger

Models and Implications

SATOSHI MATSUOKA

Department of Physiology and Biophysics, Kyoto University Graduate School of Medicine, Kyoto, 606-8501, Japan

ABSTRACT: We reevaluated the exchange stoichiometry of the Na^+/Ca^{2+} exchange current by measuring its reversal potential. The exchange current was measured from the inside-out macropatch excised from intact sarcolemma of guinea pig ventricular myocytes. This method provides more accurate control of extracellular and cytoplasmic ion concentrations and of membrane potential than is possible with a whole-cell clamped preparation. The exchange current was isolated as exchanger inhibitory peptide (XIP)-sensitive current or as cytoplasmic Na^+- and Ca^{2+}-induced current. The reversal potential of the Na^+/Ca^{2+} exchange current was, for the most part, close to the equilibrium potential of the $4Na^+:1Ca^{2+}$ exchange, although it tended to get closer to that of $3Na^+:1Ca^{2+}$ exchange at lower Na^+ concentrations. We concluded that the stoichiometry is 4 or that it may vary depending on the cytoplasmic Na^+. The $4Na^+:1Ca^{2+}$ exchange was further studied with computer modeling. A consecutive $4Na^+:1Ca^{2+}$ exchange model with two active states and two inactive states (E2 model) could not well reconstruct the current-voltage relation of the exchanger. However, a consecutive $4Na^+:1Ca^{2+}$ exchange model with 10 active states and 2 inactive states (E10 model), which included voltage-dependent Na^+ and Ca^{2+} occlusions, well simulated the current–voltage relation. Implications of $4Na^+:1Ca^{2+}$ exchange is also discussed.

KEYWORDS: stoichiometry; Na/Ca exchange; model; simulation

INTRODUCTION

The stoichiometry of Na^+/Ca^{2+} exchange, or the number of Na ions exchanged for one Ca^{2+}, is believed to be 3. $3Na^+:1Ca^{2+}$ exchange was supported by measurements of ion fluxes,[1–3] intracellular Na^+ and/or Ca^{2+} concentrations,[4–6] and reversal potential of Na^+/Ca^{2+} exchange current ($I_{Na/Ca}$).[7–9] In 1977, Mullins[10] proposed a model of $4Na^+:1Ca^{2+}$ exchange based on energy considerations and the sensitivity of Ca^{2+} fluxes to membrane potential. However, the aforementioned later studies did not support the $4Na^+:1Ca^{2+}$ exchange.

To determine stoichiometry, accurate control of both membrane potential and ion concentrations is indispensable, because activity of the Na^+/Ca^{2+} exchange primari-

Address for correspondence: Dr. Satoshi Matsuoka, Department of Physiology and Biophysics, Kyoto University Graduate School of Medicine, Kyoto, 606-8501, Japan. Voice: (81)-75-753-4357; fax: (81)-75-753-4349.

matsuoka@card.med.kyoto-u.ac.jp

ly depends on membrane potential as well as intra- and extracellular Na^+ and Ca^{2+} concentrations. In this regard, whole-cell clamp methods are not a suitable experimental system, because intracellular Na^+ and Ca^{2+} concentrations substantially change via ion flux through the exchanger, even if high concentrations of Ca^{2+} chelator (EGTA or BAPTA) are included in the pipette solution.[7]

We recently developed a method to make a large excised patch from intact guinea pig ventricular myocytes (macropatch).[11] The previously developed giant membrane patch method[12] used blebs of myocytes. By contrast, our method does not use blebs and allows us to make large inside-out patches directly from sarcolemmal membrane. This macropatch preparation enables us to more accurately control intra- and extracellular Na^+ and Ca^{2+} concentrations than does the whole-cell preparation. With this macropatch method, we presented experimental data to support the $4Na^+:1Ca^{2+}$ exchange.[11] Reversal potential of $I_{Na/Ca}$ was largely consistent with the equilibrium potential ($E_{Na/Ca}$) of $4Na^+:1Ca^{2+}$ exchange. Furthermore, stoichiometry was apparently dependent on cytoplasmic Na^+.

If the stoichiometry is 4, we need to reevaluate the ion transport mechanism of the exchanger and also the contribution of the exchanger to intracellular Ca^{2+} homeostasis. In this paper, we examine our hypothesis of the $4Na^+:1Ca^{2+}$ exchange with the *in silico* approach, a computer simulation. We specially focus on $E_{Na/Ca}$ during the action potential and Ca^{2+} transient and current-voltage (I-V) relation of the exchange.

METHODS

Isolation of Myocytes

Guinea pig ventricular cells were prepared, as previously described.[13]

Electrophysiology

Inside-out "macropatch" was formed as described in our previous study.[11] FIGURE 1 illustrates how to form the inside-out macropatch. FIGURE 1A is a picture of a pipette used for the macropatch. The tip diameter was about 7 μm, and the cell was an isolated guinea pig ventricular myocyte. Patch pipettes were prepared using borosilicate glass capillaries (O.D. 2.0 mm and I.D. 1.4 mm; Hilgenberg, Germany) and a two-step pipette puller (NARISHIGE PB-7, Japan). To facilitate seal formation, the pipette tip was slightly heat polished. The inner tip diameter was ~5–8 μm, and the electrical resistance was 0.3–0.8 MΩ when filled with pipette solutions (see below). Applying gentle negative pressure (10–15 cm H_2O) for ~5–10 minutes formed a cell-attached patch with a resistance of 2–5 GΩ (FIG.1B, 1). Lifting the pipette usually failed to establish the excised patch, and the myocyte continued to be attached. The myocyte could be blown by a stream from a capillary (theta glass) located near the cell (FIG. 1B, 2).

The concentration jump of the cytoplasmic solution was performed using the theta capillary mounted on a piezo translator.[11] The cytoplasmic solution could be changed within 30 ms.

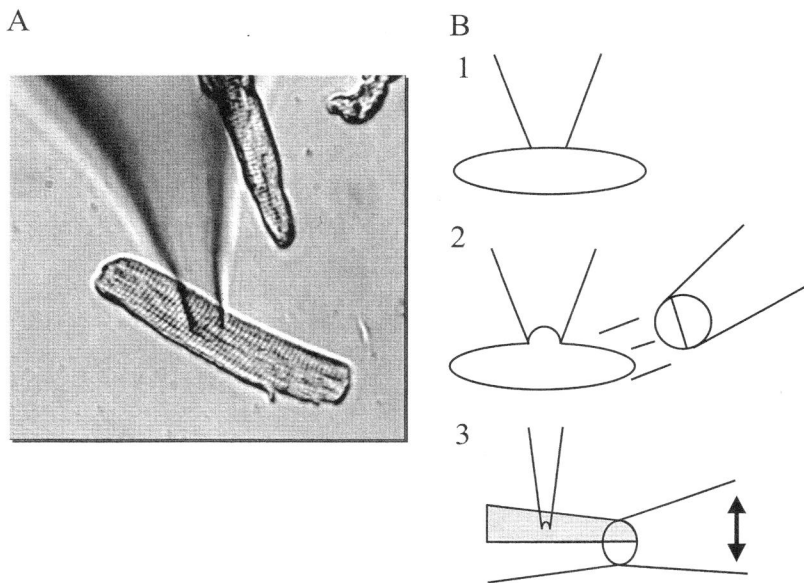

FIGURE 1. Macropatch formation. **(A)** A picture of a pipette for macropatch (inner diameter ~7 μm) and an isolated guinea pig ventricular cell. **(B)** Procedure for macropatch formation. (1) Cell-attached patch formation. (2) Lifting the pipette and cell and blowing the cell off with a stream from a θ-shaped capillary. (3) Concentration jump with the θ-shaped capillary mounted on a piezo transducer. See details in text.

The holding potential was 0 mV. The membrane current was filtered at 50–500 Hz by a low-pass filter and sampled at 50–100 Hz by an A/D converter. The I-V relationship was recorded with ramp voltage pulses (dV/dt = 0.72 V/s), as previously described,[13] and the membrane current was filtered at 500 Hz and sampled at 1.6 kHz. The temperature of the cytoplasmic solutions at the outlet of the theta capillary was 36–37°C.

Solutions and Chemicals

The pipette solution used to activate the outward $I_{Na/Ca}$ contained (in mM): N-methyl-D-glucamine (NMDG), 100; aspartate, 100; hepes, 5; TEA-Cl, 20; ouabain, 0.05; $CaCl_2$, 5; $MgCl_2$, 2; CsCl, 2; $BaCl_2$, 2; and nicardipin, 0.002 (pH = 7.4/NMDG). The pipette solution used for the inward $I_{Na/Ca}$ contained: NaOH, 100; aspartate, 100; hepes, 5; TEA-Cl, 20; ouabain, 0.05; EGTA, 0.1; $MgCl_2$, 2; CsCl, 2; $BaCl_2$, 2; and nicardipin, 0.002 (pH = 7.4/NMDG). The standard pipette solution used for recording both outward and inward $I_{Na/Ca}$ in a patch contained: NaOH, 145; aspartate, 145; hepes, 5; TEA-Cl, 20; ouabain, 0.05; $CaCl_2$, 2; $MgCl_2$, 2; CsCl, 2; $BaCl_2$, 2; and nicardipin, 0.002 (pH = 7.4/HCl).

Standard cytoplasmic solution with 100 mM Na^+ contained: NaOH, 100; EGTA, 10; hepes, 20; aspartate, 100; TEA-Cl, 20; CsCl, 20; $CaCl_2$, 8.79; and $MgCl_2$, 1.11

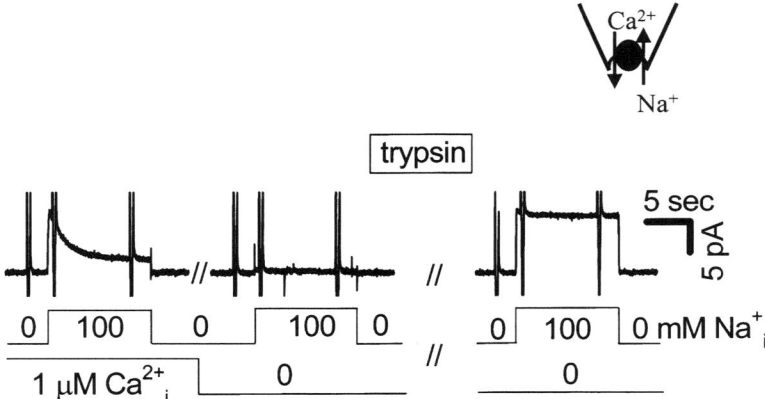

FIGURE 2. Outward $I_{Na/Ca}$ in the macropatch. The outward $I_{Na/Ca}$ was induced by applying cytoplasmic 100 mM Na^+ in the presence (*left trace*) and the absence (*middle trace*) of $Ca^{2+}{}_i$. Trypsin (0.5 mg/ml) treatment eliminated the Na^+-dependent inactivation and the Ca^{2+} activation, augmenting the current amplitude (*right trace*). Modified from Fujioka et al.[11]

(pH = 7.2/NMDG). $Na^+{}_i$ concentration was changed by replacing Na^+ with Li^+. Free Ca^{2+} and Mg^{2+} concentrations were calculated with a software developed by Bers et al.[14] and were 1.0 and 1.0 mM, respectively. The free Mg^{2+} concentration was always fixed to 1 mM when free Ca^{2+} was changed.

RESULTS

FIGURE 2 demonstrates the outward $I_{Na/Ca}$ in the macropatch excised from intact guinea pig ventricular cells. The pipette solution contained 0 mM Na^+ and 5 mM Ca^{2+}. Appling 100 mM Na^+ to the cytoplasmic side of the patch membrane induced the outward current. This current decayed during continuous application of Na_+ (Na^+-dependent inactivation). The current was dependent on cytoplasmic Ca^{2+}. In the absence of Ca^{2+}, the Na^+-induced outward current was greatly attenuated (Ca^{2+}-activation). The Na^+-dependent inactivation and the Ca^{2+}-activation were eliminated by partial proteolysis with trypsin. The outward current, thereafter, was time independent and no longer required Ca^{2+} for its activation. These properties were essentially the same as those reported previously for $I_{Na/Ca}$ in the giant membrane patch excised from the bleb of ventricular cells[15,16] and *Xenopus* oocytes expressing NCX1.[17,18] To further confirm that the Na^+-induced current was related to the Na^+-Ca^{2+} exchange, we studied the reversal potential of the current. If the current is generated by Na^+-Ca^{2+} exchange, the reversal potential should be identical to $E_{Na/Ca}$. $E_{Na/Ca}$ is determined by equation 1:

$$E_{Na/Ca} = \frac{n_x \cdot E_{Na} - 2 \cdot E_{Ca}}{n_x - 2} \tag{1}$$

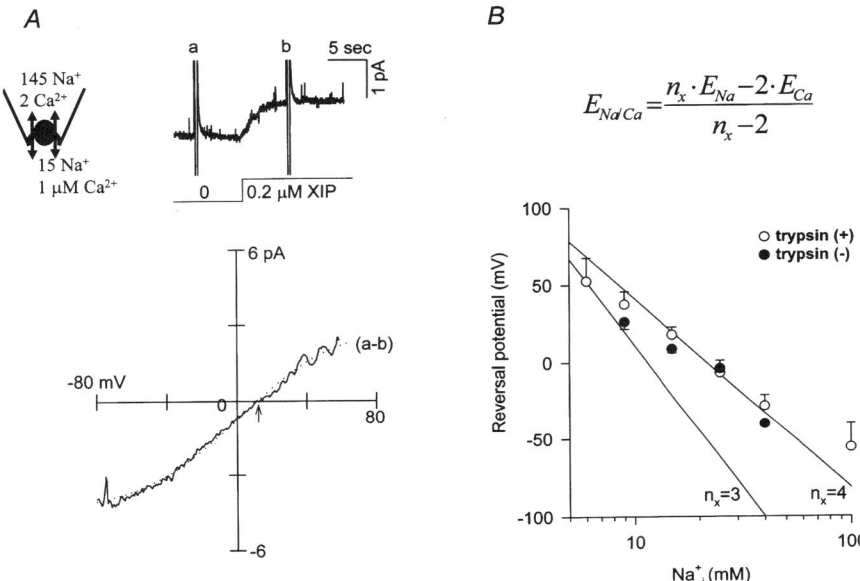

FIGURE 3. Reversal potential of XIP-sensitive $I_{Na/Ca}$. **(A)** Inhibition of $I_{Na/Ca}$ by XIP. The I-V relation was measured in the absence and the presence of 0.2 μM XIP (**a** and **b**). The I-V relation of XIP-sensitive current is shown in the lower panel. **(B)** Reversal potential-Na^+_i relation. Reversal potentials of the XIP-sensitive current were measured at various Na^+_i concentrations and plotted against Na^+_i. Data are from trypsin-treated (*open circles*) and trypsin-untreated (*closed circles*) patches. Lines are $E_{Na/Ca}$ of $3Na^+:Ca^{2+}$ exchange ($n_x = 3$) and $4Na^+ : Ca^{2+}$ exchange ($n_x = 4$). Modified from Fujioka et al.[11]

where n_x is stoichiometry (number of Na^+ exchanged with Ca^{2+}), and E_{Na} and E_{Ca} are the equilibrium potentials of Na^+ and Ca^{2+}, respectively.

$I_{Na/Ca}$ was isolated as XIP[19] (eXchanger Inhibitory Peptide)-sensitive current. XIP reversibly inhibited $I_{Na/Ca}$ without remarkable changes in background conductance.[11] FIGURE 3 illustrates the reversal potential measurements with XIP. The I-V relation was measured with the ramp pulse protocol in the absence and the presence of 0.2 mM XIP. The XIP-sensitive current crossed the x-axis at +15 mV (lower panel of FIG. 3A). In FIGURE 3B, reversal potentials measured are plotted against Na^+ concentrations. Measurements were conducted with (open circles) and without (closed circles) trypsin treatment. Deregulation by trypsin did not have a remarkable effect upon the reversal potential. To our surprise, the reversal potentials were close to the $E_{Na/Ca}$ of $3Na^+:1Ca^{2+}$ exchange only at lower Na^+ concentrations. The values of reversal potential were close to $E_{Na/Ca}$ of $4Na^+:1Ca^{2+}$ exchange at the Na^+ concentrations examined. We extended the measurements at different ionic conditions.

In FIGURE 4, $I_{Na/Ca}$ was isolated as a XIP-sensitive current (open circles) and as Na^+_i- and Ca^{2+}_i-induced currents (closed circles) in the trypsin-deregulated patch. Na^+_i dependence at lower Ca^{2+}_i (0.1 μM Ca^{2+}, FIG. 4A), Ca^{2+}_i dependence (FIG. 4B), Ca^{2+}_o dependence (FIG. 4C), and Na^+_o dependence (FIG. 4D) were studied. The majority of data were closer to $E_{Na/Ca}$ of $4Na^+:1Ca^{2+}$ exchange. Taken to-

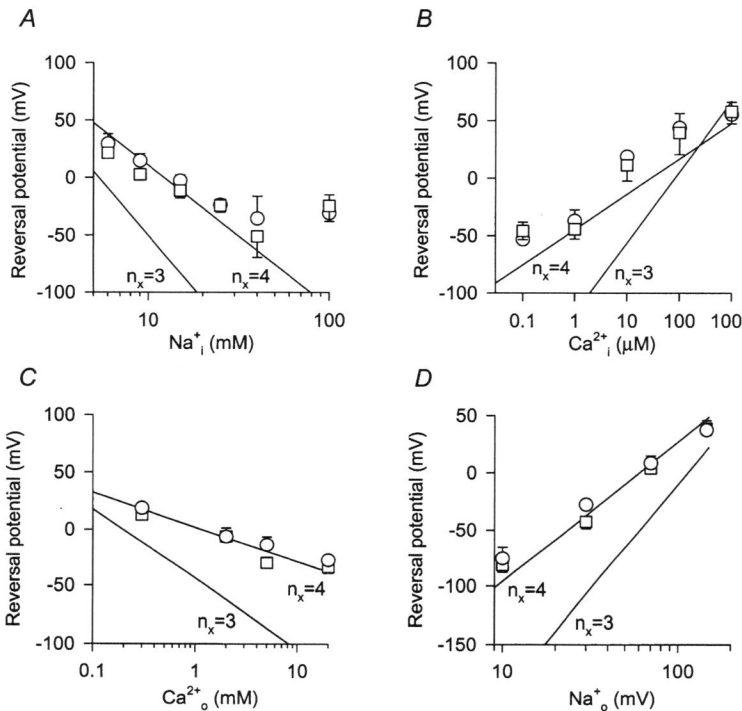

FIGURE 4. Reversal potential of $I_{Na/Ca}$. $I_{Na/Ca}$ was isolated as a XIP-sensitive current (*open circles*) and Na$^+$- and Ca^{2+}-induced currents *(closed circles)* in the trypsin-treated patch. (**A**) Na^+_i dependence at 0.1 μM Ca^{2+}. Na^+_o = 145 mM, Ca$^{2+}_o$ = 2 mM. (**B**) Ca$^{2+}_i$ dependence. Na^+_o = 145 mM, Ca$^{2+}_o$ = 2 mM, and Na^+_i = 50 mM. (**C**) Ca$^{2+}_o$ dependence. Na^+_o = 145 mM, Na^+_i = 25 mM, and Ca$^{2+}_i$ = 1 μM. (**D**) Na^+_o dependence. Ca$^{2+}_o$ = 2 mM, Na^+_i = 9 mM, and Ca$^{2+}_i$ = 1 μM. Modified from Fujioka et al.[11]

gether, we concluded that the stoichiometry is 4 rather than 3. However, it may be variable and become 3 at lower Na$^+$ concentrations and micromolar Ca$^{2+}_i$.

If the stoichiometry is 4, several aspects of exchange function must be reevaluated. FIGURE 5 demonstrates reconstructed action potentials of guinea pig ventricular cells and Ca$^{2+}_i$ transient. This model includes most known ion channels and transporters, sarcoplasmic reticulum, intracellular Na$^+$, Ca^{2+}, K$^+$, and cross-bridge formation. As shown in the upper panel, $E_{Na/Ca}$ of 4Na$^+$:1Ca^{2+} exchange is located at a more positive potential than $E_{Na/Ca}$ of 3Na$^+$:1Ca^{2+} exchange in the greater part of the action potential and resting potential. Therefore, the 4Na$^+$:1Ca^{2+} exchange would provide more driving force for Ca^{2+} extrusion (forward mode) than would the 3Na$^+$:1Ca^{2+} exchange. However, the driving force may be too strong to maintain resting Ca^{2+}. At resting potential (−80 mV, 10 mM Na^+_i, 145 mM Na^+_o, 2 mM Ca$^{2+}_o$), if Ca$^{2+}_i$ is equilibrated only with the exchanger, Ca$^{2+}_i$ would be 0.11 nM. This value is far lower than the physiological concentration (~200 nM). The ex-

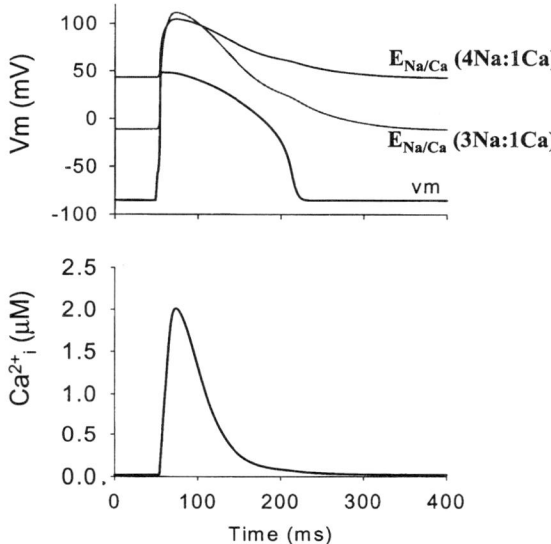

FIGURE 5. $E_{Na/Ca}$ change in simulated action potential and Ca^{2+}_i transient. A guinea pig ventricular cell model by Noma, Matsuoka, Sarai, and Kuratomi (unpublished data) was used for simulation. The prototype of the model is available at http://www.card.med.kyoto-u.ac.jp. $E_{Na/Ca}$ of $3Na^+:1Ca^{2+}$ and $4Na^+:1Ca^{2+}$ exchange are plotted. $Na^+_i = \sim 4$ mM.

changer must be inactivated at lower Ca^{2+}_i by the Ca^{2+} regulation mechanism. As reported in our previous study[20] and by Weber et al.,[21] Ca^{2+} activation and deactivation take place in several tens to several hundred ms orders.

One might speculate that the experimentally obtained I-V relation of Na^+/Ca^{2+} exchange does not support the $4Na^+:1Ca^{2+}$ exchange. In the case of $4Na^+:1Ca^{2+}$ exchange, two charges are carried per one exchange cycle, so that the I-V relation may be steep. However, in general, the I-V relation is relatively straight (FIG. 3).

To address this question, we studied the I-V relation in Na^+/Ca^{2+} exchange models. Our previous model,[20] which included two active states and two inactive states, could well reconstruct the cytoplasmic and extracellular ion dependences. FIGURE 6 is a state diagram of this "E2 model." The simulated I-V relation of the inward $I_{Na/Ca}$ (FIG. 7B) was strongly voltage dependent, and distinct from the experimentally obtained I-V relations (FIG. 7A). In this model, we assumed that the Na^+ translocation is a single voltage-dependent step. However, model consideration[9] and caged Ca^{2+} experiments[22,23] suggested that the Ca^{2+} translocation step is also voltage dependent. Moreover, the ion occlusion reaction is obviously an important reaction for determining the I-V relation.[9] Therefore, we modified the E2 model to include voltage-dependent Ca^{2+} translocation and ion occlusion steps.

FIGURE 8 is a state diagram of the modified model (E10 model). This model is largely similar to the model by Matsuoka and Hilgemann,[9] except that (1) the stoichiometry is 4; (2) the Na^+ occlusion reaction was composed of eight steps; (3) a

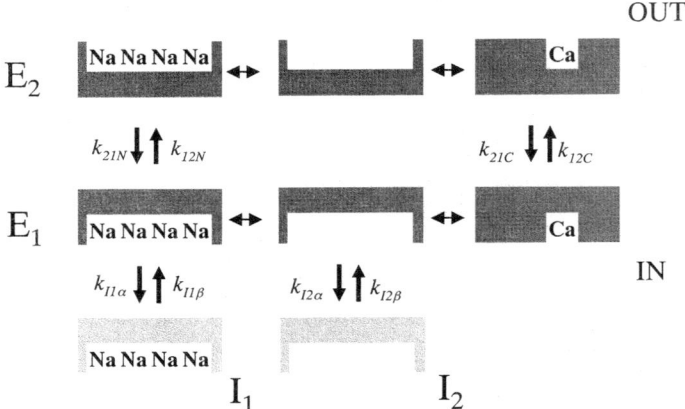

FIGURE 6. A state diagram of a two-state model (E2 model). This model is the same as our previous model.[20] Modified from Fujioka et al.[20]

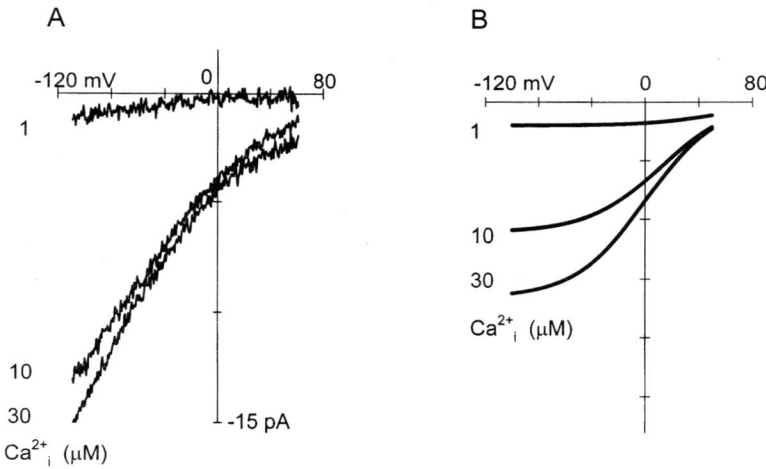

FIGURE 7. Experimentally-obtained (**A**) and simulated (**B**) I-V relation. Na^+_o = 100 mM, Ca^{2+}_o = 0 mM, and Na^+_i = 0 mM. The E2 model was used for the simulation.

state in E1 bound both the Na^+ and Ca^{2+} was eliminated for simplicity; and (4) inactive states of I_1 and I_2 were added.

FIGURES 9 and 10 demonstrate the I-V relation obtained by the macropatch experiment (A) and the E10 model (B). The E10 model can well simulate both the inward and the outward exchange currents (FIG. 9). FIGURE 10 is the I-V relation under conditions for reversal potential measurement. Reversal potentials are consistent with

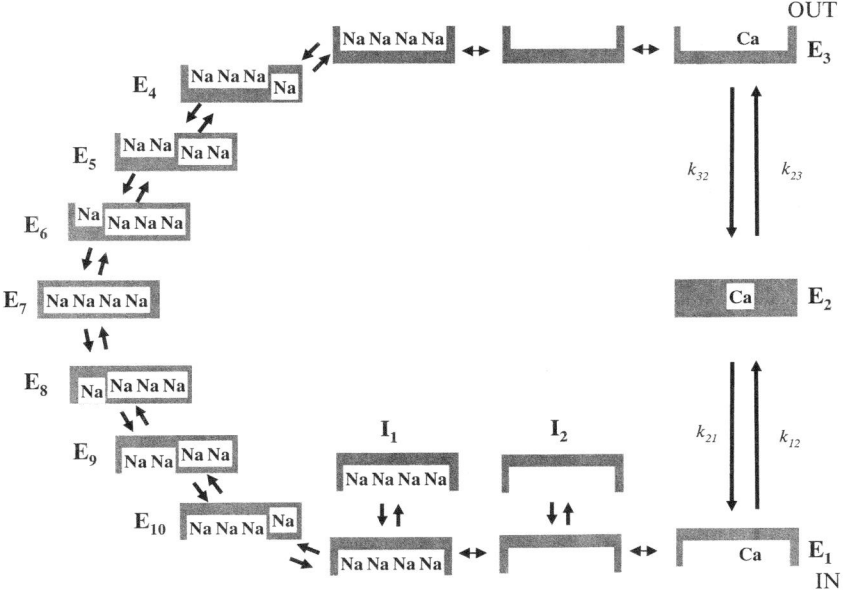

FIGURE 8. A state diagram of 10-state model (E10 model). The 8-state model by Matsuoka and Hilgemann[9] was modified. We assumed that, in total, two positive charges move per exchange cycle. Relation between charge amounts carried at each step of Na^+ occlusion (Qn), intracellular (Qci), and extracellular (Qco) half of Ca^{2+} occlusion is as follows. Qci + Qco + 8 × Qn = 2, where Qn = 0.20805, Qci = 0.6644, and Qco = −0.3288. Steady state $I_{Na/Ca}$ was calculated as follows

$$k_{12} = 1000 \times e^{\left(-Qci \times V_m \frac{F}{2 \times R \times T}\right)} \times f_{CiE1}$$

$$k_{21} = 1000 \times e^{\left(Qci \times V_m \frac{F}{2 \times R \times T}\right)}$$

$$I_{Na-Ca} = (E_2 \times k_{21} - E_1 \times k_{12}) \times \text{CaII}$$

k_{12} and k_{21} are rate constants, V_m is a membrane potential, F is a Farady constant, R is a gas constant, T is temperature, $f_{Ci/E1}$ is a fraction of Ca^{2+}_i bound exchanger in E_1 state, E_1 and E_2 are fractions of the exchanger in the E_1 and E_2 state, respectively, and CaII is an amplitude factor.

$E_{Na/Ca}$ of $4Na^+:1Ca^{2+}$ exchange. According to the aforementioned model, we concluded that stoichiometry is not the major factor for determining the I-V relation.

The reversal potential measured, for the most part, agreed with the $4Na^+:1Ca^{2+}$ exchange. However, we could not rule out the possibility that stoichiometry is variable, depending on cytoplasmic Na^+. The reversal potential tended to get closer to the $E_{Na/Ca}$ of $3Na^+:1Ca^{2+}$ exchange (FIGS. 3b and 4A). However, because the theoretical difference between the $E_{Na/Ca}$ of the $4Na^+:1Ca^{2+}$ and $3Na^+:1Ca^{2+}$ exchange

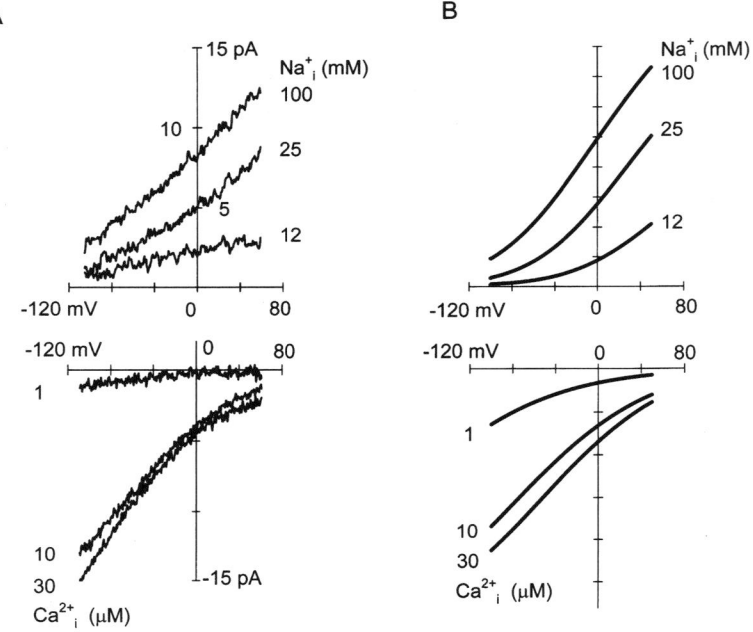

FIGURE 9. Experimentally obtained (**A**) and simulated (**B**) I-V relation. Upper traces are the outward $I_{Na/Ca}$ ($Na^+_o = 0$ mM, $Ca^{2+}_o = 5$ mM, and $Ca^{2+}_i = 1$ μM), and lower traces are inward $I_{Na/Ca}$ ($Na^+_o = 100$ mM, $Ca^{2+}_o = 0$ mM, and $Na^+_i = 0$ mM). The E10 model was used for simulation.

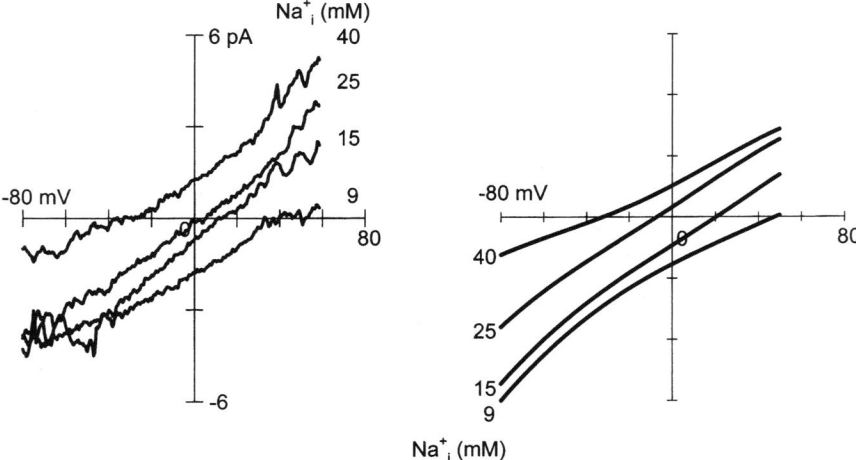

FIGURE 10. Experimentally obtained (**A**) and simulated (**B**) I-V relation under the ionic conditions for reversal potential measurement. $Na^+_o = 145$ mM, $Ca^{2+}_o = 2$ mM, and $Ca^{2+}_i = 1$ μM. The E10 model was used for simulation.

is also smaller, we could not definitely determine the stoichiometry at the lower Na^+ (<6 mM). Other methods are needed to determine the stoichiometry under those conditions.

ACKNOWLEDGMENTS

We are deeply grateful to Dr. A. Noma for his encouragement and valuable discussion. This work was supported by a Grant-in-Aid for Scientific Research from the Ministry of Education, Science and Culture and the Vehicle Racing Commemorative Foundation (to S.M.).

REFFERENCES

1. PITTS, B.J. 1979. Stoichiometry of sodium-calcium exchange in cardiac sarcolemmal vesicles. Coupling to the sodium pump. J. Biol. Chem. **254:** 6232–6235.
2. WAKABAYASHI, S. & K. GOSHIMA. 1981. Kinetic studies on sodium-dependent calcium uptake by myocardial cells and neuroblastoma cells in culture. Biochim. Biophys. Acta **642:** 158–172.
3. REEVES, J.P. & C.C. HALE. 1984. The stoichiometry of the cardiac sodium-calcium exchange system. J. Biol. Chem. **259:** 7733–7739.
4. AXELSEN, P.H. & J.H.B. BRIDGE. 1985. Electrochemical ion gradients and the Na/Ca exchange stoichiometry. Measurements of these gradients are thermodynamically consistent with a stoichiometric coefficient ≥3. J. Gen. Physiol. **85:** 471–475.
5. SHEU, S.-S. & H.A. FOZZARD. 1985. Na/Ca exchange in the intact cardiac cell. J. Gen. Physiol. **85:** 476–478.
6. CRESPO, L.M., C.J. GRANTHAM & M.B. CANNELL. 1990. Kinetics, stoichiometry and role of the Na-Ca exchange mechanism in isolated cardiac myocytes. Nature **345:** 618–621.
7. EHARA, T., S. MATSUOKA & A. NOMA. 1989. Measurement of reversal potential of Na^+-Ca^{2+} exchange current in single guinea-pig ventricular cells. J. Physiol. **410:** 227–249.
8. NOMA, A., T. SHIOYA, L.F. PAVER, et al. 1991. Cytosolic free Ca^{2+} during operation of sodium-calcium exchange in guinea-pig heart cells. J. Physiol. **442:** 257–276.
9. MATSUOKA, S. & D.W. HILGEMANN. 1992. Steady-state and dynamic properties of cardiac sodium-calcium exchange. Ion and voltage dependencies of the transport cycle. J. Gen. Physiol. **100:** 963–1001.
10. MULLINS, L.J. 1977. A mechanism for Na/Ca transport. J. Gen. Physiol. **70:** 681–695.
11. FUJIOKA, Y., M. KOMEDA & S. MATSUOKA. 2000. Stoichiometry of Na^+-Ca^{2+} exchange in inside-out patches excised from guinea-pig ventricular myocytes. J. Physiol. **523:** 339–351.
12. HILGEMANN, D.W. 1995. The giant membrane patch. In Single-Channel Recording, 2nd Ed. B. Sakmann & E. Neher, Eds. :307-327. Plenum Press. New York.
13. FUJIOKA, Y., S. MATSUOKA, T. BAN & A. NOMA. 1998. Interaction of Na^+-K^+ pump and Na^+-Ca^{2+} exchange via $[Na^+]_i$ in a restricted space of guinea-pig ventricular cells. J. Physiol. **509:** 457–470.
14. BERS, D.M., C.W. PATTON & R. NUCCITELLI. 1994. A practical guide to the preparation of Ca^{2+} buffers. Methods Cell. Biol. **40:** 3–29.
15. HILGEMANN, D.W., S. MATSUOKA, G.A. NAGEL & A. COLLINS, 1992. Steady-state and dynamic properties of cardiac sodium-calcium exchanger. Sodium-dependent inactivation. J. Gen. Physiol. **100:** 905–932.
16. HILGEMANN, D.W., A. COLLINS & S. MATSUOKA. 1992. Steady-state and dynamic properties of cardiac sodium-calcium exchanger. Secondary modulation by cytoplasmic calcium and ATP. J. Gen. Physiol. **100:** 933–961.

17. MATSUOKA, S., D.A. NICOLL, R.F. REILLY, et al. 1993. Initial localization of regulatory regions of the cardiac sarcolemmal Na^+-Ca^{2+} exchanger. Proc. Natl. Acad. Sci. USA **90:** 3870–3874.
18. MATSUOKA, S., D.A. NICOLL, Z. HE & K.D. PHILIPSON. 1997. Regulation of the cardiac Na^+-Ca^{2+} exchanger by the endogenous XIP region. J. Gen. Physiol. **109:** 273–286.
19. LI, Z., D.A. NICOLL, A. COLLINS, et al. 1991. Identification of a peptide inhibitor of the cardiac sarcolemmal Na^+-Ca^{2+} exchanger. J. Biol. Chem. **266:** 1014–1020.
20. FUJIOKA, Y., K. HIROE & S. MATSUOKA. 2000. Regulation kinetics of Na^+-Ca^{2+} exchange current in guinea-pig ventricular myocytes. J. Physiol. **529:** 611–623.
21. WEBER, C.R., K.S. GINSBURG, K.D. PHILIPSON, et al. 2001. Allosteric regulation of Na/Ca exchange current by cytosolic Ca in intact cardiac myocytes. J. Gen. Physiol. **117:** 119–131.
22. NIGGLI, E. & W.J. LEDERER. 1991. Molecular operations of the sodium-calcium exchanger revealed by conformation currents. Nature **349:** 621–624.
23. KAPPL, M., G. NAGEL & K. HARTUNG. 2001. Voltage and Ca^{2+} dependence of pre-steady-state currents of the Na-Ca exchanger generated by Ca^{2+} concentration jumps. Biophys. J. **81:** 2628–2638.

Influence of Na/Ca Exchange Stoichiometry on Model Cardiac Action Potentials

DENIS NOBLE

University Laboratory of Physiology, Parks Road, Oxford OX1 3PT, UK

> ABSTRACT: Cardiac action potential simulations were done with the stoichiometry of the Na/Ca exchanger set a 4:1. Using the Hilgemann-Noble (1987) model, this stoichiometry reduces the resting potential unless regulation by intracellular calcium is incorporated. The K_d required for such regulation is consistent with current experimental estimates of this parameter.
>
> KEYWORDS: Na/Ca exchange; stoichiometry; cardiac action potentials

Fujioka *et al.*[1] found that in inside-out patches excised from guinea pig ventricular myocytes the stoichiometry of Na/Ca exchange is closer to 4:1 than to 3:1 and depends on ion concentrations. A stoichiometry of 3:1 may still apply at $[Na^+]_i$ of less than 6 mM and a $[Ca^{2+}]_i$ higher than 1 µM. These ion concentrations are close to those that occur in physiologic conditions. It is importan, therefore, to explore the effects of stoichiometry on cellular function. Stoichiometry would be expected to have a great effect on electrical activity, because it determines the energy gradient acting on the exchanger in both resting and active conditions and may strongly influence the current-voltage relations of the exchanger. Computations, therefore, have been done using models of cardiac cell electrical activity.[2,3] These models have been very successful in accounting for the contribution of Na/Ca exchange to the plateau of the action potential and in reconstructing the time course of net calcium flux across the membrane.[4]

The equation used for sodium–calcium exchange in these models is a fairly simple one, originally developed by DiFrancesco and Noble[5] from a formulation of Mullins[6]:

$$i_{NaCa} = k(\exp(\gamma(n-2)EF/(2RT))[Na]_i^n[Ca]_o -$$

$$\exp(-(1-\gamma)(n-2)EF/(2RT))[Na]_o^n[Ca]_i)/(1 + d([Na]_i^n [Ca]_o + [Na]_o^n [Ca]_i))$$

where E, F, R, T, $[Na]_i$, $[Na]_o$, $[Ca]_i$, and $[Ca]_o$ have their usual meanings. k is a scaling parameter determining the total current carried. n is the Na/Ca stoichiometry, which was set to 3 in the original computations. This stoichiometry was sufficiently

---

Address for correspondence: Dr. Denis Noble, University Laboratory of Physiology, Parks Road, Oxford OX1 3PT, UK.
denis.noble@physiol.ox.ic.uk

large to generate an electrical current (2 corresponds to a neutral exchanger) and to enable the exchanger to maintain the calcium gradient across the cell membrane. Note that n appears in the voltage-dependent terms and therefore strongly influences the predicted iNaCa(E) relations.

γ is a voltage partition parameter that determines the way in which the energy gradient of the electrical field is distributed between the forward and the backward modes of the exchanger. To fit the current-voltage relations observed experimentally by Kimura et al.,[7] this parameter can be set in the range 0.45–0.55, that is, the voltage gradient is "felt" almost equally by the forward and back reactions. Because there is a tendency for the forward reaction to be steeper, γ is usually set to slightly less than 0.5.

d is a denominator constant. If $d \to 0$, then the denominator term can be ignored and the equation becomes even simpler. The value used in the Hilgemann-Noble model (0.0001) does influence the current and has been retained here. Similar results, however, would be obtained if d were set to zero.

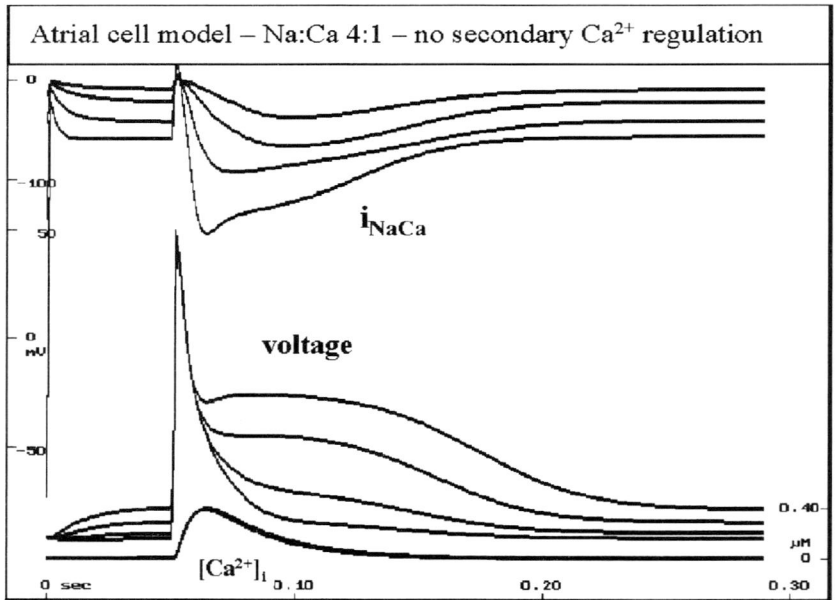

FIGURE 1. Shows Na/Ca exchange current (*top traces*), membrane potential (*middle traces*), and [Ca]$_i$ transient (*bottom traces*) for various values of k (the scaling parameter). When k was set to its smallest value (0.0000002), the resting potential was maintained, but the plateau collapsed. k was then increased to 0.0000006, 0.000002, and 0.000006. The plateau recovered, and at the highest value, a notch appeared (often observed experimentally), but there was a progressively larger membrane depolarization attributable to the larger energy gradient of the exchanger and to the steeper current-voltage relation at 4:1 compared to 3:1. The computations were done on a PC using OXSOFT HEART version 4.8.

In the computations described herein, I have explored what happens when the stoichiometry is increased to 4:1. I have used the Hilgemann-Noble atrial cell model, because the plateau in this case is almost entirely maintained by the sodium-calcium exchange current and is therefore a sensitive indicator of function. In the ventricular cell model, the contribution of the exchanger current to the plateau can be determined,[8] but the situation is then more complicated because the exchanger current overlaps in time course with calcium current activation and the activation of various potassium channels.

FIGURE 1 shows the results of setting $n = 4$. It is not possible to reproduce the contribution of exchange current to the action potential plateau without generating an unacceptably large resting current. Using these unmodified equations, therefore, a better functional result was obtained at 3:1. The resting membrane depolarization is attributable to two factors. First, at 4:1 and for the same ionic concentrations, the energy gradient acting on the exchanger was larger, so the depolarizing current carried was larger. The second is that without further manipulation of the equations, the current-voltage relation for the inward mode of the exchanger was too steep. To maintain sufficient inward current to generate the plateau, k must be increased to a value that generates an unacceptably large resting current.

This problem, however, may be avoided by including secondary calcium regulation.[9] This was done by multiplying the RHS of the equation by the term $[Ca]_i/([Ca]_i + K_d)$. FIGURE 2 shows the result obtained using the largest value of k in FIGURE 1, but with K_d set to 60 nM. This modification is sufficient to maintain both the resting potential and the plateau. It should be noted that the resting value of $[Ca]_i$ is relative-

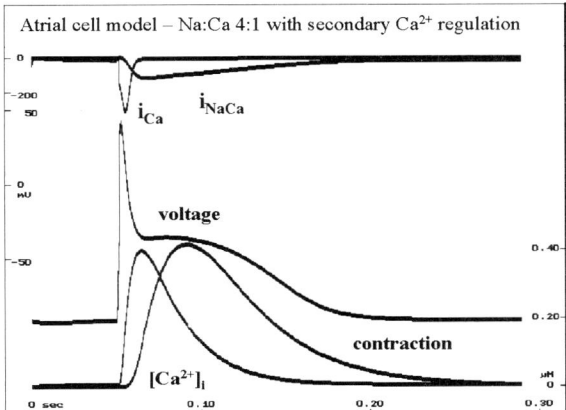

FIGURE 2. The computation is done with the largest value of k used in FIGURE 1, but including secondary calcium regulation with a $K_d = 60$ nM. With the relatively low values of resting calcium in this model, this K_d is sufficient to switch the current off at the resting potential and so prevent membrane depolarization, while still allowing the plateau to be generated.

ly low (around 20 nM instead of 50–100 nM) in the Hilgemann-Noble model. If this fault were corrected, a higher K_d (in the range of 100–150 nM) would be appropriate. This is reasonably close to the current estimates of this parameter.

These results depend critically on the steepness of the current-voltage relations predicted, because in this model increasing the stoichiometry automatically increases the voltage dependence of the current. In fact, however, Fujioka *et al.*[1] do not show such steepness in their experiments. It may be important, therefore, to develop equations for 4:1 stoichiometry while retaining current-voltage relations similar to those predicted for 3:1. This will require more complete modeling of the exchanger reactions than that used in the Hilgemann-Noble equations. At this meeting, Matsuoka[10] showed that a 10-state model could be constructed that generates the correct I(V) relations. This model has not yet been used in action potential simulations.

ACKNOWLEDGMENT

This work was supported by the British Heart Foundation, Medical Research Council, and The Wellcome Trust.

REFERENCES

1. FUJIOKA, Y., M. KOMEDA & S. MATSUOKA. 2000. Stoichiometry of Na^+/Ca^{2+} exchange in inside-out patches excised from guinea-pig ventricular myocytes. J. Physiol. **523:** 339–351.
2. HILGEMANN, D.W. & D. NOBLE. 1987. Excitation-contraction coupling and extracellular calcium transients in rabbit atrium: reconstruction of basic cellular mechanisms. Proc. Roy. Soc. B Biol. **230:** 163–205.
3. NOBLE, D., A. VARGHESE, P. KOHL & P.J. NOBLE. 1998. Improved guinea-pig ventricular cell model incorporating a diadic space, i_{Kr} and i_{Ks}, and length- and tension-dependent processes. Can. J. Cardiol. **14:** 123–134.
4. NOBLE, D., S.J. NOBLE, G.C.L. BETT, *et al.* 1991. The role of sodium-calcium exchange during the cardiac action potential. Proceeedings of 2nd international meeting on sodium-calcium exchange. Ann. N.Y. Acad. Sci. **639:** 334–353.
5. DIFRANCESCO, D. & D. NOBLE. 1985. A model of cardiac electrical activity incorporating ionic pumps and concentration changes. Philos. Trans. Roy. Soc. B **307:** 353–398.
6. MULLINS, L. 1981. Ion Transport in the Heart. Raven Press. New York.
7. KIMURA, J., S. MIYAMAE & A. NOMA. 1987. Identification of sodium-calcium exchange current in single ventricular cells of guinea-pig. J. Physiol. **384:** 199–222.
8. NOBLE, D., J.Y. LEGUENNEC & R.L. WINSLOW. 1996. Functional roles of sodium-calcium exchange in normal and abnormal cardiac rhythm. Proceedings of 3rd international meeting on sodium-calcium exchange. Ann. N.Y. Acad. Sci. **779:** 480–488.
9. HILGEMANN, D.W., A. COLLINS & S. MATSUOKA. 1992. Steady state and dynamic properties of cardiac sodium-calcium exchange. J. Gen. Physiol. **100:** 933–961.
10. MATSUOKA, S. 2002. Ann. N.Y. Acad. Sci. **976:** this volume.

Rat Heart NCX1.1 Stoichiometry Measured in a Transfected Cell System

JONATHAN LYTTON AND HUI DONG

Cardiovascular Research Group, Department of Biochemistry & Molecular Biology, University of Calgary, Calgary, AB, Canada T2N 4N1

ABSTRACT: Expression of cDNA encoding the rat heart NCX1.1 protein in HEK-293 cells provides a simple system in which properties of the expressed protein can be determined in the absence of contaminating ionic conductances or parallel transport pathways. Using this system, we have determined that NCX1.1 transports 4 Na^+ in exchange for 1 Ca^{2+} and accompanied by the movement of 2 positive charges. The measured stoichiometry is constant over a range of different intracellular Na^+ and Ca^{2+} concentrations, independent of the nature of the Ca^{2+} buffer used and not affected by the direction of the applied voltage ramp. We conclude that our measurements are unlikely to be confounded by poor control over ionic conditions close to the intracellular surface of the membrane. A stoichiometry of 4:1 for NCX1 has important implications for models of the role of NCX1 in cellular Ca^{2+} handling and suggests that in the heart the exchanger will be cardioprotective over a wider range of conditions and have a greater contribution to action potential shape than previously appreciated.

KEYWORDS: patch clamp; reversal potential

BACKGROUND

The stoichiometry with which the Na^+/Ca^{2+} exchanger binds and transports ions has become the subject of recent debate as a consequence of I_{NCX} measurements by Matsuoka and colleagues[1] in patches excised from ventricular myocytes. We decided to reinvestigate the stoichiometry issue, using NCX1.1 expressed in transfected HEK-293 cells, as this system provides a high level of protein expression in an environment that is essentially devoid of other significant membrane conductances and ion translocation pathways.[2,3] Therefore, it was possible to measure the current due to NCX1.1 function very cleanly, and to control the ionic composition on both sides of the membrane very precisely. Using whole-cell patch clamp, we could demonstrate outward currents that depended upon the presence of pipette Na^+ and perfusate Ca^{2+} that were at least 50-fold greater in NCX1.1-transfected cells than in control cells. These currents were absolutely dependent upon the presence of a small amount of regulatory Ca^{2+} on the cytoplasmic side of the membrane.[3]

Address for correspondence: Jonathan Lytton, Ph.D., Department of Biochemistry & Molecular Biology, University of Calgary Health Sciences Centre, 3330 Hospital Drive NW, Calgary, AB, Canada T2N 4N1. Voice: 403-220-2893; fax: 403-283-4841.
  jlytton@ucalgary.ca

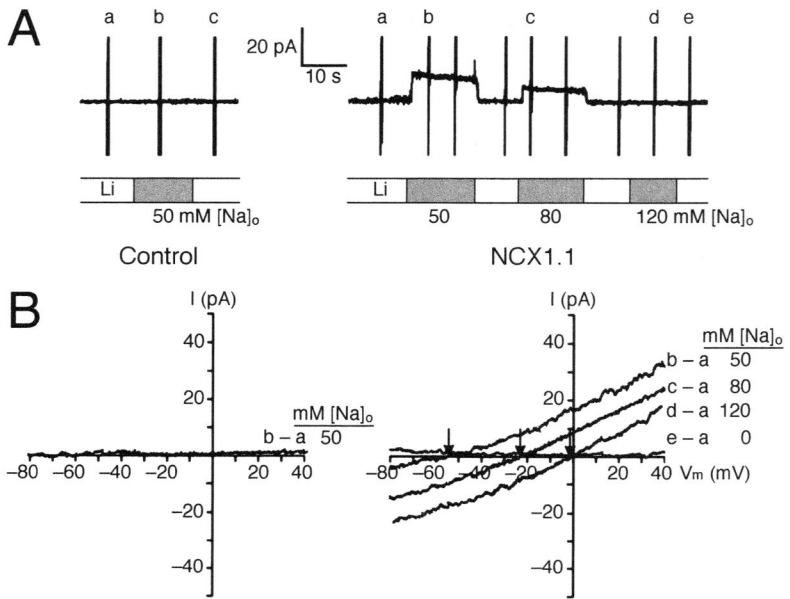

FIGURE 1. Reversal potential measurements. HEK-293 cells transfected with either vector (*control*) or rat heart NCX1.1 cDNA (*NCX1.1*) were analyzed by whole-cell patch-clamp using a holding potential of 0 mV, and a pipette solution containing (in mM) NaCl, 18; K-gluconate, 100; TEA-Cl, 20; Na$_2$ATP, 1; D-glucose, 10; EGTA, 10; CaCl$_2$, 6.4 ([Ca^{2+}]$_{free}$ = 0.3 μM); HEPES/TMA, 10, pH 7.2. Bath perfusion solutions contained (in mM): MgCl$_2$, 1; D-glucose, 10; TEA-Cl, 20; HEPES/TMA, 10, pH 7.4 and either LiCl, 125 plus EGTA, 0.5 (*Li*) or various combinations of NaCl (as indicated) and LiCl totaling 125, plus CaCl$_2$, 0.5. Voltage ramps from −100 to +60 were applied at a rate of 0.5 V/s during each perfusion as well as in LiCl/EGTA solution at the start and end of each experiment. (**A**) Representative current traces for control and NCX1.1-transfected cells. (**B**) I-V curves obtained from the indicated ramps illustrated by lowercase letters in **A**, in which the I-V curve collected at the start of the experiment (labeled *a*) has been digitally subtracted. Downward pointing *arrows* note the reversal potentials determined for the Na+ concentrations tested. Note that only the range from −80 to +40 mV of each ramp is shown, as the currents obtained at the voltage extremes were unstable or contaminated with the capacitance transient. Reproduced from Dong *et al.*[3] with permission.

STOICHIOMETRY DETERMINATIONS

Stoichiometry was determined using a thermodynamic equilibrium method in which cells were dialyzed and perfused with solutions allowing bidirectional NCX1.1 currents, and the reversal potential for I_{NCX} was measured while varying external [Na$^+$] (FIG. 1). This determination was repeated using several different pipette ionic conditions, in which [Ca^{2+}]$_i$ varied from 0.3–5 μM and [Na$^+$]$_i$ varied from 2.5–20 mM. In all cases, the relation of E_{rev} to log([Na$^+$]$_o$) had a slope from which a stoichiometry of between 1.8 and 2.2 Na$^+$ per net charge could be extracted (FIG. 2). These values were always significantly different from 3, but not from 2.

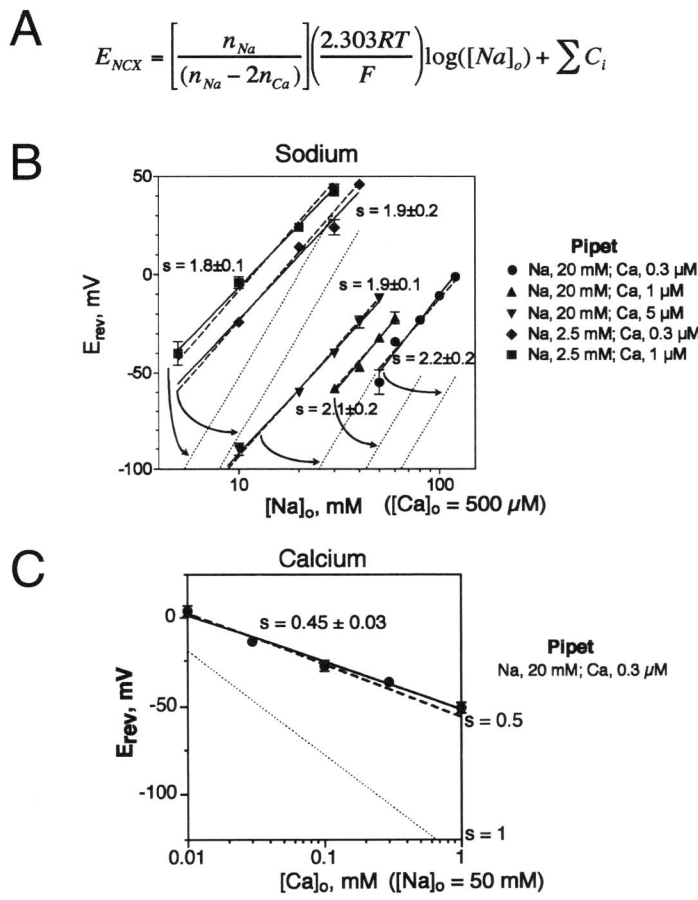

FIGURE 2. NCX1 stoichiometry determinations. (**A**) The equation that describes the relation between reversal potential and the log of the varying extracellular ion concentration (in this case [Na$^+$]). R, T, and F have their usual meanings, and the term 2.303RT/F is evaluated at 58.5 mV under the conditions used in these experiments. C_i are a series of terms, each related to one of the fixed ion species (in this case, intracellular Na$^+$, and both intracellular and extracellular Ca^{2+}). This term can be calculated explicitly from a knowledge of each fixed ion concentrations, although the value is not important in determining the stoichiometry, which is related only to the slope of the E_{rev} *versus* log([ion]) relation. (**B**) Plots of the reversal potential against log([Na$^+$]$_o$), determined as described in the legend to FIGURE 1. Bath perfusates contained 0.5 mM CaCl$_2$ and various [Na$^+$] supplemented with Li$^+$, whereas pipette solutions contained various [Na$^+$], substituted by K$^+$, and [Ca^{2+}], prepared with a 10-mM EGTA buffering system, as indicated. Data are averages ± SEM for between three and six determinations at each data point. *Solid lines* show the fit of the data to the equation in **A** above, with the calculated term s = $n_{Na}/(n_{Na}-2n_{Ca})$, the number of Na$^+$ ions moved per net charge, shown. *Dashed lines* indicate the theoretical relation, with all terms explicitly calculated, for the case where s = 2 (i.e., 4 Na$^+$ moved per two charges), whereas the corresponding relation calculated with s = 3 (i.e., 3 Na$^+$ moved per one charge) is shown

Conceptually identical experiments were then conducted in which external [Ca^{2+}] was varied. The slope of the E_{rev} versus $\log([Ca^{2+}]_o)$ plot from those experiments yielded a stoichiometry value of 0.45 Ca^{2+} per net charge, which was significantly different from a value of 1.0, but not from a value of 0.5 (FIG. 2). From these data we conclude that the transport stoichiometry for NCX1.1 operation in HEK-293 cells, measured using I_{NCX}, was 4 Na^+:1 Ca^{2+}:2 positive charges. This value is consistent with the recent determinations of Matsuoka and colleagues[1] also using I_{NCX}, but not with other published accounts, especially those that used $^{45}Ca^{2+}$-flux to measure NCX1 function.[4–9]

CONTROLS

The values we have determined could still be compatible with a 3 Na^+:1 Ca^{2+}:1 positive charge stoichiometry if during the course of our measurements, submembrane [Na^+] and [Ca^{2+}] were altered in a systematic fashion. For example, if as a consequence of the perfusion switches from solutions devoid of Na^+ and Ca^{2+} to those containing these ions, both [Na^+] and [Ca^{2+}] rose under the membrane roughly in proportion to their added concentration in the external perfusate, then a positive shift in the reversal potential would be expected, producing an E_{rev} versus log([ion]) relation shallower than predicted. Although it is difficult to formally rule out this possible artifact completely, the following pieces of data argue strongly against it. First, we saw no NCX1.1 current when internal, regulatory [Ca^{2+}] was held below 0.5 nM. Had Ca^{2+} crossed the membrane to an appreciable extent sufficient to impact our reversal potential measurements, we would have expected to see an activation of NCX1.1 under these conditions. Second, a variety of controls, particularly those that compared reversal potential measurements using ascending versus descending ramp protocols, indicate that even large fluxes through NCX1.1 itself do not influence the control of submembrane ionic composition in our dialysis solution. Third, neither using BAPTA in place of EGTA to speed-up the buffering of [Ca^{2+}] nor elevating internal [Na^+] to reduce the imposed Na^+ gradient had any impact on measured reversal potentials. Fourth, if we assume altered values of submembrane [Na^+] and [Ca^{2+}] necessary to produce our measured reversal potentials from an exchanger operating with a 3:1 stoichiometry, then we would have to correct our previous observations for stoichiometry determination of the NCKX2 exchanger[2] in the identical

---

as a *dotted line* connected to its cognate data set by a curved arrow. (**C**) As above, except the perfusion solution contained 50 mM NaCl and 75 mM LiCl, and external [Ca^{2+}] was varied in the range 10–100 μM using a 10-mM EGTA buffering system, and unbuffered above that range. The pipette solution was as described in the legend to FIGURE 1. The *solid line* shows a fit of the data to the equation shown in **A** above, with the calculated term s = $n_{Ca}/(n_{Na}-2n_{Ca})$, the number of Ca^{2+} ions moved per net charge, shown. The *dashed line* indicates the theoretical relation, with all terms explicitly calculated, for the case where s = 0.5 (i.e., 1 Ca^{2+} moved per two charges), whereas the corresponding relation calculated with s = 1 (i.e., 1 Ca^{2+} moved per one charge) is shown as a *dotted line*. All the data are clearly significantly different from an NCX1.1 stoichiometry of 3 Na^+:1 Ca^{2+}: 1 positive charge, and instead are all consistent with a stoichiometry of 4 Na^+:1 Ca^{2+}: 2 positive charges. Modified from Dong *et al.*[3] with permission.

way. Such a correction would result in values for Na$^+$ coupling of NCKX2 (2 or 3 Na$^+$ in exchange for 1 Ca^{2+} plus 1 K$^+$) that would not be compatible with its known electrogenic properties. Consequently, we conclude that alterations in submembrane [Na$^+$] and [Ca^{2+}] are not likely to confound our stoichiometry determinations.

IMPLICATIONS

We believe that a 4:1 stoichiometry for the cardiac Na$^+$/Ca^{2+} exchanger has profound implications for actual concentrations of Na$^+$ within the dyadic cleft needed to reverse the operation of the exchanger at the end of systole and for the amount of charge moved as Ca^{2+} is removed from the dyadic cleft at the beginning of systole (and the consequent influence on action potential shape).[10] We do not believe that such a 4:1 stoichiometry for NCX1.1 is likely to impact on resting [Ca^{2+}] levels, however, as these depend more on kinetic factors and the relation between pump and leak rates of various channels and transporters than they do on stoichiometry. Although a 4:1 stoichiometry has profound implications for mathematical modeling of the role of the Na$^+$/Ca^{2+} exchanger in cardiac function (see contributions in this volume by Drs. Denis Noble and Donald Bers), we do not believe that this will introduce changes that are incompatible with physiologic measurement. It does suggest, however, that the role of the Na$^+$/Ca^{2+} exchanger is likely to be cardioprotective over a wider range of conditions and that the contribution of the Na$^+$/Ca^{2+} exchanger to the shape of the later phase of the action potential will be larger than generally appreciated previously.

REFERENCES

1. FUJIOKA, Y., M. KOMEDA & S. MATSUOKA. 2000. Stoichiometry of Na$^+$/Ca^{2+} exchange in inside-out patches excised from guinea-pig ventricular myocytes. J. Physiol. (Lond.) **523 Pt 2**: 339–351.
2. DONG, H., P.E. LIGHT, R.J. FRENCH & J. LYTTON. 2001. Electrophysiological characterization and ionic stoichiometry of the rat brain K$^+$-dependent Na$^+$/Ca^{2+} exchanger, NCKX2. J. Biol. Chem. **276**: 25919–25928.
3. DONG, H., J. DUNN & J. LYTTON. 2002. Stoichiometry of the cardiac Na$^+$/Ca^{2+} exchanger NCX1.1 measure in transfected HEK cells. Biophys. J. **82**: 1943–1952.
4. BLAUSTEIN, M.P. & J.M. RUSSELL. 1975. Sodium-calcium exchange and calcium-calcium exchange in internally dialyzed squid giant axons. J. Membr. Biol. **22**: 285–312.
5. PITTS, B.J. 1979. Stoichiometry of sodium-calcium exchange in cardiac sarcolemmal vesicles. Coupling to the sodium pump. J. Biol. Chem. **254**: 6232–6235.
6. BRIDGE, J.H. & J.B. BASSINGTHWAIGHTE. 1983. Uphill sodium transport driven by an inward calcium gradient in heart muscle. Science **219**: 178–180.
7. REEVES, J.P. & C.C. HALE. 1984. The stoichiometry of the cardiac sodium-calcium exchange system. J. Biol. Chem. **259**: 7733-7739.
8. RASGADO-FLORES, H., E.M. SANTIAGO & M.P. BLAUSTEIN. 1989. Kinetics and stoichiometry of coupled Na efflux and Ca influx (Na/Ca exchange) in barnacle muscle cells. J. Gen. Physiol. **93**: 1219–1241.
9. SZERENCSEI, R.T., C.F. PRINSEN & P.P. SCHNETKAMP. 2001. Stoichiometry of the retinal cone Na/Ca-K exchanger heterologously expressed in insect cells: comparison with the bovine heart Na/Ca exchanger. Biochemistry **40**: 6009–6015.
10. BERS, D.M. 2002. Cardiac excitation-contraction coupling. Nature **415**: 198–205.

Sodium–Calcium Exchange Stoichiometry
Is The Noose Tightening?

TONG MOOK KANG, MARK STECIUK, AND DONALD WILLIAM HILGEMANN

Department of of Physiology, University of Texas Southwestern, Dallas, Texas 75390-9040, USA

ABSTRACT: The stoichiometry of cardiac sodium–calcium exchange is of profound importance to understanding its physiological function, and recent work challenges a simple 3-to-1 stoichiometry. We present a refined 3-to-1 exchange model that can explain recently measured reversal potentials that are close to those expected for a 4-to-1 exchanger. The model assumes that 1 calcium and 1 sodium ion can be transported by the exchanger, albeit more slowly than 3 sodium ions or 1 calcium ion. In this model, currents and calcium fluxes reverse at different potentials; resting free calcium would always be higher than expected for a perfect 3-to-1 exchange process. To test models such as this, we have developed new methods to study ion transport processes in giant membrane patches independent of, or in parallel with, current measurements. Briefly, ion-selective electrodes or fluorescent ion indicators are used to detect concentration changes in the pipette, close to the membrane, upon activation of transport activity. Preliminary results are presented.

KEYWORDS: sodium–calcium exchange; stoichiometry

CONSEQUENCES OF PROPOSED 4-TO-1 Na/Ca EXCHANGE COUPLING

How well do we understand the molecular and cellular functions of Na/Ca exchange and its role in cardiac excitation-contraction coupling? The answer is humbling. Our insecurities about important details have grown, not decreased, since the last meeting in this conference series. In fact, we have returned to the basic question of the exchanger's stoichiometry, which is now suggested to be 4-to-1.[1] Without a definitive resolution of this issue it seems difficult to proceed with a quantitative analysis of exchanger function or of calcium homeostasis in cardiac excitation-contraction coupling, for that matter. Therefore, this short article will (1) summarize some questions that are raised anew by this proposal, (2) present a revised model of Na/Ca exchange that might resolve the open issues in a conservative fashion, and (3) present our new approaches to measure Na/Ca exchange current-flux coupling in excised patches.

Address for correspondence: Dr. Donald William Hilgemann, Department of Physiology, University of Texas Southwestern, 5323 Harry Hines Blvd., Dallas, Texas 75390-9040, USA. Voice: 214-648-6728; fax: 214-648-8879.

hilgeman@utsw.swmed.edu

CYTOPLASMIC-FREE CALCIUM IN CARDIAC MYOCYTES

In simulations, a 3-to-1 Na/Ca exchange stoichiometry allows generation of Ca gradients in the cardiac myocyte that are nearly 10 times larger than those measured to date in resting myoctyes. Thus, when cardiac Ca homeostasis is simulated with realistic Na/Ca exchanger densities, the free resting Ca concentration is not 0.1 µM, but rather 0.01 µM.[2] This discrepancy can, in principle, be overcome by including background Ca channels in the simulations. But, we have no evidence up to now that such mechanisms exist in heart. Thus, we face a real dilemma, a "noose" in the worst case! If background Ca channels do exist in heart, they will complicate any measurements of Na/Ca exchange current near the resting conditions of the cardiac cell. If they do not exist, we cannot explain Ca homeostasis.

With a 4-to-1 Na/Ca exchange stoichiometry, the discrepancy between expected free cytoplasmic Ca and observed free Ca becomes still worse. Why is cytoplasmic free Ca 0.1 µM and not less? Possibly, free Ca next to the membrane is indeed very low, but we cannot measure it. Or does the exchanger turn itself off, for example, by the Ca-dependent regulatory reaction, when cytoplasmic free Ca decreases below 0.1 µM.[3] Two problems emerge: First, the operation of the regulatory reaction is far too slow, even in patches from fresh myocytes,[4] to operate in a beat-to-beat fashion at 2 Hz. Second, the dependence of regulation on cytoplasmic free Ca is not "steep," so that its "switch" function is quite limited as cytoplasmic free Ca increases or falls.[3] Perhaps the worst contradiction is that overexpression of Na/Ca exchangers in many different cell types has never been observed to cause a *decrease* of cytoplasmic free Ca (see other articles in this volume). Rather, cytoplasmic Ca remains unchanged or tends to increase with overexpresssion of NCX-type exchangers.

CALCIUM HOMEOSTASIS AND MEMBRANE POTENTIAL

With a 4-to-1 Na/Ca stoichiometry, the transmembrane electrical currents are twice larger than those for a 3-to-1 stoichiometry to move the same amount of Ca. Thus, to move the same amount of Ca, larger effects on membrane potential must be expected. In the sinus-atrial node, for example, the extrusion of Ca that entered through Ca channels will be expected to generate a very significant background inward current during the diastolic period. In working myocytes, extrusion of relatively little Ca would result in relatively greater depolarizing influences. Overall, the suggested 4-to-1 exchanger stoichiometry seems appropriate for a cell that must generate large gradients, but it does not seem appropriate for a cell to make large Ca fluxes in a fragile electrophysiologic environment. One possible solution might be that transmembrane Ca fluxes in intact heart are less than we presently believe from work with cardiac myocytes (e.g., because Ca channels tend to be inactivated in the intact environment).

A POSSIBLE ALTERNATIVE TO 4-TO-1 STOICHIOMETRY

The major reason that a 4-to-1 stoichiometry is presently being "reconsidered" is that reversal potentials appear more consistent with this stoichiometry than with a 3-

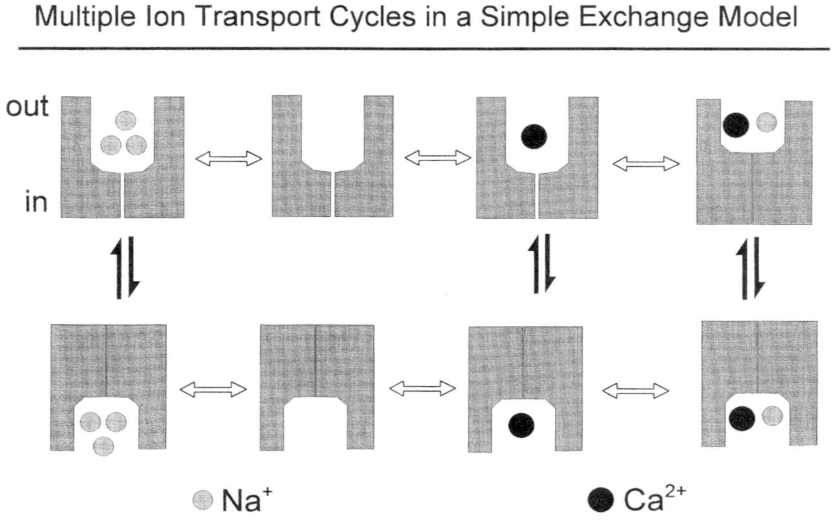

Multiple Ion Transport Cycles in a Simple Exchange Model

FIGURE 1. Cartoon of a hypothetical Na/Ca exchange system in which 3 Na are exchanged for 1 Ca. Thus, 3 Na or 1 Ca can be bound and translocated. In addition, 1 Na + 1Ca can be bound and translocated at a low rate. In this way, three transport cycles are possible. The major mechanism of transport remains 3-to-1 Na/Ca exchange. In addition, 1Ca + 1Na can be exchanged for 1 Ca, and 1Ca + 1Na can be exchanged for 3 Na. Thus, the exchanger mediates "slow" but significant 2 Na/1 Ca exchange and a background Na leak into cardiac myocytes.

to-1 stoichiometry.[1] Until now, no evidence is known to suggest that the exchanger can actually make larger Ca gradients than those dictated by a 3-to-1 stoichiometry. In fact, as already noted, the Ca gradients measured in cels are shifted from those expected for 3-to-1 toward the smaller gradients expected for 2-to-1, not toward the larger 4-to-1 gradients. In this light, we can suggest one possible alternative explanation of the data on exchanger stoichiometry, and it is described in FIGURE 1.

Both the electrical and biochemical function of cardiac Na/Ca exchange suggest that Na and Ca do not interact in a simple competitive manner, but rather that 1 Na ion and 1 Ca ion can bind together to the exchanger.[5,6] On this basis, the possibility must be considered that 1Na + 1Ca ion might be translocated by the exchanger in a relatively slow reaction, such that the perfect 3-to-1 stoichiometry of exchange is broken. Although the major portion of ion flux still can take place via the 3-to-1 ratio, new cycles of exchange are enabled. Those modes result in a small fraction of 3 Na versis "1Na + 1Ca" exchange and a small fraction of 1 Ca versus "1Ca + 1Na" exchange. The former process is effectively a 2-to-1 Na/Ca exchange process. It would physiologically generate Ca influx into the heart cell that would be counteracted by the 3-to-1 exchange mode. The latter process is effectively a Na leak pathway that requires Ca on both membrane sides, and it would physiologically contribute to a background inward current and background Na influx that would have to be compensated by Na pump activity.

Single exchanger charge- and calcium-turnover rates

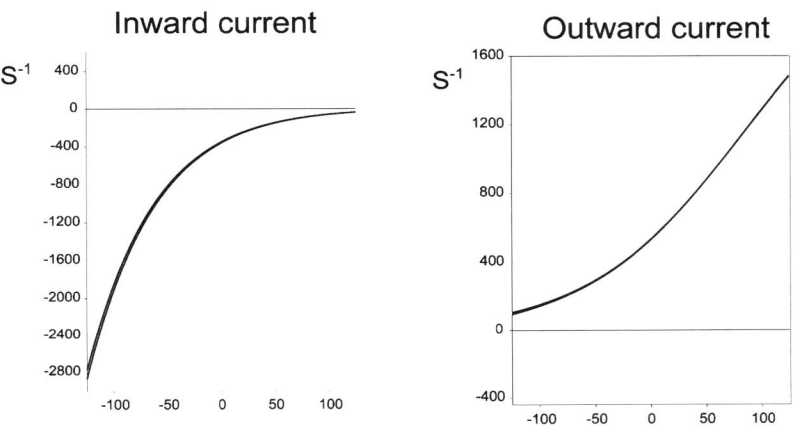

FIGURE 2. Current-voltage and Ca flux-voltage relations of the hypothetical exchanger described in FIGURE 1. Results presented are the single exchanger turnover rates, given as charge move per second and for Ca ions moved per second. The pure Ca extrusion mode is presented in the *left panel* (140 mM extracellular Na with 3 μM cytoplasmic Ca); the pure Ca influx mode is presented in the *right panel* (2 mM extracellular Ca with 60 mM cytoplasmic Na).

The most important functional consequences of this model are described in FIGURES 2 and 3. These figures show simulations of the simplest possible model that includes the necessary assumptions to make predictions of exchanger function. A simple ion binding scheme was assumed, as described in FIGURE 1. Three Na ions can bind sequentially to the empty carrier. After binding of 1 Ca ion, additional Na ion can still be bound with the same affinity as the first Na ion binds to the empty carrier (K_d, 60 mM). The translocation of 1 Ca ion is assumed to take place at 2,500 per second, 3 bound Na ions at 1,000 per second, and 1 Ca + 1 Na at 250 per second. The 3 Na and the (1Ca + 1Na) translocation steps are assumed to reflect movement of +0.6 charge through membrane field, while the translocation of Ca alone is assumed to reflect movement of −0.4 charge. (See Ref. 6 for related models.)

FIGURE 2 shows the current-voltage relations for the model under the unidirectional conditions of 140 mM extracellular Na with 3 μM cytoplasmic Ca (left panel) and with 2 mM extracellular Ca with 60 mM cytoplasmic Na (right panel). In these conditions, the exchanger behaves as a perfect 3-to-1 exchanger. That is because the movement of 1Na with 1Ca requires the presence of both ions on both sides, so that only a 3 Na-to-1 Ca exchange process is possible. Just 1 charge is moved for the net movement of each Ca ion.

FIGURE 3 shows current-voltage and flux-voltage relations in the presence of both ions on both membrane sides. The two vertical lines in each panel indicate the reversal potentials for a perfect 3-to-1 and 4-to-1 exchanger. The left panel (**A**) shows re-

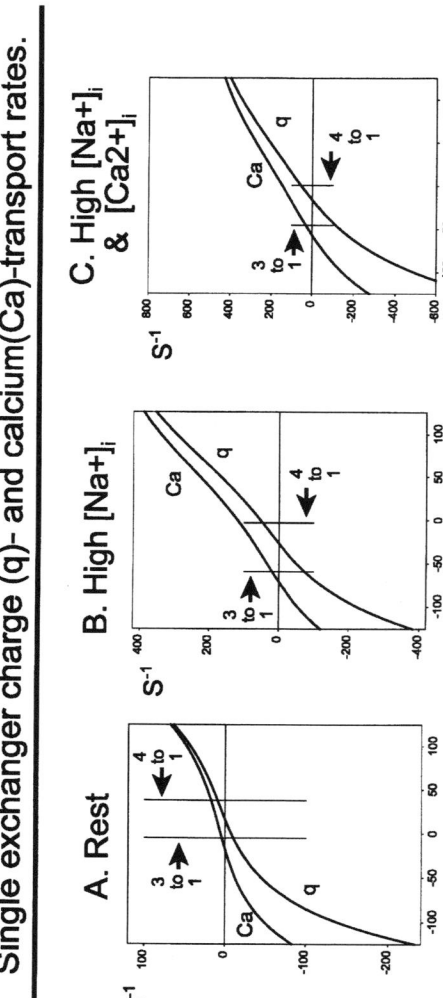

FIGURE 3. Current-voltage (q) and flux-voltage (Ca) relations for single exchangers, as described in FIGURE 1 with both Na and Ca on both membrane sides. **(A)** The resting cell condition with 150 and 1.2 mM extracellular Na and Ca, respectively, and with 7 mM and 0.1 µM cytoplasmic Na and Ca, respectively. **(B** and **C)** Conditions of increased intracellular Na 20 and 25 mM in panels **B** and **C**, respectively; 0.3 and 1 µM free cytoplasmic Ca in B and C, respectively.

sults for the expected resting conditions of a cardiac myocyte, namely, with 150 mM and 7 mM extracellular and cytoplasmic sodium, respectively, and with 1.2 mM and 0.1 µM free extracellular and cytoplasmic Ca, respectively. At a resting membrane potential of −75 mV, the exchanger would still be able to extrude Ca; that is to say, the exchanger can establish the expected resting Ca gradient. However, the reversal potential of the Ca flux (approximately −25 mV) and the net ion flux (approximately +20 mV) are not equal. This means concretely that the exchanger is always a source of background inward current that tends to load the cell with Na. An increase of the number of active exchangers would presumably lead to a positive inotropic effect via Na loading. The reversal potential of the current is shifted about half-way from the reversal expected for a 3-to-1 exchanger to that expected for a 4-to-1 exchanger.

Panels B and C of FIGURE 3 show the current- and flux-voltage relations for the condition that intracellular Na is increased to 20 and 25 mM, respectively, as might be the case with heart glycosides when Na pumps are inhibited. Cytoplasmic Ca is set to 0.3 and 1 µM, respectively, and the Ca flux-voltage relations reverse at −72 mV and −58 mV, respectively. The reversal potentials of the current (q) are shifted about half-way to the potential expected for a 4-to-1 exchanger from the reversal potentials expected for a 3-to-1 exchanger.

Overall, the simulations seem biologically plausible. (1) Resting free Ca can be maintained in the range expected to exist in myocytes. (2) There is no need for any source of background Ca influx via ion channels. (3) The exchanger contributes to the Na load of the myocyte, so that overexpression of exchangers will lead to Na loading and, therefore, an increase of resting free Ca, not a decrease.

MEASUREMENT OF CALCIUM FLUX–CURRENT RELATIONS

From the simulations presented, predictions can be made about the function of a Na/Ca exchanger with mixed transport reactions (i.e., capable of transporting 1Ca + 1Na). One of the major predictions is described in FIGURE 4. If one could measure the amount of Ca transported by the exchanger in the unidirectional condition (i.e., with only Ca on one side and only Na on the other membrane side), the relation between exchange current and Ca flux would be linear when either Ca or Na is varied on the opposite membrane sides (solid line in FIG. 4). We imagine here that the solid line in FIGURE 4 corresponds to the relation between inward exchange current with variation of cytoplasmic Ca and the amount of Ca transported per unit time (i.e., Ca flux). This linear relation will not hold when the inward exchange current is inhibited by increasing cytoplasmic Na (dotted line), as the optional transport modes become possible. With Na and Ca on the cytoplasmic side and only Ca on the extracellular side, cytoplasmic Ca can be transported to the extracellular side together with Na. This makes an electroneutral exchange mode when 3 Na are exchanged from the outside, thereby completing the pseudo-2 Na/1Ca exchange mode.

How can we test possibilities such as this? As described in FIGURE 5, it is possible to monitor Ca flux in the inward exchange mode via Ca indicators included in the pipette. In this case, 0.2 mM Fluo3 was included in the pipette with 140 mM Na and no added Ca. When inward exchange current is activated by 50 µM cytoplasmic Ca, extracellular Ca increases markedly with a time dependence that can be very closely predicted by simulations of Ca diffusion in the pipette tip. In this case, the fluores-

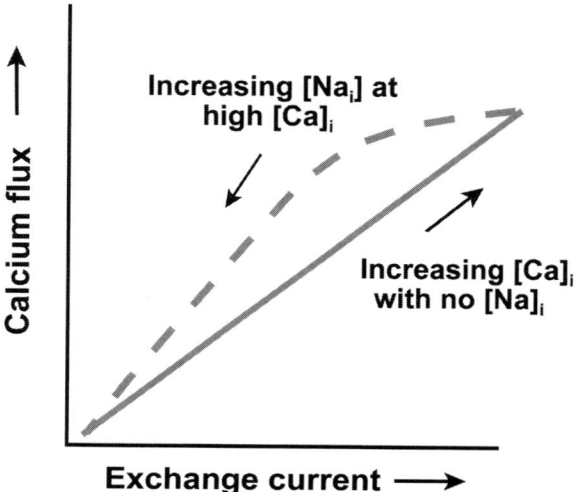

FIGURE 4. Hypothetical relations between current and Ca flux in the exchange model suggested in FIGURE 1. *Solid line* represents the expected linear relation between current and Ca flux in the unidirectional Ca influx mode of exchanger operation (i.e., with only Na on the outside and only Ca on the cytoplasmic side of the membrane). *Dotted curve* represents the expected relation between current and Ca flux as current is inhibited by increasing cytoplasmic Na in the presence of a high Ca concentration. A "2-to-1" exchange mode becomes significant, so that Ca flux is increased with respect to current.

cence signal was monitored via a CCD camera at a distance of about 10 μm from the pipette tip. The inclusion of 30 mM Na on the cytoplasmic side results in approximately 60% inhibition of both the Ca and the current signals. Thus, the stoichiometry of transport is certainly similar with and without cytoplasmic Na; there is no major discrepancy, as predicted in the previous figure.

A melted glass pipette tip is a very disadvantageous optical object, which makes the spacial resolution of these measurements very limited. Also, the dye drastically affects the free Ca in the pipette, and our ability to quantify the signal is limited. Quantification depends on our ability to saturate the dye with Ca. Given these problems of the optical method, we continued these studies using Ca-selective electrodes with tip diameters of about 3 μm. The electrodes are fabricated from quartz capillary tubing, and they can easily be placed in the giant patch clamp pipette at variable distances from the membrane. Calibration of the responses obtained is much more straightforward than that with the dye method, and it should be possible via simulations to make precise predictions about the amount of Ca transported during activation of an exchange current of known magnitude.

FIGURE 6 shows a typical extracellular Ca transient using the electrode method. In this case, the Ca electrode is positioned about 10 μm from the membrane, the patch pipette has a tip diameter of about 15 μm, and the angle of descent to the tip

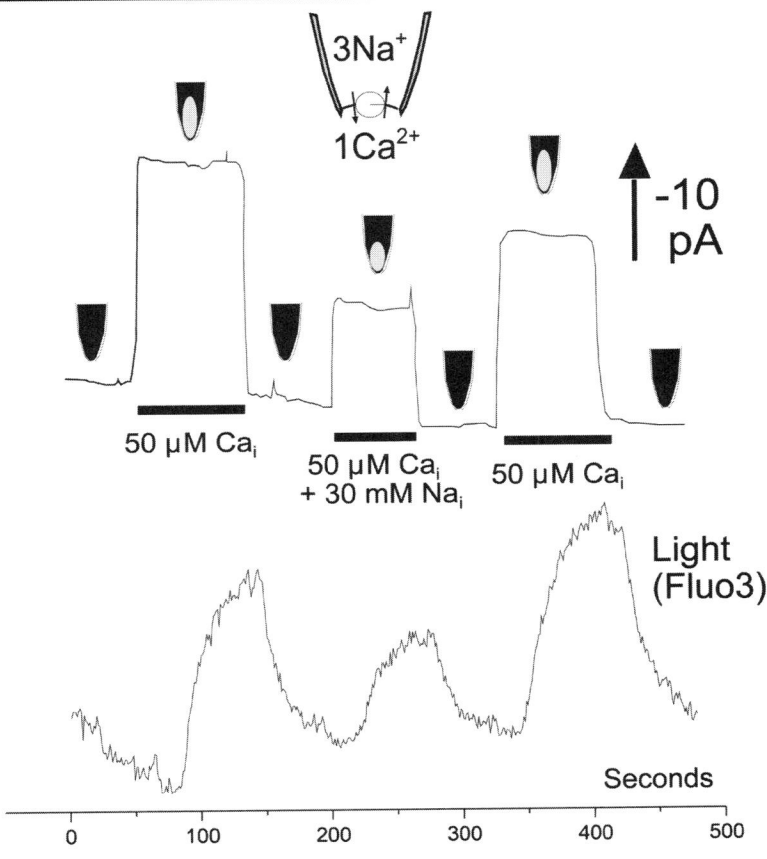

FIGURE 5. Measurement of extracellular Ca transients in a giant cardiac patch using Fluo-3 as the Ca indicator in the patch pipette (140 mM Na and no added Ca). Note the temporal relations between the current records and the Ca changes that occur with a delay. These light responses correspond to a point about 10 μ from the membrane, and the time courses are close predicted by simulations of Ca diffusion in the pipette tip (not shown). See text for details.

is about 25 degrees. The free Ca in the pipette tip rises from about 25 μM to about 75 μM within 25 seconds when an inward exchange current of about 20 pA is activated. Our initial simulation of this result (not shown) suggests that the exchange ratio is much closer to 3-to-1 than 4-to-1, although further experimentation is needed to rigorously verify this. Finally, we mention results relevant to the predictions of FIGURE 4. When the cytoplasmic Ca concentration was varied with Na in the pipette, a nearly linear relation between Ca flux and current was established. Then, in the

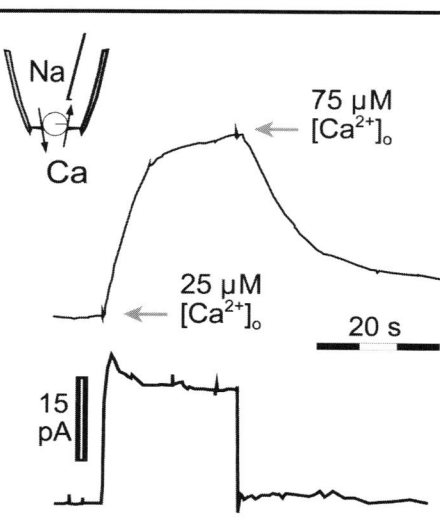

FIGURE 6. Measurement of extracellular Ca transients in a giant cardiac patch using a Ca-selective electrode placed into the pipette tip at a distance of ~10 μm from the membrane. Note the similar relation between free Ca and current as described with optical methods in the previous figure. The free Ca changes are predicted accurately in simulations by assuming a 3-to-1 Na/Ca exchange stoichiometry. See text for details.

presence of high cytoplasmic Ca (0.1 mM), nearly the same linear relation was obtained when the inward current was inhibited by adding cytoplasmic Na (data not shown). In other words, we have not obtained definitive discrepancies from a fixed stoichiometry up to now, nor have we obtained clear discrepancies from a stoichiometry of 3-to-1.

CONCLUSIONS

The stoichiometry of cardiac Na/Ca exchange is fundamental to an understanding of cardiac physiology and electrophysiolgy. Recent evidence that the stoichiometry may be 4-to-1 would force serious rethinking of numerous issues in cardiac excitation-contraction coupling, including the consequences of exchanger up- and down-regulation for cardiac excitation-contraction coupling. However, reversal potentials cannot predict flux-current coupling if stoichiometry is not fixed! Therefore, it seemed essential to develop new methods to test rigorously exchanger stoichiometry and Ca flux-current ratios in a variety of circumstances. The giant membrane patch allows such measurements using both Ca dyes and Ca-selective electrodes. Our efforts to date are not definitive. But until now, we have not obtained clear evidence for deviations from 3-to-1 stoichiometry. Definitive experiments will require further measurements of exchanger-mediated currents and Ca fluxes under conditions that are closer to the expected reversal potentials.

ACKNOWLEDGMENTS

This work was supported by HL515323 (to D.W.H.).

REFERENCES

1. FUJIOKA, Y.F., M.F. KOMEDA & S. MATSUOKA. 2000. Stoichiometry of Na^+-Ca^{2+} exchange in inside-out patches excised from guinea-pig ventricular myocytes. J. Physiol. **523:** 339–351.
2. HILGEMANN, D.W. & D. NOBLE. 1987, Excitation-contraction coupling and extracellular calcium transients in rabbit atrium: reconstruction of basic cellular mechanisms. Proc. Roy. Soc. Lond B Biol. Sci. **230:** 163–205.
3. HILGEMANN, D.W., A. COLLINS & S. MATSUOKA. 1992. Steady-state and dynamic properties of cardiac sodium-calcium exchange. Secondary modulation by cytoplasmic calcium and ATP. J. Gen. Physiol. **100:** 933–961.
4. FUJIOKA, Y.F., K.F. HIROE & S. MATSUOKA. 2000. Regulation kinetics of Na^+-Ca^{2+} exchange current in guinea-pig ventricular myocytes. J. Physiol. **529:** 611–623.
5. REEVES J.P. & J.L. SUTKO. 1983. Competitive interactions of sodium and calcium with the sodium-calcium exchange system of cardiac sarcolemmal vesicles. J. Biol. Chem. **258**(5): 3178–3182.
6. MATSUOKA, S. & D.W. HILGEMANN. 1992. Steady-state and dynamic properties of cardiac sodium-calcium exchange. Ion and voltage dependencies of the transport cycle. J. Gen. Physiol. **100:** 963–1001.

Stoichiometry
A Discussion

KENNETH PHILIPSON (*Department of Physiology/CVRL, University of California at Los Angeles, Los Angeles, CA*): What are the implications of a 4:1 stoichiometry for reverse exchange in a myocyte?

DONALD HILGEMANN (*Department of Physiology, University of Texas Southwestern, Dallas, TX*): The exchanger would probably not contribute significant calcium influx upon upstroke of the action potential, although this depends on the resting free calcium under the membrane at the time of upstroke. The resting free calcium is certainly a big problem. Even with a 3-to-1 exchange process, our problem is to explain why resting free calcium is 0.1 micromolar and not less. Up to now, we have no clear evidence that background calcium influx is significant in myocytes, be it via some type of TRP channel, Icrac, or whatever.

DONALD BERS (*Department of Physiology, Loyola University Chicago, Maywood, IL*): Our data examining calcium flux reversal over a range of membrane potential support entry of calcium at a potential much closer to that compatible with a 3:1 stoichiometry than with a 4:1 stoichiometry.

JON LEDERER (*Department of Physiology, University of Maryland, Baltimore, MD*): Don, have you looked at the rate of rise of calcium in the patch pipette with respect to the Na concentration?

HILGEMANN: That is determined by diffusion in the patch pipette rather than by transport. Simulations predict very accurately the time courses of the transients at different distances from the membrane, assuming the usual calcium diffusion coefficients.

DENIS NOBLE (*Laboratory of Physiology, Oxford University, Oxford, UK*): Don, have you considered 2° calcium regulation in your models of current and flux? What controls basal calcium under these conditions?

HILGEMANN: This is an important question, as even 3:1 models predict a lower level of calcium than observed and a 4:1 stoichiometry just makes the problem worse. It is hard to understand these values without further knowledge of the background leak or calcium channel activity.

LEDERER: It may be that slow kinetics at rest, rather than stoichiometry, dominates.

BERS: Don, have you considered the effect of protons, as reported by Niggli, in your models?

HILGEMANN: No.

MORDECAI BLAUSTEIN (*Department of Physiology, University of Maryland, Baltimore, MD*): Previous flux measurements of both sodium and calcium are clearly different from a 4:1 ratio and are very close to 3:1.

HILGEMANN: All of those measurements are under zero-trans conditions, when the model predicts perfect 3:1 stoichiometry behavior.

JUNKO KIMURA (*Department of Pharmacology, Fukushima Medical University, Fukushima, Japan*): It is important during the electrical measurements to avoid clamp artifacts.

REINALDO DIPOLO (*Centro de Biofisica y Bioquimica IVIC, Caracas, Venezuela*): Older flux measurements in dialyzed squid axons also support a 3:1 and not a 4:1 stoichiometry and were all done under zero-trans conditions.

HILGEMANN: The flux data, therefore, are all supportive of a 3:1 stoichiometry and only the current measurements at equilibrium favor a 4:1 value.

Reexamination of the Stoichiometry of Na^+/Ca^{2+} Exchange with Whole-Cell Voltage Clamp of Guinea Pig Ventricular Myocytes

MASAMITSU HINATA, HISAO YAMAMURA,[b] LIBING LI, YASUHIDE WATANABE,[a] TOMOKAZU WATANO, YUJI IMAIZUMI,[b] AND JUNKO KIMURA

Department of Pharmacology, School of Medicine,
[a]*Department of Ecology and Clinical Therapeutics, School of Nursing, Fukushima Medical University, Fukushima 960-1295, Japan*

[b]*Department of Molecular and Cellular Pharmacology, Graduate School of Pharmaceutical Sciences, Nagoya City University, Nagoya, 467-8603, Japan*

KEYWORDS: Na/Ca exchange; whole-cell voltage clamp; ventricular myocytes

INTRODUCTION

On theoretical grounds Mullins[1] proposed that the stoichiometry of Na^+/Ca^{2+} exchange (NCX) is 4:1. Subsequently, however, various workers provided experimental evidence for a $3Na^+:1Ca^{2+}$ stoichiometry of NCX.[2–6] Therefore, the stoichiometry of cardiac NCX has generally been accepted as 3:1.[7] Recently, the 3:1 stoichiometry was challenged by Fujioka *et al.*[8] who measured reversal potentials of NCX current from guinea pig ventricular cells using a giant- or macropatch method and concluded that the stoichiometry of NCX was 4:1 or variable. Therefore, the stoichiometry of NCX has become controversial once again. We reexamined the stoichiometry in guinea pig ventricular myocytes by simultaneously measuring the reversal potential of NCX current (E_{NCX}) and intracellular Ca^{2+} concentration ($[Ca^{2+}]_i$) using the whole-cell voltage clamp method and confocal microscopy, respectively.[9]

METHODS

Single ventricular cells were isolated by Langendorff perfusion of guinea pig heart with collagenase. The pipette solution contained (in mM): NaCl 20, BAPTA (1,2-bis(2-aminophenoxy)-ethane-N,N,N′,N′-tetraacetic acid) 20, $CaCl_2$ 9, 9.5, or

10 (calculated free Ca^{2+} concentration, 184, 200, or 226 nM, respectively), CsOH 120, aspartic acid 50, $MgCl_2$ 3, MgATP 5, and HEPES 20 (pH 7.2 with aspartic acid). The extracellular solution contained (in mM): NaCl 140, $CaCl_2$ 1, $MgCl_2$ 1, ouabain 0.02, nifedipine (or D600) 0.01, ryanodine 0.01, and HEPES 5 (pH 7.2 with NaOH). For measuring $[Ca^{2+}]_i$, 100 μM indo-1 was used. The indo-1 signal was calibrated using the following equation: $[Ca^{2+}]_i = K_d(R_{min} - R)/(R - R_{max})$, where K_d is the dissociation constant of indo-1 (= 217 nM), R is the fluorescence ratio, and R_{min} and R_{max} are the fluorescence ratios in the absence and saturation of Ca^{2+}, respectively.

RESULTS AND DISCUSSION

To measure E_{NCX}, we induced NCX current with 140 mM Na^+ and 1 mM Ca^{2+} in the external solution and 20 mM Na^+ and 20 mM BAPTA + 9 or 10 mM Ca^{2+} in the pipette solution and blocked I_{NCX} with 5 mM Ni^{2+}. Under these ionic conditions, the predicted E_{NCX} were −73 mV at 3:1 stoichiometry and −11 mV at 4:1 stoichiometry. Therefore, E_{NCX} was measured by setting the holding potential (HP) at −73 or −11 mV. E_{NCX} measured at the two different holding potentials were −69 ± 2 (n = 11) at −73 mV and −15 ± 1 mV (n = 15) at −11 mV HP. Thus, E_{NCX} varied depending on the HP and, furthermore, almost coincided with the HP. This indicated that intracellular Ca^{2+} and/or Na^+ concentrations might have changed because of the NCX current flow at different HPs. It also suggested that our protocol had a serious limitation for determining the stoichiometry of NCX.

To overcome the methodologic limitation, we measured $[Ca^{2+}]_i$ with 100 μM indo-1 using a confocal microscope under the whole-cell voltage clamp at the 2 different HPs. When the HP was at −11 mV, a theoretical E_{NCX} of 4:1, $[Ca^{2+}]_i$ was 374 ± 38 nM, which was significantly higher than the theoretical value of 184 nM. This was unexpected, because $[Ca^{2+}]_i$ could accumulate even in the presence of 20 mM BAPTA. In contrast, $[Ca^{2+}]_i$ was 173 ± 11 nM at −73 mV, a value close to the theoretical value of 184 nM and consistent with a 3:1 stoichiometry. Thus, the $[Ca^{2+}]_i$ measurement supported a 3:1 but not a 4:1 stoichiometry.

We examined E_{NCX} again with a different whole-cell clamp protocol. To prevent NCX current flow and Ca^{2+} accumulation, we inhibited NCX current almost completely with 5~10 mM Ni^{2+} for about 5 minutes, washed out the Ni^{2+}, and then recorded NCX current during recovery from Ni^{2+} inhibition. E_{NCX} during the recovery was measured at two different HPs, one at the theoretical E_{NCX} for 3:1 stoichiometry and the other at 4:1 stoichiometry. When the HP was at E_{NCX} for 3:1 stoichiometry, the E_{NCX} developed initially near the theoretical value for 3:1 stoichiometry and did not change with time. In contrast, when the HP was set at theoretical E_{NCX} for 4:1 stoichiometry, NCX current developed initially with E_{NCX} significantly more negative than the calculated value and then shifted with time towards the theoretical value of a 4:1 stoichiometry. These results also support a 3:1 stoichiometry.

The 4:1 or variable stoichiometry proposed by Fujioka et al.[8] could therefore be explained by Ca^{2+} accumulation on the cytoplasmic side of the membrane, even though they used a giant- or macropatch method that was theoretically devoid of Ca^{2+} accumulation. The following conditions may have allowed unexpected Ca^{2+} accumulation on the cytoplasmic side of the giant or macropatch membrane: (1) A relatively low concentration of EGTA (10 mM) as a Ca^{2+} buffer with 8.79 mM added Ca^{2+} can be calculated to yield 1 µM free Ca^{2+} on the cytoplasmic side. (2) The HP was fixed at 0 mV, which could facilitate Ca^{2+} accumulation. (3) The presence of high concentrations of Na^+ and Ca^{2+} on the cytoplasmic and extracellular sides of the membrane, respectively.

We conclude that the stoichiometry of cardiac NCX is 3:1.

ACKNOWLEDGMENT

This work was supported by Grants-in-Aid for Scientific Research from the Japan Society for the Promotion of Science (11357020 and 13670092).

REFERENCES

1. MULLINS, L.J. 1977. A mechanism for Na/Ca transport. J. Gen. Physiol. **70:** 681–695.
2. PITTS, B.J. 1979. Stoichiometry of sodium–calcium exchange in cardiac sarcolemmal vesicles. Coupling to the sodium pump. J. Biol. Chem. **254:** 6232–6235.
3. AXELEN, P.H. & J.H.B. BRIDGE. 1985. Electrochemical ion gradients and the Na/Ca exchange stoichiometry. Measurements of these gradients are thermodynamically consistent with a stoichiometric coefficient ≥3. J. Gen. Physiol. **85:** 471–475.
4. KIMURA, J., S. MIYAMAE & A. NOMA. 1987 Identification of sodium-calcium exchange current in single ventricular cells of guinea-pig. J. Physiol. **384:** 199–222.
5. EHARA, T., S. MATSUOKA & A. NOMA. 1989. Measurement of reversal potential of Na^+-Ca^{2+} exchange current in single guinea-pig ventricular cells. J. Physiol. **410:** 227–249.
6. YASUI, K. & J. KIMURA. 1990. Is potassium co-transported by the cardiac Na-Ca exchange? Pflügers Arch. **415:** 513–515.
7. BLAUSTEIN, M.P. & W.J. LEDERER. 1999. Sodium/calcium exchange: its physiological implications. Physiol. Rev. **79:** 763–854.
8. FUJIOKA, Y., M. KOMEDA & S. MATSUOKA. 2000. Stoichiometry of Na^+-Ca^{2+} exchange in inside-out patches excised from guinea-pig ventricular myocytes. J. Physiol. **523:** 339–351.
9. HINATA, M., H. YAMAMURA, L. LI, et al. 2002. Stoichiometry of Na^+/Ca^+ exchange is 3:1 in guinea-pig ventricular myocytes. J. Physiol. In press.

Simultaneous Measurement of $[Na]_i$, $[Ca]_i$, and I_{NCX} in Intact Cardiac Myocytes

KENNETH S. GINSBURG, CHRISTOPHER R. WEBER, SANDA DESPA, AND DONALD M. BERS

Department of Physiology, Loyola University Chicago, Maywood, Illinois 60153, USA

KEYWORDS: cardiac Na/Ca exchange; SBFI; Na:Ca exchange stoichiometry

To understand the role of cardiac NCX, we need to know how sodium (Na) flux, calcium (Ca) flux, and I_{NCX} are related. This relation depends on the Na:Ca exchange ratio (stoichiometry, n), which has usually been determined from the reversal potential (E_{rev}) of I_{NCX}. The value of n has widely been accepted to be 3, but recent evidence from giant membrane patches indicates n might be variable[1] and closer to 4 than 3 or, under external acidification,[2] closer to 2. The value of n is of considerable consequence, because $n = 4$ would leave any flow of outward I_{NCX} unlikely under physiologic conditions, whereas $n = 2$ would leave NCX nonelectrogenic (I_{NCX} undetectable) despite translocation of significant amounts of Na and Ca.

Measurement of n using E_{rev} requires that $[Na]_i$, $[Ca]_i$, $[Na]_o$, and $[Ca]_o$ all be constant and known, but NCX creates ionic gradients, disturbing $[Na]_i$, $[Ca]_i$, and hence E_{rev} itself whenever it translocates Na and Ca. Thus, E_{rev} measurements, especially in intact cells, may be hard to interpret. We previously used an instantaneous function to characterize the dependence of I_{NCX} on E_m, $[Ca]_i$, $[Ca]_o$, $[Na]_i$, and $[Na]_o$ in intact ventricular myocytes.[3] Although we measured $[Ca]_i$, we predicted $[Na]_i$, relying critically for this prediction on the assumption that $n = 3$ and not some other value.

To measure the E_m dependence (i-v relation) of I_{NCX} without assuming n, and to follow E_m-dependent Na and Ca fluxes via NCX, we simultaneously measured $[Na]_i$, $[Ca]_i$, and I_{NCX} in ruptured patch whole-cell voltage-clamped rabbit ventricular myocytes at 36°C using a fast wavelength scanning monochromator and synchronized photometry. Pipette solution was K^+ free and NMG based to isolate I_{NCX}, which was taken as current sensitive to 10 mM Ni. Pipettes also contained 6 mM BAPTA, 1 mM Br_2BAPTA, and 100 μM Fluo3 salt (100 nM free Ca) in order to measure $[Ca]_i$ and also to prevent it from changing rapidly, as well as 10 mM Na and 500

Address for correspondence: Donald M. Bers, Ph.D., Department of Physiology, Loyola University Chicago, Stritch School of Medicine, 2160 South First Avenue, Maywood, IL 60153. Voice: 708-216-1018; fax: 708-216-6308.

dbers@lumc.edu

Ann. N.Y. Acad. Sci. 976: 157–158 (2002). © 2002 New York Academy of Sciences.

µM SBFI salt. Na/K ATPase, I_{Ca}, Ca-activated Cl current, and SR Ca pump were all blocked.

From an initial value of −53 mV (predicted E_{rev} for the pipette conditions: $[Na]_i$ = 10 mM and $[Ca]_i$ = 100 nM with n = 3), E_{hold} was changed abruptly (e.g., to −16 mV), causing an initial large outward shift in I_{NCX}. This caused $[Ca]_i$ to rise and $[Na]_i$ to fall, as I_{NCX} relaxed towards a new equilibrium point. During and after the approach to the new steady state, we assessed E_{rev} periodically with 0.5 V/s ramps (+100 to −100 mV). These ramps were minimally perturbing because of heavy $[Ca]_i$ buffering. E_{rev} was seen to approach the new E_{hold}. Ramp data also allowed us to follow changes in the i-v relation of I_{NCX}.

Once steady states were attained at several different E_m, n was calculated from E_{rev}, $[Na]_i$, and $[Ca]_i$ by fitting $E_{rev} = n\,E_{Na} - 2\,E_{Ca}$, where $E_{Na} = (RT/F)\log_e ([Na_o]/[Na_i])$ and $E_{Ca} = (RT/2F)\log_e([Ca_o]/[Ca_i])$ with n as the parameter. The n could also be assessed by comparing Ca and Na fluxes to charge moved by I_{NCX} during transitions among various E_m. Our preliminary results are most consistent with n near 3.

REFERENCES

1. FUJIOKA, Y., K. HIROE & S. MATSUOKA. 2000. Regulation kinetics of Na^+-Ca^{2+} exchange current in guinea-pig ventricular myocytes. J. Physiol. **529:** 611–623.
2. EGGER, M. & E. NIGGLI. 2000. Paradoxical block of the Na^+-Ca^{2+} exchanger by extracellular protons in guinea-pig ventricular myocytes. J. Physiol. **523:** 353–366.
3. WEBER, C.R., K.S. GINSBURG, K.D. PHILIPSON, et al. 2001. Allosteric regulation of Na/Ca exchange current by cytosolic Ca in intact cardiac myocytes. J. Gen. Physiol. **117:** 119–131.

Electrophysiological Studies of the Cloned Rat Cardiac NCX1.1 in Transfected HEK Cells

A Focus on the Stoichiometry

HUI DONG, JEREMY DUNN, AND JONATHAN LYTTON

Cardiovascular Research Group, Departments of Biochemistry and Molecular Biology and Physiology and Biophysics, University of Calgary, Calgary, AB, Canada T2N 4N1

KEYWORDS: **patch clamp; reversal potential; NCS1; HEK cells**

INTRODUCTION

The stoichiometry of Na^+/Ca^{2+} exchanger has been of considerable interest for many years. Initial experiments suggested a coupling of 4 Na^+ to 1 Ca^{2+},[1,2] but subsequent studies converged on a stoichiometry of 3 Na^+ to 1 Ca^{2+}.[3] This was the generally accepted model for operation of the exchanger until last year, when Matsuoka and colleagues published an electrophysiological study indicating a stoichiometry of 4 Na^+ to 1 Ca^{2+} for the exchanger in macro patches from cardiac myocytes. All of the stoichiometry studies using native exchanger expressed in its endogenous environment are potentially confounded by the movement of ions through parallel pathways and uncertainty in ionic composition close to the membrane. In the current paper, we have expressed rat heart NCX1.1 in HEK-293 cells, and used patch-clamp recording to measure the thermodynamic equilibrium point for Na^+/Ca^{2+} exchanger operation under different ionic conditions. These data are consistent with a transport stoichiometry of 4 Na^+ to 1 Ca^{2+}.

METHODS

Rat heart pcDNA3.1-NCX1.1 cDNA was cotransfected together with a GFP cDNA into HEK-293 cells by Ca^{2+}-phosphate precipitation, and whole-cell patch-clamp recording of the transfected cells was performed as previously described. The bath solution used for the recording of outward currents contained (in mM): 145 LiCl; 1 $MgCl_2$; 10 glucose; 10 HEPES/TMA, pH 7.4; and either 0.5 EGTA (free $[Ca^{2+}]$ about 1 nM) or 1 $CaCl_2$. The pipette solution used to record outward currents contained (in mM): 120 NaCl; 5 KCl; 2 $MgCl_2$; 20 TEA-Cl; 1 Na_2ATP; 8 glucose; 10 HEPES/TMA, pH 7.2; and either 5 EGTA (free $[Ca^{2+}]$ < 0.5 nM), or 5 EGTA +

Address for correspondence: Jonathan Lytton, Ph.D., Department of Biochemistry & Molecular Biology, University of Calgary Health Sciences Centre, Room 2518, 3330 Hospital Drive NW, Calgary, AB, Canada T2N 4N1. Voice: 403-220-2893; fax: 403-283-4841.
  jlytton@ucalgary.ca

4.28 $CaCl_2$ (free $[Ca^{2+}] = 1$ μM). Reversal potential experiments employed solutions compatible with bidirectional operation of the NCX1.1. Bath solutions containing (in mM) various combinations of NaCl, KCl, and/or LiCl totaling 125, 1 $MgCl_2$, 20 TEA-Cl, 10 glucose, 10 HEPES/TMA, and 0.5 EGTA. External $[Ca^{2+}]$ was varied from 0.3 to 30 μM using 10 EGTA and various amounts of $CaCl_2$, or by unbuffered addition of $CaCl_2$ above this range. The pipette solution for reversal potential measurements contained (in mM) 18 NaCl and 100 potassium gluconate, plus 20 TEA-Cl; 1 Na_2ATP; 10 glucose; 10 HEPES/TMA, pH 7.2; 10 EGTA + 6.40 $CaCl_2$ (free $[Ca^{2+}] = 0.3$ (M).

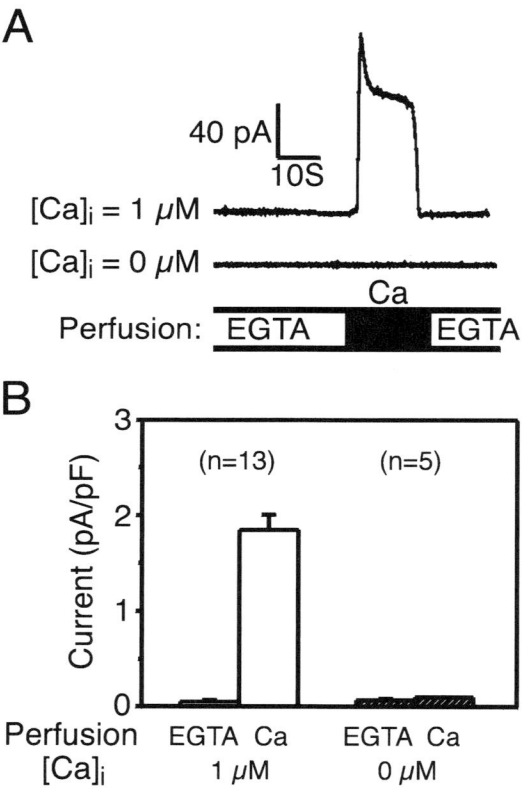

FIGURE 1. NCX1.1 outward currents recorded in transfected HEK-293 cells. **(A)** Typical traces from HEK-293 cells transfected with rat heart NCX1.1 cDNA and analyzed by whole-cell patch-clamp at a holding potential of 0 mV using a pipette solution with Ca^{2+} buffered to either <0.5 nM (0 μM) or 1 μM. **(B)** Summary data of normalized current amplitudes obtained from transfected cells perfused with either EGTA or Ca solution, and dialyzed with 0 or 1 μM $[Ca^{2+}]$, as shown in panel A. The data represent the average ± SEM from 5 (0 μM $[Ca^{2+}]_i$), or 13 (1 μM $[Ca^{2+}]_i$) cells. Modified from Dong et al.[6] with permission.

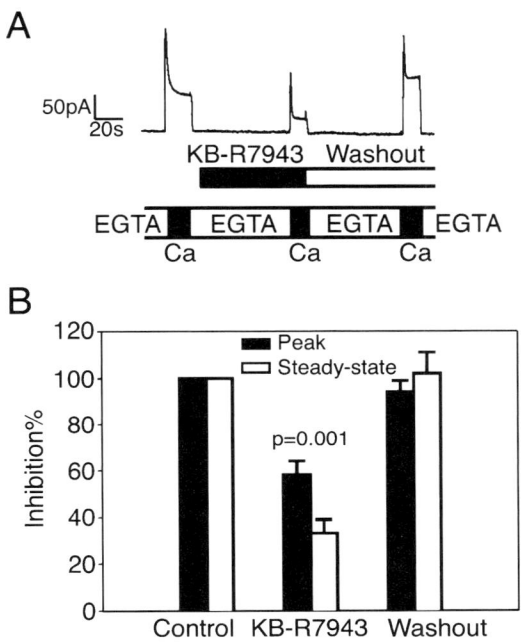

FIGURE 2. NCX1.1 outward currents are sensitive to KB-R7943. **(A)** Typical traces from HEK-293 cells transfected with rat heart NCX1.1 cDNA and analyzed by whole-cell patch-clamp at a holding potential of 0 mV as in FIGURE 1, and exposed to 1 μM KB-R7943, as illustrated. **(B)** Summary data averaged from six transfected cells.

FIGURE 3. $[Ca^{2+}]_o$ jump experiment to test the control over submembrane ionic concentration in HEK-293 cells transfected with rat heart NCX1.1 cDNA. HEK-293 cells transfected with rat heart NCX1.1 cDNA were analyzed by whole-cell patch-clamp using a holding potential of 0 mV, using pipette and bath solutions compatible with bidirectional NCX1 operation. When the bath perfusion solution changed from 125 mM LiCl and 0.5 mM EGTA to 50 mM NaCl and 0.01 mM $CaCl_2$, a small inward current was recorded. The magnitude of this inward current was not influenced by prior induction of a large outward current by a perfusion switch to 50 mM NaCl and 3 mM $CaCl_2$. The outward current will bring a significant amount of Ca^{2+} into the cell that, if not buffered properly, will induce a larger inward current when the perfusate is subsequently switched back to 50 mM NaCl and 0.01 mM $CaCl_2$.

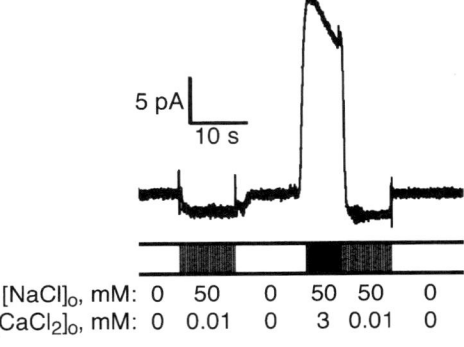

RESULTS AND DISCUSSION

FIGURE 1A illustrates the outward membrane currents, measured at a holding potential of 0 mV, elicited by a perfusion switch from EGTA- to Ca^{2+}-containing bath solution in NCX1.1, but not in control-transfected cells. These currents were regulated by cytosolic Ca^{2+} (FIG. 1) and inhibited reversibly by KB-R7943, a well-known selective inhibitor of Na^+/Ca^{2+} exchanger (FIG. 2). These results confirmed that the

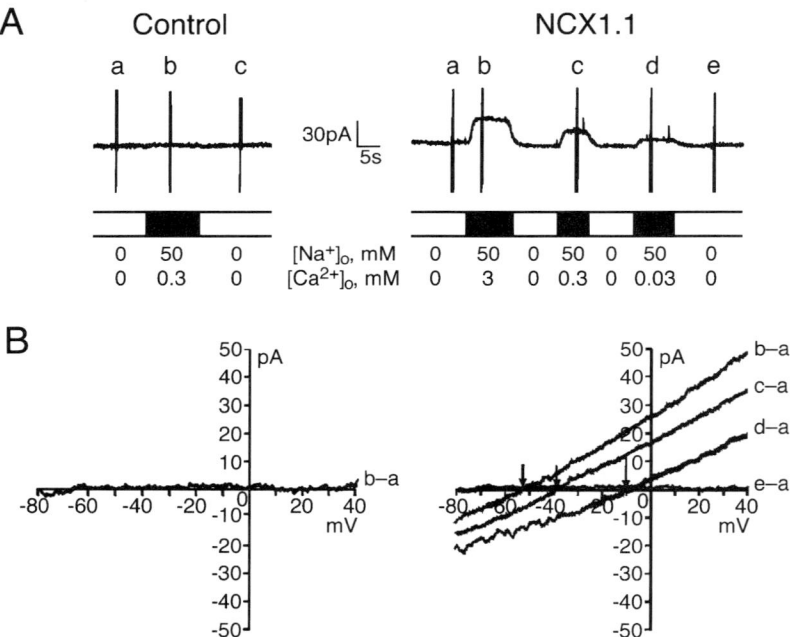

FIGURE 4. Reversal potential determinations for NCX1.1 with varying $[Ca^{2+}]_o$. HEK-293 cells transfected with either vector (control) or rat heart NCX1.1 cDNA (NCX1.1) were analyzed by whole-cell patch-clamp using solutions compatible with bidirectional NCX1.1 operation. Voltage ramps from 100 to +60 mV were applied during each Na^+/Ca^{2+} perfusion switch and in LiCl/EGTA at the start and the end of the experiment. (**A**) Representative current traces for control or NCX1.1-transfected cells subjected to perfusion switches of varying $[Ca^{2+}]_o$. (**B**) I–V curves obtained from the indicated ramps during the perfusions illustrated in A, with the I–V curve obtained in LiCl/EGTA at the start of the experiment (marked a) digitally subtracted. *Arrows* in B indicate the reversal potentials recorded for NCX1.1 under the three different $[Ca^{2+}]_o$ perfusions. Modified from Dong et al.[6] with permission.

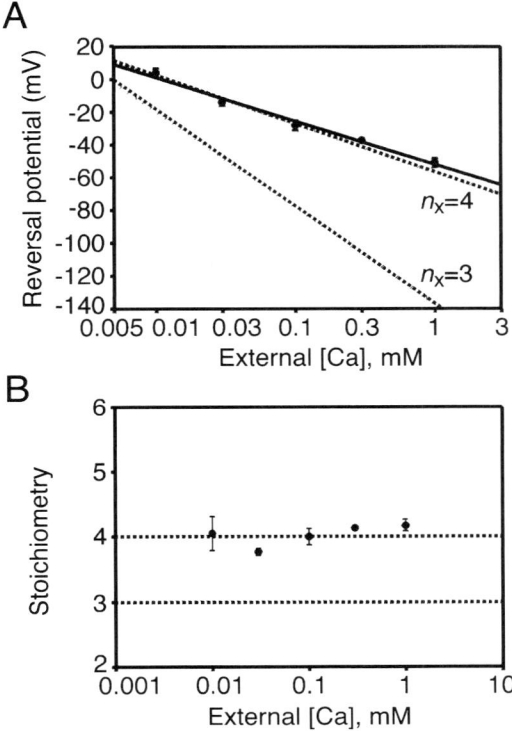

FIGURE 5. Reversal potential and stoichiometry of NCX1.1 measured by varying $[Ca^{2+}]_o$. HEK-293 cells transfected with rat heart NCX1.1 cDNA (NCX1.1) were analyzed by whole-cell patch-clamp, as described in FIGURE 4. Perfusion solutions contained 50 mM NaCl and varying $[Ca^{2+}]_o$. **(A)** Summary plot of reversal potential against log of $[Ca^{2+}]_o$. Average ± SEM for between five and seven determinations at each data point are shown. Dotted lines show theoretical E_{NCX} of 4 Na$^+$:1 Ca^{2+} and 3 Na$^+$:1 Ca^{2+} exchange, respectively, and the solid line is a fit of the data. **(B)** $[Ca^{2+}]_o$ concentration–stoichiometry relationship. Stoichiometry was calculated from the experimental data in A by using the equation: $E_{NCX} = (n_x E_{Na} - 2E_{Ca})/(n_x - 2)$. Modified from Dong et al.[6] with permission.

currents we observed in NCX1.1-transfected cells arose from the unequal movement of charge through the Na$^+$/Ca^{2+} exchanger.

The system used in the current experiment provides the advantages of a high level of NCX1 protein expression, low background ion transport levels (FIGS. 1 and 4), and excellent control over clamped voltage and ionic composition. As shown in FIGURE 3, the magnitude of inward NCX1.1 current is not influenced by prior induction of reverse-mode (outward current) operation. Thus, even high Ca^{2+} flux into the cell via exchanger operation does not alter the submembrane $[Ca^{2+}]$ significantly.

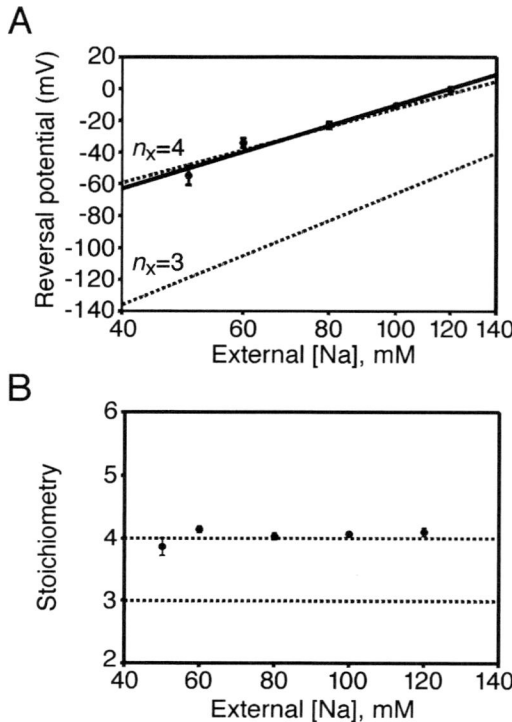

FIGURE 6. Reversal potential and stoichiometry of NCX1.1 measured by varying $[Na^+]_o$. HEK-293 cells transfected with rat heart NCX1.1 cDNA (NCX1.1) were analyzed by whole-cell patch-clamp, as described in FIGURE 4. The perfusion solutions contained various combinations of NaCl and 0.5 mM $CaCl_2$. **(A)** Summary plot of reversal potential against log of $[Na^+]_o$. Average ± SEM for between 5 and 8 determinations at each data point are shown. Dotted lines show theoretical E_{NCX} of 4 Na^+ : 1 Ca^{2+} and 3 Na^+ : 1 Ca^{2+} exchange, respectively, and the solid line is a fit of the data. **(B)** $[Na^+]_o$ concentration–stoichiometry relationship. Stoichiometry was calculated from the experimental data in A by using the equation: $E_{NCX} = (n_x E_{Na} - 2E_{Ca})/(n_x - 2)$. Modified from Dong et al.[6] with permission.

Using ionic conditions that allowed bidirectional currents, voltage ramps were employed to determine the reversal potential for NCX1.1-mediated currents (FIG. 4). Analysis of the relation between reversal potential and external $[Na^+]$ or $[Ca^{2+}]$, under a variety of intracellular conditions, yielded a stoichiometry for the NCX1.1 protein of 4 Na^+ : 1 Ca^{2+} : 2 charges moved per transport cycle, as seen in FIGURES 5 and 6.

REFERENCES

1. MULLINS, L.J. 1977. A mechanism for Na/Ca transport. J. Gen. Physiol. **70:** 681–695.
2. LEDVORA, R.F. & C. HEGYVARY. 1983. Dependence of Na^+-Ca^{2+} exchange and Ca^{2+}–Ca^{2+} exchange on monovalent cations. Biochim. Biophys. Acta **729:** 123–136.

3. BLAUSTEIN, M.P. & W.J. LEDERER. 1999. Sodium/calcium exchange: its physiological implications. Physiol. Rev. **79:** 763–854.
4. FUJIOKA, Y., M. KOMEDA & S. MATSUOKA. 2000. Stoichiometry of Na^+–Ca^{2+} exchange in inside-out patches excised from guinea-pig ventricular myocytes. J. Physiol. (London) **523**(Pt 2): 339–351.
5. DONG, H., P.E. LIGHT, R.J. FRENCH & J. LYTTON. 2001. Electrophysiological characterization and ionic stoichiometry of the rat brain K^+-dependent Na^+/Ca^{2+} exchanger, NCKX2. J. Biol. Chem. **276:** 25919–25928.
6. DONG, H., J. DUNN & J. LYTTON. 2002. Stoichiometry of the cardiac Na^+/Ca^{2+} exchanger NCX1.1 measure in transfected HEK cells. Biophys. J. **82:** 1943–1952.
7. IWAMOTO, T., T. WATANO & M. SHIGEKAWA. 1996. A novel isothiourea derivative selectively inhibits the reverse mode of Na^+/Ca^{2+} exchange in cells expressing NCX1. J. Biol. Chem. **271:** 22391–22397.
8. WATANO, T., J. KIMURA, T. MORITA & H. NAKANISHI. 1996. A novel antagonist, No. 7943, of the Na^+/Ca^{2+} exchange current in guinea-pig cardiac ventricular cells. Br. J. Pharmacol. **119:** 555–563.

Tissue-Specific Modes of Na/Ca Exchanger Regulation

LARRY V. HRYSHKO

Institute of Cardiovascular Sciences, St. Boniface General Hospital Research Centre, Winnipeg, Manitoba, Canada, R2H 2A6

ABSTRACT: Ionic regulation of Na^+/Ca^{2+} exchange describes the secondary modulating effects exerted on exchange activity by the transport substrates Na^+ and Ca^{2+}. These effects have been extensively characterized for the cardiac Na^+/Ca^{2+} exchanger, NCX1.1, primarily by the giant excised patch-clamp technique. Moreover, several studies have provided functional evidence for ionic regulation of Na^+/Ca^{2+} exchange activity in intact cellular systems. Through structure–function analyses, important protein domains involved in these regulatory processes have been identified. However, despite major progress in characterizing ionic regulation at the functional and molecular levels, the physiological importance of these processes remains unknown. In this study, we have examined Na^+/Ca^{2+} exchange activity for three members of the NCX1 family, namely NCX1.1, NCX1.3, and NCX1.4. These exchangers were expressed in *Xenopus laevis* oocytes and were characterized using the giant excised patch-clamp technique. We show that these three splice variants exhibit considerable differences in the kinetic features of their ionic regulatory profiles. Information of this type is beginning to provide insight into the physiological basis for tissue-specific expression of alternatively spliced Na^+/Ca^{2+} exchangers.

KEYWORDS: sodium–calcium exchange; ionic regulation; giant excised patch clamp; NCX1; alternative splicing

INTRODUCTION

The family of Na^+/Ca^{2+} exchangers has grown considerably since the prototypical member was cloned in 1990[1] and includes unique gene products and alternatively spliced variants.[2–4] Tissue-specific expression of distinct members has been demonstrated, and the levels of exchanger expression are known to vary considerably between different tissues and/or during development.[5] Furthermore, Na^+/Ca^{2+} exchange levels vary considerably in specific disease states,[6] including heart failure in humans.[7,8] Currently, progress is being made toward understanding the electrophysiological and electromechanical consequences of altering Na^+/Ca^{2+} exchanger expression levels. There is little to no information, however, on the physiological basis for the diversity of exchangers.

Address for correspondence: Larry V. Hryshko, Institute of Cardiovascular Sciences, St. Boniface General Hospital Research Centre, 351 Tache Avenue, Winnipeg, Manitoba, Canada, R2H 2A6. Voice: 204-235-3662; fax: 204-233-6723.
lhryshko@sbrc.ca

Exchange activity is regulated by numerous factors, including the transport substrates themselves. Na^+_i-dependent regulation describes the ability of cytoplasmic Na^+ to induce an inactive conformation of the exchanger, a process also termed I_1 inactivation.[9,10] This process leads to a prominent inactivation of outward exchanger currents at higher Na^+ concentrations. Ca^{2+}_i-dependent or I_2 inactivation describes the stimulatory effect of cytoplasmic Ca^{2+} on exchange activity.[9,11] In the absence of cytoplasmic Ca^{2+}, the exchanger enters the I_2 inactive state. Both processes have been extensively characterized using the giant excised patch-clamp technique and have also been demonstrated to operate in intact cells.[5]

The cardiac Na^+/Ca^{2+} exchanger, NCX1.1, remains the most extensively characterized member of the exchanger superfamily. Extensive structure–function analyses of NCX1.1 have provided considerable information on protein domains that play prominent roles in the ionic regulatory processes. For example, mutations within the XIP region of the NCX1.1 exchanger (encompassing amino acids 219–238) can alleviate or accelerate the I_1 inactivation process.[12] The high-affinity regulatory Ca^{2+}-binding site has been shown to reside between amino acids 371 and 508.[13,14] Mutations within this region can alter the Ca^{2+} concentration dependency of I_2 inactivation. Regulatory Ca^{2+} also strongly influences the I_1 regulatory process. For example, I_1 inactivation can be completely alleviated in the cardiac exchanger by raising regulatory Ca^{2+} to higher levels (≥ 10 μM). Almost invariably, alteration of I_1 regulation via mutagenesis alters the characteristics of I_2 regulation, and vice versa.[12,13] The physical nature of this interaction between the two regulatory processes remains unknown.

We have recently shown that the alternatively spliced NCX1.3 and NCX1.4 exchangers differ considerably in their ionic regulatory properties.[15] These exchangers are commonly referred to as the kidney (NCX1.3) and brain (NCX1.4) splice variants. Within the alternative splicing region, the kidney exchanger expresses exons B and D, whereas the brain exchanger expresses exons A and D. In general, exchangers expressing the A exon are found in excitable tissues, whereas those expressing the B exon are more widely distributed.[2] Therefore, these analyses were performed to gain insight into the function of the mutually exclusive A and B exons. Moreover, the functional differences between these isoforms may provide insight into the physiological basis for alternative splicing.

METHODS

Molecular Biology

Complementary DNAs encoding NCX1.1, NCX1.3, and NCX1.4 were linearized with *Hin*dIII (New England Biolabs Inc.) and the corresponding cRNAs were synthesized using T3 mMessage mMachine *in vitro* transcription kits (Ambion Inc.) as previously described.[15]

Oocyte Preparation

Xenopus laevis oocytes were prepared as previously described.[15] Following anesthesia with 250 mg of *p*-aminobenzoate per liter of ice water, oocytes were removed and washed in a solution containing (in mM): 88 NaCl, 15 HEPES,

2.4 NaHCO$_3$, 1.0 KCl, 0.82 MgSO$_4$, pH 7.6 at room temperature. Individual follicles were teased apart and treated with the above solution plus 16 mg/ml collagenase for 45–60 minutes with gentle agitation. Following this treatment, oocytes were washed in the above solution plus 0.41 mM CaCl$_2$, 0.3 mM Ca(NO$_3$)$_2$ and 1 mg/ml BSA (wash solution). Defolliculation was then performed by placing oocytes in 100 mM K$_2$HPO$_4$, pH 6.5, for 11–12 minutes with gentle agitation. Oocytes were washed again as above and then stored in wash solution until injection the following day. The oocytes were injected with ~20–50 ng of cRNA, and assessments of electrical activity were performed 3–6 days later.

Electrophysiology

The giant excised patch clamp technique was used to examine Na$^+$/Ca^{2+} exchange activity, as previously described.[15,16] Briefly, pipettes were pulled from borosilicate glass, tips were broken to a diameter of 30–40 μm, and subsequently fire-polished to a final diameter of 20–30 μm. Pipettes were coated with a mineral oil/Parafilm mixture to enhance patch stability and reduce pipette capacitance. Gigaohm seals to oocyte membranes were formed by gentle suction followed by excision of the patch by gently moving the pipettes with micromanipulators. For outward current measurements, pipettes contained (in mM): 100 N-methylglucamine-MES, 30 HEPES, 30 tetraethylammonium-hydroxide, 16 sulfamic acid, 8.0 CaCO$_3$, 6 KOH, 2.0 Mg(OH)$_2$, 0.25 ouabain, 0.1 niflumic acid, 0.1 flufenamic acid, pH 7.0 at 30°C. To activate outward currents, the cytoplasmic surface of the pipette (i.e. bath) was perfused with Na$^+$ vs. Li$^+$-based solutions composed of (in mM): 100 [Na + Li]-aspartate, 20 MOPS, 20 TEA-OH, 20 CsOH, 10 EGTA, 0–9.91 CaCO$_3$, 1.0–1.5 Mg(OH)$_2$, pH 7.0, at 30°C. The MAXC software program was used to calculate free Ca^{2+} and Mg^{2+} values.[17]

RESULTS

FIGURE 1 shows one type of ionic regulation observed for outward Na$^+$/Ca^{2+} exchange currents generated by the cardiac Na$^+$/Ca^{2+} exchanger, NCX1.1. This type is referred to as Na^+_i-dependent or I$_1$ regulation. In FIGURE 1A, the overlapping current traces were generated by applying different concentrations (10–100 mM) of Na$^+$ to the cytoplasmic surface of the membrane patch. Regulatory Ca^{2+} on the cytoplasmic surface (bath) and transported Ca^{2+} on the extracellular surface (pipette) were held constant at 1 μM and 8 mM, respectively. Note that both current magnitude and the extent of current inactivation increase progressively with increasing Na^+_i concentration. The graph in FIGURE 1B shows the relationship between cytoplasmic Na^+_i and F$_{SS}$, the fraction of steady-state current relative to peak current for an outward current pulse. Na^+_i-dependent or I$_1$ inactivation reflects a progressive decrease in the population of active Na$^+$/Ca^{2+} exchangers. Mutations within the XIP region of NCX1.1 (encompassing amino acids 219–238) prominently influence the characteristics of I$_1$ inactivation.[12]

In FIGURE 2A, Ca^{2+}-dependent or I$_2$ regulation of outward Na$^+$/Ca^{2+} exchange currents is shown. Note that the magnitude of outward currents increases as the concentration of regulatory Ca^{2+} is increased despite the fact that the application of reg-

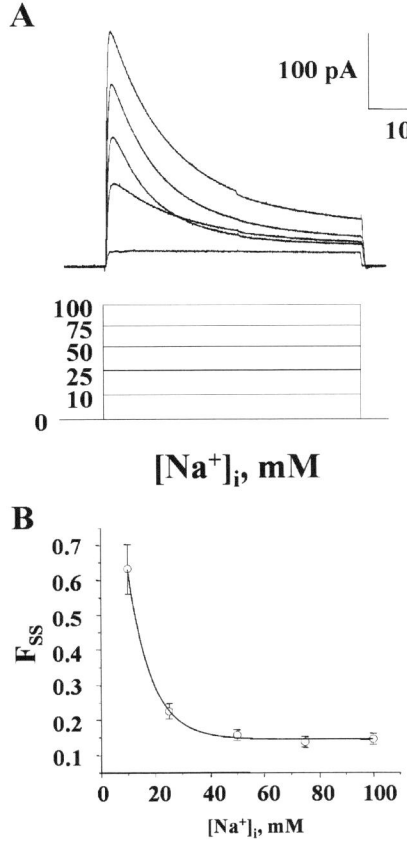

FIGURE 1. Na^+_i-dependent inactivation. **(A)** Overlapping outward Na^+/Ca^{2+} exchange current traces for the cardiac Na^+/Ca^{2+} exchanger, NCX1.1, obtained from an excised membrane patch. Outward currents were obtained by applying different concentrations of Na^+_i to the cytoplasmic surface of the patch, as indicated. Transported Ca^{2+}_o in the pipette and regulatory Ca^{2+}_i in the bath were constant at 8 mM and 1 μM, respectively. **(B)** The relationship between cytoplasmic Na^+ concentration and the extent of current inactivation for pooled results from 8 to 11 patches. Note that the extent of current inactivation increases as cytoplasmic Na^+ is increased, reflected by the progressive decease in the fraction of steady state current (F_{SS}) remaining.

ulatory Ca^{2+} decreases the electrochemical gradient favoring outward currents. At higher regulatory Ca^{2+} concentrations (e.g. ≥ 10 μM), outward current progressively declines, reflecting the progressive decrease in the electrochemical gradient favoring exchange and the competition between Na^+_i and Ca^{2+}_i at the intracellular ion-transport site. Cytoplasmic Ca^{2+} serves to recruit exchangers from the I_2-inactive state. In addition, cytoplasmic Ca^{2+} also influences the I_1-regulatory process, evident as the progressive decrease in I_1 inactivation at higher regulatory Ca^{2+} levels. The high-affinity regulatory Ca^{2+}-binding site has been identified and resides between amino acids 371 and 508 of the NCX1.1 exchanger.[14] Mutations within this region can alter the affinity of functional Ca^{2+}-dependent regulation,[13] whereas other nearby mutations can eliminate Ca^{2+} regulation altogether.[18]

As shown in FIGURE 2, the I_1 and I_2 processes show interaction. That is, increasing regulatory Ca^{2+} can alleviate the I_1 regulatory process. While this could reflect an independent effect of regulatory Ca^{2+} on the I_1 inactivation process (i.e., no in-

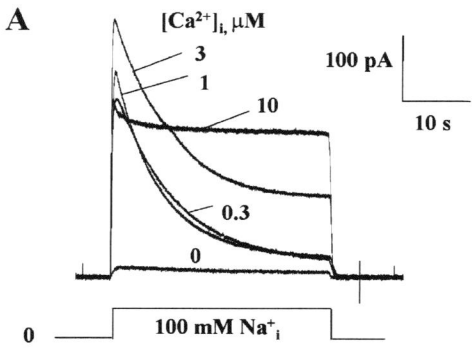
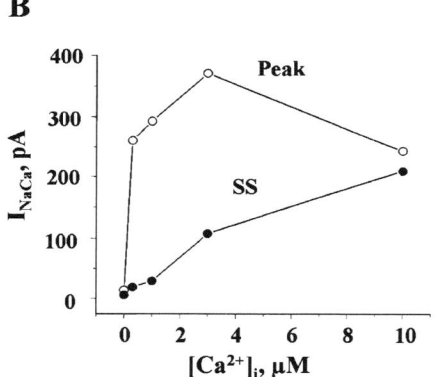

FIGURE 2. Ca^{2+}_i-dependent regulation. **(A)** The influence of cytosolic Ca^{2+} on outward Na^+/Ca^{2+} exchange currents for NCX1.1. Currents were activated by applying 100 mM Na^+ to the cytoplasmic surface of the patch, and pipette Ca^{2+} was constant at 8 mM. Regulatory Ca^{2+} levels were varied as shown. **(B)** The effects of different Ca^{2+}_i concentrations on peak and steady state currents from this same patch. Peak currents initially increase and then decrease as regulatory Ca^{2+} is increased. In contrast, steady state currents show a near linear relationship with regulatory Ca^{2+} over this concentration range, reflecting the progressive alleviation of I_1 inactivation.

teraction between I_1 and I_2), structure–function studies have shown that altering I_1 inactivation produces additional consequences on I_2 regulation and vice versa.[12,13] Furthermore, this interaction between I_1 and I_2 regulation is distinct for at least some alternatively spliced Na^+/Ca^{2+} exchangers.[15,19] FIGURE 3 shows the effects of regulatory Ca^{2+} on outward currents for NCX1.3 and NCX1.4, the splice variants commonly referred to as the kidney and brain Na^+/Ca^{2+} exchangers. Note that regulatory Ca^{2+} eliminates I_1 inactivation for the brain splice variant (NCX1.4), whereas it does not do so for the kidney splice variant (NCX1.3).

We examined the effects of kinetics of inactivation for the heart, kidney, and brain splice variants of NCX1 using rapid solution switching protocols. The intent was to more closely approximate the ionic flux rates to which these exchangers would be exposed in their physiological environments. At present, we are able to accomplish this for the typical frequencies of ion fluxes that the cardiac exchanger would be exposed to in larger mammals (i.e. 0.5–3 Hz). FIGURE 4 shows the influence of applying a 100 mM Na^+-containing solution continuously and 1 Hz to activate outward exchange currents for the cardiac exchanger NCX1.1. The results shown are from the same membrane patch using the same solutions to activate outward currents.

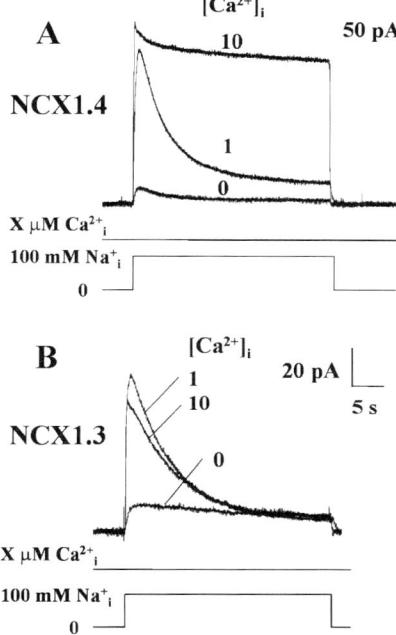

FIGURE 3. Ionic regulation of NCX1 splice variants. **(A)** Outward Na^+/Ca^{2+} exchange currents for the NCX1.4 exchanger at 0, 1, and 10 µM regulatory Ca^{2+}. Currents were activated by applying 100 mM Na^+ to the cytoplasmic surface of the patch, and pipette Ca^{2+} was constant at 8 mM. Note that increasing regulatory Ca^{2+} alleviates I_1 inactivation, similar to that observed for the cardiac exchanger, NCX1.1. In contrast, this alleviation of I_1 inactivation is not observed for the kidney exchanger, NCX1.3, as shown in (B). Here, I_1 inactivation remains prominent even at 10 µM regulatory Ca^{2+}.

Note that the extent of current inactivation at steady state is altered by changing the pattern of solution application.

To exclude the possibility that the effects of pulsatile solution application shown in FIGURE 4 were due to ionic accumulations and/or depletions or incomplete solution switching, we re-examined the above responses after proteolysis of the cytoplasmic surface of the patch with α-chymotrypsin. This treatment eliminates both forms of ionic regulation. FIGURE 5 shows that under these conditions, exchange currents are fully activated irrespective of whether or not solutions were applied continuously or in a pulsatile manner. Therefore, the influence of frequency on the extent of current inactivation reflects its influence on ionic regulation and not on ion accumulations and/or depletions.

FIGURE 6 illustrates outward Na^+/Ca^{2+} exchange currents for the 3 alternatively spliced exchangers, NCX1.1, NCX1.3, and NCX1.4, under conditions of continuous solution application and at 0.5 Hz. Note the substantial difference in the extent of current inactivation between splice variants. Specifically, at steady state, peak currents show near-complete recovery for the brain exchanger, whereas the kidney exchanger remains substantially inactivated. The cardiac exchanger shows an intermediate response. These differences in exchanger inactivation kinetics become far more prominent in response to pulsatile solution applications than is evident in response to continuous solution application.

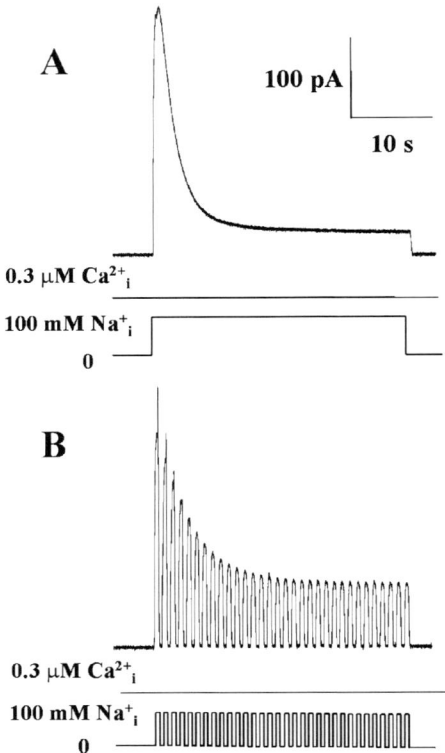

FIGURE 4. Effects of frequency on ionic regulation. Outward Na^+/Ca^{2+} exchange currents for NCX1.1 obtained from a single patch are shown. Pipette Ca^{2+} was constant at 8 mM, and regulatory Ca^{2+} was constant at 1 μM. Currents were activated by applying 100 mM Na^+ to the cytoplasmic surface of the patch either continuously (**A**) or at a frequency of 1 Hz with a duty cycle of 50% (**B**). Note that the extent of steady state current inactivation is considerably less when Na^+ is applied in a pulsatile manner.

DISCUSSION

In this report, we have examined the characteristics of ionic regulation for three splice variants of NCX1. These are often referred to as the cardiac (NCX1.1), kidney (NCX1.3), and brain (NCX1.4) Na^+/Ca^{2+} exchangers, although all three show a broader tissue distribution than these designations imply.[2] We show that these splice variants exhibit considerable differences in their patterns of ionic regulation, particularly with respect to the influence of cytoplasmic Ca^{2+} on the Na^+-dependent or I_1 inactivation process. Finally, we show that applying solutions in a pulsatile manner to giant excised membrane patches expressing different NCX1 splice variants emphasizes these differences in ionic regulation. These data demonstrate that alternative splicing alters the kinetic features of ionic regulation and may provide insight into the physiological role of these autoregulatory mechanisms.

Ionic Regulation of Na^+/Ca^{2+} Exchangers

Both of the transported ions, Na^+ and Ca^{2+}, exert secondary effects on Na^+/Ca^{2+} exchange activity through autoregulatory mechanisms. Na^+-dependent or I_1 inactivation was first described by Hilgemann[9] from measurements of Na^+/Ca^{2+} exchange

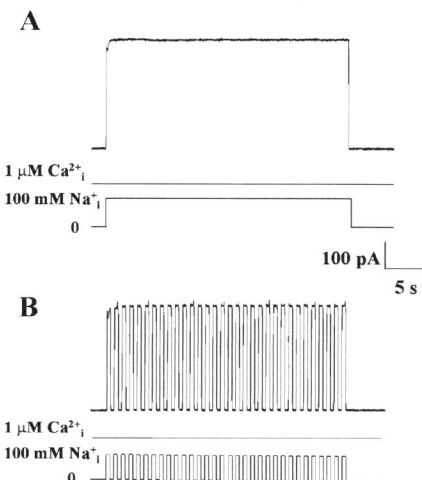

FIGURE 5. Effects of frequency of ionic regulation after α-chymotrypsin treatment. Outward Na^+/Ca^{2+} exchange currents activated by 100 mM cytoplasmic Na^+ (pipette and regulatory Ca^{2+} constant at 8 mM and 1 μM, respectively) are shown following deregulation of the exchanger by treating with 1 mg/ml α-chymotrypsin for ~1 minute. After this limited proteolysis of the cytoplasmic surface of the membrane patch, exchange currents appear fully activated and no longer exhibit I_1 and I_2 regulation.[9] Note that the effects of pulsatile solution application are no longer evident, eliminating the possibility that this phenomenon is related to ion accumulations or depletions.

currents obtained from giant excised membrane patches. In response to the application of cytoplasmic Na^+, both current magnitude and the extent of current inactivation are increased as cytoplasmic Na^+ is increased. This mechanism appears to involve a Na^+-dependent redistribution of exchangers between active and inactive states.[9,10] Although this mechanism has been demonstrated to occur in intact cardiac cells,[20] its physiological significance remains unknown. Arguing against a physiological role, it is noteworthy that I_1 inactivation is only prominent at Na^+ concentrations far in excess of those thought to occur physiologically.[20] Moreover, I_1 inactivation can be eliminated by increasing the concentration of ATP to physiological levels in excised membrane patches.[11,21] In favor of a physiological role is the fact that this mechanism is prominent and widely conserved between and across species.[22] Furthermore, the observation that alternative splicing modifies I_1 inactivation also implies an essential function for this mechanism.[15] While molecular studies have revealed a great deal about the structural counterparts responsible for I_1 inactivation, this information has yet to be applied toward an understanding of its physiological role.

Ca^{2+}-dependent or I_2 regulation describes the stimulatory effect of cytoplasmic Ca^{2+} on both inward and outward Na^+/Ca^{2+} exchange currents. This mechanism was first described in the squid axon and has been demonstrated in a variety of intact cell systems.[5] Furthermore, the kinetic features of this mechanism have been extensively studied in giant excised patches,[11] and the structural counterparts responsible for regulatory Ca^{2+} binding have been identified.[13,14] Cytoplasmic Ca^{2+} binds to a high-affinity regulatory site on the cytoplasmic surface of the exchanger, and this occupancy is essential to activate full exchange activity. Although the physiological role of Ca^{2+}-dependent regulation is not known, it seems most likely that this mechanism provides a means for matching exchange activity with cellular requirements for Ca^{2+} homeostasis. Specifically, in the presence of increased intracellular Ca^{2+} levels, exchange activity could be increased to facilitate Ca^{2+} removal. Such a mechanism would be particularly useful in cardiac muscle, where Ca^{2+} influx and efflux levels must be exquisitely matched to avoid Ca^{2+} accumulations or depletions.

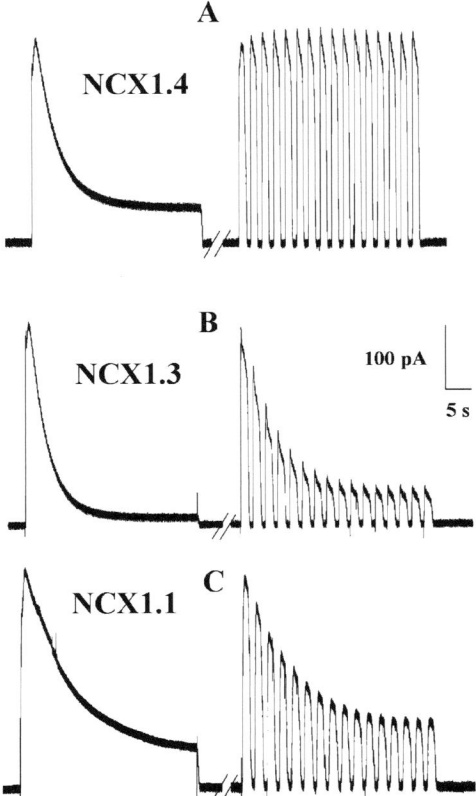

FIGURE 6. Effects of frequency on ionic regulation of NCX1 splice variants. The ionic regulatory profile of NCX1.4, NCX1.3, and NCX1.1 outward Na$^+$/Ca^{2+} exchange currents are shown. Pipette and bath Ca^{2+} were held constant at 8 mM and 1 μM, respectively. Currents were activated by applying 100 mM Na$^+$ to the cytoplasmic surface of patches, either continuously (*left panels*) or at 0.5 Hz (50% duty cycle). Note that the extent of steady state current inactivation differs considerably between NCX1 splice variants, particularly when examined using pulsatile solution application.

In this study, we show that the extent of I_1 inactivation varies considerably between the prominently expressed Na$^+$/Ca^{2+} exchanger isoforms found in brain, kidney, and cardiac muscle. Furthermore, the effects of regulatory Ca^{2+} on this process are particularly disparate between the kidney isoform and that observed for the brain and cardiac splice variants. This disparity is especially prominent when examining the behavior of Na$^+$/Ca^{2+} exchangers during pulsatile Na$^+$ application. Although a physiological role remains elusive, it is clear that alternative splicing does alter the kinetic features of ionic regulation. Information of this type makes it attractive to speculate that ionic regulation is involved in the kinetic tuning of distinct exchangers that allows them to function appropriately in their intact, cellular environments.

REFERENCES

1. NICOLL, D.A., S. LONGONI & K.D. PHILIPSON. 1990. Molecular cloning and functional expression of the cardiac sarcolemmal Na$^+$–Ca^{2+} exchanger. Science **250:** 562–565.
2. LINCK, B., Z. QIU, Z. HE, *et al.* 1998. Functional comparison of the three isoforms of the Na$^+$/Ca^{2+} exchanger (NCX1, NCX2, NCX3). Am. J. Physiol. **274:** C415–C423.

3. KOFUJI, P., W.J. LEDERER & D.H. SCHULZE. 1994. Mutually exclusive and cassette exons underlie alternatively spliced isoforms of the Na/Ca exchanger. J. Biol. Chem. **269:** 5145–5149.
4. LEE, S.L., A.S. YU & J. LYTTON. 1994. Tissue-specific expression of Na^+–Ca^{2+} exchanger isoforms. J. Biol. Chem. **269:** 14849–14852.
5. HRYSHKO, L.V. 2002. The cardiac Na^+–Ca^{2+} exchanger. *In* Handbook of Physiology. Section 2: The Cardiovascular System. V. 1: The Heart. E. Page, H.A. Fozzard & R.J. Solaro, Eds.: 388–419. Oxford University Press. Oxford, England.
6. POGWIZD, S.M., M. QI, W. YUAN, *et al.* 1999. Upregulation of Na^+/Ca^{2+} exchanger expression and function in an arrhythmogenic rabbit model of heart failure. Circ. Res. **85:** 1009–1019.
7. DIPLA, K., J.A. MATTIELLO, K.B. MARGULIES, *et al.* 1999. The sarcoplasmic reticulum and the Na^+/Ca^{2+} exchanger both contribute to the Ca^{2+} transient of failing human ventricular myocytes. Circ. Res. **84:** 435–444.
8. HOUSER, S.R. & E.G. LAKATTA. 1999. Function of the cardiac myocyte in the conundrum of end-stage, dilated human heart failure. Circulation **99:** 600–604.
9. HILGEMANN, D.W. 1990. Regulation and deregulation of cardiac Na^+–Ca^{2+} exchange in giant excised sarcolemmal membrane patches. Nature **344:** 242–245.
10. HILGEMANN, D.W., S. MATSUOKA, G.A. NAGEL & A. COLLINS. 1992. Steady-state and dynamic properties of cardiac sodium–calcium exchange. Sodium-dependent inactivation. J. Gen. Physiol. **100:** 905–932.
11. HILGEMANN, D.W., A. COLLINS & S. MATSUOKA. 1992. Steady-state and dynamic properties of cardiac sodium–calcium exchange. Secondary modulation by cytoplasmic calcium and ATP. J. Gen. Physiol. **100:** 933–961.
12. MATSUOKA, S., D.A. NICOLL, Z. HE & K.D. PHILIPSON. 1997. Regulation of cardiac Na^+–Ca^{2+} exchanger by the endogenous XIP region. J. Gen. Physiol. **109:** 273–286.
13. MATSUOKA, S., D.A. NICOLL, L.V. HRYSHKO, *et al.* 1995. Regulation of the cardiac Na^+–Ca^{2+} exchanger by Ca^{2+}. Mutational analysis of the Ca^{2+}-binding domain. J. Gen. Physiol. **105:** 403–420.
14. LEVITSKY, D.O., D.A. NICOLL & K.D. PHILIPSON. 1994. Identification of the high affinity Ca^{2+}-binding domain of the cardiac Na^+–Ca^{2+} exchanger. J. Biol. Chem. **269:** 22847–22852.
15. DYCK, C., A. OMELCHENKO, C.L. ELIAS, *et al.* 1999. Ionic regulatory properties of brain and kidney splice variants of the NCX1 Na^+–Ca^{2+} exchanger. J. Gen. Physiol. **114:** 701–711.
16. DYCK, C., K. MAXWELL, J. BUCHKO, *et al.* 1998. Structure–function analysis of CALX1.1, a Na^+–Ca^{2+} exchanger from *Drosophila*. Mutagenesis of ionic regulatory sites. J. Biol. Chem. **273:** 12981–12987.
17. BERS, D.M., C.W. PATTON & R. NUCCITELLI. 1994. A practical guide to the preparation of Ca^{2+} buffers. Methods Cell Biol. **40:** 3–29.
18. MAXWELL, K., J. SCOTT, A. OMELCHENKO, *et al.* 1999. Functional role of ionic regulation of Na^+/Ca^{2+} exchange assessed in transgenic mouse hearts. Am. J. Physiol. **277:** H2212–H2221.
19. OMELCHENKO, A., C. DYCK, M. HNATOWICH, *et al.* 1998. Functional differences in ionic regulation between alternatively spliced isoforms of the Na^+–Ca^{2+} exchanger from *Drosophila melanogaster*. J. Gen. Physiol. **111:** 691–702.
20. MATSUOKA, S. & D.W. HILGEMANN. 1994. Inactivation of outward Na^+–Ca^{2+} exchange current in guinea-pig ventricular myocytes. J. Physiol. (London) **476:** 443–458.
21. HILGEMANN, D.W. & A. COLLINS. 1992. Mechanism of cardiac Na^+–Ca^{2+} exchange current stimulation by MgATP: possible involvement of aminophospholipid translocase. J. Physiol. (London) **454:** 59–82.
22. PHILIPSON, K.D. & D.A. NICOLL. 2000. Sodium–calcium exchange: a molecular perspective. Annu. Rev. Physiol. **62:** 111–133.

NCX1 Surface Expression

A Tool to Identify Structural Elements of Functional Importance

HANNAH RAHAMIMOFF,[a,b] XIAOYAN REN,[a] CHAVA KIMCHI-SARFATY,[b] SURESH AMBUDKAR,[b] AND JUDITH KASIR[a]

[a]*Department of Biochemistry, Hebrew University-Hadassah Medical School, Jerusalem 91120, Israel*

[b]*Laboratory of Cell Biology, National Cancer Institute, National Institutes of Health, Bethesda, Maryland 20892-4255, USA*

ABSTRACT: The rat Na^+/Ca^{2+} exchanger isoforms of the NCX1 gene have 14 cysteine residues. Each of these cysteines can be mutated individually to alanine or serine without loss of functional expression in transfected HEK293 cells. Yet sequential exchange starting from the amino terminal end of three or more cysteines results in reduced transport activity and surface expression. As more and more cysteines are replaced, transport activity and surface expression decrease in parallel, and the cysteineless mutant exhibits only traces of Na^+/Ca^{2+} exchange activity and surface expression. No significant differences are detected in the amount of total cell exchanger protein between the wild-type exchanger and its functional or nonfunctional cysteine mutants. Reduced surface expression of the Na^+/Ca^{2+} exchanger NCX1 is also observed when HEK293 cells expressing the transporter are treated with cyclosporin A (CsA) or with PSC833. The reductions in transport activity and surface expression are concentration dependent and parallel. No reduction is obtained in the total amount of exchanger protein by CsA or PSC833 treatment, suggesting that the effects of these drugs on NCX1 expression are posttranslational. FK506 and rapamycin treatment of HEK293 cells expressing rat NCX1 isoforms has no effect on transport activity, surface expression, or the total amount of exchanger protein in the transfected cells.

KEYWORDS: NCX1; Na^+/Ca^{2+} exchange; cysteine mutagenesis; cyclosporin A; PSC833; immunophilins

INTRODUCTION

Membrane and secretory proteins undergo a complicated set of posttranslational processes in the ER (endoplasmic reticulum) until they mature into a functionally competent form.[1,2] Cellular quality control ensures that only properly processed and

---

Address for correspondence: Hannah Rahamimoff, Department of Biochemistry, Hebrew University–Hadassah Medical School, Box 12272, Jerusalem 91120, Israel. Voice: 972-2-6758511; fax: 972-2-6757379.

hannah@cc.huji.ac.il

folded proteins reach their target destination. Proteins that fail to mature into their functional form are retained in the ER.

In this chapter, we present two elements of the surface expression of the Na^+/Ca^{2+} exchanger NCX1: (1) the importance of the protein's cysteine residues to its functional expression and (2) the effects of the immunosupressive drug cyclosporin A and the nonimmunosupressive drug PSC833 on its functional expression.

The studies were carried out in heterologous expression systems, using the cloned (in pcDNA3.1) rat NCX1 isoforms RBE-2 (NCX1.5) and RHE-1 (NCX1.1) expressed in HEK293 cells.[3,4] Some of the experiments were also repeated in the VTF-7/HeLa cell expression systems.[5]

CYSTEINE RESIDUES ARE IMPORTANT FOR FUNCTIONAL EXPRESSION OF THE NCX1 PROTEIN

Fourteen cysteine residues are present in the rat isoforms of the NCX1 protein. Their positions along the polypeptide chain are outlined in the model presented in FIGURE 1. Four of these cysteines, C14, C20, C122, and C780 (numbering of cysteine residues as in the rat NCX1.5 isoform), are modeled to be on (or close to) the extracellular face of the membrane, two are modeled to be intramembraneous (C151 and C210), and the remaining cysteines are facing the cytoplasmic side of the membrane. The importance of cysteine residues to Na^+/Ca^{2+} exchange activity was suggested by early experiments in which NEM, DTT, MMTS, PCMB, PCMBS, and mersalyl were applied to excitable cells, and membrane preparations derived from

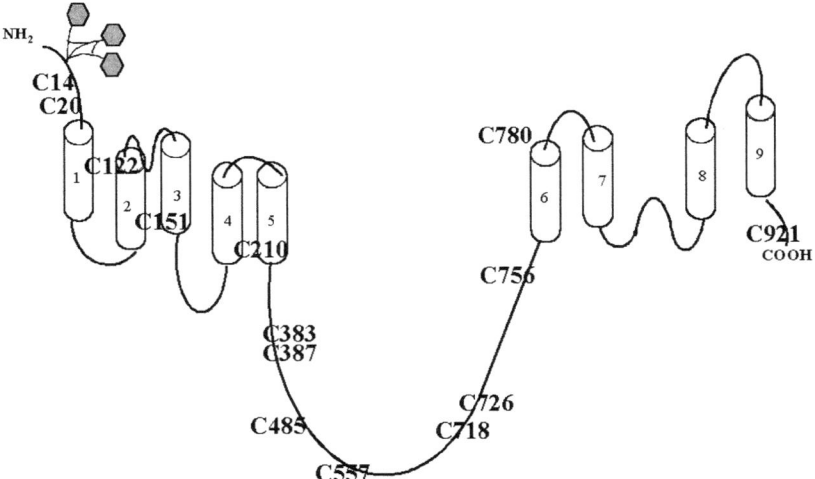

FIGURE 1. Putative topological model of the Na^+/Ca^{2+} exchanger NCX1.5 (RBE-2). The location of the cysteine residues along the polypeptide chain is shown. The model is based on Refs. 11, 33, and 34.

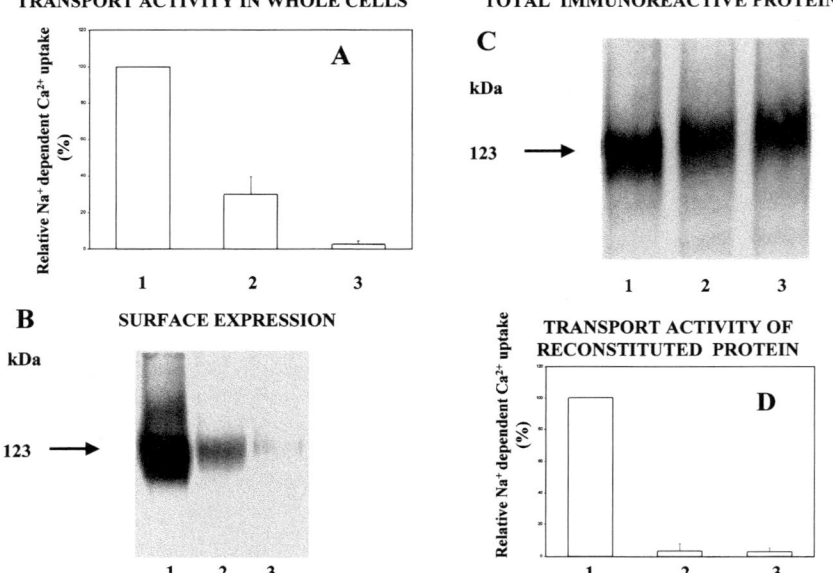

FIGURE 2. The effect of sequential cysteine substitution on functional expression of the NCX1 protein in HEK293 cells. HEK293 cells were transfected with the wild-type exchanger RBE-2 (*bar or lane 1*) and two of its cysteine mutants RBE-2/C14A/C20S/C122S (*bar or lane 2*) RBE-2/C14A/C20S/C122S/C151A/C210S/C383S/C387S (*bar or lane 3*). Twenty-four hours after transfection, Na^+ gradient-dependent Ca^{2+} uptake in whole cells (**A**), the surface expression of the protein (**B**), the total cell exchanger protein (**C**), and Na^+ gradient-dependent Ca^{2+} uptake in proteoliposomes derived from the transfected cells (**D**) were determined. (**A**) Na^+ gradient-dependent Ca^{2+} uptake in whole cells was carried out as described previously.[35] The results are expressed in relative values, the transport activity of the wild-type exchanger (which was always tested in parallel with the mutants) was taken as 100%, and the transport activity of the mutants was normalized accordingly. (**B**) Transfected cells were biotinylated *in situ* with NHS-SS-Biotin. Biotinylated proteins were isolated from cell extracts by binding to immobilized streptavidin, and Western analysis was carried out using AbO-8, as described in Refs. 33 and 35. (**C**) Western analysis of 40-μg proteins derived from cells expressing the wild-type exchanger RBE-2 and its cysteine mutants was carried out by standard procedures. AbO-8[33] was used for detection by ECL. (**D**) Transport activity of reconstituted proteins derived from HEK-293 cells expressing RBE-2 and its cysteine mutants was measured 24 hours after transfection. Cell proteins were reconstituted into proteoliposomes, and Na^+-dependent Ca^{2+} uptake was determined as described in Kasir *et al.*[35] The data are presented in relative values, the transport activity of wild-type exchanger in each experiment was taken as 100%.

them. These experiments indicated that sulfhydryl reagents have considerable effects on Na^+/Ca^{2+} exchange.[6–9] Redox modification[10] in the presence of DTT also supported an important role for cysteine residues in Na^+/Ca^{2+} exchange.

Yet examination of the effects of different MTS (methanethiosulfonate) reagents in conjunction with cysteine scanning mutagenesis (SCAM), which was carried out to study the membrane topology of the dog heart Na^+/Ca^{2+} exchanger NCX1.1 expressed in *Xenopus* oocytes,[11,12] suggested that none of the membrane-permeable or -impermeable sulfhydryl reagents tested had any significant effect on transport activity. MTSET sometimes stimulated Na^+/Ca^{2+} exchange.[11] Similar results were reported in CCL-39 cells expressing the cloned dog cardiac Na^+/Ca^{2+} with the membrane-impermeable MTSET.[13] Re-examination of the effects of sulfhydryl reagents on the Na^+/Ca^{2+} exchanger NCX1 expressed in HEK293 cells[14] suggested, however, that sulfhydryl reagents inhibited the transport activity of the rat brain and rat heart isoforms of NCX1. Moreover, our experiments suggested that intracellular cysteine residues were involved in the inhibition of transport activity of the protein by sulfhydryl reagents.[14] The importance of a disulfide bond between extracellular cysteine residues was also demonstrated for functional expression of split NCX1 exchangers.[15,16]

Single cysteine mutants of the rat heart and brain Na^+/Ca^{2+} exchangers are all functional, and their transport activity in HEK293 cells is not significantly different from that of the wild-type exchanger (not shown). Sequential exchange of cysteine residues, however, starting from the amino-terminal end, indicates that exchange of three cysteines or more results in a gradual decrease in transport activity in the expressing HEK293 cells. FIGURE 2 shows the relative transport activity (FIG. 2A), the corresponding surface expression (FIG. 2B), the total amount of immunoreactive protein (FIG. 2C), and the relative transport activity of the reconstituted preparation (FIG. 2D) of the wild-type parent exchanger RBE-2 (NCX1.5), its triple mutant RBE-2/C14A/C20S/C122S, and its mutant RBE-2/C14A/C20S/C122S/C150A/-C210S/C383S/C387S, in which the initial seven NH_2-terminal cysteines were exchanged to either alanine or serine.

It can be seen that parallel decrease in transport activity in whole cells and in the surface expression of the protein is observed when sequential cysteine replacement is carried out. Replacement of three cysteines (C14, C20, and C122) already leads to a considerable reduction in transport activity (FIG. 2A) and in surface expression (FIG. 2B) of the exchanger relative to the wild-type protein. When the initial seven NH_2-terminal cysteines (C14–C387) are replaced, only traces of transport activity and traces of the exchanger protein are detected in the surface membrane. No significant differences are detected, however, between the wild-type exchanger RBE-2 and its cysteine mutants when the total amount of immunoreactive exchanger protein in the expressing HEK293 cells is compared (FIG. 2C). Reconstitution of the transfected cell proteins into proteoliposomes (FIG. 2D) did not rescue the transport activity of the functionally impaired cysteine mutants that were retained within the transfected cells.

Because individual cysteines can be replaced without loss of transport activity, but replacement of multiple cysteines leads to reduced transport activity and surface expression, our experiments suggest that it is the interaction between the cysteines that might be important. Further studies are being done to elucidate this suggestion.

CsA AND PSC833 INHIBIT THE TRANSPORT ACTIVITY AND REDUCE SURFACE EXPRESSION OF RAT BRAIN AND HEART ISOFORMS OF THE NCX1 GENE EXPRESSED IN HEK293 CELLS

Immunophilins are ubiquitously expressed proteins comprising three families[17]: the cyclophilins, the FKBPs (FK506-binding proteins), and the bacterial parvulins[18] or their human analogues, Pin1.[19] They interact with the Ca^{2+}- and calmodulin-activated phosphatase calcineurin,[20] resulting in dephosphorylation of the transcription factor NFAT, its translocation to the nucleus, and subsequent T-cell activation. In addition to their role in T-cell activation, the immunophilins are also involved in protein folding either via their peptidyl-prolyl *cis-trans* isomerase (PPI) domain or via their chaperone domain or both.[17,22,23] Cyclosporin A (CsA) interacts with cyclophilin and FK506, and rapamycin interacts with FKBP, thereby preventing T-cell activation. CsA has been widely used for the past 15 years as an immunosupressant for transplant patients.[24] Its huge benefits in the prevention of graft rejection are accompanied by severe complications, such as hypertension and nephrotoxicity, the basis of which is not fully understood. Both side effects were linked to CsA-dependent, impaired cellular Ca^{2+} homeostasis by an unknown mechanism, which can potentially result in contraction of smooth muscle cells.[25,26] CsA was shown to modulate surface expression of membrane proteins in different ways: The drug induced the kinesis to the surface membrane of an ER-retained maturation-incompetent mutant of P-glycoprotein.[27] Since this phenomenon was also shared by other modulators of P-glycoprotein, it was probably not linked to interaction with cyclophilin but to direct interaction with the transporter. CsA was also shown to inhibit the surface expression of several membrane proteins, none of which was interacting with the drug. Among these are the homo-oligomeric acetylcholine receptor containing the α7 subunit of the homo-oligomeric 5-hydroxytryptamine type 3 receptor in *Xenopus* oocytes, the Kir2.1 potassium channel,[28] the creatinine transporter,[29] and the insulin receptor in adrenal chromaffin cells. The drug increased the surface expression of the Na^+ channels expressed in adrenal chromaffin cells, however.[31]

Addition of CsA (0.1 μM–50 μM) to HEK293 cells expressing both the rat heart and rat brain isoforms of the NCX1 gene, inhibited in a concentration-dependent manner the transport activity and reduced the surface expression of the Na^+/Ca^{2+} exchanger, without inhibiting the expression of the total amount of immunoreactive exchanger protein.[32] Similar results were obtained when the cells expressing the Na^+/Ca^{2+} exchanger were exposed to PSC833 (0.1 μM–20 μM), the nonimmunosuppressive chemical derivative of the weakly immunosuppressive CsD (cyclosporin D) analogue of CsA. FIGURE 3 shows some of these findings: In FIGURE 3A the transport activity of cells expressing the rat heart N-Flag-tagged Na^+/Ca^{2+} exchanger RHE-1 (NCX1.1) with and without 10 μM CsA or 10 μM PSC833 is shown. In FIGURE 3B the surface expression of the N-Flag-tagged Na^+/Ca^{2+} exchanger RBE-2 with and without treatment of the cells with 10 μM CsA is shown, and in FIGURE 3C the surface expression of the rat heart Na^+/Ca^{2+} exchanger RHE-1 (NCX1.1) with and without 10 μM PSC833 is shown. Exposure of the cells to these drugs did not change the expression of the Na^+/Ca^{2+} protein in the transfected cells,[32] suggesting that the reduction of transport activity and of surface expression resulted from impaired post-translational modification processes that lead to intracellular retention of the function-incompetent exchanger protein. The effects of CsA and PSC833 were specific

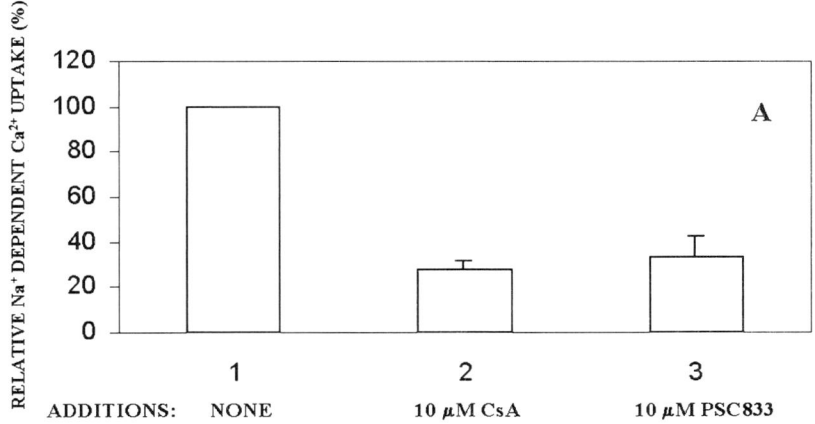

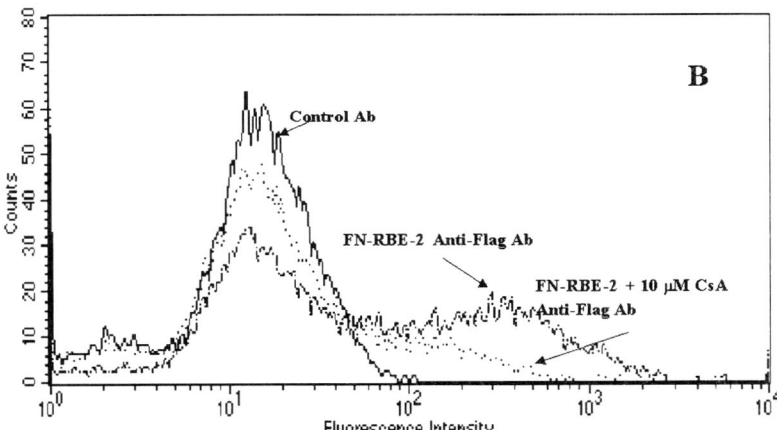

FIGURE 3. The effects of CsA and PSC833 on Na^+-dependent Ca^{2+} uptake and surface expression of the rat isoforms of NCX1 gene in HEK293 cells. **(A)** Na^+/Ca^{2+} exchange activity: Na^+ gradient-dependent Ca^{2+} uptake was measured in whole cells as previously described. CsA (10 μM) or 10 μM PSC833 were added to cells transfected with the rat heart Na^+/Ca^{2+} exchanger NCX1.1 according to the protocol described in Kimchi-Sarfaty et al.[32] The results are expressed in relative values; the transport activity without addition of any of the drugs was taken as 100%. **(B)** The effect of 10 μM CsA on the surface expression of NCX1.5: Cells expressing the amino-terminal Flag epitope–tagged rat Na^+/Ca^{2+} exchanger FN-RBE-2 were incubated with 10 μM CsA or DMSO (control cells). Twenty-four hours postransfection, the cells were analyzed for cell surface exchanger protein expression using M2, the anti-Flag antibody, or IgG1κ-control antibody. FITC-conjugated anti-mouse antibody was used for detection by FACS analysis. **(C)** The effect of 10 μM PSC833 on the surface expression of NCX1.1: The experimental protocol used was identical to that described in (B), except that the transfected cells were treated with PSC833.

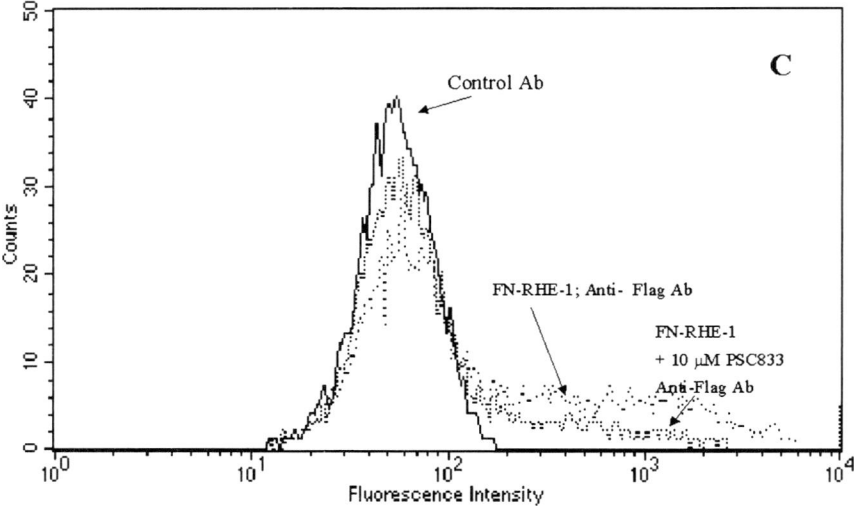

FIGURE 3C.

and not shared by the weakly immunosuppressive cyclosporin D or the immunosuppressive drugs FK506 or rapamycin. FIGURE 4A shows that addition of FK506, up to a concentration of 20 μM, to HEK293 cells transfected with NCX1 has no effect on Na^+-dependent Ca^{2+} uptake. Drug concentrations higher than 20 μM could not be used for transport studies in adherent cells due to cell pealing. Addition of up to 50 μM FK506 had no effect on the surface expression of the protein. FIGURE 4B shows the surface expression of cells transfected the N-Flag-tagged Na^+/Ca^{2+} exchanger with and without 5, 10, and 50 μM FK506 treatment. It can be seen that all the individual traces overlap, suggesting that similar amounts of exchanger protein are present in the surface membrane. Similar results are obtained when the total amount of immunoreactive exchanger protein is determined with and without FK 506 treatment of the transfected cells (FIG. 4C). The different effects of the undecapeptide and macrolide drugs on the transport activity and surface expression of the Na^+/Ca^{2+} exchanger NCX1 probably did not result from their selective expression in HEK293 cells, since cyclophilin A and B and FKBP12 and FKBP59 are all ex-

---

FIGURE 4. The effect of FK506 on transport activity, surface expression, and expression of the Na^+/Ca^{2+} exchanger in HEK293 cells. **(A)** Na^+/Ca^{2+} exchange activity: Cells were transfected with the plasmid encoding the rat heart Na^+/Ca^{2+} exchanger *fn-rhe-1* with or without exposure of the cells to different concentrations of FK506. Na^+-dependent Ca^{2+} uptake was determined 24 hours posttransfection as described. **(B)** Surface expression: Cells expressing FN-RHE-1 without or with treatment with 5, 10, and 50 μM FK506 were analyzed for surface expression using M2 monoclonal anti-Flag antibody or IgG1-control antibody. **(C)** Total immunoreactive protein expression: Experimental protocol was identical to (B) except that the cells were permeabilized before incubation with the primary antibody.

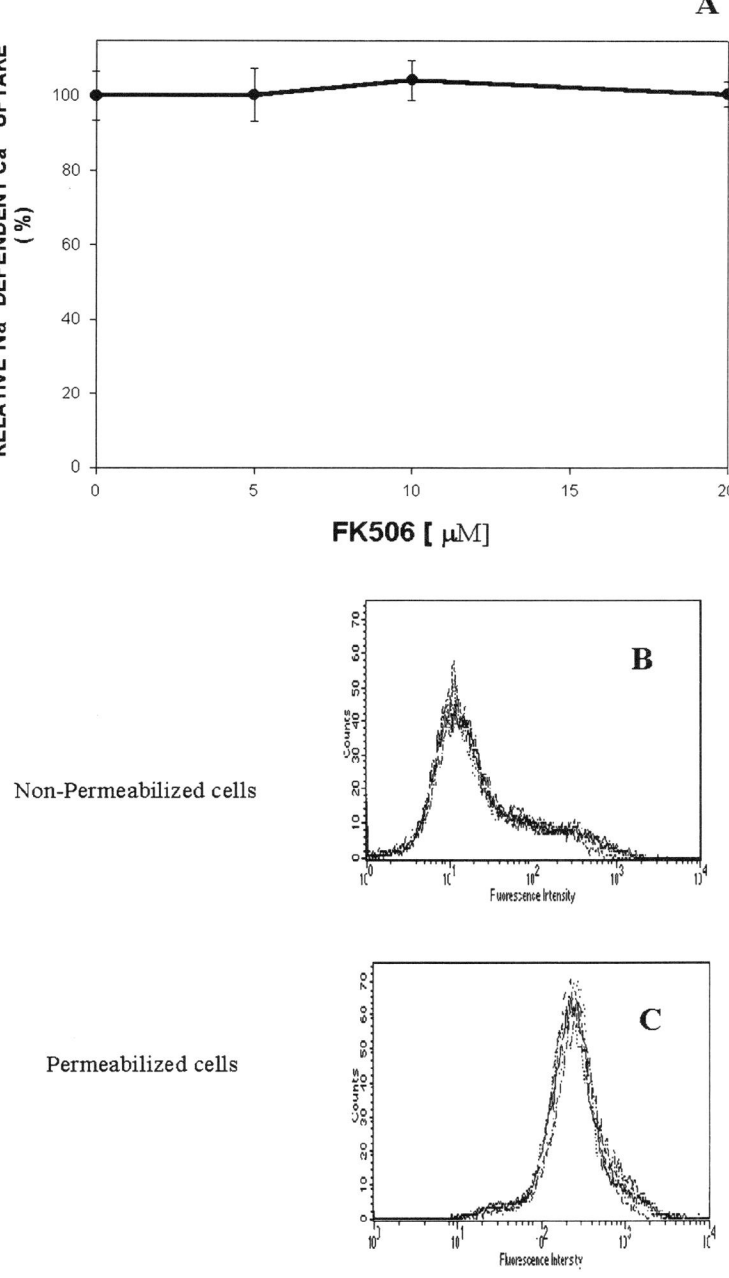

FIGURE 4. *See preceding page for legend.*

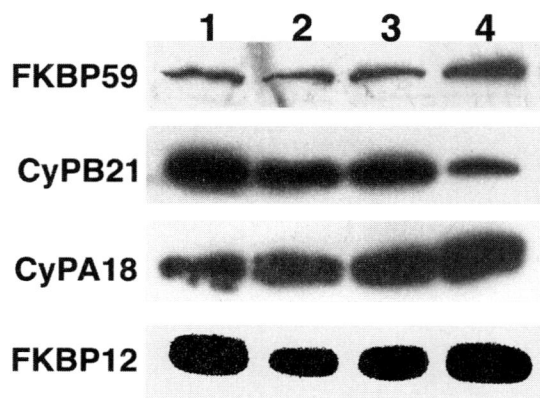

FIGURE 5. Expression of the immunophilins cyclophilin A 18, cyclophilin B 21, FKBP12, and FKBP59 in HEK293 cells: Expression of the different immunophilins was determined in HEK293 cell extracts by Western analysis. 1:1000 dilution of the appropriate anti-immunophilin antibody and 1:10000 dilution of HRP-conjugated secondary antibodies were used for detection by ECL. The same amount (20 µg) of cell extract protein was loaded in each lane. *Lane 1*: cells only; *lane 2*: mock-transfected cells (with pcDNA3.1); *lane 3*: FN-RBE-2-expressing cells; and *lane 4*: FN-RBE-2-expressing cells, treated with 10 µM CsA.

pressed in these cells. FIGURE 5 shows the expression pattern of these immunophilins in HEK293 cells with and without transfection and with CsA treatment. It can be seen that similar amounts of these proteins are expressed. Taken together, these results suggest that CsA and PSC833 interfere with one or more posttranslational processes that are important to maturation of the Na^+/Ca^{2+} exchanger NCX1 into a functional protein.

REFERENCES

1. HAMMOND, C. & A. HELENIUS. 1994. Quality control in the secretory pathway: retention of a misfolded viral membrane glycoprotein involves cycling between the ER, intermediate compartment, and Golgi apparatus. J. Cell Biol. **126:** 41–52.
2. KOPITO, R.R. 1997. ER quality control: the cytoplasmic connection. Cell **88:** 427–430.
3. FURMAN, I., O. COOK, J. KASIR & H. RAHAMIMOFF. 1993. Cloning of two isoforms of the rat brain $Na^+–Ca^{2+}$ exchanger gene and their functional expression in HeLa cells. FEBS Lett. **319:** 105–109.
4. LOW, A.M., P.J. DARBY, C.Y. KWAN & E.E. DANIEL. 1993. Effects of thapsigargin and ryanodine on vascular contractility: cross-talk between sarcoplasmic reticulum and plasmalemma. Eur. J. Pharmacol. **230:** 53–62.
5. FUERST, T., E. NILES, F. STUDIER & B. MOSS. 1986. Eukaryotic transient expression system based on recombinant vaccinia virus that synthesizes bacteriophage T& RNA polymerase. Proc. Natl. Acad. Sci. USA **83:** 8122–8126.
6. HAZAMA, S. 1983. Effects of sulfhydryl reagents on sodium–calcium exchange system in canine cardiac sarcolemmal vesicles. Hokkaido Igaku Zasshi **58:** 354–362.

7. ORLICKY, J., M. RUSCAK, O. JUHASZ & J. ZACHAR. 1987. Effects of sulfhydryl reagents on Na^+–Ca^{2+} exchange in rat brain microsomal membranes. Gen. Physiol. Biophys. **6:** 155–162.
8. PIERCE, G.N., R. WARD & K.D. PHILIPSON. 1986. Role for sulfur-containing groups in the Na^+–Ca^{2+} exchange of cardiac sarcolemmal vesicles. J. Membr. Biol. **94:** 217–225.
9. DIPOLO, R. & L. BEAUGE. 1993. In squid axons the Ca in regulatory site of Na/Ca exchanger is drastically modified by sulfhydryl blocking agents. Evidence that intracellular Ca in regulatory and transport sites are different. Biochim. Biophys. Acta **1145:** 75–84.
10. REEVES, J.P., C.A. BAILEY & C.C. HALE. 1986. Redox modification of sodium–calcium exchange activity in cardiac sarcolemmal vesicles. J. Biol. Chem. **261:** 4948–4955.
11. NICOLL, D.A., M. OTTOLIA, L. LU, et al. 1999. A new topological model of the cardiac sarcolemmal Na^+–Ca^{2+} exchanger. J. Biol. Chem. **274:** 910–917.
12. DOERING, A.E., D.A. NICOLL, J. LU, et al. 1998. Topology of a functionally important region of the cardiac Na^+–Ca^{2+} exchanger. J. Biol. Chem. **273:** 778–783.
13. IWAMOTO, T., T. Y. NAKAMURA, Y. PAN, et al. 1999. Unique topology of the internal repeats in the cardiac Na^+–Ca^{2+} exchanger. FEBS Lett. **446:** 264–268.
14. REN, X., J. KASIR & H. RAHAMIMOFF. 2001. The transport activity of the Na^+–Ca^{2+} exchanger NCX1 expressed in hek 293 cells is sensitive to covalent modification of intracellular cysteine residues by sulfhydryl reagents. J. Biol. Chem. **276:** 9572–9579.
15. QIU, Z., J. CHEN, D.A. NICOLL & K.D. PHILIPSON. 2001. A disulfide bond is required for functional assembly of NCX1 from complementary fragments. Biochem. Biophys. Res. Commun. **287:** 825–828.
16. OTTOLIA, M., S. JOHN, Z. QIU & K.D. PHILIPSON. 2001. Split Na^+–Ca^{2+} exchangers. Implications for function and expression. J. Biol. Chem. **276:** 19603–19609.
17. MARKS, A.R. 1996. Cellular functions of immunophilins. Physiol. Rev. **76:** 631–649.
18. RAHFELD, J.U., A. SCHIERHORN, K. MANN & G. FISCHER. 1994. A novel peptidyl-prolyl cis/trans isomerase from Escherichia coli. FEBS Lett. **343:** 65–69.
19. SCHIENE, C. & G. FISCHER. 2000. Enzymes that catalyse the restructuring of proteins. Curr. Opin. Struct. Biol. **10:** 40–45.
20. KLEE, C.B., H. REN & X. WANG. 1998. Regulation of the calmodulin-stimulated protein phosphatase, calcineurin. J. Biol. Chem. **273:** 13367–13370.
21. LIU, J., J.D. FARMER, JR., W.S. LANE, et al. 1991. Calcineurin is a common target of cyclophilin-cyclosporin A and FKBP-FK506 complexes. Cell **66:** 807–815.
22. IVERY, M.T. 2000. Immunophilins: switched on protein binding domains? Med. Res. Rev. **20:** 452–484.
23. GOTHEL, S.F. & M.A. MARAHIEL. 1999. Peptidyl-prolyl cis-trans isomerases, a superfamily of ubiquitous folding catalysts. Cell. Mol. Life Sci. **55:** 423–436.
24. HARIHARAN, S., C.P. JOHNSON, B.A. BRESNAHAN, et al. 2000. Improved graft survival after renal transplantation in the United States, 1988 to 1996. N. Engl. J. Med. **342:** 605–612.
25. AVDONIN, P.V., F. COTTET-MAIRE, G.V. AFANASJEVA, et al. 1999. Cyclosporine A upregulates angiotensin II receptors and calcium responses in human vascular smooth muscle cells. Kidney Int. **55:** 2407–2414.
26. FRAPIER, J.M., C. CHOBY, M.E. MANGONI, et al. 2001. Cyclosporin A increases basal intracellular calcium and calcium responses to endothelin and vasopressin in human coronary myocytes. FEBS Lett. **493:** 57–62.
27. LOO, T.W. & D.M. CLARKE. 1997. Correction of defective protein kinesis of human P-glycoprotein mutants by substrates and modulators. J. Biol. Chem. **272:** 709–712.
28. CHEN, H., Y. KUBO, T. HOSHI & S.H. HEINEMANN. 1998. Cyclosporin A selectively reduces the functional expression of Kir2.1 potassium channels in Xenopus oocytes. FEBS Lett. **422:** 307–310.
29. TRAN, T.T., W. DAI & H.K. SARKAR. 2000. Cyclosporin A inhibits creatine uptake by altering surface expression of the creatine transporter. J. Biol. Chem. **275:** 35708–35714.
30. SHIRAISHI, S., H. YOKOO, H. KOBAYASHI, et al. 2000. Post-translational reduction of cell surface expression of insulin receptors by cyclosporin A, FK506 and rapamycin in bovine adrenal chromaffin cells. Neurosci. Lett. **293:** 211–215.

31. SHIRAISHI, S., T. YANAGITA, H. KOBAYASHI, et al. 2001. Up-regulation of cell surface sodium channels by cyclosporin A, FK506, and rapamycin in adrenal chromaffin cells. J. Pharmacol. Exp. Ther. **297:** 657–665.
32. KIMCHI-SARFATY, C., J. KASIR, S. AMBUDKAR & H. RAHAMIMOFF. 2002. Transport activity and surface expression of the Na^+–Ca^{2+} exchanger NCX1 is inhibited by the immunosuppressive agent cyclosporin A and the non-immunosupressive agent PSC833. J. Biol. Chem. **277:** 2505–2510.
33. COOK, O., W. LOW & H. RAHAMIMOFF. 1998. Membrane topology of the rat brain Sodium-Calcium exchanger. Biochim. Biophys. Acta **1371:** 40–52.
34. IWAMOTO, T., A. UEHARA, I. IMANAGA & M. SHIGEKAWA. 2000. The Na/Ca exchanger NCX1 has oppositely oriented reentrant loop domains that contain conserved aspartic acids whose mutation alters its apparent Ca^{2+} affinity. J. Biol. Chem. **275:** 38571–38580.
35. KASIR, J., X. REN, I. FURMAN & H. RAHAMIMOFF. 1999. Truncation of the C-terminal of the rat brain Na^+–Ca^{2+} exchanger RBE-1 (NCX1.4) impairs surface expression of the protein. J. Biol. Chem. **274:** 24873–24880.

Functional Regulation of Alternatively Spliced Na$^+$/Ca^{2+} Exchanger (NCX1) Isoforms

D.H. SCHULZE, S.K. POLUMURI, T. GILLE, AND A. RUKNUDIN

Department of Microbiology and Immunology, University of Maryland, Baltimore, Maryland 21201, USA

ABSTRACT: Alternative splicing of RNA transcripts is a general characteristic for NCX genes in mammals, mollusks, and arthropods. Among the family of three NCX genes in mammals, the NCX1 gene contains six exons, namely, A, B, C, D, E, and F, that make up the alternatively spliced region. Studies of the NCX1 gene transcripts suggested that 16 distinct gene products can be produced from the NCX1 gene. The exons A and B are mutually exclusive when expressed. Generally, exon A–containing transcripts are predominantly found in excitable cells like cardiomyoctes and neurons, whereas exon B–containing transcripts are mostly found in nonexcitable cells like astrocytes and kidney cells. Other alternatively spliced exons (C–F) appear to be cassette-type exons and are found in various combinations. Interestingly, exon D is present in all characterized transcripts. The alternatively spliced isoforms of NCX1 show tissue-specific expression patterns, suggesting functional adaptation to tissues. To investigate functional differences among alternatively spliced isoforms of NCX1, we expressed an exon A–containing transcript present in cardiac tissue (NCX1.1) and an exon B–containing transcript found in the kidney (NCX1.3) in *Xenopus* oocytes. We demonstrated that the Na$^+$/Ca^{2+} exchangers expressed by exon A– and exon B–containing transcripts display differences in activation by PKA and by [Ca^{2+}]$_i$. We also observed that these two isoforms show differences in voltage dependence. Suprisingly, the alternatively spliced isoforms of NCX1 display greater functional differences among themselves than the products of different gene loci, NCX1, NCX2, and NCX3.

KEYWORDS: sodium–calcium exchanger; alternative splicing; NCX1-3

INTRODUCTION

The cloning and characterization of the canine cardiac Na$^+$/Ca^{2+} exchanger by Nicoll *et al.*[1] provided the initial template against which subsequent NCX sequences would be compared. When our sequence of the human cardiac Na$^+$/Ca^{2+} exchanger cDNA was compared to the canine sequence,[2] we were quite surprised at the level of conservation between the two deduced amino acid sequences (greater than 98% sequence identity). Although the level of conservation was quite high, most of the

---

Address for correspondence: Dan H. Schulze, Department of Microbiology and Immunology, University of Maryland School of Medicine, 655 W. Baltimore Street, Baltimore, MD 21201. Voice: 410-706-5180; fax: 410-706-2129.

dschulze@umaryland.edu

Differences between Human and Dog Cardiac NCX1

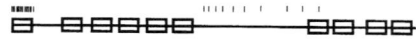

Differences between Human Cardiac and Rabbit Kidney NCX1

FIGURE 1. Identification of amino acid sequence differences within the NCX1.1 gene sequences. The boxes identify the positions of predicted hydrophobic/transmembrane regions. Ticks above the sequence identify the position of the differences in the comparisons. Break in the bottom line identifies the 36-amino-acid deletion present in the rabbit kidney sequence.

amino acid substitutions were present in signal sequence and isolated differences within the large intracellular loop of the Na^+/Ca^{2+} exchanger. (See FIG. 1, top panel.)

After the cloning of the human NCX1,[2,3] Reilly and Shugrue published the sequence of the rabbit kidney Na^+/Ca^{2+} exchanger.[4] Although the overall sequence identity for the rabbit sequence was similar to the dog heart (94% amino acid sequence identity), the position of differences and a deletion within the intracellular loop portion of the sequences suggested important differences between the rabbit kidney sequence and the other published NCX1 sequences. In addition to the differences seen in the signal sequence, there were 12 AA differences clustered within a short stretch of the intracellular loop, a pattern not seen in the human/canine comparison (FIG. 1). In addition to these differences, a 36-amino-acid deletion was present in the rabbit kidney sequence when compared to the canine or human cardiac exchangers. We hypothesized that either the sequence of the rabbit was surprisingly divergent or the source of the tissue (kidney) accounted for these differences.

cDNAs prepared from the rabbit cardiac and kidney tissues showed that the restricted region that contained these differences within the putative intracellular loop was variable in size and the sequence difference depended on the tissue source.[5] The comparison of the rabbit cardiac sequence with the dog or human cardiac sequences revealed 98% sequence identity with no deletions. This observation suggested that alternative splicing of the RNA was the cause for the differences noted in rabbit kidney sequence. We screened a genomic rabbit library with probes containing the regions of divergence in the predicted loop region of the exchanger NCX1 and were able to identify and describe the exons within the genomic clone.[6]

Analysis of genomic sequence led to the discovery of 6 exons (named A through F based on their position) that are used in various combinations to assemble a 76-amino-acid region in the loop of the Na^+/Ca^{2+} exchanger.[6] It was clear from the se-

TABLE 1. Sequence alignment of exons A and B

Rabbit exon A	KTISVKVIDDEEYEKNKTFFLEIGEPRLVEMSEKK
	| |:::: |:||||| :| | : ||: : |
Rabbit exon B	KIITIRIFDREEYEKECSFSLVLEEPKWIRRGMK

NOTE: Perpendicular line (|) between two sequences indicates sequence identity and a colon (:) indicates similar amino acid pairs.

quence that either exon A or B would be required; otherwise, the rest of the transcript would be out of frame. Another interesting aspect of exons A and B is that they contain similar amino acid sequences, suggesting that these two segments are the product of an ancestral sequence duplication (see TABLE 1). The remaining exons were considered to be cassette-type, in that any combination could potentially produce a functional transcript. We also have noted that all functional alternatively spliced transcripts contain exon D, resulting in 16 possible alternatively spliced isoforms many of which have been described.

SPLICING PATTERN DEPENDS ON THE TISSUE

Once alternative splicing had been confirmed by genomic analysis, it was important to identify the distribution of alternatively spliced transcripts in various tissues. Tissue distribution for the alternatively spliced isoforms was demonstrated independently by several laboratories using either sequence analysis of cDNAs,[6–8] hybridization approaches using exon-specific probes,[9] or by use of RT-PCR techniques.[9] An example of the latter technique uses oligonucleotide primers that flank the region that undergoes alternative splicing to amplify NCX1 cDNAs. When one of the primers is end-labeled with ^{32}P and the PCR products can be separated on sequencing gels, the size of the PCR products can be used to identify the exon combinations that are contained in the transcript, and the intensity of the band reflects the relative amount of the transcript in the sample. We demonstrated in cDNAs prepared from astrocytes that three B-containing isoforms (BDEF, BDF, and BD) are expressed at comparable levels. Interestingly, in cultured hippocampal neurons the two predominant isoforms are A exon–containing transcripts (ADF and AD).[9] In combining results for tissue distribution of NCX1 alternatively spliced isoforms, a picture emerges suggesting that cells that routinely experience large changes in membrane potential, such as cardiac cells and neurons, express A exon–containing transcripts; whereas those cells that do not show much variation in membrane potential express B exon–containing transcripts.[9]

REGULATION OF ALTERNATIVELY SPLICED ISOFORM FUNCTION

Because tissues may express multiple isoforms of NCX1, it is important to study function in a system in which a single alternatively spliced isoform can be studied. The alternatively spliced region is present on the large intracellular loop of the transporter. Using constructs that have the intracellular loop deleted or methods to disrupt the integrity of the intracellular loop, it has been demonstrated that the entire loop is involved in regulation of transport.[10,11] Disruption of the intracellular loop results in a deregulation of Na^+/Ca^{2+} exchanger function.

To determine the effect of the alternative splicing region, we compared transport function of the Na^+/Ca^{2+} exchanger cardiac isoform (containing exons ACDEF, NCX1.1) with the kidney/astrocyte isoform (containing exons BD, NCX1.3). These two isoforms were chosen because they provided the greatest possible difference in sequence between alternatively spliced isoforms. We expressed these two isoforms in *Xenopus* oocytes by injection of cRNA for either of the two isoforms, and they were studied 2 to 4 days after injection.[12]

REGULATION BY VOLTAGE

Although there is controversy concerning the stoichiometry of the Na^+/Ca^{2+} exchanger,[13,14] it has been clearly established that when the exchanger functions a positive charge is translocated across the membrane. Oocytes expressing either the cardiac (containing exons ACDEF, NCX1.1) or kidney (containing exons BD, NCX1.3) alternatively spliced isoforms were studied for their voltage dependence using two-electrode clamp techniques. The cell was held at a potential of −70 mV, and then a series of voltage pulses from −80 mV to +60 mV are given with 20-mV increments and the current was measured. In previous studies we have shown that there was a significant difference in the current/voltage relationship when the two isoforms were normalized and averaged over multiple oocytes.[12] The data were con-

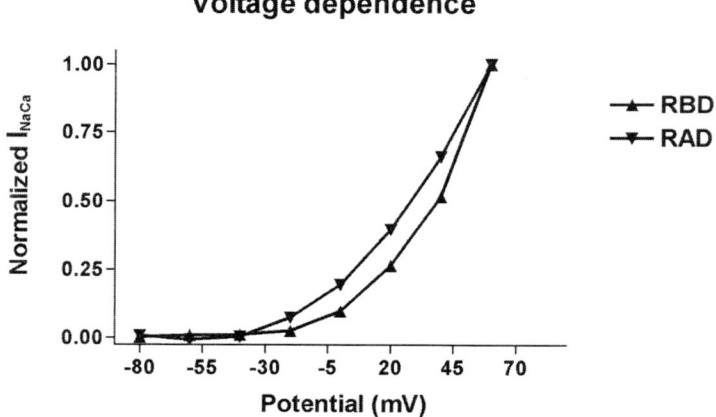

FIGURE 2. Na^+/Ca^{2+} exchanger current produced by alternatively spliced isoforms for NCX1.4 and NCX1.3. Averaged results for voltage dependency of normalized currents of the exon A–containing isoform (AD) and exon B–containing isoform (BD) Na^+/Ca^{2+} exchangers studied in *Xenopus* oocytes.

sistent with the kidney isoform doubling the current every 28.5 mV, whereas the cardiac isoform doubled every 19.5 mV. FIGURE 2 shows that the NCX1.4 isoform (containing exons AD) similarly functioned more efficiently in the lower voltage range than the NCX1.3 isoform. For example, nearly 25% of the maximum current could be produced below 0 mV for NCX1.4, whereas only about 10% of the maximum current for NCX1.3 was produced under similar conditions. These results demonstrate that two exon A–containing isoforms (ACDEF, NCX1.1 and AD, NCX1.4) function more efficiently over a wider voltage range than NCX1.3 (BD). When voltage dependence was studied using the excised giant patch from oocytes expressing the same isoforms, AD (NCX1.4) or BD (NCX1.3), no voltage-dependant differences were noted.[15] The ability to detect voltage-dependent differences between alternatively spliced isoforms of NCX1 depends on the methods of study.

REGULATION BY INTRACELLULAR Ca^{2+}

From early studies of the Na^+/Ca^{2+} exchanger using the excised giant patch, it was noted that besides the Na^+ gradient, $[Ca^{2+}]_i$ was required to initiate Na^+/Ca^{2+} exchange current. This current is seen in the presence of "regulatory" Ca^{2+}, and this Ca^{2+} was shown to bind to specific residues in the intracellular loop.[16] We wanted to study how intracellular Ca^{2+} would affect Na^+/Ca^{2+} exchanger regulation in oocytes expressing either alternatively spliced cardiac isoform ACDEF (NCX1.1) or kidney isoform BD (NCX1.3). We studied the Na^+/Ca^{2+} exchanger function using reverse mode, in which Ca^{2+} can enter *Xenopus* oocytes. The experimental protocol was a series of eight depolarizing pulses from a holding potential of –70 mV to +60 mV with current measurement after each depolarizing pulse. The intervals between pulses were changed between 1 sec and 16 sec. When intervals between depolarizing pulses were more than 5 sec, there was no significant increase in current generated from subsequent pulses. As the interval decreased to 1 sec, significant differences in current could be measured (FIG. 3A). In plotting the current after each pulse, we only observed increases for each isoform when there was a 1-sec interval between pulses. Analysis revealed that the NCX1.3 is more responsive to the effects of regulatory Ca^{2+} than the NCX1.1 isoform.[12]

What we think is happening using the repeated pulse protocol is after the initial depolarizing pulse, Ca^{2+} will enter the oocyte. If the second pulse is given at a short enough interval (1 sec), the Ca^{2+} does not have a chance to diffuse away from the membrane (and the Na^+/Ca^{2+} exchanger). Under these conditions the increase in current after subsequent depolarizing pulses observed was presumably due to the regulatory Ca^{2+} on the exchanger. If longer intervals are used between pulses (16 sec) there would be time for the local Ca^{2+} to diffuse from the membrane (and the exchanger) and no increase will be seen.

To test this hypothesis, we replaced the extracellular Ca^{2+} with Ba^{2+} that can be transported via the Na^+/Ca^{2+} exchanger instead of Ca^{2+}. When Ba^{2+} was used to replace Ca^{2+}, there was no significant difference between the two alternatively spliced isoforms of the Na^+/Ca^{2+} exchanger (FIG. 3B). This result confirmed our hypothesis that local Ca^{2+} was causing the difference in isoform function.

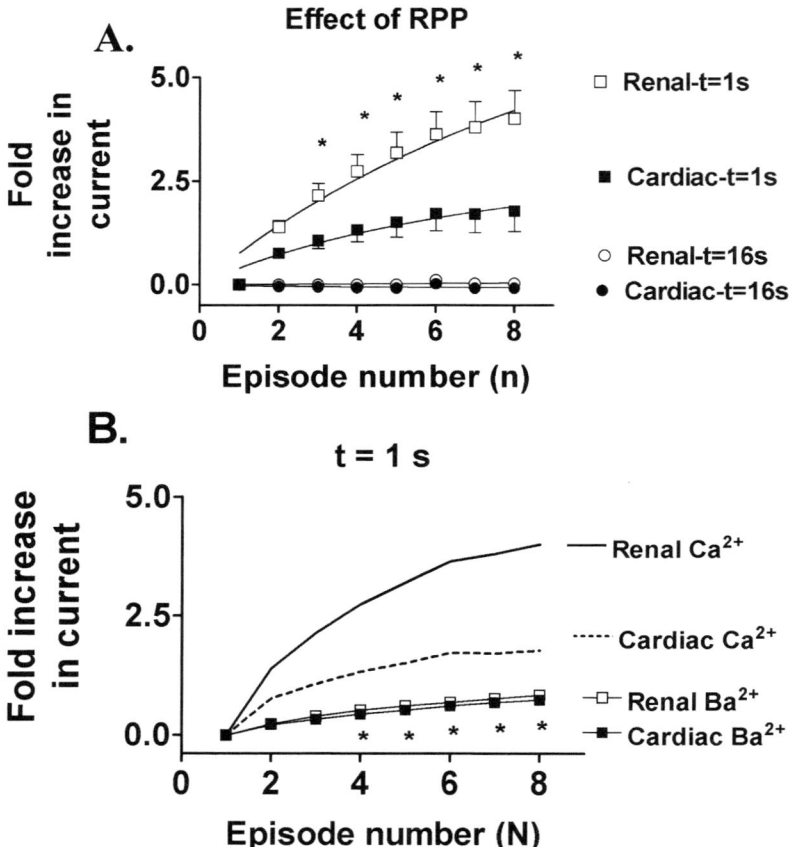

FIGURE 3. Repeated voltage pulse protocol affects cardiac and renal alternatively spliced isoforms of NCX1 differently. **(A)** Averaged results for the increase in current as a fractional increase from $n = 1-8$ pulses are shown for the cardiac (*filled symbols*, $n = 11$) and the renal (*open symbols*, $n = 10$) Na^+/Ca^{2+} exchanger. **(B)** The fold increase in current seen in (A) for 1-sec pulse in the presence of Ca^{2+} is presented as lines for both cardiac and renal NCX1 alternatively spliced isoforms. Increase in current for the two isoforms in the presence of extracellular Ba^{2+} is presented for $n = 1-8$ for the cardiac (*filled symbols*, $n = 8$) and for the renal (*open symbols*, $n = 8$) isoforms. Asterisks denote significant difference between the current in Ca^{2+} and the current in the presence of Ba^{2+} ($P > 0.001$). Modified from Ruknudin et al.[12]

REGULATION BY PHOSPHORYLATION

To determine whether different alternatively spliced isoforms of NCX1 differed in their response to PKA stimumation, oocytes expressing various isoforms were treated to activate the PKA pathway in oocytes. Expression of exon A–containing isoforms increase their expression between 30 and 50% after activation of the PKA pathway in *Xenopus* oocytes.[12] For example, oocytes expressing NCX1.1 or NCX1.4 increase their function after treatment with a cocktail that activates and stabilizes PKA activity in oocytes. The increase in function is not seen when activation with PKA cocktail and an inhibitor of PKA were administered at the same time. The increase was not observed upon activation with PKA when exon B–containing isoform (NCX1.3) expressing oocytes are studied.[12] These results suggest that there is a difference in response to activation of the PKA pathway depending on the alternatively spliced isoform expressed.

To determine whether the Na^+/Ca^{2+} exchanger is directly phosphorylated, the exchanger protein was immunoprecipitated from expressing oocytes. We show that the precipitated protein from oocytes expressing NCX1.1 isoform can be phosphorylated to a greater extent than the same amount of protein from NCX1.3 expression oocytes in the presence of ^{32}P and the active subunit of PKA *in vitro*.[12] Currently, site-directed mutations within exon A are being studied to identify the residues that are responsible for the PKA-specific increase in function.

FUNCTION OF THREE NCX GENES IS SIMILAR

NCX1 is part of a multigene family of three NCX genes (NCX1–3)[1,17,18] present in mammals and at least three NCKX genes (NCKX1–3) that have been described to date.[19] Commonalities between these proteins include sequence homology, the presence of an NH_2-terminal signal sequence, two sets of multiple transmembrane alpha helices near the ends of the protein, and a large intracellular loop. Contained in the transmembrane regions for each of the proteins are two shared sequence motifs (alpha repeats), which are thought to be important in the transport processes for these proteins.[20] The three NCX genes (NCX1–3) display moderate sequence identity. The three genes share between 68 and 75% sequence identity between any combination of the three genes. Interestingly, these three genes are each about 45% identical to the orthologous NCX sequence cloned from *Drosophila*.[21]

When the functioning of these three NCX genes were studied using multiple techniques, interestingly only minor differences in function were noted. NCX1, NCX2, and NCX3 function quite similarly in their apparent affinities for both Na^+ and Ca^{2+}, pH, inhibition by XIP peptide, inhibition by isothiourea, effects of regulatory Ca^{2+} and after intracellular application of chymotrypsin.[22] The only differences between the NCX1, NCX2, and NCX3 gene products are in more subtle aspects of regulation of function. ATP depletion inhibited NCX1 and NCX2 significantly but not NCX3. Activation of the PKA or PKC pathways increased both NCX1 and NCX 3 but not NCX2.[23] For NCX3, Ni^{2+} is 10-fold less effective in blocking Ca^{2+} influx when compared to NCX1 or NCX2. In addition, an isothiourea-family compound, KB-R7943 was threefold more inhibitory for NCX3 than the other members of the gene family.[24]

These functional similarities between the three independent NCX genes highlight an interesting paradox. The three NCX genes found on different chromosomes differ more significantly in sequence when compared to the spliced isoforms of NCX1, but they are more similar in their function than the alternatively spliced isoforms of the NCX1 gene.

WHY IS THERE FUNCTIONAL SIMILARITY AND SEQUENCE DISPARITY IN NCX 1, 2, AND 3?

The pattern of expression of the three NCX genes is different. NCX1 is expressed in all mammalian tissues studied. The amount of NCX1 can vary dramatically depending on the tissue. Heart ventricle contains the highest levels of transcripts and protein, while lymphocytes and lung expression is low in both message and protein. NCX2 and NCX3 are only expressed in brain and skeletal muscle.[17,18] In studies of brain development, each region of the brain contains transcripts for all the three NCX genes. In the neonatal brain tissues studied, NCX1 levels are the highest (about 70% of the total NCX transcripts), NCX3 next, and NCX2 being the lowest in abundance.[25] As the brain develops, we observed that NCX transcript levels for the three genes can differ depending on the region of the brain studied. For example, after birth the telecephalon showed marked increase in the amount of the NCX2 transcripts.[25] The increase in NCX2 expression during development of the rat brain was consistent with recent studies of embryonic and adult brain Na^+/Ca^{2+} exchanger expression.[26,27] Other parts of the brain such as the spinal cord appear unchanged with respect to NCX1-3 expression levels.[25]

Sequence analyses of the three NCX genes, when compared to the orthologous *Drosophila* sequence, are equally divergent. This observation is consistent with the possibility that the NCX gene family could have expanded through genomic duplication events that have been postulated during vertebrate radiation.[28] The genomic duplication during vertebrate divergence has been shown in a large number of cases to expand a single gene present in *Drosophila,* to gene families with as many as four distinct mammalian genes.[29]

With the development of the brain in vertebrates, dramatic changes in cell specialization, metabolic needs, and sheer growth, the three distinct NCX genes could have been developmentally regulated to balance Ca^{2+} requirements in brain tissue. In other tissues that were not "expanded" to the extent of the brain, we could possibly use alternative splicing mechanisms to more specifically "tailor" NCX function in those cells. For example, in heart tissue, while there have certainly been major changes in development during vertebrate evolution, the diversification of cell types in heart has not changed to same extent as in brain. Cardiac cells could then choose to alter the splice preference to use exon A–containing alternatively spliced transcripts, which functions over a wide range of depolarizing voltages. Brain cells could use different genes to regulate the level of Na^+/Ca^{2+} exchanger activity depending on the cell type. Future experiments elucidating the mechanisms of NCX gene regulation will help to understand this process more fully.

REFERENCES

1. NICOLL, D.A., S. LONGONI & K.D. PHILIPSON. 1990. Molecular cloning and functional expression of the cardiac sarcolemmal Na–Ca exchanger. Science **250:** 562–565.
2. KOFUJI, P., R.W. HADLEY, R.S. KIEVAL, *et al.* 1992. Expression of the Na^+/Ca^{2+} exchanger in diverse tissues: a study using the cloned human cardiac Na^+/Ca^{2+} exchanger. Am. J. Physiol. (Cell Physiol.) **263:** C1241–1249.
3. KOMURO, I., K.E. WENNINGER, K.D. PHILIPSON & S. IZUMO. 1992. Molecular cloning and characterization of the human cardiac Na/Ca exchanger cDNA. Proc. Natl. Acad. Sci. USA **89:** 4769–4773.

4. REILLY, R. & C.A. SHUGRUE. 1992. cDNA cloning of a renal Na^+–Ca^{2+} exchanger. Am. J. Physiol. **262:** F1105–F1109.
5. KOFUJI, P., W.J. LEDERER & D.H. SCHULZE. 1993. Na^+/Ca^{2+} exchanger isoforms expressed in heart and kidney. Am. J. Physiol. **265:** F598–F603
6. KOFUJI, P., W.J. LEDERER & D.H. SCHULZE. 1994. Mutually exclusive and cassette exons underlie alternatively spliced isoforms of the Na^+/Ca^{2+} exchanger. J. Biol. Chem. **269:** 5145–5149.
7. QUEDNAU, B.D., D.A. NICOLL & K.D. PHILIPSON. 1997. Tissue specificity and alternative splicing of Na^+–Ca^{2+} exchanger isoforms NCX1, NCX2 and NCX3 in rat. Am. J. Physiol. (Cell Physiol.) **41:** C1250–C1261.
8. LEE, S.L., A.S. YU & J. LYTTON. 1994. Tissue-specific expression of Na^+–Ca^{2+} exchanger isoforms. J. Biol. Chem. **269:** 14849–14852.
9. HE, S., A. RUKNUDIN, L.L. BAMBRICK, et al. 1998. Isoform-specific regulation of the Na^+/Ca^{2+} exchanger in rat astrocytes and neurons by PKA. J. Neurosci. **18:** 4833–4941.
10. NICOLL, D.A. & K.D. PHILIPSON. 1991. Molecular studies of the cardiac sarcolemmal sodium calcium exchanger. Ann. N.Y. Acad. Sci. **639:** 181–188.
11. MATSUOKA, S., D.A. NICOLL, R.F. REILLY, et al. 1993. Initial localization of regulatory regions of the cardiac sarcolemmal Na^+–Ca^{2+} exchanger. Proc. Natl. Acad. Sci. USA **90:** 3870–3874.
12. RUKNUDIN, A., S. HE, W.J. LEDERER & D.H. SCHULZE. 2000. Functional differences between the cardiac and renal isoforms of the rat Na^+/Ca^{2+} NCX1 expressed in *Xenopus* oocytes. J. Physiol. **529:** 599–610.
13. FUJIOKA, Y., M. KOMEDA & S. MATSUOKA. 2000. Stoichiometry of Na^+–Ca^{2+} exchange in inside-out patches from guinea-pig ventricular myocytes. J. Physiol. **523:** 339–351.
14. DONG, H. & J. LYTTON. 2002. Stoichiometry of the cloned rat cardiac Na^+/Ca^{2+} exchanger, NCX1.1. Ann. N.Y. Acad. Sci. This volume.
15. DYCK, C., A. OMELCHENKO, C.L. ELIAS, et al. 1999. Ionic regulation properties of brain and kidney splice variants of the NCX1 Na^+–Ca^{2+} exchanger. J. Gen. Physiol. **114:** 701–711.
16. LEVITSKY, D.O., D.A. NICOLL & K.D. PHILIPSON. 1994. Identification of the high affinity Ca^{2+} binding domain of the cardiac Na^+–Ca^{2+} exchanger. J. Biol. Chem. **269:** 22847–22852.
17. LI, Z., S. MATSUOKA, L.V. HRYSHKO, et al. 1994. Cloning of the NCX2 isoform of the plasma membrane Na^+–Ca^{2+} exchanger. J. Biol. Chem. **269:** 17434–17439.
18. NICOLL, D.A., B.D. QUEDNAU, Z. QUI, et al. 1996. Cloning of a third mammalian Na^+–Ca^{2+} exchanger, NCX3. J. Biol. Chem. **271:** 24914–24921.
19. KRAEV, A., B.D. QUEDMAU, S. LEACH, et al. 2001. Molecular cloning of a third member of the potassium-dependent sodium–calcium exchanger gene family, *NCKX3*. J. Biol. Chem. **276:** 23161–23172.
20. PHILIPSON, K.D. & D.A. NICOLL. 2000. Sodium–calcium exchange: a molecular perspective. Ann. Rev. Physiol. **62:** 111–133.
21. RUKNUDIN, A., C. VALDIVIA, P. KOFUJI, et al. 1997. Na^+/Ca^{2+} exchanger in *Drosophila*: cloning, expression and transport differences. Am. J. Physiol. (Cell Physiol.) **273:** C257–C265.
22. LINCK, B., Z. QUI, Z. HE, et al. 1998. Functional comparison of the three isoforms of the Na^+/Ca^{2+} exchanger (NCX1, NCX2 and NCX3). Am. J. Physiol. (Cell Physiol) **43:** C415–C423.
23. IWAMOTO, T., Y. PAN, T.Y. NAKAMURA, et al. 1998. Protein kinase C dependent regulation of Na^+/Ca^{2+} exchanger isoforms and NCX3 does not require their direct phosphorylation. Biochemistry **31:** 117230–17238.
24. IWAMOTO, T. & M. SHIGEKAWA. 1998. Differential inhibition of Na^+/Ca^{2+} exchanger isoforms by divalent cations and isothiourea derivative. Am. J. Physiol. (Cell Physiol.) **275:** C423–C430.
25. POLUMURI, S.K., A. RUKNUDIN, M.M. MCCARTHY, et al. 2002. Sodium–calcium exchanger NCX1, NCX2, and NCX3 transcripts in developing rat brain. Ann. N.Y. Acad. Sci. This volume.

26. YU, L. & R.A. COLVIN. 1997. Regional differences in expression of transcripts for Na^+/Ca^{2+} exchanger isoforms in rat brain. Brain Res. Mol. Brain Res. **15:** 285–292.
27. SAKAUE, M., H. NAKAMURA, I. KANEKO, *et al.* 2000. Na^+/Ca^{2+} exhanger isoforms in rat neuronal preparations: different changes in their expression during postnatal development. Brain Res. **881:** 212–216.
28. KASHARA, M., J. NAKAYA, Y. SATTA & N. TAKAHATA. 1997. Chromosomal duplication and the emergence of the adaptive immune system. Trends Genet. **13:** 90–92.
29. SPRING, J. 1997. Vertebrate evolution by interspecific hybridisation: are we polyploid? FEBS Lett. **400:** 2–8.

The Cardiac Sodium–Calcium Exchanger Associates with Caveolin-3

JULIE BOSSUYT,[a,b] BONNIE E. TAYLOR,[a] MARILYN JAMES-KRACKE,[c] AND CALVIN C. HALE[a,b]

[a]*The Dalton Cardiovascular Research Center,* [b]*Department of Biomedical Sciences,* [c]*Department of Pharmacology, University of Missouri, Columbia, Missouri 65211, USA*

ABSTRACT: The cardiac Na/Ca exchanger's (NCX1) role in calcium homeostasis during myocardial contractility makes it a possible target of signaling factors regulating inotropy. Caveolae, structured invaginations of the plasmalemma, are known to concentrate a wide variety of signaling factors. The predominant coat proteins of caveolae, caveolins, dock to and regulate the activity of these signaling factors and other proteins through interaction with their scaffolding domain. In this study we investigated the interaction of NCX1 with caveolin proteins. Western blots of bovine cardiac sarcolemmal vesicles revealed the presence of caveolin-1, -2, and -3. Immunoprecipitation of detergent-solubilized vesicle proteins with either NCX1 or caveolin-3 antibodies indicated that NCX1 coprecipitates with caveolin-3, but not with caveolin-1 and -2. Functional disruption of caveolae, by β-cyclodextrin treatment of vesicles, diminished coprecipitation of caveolin-3 and NCX1 activity. NCX1 has five potential caveolin-binding motifs, two of which are in the transporter's exchange inhibitory peptide (XIP) domain. The presence of 50 mM XIP peptide enhanced coprecipitation of caveolin-3 with NCX1 independent of calcium concentration. We conclude that NCX1 associates specifically with caveolin-3. Partitioning of NCX1 in caveolae has implications for temporal and spatial regulation of excitation-contraction and -relaxation coupling in cardiac myocytes.

KEYWORDS: caveolae; sodium–calcium exchange; exchange inhibitory peptide; caveolin-3

INTRODUCTION

Modulation of Na/Ca exchange (NCX1) by signaling factors regulating the inotropic activity of the heart may be important because of NCX1's role in myocardial Ca^{2+} homeostasis. At present it is not known whether NCX1 is located in caveolae, which are a subclass of lipid rafts and act as a platform where many signaling factors can congregate.[1–3] Although caveolae composition is cell specific, they characteristically contain caveolin protein isoforms (caveolin-1, -2, and -3). Caveolins dock to

Address for correspondence: Calvin C. Hale, Dalton Cardiovascular Research Center, University of Missouri, Columbia, 134 Research Park Drive, Columbia, MO 65211. Voice: 573-882-3244; fax: 573-884-4232.
chale@Missouri.edu

and regulate the activity of many signaling pathway components through their "scaffolding" domain. It appeared worthwhile to test whether NCX1 was associated with caveolae or, more specifically, caveolin isoforms in the cardiac myocytes.

Recent studies have reported an increase or upregulation of NCX1 in heart failure, although there are still few reports on NCX1 function in heart failure.[4] Of interest for the present study was that the muscle-specific caveolin-3 isoform has also been shown to increase in heart failure.[5] This increase in caveolin-3 and sarcolemmal caveolae during heart failure has been linked to enhanced NO signaling. In endothelial cells, nitric oxide synthase (NOS) copurified with caveolin-1 in sucrose density gradients.[6] That study also reported that endothelial NCX tended to be concentrated in fractions enriched in caveolin-1 and NOS, although the association seemed less specific as NCX tended to be present in all fractions. So far there have been no reports of such an association in cardiac myocytes.

MATERIALS AND METHODS

Bovine cardiac sarcolemmal (bsl) vesicles were prepared, and Na/Ca exchange activity was assayed as previously described.[7] For the evaluation of β-cyclodextrin treatment on activity, vesicles were exposed for 1 hour on ice before being assayed. The sucrose density gradient fractionation[6] for caveolin proteins was performed as previously described on bovine myocardial tissue. The adaptation in the procedure consisted of a centrifugation at $2,500 \times g$ for 15 min to remove large debris.

Immunoprecipitation experiments were performed with antibodies immobilized to CNBr 4B beads (coupled according to manufacturer's instructions). Antibody-bound beads were then incubated with detergent-extracted sarcolemmal proteins. The beads were extensively washed to remove nonspecifically associated proteins. The final washed bead pellet was then subjected to Western blotting. Images of the Western blots were captured and analyzed using the NIH Image program.

RESULTS

Caveolin-Binding Motifs in NCX1

Two related caveolin-binding motifs are present on caveolin-associated proteins.[2] The motifs consist of either φ x φ x x x x x φ or φ x x x x x φ x x φ, where x is any amino acid and φ is aromatic amino acid Trp, Phe, or Tyr. Five such potential caveolin binding motifs can be located on NCX1 at amino acids 171–179, 223–231, 224–231, 620–627, and 875–883 (FIG. 1). Three of these motifs are therefore located at regions of special interest on the exchanger. Two (the motifs at 223–231 and 224–231) are nestled within the endogenous XIP domain, while the motif at 620–627 is located at an alternative splicing exon found in NCX1, but not other exchanger isoforms.[8] The motifs at amino acids 171–179 and 875–883, modeled to be located near the extracellular side of the protein, seem less likely candidates for actual binding sites since the caveolin scaffolding domain is located at or near the cytoplasmic side of the membrane.

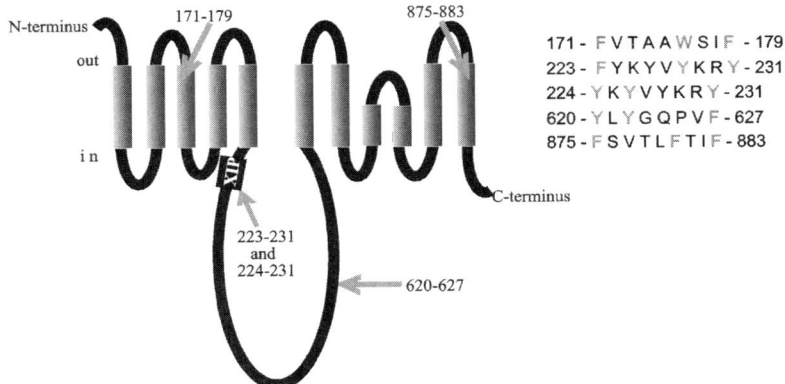

FIGURE 1. Predicted location of NCX1 caveolin-binding motifs. The five NCX1 amino acid segments containing caveolin-binding motifs are listed, and their relative positions on NCX1 are shown. The gray-colored amino acids indicate the N position of the binding motif (see text for additional details). Two (amino acids 171–179 and 875–883) are modeled to be in transmembrane helices near the external face of the sarcolemmal membrane. Amino acids 223–232, 224–231, and 620–627 are all modeled to exist on the protein's large cytoplasmic loop with the first two segments actually overlapping within the endogenous XIP domain.

Presence of Caveolin Protein in Bovine Sarcolemmal Vesicle Proteins

The presence of caveolin proteins in the "standard" bovine sarcolemmal (bsl) vesicle preparation was detected by Western blot analysis with isoform-specific antibodies (FIG. 2a). All three caveolin protein isoforms (22, 20, and 18 kDa) were present in bsl vesicles, although caveolin-2 was barely detectable. In support of this result, all three isoforms were also detected in canine and porcine sarcolemmal vesicle preparations (not shown). It has been reported that differentiated muscle cells express mainly caveolin-3, so much if not all caveolin-1 and -2 may be from endothelial cells present in the myocardial tissue source.[1]

Cofractionation of Caveolin and NCX1

Myocardial tissue was fractionated using a slightly adapted multistep gradient method previously reported to fraction caveolar proteins.[6] FIGURE 3 shows a representative result of the sucrose gradient fractionation of total cellular (soluble and membrane) protein from myocardial tissue. It is clear that the light membrane fractions are separated not only from the denser membranes but also from the majority of the total membrane and cytosolic proteins. The peak of NCX1 and caveolin-1 and -3 protein cofractionated in fractions 4 and 5. This suggests that NCX1 and caveolin proteins are located in the same light-membrane gradient fractions. Similar results were obtained for gradients containing only total cellular membrane proteins.

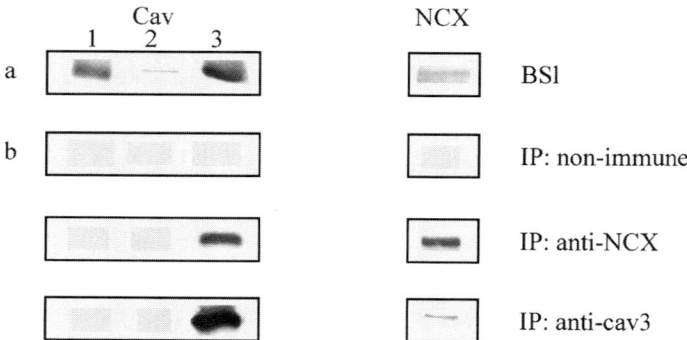

FIGURE 2. Western blot analysis of caveolin and NCX1 proteins. (**a**) Blots of sarcolemmal vesicle proteins separated in 10% polyacrylamide gels were probed with antibodies against caveolin-1, caveolin-2, and caveolin-3 or NXC1. All three caveolin isoforms were present in the vesicle preparations; however, the observed level of caveolin-2 was low. (**b**) Blots of immunoprecipitated proteins. Detergent-solubilized BSl proteins were incubated with Sepharose bead–coupled nonimmune antibody, anti-NCX1 antibody, or anti-caveolin-3 antibody. Blots were then probed with anti-caveolin-1, anti-caveolin-2, and anti-caveolin-3 antibodies or anti-NXC1 antibody as indicated. When immunoprecipitating with caveolin-3 antibody, sixfold more sample was loaded on gels to visualize NCX1 compared to the sample volume used to visualize caveolin-3.

Coprecipitation of Caveolin-3 and NCX1

Co-immunoprecipitation experiments were performed to determine whether any of the caveolin protein isoforms associate with NCX1. Monoclonal antibodies against NCX1, caveolin-3, or nonimmune IgG were covalently coupled to Sepharose beads and used as immunoprecipitant. The results of the immunoprecipitation of detergent-extracted sarcolemmal proteins are shown in FIGURE 2b. Following nonimmune IgG immunoprecipitation, neither NCX1 nor caveolin isoforms could be detected. Both caveolin-3 and NCX1 were detected on blots for NCX1 immunoprecipitation. Caveolin-1 and -2, while present in the sarcolemmal vesicles, did not co-immunoprecipitate with NCX1. NCX1 and caveolin-3 were also detected following precipitation with immobilized anti-caveolin-3. The results shown in FIGURE 2b suggest that NCX1 specifically associates with caveolin-3, which is the muscle-specific isoform.

Cholesterol Depletion of Caveolae and Coprecipitation

Function and structural integrity of caveolae are dependent on a critical level of cholesterol in the plasma membrane and the caveolae themselves.[1] β-Cylcodextrin, by binding cholesterol, has been shown to disrupt function and structure of caveolae, as caveolin requires cholesterol to remain in the membrane. The results in TABLE 1 indicate that the amount of caveolin-3 coprecipitated diminished upon β-cyclodextrin treatment (unlike the amount of NCX1). These data suggest that functional integrity of caveolae may be required for NCX1/caveolin-3 interaction. In oth-

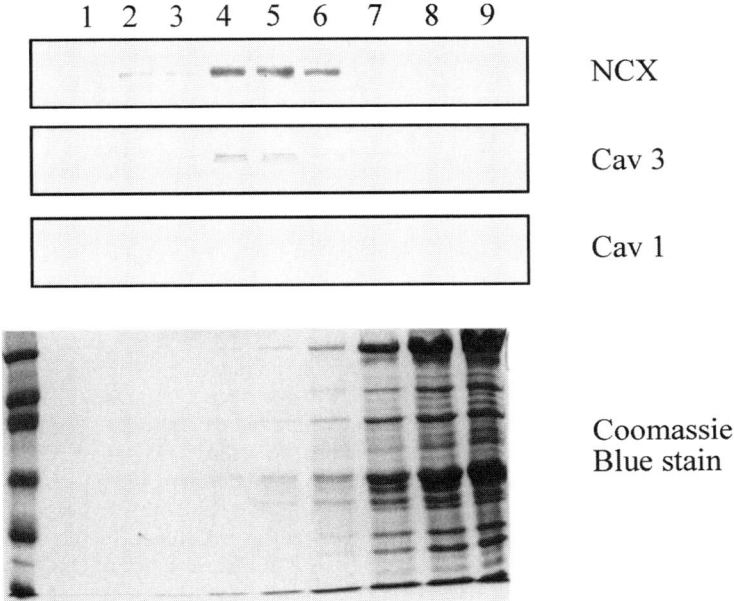

FIGURE 3. Sucrose gradient fractionation of myocardial membranes. A crude membrane preparation of bovine ventricular tissue was fractionated on a multistep sucrose gradient. *Upper*: Western blot analysis of gradient fractions. *Lower*: Coomassie blue staining pattern of sucrose gradient fractions. Caveolin-1, which appeared in low abundance in crude membrane fractions and was difficult to detect, peaked in fractions 5 and 6.

TABLE 1. The effect of β-cyclodextrin on coprecipitation of caveolin-3 with NCX1 and NCX1 activity

β-Cyclodextrin	Percent control IP	Percent control activity	Specific activity (nmol/mg/sec)
0	100	100	0.080 ± 0.013
0.02	nd	83.1	0.065 ± 0.013
0.2	58	62.6	0.049 ± 0.020
2.0	46	59.6	0.044 ± 0.016

NOTE: Cardiac sarcolemmal vesicles were treated with β-cyclodextrin as indicated for 30 min. Treated vesicles were either solubilized and subjected to immunoprecipitation (IP) with immobilized anti-NCX1 antibody or assayed for exchange activity ($n = 5$). Coprecipitation of caveolin-3 with NCX1 was quantified densitometrically. Specific activity is presented showing standard error.

er experiments, a dose-dependent decrease of NCX1 transport activity was observed following β-cyclodextrin treatment, which paralleled the decrease in coprecipitated caveolin-3. Passive $^{45}Ca^{2+}$ equilibration in β-cyclodextrin-treated vesicles was not different from control (untreated), suggesting that the treatment did not cause the vesicles to become leaky. In these experiments both coprecipitation of caveolin-3 and NCX1 activity were diminished, but not abolished, even at the highest concentration of β-cyclodextrin (2%).

Effect of XIP on Co-immunoprecipitation

As two potential caveolin-binding motifs were identified as being nestled within the endogenous XIP region, we tested the effect of exogenous XIP peptide on the NCX1/caveolin-3 interaction. FIGURE 4a indicates that the presence of 50 µM XIP peptide greatly enhanced the amount of caveolin-3 coprecipitated with NCX1. Having expected the opposite result, we confirmed the presence of XIP peptide in the immunoprecipitate by Coomassie blue staining of SDS-PAGE (data not shown). Densitometric analysis was used to quantify the XIP-mediated enhanced caveolin-3 coprecipitation. In this analysis ($n = 7$), the amount of caveolin-3 coprecipitated was normalized to the immunoprecipitated NCX1 observed. The results indicated a slightly greater than twofold increase in coprecipitated caveolin-3.

An earlier report suggested a partial overlap between the putative XIP-binding site and regulatory calcium-binding site.[9] Therefore, calcium could potentially modulate the effect of XIP on co-immunoprecipitation. We tested this hypothesis in another set of experiments (FIG. 4b). We observed no difference in the amount of caveolin-3 coprecipitated in the presence of 2 mM EGTA (absence of Ca^{2+}) or 2 mM

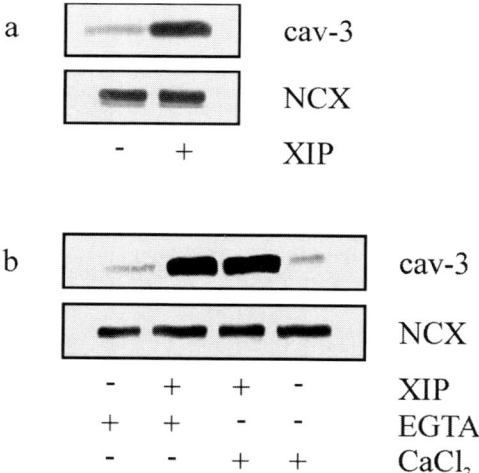

FIGURE 4. The effect of exogenous XIP peptide on caveolin-3 coprecipitation. Immobilized anti-NCX1 antibody was used to immunoprecipitate NCX1 and coprecipitate caveolin-3 as in FIGURE 2. Where indicated, 50 µM XIP peptide, 2 mM EGTA, and/or 2 mM $CaCl_2$ were incubated with solubilized BSl proteins. **(a)** *Upper*: precipitate was probed with anti-caveolin-3 antibody. Densitometric analysis indicated that, when normalized to NCX1 immunoprecipitate, XIP enhanced caveolin-3 coprecipitation slightly greater than twofold ($n = 7$). *Lower*: precipitate was probed with anti-NCX1 antibody. **(b)** The effect of Ca^{2+} and EGTA on coprecipitation of caveolin-3 with or without exogenous XIP peptide.

$CaCl_2$. Calcium therefore appears to have no effect on the XIP-mediated increase in coprecipitated caveolin-3.

DISCUSSION

In this report we provide evidence for the association of NCX1 with the muscle-specific caveolin-3 isoform. Whereas all three known caveolin isoforms were present in sarcolemmal vesicle preparations, only caveolin-3 coprecipitated with NCX1 (FIG. 2b). The ability of caveolin-3 to associate with a wide variety of proteins besides NCX1 is probably reflected in the low level of NCX1 coprecipitated with caveolin-3. These observations strongly suggest the presence and partitioning of NCX1 in myocyte caveolae. Further support for this idea comes from the observations that NCX1 and caveolin-3 cofractionate on sucrose gradients (FIG. 2) and that cholesterol depletion of caveolae with β-cyclodextrin diminishes coprecipitation of caveolin-3 with NCX1 (TABLE 1). It is therefore possible that the recently reported increased expression of caveolin-3 and NCX1 protein during heart failure, may not be casually unrelated events.[4,5]

Of the five potential caveolin-binding motifs, it is possible that the endogenous XIP domain is the functionally relevant interacting region with caveolin-3. This suggestion is based on the following: (1) the XIP domain contains two nestled motifs; (2) the motifs at amino acids 171–179 and 875–883 are less likely candidates, since their positions are modeled near the external side of the membrane in contrast to the caveolin scaffolding domain; and (3) XIP peptide significantly enhances coprecipitation of caveolin-3 with NCX1, suggesting a relationship exists. We are currently addressing the question as to which caveolin motif(s) is crucial for caveolin-3 association by mutagenesis experiments, and at this time the functional relevance of the XIP region or AA 620–627 cannot be ruled in or out.

An earlier model of XIP's regulatory function predicted that interaction of the domain with either another portion of the exchanger or with anionic phospholipids would result in inhibition or activation of transport, respectively.[10] Given these predictions, we suggest that endogenous XIP domain association with caveolin-3 might result in an active transporter. The reasoning here is that the XIP domain, when bound to caveolin-3, would be unable to associate with the binding site on the exchanger, which would result in inhibition. Other constituents of caveolae may also contribute to the activity of the transporter. Our suggestion that the transporter in caveolae is active is supported by our observation that β-cyclodextrin both diminishes coprecipitation and transport activity and is consistent with recent reports indicating upregulation of caveolin-3 and NCX1 protein during heart failure.[4,5]

The observation that XIP enhances the coprecipitation of caveolin-3 with NCX1 could be explained by the creation of an extra interaction site for the caveolin-3 protein: the XIP peptide bound to the transporter may provide an additional interaction site for caveolin-3 (1), or a conformational change, induced by the addition of XIP, could result in an additional interaction site on the exchanger itself (2). The twofold increase in caveolin-3 co-immunoprecipitation (as revealed by densitometry), suggests a stochiometric doubling of the amount of caveolin-3 bound to each transporter.

A previous report suggested that the putative XIP-binding site partially overlaps with a regulatory calcium-binding site.[9] In this study no effect of calcium on the association of NCX1 with caveolin-3 was observed. This finding presents a possible contradiction to the earlier study if the XIP domain is indeed the relevant interacting region. In that case the finding could be interpreted as follows: the XIP peptide and regulatory Ca^{2+} do not bind to the same site or both bind to a portion of this region without overlap.

Partitioning of NCX1 in caveolae has other potential ramifications in terms of exchanger function beyond Ca^{2+} homeostasis in myocardial contractility. In cardiac myocytes it remains to be shown whether NCX1 has additional functions beyond bulk cytosolic Ca^{2+} homeostasis related to contractility, but the possibility is intriguing.

REFERENCES

1. SMART, E.J., et al. 1999. Caveolins, liquid-ordered domains, and signal transduction. Mol. Cell. Biol. **19**: 7289–7304.
2. OKAMOTO, T., et al. 1998. Caveolins, a family of scaffolding proteins for organizing "preassembled signaling complexes" at the plasma membrane. J. Biol. Chem. **273**: 5419–5422.
3. GALBIATI, F., B. RAZANI & M.P. LISANTI. 2001. Emerging themes in lipid rafts and caveolae. Cell **106**(4): 403–411.
4. HOBAI, I.A. & B. O'ROUKE. 2000. Enhanced Ca^{2+}-activated Na^+–Ca^{2+} exchange activity in canine pacing-induced heart failure. Circ. Res. **87**: 690–698.
5. HARE, J.M., et al. 2000. Contribution of caveolin protein abundance to augmented nitric oxide signaling in conscious dogs with pacing-induced heart failure. Circ. Res. **86**: 1085–1092.
6. TEUBL, M., et al. 1999. Na^+/Ca^{2+} exchange facilitates Ca^{2+}-dependent activation of endothelial nitric-oxide synthase. J. Biol. Chem. **274**: 29529–29535.
7. KLEIBOEKER, S.B., M.A. MILANICK & C.C. HALE. 1992. Interactions of the exchange inhibitory peptide with Na–Ca exchange in bovine cardiac sarcolemmal vesicles and ferret red cells. J. Biol. Chem. **267**: 17836–17841.
8. KOFUJI, P., W.J. LEDERER & D.H. SCHULZE. 1994. Mutually exclusive and cassette exons underlie alternatively spliced isoforms of the Na/Ca excahanger. J. Biol. Chem. **269**: 14894–14852.
9. HALE, C.C., et al. 1997. Localization of an exchange inhibitory peptide (XIP) binding site on the cardiac sodium–calcium exchanger. Biochem. Biophys. Res. Commun. **236**: 113–117.
10. SHANNON, T.R., C.C. HALE & M.A. MILANICK. 1994. Interaction of cardiac Na–Ca exchanger and exchange inhibitory peptide with membrane phospholipids. Am. J. Physiol. **266**: C1350–C1356.

Kinetic Model of Transport and Regulation of the Cardiac, Brain, and Kidney Isoforms of NCX1

ALEXANDER OMELCHENKO, CHADWICK L. ELIAS, ROBERT SCATLIFF, PLATON CHOPTIANY, MARK HNATOWICH, AND LARRY V. HRYSHKO

Institute of Cardiovascular Sciences, St. Boniface General Hospital Research Centre, University of Manitoba, Winnipeg, Canada, R2H 2A6

Na^+/Ca^{2+} EXCHANGE, ALTERNATIVELY SPLICED ISOFORMS, IONIC REGULATION

We have previously demonstrated pronounced differences in the ionic regulatory profiles of alternatively spliced isoforms of the NCX1 subtype of Na^+–Ca^{2+} exchanger expressed in *Xenopus* oocytes.[1,2] Using cardiac (i.e., NCX1.1), brain (i.e., NCX1.4), and kidney (i.e., NCX1.3) splice variants, and the giant excised patch clamp technique,[1–3] we found that peak outward exchange current for each isoform progressively rises with increasing Na^+_i (10–100 mM) and each exhibits sodium-dependent (i.e., I_1) inactivation,[3,4] but that only NCX1.1- and NCX1.4-mediated steady-state outward currents correspondingly rise with increasing Na^+_i. In the case of NCX1.3, we showed that steady-state outward currents were mainly insensitive to elevated Na^+_i, apparently being offset by extensive I_1 inactivation, as indicated by the observation that this behavior could be abolished by treatment of the patch with α-chymotrypsin, a procedure known to deregulate Na^+–Ca^{2+} exchangers.[3] With respect to Ca^{2+}-dependent (i.e., I_2) regulation,[5] outward mode Na^+–Ca^{2+} exchange activity was stimulated by application of increasing Ca^{2+}_i for all three isoforms. However, high (i.e., 10 µM) regulatory Ca^{2+}_i was effective in alleviating I_1 inactivation only in the cases of NCX1.1 and NCX1.4. With NCX1.3, the extent of I_1 was not appreciably relieved by this maneuver.

In other experiments aimed at providing greater insight into the differences between NCX1 splice variants, we examined the influence of regulatory Ca^{2+}_i added either simultaneously with, or 32 sec before, application of 100 mM Na^+_i to activate outward current. With NCX1.3, a marked *inhibition* (up to ≈ twofold) of peak current was observed when regulatory Ca^{2+}_i was applied before current activation, as compared with simultaneous application of Na^+_i and Ca^{2+}_i (FIG. 1A). In contrast, NCX1.1-mediated peak currents were less if Na^+_i and Ca^{2+}_i were added simultaneously, as compared with prior application of Ca^{2+}_i; whereas NCX1.4-mediated

---

Address for correspondence: Larry V. Hryshko, St. Boniface General Hospital Research Centre, University of Manitoba, Cardiovascular Sciences, 351 Tache Avenue, R3032, Winnipeg R2H 2A6, Canada. Voice: 204-235-3662; fax: 204-233-6723.
lhryshko@sbrc.ca

Ann. N.Y. Acad. Sci. 976: 205–208 (2002). © 2002 New York Academy of Sciences.

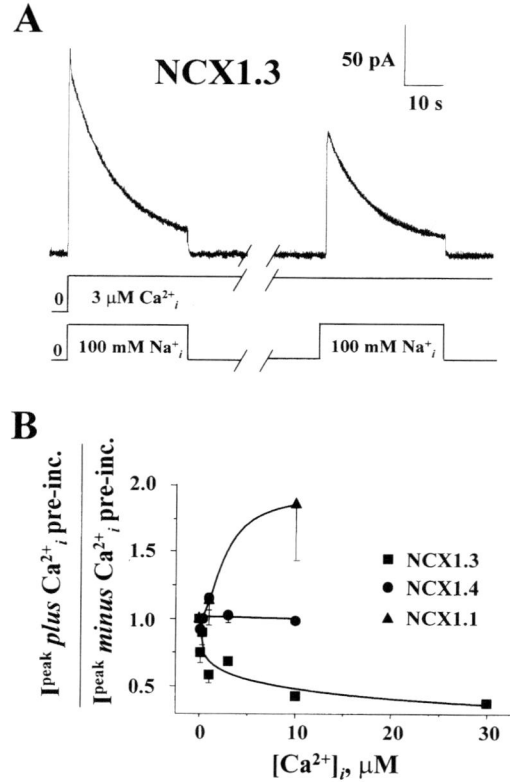

FIGURE 1. (**A**) shows representative NCX1.3-mediated outward current traces generated by rapid application of 100 mM Na^+_i to the cytoplasmic side of a giant patch without (*left trace*) and with (*right*) regulatory Ca^{2+}_i preincubation for 32 s. (**B**) illustrates the regulatory Ca^{2+}_i-dependence for "offsetting" peak outward current in response to Ca^{2+}_i preincubation, versus simultaneous addition with Na^+_i, for NCX1.1, NCX1.3, and NCX1.4.

currents were equivalent whether or not the patches were preincubated with regulatory Ca^{2+}_i. For both NCX1.1 and NCX1.3, these "offsets," although differing in sign, increased with increasing Ca^{2+}_i (FIG. 1B).

To account for these peculiarities, we devised a modified model of Na^+–Ca^{2+} exchange in which transport can proceed via "slow" and "fast" transport cycles (FIG. 2). The slow cycle corresponds to the regulatory Ca^{2+}_i-unbound form (i.e., the I_2 conformation) of the molecule, whereas the regulatory Ca^{2+}_i-bound form constitutes the fast transport configuration. It is assumed that these transitions can occur from any transport configuration of the exchanger and, in particular, even in the absence of Na^+_i when the exchanger is idle. For each isoform, I_1 inactivation is seen to originate from the 3 Na^+_i-loaded exchanger state, as originally modeled by Hilgemann *et al.*[4]; whereas we postulate a new form of regulation, termed I_3, that arises

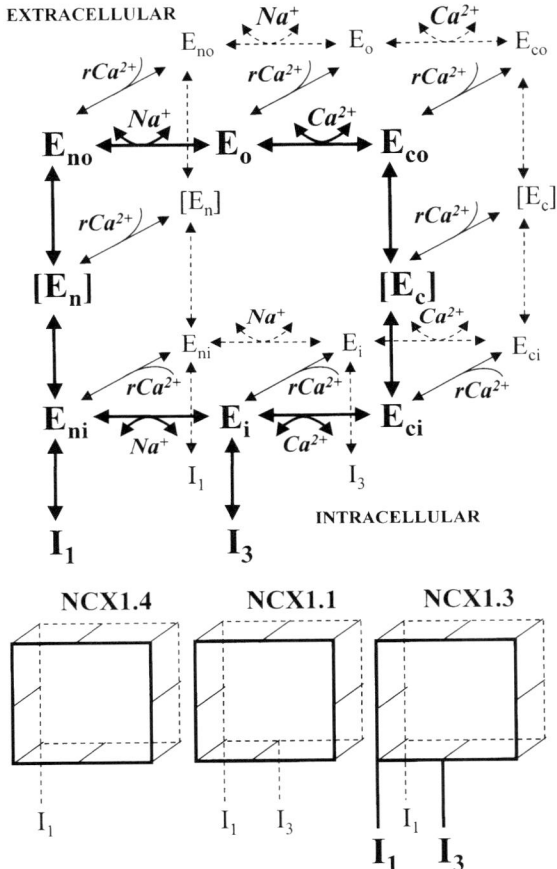

FIGURE 2. The *upper panel* shows our hypothetical general model accounting for the ionic regulatory behavior of NCX1 splice variants. The representation depicts I_1 and I_3 inactive states in the slow and fast transport configurations. Slow and fast cycles are stabilized by the absence (i.e., I_2) and presence, respectively, of regulatory Ca^{2+}_i (rCa^{2+}). Transition between slow and fast cycles is modeled to occur from any transport configuration of the generic exchanger, and *arrows* designate kinetic transitions between specific states (the corresponding rate constants have been omitted for brevity). E_i, E_{ni}, E_{ci}, and E_o, E_{no}, E_{co} designate empty (i.e., ion unbound), 3 Na^+-loaded, and Ca^{2+}-loaded transport states of the exchanger facing intra- and extracellularly, respectively. $[E_n]$ and $[E_c]$ designate the occluded 3 Na^+- and Ca^{2+}-loaded states of the exchanger. Letters in boldface represent transport and inactive states belonging to the fast cycle. The *bottom panel* shows abbreviated versions of the general model that highlights the specific ionic regulatory characteristics of NCX1.4, NCX1.1, and NCX1.3.

from the empty state of the exchanger facing the cytoplasmic surface of the membrane (i.e., E_i; FIG. 2). We suggest that the distinctive features of ionic regulation observed with each isoform are solely dependent on the relative values of the specific rate constants governing their I_1- and I_3-inactivation mechanisms. Our results are consistent with the following suppositions: (1) NCX1.4 forms an I_1 state in its slow transport cycle only; (2) NCX1.1 forms I_1 and I_3 states in its slow cycle only, and (3) NCX1.3 forms an I_1 configuration in both its slow and fast cycles, and an I_3 state in its fast cycle only (FIG. 2). Results from computer simulations of this model closely approximate our experimental data.

ACKNOWLEDGMENTS

This work was supported by a grant from the Canadian Institutes of Health Research.

REFERENCES

1. TRAC, M., C. DYCK, M. HNATOWICH, et al. 1997. Transport and regulation of the cardiac Na^+–Ca^{2+} exchanger, NCX1. Comparison between Ca^{2+} and Ba^{2+}. J. Gen. Physiol. **109:** 361–369.
2. DYCK, C., A. OMELCHENKO, C.L. ELIAS, et al. 1999. Ionic regulatory properties of brain and kidney splice variants of the NCX1 Na^+–Ca^{2+} exchanger. J. Gen. Physiol. **114:** 701–711.
3. HILGEMANN, D.W. 1990. Regulation and deregulation of cardiac Na^+–Ca^{2+} exchange in giant excised sarcolemmal membrane patches. Nature **344:** 242–245.
4. HILGEMANN, D.W., S. MATSUOKA, G.A. NAGEL & A. COLLINS. 1992. Steady-state and dynamic properties of cardiac sodium–calcium exchange: sodium-dependent inactivation. J. Gen. Physiol. **100:** 905–932.
5. HILGEMANN, D.W., A. COLLINS & S. MATSUOKA. 1992. Steady-state and dynamic properties of cardiac sodium–calcium exchange: secondary modulation by cytoplasmic calcium and ATP. J. Gen. Physiol. **100:** 933–961.

Phosphorylation of the Na$^+$/Ca^{2+} Exchangers by PKA

A. RUKNUDIN AND D. H. SCHULZE

Department of Microbiology and Immunology, University of Maryland, Baltimore, Maryland 21201, USA

KEYWORDS: sodium–calcium exchangers; protein kinases; protein phosphorylation; NCX1

INTRODUCTION

Protein phosphorylation plays a central role in the regulation of cellular function and signal transduction in cells. Phosphorylation of the proteins occurs as a result of the activation of the protein kinases like PKA and PKC, and the phosphorylated proteins are altered in their functional properties. In many instances, the phosphorylated proteins are dephosphorylated by protein phosphatases, thereby restoring the particular phosphorylation system to its unphosphorylated or basal state. Thus, the function of various cellular proteins can be regulated by activating either a protein kinase or a phosphatase. Phosphorylation of the Na$^+$/Ca^{2+} exchanger and its effect on function has been recently demonstrated for the Na$^+$/Ca^{2+} exchanger proteins.

Na$^+$/Ca^{2+} exchange was initially described in giant squid axons,[1] and the regulation of the Na$^+$/Ca^{2+} exchange activity by ATP[2] was also first observed in squid axons.[3,4] Later it was suggested that ATP phosphorylates the Na$^+$/Ca^{2+} exchanger protein, resulting in an increase in the activity of the squid axon Na$^+$/Ca^{2+} exchanger.[5] In mammals, Na$^+$/Ca^{2+} exchange plays a critical role during systole and diastole in heart. The Na$^+$/Ca^{2+} exchanger in heart has been extensively studied, and these studies have contributed to the general understanding of the function and regulation of the Na$^+$/Ca^{2+} exchange itself. Early studies of heart Na$^+$/Ca^{2+} exchange using sarcolemmal vesicles showed an increased activity in the presence of ATP.[6] Na$^+$/Ca^{2+} exchanger function was enhanced after ATP was applied to the cytoplasmic surface in the excised giant patch, even though the application of protein kinases in giant patches did not increase the activity of the cardiac Na$^+$/Ca^{2+} exchange in the presence of ATP.[7–9] Recent studies confirmed that ATP depletion reduced the Na$^+$-dependent Ca^{2+} uptake in Na$^+$-loaded rat cardiomyocytes.[10] Although squid axons provided strong evidence for the involvement of protein phosphorylation in the ATP-dependent stimulation of Na$^+$/Ca^{2+} exchange, biochemical evidence was still lacking. Shigekawa *et al.* discovered that NCX1 from rat aorta is phosphorylated and

---

Address for correspondence: Abdul Ruknudin, Ph.D., Department of Microbiology and Immunology, School of Medicine, University of Maryland, 655 W. Baltimore Street, Baltimore, MD 21201. Voice: 410-706-1379; fax: 410-706-2129.
    aruknudi@umaryland.edu

concomitantly activated in response to growth factors in primary cultures of rat aorta smooth muscle cells.[11] A subsequent study from the same laboratory demonstrated that the cardiac isoform of the Na^+/Ca^{2+} exchanger, NCX1, is regulated by PKC-catalyzed protein phosphorylation.[12]

Phosphorylation of the cardiac Na^+/Ca^{2+} exchanger NCX1 by a PKA-dependent pathway had not been demonstrated until recently. PKA-dependent phosphorylation is very important since the heart rate is regulated by catecholamines through β-adrenergic receptors. When β-adrenergic receptors are activated, calcium channels are shown to be phosphorylated by a PKA-dependent pathway, resulting in increased calcium influx through phosphorylated calcium channels in the sarcolemma. Because the Na^+/Ca^{2+} exchanger is the main Ca^{2+} efflux mechanism in cardiomyocytes, the Na^+/Ca^{2+} exchanger is possibly activated by phosphorylation to increase its activity. If the Ca^{2+} influx is not balanced by Ca^{2+} efflux during the beat-to-beat cycle, it may lead to an increase in diastolic Ca^{2+} or in Ca^{2+} overload in the sarcoplasmic reticulum, resulting in cardiac arrhythmia. One may argue that increased Ca^{2+} influx could enhance the activity of the Na^+/Ca^{2+} exchanger by catalytic effect of $[Ca^{2+}]_i$, but it has not been demonstrated unequivocally that catalytic $[Ca^{2+}]_i$ enhances the Na^+/Ca^{2+} exchanger activity to compensate for the increased Ca^{2+} influx under these conditions.

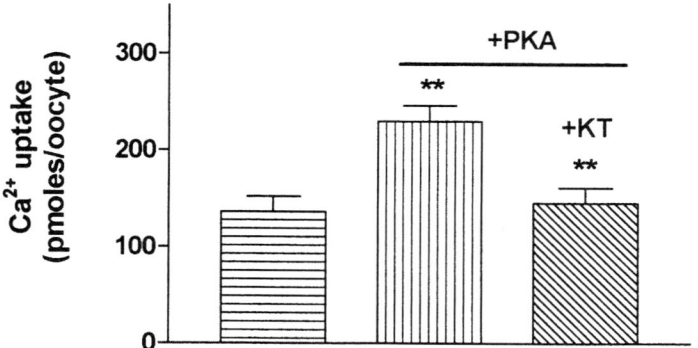

FIGURE 1. Bar graph showing the increase in Ca^{2+} uptake after PKA activation. ^{45}Ca uptake was measured in Na^+-loaded oocytes expressing the cardiac isoform of NCX1 after reducing $[Na^+]_o$ to 0 mM. PKA activation was obtained with a cocktail of IBMX, dibutryl cAMP, and forskolin. KT5720, an inhibitor of PKA, was added to the cocktail where indicated (KT). Asterisks indicate level of significance: $P < 0.001$. (Modified from Ruknudin et al.[13])

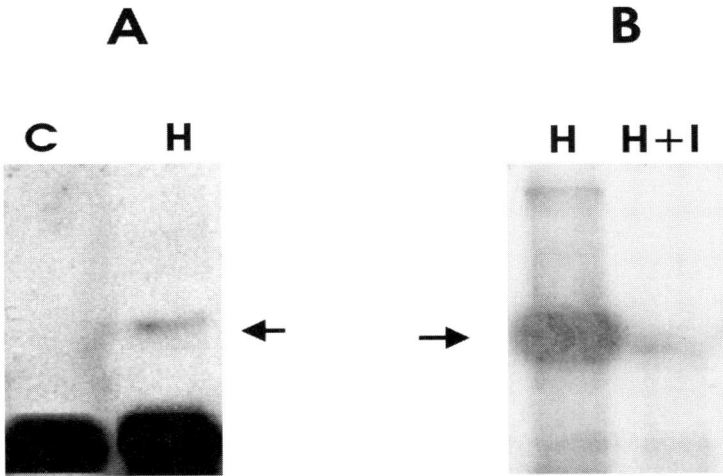

FIGURE 2. Western and autoradiograph showing the NCX1.1 protein and its phosphorylation. (A) Western blot of the Na^+/Ca^{2+} exchanger protein immunoprecipitated by NCX1 antibody. The 110-kDa fraction (H) is identified by the VSV antibody in lane 2 (*arrow*). The immunoglobulins appear as a thick band at 50 kDa. The control oocytes (c) did not show any Na^+/Ca^{2+} exchanger protein in lane 1. (B) Autoradiograph of the Na^+/Ca^{2+} exchanger protein showing the phosphorylation by PKA *in vitro* (H). The inclusion of the pseudo-substrate for PKA enzyme prevented the *in vitro* phosphorylation of the exchanger (H + I). (Modified from Ruknudin *et al.*[13])

RESULTS AND DISCUSSION

We investigated the effect of PKA on the cloned rat Na^+/Ca^{2+} exchanger, NCX1, expressed in *Xenopus* oocytes. We found that the activity of the cardiac isoform of NCX1 (NCX1.1) was increased by about 40% after activating the PKA pathway in these cells (FIG. 1). We also observed that the NCX1 protein was phosphorylated under these conditions (FIG. 2). The cardiac isoform of NCX1 protein containing a VSV-epitope tag was precipitated by the NCX antibody and could be identified by the VSV antibody as a 110-kDa band in the Western blots (FIG. 2B). The NCX1 protein immunoprecipitated by NCX antibody was heavily phosphorylated in the presence of $\gamma^{32}PATP$ and the PKA-catalytic subunit. Moreover, inclusion of the pseudo-substrate for the enzyme eliminated the phosphorylation (FIG. 2B), indicating the specificity of the PKA reaction. The increased activity of the Na^+/Ca^{2+} exchanger after PKA activation was also demonstrated in rat cardiomyocytes.[13] Moreover, Perchenet *et al.* have demonstrated that the β-adrenergic stimulator isoprenaline mediated upregulation of Na^+/Ca^{2+} exchange activity in guinea pig ventricular myocytes.[14]

To investigate whether the effect of PKA can be demonstrated in other Na^+/Ca^{2+} exchangers, we examined the effect of PKA on the Na^+/Ca^{2+} exchanger of *Drosophila*. We have cloned the Na^+/Ca^{2+} exchanger from *Drosophila*, *Dmel/Ncx*, by homology screening using the human heart Na^+/Ca^{2+} exchanger. *Dmel/Ncx* showed a

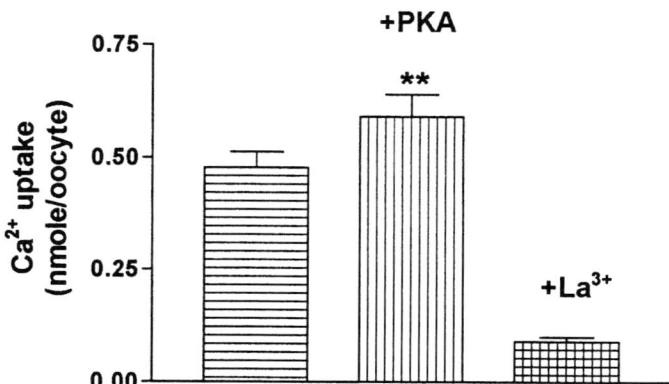

FIGURE 3. Bar graph showing increased Ca^{2+} uptake after PKA activation. ^{45}Ca uptake over a 10-min period was measured in Na^+-loaded oocytes expressing $Dmel/Ncx$ exchanger after reducing $[Na^+]_0$ to 0 mM. PKA activity was enhanced as described in FIGURE 1. Asterisks indicate significance: $P < 0.001$; $n = 13$. La^{3+} blocked the Na^+-dependent Ca^{2+} uptake.

54% difference from the mammalian Na^+/Ca^{2+} exchangers at the amino acid level.[13] There are also functional differences between the insect and mammalian Na^+/Ca^{2+} exchangers.[16–18] Moreover, the sequence of the $Dmel/Ncx$ exchanger showed consensus PKA sites. We investigated the effect of PKA on $Dmel/Ncx$ exchanger in heterologous expression system. We expressed the $Dmel/Ncx$ exchanger in $Xenopus$ oocytes by injecting the cRNA and found that the activity of the $Dmel/Ncx$ exchanger was also increased after activation of the PKA pathway (FIG. 3).

CONCLUSION

We have shown the PKA-induced phosphorylation of the mammalian cardiac Na^+/Ca^{2+} exchanger NCX1. The insect Na^+/Ca^{2+} exchanger is similarly regulated by the PKA-dependent phosphorylation.

REFERENCES

1. BAKER, P.F. & M.P. BLAUSTEIN. 1968.Sodium-dependent uptake of calcium by crab nerve. Biochim. Biophys. Acta **150:** 167–170.
2. DIPOLO, R. 1974. Effect of ATP on the calcium efflux in dialyzed squid giant axons. J. Gen. Physiol. **64:** 503–517.

3. DIPOLO, R. 1977. Characterization of the ATP-dependent calcium efflux in dialyzed squid giant axons. J. Gen. Physiol. **69**: 795–813.
4. DIPOLO, R. & L. BEAUGE. 1987. In squid axons, ATP modulates Na^+–Ca^{2+} exchange by a $[Ca^{2+}]_i$-dependent phosphorylation. Biochim. Biophys. Acta **897**: 347–354.
5. DIPOLO, R. & L. BEAUGE. 1994. Effects of vanadate on MgATP stimulation of Na–Ca exchange support kinase-phosphatase modulation in squid axons. Am. J. Physiol. **266**: C1382–C1391.
6. CARONI, P. & E. CARAFOLI. 1983. The regulation of the Na^+–Ca^{2+} exchanger of heart sarcolemma. Eur. J. Biochem. **132**: 451–460.
7. HILGEMANN, D.W. 1990. Regulation and deregulation of cardiac Na^+–Ca^{2+} exchange in giant excised sarcolemmal membrane patches. Nature **344**: 242–245.
8. HILGEMANN, D.W., A. COLLINS & S. MATSUOKA. 1992. Steady-state and dynamic properties of cardiac sodium–calcium exchange. Secondary modulation by cytoplasmic calcium and ATP. J. Gen. Physiol. **100**: 933–961.
9. MATSUOKA, S. & D.W. HILGEMANN. 1994. Inactivation of outward Na^+–Ca^{2+} exchange current in guinea-pig ventricular myocytes. J. Physiol. (London) **476**: 443–458.
10. HAWORTH, R.A. & A.B. GOKNUR. 1996. Regulation of sodium–calcium exchange in intact myocytes by ATP and calcium. Ann. N.Y. Acad. Sci. **779**: 464–479; 464–479.
11. SHIGEKAWA, M., T. IWAMOTO & S. WAKABAYASHI. 1996. Phosphorylation and modulation of the Na^+–Ca^{2+} exchanger in vascular smooth muscle cells. Ann. N.Y. Acad. Sci. **779**: 249–257; 249–257.
12. IWAMOTO, T., Y. PAN, S. WAKABAYASHI, et al. 1996. Phosphorylation-dependent regulation of cardiac Na^+/Ca^{2+} exchanger via protein kinase C. J. Biol. Chem. **271**: 13609–13615.
13. RUKNUDIN, A., S. HE, W.J. LEDERER & D.H. SCHULZE. 2000. Functional differences between cardiac and renal isoforms of the rat Na^+–Ca^{2+} exchanger NCX1 expressed in *Xenopus* oocytes. J. Physiol. **529**(Pt. 3): 599–610.
14. PERCHENET, L., A.K. HINDE, K.C. PATEL, et al. 2000. Stimulation of Na/Ca exchange by the beta-adrenergic/protein kinase A pathway in guinea-pig ventricular myocytes at 37 degrees C. Pflugers Arch. **439**: 822–828.
15. RUKNUDIN, A., C. VALDIVIA, P. KOFUJI, et al. 1997. Na^+/Ca^{2+} exchanger in *Drosophila*: cloning, expression, and transport differences. Am. J. Physiol. **273**: C257–C265.
16. DYCK, C., K. MAXWELL, J. BUCHKO, et al. 1998. Structure–function analysis of CALX1.1, a Na^+–Ca^{2+} exchanger from *Drosophila*. Mutagenesis of ionic regulatory sites. J. Biol. Chem. **273**: 12981–12987.
17. HRYSHKO, L.V., S. MATSUOKA, D.A. NICOLL, et al. 1996. Anomalous regulation of the *Drosophila* Na^+–Ca^{2+} exchanger by Ca^{2+}. J. Gen. Physiol. **108**: 67–74.
18. SCHWARZ, E.M. & S. BENZER. 1997. Calx, a Na–Ca exchanger gene of *Drosophila melanogaster*. Proc. Natl. Acad. Sci. USA **94**: 10249–10254.

Cellular Regulation of Sodium–Calcium Exchange

MADALINA CONDRESCU, KWABENA OPUNI, BASIL M. HANTASH, AND
JOHN P. REEVES

Department of Pharmacology and Physiology, University of Medicine and Dentistry of New Jersey, New Jersey Medical School and Graduate School of Biomedical Sciences, Newark, New Jersey 07103, USA

ABSTRACT: Na^+/Ca^{2+} exchange activity was studied in transfected Chinese hamster ovary (CHO) cells expressing the wild-type cardiac exchanger (NCX1.1) or mutants created by site-directed mutagenesis. The activity of the wild-type exchanger, but not exchanger mutants deficient in Ca^{2+}-dependent activation, was inhibited by sphingolipids such as ceramide and sphingosine. We propose that sphingolipids interfere with the regulatory activation of exchange activity by Ca^{2+} and suggest that this interaction provides a means for monitoring and regulating diastolic Ca^{2+} levels in beating cardiac myocytes. Exchange activity in CHO cells was also linked, through a poorly understood feedback mechanism, to Ca^{2+} accumulation within internal stores such as the endoplasmic reticulum and the mitochondria. Finally, the F-actin cytoskeleton was shown to modulate exchange activity through interactions involving the exchanger's central hydrophilic domain. We conclude that regulation of exchange activity in intact cells involves multiple interactions with various lipid species, cytosolic Ca^{2+}, organellar Ca^{2+} stores, and the cytoskeleton.

KEYWORDS: Na^+/Ca^{2+} exchange; NCX1.1; sphingolipids; cardiac myocytes; endoplasmic reticulum; mitochondria; F-actin cytoskeleton

INTRODUCTION

The Na^+/Ca^{2+} exchanger is an important regulator of cardiac contraction, and it is reasonable to expect that this activity would itself be tightly regulated. As attested by many papers in this volume, *in vitro* studies have shown that exchange activity can be modulated by such factors as ATP, Ca^{2+} (acting at regulatory sites), phosphorylation, and various lipids, especially phosphatidylinositol-4,5-*bis*phosphate (PIP2) and sphingolipids (see below). However, the contribution of these processes to regulating exchange activity under physiological conditions remains to be established. Indeed, there is no compelling evidence that exchange activity is in fact regulated in a normal physiological context. On the other hand, more than half of the exchange

---

Address for correspondence: John P. Reeves, Department of Pharmacology and Physiology, University of Medicine and Dentistry of New Jersey, New Jersey Medical School and Graduate School of Biomedical Sciences, 185 South Orange Avenue, Newark, NJ 07103. Voice: 973-972-3890; fax: 973-972-7950.

reeves@umdnj.edu

protein is composed of a hydrophilic domain of 546 amino acids that can be proteolyzed or deleted by mutation without altering the fundamental kinetics of the exchange process.[1] Portions of this domain have been shown to be essential for the various *in vitro* modes of regulation mentioned above. It is logical to infer that this domain serves an important regulatory function—the task confronting investigators in the exchange field is to identify that regulatory function and to demonstrate its physiological importance.

In this article, we shall briefly review current concepts of exchanger regulation, based primarily on exchange current measurements with excised membrane patches and then summarize some recent results from our laboratory on the factors modulating exchange activity in transfected Chinese hamster ovary (CHO) cells.

IN VITRO MODES OF REGULATION

Early studies of exchange activity with squid giant axons revealed two general modes of exchanger regulation dependent on ATP and cytosolic Ca^{2+}.[2–4] ATP reduced the K_m values for transport of cytosolic Ca^{2+} and extracellular Na^+, whereas cytosolic Ca^{2+} appeared to be a necessary cofactor for exchanger operation, even when operating in reverse, that is, Na^+_i-dependent Ca^{2+} influx. The evidence with squid axons favored a phosphorylation mechanism for ATP-dependent regulation.[5]

Later studies with excised "giant" patches from cardiac myocytes revealed more complex modes of regulation.[6] Inward exchange currents were found to decay from a peak toward a steady-state value following application of Na^+ to the cytosolic surface of the membrane patch. The decay process was postulated to reflect the time-dependent entry of exchangers into an inactive state. This process was termed "Na^+-dependent inactivation" since it was initiated by Na^+ ions binding to the cytosolically disposed transport sites. ATP prevented Na^+-dependent inactivation, and a landmark report by Hilgemann and Ball[7] showed that ATP's effects were due to the formation of PIP2 within the membrane patch (cf. the article by Hilgemann in this volume). PIP2 appeared to interact with the exchanger's hydrophilic, regulatory domain, most likely with the "XIP" region,[8] an amphipathic, positively charged segment of 20 amino acids at the NH_2-terminal portion of the hydrophilic domain. It has been suggested that PIP2 blocks Na^+-dependent inactivation by preventing an inhibitory interaction of the XIP region with another, as yet unidentified, segment of the exchanger; this inhibitory interaction is presumably facilitated by the binding of Na^+ ions to the transport sites.

Philipson and his colleagues identified two short, acidic segments within the hydrophilic domain that bind Ca^{2+} with high affinity.[9] Certain mutations within these segments altered Ca^{2+} binding and also altered Ca^{2+}-dependent activation, suggesting that these were the "regulatory" Ca^{2+} binding sites.[10] Mutations outside of this region also disrupted Ca^{2+}-dependent activation. Indeed, merely deleting 6 amino acids in one portion of the hydrophilic domain abrogated both Ca^{2+}- and Na^+-dependent modes of regulation,[11] suggesting that regulation depends on precise conformational alterations within the hydrophilic domain.

Are these modes of regulation physiologically important? Na^+-dependent inactivation is most prominent under conditions that are likely to be pathological rather than physiological, that is, high cytosolic $[Na^+]$, low ATP and/or low PIP2. As for

Ca^{2+}-dependent activation, studies with intact cells (as opposed to patches) have reported K_D values for "regulatory" Ca^{2+} ranging from 20 to 125 nM,[12–14] suggesting that under resting conditions, the exchanger is nearly fully activated by Ca^{2+}. It appears that neither of these modes of regulation is poised to modulate exchange activity over a significant range under normal physiological conditions.

To better understand the role of exchange activity in a cellular environment, we launched a series of studies of exchange activity in transfected CHO cells. These studies have revealed new and unexpected modes of regulation of exchange activity and have underscored the importance of understanding the role of exchange activity within the broader context of cellular Ca^{2+} homeostasis.

Na$^+$/Ca^{2+} EXCHANGE IN TRANSFECTED CHO CELLS

Transfected cells provide a potentially powerful approach to the study of exchanger regulation. CHO cells, for example, are without endogenous exchange activity, are easily transfected, and have few endogenous Ca^{2+} or Na$^+$ channels that could confound the interpretation of data. On the other hand, the fact that CHO cells are not heart cells means that they might be lacking structural or biochemical elements that are essential for normal exchanger regulation.

Early studies[15] on the regulation of the bovine cardiac Na$^+$/Ca^{2+} exchanger expressed in CHO cells have led to the conclusions listed below:

- exchange activity is activated in a regulatory fashion by cytosolic Ca, with a K_D of ~50 nM;

- F-actin filaments modulate the kinetics of the exchange process through an interaction involving the hydrophilic domain of the exchanger;

- exchange activity is not affected by variations in total cellular PIP2 levels over a two- to threefold range, or by activation or inhibition of PKA or PKC.

Thus, cytosolic Ca and the cytoskeleton appear to be the main regulators of exchange activity in CHO cells. However, neither mode of regulation is straightforward. The K_D for Ca-dependent regulation is below normal resting levels of [Ca^{2+}]$_i$, leaving little opportunity for meaningful regulation of exchange activity under physiological conditions. Cytoskeletal regulation appears to be important under some assay conditions, but not others,[16] suggesting that the linkage between the exchange activity and the cytoskeleton involves aspects of Ca^{2+} homeostasis that are poorly understood.

SPHINGOLIPIDS—PHYSIOLOGICAL REGULATORS OF EXCHANGE ACTIVITY?

Recent results have demonstrated that exchange activity in CHO cells is markedly inhibited by the sphingolipids ceramide and sphingosine.[17] FIGURE 1 shows the results of a typical Ba^{2+} influx assay for measuring exchange activity in CHO cells. In these experiments, the cells were treated with gramicidin, a channel-forming antibiotic, in a medium containing 20 mM NaCl and 120 mM KCl; the gramicidin clamps cytosolic Na$^+$ at 20 mM, a concentration that supports a high level of exchange ac-

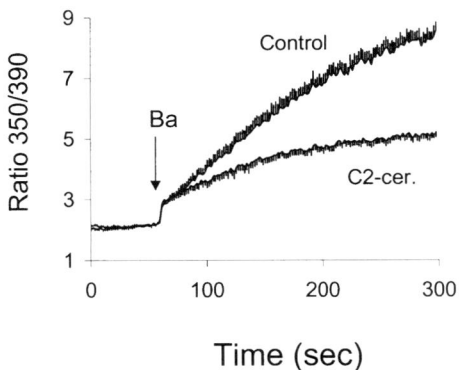

FIGURE 1. Inhibition of Na^+/Ca^{2+} exchange by C2-ceramide. CHO cells expressing NCX1.1 were loaded with fura-2 and preincubated for 5 min in Na-PSS + 1 mM $CaCl_2$. The cells were assayed for Na^+/Ca^{2+} exchange activity (Ba^{2+} influx) in 20/120 Na/K-PSS + 0.3 mM EGTA containing 1 µg/ml gramicidin with or without 15 µM C2-ceramide, as indicated. $BaCl_2$ (1 mM) was added, as indicated by the *arrow*. An increase in the ratio of fura-2 fluorescence for excitation at 350 nm versus 390 nm (ratio 350/390) indicates Ba^{2+} entry into the cytosol. Na-PSS contained, in mM: 140 NaCl, 5 KCl, 1 $MgCl_2$, 10 glucose, and 20 Mops, adjusted to pH 7.4 with Tris. The composition of 20/120 Na/K-PSS was the same except that the NaCl and KCl concentrations were 20 mM and 120 mM, respectively. C2-ceramide: *N*-acetylsphingosine. Figure reprinted from Condrescu and Reeves[17] with permission.

tivity. Ba^{2+} enters the cell in exchange for internal Na^+, and the rate of Ba^{2+} entry was monitored using fura-2. As shown, the addition of a short-chain analogue of ceramide (C2-ceramide; 15 µM), reduced the rate of Ba^{2+} influx. Exchange activity was not inhibited when a biologically inactive ceramide analogue (C2-dihydroceramide) was used in place of C2-ceramide. Importantly, C2-ceramide was ineffective in cells expressing exchanger mutants in which segments of the central "regulatory" domain were deleted, that is, $\Delta(241-680)$ and $\Delta(680-685)$; these mutants were defective in Ca^{2+}-dependent activation. In contrast, C2-ceramide *did* inhibit a mutant exchanger with alterations in the XIP region; this mutant showed normal Ca^{2+} activation but did not show Na^+-dependent inactivation. Finally, C2-ceramide also inhibited exchange activity operating in the Ca^{2+} efflux mode. A similar pattern of results was obtained with 2.5 µM sphingosine in place of C2-ceramide, and also when cells were pretreated with an agent that blocked ceramide utilization pathways so that endogenous ceramide accumulated within the cells.

These results indicate that sphingolipids inhibit exchange activity in a manner that is linked to the regulatory properties of the exchanger. The differential effects of ceramide on the mutants described above led us to propose that ceramide interferes with the regulatory activation of the exchanger by cytosolic Ca. These considerations have led to a more general hypothesis, called the "diastolic trap" hypothesis, in which sphingolipids and the exchanger act together to regulate diastolic $[Ca^{2+}]_i$ in cardiac myocytes. As shown in FIGURE 2, during diastole a fraction of the exchanger population would be expected to lose Ca^{2+} from their regulatory sites and enter an inactive (I_2) state. If, as suggested by our results, endogenous sphingolipids prevent

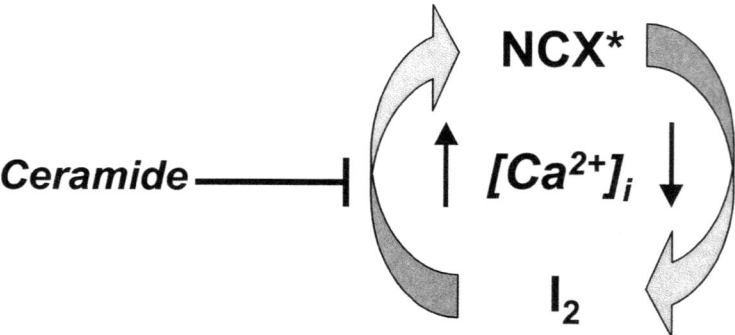

FIGURE 2. The diastolic trap hypothesis. When Ca^{2+} dissociates from regulatory binding sites in the exchanger's hydrophilic domain, a conformational transition from an active state (NCX*) to an inactive state (I_2) occurs. The designation I_2 distinguishes this inactive state from the one produced by Na^+-dependent inactivation (I_1). Elevation of $[Ca^{2+}]_i$ (*left arrow*) leads to refilling of the regulatory Ca^{2+}-binding sites and re-establishment of the exchanger's active state. We propose that ceramide stabilizes the I_2 state and thereby interferes with the $I_2 \rightarrow NCX*$ transition. In beating myocytes, dissociation of Ca^{2+} from regulatory sites during diastole would act in conjunction with endogenous ceramide to govern the distribution of exchangers between the NCX* and I_2 states. Figure reprinted from Condrescu and Reeves[17] with permission.

a portion of these exchangers from being reactivated during the systolic rise in $[Ca^{2+}]_i$, a population of inactive exchangers would be "trapped" in the I_2-inactive state. Over time, a steady-state population of inactive exchangers would build up, governed by the relative rates of entry and exit of exchangers into the I_2 state. Entry of exchangers into the I_2 state would be determined by diastolic levels of $[Ca^{2+}]_i$, and sphingolipids would control the rate of exit from the I_2 state to the active state. Thus, if diastolic $[Ca^{2+}]_i$ were to decline for some reason, the interplay between endogenous sphingolipids and the increased number of exchangers entering the I_2 state would increase the population of inactive exchangers, thereby inhibiting exchange-mediated Ca^{2+} efflux and restoring diastolic $[Ca^{2+}]_i$ to an appropriate level over subsequent beats. While highly speculative at this point, the diastolic trap hypothesis provides a welcome conceptual framework for understanding the physiological role of Ca^{2+}-dependent activation in the context of the exchanger's surprisingly high affinity for regulatory Ca^{2+}.

MITOCHONDRIAL REGULATION OF EXCHANGE ACTIVITY

An entirely unexpected mode of exchanger regulation was revealed during a simple two-pulse procedure for examining exchange activity in transfected cells.[18] The procedure is illustrated in FIGURE 3. Here, fura-2-loaded CHO cells expressing NCX1.1 are subjected to two 5-min intervals in 20 mM extracellular $[Na^+]$, separat-

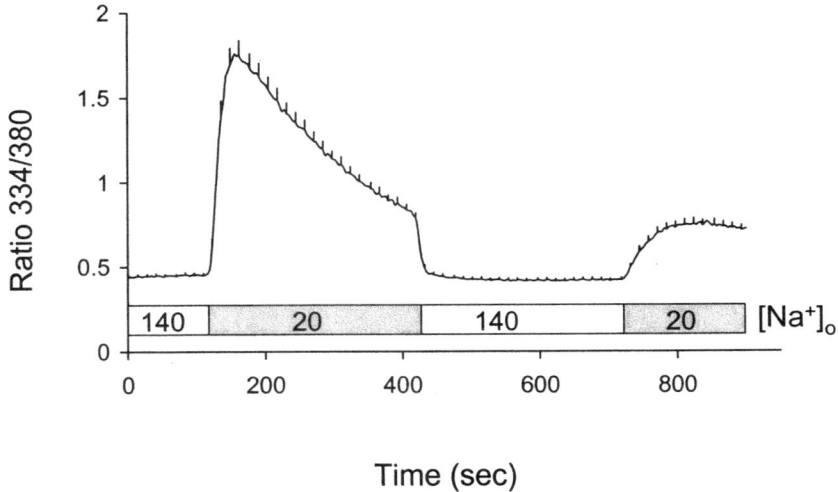

FIGURE 3. Two-pulse protocol for Na^+/Ca^{2+} exchange. Transfected CHO cells were loaded with fura-2 and 20–70 individual cells were monitored per cover slip using an Attofluor digital imaging apparatus. The external medium was changed from Na-PSS + 1 mM $CaCl_2$ to 20/120 Na/K-PSS + 1 mM $CaCl_2$ for the intervals indicated by the bar below the trace. An increase in the 340/380 ratio indicates an increase in the cytosolic Ca^{2+} concentration. Results are the means ± SE of data obtained with 29 separate coverslips. Figure reprinted from Opuni and Reeves[18] with permission.

ed by a 5-min recovery period in 140 mM Na^+. As shown in FIGURE 3, Ca^{2+} entered the cells by reverse-exchange activity during the first interval in 20 mM Na^+, leading to a rise in $[Ca^{2+}]_i$. During the second interval, however, the rise in $[Ca^{2+}]_i$ was only 25% of that observed during the first interval. Various control experiments established that the reduced Ca^{2+} influx during the second interval was due to an inhibition of exchange activity rather than loss of cytosolic Na^+ or increased organellar sequestration of cytosolic Ca^{2+}. Recent preliminary results indicated that the Δ(241–680) deletion mutant, missing a large portion of the exchanger's "regulatory" domain, did not exhibit reduced exchange activity during the second pulse in low Na^+.

The Ca^{2+} that entered the cell during the first interval in low Na^+ was accumulated by mitochondria, as determined using rhod-2, a Ca^{2+}-indicating fluorescent dye that selectively labeled the mitochondria under our loading conditions (see FIG. 4). The following evidence indicates that mitochondrial Ca^{2+} accumulation is necessary for the observed inhibition of exchange activity during the second interval in low Na^+: Uncouplers such as Cl-CCP blocked mitochondrial Ca^{2+} accumulation and abolished the inhibition of exchange activity. Conversely, diltiazem, which inhibits Ca^{2+} efflux from mitochondria,[19] promoted increased mitochondrial Ca^{2+} accumulation and exacerbated the inhibition of exchange activity. Finally, nocodazole, an agent that disrupts microtubules, blocked mitochondrial Ca^{2+} accumulation under the conditions of our experiments and abolished inhibition of exchange activity dur-

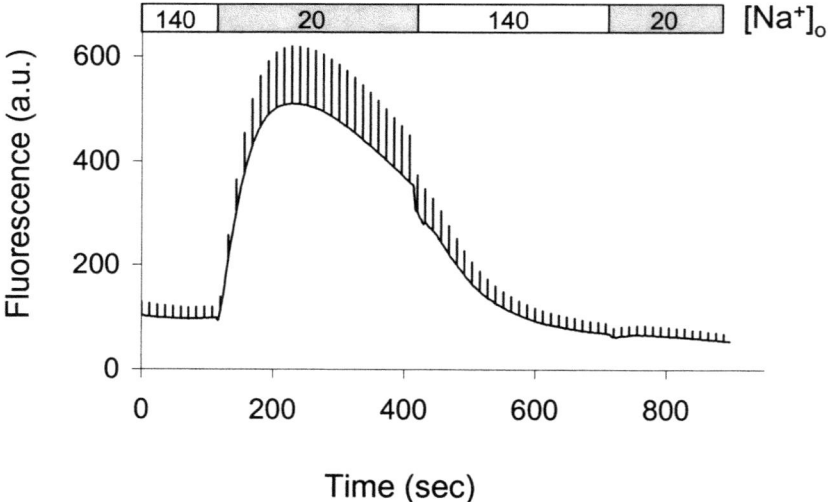

FIGURE 4. Mitochondrial Ca^{2+} uptake during the 2-pulse protocol. Transfected CHO cells were loaded for 12 min with rhod-2/AM, a Ca^{2+}-sensitive fluorescent probe that preferentially labels mitochondria. The 2-pulse protocol described in FIGURE 3 was then applied to the rhod-2-labeled cells. Rhod-2 fluorescence in the mitochondrion-rich perinuclear regions of the cells was monitored with a computer-controlled digital imaging apparatus. An increase in rhod-2 fluorescence corresponds to an increase in mitochondrial Ca concentration. Note that rhod-2 fluorescence increased during the first interval in low Na^+, but not the second, consistent with an inhibition of exchange activity during the second interval. Results are the means ± SE (for every fourth data point) of rhod-2 fluorescence (arbitrary units) for 6 coverslips, with 20–60 individual cells monitored per coverslip. For full experimental details, see Ref. 18. Figure reprinted from Opuni and Reeves[18] with permission.

ing the second pulse of low Na^+. The result with nocodazole is poorly understood but suggests that microtubules may be involved in conveying Ca^{2+} from the plasma membrane to the mitochondria.

We conclude that mitochondrial Ca^{2+} accumulation leads to the elaboration of a second messenger, or induces some other alteration, that inhibits Na^+/Ca^{2+} exchange activity. Despite much experimental effort, the nature of the inhibitory signal remains unknown. More recently, we have found that not only mitochondria, but also a thapsigargin-sensitive internal Ca^{2+} store, mostly likely the endoplasmic reticulum (ER), is involved in this process. In cells treated with thapsigargin, an inhibitor of the ER Ca^{2+}-transporting ATPase,[20] the rates of Ca^{2+} influx during the first and second intervals are equal, although mitochondrial Ca^{2+} accumulation is unaffected. Moreover, when the combination of extracellular ATP (a purinergic agonist) and thapsigargin was applied to release Ca^{2+} from internal stores following the first interval in low Na^+, Ca^{2+} influx during the second interval was no longer inhibited. We conclude that mitochondria and the ER collaborate in some way to inhibit exchange activity.

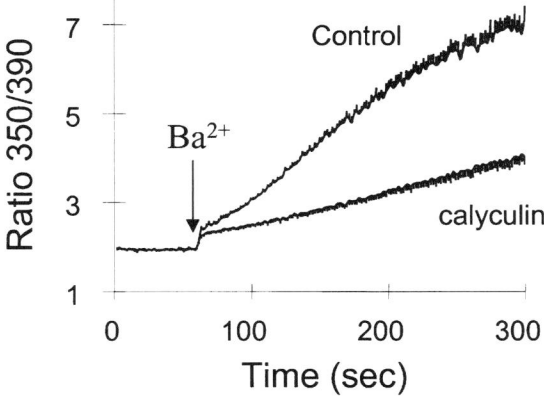

FIGURE 5. Effect of calyculin A, a protein phosphatase inhibitor, on Na^+/Ca^{2+} exchange. Transfected CHO cells were assayed for Ba^{2+} influx as described in FIGURE 1. Before assay, the cells were preincubated for 5 min in Na-PSS + 1 mM $CaCl_2$ with or without 100 nM calyculin A, as indicated. Figure reprinted from Condrescu et al.[21] with permission.

PROTEIN PHOSPHATASE INHIBITORS, THE F-ACTIN CYTOSKELETON, AND EXCHANGE ACTIVITY[21]

A dramatic illustration of cellular effects on exchange activity is presented in FIGURE 5. In this experiment, transfected CHO cells were treated for 5 min with calyculin A, an agent that inhibits protein phosphatases 1 and 2A. As shown, exchange-mediated Ba^{2+} influx was markedly inhibited by this treatment. With okadaic acid, another protein phosphatase inhibitor, we observed nearly a complete suppression of exchange activity following 60 min of treatment. Curiously, the inhibition of activity was observed only for the Ca^{2+} influx mode of exchange activity—little or no inhibition of Na^+-dependent Ca^{2+} efflux was observed following calyculin A treatment. The protein phosphatase inhibitors produced a breakdown in the major cytoskeletal filament systems, microfilaments, microtubules, and intermediate filaments, and dramatically altered cellular morphology. Thus, their effects on exchange activity may be secondary consequences of these cytoskeletal alterations.

Two additional findings strongly implicate the actin cytoskeleton in the response of exchange activity to protein phosphatase inhibitors. First, treatment of the cells with cytochalasin D, or latrunculin A, agents that lead to a breakdown in the actin cytoskeleton, partially protected exchange activity against the inhibitory effects of calyculin A. Similar protective effects of actin filament breakdown have been reported for calyculin's effects on store-dependent Ca^{2+} entry.[22] More dramatically, as shown in FIGURE 6, when cells were treated with jasplakinolide, an agent that binds to F-actin filaments and stabilizes them against breakdown,[23] nearly complete protection against calyculin A was observed. The data in FIGURE 6 also show that jas-

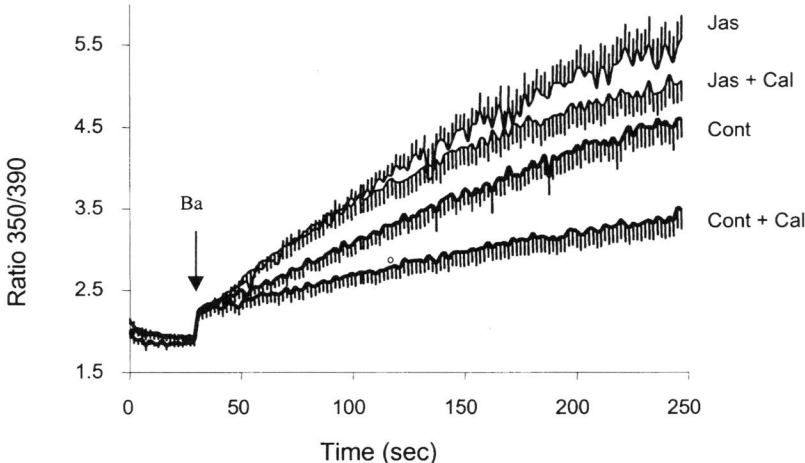

FIGURE 6. Effect of jasplakinolide treatment on Na^+/Ca^{2+} exchange and response to calyculin. Transfected CHO cells were treated for 60 min with (Jas) or without (Control) 1 µM jasplakinolide, an agent that stabilizes F-actin filaments. Following fura-2 loading, the cells were preincubated for 5 min in Na-PSS + 1 mM $CaCl_2$ with (Cal) or without 100 nM calyculin A and assayed for Na^+/Ca^{2+} exchange as described in FIGURE 1.

plakinolide treatment *stimulates* exchange activity in comparison to untreated cells. This is the first agent we have found that stimulates exchange activity in CHO cells. Importantly, jasplakinolide did *not* stimulate the activity of the Δ(241–680) deletion mutant, which is missing much of the "regulatory" domain.

These findings offer further support for our early conclusion, based on $^{45}Ca^{2+}$ uptake studies, that an interaction between the exchanger and F-actin filaments is an important determinant of exchange activity and that the exchanger's central hydrophilic domain is essential for this interaction. Additional support for this conclusion is provided by recent immunofluorescence imaging results from our laboratory, showing that the wild-type exchanger, but not the Δ(241–680) deletion mutant, interacts with F-actin stress fibers in CHO-T cells (unpublished observations).

CONCLUDING REMARKS

The results summarized above indicate that the regulation of exchange activity in intact cells involves important interactions with internal organellar Ca^{2+} stores (mitchondria, endoplasmic reticulum) and the cytoskeleton. Thus, the regulatory behavior of the exchanger in intact cells is vastly more complicated than the *in vitro* studies of PIP2 and Ca^{2+} would suggest. The emerging functional and physical evidence for F-actin involvement in exchange activity will provide a focus for future research in this important area.

REFERENCES

1. MATSUOKA, S., D.A. NICOLL, R.F. REILLY, et al. 1993. Proc. Natl. Acad. Sci. USA **90:** 3870–3874.
2. DIPOLO, R. 1979. J. Gen. Physiol. **73:** 91–113.
3. DIPOLO, R. 1976. Fed. Proc. **35:** 2579–2582.
4. BLAUSTEIN, M.P. & E.M. SANTIAGO. 1977. Biophys. J. **20:** 79–111.
5. DIPOLO, R. & L. BEAUGE. 1991. Ann. N.Y. Acad. Sci. **639:** 100–111.
6. HILGEMANN, D.W. 1996. Ann. N.Y Acad. Sci. **779:** 136–158.
7. HILGEMANN, D.W. & R. BALL. 1996. Science **273:** 956–959.
8. HE, Z., S. FENG, Q. TONG, et al. 2000. Am. J. Physiol. Cell. Physiol. **278:** C661–C666.
9. LEVITSKY, D.O., D.A. NICOLL & K.D. PHILIPSON. 1994. J. Biol. Chem. **269:** 22847–22852.
10. MATSUOKA, S., D.A. NICOLL, L.V. HRYSHKO, et al. 1995. J. Gen. Physiol. **105:** 403–420.
11. MAXWELL, K., J. SCOTT, A. OMELCHENKO, et al. 1999. Am. J. Physiol. **277:** H2212–H2221.
12. MIURA, Y. & J. KIMURA. 1989. J. Gen. Physiol. **93:** 1129–1145.
13. FANG, Y., M. CONDRESCU & J.P. REEVES. 1998. Am. J. Physiol. **275:** C50–C55.
14. WEBER, C.R., K.S. GINSBURG, K.D. PHILIPSON, et al. 2001. J. Gen. Physiol. **117:** 119–131.
15. REEVES, J.P. 1998. J. Bioenerg. Biomembr. **30:** 151–160.
16. CONDRESCU, M., J.P. GARDNER, G. CHERNAYA, et al. 1995. J. Biol. Chem. **270:** 9137–9146.
17. CONDRESCU, M. & J.P. REEVES. 2001. J. Biol. Chem. **276:** 4046–4054.
18. OPUNI, K. & J.P. REEVES. 2000. J. Biol. Chem. **275:** 21549–21554.
19. HANNUN, Y.A., A.H. MERRILL, JR. & R.M. BELL. 1991. Methods Enzymol. **201:** 316–328.
20. LYTTON, J., M. WESTLIN & M.R. HANLEY. 1991. J. Biol. Chem. **266:** 17067–17071.
21. CONDRESCU, M., B.M. HANTASH, Y. FANG & J.P. REEVES. 1999. J. Biol. Chem. **274:** 33279–33286.
22. PATTERSON, R.L., D.B. VAN ROSSUM & D.L. GILL. 1999. Cell **98:** 487–499.
23. BUBB, M.R., A.M. SENDEROWICZ, E.A. SAUSVILLE, et al. 1994. J. Biol. Chem. **269:** 14869–14871.

Intracellular Ionic and Metabolic Regulation of Squid Nerve Na^+/Ca^{2+} Exchanger

REINALDO DIPOLO[a,c] AND LUIS BEAUGÉ[b,c]

[a]*Laboratorio de Permeabilidad Iónica, Centro de Biofísica y Bioquímica, IVIC, Caracas 1020-A, Venezuela*

[b]*Laboratorio de Biofísica, Instituto de Investigación Médica M. y M. Ferreyra, Casilla de Correo 389, 5000 Córdoba, Argentina*

[c]*Marine Biological Laboratory, Woods Hole, Massachusetts 02543, USA*

ABSTRACT: Intracellular Na^+ and H^+ synergistically inhibit the squid Na^+/Ca^{2+} exchanger by reducing the affinity for Ca^{2+} of its regulatory site. MgATP antagonizes H^+_i and Na^+_i inhibition; this effect must occur through a phosphorylation–dephosphorylation process, because exogenous protein phosphatases prevent MgATP activation of the exchanger. Protection by ATP against H^+_i and Na^+_i inhibition happens by decreasing the apparent affinity for the synergistic binding of these cations to the carrier. In this way ATP modifies the apparent affinity for Ca^{2+} of its regulatory site. Mg^{2+} ions play an important role in the process because they are essential for ATP activation of Na^+/Ca^{2+} exchange but can also promote deactivation of the ATP upregulated exchanger. At constant [ATP], activation at low $[Mg^{2+}]_i$ is followed by deactivation as $[Mg^{2+}]_i$ is increased. The most likely explanation for deactivation is stimulation of endogenous phosphatases. We developed a kinetic model that predicts all H^+_i, Na^+_i, and MgATP described above. This scheme includes the following conditions: (i) The binding of Ca^{2+} to the regulatory site is essential for the binding of Na^+_i or Ca^{2+}_i to the transporting sites. (ii) The binding of a first H^+_i to the carrier displaces Ca^{2+}_i from its regulatory site and allows binding of one Na^+ forming a $H.E_1.Na$ complex. The $H.E_1.Na$ complex can bind a second H^+_i forming a dead-end inhibitory $H_2.E_1.Na$ complex. (iii) MgATP, through an unspecified phosphorylation process, decreases the apparent affinity for the synergistic H^+_i and Na^+_i binding to the carrier.

KEYWORDS: Na^+/Ca^{2+} exchanger; H^+ and Na^+ inhibition; MgATP; Mg^{2+} ions; squid nerve

INTRODUCTION

The Na^+/Ca^{2+} exchanger is subjected to at least two main types of regulation on the intracellular side of the molecule, most of which occur at the large intracellular loop of the exchange protein. One type of regulation is exerted by ions, among them

Address for correspondence: Reinaldo DiPolo, Laboratorio de Permeabilidad Ionica, Centro de Biofísica y Bioquímica, IVIC, Apartado Postal 21827, Caracas 1020-A, Venezuela. Voice: 58-212-504-1230; fax: 58-212-504-1093. Also: Luis Beaugé, Instituto de Investigacion Médica M.Y.M. Ferreyra, Casilla de Correo 389, 5000 Córdoba, Argentina.
rdipolo@ivic.ve and lbeauge@immf.uncor.edu

Ann. N.Y. Acad. Sci. 976: 224–236 (2002). © 2002 New York Academy of Sciences.

Na^+ inactivation, Ca^{2+} stimulation, and H^+ modulation. In the other, the metabolic regulation is related to compounds like ATP and, in the squid, also to the phosphagen, phosphoarginine.[1] These two modes of regulation are intimately interrelated. For instance, in squid nerve axons and mammalian heart, intracellular Ca^{2+} and ATP antagonize Na^+_i inhibition,[2,3] whereas in the mammals Na^+_i increases H^+_i inhibition of Na^+/Ca^{2+} exchange.[4] In addition, in squid axons proton inhibition takes place only at the intracellular side.[5] On the other hand, another cation must be taken into account, and that is intracellular Mg^{2+}. The available experimental evidence indicates that all ATP and phosphagen regulation of the exchanger involves the phosphorylation-dephosphorylation processes, although their exact nature is, in some cases like the squid, unknown.[6] In other words, kinases and phosphatases have a role to perform. All kinase-dependent phosphorylations and phosphatase dephosphorylations require Mg^{2+} ions as essential cofactors.[7,8] Therefore, Mg^{2+}_i also becomes bound to this interplay of ionic and metabolic regulation of the Na^+/Ca^{2+} exchanger and deserves to be investigated. Finally, not much information is available regarding the effects of ATP on H^+_i inhibition and its interaction with Na^+_i. In this paper we show experiments with dialyzed squid giant axons, a preparation that, by allowing accurate control of the intracellular biochemical composition of these nerves, provides an excellent way to study how H^+_i–Na^+_i–Ca^{2+}_i interactions are influenced by ATP and the role of Mg^{2+} ions in these processes.

METHODS

The preparation used in most experiments was the giant axon of the squid subjected to internal dialysis. Two squid species were used: *Loligo pealei*, at the Marine Biological Laboratory, Woods Hole, Massachussetts, and *Loligo plei*, at the Instituto Venezolano de Investigaciones Científicas; Fundaciencia–IVIC. Dissected axons were mounted into the dialysis chamber and dialyzed using capillaries of regenerated cellulose fibers, which is permeated by solutes (MW cutoff of 18 kDa) but little water (210 μm OD; 200 μm ID; Spectrapor number 132226; Spectrum, Houston, Texas).[9] The usual dialysis solution contained 385 mM Tris-Mops, 45 mM NaCl, 2 mM ionized Mg^{2+}, 285 mM glycine 285, 1 mM Tris-EGTA (or 1–3 mM BAPTA [1,2-*bis*(*O*-aminophenoxy)ethane-*N*,*N*,*N''*,*N'''*-tetraacetic acid] or dibromo BAPTA in the experiments where pH was modified between 6.9 and 8.8 units) and a pH of 7.3 (considered normal for these nerves); temperature between 17 and 18°C. The solution bathing the external surface of the axons usually had 440 mM NaCl, 0.3 mM $CaCl_2$, 60 mM $MgCl_2$, and 10 mM Tris-Cl (pH 7.6). Osmolarity was adjusted to 940 mosmols per liter. Calculations of ionized $[Ca^{2+}]$ were performed with the WinMaxc computer program (Version 2.00, 1999; Chris Patton Hopkins Marine Station, CA). BAPTA was used to buffer $[Ca^{2+}]_i$ from 0.3 to 0.7 μM, while dibromo BAPTA was the Ca^{2+} chelator used for $[Ca^{2+}]_i$ from 1.2 to 10 μM. Higher $[Ca^{2+}]_i$ was considered as equal to the $CaCl_2$ added in excess to the amount required to attain 10 μM in dibromo BAPTA-containing solutions. Lithium compensated for the removal of extracellular sodium. The production of ATP was prevented by including 1 mM NaCN in the external medium. Fluxes of Na^+ through Na^+ channels were blocked with outside 100 nM TTX. Ca^{2+} efflux through the Ca^{2+} pump and the Na^+/K^+ pump were halted by adding 100 μM vanadate to the dialysis medium. In the pH experiments, axons were predialyzed for about 45 min with the standard medium containing 0.2 mM EGTA and no calcium or ATP. Addition of ATP (3 mM) to the dialysis medium was done at a constant free $[Mg^{2+}]_i$ of 1 mM. Steady-state $^{45}Ca^{2+}$ or $^{22}Na^+$ effluxes were measured before and after a given experimental condition; therefore, each axon serves as its own control. BAPTA and dibromo BAPTA were from Molecular Probes (Molecular Probes, Eugene, OR). All other reagents were from SIGMA (St. Louis, MO).

The preparation of membrane vesicles of squid optical nerves, the partial purification of the required cytosolic soluble protein factor, and the technique for fluxes measurements in vesicles are all described in detail elsewhere.[10]

RESULTS AND DISCUSSION

Effects of H^+_i, Na^+_i, and ATP with the Squid Na^+/Ca^{2+} Exchange

FIGURE 1 illustrates experiments in three different axons in which the Na^+_o-dependent Ca^{2+} efflux (forward Na^+_o/Ca^{2+}_i exchange) was measured as a function of the internal Na^+ at three different pH_is: 6.9 (FIG. 1A), 7.3 (FIG. 1B), and 8.8 (FIG. 1C). As the concentration of Na^+_i increased, a progressive inhibition of Ca^{2+} efflux was observed at every pH_i investigated. However, Na^+_i became a more powerful inhibitor as pH_i was reduced; that is, Na^+_i inhibition was more noticeable at higher $[H^+]_i$. Another important result illustrated in this figure is given by the levels of fluxes at zero Na^+_i, which went from about 23 fmole $\cdot$ cm^{-2} $\cdot$ sec^{-1} at pH_i 6.9 to 60 fmole. cm^{-2} $\cdot$ sec^{-1} at pH_i 7.3 and around 600 fmole $\cdot$ cm^{-2} $\cdot$ sec^{-1} at pH_i 8.8. In other words, the presence of internal Na^+ is not an absolute requirement for H^+_i inhibition of the squid Na^+/Ca^{2+} exchanger. That we are really looking at forward Na^+/Ca^{2+} exchange fluxes is evidenced by the fact that, when extracellular Na^+ was removed (in this case in FIG. 1B), the fluxes dropped to practically background levels. The power of H^+_i to inhibit the exchanger in the absence of Na^+_i and ATP, at a constant $[Ca^{2+}]_i$ of 1.2 µM, is shown even better when comparing the relative values of the fluxes. Thus varying the pH_i from its physiological value of 7.3 to 6.9, a range of only 0.4 pH_i units, causes about 70% inhibition of the forward Na^+/Ca^{2+} exchange, while raising the pH_i to 8.8 increases the exchange rate to a level more than 20 times higher than that seen at pH 6.9.

One obvious question is whether there is any interaction between Ca^+_i and H^+_i. To answer it, we followed the effects of different $[H^+]_i$ of the forward Na^+/Ca^{2+} exchange as a function of $[Ca^+]_i$, in (Na^+_i + ATP)-free solutions. FIGURE 2A shows that between 0.7 µM and 10 µM $[Ca^{2+}]_i$ intracellular protons have always had an inhibitory effect on the forward Na^+/Ca^{2+} exchange. However, the fractional proton inhibition is reduced as $[Ca^{2+}]_i$ is increased. At a saturating $[Ca^{2+}]_i$ of 1000 µM, H^+_i becomes ineffective. This indicates that, besides Na^+_i not being essential for H^+_i inhibition, inhibition is antagonized by Ca^{2+}_i. The actual location at which this antagonism takes place is not indicated by these experiments. Other experiments, however, some of which are shown here, point to the intracellular Ca^{2+} regulatory site. On that basis we developed the model described below and performed simulations for all the experimental conditions investigated. One of these simulations, which has to do with the experiment in FIGURE 2A, is illustrated in FIGURE 2B. Ex-

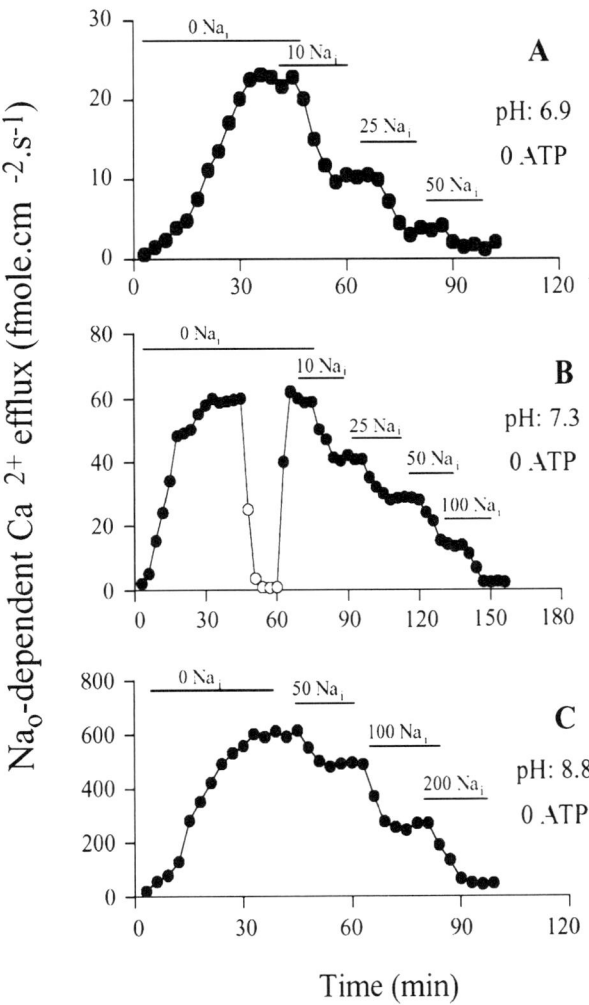

FIGURE 1. Effect of Na^+_i on the absolute values of forward Na^+/Ca^{2+} exchange at different pH_is in axons dialyzed without ATP. **(A)** Na^+_o-dependent Ca^{2+} efflux at pH 6.9. **(B)** Na^+_o-dependent Ca^{2+} efflux at pH 7.3. *Filled circles*: in the presence 440 mM extracellular Na^+_o. *Open circles*: in the absence of Na^+_o (presence of 440 mm Li^+). **(C)** Na^+_o-dependent Ca^{2+} efflux at pH 8.8. All concentrations are given in millimolar units. Notice the marked synergism between Na^+_i and H^+_i in inhibiting the exchanger. The average temperature was 17°C. (Reproduced with permission of the *Journal of Physiology*.[12])

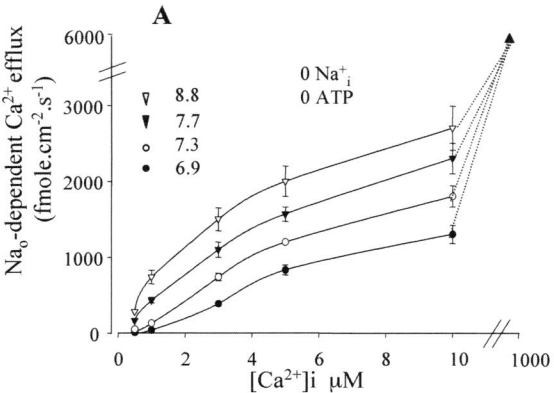

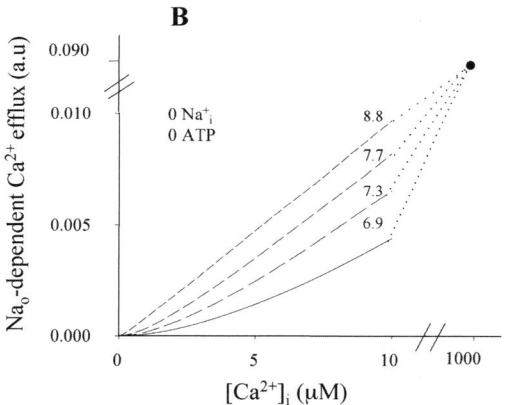

FIGURE 2. Effects of $[Ca^{2+}]_i$ on the forward Na^+/Ca^{2+} exchange flux at different pH_is in the absence of Na^+_i and ATP. **(A)** Na^+_o-dependent Ca^{2+} efflux at different $[Ca^{2+}]_i$ at pHs 6.9, 7.3, 7.7, and 8.8. The error bars indicate SEM. The mean temperature was 17°C. **(B)** Simulation using the model illustrated in FIGURE 8. Notice that the Ca^{2+}_i–H^+_i antagonism as well as the lack of effect of protons on the exchanger at saturating $[Ca^{2+}]_i$ (1000 µM), found experimentally, are predicted by the model. (Redrawn with permission of the *Journal of Physiology*.[12])

cept for the lack of the initial saturation kinetics seen in FIGURE 1A, the model predicts the main observation of FIGURE 2A: the inhibition by H^+_i at limiting $[Ca^{2+}]_i$ is overcome when Ca^+_i concentration is saturating.

FIGURE 3A summarizes the results of a number of axons where Na^+_i inhibition of forward Na^+_o/Ca^{2+}_i exchange was explored from 0 to 200 mM $[Na^+]_i$ at pH 6.9, 7.3, and 8.8. The data are plotted as a percentage of the fluxes obtained at zero Na^+_i. The open symbols correspond to ATP-depleted axons, and filled circles represent axons dialyzed with 3 mM MgATP at pH 6.9. FIGURE 3A shows clearly that H^+_i and Na^+_i

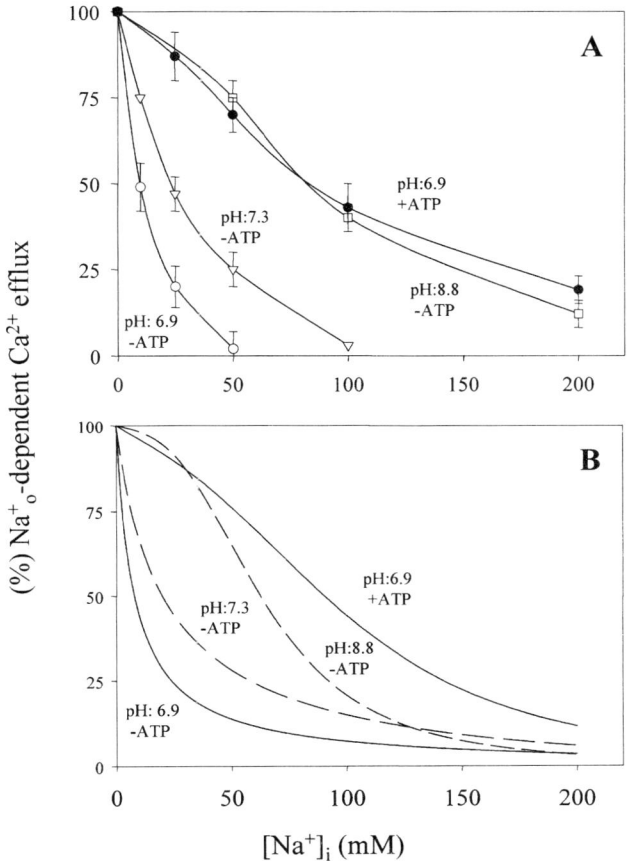

FIGURE 3. Na^+_i-dependent inhibition of forward Na^+/Ca^{2+} exchange at different pH_is in the absence and presence of 3 mM ATP. **(A)** Summary of the results obtained in several axons plotted as percentage of the Na_o-dependent Ca^{2+} efflux as a function of the $[Na^+]_i$ in the presence and absence of ATP. *Open symbols*: absence of ATP. *Filled circles*: presence of 3 mM MgATP. Ordinate: percentage of Na^+_o/Ca^{2+}_i exchange. Abscissa: Intracellular Na^+ in millimolar. The error bars indicate SEM. The mean temperature was 17°C. **(B)** Simulations using the model depicted in FIGURE 8. Notice the exquisite sensitivity of the exchange activity to Na^+_i at the acidic pH and the similarity between experimental and simulated data. (Redrawn with permission of the *Journal of Physiology*.[12])

act synergistically. In the absence of ATP, the apparent affinity for Na^+_i, taken as the $K_{0.5}$, was 10 mM, 40 mM, and 90 mM at pH 6.9, 7.3, and 8.8, respectively. In this regard our results agree with those of Doering and Lederer,[4] who found that in giant excised patches of guinea-pig cardiac sarcolemma, intracellular Na^+ acts as a cofactor in proton inhibition of the Na^+/Ca^{2+} exchanger. In addition, at pH_i 8.8, the curve for Na^+_i inhibition suggests that more than one Na^+ ion is involved, whereas at pH_i 7.3 and 6.9 the curves seem to follow single-ion kinetics. The role of ATP in the synergistic Na^+_i–H^+_i inhibition of the exchanger is evident in the fluxes represented by the filled circles. FIGURE 3A shows that upon the addition of 3 mM MgATP, the inhibition by Na^+ ions seen at pH 6.9 is markedly reduced; actually, the values of Ca^{2+} efflux observed at the different $[Na^+]_i$ investigated are very close to those seen at pH 8.8 in the absence of ATP. In other experiments not shown here, we found that in Na^+_i-free dialysis solutions at pH_i 6.9, MgATP enhances the Na^+_o-dependent Ca^{2+} efflux by fourfold, whereas it has no effect at pH 8.8. When 40 mM Na^+_i was included in the intracellular medium, MgATP increased the fluxes by sixfold at pH 6.9 and by less than 10% at pH 8.8. Therefore, and very interestingly, both ATP and alkalinization antagonize Na^+_i inhibition of the squid Na^+/Ca^{2+} exchanger. FIGURE 3B shows simulations of the experimental conditions described in FIGURE 3A on the basis of the model described below. As can be seen, the simulated curves are almost exact replicas of those obtained experimentally.

The Role of the Ca_i Regulatory Site on H^+_i–Na^+_i–ATP Interactions

The experiments described in FIGURES 1, 2, and 3 show that Na^+_i favors, whereas ATP and Ca^{2+}_i antagonize, H^+_i inhibition of the Na^+/Ca^{2+} exchanger estimated in its forward mode. Nevertheless, even with the appropriate simulation of FIGURE 1B, we cannot establish on which sites, or at which part of the reaction cycle, the antagonism of Ca^{2+} ions takes place. And this is simply because when measuring the forward exchange mode, the regulatory and transport Ca^{2+} sites cannot be distinguished. We already know, however, that in squid axons at limiting $[Ca^{2+}]_i$, ATP stimulation of the Na^+/Ca^{2+} exchanger occurs via an increase in the affinity of the intracellular Ca^{2+}_i regulatory site.[6] With these considerations in mind, we decided to carry out experiments in which the effects of Ca^{2+}_i on the Ca_i regulatory and transport sites could be dissociated. With that aim, the effects of H^+_i, ATP, and Ca^{2+}_i on the Na^+_o/Na^+_i partial reaction of the exchanger were followed in which the binding of Ca^{2+} to its intracellular regulatory site is required, but this cation is not transported. The Na^+ exchange rates were estimated on the basis of the Ca^{2+}_i-dependent, Na^+_o-stimulated [^{22}Na]Na^+ efflux being at a constant 40 mM $[Na^+]_i$. FIGURE 4A shows that in the absence of ATP (open symbols), the apparent affinity of the regulatory site for Ca^{2+}_i was strongly influenced by pH_i, with the $K_{0.5}$ for Ca^{2+} ranging from around 20 µM at pH_i 6.9 to 2 µM at pH_i 7.3 and 0.3–0.5 µM Ca^{2+}_i at pH_i 8.8. Interestingly, a fall of 0.4 pH_i units around the physiological axonal pH (from 7.3 to 6.9) reduced the $K_{0.5}$ for Ca^{2+}_i stimulation of Na^+_o/Na^+_i exchange about 10-fold. In addition, it is also remarkable that a similar affinity for Ca^{2+}_i can be obtained with pH_i 7.3 in the absence of ATP as with 3 mM ATP at an internal pH of 6.9 (filled circles). FIGURE 4B illustrates the excellent agreement between the experimental data and the simulations produced with the model described below where the Ca^{2+}_i regulatory site is central to the interaction of the squid Na^+/Ca^{2+} exchanger H^+_i, Na^+_i and ATP.

The Metabolic Role of ATP and the Relevance of Mg^{2+} Ions

The results presented above indicate that ATP stimulation of the squid Na^+/Ca^{2+} exchanger is consequence of the removal of the inhibition exerted by intracellular H^+ and Na^+ ions, which in turn prevent the binding of Ca^{2+} to its regulatory site. The experiments that follow explore the role played by Mg^{2+} ions in these interactions.

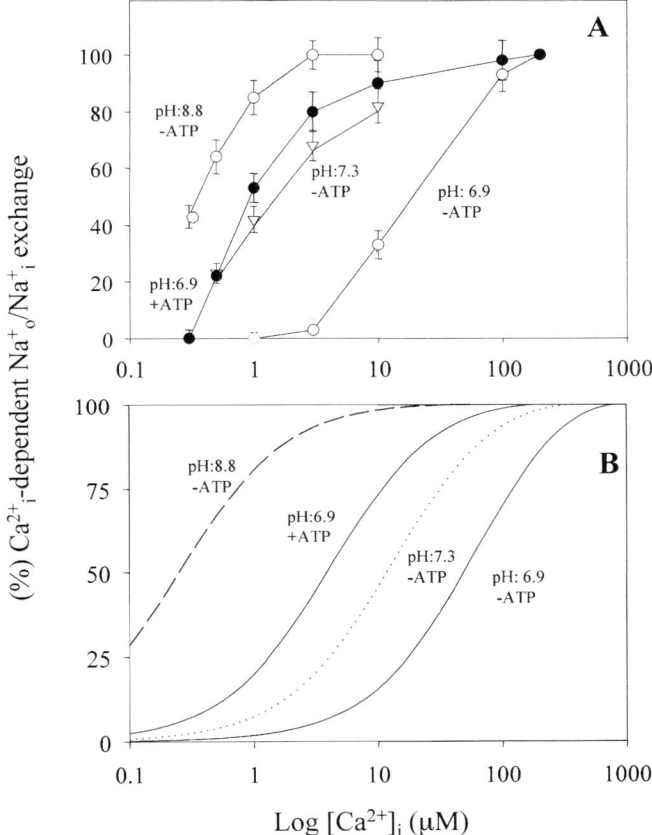

FIGURE 4. Effect of pH_i on the Ca^{2+}_i-dependent Na^+_o/Na^+_i exchange in the absence and presence of ATP. **(A)** Steady-state Na^+_o-dependent Na^+ efflux at three different pH_is as a function of $[Ca^{2+}]_i$, plotted as percentage of the maximum fluxes at saturating concentrations of calcium (1000 µM). *Open symbols*: solutions free of ATP at pH 6.9, 7.3, and 8.8. *Filled circles*: fluxes obtained at pH 6.9 in the presence of 3 mM MgATP. The error bars indicate SEM. The mean temperature was 17.5°C. Notice that (i) there is a large change in the apparent affinity of the Na^+_o/Na^+_i exchange for Ca^{2+}_i between acid and alkaline pH in the absence of ATP and (ii) the addition of 3 mM MgATP brings the Ca^{2+}_i affinity close to that seen at pH 7.3 without ATP. **(B)** Simulation using the model described in FIGURE 8. Note the agreement between experimental and simulated data. (Redrawn with permission of the *Journal of Physiology*.[12])

FIGURE 5 depicts the intracellular Mg^{2+} dependence of the ATP stimulation of Na^+/Ca^{2+} exchange in experiments where the magnitude of the ATP stimulation of the Na^+_o-dependent Ca^{2+} efflux was followed at a constant ATP concentration of 4 mM. The concentration of ionized Mg^{2+}_i was varied between 0.5 and 10 mM. As illustrated in FIGURE 5, the response is biphasic: stimulation up to about 1 mM Mg^{2+}_i followed by a progressive decline, which at 10 mM Mg^{2+}_i reaches the values of the

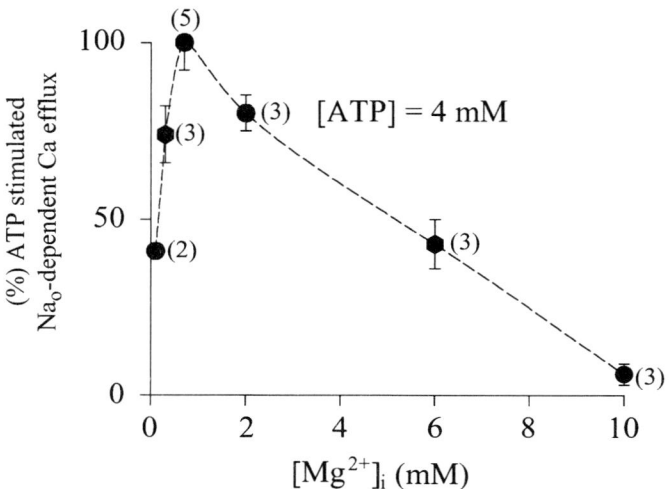

FIGURE 5. Dose–response relationship between the ATP-stimulated, Na^+_o-dependent Ca^{2+} efflux and intracellular ionized $[Mg^{2+}]$ at constant [ATP]. Na^+_o-dependent Ca^{2+} efflux in the presence of 4 mM ATP was measured in five different dialyzed squid axons in which $[Mg^{2+}]$ in the dialysis was varied between 0.2 and 10 mM. The points correspond to the mean ± SEM, normalized to the Ca^{2+} efflux values obtained with 1 mM $[Mg^{2+}]$. The number in parentheses referred to different axons. Notice the bell shape of the ATP-stimulated Na^+/Ca^{2+} exchange fluxes. Temperature: 17.5°C. (Reproduced with permission from the *American Journal of Physiology*.[9])

nonstimulated Ca^{2+} fluxes. The rising phase of the curve likely indicates the need for Mg^{2+} in order to promote phosphorylation by ATP through some kinase system. This proposition is supported by additional observations: (i) ATP-γ-S, a substrate for kinases but not for ATPases, also stimulates the exchanger and (ii) CrATP, inhibitor of most kinases, prevents the ATP stimulation.[6] Unfortunately, and despite all our efforts, we have been unable to individualize the enzyme involved, but we have established that a cytosolic protein of low molecular weight is essential for this process.[6,10] If a kinase-dependent phosphorylation is part of the Mg stimulation part,[7] it is likely that a phosphatase (or phosphatases), which is Mg^{2+} dependent,[8] is responsible for the deactivation at high Mg^{2+}_i concentrations. We have tried several phosphatase inhibitors but, again, were unable to identify one that works. However, the experiments described in the two following figures point to these enzymes as responsible for the decline in fluxes observed in FIGURE 5. FIGURE 6A describes an experiment in which an axon was dialyzed, always in the presence of 2 mM free Mg^{2+}. The addition of ATP produces the usual stimulation of the Na^+_o-dependent Ca^{2+} efflux (evidenced by its disappearance upon removal of external Na^+), and that stimulation ceases when ATP is taken away from the dialysis solution. FIGURE 6B shows that removal of ATP *per se* is not the main factor, because deactivation following removal of ATP is prevented when the intracellular medium has been previously depleted of Mg^{2+} ions. In other experiments (not shown here), deactivation of the ATP-

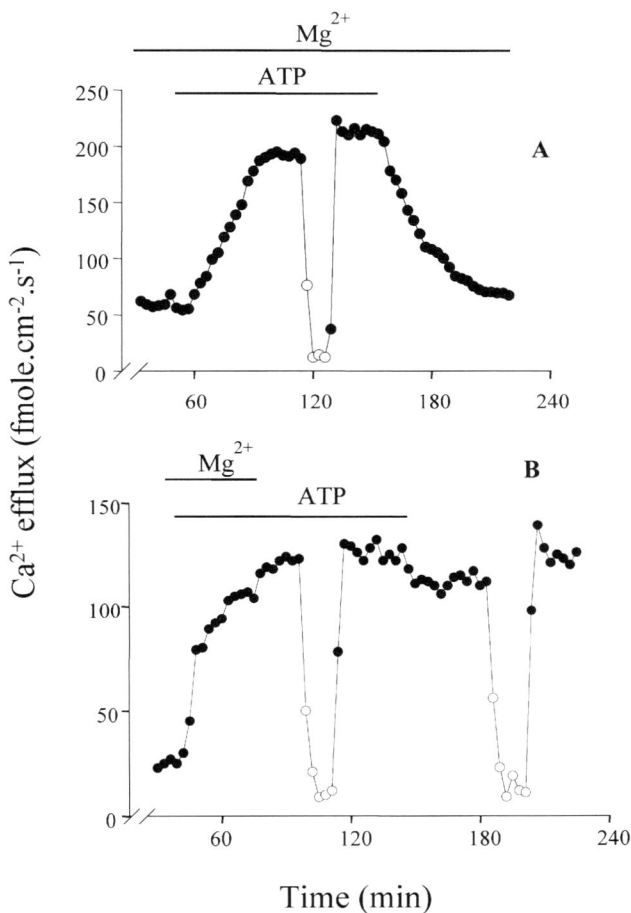

FIGURE 6. Mg^{2+}-ATP interactions in the upregulation of the Na^+/Ca^{2+} exchanger in dialyzed squid axons. *Filled circles*: Ca^{2+} efflux in the presence of external Na^+. *Open circles*: Ca^{2+} efflux in the absence of external Na^+. Vanadate (0.1 mM) was included in all dialysis solutions to inhibit the Ca^{2+} pump. Notice that all Ca^{2+} efflux is Na^+_o-dependent forward Na^+/Ca^{2+} exchange. **(A)** Reversibility of the MgATP stimulation of forward Na^+/Ca^{2+} exchange upon the removal of ATP in the presence of 2 mM intracellular Mg^{2+}. **(B)** Removal of Mg^{2+} in the presence of ATP or removal of ATP in the absence of Mg^{2+} keeps the exchanger in the upregulated state. Notice that the exchanger can be maintained in this activated state for more than 2 hours. (Reproduced with permission from the *American Journal of Physiology*.[9])

upregulated Na^+/Ca^{2+} exchanger can be achieved by re-addition of ionized Mg^{2+} in the absence of ATP. These results are consistent with the idea that Mg^{2+} stimulates a dephosphorylation process (a phosphatase[8]). This possibility is strengthened even further by the experiment illustrated in FIGURE 7 performed in membrane vesicles of squid optical nerves. In this case, addition of 50 units/ml of an alkaline phosphatase

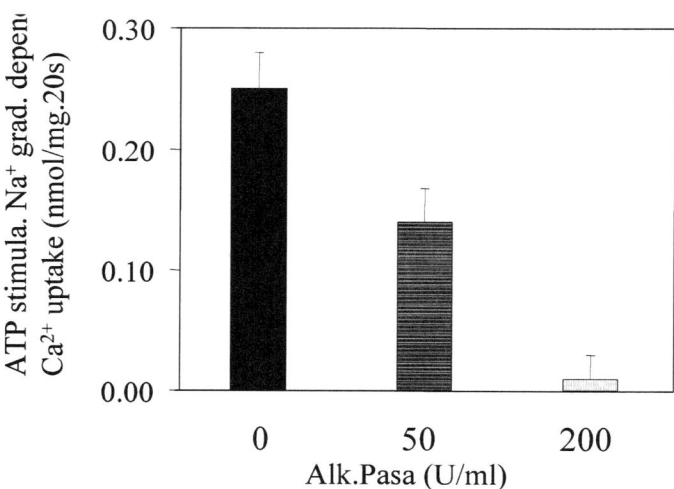

FIGURE 7. Effect of alkaline phosphatase on ATP stimulation of Na^+/Ca^{2+} exchange in squid optic nerve vesicles. Uptake of $[^{45}Ca]Ca^{2+}$ was estimated without or with 50 or 200 U/ml of alkaline phosphatase. Before starting ^{45}Ca uptake, concentrated membrane vesicles were incubated for 2 min at 20°C in the absence of a Na^+ gradient with no Ca^{2+} but with Mg^{2+} and ATP in the presence and absence of alkaline phosphatase. All experiments were done in the presence of the soluble cytosolic protein required for MgATP stimulation of the Na^+/Ca^{2+} exchanger in the squid.[6] Values are the mean ± SEM of triplicate determinations. (Reproduced with permission from the *American Journal of Physiology*.[9])

reduced to half, and 200 units/ml of that phosphatase completely prevented, the MgATP stimulation of the Na^+/Ca^{2+} exchange fluxes.

A Kinetic Model for H^+_i–Na^+_i–ATP Interactions with the Squid Na^+/Ca^{2+} Exchanger

FIGURE 8 shows the basic features of a model of the Na^+/Ca^{2+} exchanger in which for simplicity only the intracellular ionic interactions related to the Ca^{2+} regulatory site and associated with H^+ and Na^+ inhibition have been taken into account; that is, the forms loaded with Na^+ and Ca^{2+} into their transporting sites are ignored. Therefore, the forms of the carrier considered here are the following: X, free of ligands; $Ca_r.X$, loaded with Ca^{2+} at the regulatory site; and H.X, H.X.Na, and $H_2.X.Na$, which are carriers binding H^+ and Na^+ at their inhibitory sites. The basic assumptions are: (i) proton inhibition occurs even in the absence of Na^+ (by the formation of the dead-end H.X complex); (ii) H.X can bind one Na^+ to a nontransporting site, and the H.X.Na complex allows the binding of a second proton forming the dead-end $H_2.X.Na$ complex; (iii) the binding of the first proton excludes that of Ca^{2+} to its regulatory site, which is essential for Ca^{2+} and Na^+ binding to their transporting sites but not for cation translocation; (iv) Na^+ and Ca^{2+}_i compete for the intracellular transporting sites; (v) MgATP, through an unknown phosphorylation process, antagonizes inhibition by protons in the absence, but more conspicuously, in the presence of Na^+_i by decreasing their apparent affinity for binding to the carrier; and

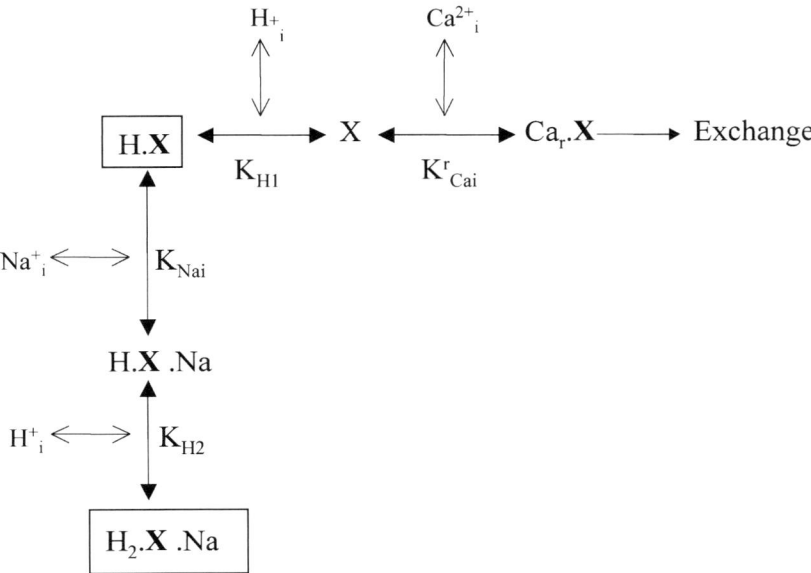

FIGURE 8. Kinetic model simulations for the Na^+_i, H^+_i, Ca^{2+}_i, and ATP interactions in the regulation of the squid Na^+/Ca^{2+} exchanger. Notice that with the values of the constant used, at physiological pH, Ca^{2+}_i, and Na^+_i, the fraction of carriers available for translocations is quite small. (Reproduced with permission of the *Journal of Physiology*.[12])

(vi) the binding of all ligands is at rapid random equilibrium and 3 Na^+ do so simultaneously.

On the basis of the following constants:

- K_{H10}, true affinity for the first proton site, 1×10^{-9} M
- K_{H20}, true affinity of the $H_2.X.Na$ for the second proton, 1×10^{-8} M
- K_{Naio}, true affinity for Na^+_i of the H.X complex, 1×10^{-1} M
- K_{ATP}, K_m for the effects of MgATP, 200 µM
- Factor_ATP_K_{H1}, decrease by ATP of the apparent affinity for H^+_i binding to X = 2
- Factor_ATP_K_{H2}, decrease by ATP of the apparent affinity for H^+_i binding to H.X.Na = 2
- Factor_ATP_K_{Nai}, decrease by ATP of the apparent affinity for Na^+_i binding to H.X = 10
- $K_{H1} = K_{H10} \times (1 + (\text{factor\_ATP\_}K_{H1} \times [ATP]/(K_{ATP} + [ATP]))$
- $K_{H2} = K_{H20} \times (1 + (\text{factor\_ATP\_}K_{H2} \times [ATP]/(K_{ATP} + [ATP]))$
- $K_{Nai} = K_{Naio} \times (1 + (\text{factor\_ATP\_}K_{Nai} \times [ATP]/(K_{ATP} + [ATP]))$

we can reproduce all the experimental results shown above.

This work permits us to conclude that the intracellular Ca^{2+} regulatory site plays a central role in the synergistic ($H^+_i + Na^+_i$) inhibition of the Na^+/Ca^{2+} exchanger. In addition, ATP modulation of the Na^+/Ca^{2+} exchange occurs with protection from intracellular proton and sodium inhibition. At physiological $[Na^+]_i$, this effect occurs primarily in the pH_i range from 6.9 to 7.3. These effects of ATP are linked to the interplay of phosphorylation–dephosphorylation processes. A simple kinetic model for Na^+/Ca^{2+} exchange that takes into account H^+_i–Na^+_i–ATP interactions by influencing the binding of Ca^{2+} to its intracellular regulatory site satisfactory reproduces the experimental results.

ACKNOWLEDGMENTS

This work was supported by grants from National Science Foundation (USA) (IBN-9631107); Consejo Nacional de Investigaciones Científicas y Tecnólogicas (CONICIT-Venezuela S1-99000946); Fundación Polar (Venezuela); Fundaciencias-IVIC, Consejo Nacional de Investigaciones Científicas y Técnicas (CONICET-Argentina 4904/97); Agencia Nacional de Promoción Científica y Tecnológica-FONCYT-Argentina (PICT99 05-05158); Agencia Córdoba Ciencia, Argentina (181/01); and Beca Carrillo-Oñativia, Ministerio de Salud de la Nación, Argentina.

REFERENCES

1. BLAUSTEIN, M.P. & W.J. LEDERER. 1999. Sodium/calcium exchange: its physiological implications. Physiol. Rev. **79:** 764–830.
2. REQUENA, J. 1978. Calcium efflux from squid axons under constant sodium electrochemical gradient. J. Gen. Physiol. **72:** 443–470.
3. HILGEMANN, D.W. & S. MATSUOKA. 1992. Steady-state and dynamic properties of cardiac sodium–calcium exchanger. Secondary modulation by cytoplasm calcium and ATP. J. Gen. Physiol. **100:** 933–961.
4. DOERING, A.E. & W.J. LEDERER. 1994. The action of Na^+ as a cofactor in the inhibition by cytoplasmic protons of the cardiac Na^+/Ca^{2+} exchanger in the guinea-pig. J. Physiol. **480:** 9–20.
5. DIPOLO, R. & L. BEAUGÉ. 1982. The effect of pH on Ca extrusion mechanism in dialyzed squid axons. Biochim. Biophys. Acta **688:** 237–245.
6. DIPOLO, R. & L. BEAUGÉ. 1999. Metabolic pathways in the regulation of invertebrate and vertebrate Na^+/Ca^{2+} exchange. Biochim. Biophys. Acta **1422:** 57–71.
7. DUNAWAY-MARIANO, D. & W.W. CLELAND. 1980. Investigations of substrate specificity and reaction mechanism of several kinases using chromium (III) adenosine 5-triphosphate and chromium (III) adenosine 5 diphosphate. Biochemistry **19:** 1506–1515.
8. COHEN, P.T. 1989. The structure and regulation of protein phosphatases. Annu. Rev. Biochem. **58:** 453–508.
9. DIPOLO, R., G. BERBERIÁN & L. BEAUGÉ. 2001. In squid nerves intracellular Mg promotes deactivation of the ATP-upregulated Na/Ca exchanger. Am. J. Physiol. **279:** C1631–C1639.
10. DIPOLO, R., G. BERBERIAN, D. DELGADO, et al. 1997. A novel 13 kD cytoplasmic soluble protein is required for the nucleotide modulation of the Na/Ca exchange in squid nerve fibers. FEBS Lett. **401:** 6–10.
11. DIPOLO, R. & L. BEAUGÉ. 1987. In squid axons, ATP Modulates Na/Ca exchange by a Ca^{2+}-dependent phosphorylation. Biochim. Biophys. Acta **897:** 347–354.
12. DIPOLO, R. & L. BEAUGÉ. 2002. MgATP counteracts intracellular proton inhibition of the sodium–calcium exchange in dialysed squid axons. J. Physiol. **539:** 791–803.

Pathways Regulating Na$^+$/Ca^{2+} Exchanger Expression in the Heart

DONALD R. MENICK,[a] LIN XU,[a] CHRISTIANA KAPPLER,[a] WENJING JIANG,[a] PATRICK R. WITHERS,[a] NEAL SHEPHERD,[b] SIMON J. CONWAY,[c] AND JOACHIM G. MÜLLER[a]

[a]*Gazes Cardiac Research Institute, Medical University of South Carolina, Charleston, South Carolina 29425, USA*

[b]*Department of Veterans Affairs, Durham, North Carolina, USA*

[c]*Medical College of Georgia, Augusta, Georgia, USA*

ABSTRACT: The Na$^+$/Ca^{2+} exchanger (NCX1) is regulated at the transcriptional level in cardiac hypertrophy, ischemia, and failure. Following pressure overload, activation of MAPKs coincides with the kinetics of NCX1 gene upregulation in adult cardiocytes. Using adenoviral gene delivery, we begin to identify the molecular pathways responsible for upregulation of the exchanger gene. Inhibition of ERK with the MEK inhibitor UO126, the ERK protein phosphatase MKP-3, inhibited ERK activation, but only inhibited α-adrenergic-induced NCX1 upregulation by 30%. Overexpression of DN-JNK lowered basal NCX1 expression. Overexpression of activated MKK-3 was sufficient for α-adrenergic-stimulated upregulation of the reporter gene. Together, this data indicates that (1) JNK mediates basal cardiac expression of the NCX1 gene, (2) ERK and p38 play a role in α-adrenergic-stimulated NCX1 upregulation, and (3) p38 activation alone is sufficient for NCX1 upregulation.

KEYWORDS: gene expression; regulation; signal transduction; integrins; MAP kinases; hypertrophy; transgenic mice

INTRODUCTION

The upregulation or downregulation of genes, several of which are normally expressed in the embryonic heart, is associated with distinct pathophysiological phenotypes in the hypertrophying and failing heart. One of those genes is the one for the Na$^+$/Ca^{2+} exchanger (NCX1). There is a rapid upregulation of NCX1 mRNA in response to pressure overload.[1–3] The exchanger is also upregulated in end-stage heart failure.[3–7] Importantly, variations in the concentration of the Na$^+$/Ca^{2+} exchanger protein among myocytes from failing human hearts are inversely related to variations in the frequency-dependent increase in diastolic calcium, which results in diastolic dysfunction.[8] Therefore, the upregulation of the exchanger has been

Address for correspondence: Donald R. Menick, Ph.D., Gazes Cardiac Research Institute, Medical University of South Carolina, 114 Doughty Street, Rm 203, Charleston, SC 29425. Voice: 843-876-5045; fax: 843-792-4762.
menickd@musc.edu

proposed as a compensatory mechanism for the decrease in SERCA level. The increase in exchanger activity would increase calcium extrusion and act to preserve low diastolic calcium levels. But others have proposed that any change in either the abundance or activity of the exchanger or SERCA would lead to imbalances in Ca^{2+} homeostasis.[9] It has recently been demonstrated that Ca^{2+} extrusion via the exchanger produces a transient inward, depolarizing current I_{ti}, which can produce a delayed depolarization and brings the diastolic membrane potential closer to the threshold for triggering an inappropriately timed action potential.[10] With the increased expression of the exchanger resulting in increased activity, any given amount of SR Ca^{2+} release will result in greater inward I_{ti} and increased probability that a triggered arrhythmia will result.[11] Although still controversial, there is growing evidence for an important role for the exchanger in altered E–C coupling and arrhythmogenesis in the context of cardiac hypertrophy and failure. Unfortunately, very little is known about the genetic elements and transcription factors that regulate the level of NCX1 expression in the heart.

NCX1H1 PROMOTER REGULATES CARDIAC–SPECIFIC EXPRESSION IN A TRANSGENIC MOUSE MODEL

The multiple tissue-specific variants of the Na^+/Ca^{2+} exchanger are the result of alternative promoter usage and splicing.[12–15] The NCX1 gene contains three promoters adjacent to three 5′-UTR exons located over a region of 35 kb, approximately 30 kb 5′ of the first coding exon. The NCX1 gene has been cloned in cat, rat, and human,[12,16,17] each having three promoters with nearly identical structure. On the basis of 5′-RACE results together with genomic mapping, we proposed that the most distal promoter was cardiac-specific, the promoter 5′ of the center 5′-UTR exon-regulated expression in the kidney, and the most proximal promoter was responsible for low-level ubiquitous expression as well as higher levels of expression in the brain. Therefore, the three NCX1 5′-UTR exons were designated H1, K1, and Br1; and the corresponding promoter regions distal to each designated cardiac, kidney,

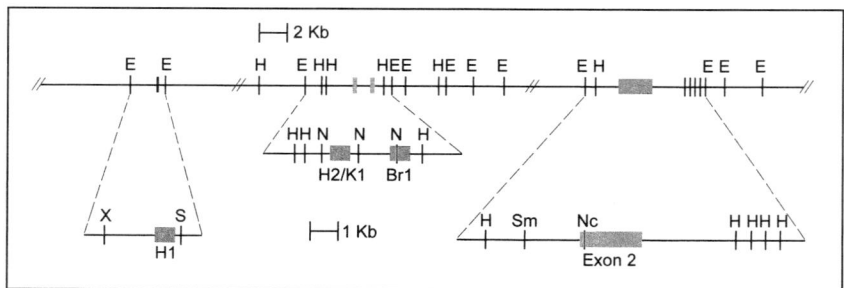

FIGURE 1. Structure of the 5′-end of the feline NCX1 gene. A partial restriction map of this region is shown, with closed boxes depicting the exons with noncoding regions H1, K1, and Br1 and the first exon (exon 2), encoding translated regions. (Modified from Barns et al.[12] Reprinted by permission from *Journal of Biological Chemistry*.)

and brain (FIG. 1). Alternatively, Scheller proposed that no single transcription start can be called tissue specific.[17] In this model, the Br1 promoter drives ubiquitous expression of NCX1, and the H1 and K1 promoters serve as auxiliary promoters augmenting expression in various tissues. But transfection data from our work and that of Susan Nicolas[16] was consistent with the distal promoter being important for cardiac expression. Our construct of the feline distal promoter containing 184 bases of the 5′-flanking region, the H1 exon, and 67 bases of the first intron was sufficient for expression and α-adrenergic stimulation of the luciferase reporter gene in neonatal cardiocytes.[18] Furthermore, the feline NCX1 heart promoter did not drive expression in mouse L cells, human CHO, or 293 cells.

In order to directly address the role of the NCX1H1 promoter in the adult heart, we produced transgenic mouse lines containing a luciferase reporter driven by the H1 promoter of the feline NCX1 gene. The NCX1H1 promoter was sufficient for driving the normal spatio-temporal pattern of NCX1 expression in cardiac development.[19] The luciferase reporter gene was expressed in a heart-restricted pattern in both the early embryo (E8–E14 dpc) and in later embryos (after E14 dpc) when NCX1 was also expressed in other tissues. In the adult, no luciferase activity was detected in the kidney, liver, spleen, uterus, or skeletal muscle; minimal activity in the brain; but very high levels of luciferase expression were detected in heart. Transverse aortic constriction-operated mice showed a significantly increased left ventricle mass after 7 days. In addition, there is a twofold upregulation of NCX1H1 promoter activity in the LV after 7 days of pressure overload compared to both control and sham-operated animals. This work demonstrates that the NCX1H1 promoter directs cardiac-specific expression of the exchanger in both the embryo and adult and is also sufficient for the upregulation of NCX1 in response to pressure overload.[19]

This work shows for the first time that the distal NCX1 promoter region, including −1831 bp to +67 bp of intron 1, encompassing all of exon H1, is responsible for heart-specific expression of the exchanger in both the embryo and adult and also is responsible for the upregulation of NCX1 in response to pressure overload.

IDENTIFICATION OF *CIS*-ELEMENTS AND NUCLEAR FACTORS MEDIATING CARDIAC EXPRESSION AND UPREGULATION OF NCX1

Using the full-length (1831-bp) construct, we introduced site-specific point mutations into each of the consensus elements, and the activity of each of the mutants was compared with the wild-type full-length construct to determine its contribution to cardiac-specific expression.[18] This mutational analysis revealed that both the CArG element at −80 and the GATA element at −50 were required for expression in neonatal cardiocytes (FIG. 2). To analyze the potential interactions of the NCX1 cardiac elements with *trans*-acting factors, we conducted electrophoretic mobility shift assays. Supershift analysis demonstrated that the serum response factor binds to the −80 CArG element. Supershift analysis also showed that GATA-4, but not GATA-6, binds to the −50 GATA element.[18] Stimulation of the α-adrenergic receptor results in a hypertrophic response characterized by an increase in cell size, increased protein synthesis, and the transcriptional upregulation of a number of cardiac genes, including NCX1. We have identified two elements, the E-box at −172 and a novel element

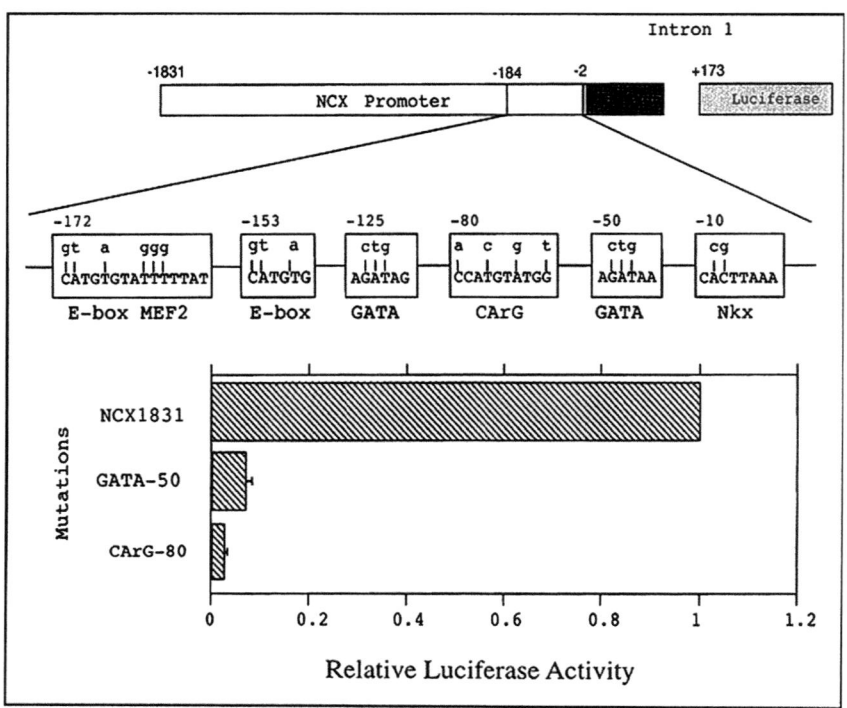

FIGURE 2. The GATA and CArG elements are essential for cardiac expression of the NCX1 gene. Effect of transcriptional element mutations on expression of the NCX1 gene. Point mutations were created in the full-length, 1831-bp NCX1 promoter–luciferase construct. Relative luciferase values for the mutant constructs are shown as a percentage of the wild type. A CMV promoter β-gal fusion vector was cotransfected to normalize transfection efficiency.

at +104, that mediate upregulation of the NCX1 gene in response to α-adrenergic stimulation in neonatal cardiocytes.[18] Analysis of the NCX1 rat promoter revealed several GATA-binding sequences,[20] two of which correspond in position to −125 and −50 sites that we had examined in the feline promoter. Mutation of the rat NCX1 proximal GATA site (−75) abolishes GATA-4 binding and reduces promoter activity by 60%, whereas mutation of the −145 GATA site reduces activity by 30%.[21] Again, this is consistent with what we observed with the feline promoter.

CONSTRUCTION OF THE NCX1H1 PROMOTER LUCIFERASE ADENOVIRUS

Most of the signal transduction and gene regulation studies to date (including what is discussed above) have used the rat neonatal cardiocyte as a model for adult cardiac hypertrophy. Neonatal cardiocytes are in a state of growth by their very nature such that conceptual extension of what is found in the neonatal cardiocyte to hypertrophy in the terminally differentiated adult cardiocyte can be misleading. Furthermore, the advent of adenoviral vectors now allows for efficient gene delivery to the adult cardiocyte and permits us to carry out these studies in a physiologically more relevant model system.

A construct containing 1831 bases of the promoter region, the first exon and 67 bases of the first intron, was fused to the luciferase reporter gene, and the fragment digested out with *Bam*H1 and ligated into the *Bgl*II site of the promotorless pAdTrack vector. The pAdTrack vector contains the green fluorescent protein (GFP) gene in the viral backbone. GFP expression is driven by a cytomegalovirus (CMV) promoter.[22] This allows for the direct assessment of the efficiency of viral infection and even the normalization of infection when using more than one NCX1 promoter-luciferase construct. The resultant plasmid was linearized with *Pme1* and cotransformed into *E. coli* BJ5183 cells with the adenoviral backbone plasmid pAdEasy-1. Recombinants were selected and high-titer viral supernatants prepared.

Adult feline cardiocytes were plated on laminin-coated dishes and immediately infected with the NCX1 promoter–luciferase adenovirus. Experiments were performed to determine the optimum multiplicity of infection (MOI), infection time, and the kinetics of luciferase expression post infection. We found that MOIs of ~1.5 viruses/adult cardiocyte with infection times between 4 and 12 h were optimum. GFP expression demonstrated that 80–90% of the adult cardiocytes were infected. Luciferase activity could be detected by 6–12 h post infection but reached steady-state levels of expression around 48 h post infection (data not shown). Importantly, there are no dramatic differences in morphology between the infected cells and non-infected cells from the same isolation. The infected cardiomyocytes contract when electrically stimulated and have completely normal morphology and normal resting sarcomere lengths.

ROLE OF MAP KINASES IN α-ADRENERGIC-STIMULATED UPREGULATION OF THE EXCHANGER

There are three subfamilies of the MAPK: ERKs, JNKs, and the p38 MAPK. MAPKs are activated in both *in vitro* and *in vivo* models of hypertrophy. Stimulation of the α-adrenergic receptor results in a hypertrophic response characterized by an increase in cell size, increased protein synthesis, activation of MAPKs, and the transcriptional upregulation of a number of cardiac genes, including NCX1. Sugden has demonstrated that activation of MAP kinases results in the overexpression of several hypertrophic gene markers.[23] Here we test whether MAPKs play a role in the upregulation of the exchanger in adult cardiocytes.

When adult cardiocytes are co-infected with our NCX1H1 promoter-luciferase and the ERK-specific MAP kinase phosphatase adenovirus (MKP-3), 40% of the PE-stimulated upregulation of NCX1 gene expression is inhibited. In addition, when cardiocytes are treated with the MEK1/2 inhibitor, UO126 (5 µM), PE stimulation is inhibited by 40% (Fig. 3). Importantly, the treatment with UO126 or overexpression of MKP-3 completely inhibits the α-adrenergic activation of ERK. Therefore, α-adrenergic-stimulated upregulation of the exchanger is mediated in part by ERK1/2.

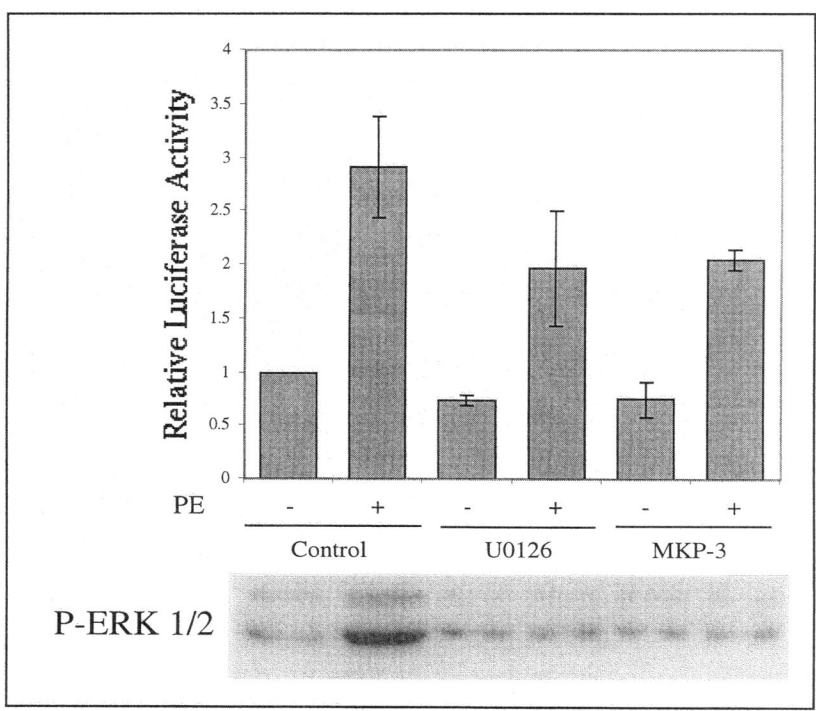

FIGURE 3. ERK mediates part of the α-adrenegric stimulated upregulation of the exchanger. Adult cardiocytes were infected with the NCX1 promoter–luciferase adenovirus or co-infected with NCX1 promoter–luciferase and MKP-3 adenovirus. Twelve hours post infection, medium was changed, and the cell was treated with 10 μM PE for 48 h. Luciferase activities were normalized to GFP fluorescence and expressed as a percent of control. Where indicated, 5 μM UO126 was added with PE.

We next looked at the possible contributions of JNK and p38 to NCX1 upregulation. FIGURE 4 demonstrates that when adenoviruses expressing a dominant-negative (dn) JNK are co-infected with the NCX1-promoter-luciferase adenovirus, basal NCX1 expression is inhibited. dn JNK also inhibited PE stimulation of NCX1. We are currently determining whether the inhibition of stimulated activity is only a reflection of the effect of dn JNK on NCX1 basal expression.

In order to determine whether p38 contributes to the α-adrenergic upregulation of NCX1, we used the p38α/β inhibitor, SB202190. FIGURE 5 clearly shows the treatment of cardiocytes with SB202190 (1 μM) inhibits the PE stimulation of NCX1 luciferase activity but has no effect on basal expression of the exchanger. These results were verified by using an adenovirus expressing dn p38. Overexpression of the dn p38 at levels equal to or greater than that of wild-type p38 had very little effect on basal NCX1 expression. At these same levels, PE-induced upregulation was dramatically inhibited. Importantly, expression of an activated MKK3 re-

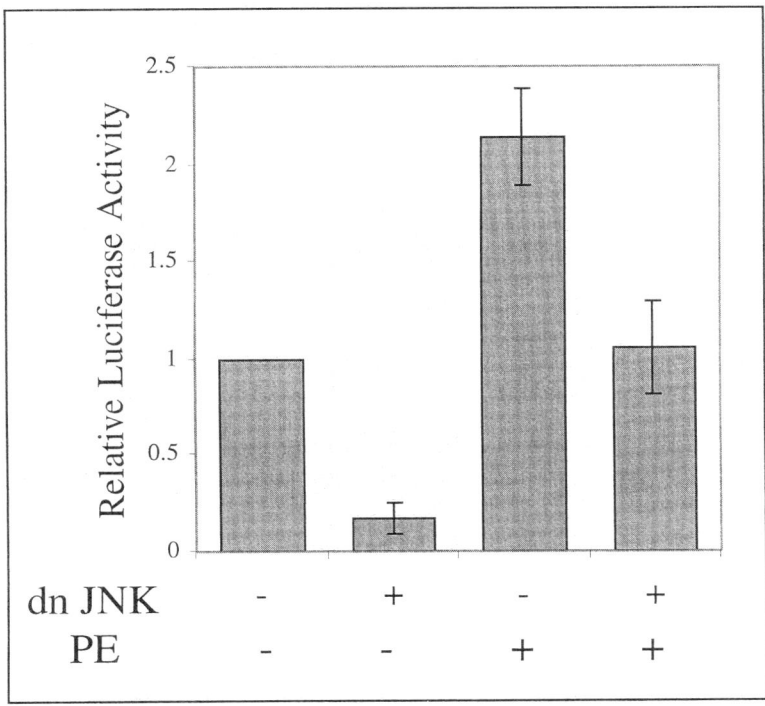

FIGURE 4. JNK mediates NCX1 basal expression. Adult cardiocytes were plated on laminin-coated dishes and co-infected with the NCX1 promoter–luciferase (MOI 1.5) and the dn JNK adenovirus (MOI 5). Twelve hours post infection, medium was changed, and cells were treated with 10 μM PE for 48 h. Luciferase activities were normalized to GFP fluorescence and expressed as a percent of control. Levels of the dn JNK were compared to endogenous JNK by immunoblotting with anti-JNK antibodies (not shown).

sults in the upregulation of NCX1, whereas overexpression of wild-type p38 alone had no effect (data not shown).

ROLE OF INTEGRIN ACTIVATION IN THE UPREGULATION OF NCX1 IN ADULT CARDIOCYTES

Integrins bind to extracellular matrix molecules, including fibronectin, laminin, and collagen. Individual integrins often bind to more than one ligand, and individual matrix molecules bind to more than one integrin. These ligands bind to the integrins via a tripeptide RGD motif. The "RGD" motif of fibronectin and vitronectin is critical for their interaction with their respective integrins (α5β1 and αvβ3). Unfortunately, this motif is masked in the native conformation of these ECM proteins[24] when they are in solution. Therefore, we utilized a synthetic penta-peptide that has a sequence homology to part of the type III-10 domain repeat of fibronectin

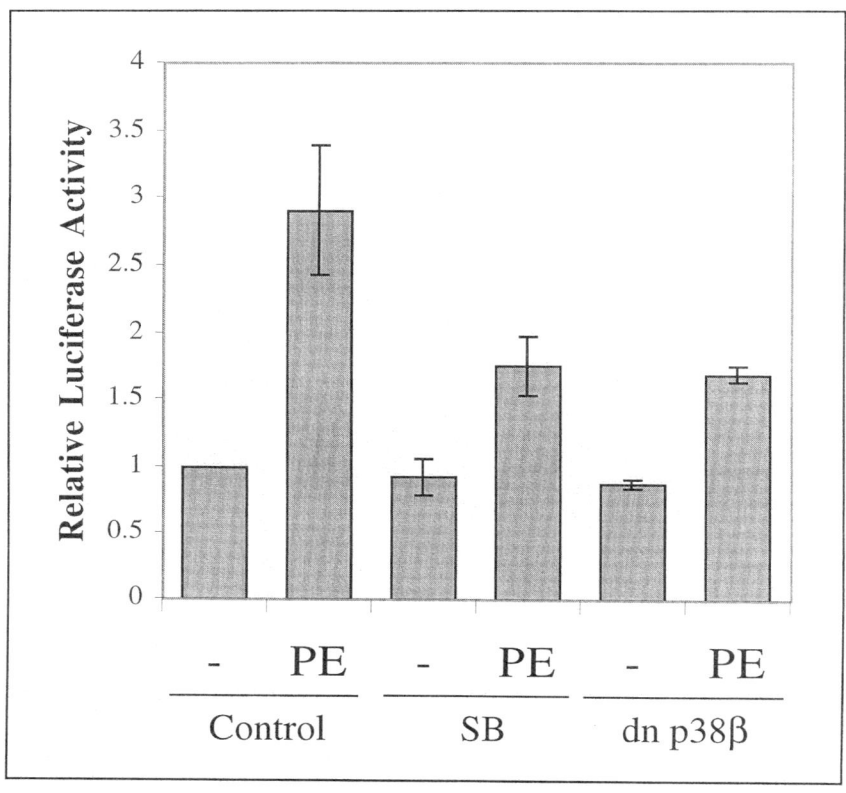

FIGURE 5. Effects of p38 inhibitor and expression of dn p38 on NCX1 basal and α-adrenergic-stimulating expression. Adult cardiocytes were infected with the NCX1 promoter-luciferase adenovirus at a MOI of 1.5 and, where indicated, co-infection with dn p38β adenovirus (MOI 5). Twelve hours post infection, medium was changed, and cells were treated with 10 μM PE or with 10 μM PE in the presence of 1 μM SB 202190 for 48 h. Luciferase activities were normalized to GFP fluorescence and expressed as a percent of control. Total p38 and activated p38 were measured by Western blotting (not shown).

GRGDS. Peptides containing the RGD motif can be used at low concentrations to block integrin binding to extracellular ligands, but at higher concentrations they can stimulate integrin signaling. The **GRGES** peptide, which does not interact with integrins, was used as a control peptide. On day 1 cardiocytes infected with the NCX1 promoter-luciferase adenovirus were left untreated or treated with RGD (2.5 mM) or RGE (2.5 mM) for 48 h. Activation of integrins with RGD induced an approximately twofold increase in luciferase activity over that of untreated or RGE-treated cardiocytes (FIG. 6). Importantly, treatment with the MEK1/2 inhibitor, U0126 (5 μM), inhibited >60% of the RGD-stimulated upregulation. Further treatment with the PKC

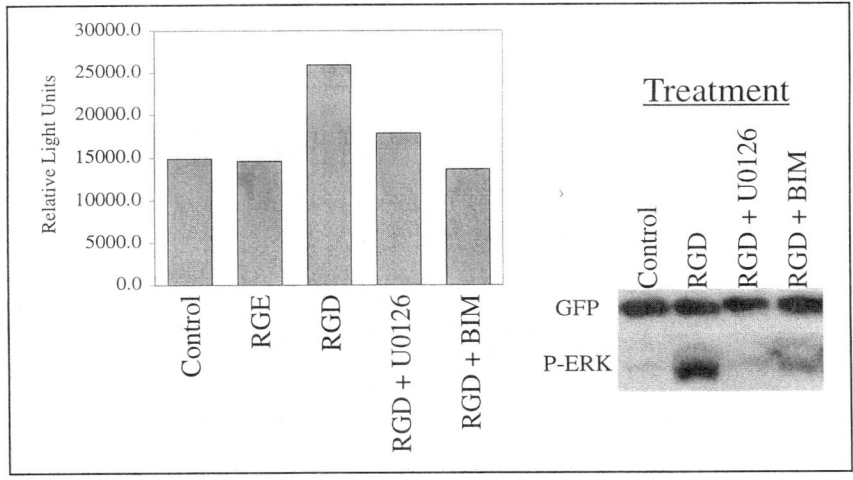

FIGURE 6. Integrin activation in adult cardiocytes results in the upregulation of the NCX1 gene. Adult cardiocytes were plated on laminin-coated dishes and infected with the NCX1 promoter–luciferase adenovirus at a MOI of 1.5. Twelve hours post infection, medium was changed, and cells were treated with RGE (2.5 mM), RGD (2.5 mM), or RGD (2.5 mM) in the presence of the MEK inhibitor 5 μM UO126 or PKC inhibitor 1.4 μM BIM. Relative luciferase values are shown for each treatment. Western blot was probed with anti-GFP and anti-phospho-ERK1/2.

inhibitor bisindoylmaleimide (BIM) at 1.4 μM completely inhibited integrin-stimulated NCX1 upregulation.

These data demonstrate that RGD stimulation of integrins results in activation of ERK1/2, which mediates in part the upregulation of the NCX1 gene in isolated adult cardiocytes.

In summary, the distal NCX1 promoter region, including −1831 bp to +67 bp of intron 1 and encompassing all of exon H1, is responsible for heart-specific expression of the exchanger in both the embryo and adult and also is responsible for the upregulation of NCX1 in response to pressure overload. Furthermore, the results presented here from several studies provide evidence that MAP kinases play an important role in regulating the expression of NCX1 in adult cardiocytes. JNK does not appear to be part of either α-adrenergic upregulation or integrin-mediated upregulation of the exchanger but is important for basal expression. P38 is very important for α-adrenergic-stimulated NCX1 upregulation but does not play a role in integrin-mediated upregulation. Finally, ERK1/2 plays a major role in integrin-mediated upregulation of NCX1 upregulation and also contributes to α-adrenergic upregulation. Although many questions remain to be answered, this work demonstrates the feasibility of delineating the components and molecular mechanisms of signaling pathways responsible for changes in gene expression in adult cardiocytes using adenoviral gene delivery.

ACKNOWLEDGMENTS

This work is supported in part by National Institutes of Health Program Project HL48788 (Project 3 to D.R.M.); NIH R01 HL66223 (D.R.M.), HL60104, and HL60714 (S.J.C.); a postdoctoral research fellowship from Deutsche Forschungsgemeinschaft Mu 1454/1-1 (J.G.M.); and Department of Veterans Affairs Merit funding (N.S.).

REFERENCES

1. KENT, R.L., J.D. ROZICH, P.L. MCCOLLAM, et al. 1993. Rapid expression of the Na^+–Ca^{2+} exchanger in response to cardiac pressure overload. Am. J. Physiol. **265**: H1024–1029.
2. MENICK, D.R., K.V. BARNES, U.F. THACKER, et al. 1996. The exchanger and cardiac hypertrophy. Ann. N.Y. Acad. Sci. **779**: 489–501.
3. STUDER, R., H. REINECKE, R. VETTER, et al. 1997. Expression and function of the cardiac Na^+/Ca^{2+} exchanger in postnatal development of the rat, in experimental-induced cardiac hypertrophy and in the failing human heart. Basic Res. Cardiol. **92**: 53–58.
4. HASENFUSS, G., H. REINECKE, R. STUDER, et al. 1994. Relation between myocardial function and expression of sarcoplasmic reticulum Ca^{2+}-ATPase in failing and non-failing human myocardium. Circ. Res. **75**: 434–442.
5. HASENFUSS, G., M. MEYER, W. SCHILLINGER, et al. 1997. Calcium handling proteins in the failing human heart. Basic Res. Cardiol. **92**: 87–93.
6. HASENFUSS, G., M. PREUSS, S. LEHNART, et al. 1996. Relationship between diastolic function and protein levels of sodium–calcium exchanger in end-stage failing human hearts. Circulation **94**: 1–443.
7. STUDER, R., H. REINECKE, J. BILGER, et al. 1994. Gene expression of the cardiac Na^+–Ca^{2+} exchanger in end-stage human heart failure. Circ. Res. **75**: 443–453.
8. HASENFUSS, G., W. SCHILLINGER, S.E. LEHNART, et al. 1999. Relationship between Na^+–Ca^{2+} exchanger protein levels and diastolic function of failing human myocardium. Circulation **99**: 641–648.
9. GAUGHAN, J.P., S. FURUKAWA, V. JEEVANANDAM, et al. 1999. Sodium/calcium exchange contributes to contraction and relaxation in failed human ventricular myocytes. Am. J. Physiol. **277**: H714–724.
10. POGWIZD, S.M., M. QI, W. YUAN, et al. 1999. Upregulation of Na^+/Ca^{2+} exchanger expression and function in an arrhythmogenic rabbit model of heart failure. Circ. Res. **85**: 1009–1019.
11. SCHLOTTHAUER, K. & D.M. BERS. 2000. Sarcoplasmic reticulum Ca^{2+} release causes myocyte depolarization : underlying mechanism and threshold for triggered action potentials. Circ. Res. **87**: 774–780.
12. BARNES, K.V., G. CHENG, M.M. DAWSON, et al. 1997. Cloning of cardiac, kidney, and brain promoters of the feline *ncx1* gene. J. Biol. Chem. **272**: 11510–11517.
13. LEE, S.L., A.S. YU & J. LYTTON. 1994. Tissue-specific expression of Na^+–Ca^{2+} exchanger isoforms. J. Biol. Chem. **269**: 14849–14852.
14. KOFUJI, P., W.J. LEDERER & D.H. SCHULZE. 1993. Na/Ca exchanger isoforms expressed in kidney. Am. J. Physiol. **265**: F598–603.
15. QUEDNAU, B.D., D.A. NICOLL & K.D. PHILIPSON. 1997. Tissue specificity and alternative splicing of the Na^+/Ca^{2+} exchanger isoforms NCX1, NCX2, and NCX3 in rat. Am. J. Physiol. **272**: C1250–1261.
16. NICHOLAS, S.B., W. YANG, S.L. LEE, et al. 1998. Alternative promoters and cardiac muscle cell-specific expression of the Na^+/Ca^{2+} exchanger gene. Am. J. Physiol. **274**: H217–232.
17. SCHELLER, T., A. KRAEV, S. SKINNER, et al. 1998. Cloning of the multipartite promoter of the sodium-calcium exchanger gene NCX1 and characterization of its activity in vascular smooth muscle cells. J. Biol. Chem. **273**: 7643–7649.

18. CHENG, G., T.P. HAGEN, M.L. DAWSON, et al. 1999. The role of GATA, CArG, E-box, and a novel element in the regulation of cardiac expression of the Na^+–Ca^{2+} exchanger gene. J. Biol. Chem. **274:** 12819–12826.
19. MÜLLER, J.G., Y. ISOMATSU, S.V. KOUSHIK, et al. 2001. Cardiac-specific expression and hypertrophic upregulation of the feline Na^+/Ca^{2+} exchanger gene H1-promoter in a transgenic mouse model. Circ. Res. **90:** 158–164.
20. KOBAN, M.U., A.F. MOORMAN, J. HOLTZ, et al. 1998. Expressional analysis of the cardiac Na-Ca exchanger in rat development and senescence. Cardiovasc. Res. **37:** 405–423.
21. NICHOLAS, S.B. & K.D. PHILIPSON. 1999. Cardiac expression of the $Na^{(+)}/Ca^{(2+)}$ exchanger NCX1 is GATA factor dependent. Am. J. Physiol. **277:** H324–330.
22. HE, T.C., S. ZHOU, L.T. DA COSTA, et al. 1998. A simplified system for generating recombinant adenoviruses. Proc. Natl. Acad. Sci. USA **95:** 2509–2514.
23. GILLESPIE-BROWN, J., S.J. FULLER, M.A. BOGOYEVITCH, et al. 1995. The mitogen-activated protein kinase kinase MEK1 stimulates a pattern of gene expression typical of the hypertrophic phenotype in rat ventricular cardiomyocytes. J. Biol. Chem. **270:** 28092–28096.
24. ERICKSON, H.P. & N.A. CARRELL. 1983. Fibronectin in extended and compact conformations. Electron microscopy and sedimentation analysis. J. Biol. Chem. **258:** 14539–14544.

Multiple Modes of Regulation of Na$^+$/H$^+$ Exchangers

HISAYOSHI HAYASHI,[a] KATALIN SZÁSZI,[a] AND SERGIO GRINSTEIN

Cell Biology Programme, The Hospital for Sick Children,
Toronto, Ontario, M5G 1X8, Canada

ABSTRACT: Mammalian Na$^+$/H$^+$ exchangers (NHE) mediate electroneutral countertransport of H$^+$ for Na$^+$ across the plasmalemmal and organellar membranes. They contribute to cellular and organellar pH and volume regulation and transepithelial Na$^+$ transport. The aim of this review is to illustrate the complex regulation of these transporters by focusing on the multiple mechanisms controlling the epithelial isoform, NHE3. A variety of agents and conditions (e.g., hormones, growth factors, cellular pH, and medium osmolarity) act in concert to achieve short-term and long-term regulation of this isoform. The underlying mechanism involves changes in the number of transporters on the cell surface and/or altered activity of the individual exchangers due to allosteric activation by intracellular protons, phosphorylation and interaction with accessory proteins and the cytoskeleton. A similar regulatory versatility probably applies to other NHE isoforms, and the lessons learned from studying members of the NHE family could serve as a useful reference when exploring the modes and levels of regulation of other transporters.

KEYWORDS: NHE; epithelial Na$^+$/H$^+$ exchanger; intracellular pH; transport regulation

INTRODUCTION: THE Na$^+$/H$^+$ EXCHANGER FAMILY

Intracellular as well as systemic pH, Na$^+$ concentration, and fluid volume are vital parameters that have to be tightly regulated. Members of the mammalian Na$^+$/H$^+$ exchanger (NHE) family participate in the regulation of these parameters at both cellular and systemic levels. (For a review, see Refs. 1–3.) The plasmalemmal Na$^+$/H$^+$ exchangers catalyze the one-for-one exchange of intracellular H$^+$ for extracellular Na$^+$. To date seven NHE isoforms have been identified and cloned. All of them consist of ≈800 amino acids and share approximately 40% homology among them. They are all thought to span the membrane 10–12 times, with both N- and C-termini in the cytosol. The C-terminal domains of the different isoforms show the lowest degree of homology. Unlike the N-terminal half, which mediates cation exchange, the C-terminus is thought to be involved in transport regulation.[1,2]

---

[a]Contributed equally to the study.
Address for correspondence: Dr. Sergio Grinstein, Cell Biology Programme, The Hospital for Sick Children, 555 University Ave, Toronto, Ontario M5G1X8, Canada. Voice: 416-813-5727; fax: 416-813-5028.
sga@sickkids.on.ca

Members of the NHE family show unique tissue distribution. NHE1 is expressed in the plasma membrane of most, if not all, mammalian cells. It is often referred to as the "housekeeping" isoform, as it is responsible for regulation of the cytosolic pH and cell volume. By contrast, the tissue distribution of NHE2–5 is more restricted, indicating that these isoforms serve more specialized functions. NHE2, 3, and 4 are present in the gastrointestinal tract and the kidney, whereas NHE5 was shown to be abundant in the brain. Recent studies identified another group of NHE isoforms, namely, NHE6 and 7, that are localized exclusively in intracellular organelles. NHE6 is the mitochondrial exchanger,[4] whereas NHE7 was found in the *trans*-Golgi network.[5] These isoforms are likely to be involved in maintaining the alkali cation and/or pH homeostasis of these subcellular organelles.

The focus of this chapter is the versatile regulation of transport by NHE. To illustrate this issue, we concentrate on the epithelial isoform NHE3, which is controlled by multiple agents and conditions, including hormones, growth factors, and medium osmolarity.

PHYSIOLOGIC ROLE OF THE EPITHELIAL ISOFORM NHE3

Polarized epithelial cells such as those of the gastrointestinal tract or kidney tubules express more than one NHE isoform.[6] Na^+/H^+ exchanger activity has been shown to exist on both the apical and the basolateral membrane, although the functional properties of the transporters differ. The basolateral isoform, responsible for overall cellular pH regulation, proved to be NHE1. By contrast, the apical membrane expresses NHE3 (and in some cells NHE2 or 4). NHE3 is an important contributor to the regulation of pH, Na^+, and water homeostasis of the organism, as it catalyzes Na^+ and fluid (re)absorption across epithelia. NHE3 knockout mice show impaired systemic fluid homeostasis and disturbed acid-base balance. The major changes detected include decreased reabsorption of Na^+ and HCO_3^- by the kidney, reduced blood pressure, mild acidosis, and diarrhea due to absorptive abnormalities in the intestine.[7]

Fine control of the activity of NHE3 allows adaptation to both short- and long-term changes. Moreover, altered activity and/or expression of NHE3 is likely to be associated with certain pathologic conditions. In the following sections we discuss molecular aspects of the complex regulation of this exchanger.

REGULATION OF NHE3

A variety of hormones and other factors have been identified as regulators of NHE3.[1,8] Some of these, including α-adrenergic receptor agonists,[9] endothelin,[10] and growth factors (epidermal growth factor[11] and fibroblast growth factor[12]), stimulate NHE3. By contrast, hormones that elevate cAMP or cGMP inhibit the exchanger.[13–15] NHE3 is also sensitive to some physical parameters. Thus, cell shrinkage induced by hyperosmolarity was shown to downregulate transport.[16,17] Although in recent years considerable progress has been made in understanding the regulation of NHE3, the exact underlying mechanisms are still only partially understood. It is clear that more than one mechanism accounts for the regulation, but we lack a detailed understanding of these processes and of how the various factors act in concert to achieve the fine control of transport.

The coexistence of multiple isoforms in epithelial cells complicates the study of NHE3. The characterization of its functional properties and regulation was therefore greatly facilitated by expressing NHE3 in antiporter-deficient cell lines, such as AP-1 and PS120 fibroblasts.[18,19] Although many of the properties and regulation of the exchanger are recapitulated in these cells, it should be noted that some important factors specific to epithelial cells might be missing in these heterologous expression systems. The following information has emerged from studies of NHE3 function in isolation. First, the exchanger is sensitive to the intra- and extracellular concentration of its substrates. At physiologic intracellular pH (around 7.2) it is dormant, but decreasing cytosolic pH sharply increases its activity. As discussed later, this activation probably involves a specific proton modifier site on the intracellular aspect of the exchanger. Mechanisms implicated in the acute regulation by hormones and growth factors include phosphorylation of the protein and/or accessory factors, alterations in the surface expression of the exchangers, and changes mediated through the cytoskeleton. Long-term regulation, in turn, is achieved by altered expression of the protein.

SHORT-TERM REGULATION BY THE SUBSTRATES OF THE EXCHANGER

Regulation by the Intracellular Proton Concentration. For an electroneutral Na^+/H^+ exchanger, thermodynamic equilibrium will be attained when $[H^+]_i/[H^+]_o = ([Na^+]_i/[Na^+]_o) \times K_e$, where K_e is the equilibrium constant and the subscripts i and o denote intra- and extracellular, respectively. In almost all cells, a steep inwardly directed Na^+ gradient is maintained across the plasma membrane due to extrusion of Na^+ by the Na^+-K^+ ATPase. Assuming that $[Na^+]_i = 14$ mM and $[Na^+]_o = 140$ mM, it is clear that the exchangers could drive pH_i up to 1 unit above pH_o. However, under physiologic conditions, intracellular pH remains around 7.2. This indicates that a mechanism exists to prevent the system from reaching thermodynamic equilibrium, a situation that would dangerously alkalinize the cytoplasm.

The concentration dependence on extracellular Na^+ obeys Michaelis-Menten kinetics, suggesting the existence of a single transport site. However, the kinetics for intracellular H^+ does not conform to Michaelis-Menten kinetics, revealing a Hill coefficient of ~2.[19–21] It is believed that the protein contains an additional H^+ site(s) that serves as an allosteric modifier. Occupancy of this site by H^+ causes an increase in the exchange activity.[20]

Although the regulation of NHE3 activity has been extensively studied, little attention has been paid to the initial events mediating the activation of NHE3 by intracellular protons. In the course of experiments using brush border membrane vesicle at subphysiologic temperatures, Kinsella *et al.*[22] demonstrated hysteresis in the inactivation of NHE3. This slow responsiveness cannot easily be reconciled with simple protonation of a side chain of the exchanger. We therefore decided to investigate whether slow activation and/or inactivation transitions are observed in intact cells and to examine the underlying mechanism.

Mechanism of the Activation by Intracellular Proton. We first studied the time dependence of the activation of NHE3 in LLC-PK$_1$ cells by monitoring the Na$^+$-induced pH$_i$ recovery following an acid load. Cytosolic pH was measured by microphotometry, and an acid load was imposed on the cells using NH$_4$Cl. When the cells were acidified for 1 minute, Na$^+$-dependent pH$_i$ recovery was slow. However, the Na$^+$-dependent pH$_i$ recovery rate increased gradually as the period of acidification was lengthened. When the cells were acidified for 5 minutes, the rate of alkalinization was over 2-fold greater than the rate measured at 1 minute. This result likely reflects increased activity of NHE3. However, it is conceivable that the cytosolic buffer power diminishes over time after acidification. To discount this possibility, we measured the pH$_i$ recovery induced by nigericin/K$^+$. The pH$_i$ recovery was similar whether nigericin was added 1 or 5 minutes after acidification was imposed, suggesting that the intracellular buffer capacity does not decrease during acidification.

Imposition of an acid load of variable duration required incubation in Na$^+$-free medium for different periods of time. Therefore, it was conceivable that the observed activation of NHE3 was secondary to the intracellular Na$^+$ depletion of the cells, because NHE activity is affected by the gradient of Na$^+$. However, similar results were obtained under conditions in which intracellular Na$^+$ remained constant.

In native as well as in heterologous expression systems, NHE3 is present in at least two subcellular compartments: the plasma membrane and an endomembrane vesicular compartment, which includes subapical/recycling endosomes.[23] It was shown that chronic acidification increases the activity of NHE3 by mobilization of the endomembrane pool to the cell surface.[24] By analogy, it is conceivable that the density of surface NHE3 may change during acute intracellular acidification. To assess this possibility, we quantified the abundance of exofacial NHE3 molecules. When the surface exposure of an epitope-tagged NHE3 was assessed, we found that acidification did not alter this parameter, implying that the pH change altered the intrinsic activity of a constant number of NHE3 instead.

It is well established that NHE3 can become phosphorylated on serine residues and that phosphorylation is accompanied by change in transport activity. In addition, tyrosine phosphorylation may play a role in NHE3 regulation.[25] We therefore also determined whether phosphorylation of NHE3 changes during acidification, using [^{32}P] orthophosphate labeling and immunoprecipitation. No changes in total phosphorylation of NHE3 were detected during acidification.

The molecular nature of the event mediating the slow activation of NHE3 remains to be completely defined. Nevertheless, our results indicate that activation of NHE3 is not due to increased density at the cell surface or is it accompanied by enhanced phosphorylation. We tentatively suggest that upon cytosolic acidification NHE3 undergoes a slow conformational change that activates the intrinsic activity of NHE3.

MECHANISMS INVOLVED IN HORMONAL CONTROL OF NHE3

Regulation by Phosphorylation. The primary structure of the protein indicates the presence of potential phosphorylation sites for both protein kinases C (PKC) and A

(PKA). Indeed, the exchanger is phosphorylated under basal conditions. Moreover, acute activation of both PKA and PKC has been shown to increase phosphorylation of the protein.[14,15,26,27] Despite these studies, however, controversy still exists regarding the precise role of phosphorylation in transport regulation and the sites involved. Mutation of the primary target sites of PKA or PKC phosphorylation markedly decreases, but it does not abolish the inhibitory effect of cAMP or phorbol esters, respectively.[15,27] Moreover, increased phosphorylation does not seem to be involved in mediating the effects of growth factors.[28] These findings suggest that direct phosphorylation is an important, but not exclusive mode of regulation.

Changes in the Surface Expression of NHE3. A fast and effective way to control the function of cell surface proteins is by altering the number of available molecules at the plasma membrane. Recent evidence obtained by electron microscopy and confocal immunofluorescence microscopy indicates that NHE3, in addition to being present at the surface, is detectable in intracellular vesicles.[23,29,30]

Several studies have provided evidence for modulation of the dynamic equilibrium between surface and endomembrane transporters. NHE3 undergoes constitutive uptake into clathrin-coated vesicles[31] and is recycled back to the plasma membrane in a phosphatidylinositol 3-kinase–dependent manner.[32] Inhibition of this kinase by wortmannin or LY294002 blocks exocytosis of NHE3, depleting the plasmalemmal pool of exchangers and causing a marked drop in the transport rate.[12,32] Moreover, transfection of NHE3-expressing PS120 cells with constitutively active phosphatidylinositol 3-kinase or AKT stimulates the exchanger and increases the percentage of NHE3 present on the plasma membrane.[33] Also, a phosphatidylinositol 3-kinase-dependent increase in surface NHE3 was shown after stimulation of cells with basic fibroblast growth factor[12] and EGF.[11] Conversely, decreased surface expression was associated with inhibition by PKC[30] or, under certain conditions, by PKA[34] or by the cAMP-elevating hormone dopamine.[35] Collectively, these studies imply that redistribution of NHE3 between subcellular compartments is an effective means of transport regulation.

Role of Accessory Proteins in NHE3 Regulation. Changes in the transport rate can also occur without changes in the surface density of NHE3. Acute inhibition by cAMP elevation does not involve alteration of the surface expression in AP-1[36] or OK cells.[37] Moreover, inhibition of NHE3 in renal cells by the cAMP-elevating parathyroid hormone does not parallel the course of disappearance of the exchangers.[34] These data point to the existence of alternative regulatory mechanisms that modulate the intrinsic transport activity of NHE3 as opposed to the number of available exchangers.

A search for accessory proteins involved in the regulation of NHE3 led to the identification of Na^+/H^+ exchanger regulatory factors (NHERF). Currently, two homologous proteins associating with NHE3 have been described: NHERF1 (also known as ezrin-binding protein 50 or EBP50) and NHERF2 (also known as E3KARP/TKA1; for a review see Refs. 38 and 39). These PDZ domain-containing proteins were suggested to mediate specific protein-protein interactions and thereby serve as adaptors. Indeed, NHERF also binds ezrin and merlin, members of the ERM (ezrin/radixin/moesin) family of cytoskeletal proteins. The ERM family members link membrane proteins with the actin skeleton. In addition, ezrin is also a protein kinase A-binding protein (AKAP). Studies in PS120 cells showed that formation of a complex between NHE3, NHERF, and ezrin is essential for PKA-mediated phosphorylation and inhibition of NHE3 activity.[38] A model based on these data postulates that NHE3, NHERF, and ezrin form a signaling complex, mediating the regulation of Na^+/H^+ exchange. Recently, the model developed in fibroblasts has been extended to epithelial cells.[40] Using specific antibodies, NHERF was identified in the renal proximal tubule, where it colocalized with ezrin and NHE3.[41] The two proteins were shown to coimmunoprecipitate, and expression of a truncated NHERF lacking the ezrin-binding domain blunted the inhibitory response to cAMP in OK cells.[37] Interestingly, both basal transport and cell shrinkage-induced inhibition of the exchanger were unaltered upon expression of the mutant NHERF.

Regulation by the Cytoskeleton. The actin-based cytoskeleton is a highly dynamic structure. Many actin-binding proteins are involved in regulating the assembly, length, and stability of filamentous (F) actin. Several transporters, including NHE3, have been shown in recent years to be associated with and regulated by the cytoskeleton. Our studies provided evidence that optimal NHE3 function requires an intact and dynamic actin cytoskeleton.[42] Disruption of F-actin by drugs that interfere with the normal organization of the cytoskeleton, such as cytochalasin D and latrunculin B, resulted in a drastic inhibition of exchanger activity.[43] Interestingly, the effect proved to be isoform specific, as NHE1 was not inhibited by cytochalasin treatment. Analysis of truncated mutants of NHE3 and chimeras constructed from NHE3 and NHE1 revealed that cytoskeletal inhibition requires the cytosolic tail of the protein. The domain encompassed by residues 650 and 684 of NHE3 proved to be essential for the cytoskeletal control of this transporter.[43] Interestingly, this region overlaps with that defined to mediate the interaction of NHE3 with NHERF2[44] and contains a putative ezrin-binding sequence (RKRL, residues 656-659). It is possible, yet not proven, that ERM proteins mediate anchorage of NHE3 to the cytoskeleton.

The small GTP-binding protein RhoA, a major organizer of the actin skeleton,[45] proved to be a regulator of NHE3. Dominant-negative mutants of RhoA as well as of one of its downstream effectors, Rho-kinase, markedly decreased NHE3 activity.[46] Rho-kinase is a potent regulator of the phosphorylation state of myosin, which in turn determines the organization of actin and influences cell morphology and contractility.[47] It is therefore conceivable that inhibition of the Rho signaling pathway affects the exchanger by altering the phosphorylation state of myosin. Phosphorylation of myosin seems to be a requisite for optimal function of NHE3, as its dephosphorylation induced by inhibition of myosin light chain-kinase or Rho-kinase diminishes the exchanger activity.[46] Moreover, our recent studies provided evidence that PKA might exert its effect partially by inducing myosin dephosphorylation, thus altering the organization of the cytoskeleton.[36]

In nonepithelial cells heterologously transfected with NHE3, cytoskeleton disruption does not seem to alter the surface density of NHE3.[43,46] Thus, the cytoskeleton must influence the intrinsic activity of the transporter itself. The molecular mechanisms of this regulation remain incompletely understood. Physical linkage to the cytoskeleton might be essential to maintain the active conformation of the protein. Another possibility is that NHE3 might be fully active only in special subdomains of the cell, and its localization to these regions may depend on the cytoskeleton. This may be of particular relevance to epithelial cells, where the apical membrane is rich in cholesterol, likely assembled in specialized membrane domains known as rafts. Experiments are underway to determine whether partitioning into rafts is essential for optimal NHE3 activity.

LONG-TERM REGULATION

The activity of NHE3 in renal or intestinal epithelial cells is chronically regulated by a variety of hormones involved in systemic Na^+ and blood volume regulation (e.g., mineralo- and glucocorticoids), thyroid hormones, and endothelin and in conditions such as chronic acidosis or inflammation. This regulation is achieved by altering the expression levels of NHE3 mRNA and protein.

Chronic Upregulation of NHE3. Several hormones and factors induce an increase in the level of NHE3. Cloning and characterization of the rat NHE3 gene[48,49] not only provided insights into the structure and tissue-specific expression of the exchanger, but also clarified aspects of its transcriptional regulation. Sequence analysis of the 5′-flanking promoter region revealed the presence of putative *cis*-acting elements recognized by various transcription factors (e.g., AP-1, AP-2, C/EBP, NF-I, OCT-1/OTF-1, PEA3, Sp1, glucocorticoid, and thyroid hormone receptors).

Glucocorticoids elevate NHE3 activity and/or mRNA levels both in the ileum[50] and in the renal proximal tubule.[51] Aldosterone was also shown to stimulate NHE3 expression, but this effect was detected only in the proximal colon.[52] Thyroid hormone, which causes a general increase in metabolism, was also demonstrated to stimulate transcription of the NHE3 gene.[53]

Butyrate, a short chain fatty acid that acts as a differentiation and trophic factor in the mucosa of the large intestine, potently stimulates sodium and water absorption by elevating NHE3 activity.[54] Recent studies provided evidence that butyrate regulates the NHE3 gene promoter in Caco2 cells.[55]

Similarly, chronic extracellular acidosis has been known for years to increase NHE3 activity in cultured lines[56,57] and in the rabbit renal proximal tubule.[58] This stimulation is similarly associated with an increase in NHE3 mRNA abundance.[57] The mechanisms responsible for the translational regulation of NHE3 by chronic acidosis remain largely unexplored. In OKP cells, the increased NHE3 activity noted after chronic extracellular acidosis is independent of PKC stimulation, yet appears to be associated with nonreceptor tyrosine kinases of the *src* family.[25] As intracellular acidosis activates *src*, this kinase might be involved in signaling enhanced transcription. However, the pathways(s) linking *src* stimulation to NHE3 mRNA generation remain mysterious.

It is noteworthy that the acid-induced increase in NHE3 activity is not completely blocked by protein synthesis inhibitors, suggesting that posttranslational regulation also contributes to the effect. Indeed, it was recently reported that acid incubation also caused an increase in exocytic insertion of NHE3 into the apical membrane.[24]

Chronic Downregulation of NHE3. Whereas a large number of studies have focused on the chronic stimulation of NHE3, relatively little is known about the chronic downregulation of the exchanger. The importance of this phenomenon is probably underestimated, as it could potentially be associated with certain diseases. This is highlighted by the finding that Na^+ absorptive capacity is diminished in colonic mucosa in patients with inflammatory bowel disease (IBD).[59] Recently, it was shown that gamma interferon, which is present in high levels in IBD tissues, can significantly downregulate NHE3 protein and mRNA expression without changing mucosal histology in rat intestine and Caco-2 cells.[60]

It is also conceivable that some bacteria that cause diarrhea might secrete toxins affecting NHE3 activity in the intestine. Although this area is largely unexplored, recent studies have identified the first inherited disease that probably alters apical membrane trafficking selectively. Microvillus inclusion disease (also known as familial microvillous atrophy) is a rare genetic disorder characterized by continuous diarrhea from birth. Histologic analysis shows hypoplastic villus atrophy in the intestine, which is not accompanied by inflammatory changes. Interestingly, immunofluorescence analysis revealed that several apical proteins, including NHE3, are absent from the membrane.[61]

CONCLUDING REMARKS

As the preceding overview suggests, NHE3 is regulated by multiple agents through a variety of acute as well as chronic mechanisms. Similar functional and regulatory versatility has been demonstrated for NHE1 and is likely to apply to other exchanger isoforms. The lessons learned from studying members of the NHE family should serve as a useful reference when exploring the modes and levels of regulation of Na^+/Ca^{2+} exchangers and other transporters.

ACKNOWLEDGMENTS

This work was supported in part by a Fellowship from the Canadian Arthritis Network (to H.H.) and a Canadian Institutes of Health Research fellowship (to K.S.). One of the authors (S.G.) is an International Scholar of the Howard Hughes Medical Institute, is the current holder of the Pitblado Chair in Cell Biology at The Hospital for Sick Children, and was cross-appointed to the Department of Biochemistry, University of Toronto.

REFERENCES

1. WAKABAYASHI, S., M. SHIGEKAWA & J. POUYSSEGUR. 1997. Molecular physiology of vertebrate Na^+/H^+ exchangers. Physiol. Rev. **77:** 51–74.
2. ORLOWSKI, J. & S. GRINSTEIN. 1997. Na^+/H^+ exchangers of mammalian cells. J. Biol. Chem. **272:** 22373–22376.
3. COUNILLON, L. & J. POUYSSEGUR. 2000. The expanding family of eucaryotic Na^+/H^+ exchangers. J. Biol. Chem. **275:** 1–4.
4. NUMATA, M. et al. 1998. Identification of a mitochondrial Na^+/H^+ exchanger. J. Biol. Chem. **273:** 6951–6959.
5. NUMATA, M. & J. ORLOWSKI. 2001. Molecular cloning and characterization of a novel $(Na^+,K^+)/H^+$ exchanger localized to the trans-Golgi network. J. Biol. Chem. **276:** 17387–17394.
6. BIEMESDERFER, D. et al. 1993. NHE3: a Na^+/H^+ exchanger isoform of renal brush border. Am. J. Physiol. **265:** F736–742.
7. SCHULTHEIS, P.J. et al. 1998. Renal and intestinal absorptive defects in mice lacking the NHE3 Na^+/H^+ exchanger. Nat. Genet. **19:** 282–285.
8. NOEL, J. & J. POUYSSEGUR. 1995. Hormonal regulation, pharmacology, and membrane sorting of vertebrate Na^+/H^+ exchanger isoforms. Am. J. Physiol. **268:** C283–296.
9. LIU, F. & F.A. GESEK. 2001. Alpha(1)-adrenergic receptors activate NHE1 and NHE3 through distinct signaling pathways in epithelial cells. Am. J. Physiol. Renal Physiol. **280:** F415–425.

10. PENG, Y. *et al.* 2001. ET(B) receptor activation causes exocytic insertion of NHE3 in OKP cells. Am. J. Physiol. Renal Physiol. **280:** F34–42.
11. DONOWITZ, M. *et al.* 2000. Short-term regulation of NHE3 by EGF and protein kinase C but not protein kinase A involves vesicle trafficking in epithelial cells and fibroblasts. Ann. N.Y. Acad. Sci. **915:** 30–42.
12. JANECKI, A.J. *et al.* 2000. Basic fibroblast growth factor stimulates surface expression and activity of Na^+/H^+ exchanger NHE3 via mechanism involving phosphatidylinositol 3-kinase. J. Biol. Chem. **275:** 8133–8142.
13. MCSWINE, R.L. *et al.* 1998. Regulation of apical membrane Na^+/H^+ exchangers NHE2 and NHE3 in intestinal epithelial cell line C2/bbe. Am. J. Physiol. **275:** C693–701.
14. MOE, O.W., M. AMEMIYA & Y. YAMAJI. 1995. Activation of protein kinase A acutely inhibits and phosphorylates Na/H exchanger NHE-3. J. Clin. Invest. **96:** 2187–2194.
15. KURASHIMA, K. *et al.* 1997. Identification of sites required for down-regulation of Na^+/H^+ exchanger NHE3 activity by cAMP-dependent protein kinase. J. Biol. Chem. **272:** 28672–28679.
16. KAPUS, A. *et al.* 1994. Functional characterization of three isoforms of the Na^+/H^+ exchanger stably expressed in Chinese hamster ovary cells. ATP dependence, osmotic sensitivity, and role in cell proliferation. J. Biol. Chem. **269:** 23544–23552.
17. SOLEIMANI, M. *et al.* 1994. Effect of high osmolality on Na^+/H^+ exchange in renal proximal tubule cells. J. Biol. Chem. **269:** 15613–15618.
18. ORLOWSKI, J. & R.A. KANDASAMY. 1996. Delineation of transmembrane domains of the Na^+/H^+ exchanger that confer sensitivity to pharmacological antagonists. J. Biol. Chem. **271:** 19922–19927.
19. LEVINE, S.A. *et al.* 1993. Kinetics and regulation of three cloned mammalian Na^+/H^+ exchangers stably expressed in a fibroblast cell line. J. Biol. Chem. **268:** 25527–25535.
20. ARONSON, P.S., J. NEE & M.A. SUHM. 1982. Modifier role of internal H^+ in activating the Na^+–H^+ exchanger in renal microvillus membrane vesicles. Nature **299:** 161–163.
21. Orlowski, J. 1993. Heterologous expression and functional properties of amiloride high affinity (NHE-1) and low affinity (NHE-3) isoforms of the rat Na/H exchanger. J. Biol. Chem. **268:** 16369–16377.
22. KINSELLA, J.L., P. HELLER & J.P. FROEHLICH. 1998. Na^+/H^+ exchanger: proton modifier site regulation of activity. Biochem. Cell Biol. **76:** 743–749.
23. D'SOUZA, S. *et al.* 1998. The epithelial sodium-hydrogen antiporter Na^+/H^+ exchanger 3 accumulates and is functional in recycling endosomes. J. Biol. Chem. **273:** 2035–2043.
24. YANG, X. *et al.* 2000. Acid incubation causes exocytic insertion of NHE3 in OKP cells. Am. J. Physiol. Cell Physiol. **279:** C410–419.
25. YAMAJI, Y. *et al.* 1995. Overexpression of csk inhibits acid-induced activation of NHE-3. Proc. Natl. Acad. Sci. USA **92:** 6274–6278.
26. ZHAO, H. *et al.* 1999. Acute inhibition of Na/H exchanger NHE-3 by cAMP. Role of protein kinase a and NHE-3 phosphoserines 552 and 605. J. Biol. Chem. **274:** 3978–3987.
27. WIEDERKEHR, M.R., H. ZHAO & O.W. MOE. 1999. Acute regulation of Na/H exchanger NHE3 activity by protein kinase C: role of NHE3 phosphorylation. Am. J. Physiol. **276:** C1205–1217.
28. YIP, J.W. *et al.* 1997. Regulation of the epithelial brush border Na^+/H^+ exchanger isoform 3 stably expressed in fibroblasts by fibroblast growth factor and phorbol esters is not through changes in phosphorylation of the exchanger. J. Biol. Chem. **272:** 18473–18480.
29. BIEMESDERFER, D. *et al.* 1997. Monoclonal antibodies for high-resolution localization of NHE3 in adult and neonatal rat kidney. Am. J. Physiol. **273:** F289–299.
30. JANECKI, A.J. *et al.* 1998. Subcellular redistribution is involved in acute regulation of the brush border Na^+/H^+ exchanger isoform 3 in human colon adenocarcinoma cell line Caco-2. Protein kinase C-mediated inhibition of the exchanger. J. Biol. Chem. **273:** 8790–8798.
31. CHOW, C.W. *et al.* 1999. The epithelial Na^+/H^+ exchanger, NHE3, is internalized through a clathrin-mediated pathway. J. Biol. Chem. **274:** 37551–37558.

32. KURASHIMA, K. et al. 1998. Endosomal recycling of the Na^+/H^+ exchanger NHE3 isoform is regulated by the phosphatidylinositol 3-kinase pathway. J. Biol. Chem. **273**: 20828–20836.
33. LEE-KWON, W. et al. 2001. Constitutively active phosphatidylinositol 3-kinase and AKT are sufficient to stimulate the epithelial Na^+/H^+ exchanger 3. J. Biol. Chem. **276**: 31296–31304.
34. COLLAZO, R. et al. 2000. Acute regulation of Na/H exchanger NHE3 by parathyroid hormone via NHE3 phosphorylation and dynamin-dependent endocytosis. J. Biol. Chem. **275**: 31601–31608.
35. HU, M.C. et al. 2001. Dopamine acutely stimulates Na^+/H^+ exchanger (NHE3) endocytosis via clathrin-coated vesicles: dependence on protein kinase A-mediated NHE3 phosphorylation. J. Biol. Chem. **276**: 26906–26915.
36. SZASZI, K. et al. 2001. Role of the cytoskeleton in mediating cAMP-dependent protein kinase inhibition of the epithelial Na^+/H^+ exchanger NHE3. J. Biol. Chem. **276**: 40761–40768.
37. WEINMAN, E.J. et al. 2001. Ezrin binding domain-deficient NHERF attenuates cAMP-mediated inhibition of Na^+/H^+ exchange in OK cells. Am. J. Physiol. Renal Physiol. **281**: F374–380.
38. WEINMAN, E.J. et al. 2000. NHERF associations with sodium-hydrogen exchanger isoform 3 (NHE3) and ezrin are essential for cAMP-mediated phosphorylation and inhibition of NHE3. Biochemistry **39**: 6123–6129.
39. SHENOLIKAR, S. & E.J. WEINMAN. 2001. NHERF: targeting and trafficking membrane proteins. Am. J. Physiol. Renal Physiol. **280**: F389–395.
40. WEINMAN, E.J., D. STEPLOCK & S. SHENOLIKAR. 2001. Acute regulation of NHE3 by protein kinase A requires a multiprotein signal complex. Kidney Int. **60**: 450–454.
41. WADE, J.B. et al. 2001. Differential renal distribution of NHERF isoforms and their colocalization with NHE3, ezrin, and ROMK. Am. J. Physiol. Cell Physiol. **280**: C192–198.
42. SZASZI, K. et al. 2000. Regulation of the epithelial Na^+/H^+ exchanger isoform by the cytoskeleton. Cell Physiol. Biochem. **10**: 265–272.
43. KURASHIMA, K. et al. 1999. The apical Na^+/H^+ exchanger isoform NHE3 is regulated by the actin cytoskeleton. J. Biol. Chem. **274**: 29843–29849.
44. YUN, C.H. et al. 1998. NHE3 kinase A regulatory protein E3KARP binds the epithelial brush border Na^+/H^+ exchanger NHE3 and the cytoskeletal protein ezrin. J. Biol. Chem. **273**: 25856–25863.
45. HALL, A. 1998. Rho GTPases and the actin cytoskeleton. Science **279**: 509–514.
46. SZASZI, K. et al. 2000. RhoA and rho kinase regulate the epithelial Na^+/H^+ exchanger NHE3. Role of myosin phosphorylation. J. Biol. Chem. **275**: 28599–28606.
47. KIMURA, K. et al. 1996. Regulation of myosin phosphatase by Rho and Rho-associated kinase (Rho-kinase). Science **273**: 245–248.
48. KANDASAMY, R.A. & J. ORLOWSKI. 1996. Genomic organization and glucocorticoid transcriptional activation of the rat Na^+/H^+ exchanger Nhe3 gene. J. Biol. Chem. **271**: 10551–10559.
49. CANO, A. 1996. Characterization of the rat NHE3 promoter. Am. J. Physiol. **271**: F629–636.
50. YUN, C.H. et al. 1993. Glucocorticoid stimulation of ileal Na^+ absorptive cell brush border Na^+/H^+ exchange and association with an increase in message for NHE-3, an epithelial Na^+/H^+ exchanger isoform. J. Biol. Chem. **268**: 206–211.
51. BAUM, M. et al. 1994. Effect of glucocorticoids on renal cortical NHE-3 and NHE-1 mRNA. Am. J. Physiol. **267**: F437–442.
52. CHO, J.H. et al. 1998. Aldosterone stimulates intestinal Na^+ absorption in rats by increasing NHE3 expression of the proximal colon. Am. J. Physiol. **274**: C586–594.
53. CANO, A., M. BAUM & O.W. MOE. 1999. Thyroid hormone stimulates the renal Na/H exchanger NHE3 by transcriptional activation. Am. J. Physiol. **276**: C102–108.
54. MUSCH, M.W. et al. 2001. SCFA increase intestinal Na absorption by induction of NHE3 in rat colon and human intestinal C2/bbe cells. Am. J. Physiol. Gastrointest. Liver Physiol. **280**: G687–693.

55. KIELA, P.R. *et al.* 2001. Regulation of the rat NHE3 gene promoter by sodium butyrate. Am. J. Physiol. Gastrointest. Liver Physiol. **281:** G947–956.
56. IGARASHI, P. *et al.* 1992. Effects of chronic metabolic acidosis on Na^+-H^+ exchangers in LLC-PK1 renal epithelial cells. Am. J. Physiol. **263:** F83–88.
57. AMEMIYA, M. *et al.* 1995. Expression of NHE-3 in the apical membrane of rat renal proximal tubule and thick ascending limb. Kidney Int. **48:** 1206–1215.
58. HORIE, S. *et al.* 1990. Preincubation in acid medium increases Na/H antiporter activity in cultured renal proximal tubule cells. Proc. Natl. Acad. Sci. USA **87:** 4742–4745.
59. SANDLE, G.I. *et al.* 1990. Cellular basis for defective electrolyte transport in inflamed human colon. Gastroenterology **99:** 97–105.
60. ROCHA, F. *et al.* 2001. IFN-gamma downregulates expression of Na^+/H^+ exchangers NHE2 and NHE3 in rat intestine and human Caco-2/bbe cells. Am. J. Physiol. Cell Physiol. **280:** C1224–1232.
61. AMEEN, N.A. & P.J. Salas. 2000. Microvillus inclusion disease: a genetic defect affecting apical membrane protein traffic in intestinal epithelium. Traffic **1:** 76–83.

Cyclosporin A Regulates Sodium–Calcium Exchanger (NCX1) Gene Expression *in Vitro* and Cardiac Hypertrophy in NCX1 Transgenic Mice

MARIA C. JORDAN,[a] BEATE D. QUEDNAU,[a] KENNETH P. ROOS,[a] ROBERT S. ROSS,[a] KENNETH D. PHILIPSON,[a] AND SUSANNE B. NICHOLAS[b]

[a]*Departments of Physiology and Medicine and the Cardiovascular Research Laboratory, University of California, Los Angeles (UCLA), Los Angeles, California 90095, USA*

[b]*Department of Medicine, Divisions of Nephrology and Endocrinology, UCLA, Los Angeles, California 90095, USA*

ABSTRACT: The cardiac-specific sodium-calcium exchanger (NCX1) is a GATA-4 dependent gene that is upregulated during cardiac hypertrophy and heart failure. To date, lack of an appropriate inhibitor of NCX1 and embryonic lethality of NCX1 knockout mice have slowed investigation of the relation between NCX1 upregulation and cardiac hypertrophy. Recently, *in vitro* studies have shown that cyclosporin A (CSA), a calcineurin inhibitor, significantly downregulated expression of the hypertrophic genes atrial natriuretic factor and β-myosin heavy chain and protected against cardiac hypertrophy and heart failure in calcineurin overexpressing mice. This suggested that CSA might play an important role in the treatment of hypertrophy and heart failure. In an *in vitro* model of cardiac hypertrophy, we showed that CSA is a potent inhibitor of NCX1 basal expression and NCX1 promoter activity. Female homozygous transgenic mice that overexpress NCX1 develop heart failure and die prematurely after two or more pregnancies. Others have demonstrated that pressure overloaded wild-type mice treated with CSA do not develop cardiac hypertrophy and downregulate expression of NCX1. We investigated the effect of CSA on NCX1 expression and transverse aortic constriction-induced cardiac hypertrophy in NCX1 overexpressing mice. We found that CSA blunted these responses.

KEYWORDS: cyclosporin A; sodium–calcium exchanger; NCX1; cardial hypertrophy; transgenic mice

---

Address for correspondence: Susanne B. Nicholas, M.D., Ph.D., University of California, Los Angeles, 900 Veteran Avenue, Suite 24-130, Los Angeles, CA 90095. Voice: 310-794-7555; fax: 310-794-7654.

sunicholas@mednet.ucla.edu

INTRODUCTION

The sarcolemmal Na^+/Ca^{2+} exchanger (NCX1) plays an important role in maintaining intracellular Ca^{2+} homeostasis in cardiomyocytes. The protein mediates the electrogenic exchange of one intracellular Ca^{2+} ion for three extracellular Na^+ ions in each reaction cycle. During the contraction and relaxation cycle of the heart the exchanger extrudes cellular Ca^{2+} at the onset of diastole and thus is a major modulator of contractile function.[1]

In recent years it has increasingly been recognized that alterations in myocardial Ca^{2+} levels play a central role in the pathophysiology of cardiac hypertrophy, which is associated with significant changes in myocardial contractility as well as in arrhythmias associated with congestive heart failure.[2,3] In myocytes from failing hearts, intracellular Ca^{2+} homeostasis is no longer maintained, and increased levels of diastolic Ca^{2+} concentrations along with decreased amplitudes of systolic whole-cell Ca^{2+} transients have been observed.[2] These changes are accompanied by alterations in the expression profiles of calcium-regulatory proteins. Sarcoplasmic reticulum (SR) Ca^{2+} ATPase mRNA levels are significantly lowered in animal models of heart failure and in terminal human heart failure.[4] In addition, L-type Ca^{2+} channels have been reported to be downregulated in various animal models of heart failure, whereas reports on the human failing heart are still controversial.[4]

On the other hand, it has been shown that NCX1 expression levels[5–7] and activity[3,8–11] are increased in a variety of animal models of heart failure as well as in the failing human heart. These results are interpreted as an adaptive response to preserve Ca^{2+} homeostasis and myocardial function. Compared to terminal heart failure, Ca^{2+} homeostasis and signaling in hypertrophic cardiomyocytes appear to be different.[12] In terms of sodium–calcium transport, less is known about the role of NCX1 in the pathophysiology of cardiac hypertrophy. There are few reports to indicate that NCX1 transcript levels[13] and protein expression[14] are upregulated in a pressure overload model of cardiac hypertrophy. As a result, the molecular basis of the transition from an initially beneficial hypertrophic response of the myocardium to terminal heart failure is the subject of intense investigations.

Numerous cellular signaling pathways have been implicated in the regulation of cardiac hypertrophy.[15] Recently, calcium–calmodulin-dependent serine-threonine phosphatase calcineurin in conjunction with the nuclear factor of activated T cells (NFAT) transcription factors has been demonstrated to play a central regulatory role in the hypertrophic response.[16] In addition, the immunosuppressive drugs cyclosporin A (CSA) and FK506, which specifically inhibit calcineurin, have been shown to attenuate cardiac hypertrophy *in vivo* and *in vitro*.[17] However, studies on the beneficial effect of CSA on experimental pressure-overload hypertrophy have shown contrasting results.[17] The proposal that a calcium-activated signal transduction pathway may be a central mediator of cardiomyocyte hypertrophy invites further studies of cardiac calcium-regulatory proteins in general and of the Na^+/Ca^{2+} exchanger in particular.

In this work, we demonstrate that the cardiac NCX1 promoter is stimulated by the α-adrenergic agent phenylephrine and may be regulated through a calcineurin-dependent transcriptional pathway *in vitro*. Furthermore, we show that homozygous mice overexpressing the cardiac Na^+/Ca^{2+} exchanger develop hypertrophy when

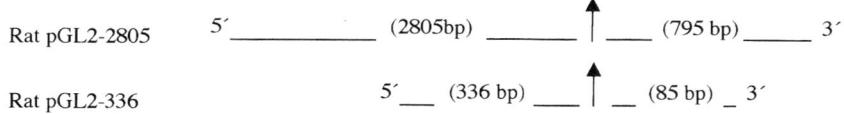

FIGURE 1. Cardiac Na^+/Ca^{2+} exchanger gene promoter expression constructs used in transient transfection experiments of cultured rat primary neonatal cardiomyocytes. pGL-2805, cardiac NCX1 full-length promoter construct in vector pGL2B; pGL-336, cardiac NCX1 truncated promoter construct. The transcriptional start site is indicated by an *arrow*. Numbers downstream of the transcriptional start site represent the various lengths of 5'-untranslated sequence in the construct. For a detailed description of the constructs see Ref. 18.

subjected to transverse aortic constriction (TAC) and that this hypertrophic response can be blunted by treating the animals with cyclosporin A.

METHODS

The following expression constructs were used for this study: the rat cardiac-specific NCX1 full-length (-2805) and truncated promoter (-336)[18] (FIG. 1) and the rat atrial natriuretic factor (-3003) promoter in the pGL2B promoterless vector, which contains the firefly luciferase reporter gene (Promega, Madison, WI). The constructs were transiently transfected into cultured rat primary neonatal cardiac myocytes, as previously described.[18] The pTK vector containing the thymidine kinase promoter and the *Renilla* luciferase reporter gene was used to standardize transfection efficiency and to facilitate measurement of promoter activity by a dual luciferase reporter gene assay (Promega). Cultured cardiomyocytes were treated with 75 µM phenylephrine or pretreated with cyclosporin A (1.0, 3.0, or 5.0 µM; Sigma Chemical Co.) for 48 hours before promoter activity was measured. Promoter activity data from independent experiments are presented as means ±SEM. Statistical evaluation of data was done using Student's *t* test. A *P* value of <0.05 was chosen as the limit of statistical significance.

Age-matched, homozygous, NCX1-overexpressing mice were bred from heterozygous backgrounds.[19] Mice received subcutaneous injections of CSA 25 mg/kg, twice daily for 2 days followed by TAC[20] and CSA for an additional 2 weeks. Animals were divided into 4 groups: CSA-TAC, $n = 6$; vehicle-TAC, $n = 6$; surgical sham-CSA, $n = 3$; and surgical sham-vehicle, $n = 3$. All mice for this study were evaluated by ultrasound echocardiography 1–5 days prior to initiation of the 16-day treatment protocol and at the end of the study just prior to terminal hemodynamic assessment. Ultrasound imaging was performed according to Tanaka *et al.*[21] Dimensions of the left ventricular cavity (EDD and ESD) and its wall thickness (PWT and VST) during systole and diastole were determined with SigmaScan V3.0 software (Systat Inc.), and left ventricular shortening (%LVFS) was calculated according to Pollick *et al.*[22] At the conclusion of the study, each animal's total body weight, heart weight, and tibia length were obtained. Data were compiled and are shown as means and standard error. Data were statistically evaluated using two-tailed *t* tests with Instat V3.05 software (GraphPad Inc., San Diego, CA). A P value of 0.05 was chosen as the limit of statistical significance.

RESULTS AND DISCUSSION

In Vitro *Regulation of the NCX1 Promoter by Phenylephrine and Cyclosporin A*

The Na^+/Ca^{2+} exchanger gene promoter consists of three tissue-specific, alternative promoters[18,23] that regulate tissue-exclusive transcription in heart and kidney as well as transcription in brain. The brain NCX1 promoter has also been shown to drive ubiquitous NCX1 expression.[18] The tissue-specific activity of the cardiac NCX1 promoter is mediated by heart-specific transcription factors, including members of the GATA family of transcription factors.[18,23,24] GATA-4 has been identified in the rat cardiac NCX1 promoter.[25]

The serine-threonine phosphatase calcineurin directly dephosphorylates the transcription factor NFAT3 in the cytoplasm. NFAT3 then translocates to the nucleus and activates the transcription of specific genes by binding to its regulatory elements in conjunction with other transcription factors, such as GATA-4.[16,26] Because the cardiac NCX1 promoter is GATA4-factor dependent, we hypothesized that a specific block of the calcineurin-signaling pathway may result in a decrease of NCX1 promoter function. We transfected rat neonatal primary cardiomyocytes with NCX1 promoter constructs and treated the cells with increasing concentrations of cyclosporin A (CSA). As shown in FIGURE 2, the activity of the NCX1 cardiac promoter, measured by dual luciferase reporter gene assays, is downregulated by CSA in a dose-dependent manner, suggesting that a calcineurin-specific signaling pathway

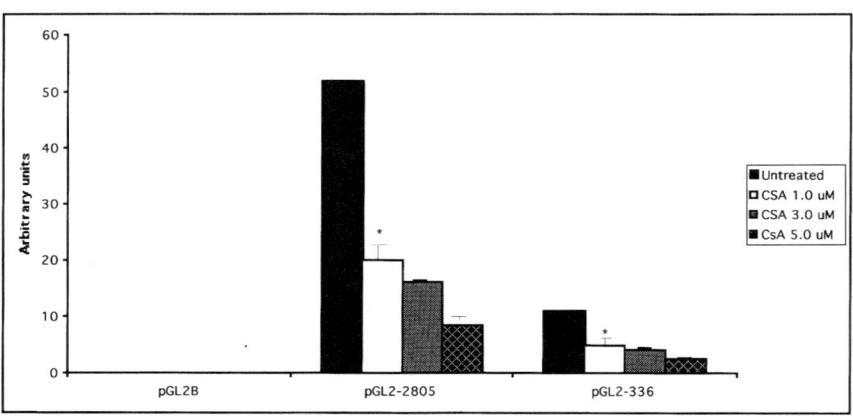

FIGURE 2. Dose-dependent downregulation of NCX1 promoter activity by cyclosporin A (CSA) *in vitro*. The NCX1 full-length (pGL2-2805) and truncated (pGL2-336) promoter constructs along with promoterless vector (pGL2B) and the luciferase reporter gene and the pTK vector containing the *Renilla* luciferase reporter gene were transiently transfected into rat primary neonatal cardiomyocytes. Cells were treated with CSA for 48 hours. Promoter activity was determined by dual luciferase reporter gene assay ($n = 4$; *P <0.05).

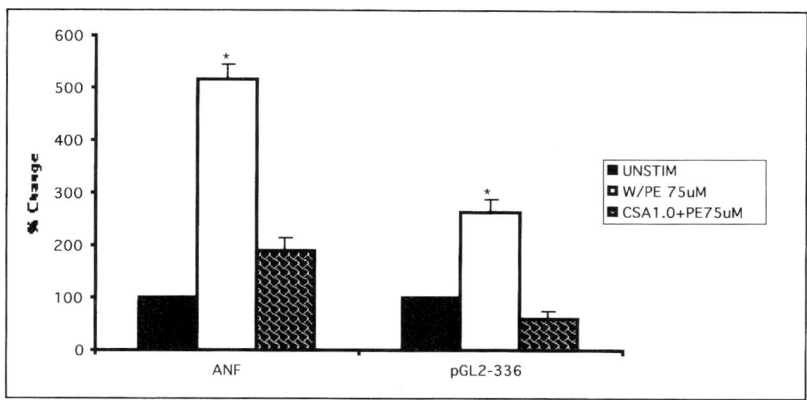

FIGURE 3. The phenylephrine-induced activity of the NCX1 and ANF promoters is significantly downregulated by cyclosporin A (CSA). The NCX1 truncated promoter (pGL2-336), the rat ANF promoter construct, pANF3003, and the pTK vector were transiently transfected into cultured rat neonatal cardiomyocytes. The cells were treated with 75 µM of phenylephrine for 48 hours or pretreated with 1.0 µM CSA for 30 minutes. Promoter activity was measured by dual luciferase assay ($n = 4$; *$P <0.05$).

may be involved. The results in FIGURE 2 demonstrate that the CSA-induced downregulation of the promoter activity is observed in the full-length (pGL2-2805) as well as in the truncated (pGL2-336) NCX1 promoters. This implies that the binding site(s) for the transcription factor(s) involved in this signaling pathway lies within the first 336 bp upstream of the transcriptional start site.

It was reported earlier that the activity of the cardiac-specific promoter from cat can be stimulated by the α-adrenergic agent phenylephrine in an *in vitro* model of cardiac hypertrophy.[23] We treated cultured cardiomyocytes transiently transfected with the -336 rat NCX1 cardiac promoter with 75 µM phenylephrine and observed a significant 2- to 3-fold stimulation of NCX1 promoter activity (FIG. 3). This increase in activity was significantly blunted following pretreatment with 1.0 µM CSA. Phenylephrine is known to induce changes in cultured cardiomyocytes that are associated with a cellular hypertrophic response. This includes the upregulation of certain fetal genes such as atrial natriuretic factor (ANF) and β-myosin heavy chain.[27] As shown in FIGURE 3, activity of the rat ANF promoter was upregulated when the cultured cardiomyocytes were stimulated with phenylephrine, and this stimulatory effect was also dramatically reduced when the cells were treated with CSA.

The results presented here show that the NCX1 cardiac promoter behaves in a way similar to the hypertrophic gene ANF in an *in vitro* model of cardiac hypertrophy and that this response can be blocked by the inhibition of calcineurin. We have additional evidence (data not shown) that NCX1 protein levels are significantly upregulated in the hearts of transgenic mice overexpressing a constitutively active form of calcineurin,[16] further supporting the observation that NCX1 expression may be regulated by a calcineurin-dependent transcriptional pathway.

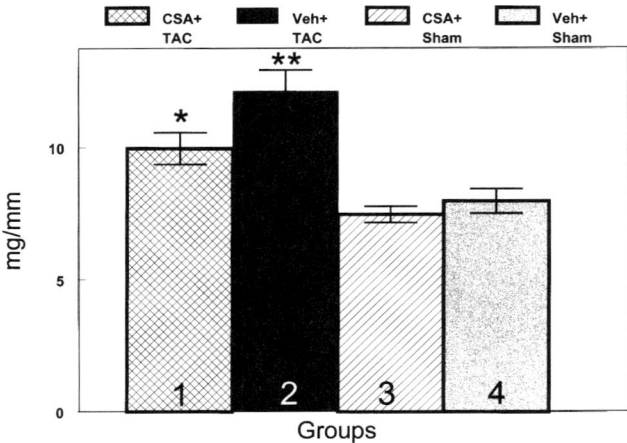

FIGURE 4. Heart weight-to-tibia-length ratio in experimental animals. Fourteen days postsurgery, TAC-CSA animals (Group 1*) showed increased heart weight-to-tibia-length ratio indicative of the manifestation of cardiac hypertrophy when compared to surgical sham-CSA (Group 3) and surgical sham-vehicle (Group 4; P <0.025) animals. The hypertrophy in vehicle-TAC animals (Group 2**) was also significant compared to that of TAC animals treated with CSA (Group 1; P <0.05), surgical sham-CSA (Group 3; P <0.01), and surgical sham-vehicle (Group 4; P <0.01).

Cyclosporin A May Attenuate Cardiac Hypertrophy in NCX1 Transgenic Mice

We were also interested in elucidating the role of a calcineurin-dependent signaling pathway in an *in vivo* model of hypertrophy involving NCX1 overexpression. When undergoing cardiovascular stress, such as pregnancy, homozygous, NCX1 overexpressing mice develop cardiac hypertrophy and die prematurely due to heart failure.[28] We subjected NCX1 overexpressing transgenic (NCX-TG) mice to TAC in order to establish a pressure overload in never-pregnant mice, and we compared those animals to surgical sham control mice. In addition, an equal number of the surgical sham and TAC animals were given CSA for 2 weeks. As shown in FIGURE 4, the heart weight-to-tibia-length ratio (mean ± SEM, mg/mm) was significantly different between surgical sham-vehicle (Group 4; 7.97 ± 0.81) and TAC-vehicle animals (Group 2; 11.51 ± 2.19; P <0.01), indicating that the TAC-treated mice manifested cardiac hypertrophy. However, this number was decreased in TAC animals treated with CSA (Group 1; 9.52 ± 1.61; P <0.05) compared to TAC-vehicle (Group 2). The heart weight-to-tibia-length ratio in surgical sham-CSA animals was similar to those in surgical sham-vehicle–treated animals. These results show that the hypertrophic response in TAC-treated animals could be blunted by CSA.

We further measured left ventricular fractional shortening (LV%FS) in all experimental groups as an index of ventricular function (FIG. 5). The data demonstrate a significant decrease in left ventricular function in the TAC-vehicle animals (Group

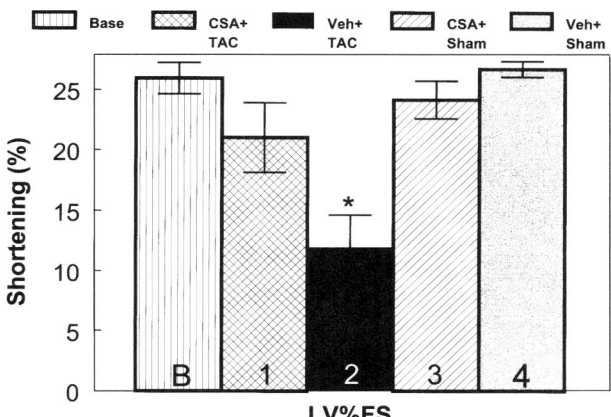

FIGURE 5. Echo-derived left ventricular fractional shortening (LV%FS) as a measurement of ventricular function. Left ventricular fractional shortening was determined by echocardiography for Group 1 (TAC-CSA), Group 2 (TAC-vehicle; $P <0.0001$), Group 3 (surgical sham-CSA), and Group 4 (surgical sham-vehicle) and compared to baseline recordings (*Base*) 14 days postsurgery.

2) compared to baseline (B), surgical sham-CSA (Group 3), and surgical sham-vehicle animals (Group 4) after 14 days ($P <0.0001$). However, no statistically significant difference in LV%FS was noted between TAC animals treated with CSA (Group 1) and baseline (B) at day 14, indicating that cyclosporin A protected against the TAC-induced decrease in left ventricular function associated with cardiac hypertrophy.

Preliminary echocardiographic data also demonstrated increases in both end systolic and end diastolic chamber dimensions in TAC-vehicle animals that were blunted with CSA treatment (data not shown). Similar improvements were observed in other cardiac parameters, such as ventricular septal thickness and posterior wall thickness (data not shown).

We have demonstrated that homozygous NCX1 overexpressing mice develop cardiac hypertrophy when subjected to transverse aortic constriction and thus represent another model for studying the effects of CSA on hypertrophied hearts. Furthermore, our findings are consistent with reports showing that the calcineurin inhibitor CSA has a beneficial effect in various *in vivo* animal models of cardiac hypertrophy.[17] However, many studies also report no beneficial effect of CSA in animals with experimental cardiac hypertrophy.[17] The nature of this discrepancy remains to be uncovered, but it could, for instance, be based on the genetic background of the animals used in the different experiments.

So far, very few studies are investigating the expression profile of NCX1 in experimentally induced cardiac hypertrophy. Kent *et al.*[13] documented a rapid increase of NCX1 mRNA and protein levels in pressure-overloaded right ventricles from cats. A study by Wang *et al.*[14] recently showed that NCX1 transcripts as well as protein levels were upregulated in the hypertrophied hearts of wild-type mice that were subjected to thoracic aortic banding. This upregulation as well as the hypertrophied response in general was blunted by treating the animals with CSA. However, NCX1 activity was described as being diminished. The discrepancy of this finding is unclear. The field of NCX1 research in cardiac hypertrophy is still in its beginning, and more detailed investigations are needed to elucidate the role of the Na^+/Ca^{2+} exchanger in the pathogenesis of this disease.

CONCLUSION

The Na^+/Ca^{2+} exchanger gene NCX1, like other hypertrophic genes, is significantly upregulated in an *in vitro* model of cardiac hypertrophy. The calcineurin inhibitor cyclosporin A effectively inhibits NCX1 promoter activity in a dose-dependent manner *in vitro*, suggesting that NCX1 expression may be regulated by a calcineurin-dependent signaling pathway. Preliminary results with calcineurin overexpressing mice suggest an increase in NCX1 transcript levels in the heart of the animals.

We further showed that homozygous mice overexpressing NCX1 develop cardiac hypertrophy when subjected to transverse aortic constriction and that this hypertrophic response can be blunted by treating the animals with cyclosporin A. We will use this model to further investigate how a calcineurin-dependent signaling pathway affects NCX1 expression, which may ultimately identify the role of the Na^+/Ca^{2+} exchanger in cardiac hypertrophy *in vivo*.

ACKNOWLEDGMENTS

This research was conducted while Susanne B. Nicholas was a Pfizer scholar. The study was further supported by National Institutes of Health Grant HL48509 (to K.D.P.) and The Laubisch Endowment (to K.P.R.)

REFERENCES

1. BERS, D.M. 2000. Circ. Res. **87:** 275–281.
2. BALKE, C.W. & S.R. SHOROFSKY. 1998. Cardiovasc. Res. **37:** 290–299.
3. POGWIZD, S.M., M. QI, W. YUAN, *et al.* 1999. Circ. Res. **85:** 1009–1019.
4. HASENFUSS, G. 1998. Cardiovasc. Res. **37:** 279–289.
5. STUDER, R., H. REINECKE, J. BILGER, *et al.* 1994. Circ. Res. **75:** 443–453.
6. HASENFUSS, G., W. SCHILLINGER, S. E. LEHNART, *et al.* 1999. Circulation **99:** 641–648.
7. FLESCH, M., R.H.G. SCHWINGER, F. SCHIFFER, *et al.* 1996. Circulation **94:** 992–1002.
8. O'ROURKE, B., D.A. KASS, G.F. TOMASELLI, *et al.* 1999. Circ. Res. **84:** 562–570.
9. HATEM, S.N., I.S.K. SHAM & M. MORAD. 1994. Circ. Res. **74:** 253–261.
10. REINECKE, H., R. STUDER, R. VETTER, *et al.* 1996. Cardiovasc. Res. **31:** 48–54.
11. HOBAI, I.A. & B. O'ROURKE. 2000. Circ. Res. **87:** 690–698.
12. TOMASELLI, G.F. & E. MARBÁN. 1999. Cardiovasc. Res. **42:** 270–283.
13. KENT, R.L., J.D. ROZICH, P.L. MCCOLLAM, *et al.* 1993. Am. J. Physiol. **265:** H1024–H1029.
14. WANG, Z., B. NOLAN, W. KUTSCHKE & J.A. HILL. 2001. J. Biol. Chem. **276:** 17706–17711.
15. MOLKENTIN J.D. & G.W. DORN, II. 2001. Annu. Rev. Physiol. **63:** 391–426.
16. MOLKENTIN J.D., J.-R. LU, C.L. ANTOS, *et al.* 1998. Cell **93:** 215–228.
17. MOLKENTIN, J.D. 2000. Circ. Res. **87:** 731–738.

18. NICHOLAS, S.B., W. YANG, S.-L. LEE, et al. 1998. Am. J. Physiol. **274:** H217–H232.
19. ADACHI-AKAHANE, S., L. LU, Z. LI, et al. 1997. J. Gen. Physiol. **109:** 717–729.
20. ROCKMAN, H.A., R.S. ROSS, A.N. HARRIS, et al. 1991. Proc. Natl. Acad. Sci. USA **88:** 8277–8281.
21. TANAKA, N., N. DALTON, L. MAO, et al. 1996. Circulation **94:** 1109–1117.
22. POLLICK, C., S.L. HALE & R.A. KLONER. 1995. J. Am. Soc. Echocardiogr. **8:** 602–610.
23. BARNES, K.V., G. CHENG, M.M. DAWSON & D.R. MENICK. 1997. J. Biol. Chem. **272:** 11510–11517.
24. CHENG, G., T.P. HAGEN, M.L. DAWSON, et al. 1999. J. Biol. Chem. **274:** 12819–12826.
25. NICHOLAS, S.B. & K.D. PHILIPSON. 1999. Am. J. Physiol. **277:** H324–H330.
26. RAO, A., C. LUO & P.G. HOGAN. 1997. Annu. Rev. Immunol. **15:** 707–747.
27. CHIEN, K.R., H. ZHU, K.U. KNOWLTON, et al. 1993. Annu. Rev. Physiol. **55:** 77–95.
28. ROOS, K.P., M.C. JORDAN, L. LU, et al. 2000. FASEB J. **14:** A696 (Abstr.).

Role of Sodium–Calcium Exchanger (*Ncx1*) in Embryonic Heart Development

A Transgenic Rescue?

SIMON J. CONWAY,[a] AGNIESZKA KRUZYNSKA-FREJTAG,[a] JIAN WANG,[a] RHONDA ROGERS,[a] PAIGE L. KNEER,[a] HONGMEI CHEN,[a] TONY CREAZZO,[a,b] DONALD R. MENICK,[c] AND SRINAGESH V. KOUSHIK[a]

[a]*Institute of Molecular Medicine and Genetics and Department of Cell Biology and Anatomy, Medical College of Georgia, Augusta, Georgia 30912, USA*

[b]*Department of Pediatrics, Duke University Medical Center, Durham, North Carolina 27710, USA*

[c]*Gazes Cardiac Research Institute, Medical University of South Carolina, Charleston, South Carolina 29425, USA*

ABSTRACT: Na^+/Ca^{2+} exchanger (*Ncx-1*) is highly expressed in cardiomyocytes, is thought to be required to maintain a low intracellular Ca^{2+} concentration, and may play a role in excitation-contraction coupling. Significantly, targeted deletion of *Ncx-1* results in *Ncx1*-null embryos that do not have a spontaneously beating heart and die *in utero*. Ultrastructural analysis revealed gross anomalies in the *Ncx1*-null contractile apparatus, but physiologic analysis showed normal field-stimulated Ca^{2+} transients, suggesting that *Ncx-1* function may not be critical for Ca^{2+} extrusion from the cytosol as previously thought. Using caffeine to empty the intracellular Ca^{2+} stores, we show that the sarcoplasmic reticulum is not fully functional within the 9.5-dpc mouse heart, indicating that the sarcoplasmic reticulum is unlikely to account for the unexpected maintenance of intracellular Ca^{2+} homeostasis. Using the *Ncx1-lacZ* reporter, our data indicate restricted expression patterns of *Ncx1* and that *Ncx1* is highly expressed within the conduction system, suggesting *Ncx1* may be required for spontaneous pacemaking activity. To test this hypothesis, we used transgenic mice overexpressing one of the two known adult *Ncx1* isoforms under the control of the cardiac-specific α-myosin heavy chain promoter to restore *Ncx1* expression within the *Ncx1*-null hearts. Results indicate that the transgenic reexpression of one *Ncx1* isoform was unable to rescue the lethal null mutant phenotype. Furthermore, our *in situ* results indicate that both known adult *Ncx1* isoforms are coexpressed within the embryonic heart, suggesting that effective transgenic rescue may require the presence of both isoforms within the developing heart.

KEYWORDS: cardiac Na^+/Ca^{2+} exchanger; knockout mouse mutant; transgenic rescue; gene expression; embryonic heartbeat

Address for correspondence: Simon J. Conway or Srinagesh V. Koushik, Institute of Molecular Medicine and Genetics, Medical College of Georgia, CB2803, 1120 15th St., Augusta, GA 30912. Voice: 706-721-8775; fax: 706-721-8685.
sconway@mail.mcg.edu or skoushik@immagene.mcg.edu

INTRODUCTION

The Na^+/Ca^{2+} exchanger (*Ncx1*), an ion transport protein present in virtually all adult animal cells, is a plasma membrane-bound protein that catalyzes the electrogenic exchange of 1 intracellular Ca^{2+} ion for 3 extracellular Na^+ ions.[1,2] In addition to being highly expressed in cardiomyocytes and being one of the earliest functional genes to be expressed in the developing mouse heart,[3] there is rapid upregulation of human *Ncx1* in response to pressure overload,[4,5] end-stage heart failure,[6,7] and senescence.[8] The primary role of *Ncx1* is thought to be as a housekeeping gene involved in maintaining a low intracellular Ca^{2+} concentration, and it may play a role in sustaining excitation-contraction (EC) coupling (although this is still contentious). Significantly, *Ncx1* is one of the earliest functional genes to be expressed in the developing heart,[3] making this gene one of the earliest markers of cardiac differentiation. However, little is known about the actual role of *Ncx1* in the embryonic heart and in cardiac function.[2]

To determine the precise role of *Ncx1*, we generated a *β-galactosidase* reporter-gene *lacZ* knockin mutation into the *Ncx1* locus, resulting in null embryos that are embryonic lethal.[9] Analysis of mutants revealed that the reporter faithfully recapitulated the previously reported expression of endogenous *Ncx1*.[3] Furthermore, heterozygous mice are normal and fertile, whereas *Ncx1*-null embryos are growth retarded, survive to 11.0 days postcoitum (dpc), but lack a spontaneously beating heart (normal embryonic heart starts beating at ~8.25 dpc). Significantly, as endogenous *Ncx1* mRNA (which is ubiquitously expressed in adults) is only expressed within the embryonic heart prior to lethality, the *Ncx1*-null embryos essentially represent a "heart-specific" knockout. Surprisingly, the *Ncx1*-null embryos displayed transient Ca^{2+} cycling when electrically stimulated. This suggests that the Ca^{2+} delivery mechanism was fundamentally intact and that *Ncx1*-null cardiomyocytes can regulate intracellular Ca^{2+} concentrations despite the absence of *Ncx1*. Given that embryonic EC coupling is mainly thought to involve Ca^{2+} entering via the plasma membrane Ca^{2+} channel and exiting via *Ncx1* and that there is not thought to be any functional sarcoplasmic reticulum (SR) in these early embryonic hearts[1,10] — the presence of these electrically produced transients in the absence of *Ncx1* — strongly suggest that our understanding of embryonic EC coupling and the genes involved is incomplete, as Na^+/Ca^{2+} exchange may not be the primary mechanism for removal of Ca^{2+} within embryonic hearts.

The presence of these relatively normal electrically stimulated Ca^{2+} transients suggested an additional, yet novel, role for *Ncx1* in embryonic hearts, in that *Ncx1* may be involved within the actual generation of the heartbeats and that *Ncx1* may function as part of the pacemaker during normal cardiac function.[11] Supporting these suggestions are recent data that the sinoatrial nodal cell ryanodine receptor and *Ncx1* act as molecular partners in pacemaker regulation, as localized SR Ca^{2+} release and Ca^{2+} activation of *Ncx1* may be a critical element in a chain of molecular interactions that permits the heartbeat to occur and determines its beating rate.[12] Similarly, using a distant upstream region of the rat multipartite *Ncx1* promoter, which is sufficient to confer cardiac-specific expression, rat promoter activity is significantly reduced, except in the sinoatrial/atrioventricular node and atrioventricular bundles of the conduction system.[13] Given that the lack of heartbeat during embryonic development leads to numerous defects such as growth retardation, apoptosis,[9,11,14,15] and vascular defects[14] prior to lethality of the embryo and that *Ncx1* is only expressed in the developing embryonic heart, we investigated whether transgenic reexpression of *Ncx1* within mutant cardiac myocytes could rescue these lethal defects.

Transgenic mice overexpressing canine *Ncx1* under the control of the cardiac-specific α-myosin heavy chain (α-MHC) promoter[16] were used to restore *Ncx1* expression within the homozygous null mutant hearts. Our results indicate that the transgenic reexpression of *Ncx1* was unable to rescue the lethal null mutant phenotype. Furthermore, our results indicate that both known adult *Ncx*1 isoforms are coexpressed within the embryonic heart, suggesting that effective transgenic rescue may require both isoforms to be present within the developing heart.

MATERIAL AND METHODS

Transgenic Rescue. A breeding pair of transgenic mice that express canine *Ncx1* (*Ncx1 Tg*) under the control of α-MHC[16] (generously provided by Dr. Kenneth Philipson, UCLA) were crossed with *Ncx1-lacZ* heterozygous mice[9] to obtain $Ncx1^{lacZ}$/+ transgenic positive (Tg) heterozygous mice. Null ($Ncx1^{lacZ}/Ncx1^{lacZ}$) embryos were generated that either carried the transgene and expressed canine *Ncx1* in the null heart or were negative. Tail biopsies/limb buds were subjected to polymerase chain reaction (PCR) genotyping.[9,16]

Ventricular Ca^{2+} Transient Measurement and Caffeine Stimulation. Ca^{2+} transients were measured from 9.5 and 10.5 dpc. as previously described.[9] Caffeine (20 mM) stimulation and either thapsigargin or cyclopiazonic acid exposure was carried out on fura-2–loaded whole heart tubes and isolated myocytes. Experiments using heart tubes were at 32°C and isolated myocytes were at room temperature.

Immunohistochemistry. Sections were postfixed in either 4% paraformaldehyde or acetone and subsequently stained with 5A5 antibody (1:200 dilution; Developmental Studies Hybridoma Bank) that detects highly polysialylated NCAM; connexin 40 rabbit polyclonal antibody (1:1000 dilution; a gift from Dr. R. Gourdie, MUSC) and mouse monoclonal antibody that detect ventricular myosin heavy chains (1:2000 dilution; Accurate Chemical & Scientific), which are molecular markers of the developing cardiac conduction system.[17] Biotinylated secondary antibodies (Southern Biotechnology) were used with the ABC Complex kit (Vector).

X-gal Staining, Northern Blotting, and Radioactive Section in Situ Hybridization. Embryos from $Ncx1^{lacZ}$/+ Tg matings were genotyped and stained with *X-gal* as whole embryos. Northern blotting using the *Ncx1* exon 2-specific probe[3] and the GAPDH housekeeping gene was performed on 30 µg total RNA 10.5-, 13.5-dpc, and isolated adult heart samples. Radioactive section *in situ* hybridization was performed on wax sections, as previously described,[18] using either *Ncx1* exon 2 probe[3] or the PCR generated 3' untranslated region *Ncx1-AF* and *Ncx1-U* probes (see below for details).

RT-PCR from cDNA. RNA was isolated from either whole embryos at 8.5 and 11.5 dpc or isolated hearts from 15.5-dpc embryos or adults using previously described protocols.[3] cDNA was synthesized using the first-strand cDNA synthesis kit (Invitrogen). The PCR was designed to specifically amplify the two different published isoforms of *Ncx1* that varied in the COOH terminus (Genbank AF004666[19] [called *Ncx1-AF*] and U70033[20] [called *Ncx1-U*] isoforms). *Ncx1-AF* specific primers (sense 5'-CCATCTTCGGAATGTCAATG-3' and antisense 5'-GCCTTTTTG-GTCTTTATTTC-3'), and *Ncx1-U* specific primers (sense 5'-GTGAGCTGGGAGGGA-3' and antisense 5'-ATGGTTTCTGACTTGGC-3') were designed within the 3' untranslated region of each isoform (FIG. 6C, arrow).

Cloning of Isoform-Specific in Situ *Probes.* Both 288-bp *Ncx1-U* and 448-bp *Ncx1-AF*–specific fragments were amplified from adult heart cDNA and cloned using a TOPO TA cloning kit (Invitrogen), sequenced to confirm the identity and used for *in situ* analysis.

RESULTS AND DISCUSSION

Restricted Expression Patterns of Ncx1 *within Both the Embryo and the Adult Mouse.* *Ncx1* is generally thought to be widely expressed within adult mammalian tissues,[21,22] but several recent gene expression studies[3] and transgenic studies[9,13,23] have shown that *Ncx1* is restricted to the embryonic heart during early development (FIG. 1a). Our knockin of a *lacZ* reporter into the *Ncx1* genomic locus has enabled us to demonstrate unequivocally that *Ncx1* mRNA expression is confined to the early embryonic cardiomyocytes and is absent from the rest of the developing embryo and extraembryonic tissues. Thus, the placental and vascular defects, growth retardation, and ultimate lethality associated with the targeted deletion of *Ncx1*[9,14,15] are all secondarily due to abnormal cardiomyocyte morphogenesis, cardiac function,

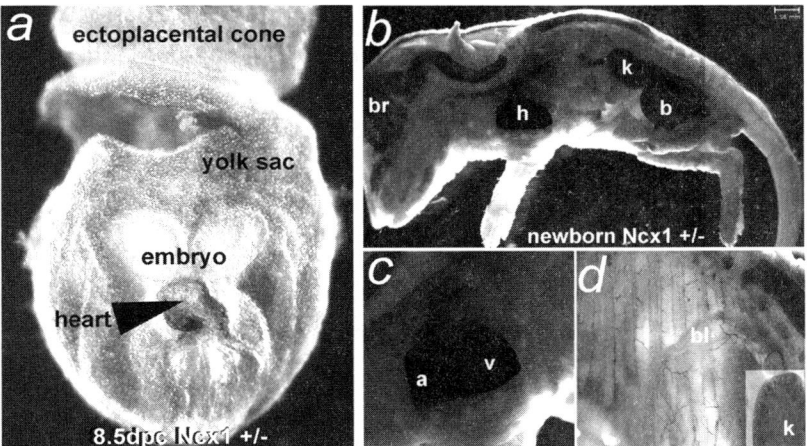

FIGURE 1. See Plate 2 in color section. Spaciotemporal expression of *Ncx1* within the cardiovascular system. (**a**) Whole-mount *lacZ* staining of 8.5-dpc embryo, showing that *Ncx1^{lacZ}* is only present within the primitive "linear" heart. Note no *lacZ* staining within yolk sac or ectoplacental cone. (**b-d**) *Ncx1^{lacZ}* staining of heterozygous newborn mouse. Note that *lacZ* staining is present within the heart (h), brain (br), kidney (k), and bladder (b). Also note that *Ncx1^{lacZ}* staining is restricted to the blood vessel (bl) and nerves and to the cortex and not the medulla of the newborn kidney (*inset*).

and pacing of the heart. Additionally, *lacZ* staining of newborn animals indicates that *Ncx1* mRNA is also not as widespread as previously thought. The highest levels of expression were within the newborn heart, followed by the kidney/brain/bladder and spinal cord (FIG. 1b and c). Although *Ncx1 lacZ*–positive staining could be seen within most tissues of the body, staining was confined to the smooth muscle surround in the blood vessels or the major nerves (FIG. 1d).

SR is not fully functional within the 9.5-dpc mouse heart. The SR is thought to be poorly developed within the embryonic heart, and thus embryonic EC coupling is thought to depend heavily on trans-sarcolemmal Ca^{2+} fluxes provided by the sarcolemmal Ca^{2+} channel and *Ncx1*.[1,10] However, the presence of electrically produced excitable Ca^{2+} transients and the unexpectedly rapid decay of these Ca^{2+} transients in our *Ncx1*-null embryos suggested that upregulation of the sarcolemmal Ca^{2+}-ATPase, increased SR, or some other novel mechanism is responsible for the slightly slower Ca^{2+} removal within the *Ncx1* null embryos. This raises the intriguing possibility that contrary to current understanding, Ca^{2+} extrusion through the surface membrane via *Ncx1* may not be the primary mechanism for Ca^{2+} removal in early mouse embryos during EC coupling.

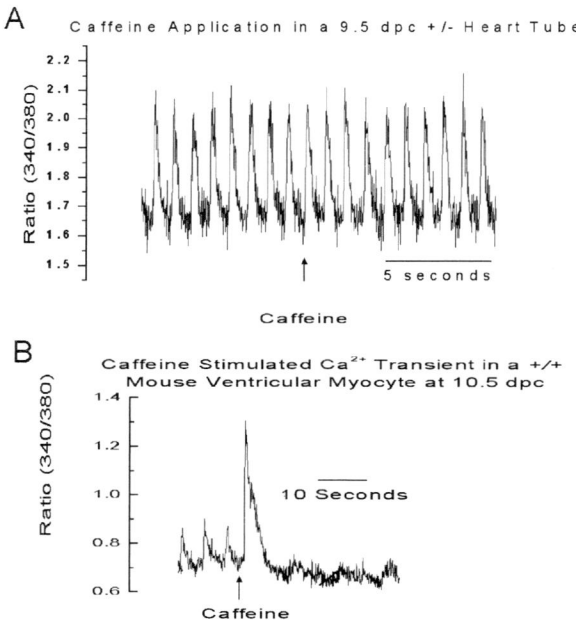

FIGURE 2. Effect of caffeine on Ca^{2+} homeostasis in early embryonic heart development in mouse. (**a**) Ventricular Ca^{2+} transients in a spontaneously beating heart tube from a $Ncx1^{lacZ}/+$ embryo at 9.5 dpc. Caffeine (20 mM) has no effect. Results are typical of data from 5 heart tubes (2 +/+ and 3 $Ncx1^{lacZ}/+$ embryos). (**b**) Caffeine-stimulated Ca^{2+} transients were apparent in +/+ embryos by 10.5 dpc, as shown in this example of an isolated ventricular myocyte, indicating the presence of an SR Ca^{2+} store. The myocyte in this example was stimulated at 0.2 Hz.

To determine if the SR is functional, spontaneous Ca^{2+} transients were measured in either +/+ or +/− spontaneously beating heart tubes or in +/+, $Ncx1^{lacZ}$/+, and $Ncx1^{lacZ}$/$Ncx1^{lacZ}$ 9.5- and 10.5-dpc embryos in which Ca^{2+} transients were elicited by electrical field stimulation. Application of caffeine to empty the intracellular stores of Ca^{2+} had no effect at 9.5 dpc (FIG. 2a), but it had a dramatic effect on 10.5-dpc hearts (FIG. 2b). Additionally, attempts to block SR Ca^{2+}-ATPase with either thapsigargin or cyclopiazonic acid had no effect on the magnitude of spontaneous Ca^{2+} transients in all three genotypes (data not shown), suggesting that the SR is nonfunctional within 9.5-dpc normal heart tubes and that there is no increased SR within the Ncx1-null embryos that could account for the unexpected maintenance of intracellular Ca^{2+} homeostasis.

Ncx1 *Is Highly Expressed within the Conduction System.* The most striking finding was lack of spontaneous Ca^{2+} transient activity in Ncx1-null hearts, which was always observed in wild-type and heterozygous littermate embryos. A novel interpretation of these results is that the pacemaking (excitation) mechanism is also defective in Ncx1–null mutants. Thus, lack of spontaneous rhythmic activity in these hearts could be due to the absence of the diastolic pacemaker potential, implying that the inward current generated by electrogenic Ncx1 in extruding Ca^{2+} during the relaxing phase of the preceding heartbeat leads to subsequent diastolic depolarization.

Pacemaker and conducting systems coordinate the heartbeat.[24,25] However, the existence of a pacemaker/conduction system within the embryo is a highly contentious issue, as the components are not morphologically recognizable, but heart tubes have an adult-like ECG.[25] Fate studies indicated that adult pacemaker/conduction systems arise from rhythmically beating embryonic precursors,[24] with the whole primary myocardium acting as conducting tissue.[26] Supporting the idea that Ncx1 expression is essential for generating rhythmic activity, both $Ncx1^{lacZ}$ reporter and Ncx1 mRNA are present within the early fetal atrioventricular conduction axis and coexpressed with known conduction system markers[24,17] (FIG. 3a-e) and later within the adult conduction system (FIG. 3f). In fact, the highest level of endogenous Ncx1 expression within the adult heart is within the sinoatrial node, a finding similar to those using a transgenic reporter system.[13] The heartbeat is thought to originate from the sequential activation of ionic currents in pacemaker cells,[27] but recently it was suggested that it may be influenced by transient changes in cytosolic Ca^{2+} concentrations.[28,29] Thus, colocalization of Ncx1 in adult pacemaker cells and the lack of spontaneous Ca^{2+} transients suggest that Ncx1 expression/function may be required for generation of the embryonic heartbeat and spontaneous pacemaking activity.

Transgenic Reexpression of Canine Ncx1 *within the Lethal* Ncx1-*Null Mutant.* Overexpression of canine Ncx1 transgene (Tg) does not adversely affect the survival and/or mating performance of Ncx1-lacZ heterozygous embryos or adults (FIG. 4). However, when the offspring from $Ncx1^{lacZ}$/+ Tg-positive matings were genotyped, only +/+ and $Ncx1^{lacZ}$/+ (either Tg-positive or Tg-egative) mice were present, indicating that even Ncx1-null mutants that express canine Ncx1 mRNA are *in utero* lethal and that there is no difference in the survival rate of the mutants. Additionally, when 13.5-dpc fetuses resulting from $Ncx1^{lacZ}$/+ Tg-positive matings were genotyped, no Ncx1-null Tg-positive fetuses were present. However, when 9.5- and 10.5-dpc embryos resulting from $Ncx1^{lacZ}$/+ Tg-positive matings were genotyped, Ncx1-null Tg-positive embryos were present at the expected frequencies (TABLE 1), but they were malformed and lacked a vigorous heartbeat. Thus, this particular trans-

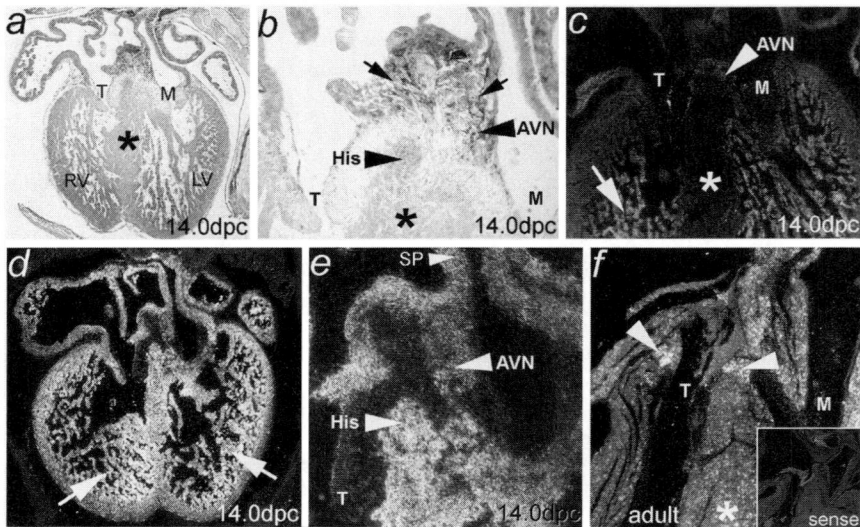

FIGURE 3. See Plate 3 in color section. Overlapping expression of *Ncx1* and cardiac conduction system markers PSA-NCAM and connexin 40. (**a**) Low-power transverse section through the mitral (M) and tricuspid (T) valve level of a 14.0-dpc heterozygous embryo, showing both $Ncx1^{lacZ}$ staining (blue) and PSA-NCAM protein expression (light brown) within the heart. (**b**) Higher power image of atrioventricular junction region of heart shown in **b**. Note the colocalization of PSA-NCAM antibody and $Ncx1^{lacZ}$ staining within the subendocardial myocardial lining of the atrioventricular junction (indicated by *small black arrows*) and within the atrioventricular node of the conduction system (AVN). $Ncx1^{lacZ}$ staining is also present within the His bundle (His). (**c**) Connexin 40 antibody staining (viewed under dark-field illumination the brown DAB staining appears bright blue) of serial section of the heart, shown in **a**, shows that $Ncx1^{lacZ}$ staining is identical to connexin 40 expression (indicated by *small black arrows*). (**d**) Low-power image of 14.0-dpc wild-type embryo following radioactive *in situ* hybridization. Note *Ncx1* mRNA is particularly highly expressed within the trabecular ventricular compartment (indicated by *white arrows*) where propagation of the impulse is preferentially achieved. (**e**) Higher power image of atrioventricular junction/septum primum (SP) region of heart shown in **d**. Note that *Ncx1* mRNA expression is present within the atrioventricular node (AVN) and is particularly strongly expressed within the His bundle. (**f**) High-power transverse section through the mitral and tricuspid valve level of an adult heart following radioactive *in situ* hybridization analysis of *Ncx1* mRNA expression. Note that *Ncx1* mRNA expression is present throughout the adult heart and is highly expressed within the atrioventricular fascicles (indicated by *white arrowheads*). *Inset*, *in situ* hybridization negative control using an *Ncx1* sense probe.

gene is not sufficient to rescue the *Ncx1*-null phenotype. Nevertheless, closer comparison of the 10.0-dpc *Ncx1*-null Tg-positive and Tg-negative embryos shows that the *Ncx1*-null Tg-positive embryos have pericardial effusions surrounding the developing heart (FIG. 5). Pericardial effusions are thought to be due to osmotic imbalances that arise from improper development of the cardiovascular system[30] and are usually associated with a failing, yet beating, embryonic heart. This indicates that the *Ncx1*-null Tg-positive hearts are attempting to beat, but that the canine *Ncx1* transcripts are not sufficient for *in utero* survival.

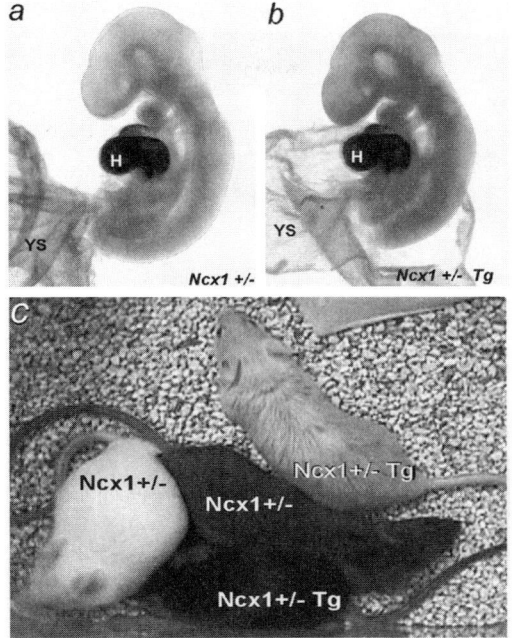

FIGURE 4. See Plate 4 in color section. Expression of $Ncx1^{lacZ}$ in transgenic animals. Whole-mount *lacZ* staining of 9.5-dpc $Ncx1^{lacZ}/+$ (**a**) and $Ncx1^{lacZ}/+$ transgenic positive (**b**) embryos, showing that $Ncx1^{lacZ}$ is only present within the heart and that the presence of the transgene does not affect either normal development or endogenous *Ncx1* expression. (**c**) Four adult $Ncx1^{lacZ}/+$ healthy mice are shown; two carry the transgene.

TABLE 1. Transgenic rescue data

	+/+	+/–	–/–	+/+ Tg	+/– Tg	–/– Tg
13.5 dpc (Litters $n = 3$)	3	6	—	5	10	—
10.5 dpc (Litters $n = 4$)	4	12	[5*]	3	9	[4*]
9.5 dpc (Litters $n = 6$)	6	18	[8*]	5	11	[6*]

n, numbers of individual litters analyzed at each age are listed below each age group.
*Abnormal embryos (growth retarded and lacking heartbeat).
[ ] Numbers of abnormal mutants.

Why No Transgenic Rescue? The lack of complete rescue of the null mutant phenotype is intriguing, as it suggests that either one or some combination of the following reasons may act to impede complete rescue of the mutant phenotype. Firstly, the canine protein may not sufficiently complement the lack of mouse *Ncx1* gene. Although canine and mouse genes are 95% identical, at the protein level, subtle differences in functional properties of the canine transgene may prevent normal localization/function. Previous reports have indicated that the overexpression of canine Ncx1 protein may cause it to be functionally excluded from the Ca^{2+} microdomains

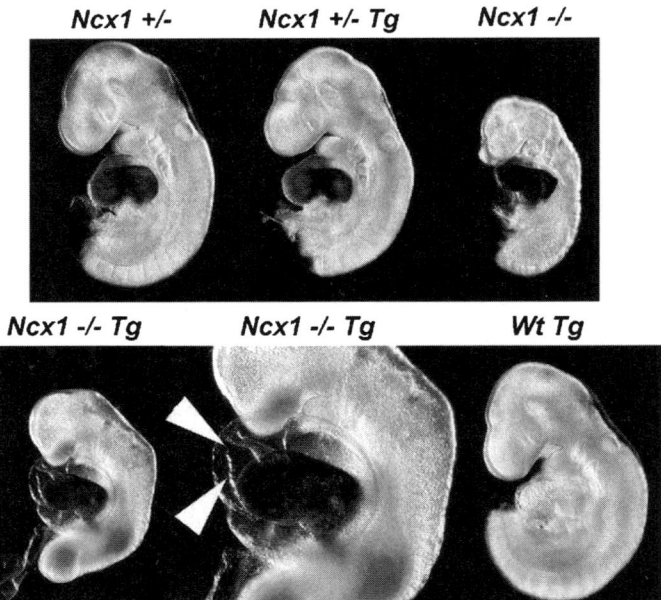

FIGURE 5. See Plate 5 in color section. Reexpression of the canine *Ncx1* transgene does not rescue embryonic development. Five 10.0-dpc embryos are shown following *lacZ* staining. Note that the heterozygous $Ncx1^{lacZ}/+$ (both transgenic (Tg)-positive and Tg-negative) embryos are normal (identical to wild-type nonstained embryo) and have a spontaneously beating heart. Both the *Ncx1*-null embryos (both Tg-positive and Tg-negative) are abnormal and lack a spontaneously beating heart. However, the Tg-positive embryos have significant pericardial effusion, indicating some cardiac function (illustrated by *arrowheads* in enlarged image).

surrounding the ryanodine receptor complex and causes SR Ca^{2+} overload and upregulation of SR Ca^{2+} ATPase function.[16,31] Secondly, the spatiotemporal expression driven by α-MHC promoter may not be recapitulating the endogenous *Ncx1* expression. Although the α-MHC gene is expressed early during heart development, a thorough examination of the spatiotemporal expression pattern of the canine *Ncx1* mRNA driven by the α-MHC promoter in this particular transgenic line has not been performed. The inability of this transgene to rescue the mutant phenotype could be due to either differences in spatiotemporal expression or levels of expression. Thirdly, because the transgene expresses only a single isoform, this may not be sufficient to restore normal Ncx1 protein activity. There are numerous reports on alternative splicing isoforms of *Ncx1* that are generated using mutually excluded exons labeled A-F that have been shown to be functional[1,2] in different tissues. However, there are two reports that describe the cloning of mouse cDNAs that encode for a 970-amino acid (AA) protein[20] (Genbank accession #U70033, which was similar to the canine cardiac *Ncx-1*[32]) used to construct the transgenic mouse[16] and the other for a truncated 940AA[19] (Genbank accession #AF004666) that lacks the last transmembrane segment of *Ncx-1*, which seem to be generated by alternative splicing at the 3' end of the protein. The truncated *Ncx1-AF* isoform has an AATAAA polyadenylation

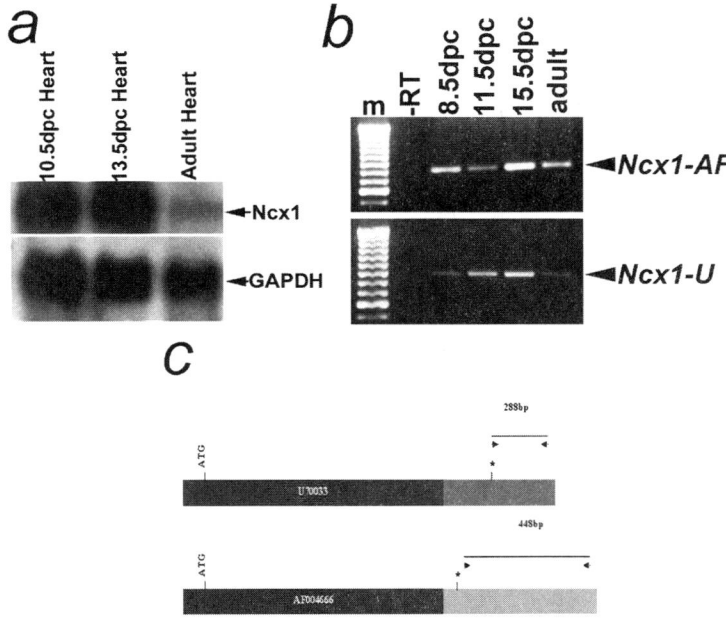

FIGURE 6. Identification of *Ncx1* expression within the cardiovascular system. (**a**) Northern analysis of *Ncx1* expression within the 10.5-, 13.5-dpc, and adult isolated hearts. A single band of ~7 kb is present at all three ages, and *Ncx1* is significantly diminished within the adult heart. (**b**) RT-PCR analysis of the relative levels and timing of both *Ncx1-AF* and *Ncx1-U* isoforms during development and within the adult heart. Note that both isoforms are present at all ages and that *Ncx1-AF* is expressed at higher levels. (**c**) Schematic showing the different isoforms and the positions of the isoform-specific primers and *in situ* probes. *Asterisk* indicates stop codons and the red/blue sections have no sequence similarity.

signal in the 3′ untranslated region that caused the 3344-bp clone to stop at a premature termination site and encode for a shorter protein. This is in contrast to the other isoform, which encodes a longer protein. Comparing the predicted AA sequence of *Ncx1* isoforms by hydropathic plot, it was found that *Ncx1-AF* isoform protein has only 11 extracellular domains, missing 1 domain at the COOH terminal, whereas *Ncx1-U* isoform protein has 12 extracellular domains.[20] Although the presence of these two isoforms was known for a while, it is unclear as to what are the spatiotemporal expression patterns of these proteins in a developing embryo.

Both Ncx1-AF *and* Ncx1-U *Cardiac Isoforms Are Expressed within the Normal Heart.* Cardiac Na^+/Ca^{2+} exchange activity is highest during *in utero* development and decreases rapidly after birth, mirroring the neonatal development of the SR and T-tubules.[33] The mouse *Ncx1* gene contains a cluster of six exons (A, B, C, D, E, and F) that encode a variable region in the large intracellular loop of the protein. Several reports have noted novel splice variants, suggesting that an enlarged repertoire of *Ncx1* isoforms may be present in the heart.[19,20,22] However, the significance of these various isoforms is currently unknown. To verify what sequences are ex-

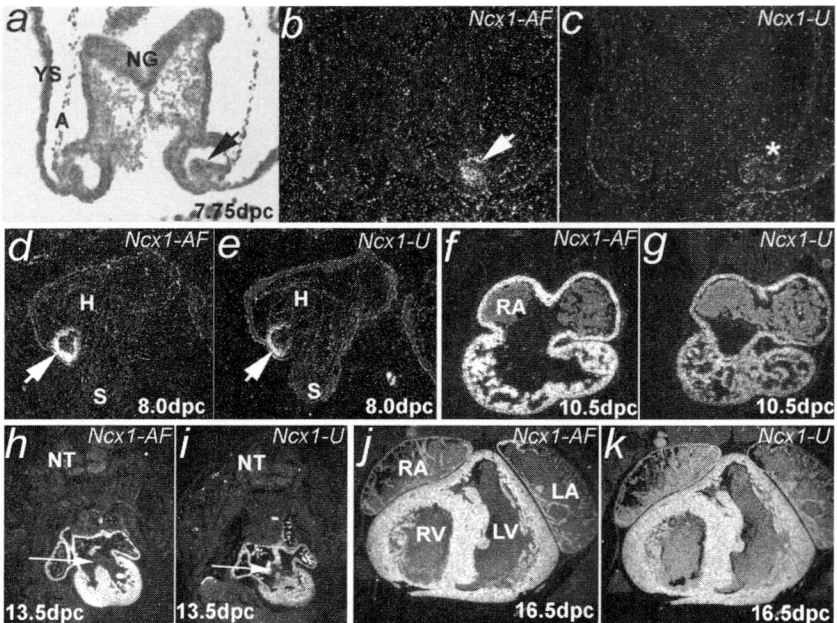

FIGURE 7. Spaciotemporal expression of the *Ncx1-AF* and *Ncx1-U* isoforms within the cardiovascular system. (**a**) H&E staining of 7.75-dpc wild-type transverse section. *Arrow* indicates the cardiogenic plate, and note the yolk sac (ys), amnion (a), and neural grove (ng). (**b**) Radioactive *in situ* hybridization analysis of *Ncx1-AF* (**b**) and *Ncx1-U* (**c**) mRNA expression within embryo in **a**. Note that silver grains are evident in the cardiogenic plate in only the *Ncx1-AF*–probed sections (*white arrow*). 8.0-dpc embryos (**d** and **e**), 10.5-dpc (**f** and **g**), and 13.5-dpc (**h** and **i**) embryos probed with both *Ncx1-AF* and *Ncx1-U*. Note that at all ages, *Ncx1-AF* and *Ncx1-U* are coexpressed and that *Ncx1-AF* is expressed at slightly higher levels. However, by 16.5 dpc (**j** and **k**) the expression patterns and levels are identical. Both isoforms are restricted to the myocardial layer within the developing heart until 10.5 dpc, and both isoforms start to become weakly expressed within the 13.5-dpc neural tube (nt).

pressed within the developing heart, we used Northern blotting to determine how many bands are present. Using our exon 2-specific probe (which is present within all known isoforms), only a single band (~7.0 kb) was present and, as expected, *Ncx1* expression diminished within the adult heart (FIG. 6a).

Currently, at least two different isoforms are thought to be found in the heart using 5′ rapid amplification of cDNA ends (RACE)[34] and up to five using RT-PCR amplification.[20] Although the presence of these two isoforms was known for a while, it is unclear if they have different functional properties and what the spatiotemporal expression patterns of these isoforms are within an embryo.

Using 3′-RACE on 13.5-dpc embryonic heart samples (data not shown), we could only detect two bands that, when sequenced, recognized two different isoforms in Genbank (*Ncx1-AF*, AF004666; *Ncx1-U*, U70033). To verify the presence of both isoforms and determine the spatiotemporal patterns, RT-PCR analysis was carried out using isoform-specific primers (FIG. 6b and c). Both isoforms are present within the 8.5- and 11.5-dpc embryo (*Ncx1* expression is restricted to the heart at these stages) and in the isolated 15.5-dpc fetal and adult hearts, but the *Ncx1-AF* is expressed at a significantly higher level than the *Ncx1-U* isoform. Additionally, the reduced level of the *Ncx1-U* isoform was also seen when *in situ* hybridization analysis was performed using isoform-specific probes (FIG. 7). The *in situ* analysis also suggested that there may be a slight delay in onset of *Ncx1-U*, but that both isoforms are coexpressed from 8.0 to 16.5 dpc. Thus, both isoforms are expressed in the developing embryonic heart, and as the α-MHC–canine transgenic construct only contains the *Ncx1-U* isoform, transgenic failure may also be due to failure to coexpress *Ncx1-AF*. This suggests that to achieve transgenic rescue of the null mutant phenotype, we may need to express both isoforms of *Ncx1*.

SUMMARY

These data demonstrate that *Ncx1* is a Ca^{2+}-handling gene that is essential for normal cardiomyocyte development and function and may serve as an interesting animal model for understanding functionally related human congenital heart defects. Additionally, the presence of an *Ncx1*-null animal model should enable us to determine what role, if any, the different isoforms play in Na^+/Ca^{2+} exchange, generation of the heartbea, and maintenance of normal intracellular Ca^{2+} homeostasis.

ACKNOWLEDGMENTS

This work is supported by American Heart Association SE-Affiliate Postdoctoral Fellowship 0020402B (to S.V.K) and by National Institutes of Health Grants HL48788/HL66223 (to D.R.M), HL36059/HL58861 (to T.C), and HL60104 (to S.J.C).

REFERENCES

1. PHILIPSON, K.D. & D.A. NICOLL. 2000. Sodium-calcium exchange: a molecular perspective. Annu. Rev. Physiol. **62:** 111–133.
2. BLAUSTEIN, M.P. & W.J. LEDERER. 1999. Sodium/calcium exchange: its physiological implications. Physiol. Rev. **79:** 763–854.
3. KOUSHIK, S.V., J. BUNDY & S.J. CONWAY. 1999. Sodium-calcium exchanger is initially expressed in a heart-restricted pattern within the early mouse embryo. Mech. Dev. **88:** 119–122.
4. KENT, R.L. *et al.* 1993. Rapid expression of the Na(+)-Ca2+ exchanger in response to cardiac pressure overload. Am. J. Physiol. **265:** H1024–1029.
5. STUDER, R. *et al.* 1997. Expression and function of the cardiac Na^+/Ca^{2+} exchanger in postnatal development of the rat, in experimental-induced cardiac hypertrophy and in the failing human heart. Basic Res. Cardiol. **92**(Suppl 1): 53–58.
6. STUDER, R. *et al.* 1994. Gene expression of the cardiac Na(+)-Ca2+ exchanger in end-stage human heart failure. Circ. Res. **75:** 443–453.
7. HASENFUSS, G. *et al.* 1997. Calcium handling proteins in the failing human heart. Basic Res. Cardiol. **92** (Suppl 1): 87–93.
8. KOBAN, M.U. *et al.* 1998. Expressional analysis of the cardiac Na-Ca exchanger in rat development and senescence. Cardiovasc. Res. **37**(2): 405–423.

9. KOUSHIK, S.V. *et al.* 2001. Targeted inactivation of the sodium-calcium exchanger (*Ncx1*) results in the lack of a heartbeat and abnormal myofibrillar organization. FASEB J. 10.1096/fj.00–069fje.
10. BOERTH, S.R., D.B. ZIMMER & M. ARTMAN. 1994. Steady-state mRNA levels of the sarcolemmal Na(+)-Ca2+ exchanger peak near birth in developing rabbit and rat hearts. Circ. Res. **74:** 354–359.
11. KOUSHIK, S.V. *et al.* 2001. Targeted inactivation of the sodium-calcium exchanger (*Ncx1*) results in the lack of a heartbeat and abnormal myofibrillar organization. FASEB J. (Summary) **15:** 1209–1211.
12. BOGDANOV, K.Y., T.M. VINOGRADOVA & E.G. LAKATTA. 2001. Sinoatrial nodal cell ryanodine receptor and Na(+)-Ca2+ exchanger: molecular partners in pacemaker regulation. Circ. Res. **88**(12): 1254–1258.
13. KOBAN, M.U. *et al.* 2001. A distant upstream region of the rat multipartite Na(+)-Ca2+ exchanger NCX1 gene promoter is sufficient to confer cardiac-specific expression. Mech. Dev. **109**(2): 267–279.
14. WAKIMOTO, K. *et al.* 2000. Targeted disruption of Na(+)-Ca2+ exchanger gene leads to cardiomyocyte apoptosis and defects in heartbeat. J. Biol. Chem. **275**(47): 36991–36998.
15. CHO, C.H. *et al.* 2000. The Na+ -Ca2+ exchanger is essential for embryonic heart development in mice. Mol. Cells **10**(6): 712–722.
16. ADACHI-AKAHANE, S. *et al.* 1997. Calcium signaling in transgenic mice overexpressing cardiac Na(+)-Ca2+ exchanger. J. Gen. Physiol. **109**(6): 717–729.
17. CHUCK, E.T. & M. WATANABE. 1997. Differential expression of PSA-NCAM and HNK-1 epitopes in the developing cardiac conduction system of the chick. Dev. Dyn. **209:** 182–195.
18. CONWAY, S.J. 1996. *In situ* hybridization of cell and tissue sections. *In* Methods in Molecular Medicine, Molecular Diagnosis of Cancer. Chapt. **15:** 193–206. F.E. Cotter, Ed. Humana Press Inc. Totowa, NJ.
19. SHI, S., B. CHANG & S.R. BRUNNERT. 1998. Identification and cloning of a truncated isoform of the cardiac sodium-calcium exchanger in the BALB/c mouse heart. Biochem. Genet. **36**(3-4): 119–135.
20. KIM, I. & C.O. LEE. 1996. Cloning of the mouse cardiac Na(+)-Ca2+ exchanger and functional expression in *Xenopus* oocytes. Ann. N.Y. Acad. Sci. **779:** 126–128.
21. KRAEV, A., I. CHUMAKOV & E. CARAFOLI. 1996. The organization of the human gene NCX1 encoding the sodium-calcium exchanger. Genomics **37**(1): 105–112.
22. WAKIMOTO, K. *et al.* 2000. Isolation and characterization of Na+/Ca2+ exchanger gene and splicing isoforms in mice. DNA Seq. **11**(1-2): 75–81.
23. MÜLLER, J.G. *et al.* 2002. Cardiac specific expression and hypertrophic upregulation of the feline Na+/Ca2+ exchanger gene H1-promoter in a transgenic mouse model. Circ. Res. **90:** 158–164.
24. GOURDIE, R.G., S. KUBALAK & T. MIKAWA. 1999. Conducting the embryonic heart: orchestrating development of specialized cardiac tissues. Trends Cardiovasc. Med. **9:** 18–26.
25. MOORMAN, A.F. *et al.* 1998. Development of the cardiac conduction system. Circ. Res. **82:** 629–644.
26. PATTEN, B.M. 1956. The development of the sinoventricular conduction system. Univ. Mich. Med. Bull. **22:** 1–21.
27. IRISAWA, H., H.F. BROWN & W. GILES. 1993. Cardiac pacemaking in the sinoatrial node. Physiol. Rev. **73:** 197–227.
28. HUSER. J., L.A. BLATTER & S.L. LIPSIUS. 2000. Intracellular Ca2+ release contributes to automaticity in cat atrial pacemaker cells. J. Physiol. (Lond.) **524:** 415–422.
29. TERRAR, D. & L. RIGG. 2000. What determines the initiation of the heartbeat? J. Physiol. **524:** 16.
30. COPP, A.J. 1995. Death before birth: clues from gene knockouts and mutations. Trends Genet. **11**(3): 87–93.
31. TERRACCIANO, C.M. *et al.* 1998. Na+-Ca2+ exchange and sarcoplasmic reticular Ca2+ regulation in ventricular myocytes from transgenic mice over expressing theNa+-Ca2+ exchanger. J. Physiol. **512**(Pt 3): 651–667.

32. NICOLL D.A., S. LONGONI & K.D. PHILIPSON. 1990. Molecular cloning and functional expression of the cardiac sarcolemmal Na(+)-Ca2+ exchanger. Science **250**(4980): 562–565.
33. ARTMAN, M. *et al.* 1995. Na+/Ca2+ exchange current density in cardiac myocytes from rabbits and guinea pigs during postnatal development. Am. J. Physiol. **268**(4 Pt 2): H1714–1722.
34. MENICK, D.R. *et al.* 1996. Gene expression of the Na-Ca exchanger in cardiac hypertrophy. J. Cardiol. Fail. **2**(4 Suppl): S69–76.

The Gene Promoter of Human Na^+/Ca^{2+} Exchanger Isoform 3 (SLC8A3) Is Controlled by cAMP and Calcium

NADIA GABELLINI,[a] STEFANIA BORTOLUZZI,[b] GIAN A. DANIELI,[b] AND ERNESTO CARAFOLI[a]

[a]*Department of Biological Chemistry and* [b]*Department of Biology, University of Padua, 35121 Padua, Italy*

KEYWORDS: isoforms; splicing; calcium; cAMP; retinoic acid; BDNF; promoter; regulation; expression

Three genes (SLC8A1, 2, and 3) encoding highly conserved Na^+/Ca^{2+} exchanger isoforms (NCX1, 2, and 3) were found to be located on human chromosomes 2p23-p22, 19, and 14q241,[1–3] respectively. Only the gene and promoter structure of isoform 1 has been described. It consists of alternative promoters directing tissue-specific transcription in heart, kidney, brain, and skeletal muscles. Expression of the three isoforms during the development of rat cerebellar granule neurons in culture has been studied. The study has shown that the levels of NCX2 mRNA and protein were downregulated by Ca^{2+}. An opposite effect was observed on NCX3. The downregulation by Ca^{2+} was mediated by the Ca^{2+}-dependent phosphatase calcineurin.[4] The human genomic DNA region, including the SLC8A3 promoter, has been identified by computer program analysis. The promoter sequence had indicated the presence of a cyclic AMP-responsive element (CRE) that mediates the transcriptional responses to second messengers cAMP and Ca^{2+} (by protein kinase A and Ca^{2+}-calmodulin kinases CaMKII and IV) phosphorylation of transcription factor CREB. The study has explored the responses of the newly identified SLC8A3 promoter to cAMP and Ca^{2+}. The former induces promoter activity, whereas the latter inhibits it.

The human SLC8A3 gene was identified by a similarity search (BLAST Human Genome). The gene consists of 9 exons numbered according to the homologous NCX1 exons;[1] however, the NCX3 sequence lacks the small alternative exons 6, 7, and 8. The human sequence exhibits 93% and 97% nucleotide and amino acid identities, respectively, with the corresponding rat sequence.[5] The alternative splicing in the region encoding the large hydrophilic loop of NCX3 mRNA was analyzed by reverse transcriptase-polymerase chain reaction (RT-PCR) with primers matching in the vicinity of the 3′ end of exon 2 and exon 9. In human skeletal muscle, three DNA

Address for correspondence: Dr. Nadia Gabellini, Department of Biological Chemistry, University of Padua, Via G. Colombo, 3, 35121 Padua, Italy.
nadia.gabellini@unipd.it

fragments were obtained. Their sequence indicated one isoform including exons 4 and 5, a second containing only exon 4, and a third in which exon 4 was also skipped. Densitometric analysis of the intensities of the amplified bands indicated that for each copy of the shortest isoform in which exon 2 was connected to exon 9, 0.8 copies of the two largest isoforms were amplified. These were estimated together because they were not well resolved on agarose gels, inasmuch as they differed only in the presence of 18 bp of exon 5. Analysis was also performed on human total brain cDNA, in which case only the isoform including exon 4 was found. The same result was also obtained on human cell line SH-SY5Y.

The homology with the rat cDNA sequence[4] at the 3' end of exon 1 allowed the location of the promoter region; the transcription start site (TSS) was predicted by the Proscan program upstream of the conserved sequence. A large genomic region upstream of the predicted TSS was analyzed by several programs for the presence of promoter sequences. The most likely promoter region was indicated in the 250 bp upstream of the TSS. This included a TATA box and several Sp-1 elements (GC boxes), indicating constitutive transcription. The promoter also suggested inducible transcription by elements binding transcription factors CREB/ATF, AP1, KROX-24, and EGR-1. It also included binding sites for the brain-specific AP-2 transcription factor and for muscle-specific MyoD and GATA 2/3. In agreement with these findings, the SLC8A3 gene has been shown to be expressed dominantly in brain and skeletal muscle. The ability of the indicated sequence to direct transcription in HEK293 and SH-SY5Y cells was tested in transient transfection assays of reporter constructs. Two plasmids carrying 1.6 kb (pGLN3-1.8 kb) and 258 bp (pGLN3-0.5kb) upstream of the TSS directed similar rates of luciferase synthesis. Thus, it is likely that the minimal promoter resides in the 258-bp region upstream of the TSS, as indicated by the Proscan program.

The luciferase expression driven by the SLC8A3 promoter was followed during the neuronal differentiation of SH-SY5Y cells promoted by retinoic acid (RA) and the brain-derived neurotrophic factor (BDNF).[6] Cells were cotransfected with pGLN3-0.5kb and pSV-βgalactosidase. In untreated proliferating cells the luciferase and the βgalactosidase activities decreased progressively during the initial 4 days following transfection. Treatment with RA (10 µM) enhanced the SLC8A3 promoter activity to a maximum 48 hours after the addition of RA, whereas the βgalactosidase activity declined progressively. Subsequent stimulation with BDNF (50 ng/ml) to promote full neuronal differentiation rapidly induced promoter activity that became maximal 4–5 hours after the addition of BDNF. The results suggest that the minimal SLC8A3 promoter sequence directs neuronal specific expression. In proliferating and RA-treated cells luciferase expression from the SLC8A3 promoter was induced about 10-fold by rising cAMP for 16 hours, using the membrane-permeable derivative Bt_2cAMP. The strong inducibility by cAMP was likely to be mediated by the CRE sequence in the SLC8A3 promoter. By contrast, the partial depolarization of the plasma membrane by increasing the concentration of KCl in the medium to 30 mM caused the downregulation of promoter activity by 50%. The depolarizing treatment caused an increase in intracellular Ca^{2+}. Most probably, the inhibition depended on the phosphorylation of transcription factor CREB by CaMKII, because the specific kinase inhibitor KN-93 (5 µM) partially reversed the inhibition. The calcineurin inhibitor FK506 failed to affect the promoter inhibition. This indicates that the inhibition by Ca^{2+} was mediated by CaMKII-directed phosphorylation of CREB.

Transcription factor CREB can be activated by CaMKIV-directed phosphorylation of Ser-133, whereas CaMKII phosphorylates CREB at a second position (Ser-142), causing an inhibitory effect on transcription. Western blotting of whole cells under the conditions employed for the luciferase measurements indicated that the amount of phospho-CREB was consistently increased following cAMP elevation, but it was significantly reduced in the presence of 30 mM KCl. Cells were treated with BDNF for 8 hours and then were submitted to procedures increasing cAMP and/or Ca^{2+} for an additional 16 hours. The luciferase activity increased about 4-fold in response to the rise in cAMP; however, the Ca^{2+} increase failed to influence the promoter activity also when it was applied concomitantly with the cAMP. Monitoring of the activating phosphorylation of the ATF and CREB transcription factor family (Ser-133 in CREB) indicated that following BDNF stimulation ATF became phosphorylated instead of CREB.

REFERENCES

1. KRAEV, A., I. CHUMAKOV & E. CARAFOLI. 1996. Genomics **37:** 105–112.
2. KIKUNO, R., T. NAGASE, K. ISHIKAWA, et al. 1999. DNA Res. **6:** 197–205.
3. GABELLINI, N., S. BORTOLUZZI, G.A. DANIELI & E. CARAFOLI. 2002. Gene, in press.
4. LI, L., D. GUERINI & E. CARAFOLI. 2000. J. Biol. Chem. **275:** 20903–20910.
5. NICOLL, D.A., B.D. QUEDNAU, Z. QUI, et al. 1996. J. Biol. Chem. **271:** 24914–24921.
6. ENCINAS, M., M. IGLESIAS, Y. LIU, et al. 2000. J. Neurochem. **75:** 991–1003.

Role of MAP Kinases in the Na^+/Ca^{2+} Exchanger Gene Expression in Feline Adult Cardiocytes

LIN XU, JOACHIM G. MULLER, PATRICK R. WITHERS, CHRISTIANA KAPPLER, WENJING JIANG, AND DONALD R. MENICK

Gazes Cardiac Research Institute, Medical University of South Carolina, Charleston, South Carolina 29425, USA

KEYWORDS: gene expression; regulation; signal transduction; MAP kinases; hypertrophy

The cardiac hypertrophic response is characterized by an increase in cardiac mass, protein content, and size of the individual cardiocyte. This results in increased myofilaments and improved contractile function. However, if the pathologic stimulus is prolonged or sufficiently severe and the increase in mass is insufficient to normalize ventricular wall stress, decompensated hypertrophy or heart failure will occur. The upregulation or downregulation of genes is associated with distinct pathophysiologic phenotypes in the hypertrophying and failing heart. The gene, encoding the Na^+/Ca^{2+} exchanger (NCX1), is upregulated at the transcriptional and translational levels in cardiac hypertrophy.[1,2] NCX1 is most abundant in the heart, where it regulates Ca^{2+} fluxes across the sarcolemma and serves a critical role in the maintenance of the cellular calcium balance for excitation-contraction coupling.[3] The mitogen-activated protein kinase (MAPK) superfamily of protein Ser/Thr kinase is a widely distributed group of enzymes that have been highly conserved through evolution. MAPK is activated within 30 minutes of pressure overload and remains active through most of the period of ventricular growth, returning to basal activity within 7 days. Because this coincides with the kinetics of NCX1 gene upregulation in our model, we examined the role of MAP kinases in exchanger regulation.

Stimulation of the α–adrenergic receptor results in a hypertrophic response characterized by an increase in cell size, an increase in protein synthesis, and the transcriptional upregulation of a number of cardiac genes, including NCX1. When adult cardiocytes are coinfected with our NCX1H1 promoter-luciferase and the ERK-specific MAP kinase phosphatase-3 adenovirus (MKP-3), 40% of the PE-stimulated upregulation of NCX1 gene expression is inhibited. In addition, when cardiocytes are treated with the MEK1/2 inhibitor UO126 (5 μM), PE stimulation is inhibited by

Address for correspondence: Donald R. Menick, Ph.D., Gazes Cardiac Research Institute, Medical University of South Carolina, 114 Doughty Street, Rm 203, Charleston, SC 29425. Voice: 843-876-5045; fax: 843-792-4762.
menickd@musc.edu

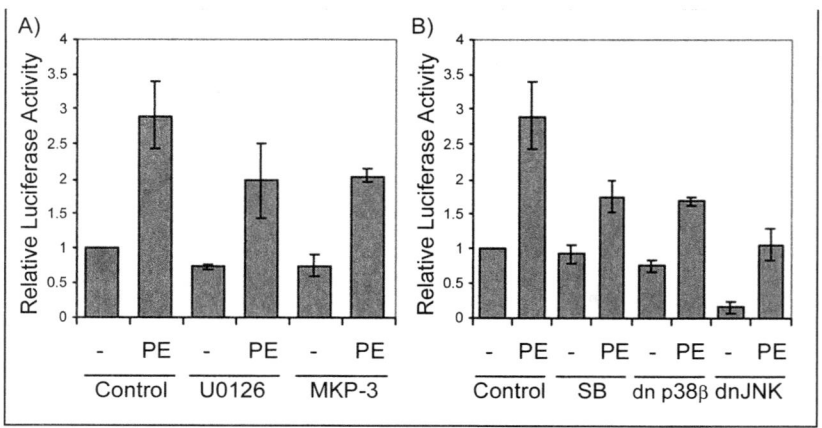

FIGURE 1. (**A**) Adult cardiocytes were infected with the NCX1 promoter-luciferase adenovirus or coinfected with NCX1 promoter-luciferase and MKP-3. Twelve hours post-infection, medium was changed and cells were treated with 10 μM PE for 48 hours. Luciferase activities were normalized to GFP fluorescence and expressed as a percentage of control. Where indicated, 5 μM U0126 was added with the PE. (**B**) Same conditions as in **A**, but cells were treated with the p38α/β inhibitor SB202190 (1 μM) or coinfected with dn JNK or dn p38 adenovirus.

40% (FIG. 1A). Importantly, treatment with UO126 or overexpression of MKP-3 completely inhibits the α-adrenergic activation of ERK. α–Adrenergic stimulation can also activate MEKK1, which can lead to activation of the c-Jun N-terminal protein kinase (JNK) and p38. Therefore, we looked at the possible contributions of JNK and p38 to NCX1 upregulation. When adenoviruses expressing a dominant-negative JNK are coinfected with the NCX1-promoter-luciferase adenovirus, basal NCX1 expression is inhibited. Dominant-negative JNK also inhibited PE stimulation of NCX1 (FIG. 1B). We are currently determining if inhibition of stimulated activity is only a reflection of the effect of dn JNK on NCX1 basal expression.

To determine if p38 contributes to the α-adrenergic upregulation of NCX1, we used the p38α/β inhibitor SB202190. Treatment of cardiocytes with SB202190 inhibits the PE stimulation of NCX1 luciferase activity, but it has no effect on basal expression of the exchanger (FIG. 1B). These results were verified by using an adenovirus expressing dn p38. Overexpression of dn p38 at levels equal to or greater than that of wild-type p38 had very little effect on basal NCX1 expression. At these same levels, PE-induced upregulation was dramatically inhibited. Therefore, our results demonstrate that JNK mediates the basal cardiac expression of the NCX1 gene, whereas ERK and p38 play a role in α-adrenergic–stimulated NCX1 upregulation.

ACKNOWLEDGMENTS

This work was supported in part by National Institutes of Health Program Project HL48788 (Project 3 to D.R.M.), National Institutes of Health R01 HL66223 (to

D.R.M.), and a postdoctoral research fellowship from Deutsche Forschungsgemeinschaft Mu 1454/1-1 (to J.G.M.).

REFERENCES

1. KENT, R.L., J.D. ROZICH, P.L. MCCOLLAM, et al. 1993. Rapid expression of the Na^+-Ca^{2+} exchanger in response to cardiac pressure overload. Am. J. Physiol. **265:** H1024–1029.
2. MENICK, D.R., K.V. BARNES, U.F. THACKER, et al. 1996. The exchanger and cardiac hypertrophy. Ann. N.Y. Acad. Sci. **779:** 489–501.
3. BERS, D. & J. BRIDGE. 1989. Relaxation of rabbit ventricular muscle by Na^+-Ca^{2+} exchanger and sarcoplasmic reticulum calcium pump. Ryanodine and voltage sensitivity. Circ. Res. **65:** 334–342.

Regulation of Phosphatidylinositol-4,5-Biphosphate Bound to the Bovine Cardiac Na^+/Ca^{2+} Exchanger

LUIS BEAUGÉ, CARLA ASTEGGIANO, AND GRACIELA BERBERIÁN

Laboratorio de Biofísica, Instituto de Investigación Médica Mercedes y Martín Ferreyra, 5000 Córdoba, Argentina

ABSTRACT: Western blot and cross immunoprecipitation analysis with specific antibodies demonstrate that in bovine heart sarcolemmal vesicles phosphatidylinositol-4,5-biphosphate (PtdIns-4,5-P_2) binds strongly to the Na^+/Ca^{2+} exchanger (NCX1). This binding is modulated by ATP, Ca^{2+}, vanadate, exchanger inhibitory peptide (XIP), and PLC-PtdIns specific in a way resembling the ATP regulation of the exchange fluxes. With 1 µM Ca^{2+}, 3 mM Mg^{2+}, and 0.4 mM vanadate, 1 mM ATP increased about twofold the bound PtdIns-4,5-P_2, reaching a steady state in 3–5 s at 37°C. With 100 µM Ca^{2+}, ATP had no effect on the PtdIns-4,5-P_2 bound to NCX1 or on the exchange fluxes. Without vanadate the bound PtdIns-4,5-P_2 was largely reduced; under this condition ATP failed to increase it and did not stimulate the exchanger. XIP inhibits the exchanger, more noticeable in the absence of ATP. With XIP, ATP does not modify the levels of bound PtdIns-4,5-P_2; however there is a small but distinct ATP stimulation of the exchanger. Vesicles pretreated with PtdIns-PLC, showed no *de novo*, [^{32}P]ATP-induced, production of PtdIns-4,5-P_2, but some ATP-stimulated increase in the bound PtdIns-4,5-P_2 was detected; however, that increase did not exceed the levels found with vanadate and no ATP. These results indicate that in bovine heart sarcolemmal vesicles, ATP upregulation of NCX1 is related to PtdIns-4,5-P_2 bound to the exchanger, perhaps over a "threshold" or "unspecific" amount. In addition, vanadate could influence the amount of detected PtdIns-4,5-P_2 either by inhibiting phosphoinositide-specific phosphatases and/or by inducing a redistribution of PtdIns-4,5-P_2 molecules associated with the Na^+/Ca^{2+} exchanger.

KEYWORDS: Na^+/Ca^{2+} exchanger; bovine heart; phosphoinositides; metabolic regulation; ATP; XIP; vanadate

INTRODUCTION

In practically all cell preparations the Na^+/Ca^{2+} exchange is stimulated by ATP.[1,2] The fact that Mg^{2+} ions are essential indicates that phosphorylation–dephosphorylation processes are involved.[2] A primary division among these processes occurs be-

Address for correspondence: Luis Beaugé, Instituto de Investigación Médica Mercedes y Martín Ferreyra, Casilla de Correo 389, 5000 Córdoba, Argentina. Voice: (54-351) 468-1465; fax: (54-351) 469-5163).
lbeauge@immf.uncor.edu

tween those that involve phosphorylation of the carrier protein and those that do not. Direct phosphorylation of serine residues was observed in cells transfected with canine exchanger, in isolated rat cardiomyocytes in culture, in this case by a protein kinase C (PKC)-related pathway,[3] and in aortic smooth muscle of the rat, which was induced by a growth factor and phorbol.[4] Structures other than the exchanger are targets for phosphorylation in squid nerve and mammalian heart. The actual target for phosphorylation in the squid is unknown; however, MgATP stimulation requires the simultaneous presence of a cytosolic soluble protein of low molecular weight.[5,6] In the mammalian heart, stimulation of the Na^+/Ca^{2+} exchanger by MgATP takes place via the phosphoinositide cycle, mainly through the synthesis of $PtdIns-4,5-P_2$.[7–9] Specifically, in bovine heart sarcolemmal vesicles, MgATP stimulation of the Na^+/Ca^{2+} exchange involves the synthesis of PtdIns-4,5-P2 from PtdIns through a fast phosphorylation chain: Ca^{2+}-independent formation of PtdIns-4-P followed by Ca^{2+}-dependent synthesis of $PtdIns-4,5-P_2$.[9] Although the actual mechanism of the stimulation was unknown, an intimate association of $PtdIns-4,5-P_2$ with the carrier seemed necessary, because this *de novo* synthesized $PtdIns-4,5-P_2$ and the exchanger coimmunoprecipitate.[9] In this work we describe experiments indicating that, in general, this is the case.

METHODS

The techniques used in this work were previously described in detail.[9,10] Therefore, only the more relevant points or new information will be dealt with here.

Sarcolemmal vesicles were made from bovine hearts by differential centrifugation (as indicated in Ref 11), loaded, and stored in 160 mM NaCl, 20 mM Mops/Tris (pH 7.4) and 0.1 mM EDTA. This preparation practically lacked any intracellular organelle contamination: (1) More than 90% gave positive results to the 5-nucleotidase activity assays, indicating that at least that fraction consisted of sarcolemmal membranes. (2) Compared with pure endoplasmic reticulum, less than 5% reacted with the anti-calnexin antibody. (3) Compared with pure lisosomal fraction, it had no detectable reactivity with the anti-Lamp-1 antibody.

The transport of Ca^{2+} through the Na^+/Ca^{2+} exchanger was defined as the difference in $[^{45}Ca]Ca^{2+}$ uptake between low (20 mM) and high (160 mM) Na^+-containing solutions. Routinely, the vesicles were incubated for 15–20 seconds at 37°C in the presence of 20 or 160 mM NaCl, 1 µM Ca^{2+}, 3 mM free Mg^{2+}, 0.3 mM vanadate, and 20 mM Mops/Tris (pH 7.4 at 37°C) with and without 1 mM ATP. Extravesicular Na^+ was replaced with the equivalent of 260 mOsm NMG-Cl.[9]

For the immunologic quantification of PtdIns-4,5-P2 bound to Na^+/Ca^{2+} exchanger the vesicles were incubated at 37°C in media that, unless otherwise stated, had similar composition to those used for Ca^{2+} uptake experiments. The reaction was stopped by adding Laemmli SDS-sample buffer at 5 times the final concentration; the samples were heated for 10 minutes at 37°C, run in discontinuous SDS-PAGE,[12] and then electrotransferred to PVDF (polyvinylidene difluoride) membranes for Western blots. The same sample was used first to distinguish the PtdIns-4,5-P2 (mouse monoclonal antibody against PtdIns-4,5-P2 from Per Septive Biosystems or Assay Designs, Inc.) and then, after stripping, to detect the Na^+/Ca^{2+} exchanger protein by means of a primary guinea pig polyclonal antibody against the

N-terminal portion of the bovine cardiac Na^+/Ca^{2+} exchanger (kindly provided by Dr. J.P. Reeves) or a commercial monoclonal antibody against the NCX1 (Novus Biologicals). The bands recognized by specific antibodies were detected with the ECF Kit (Amersham Pharmacia) and quantified by using a Storm 840 image analyzer (Molecular Dynamics) with a Scion PC (NIH) software. The band densities obtained with the anti PtdIns-4,5-P2 antibody (in arbitrary units) were divided by those obtained from the anti-NCX1 antibody (in arbitrary units). The different experiments were compared by assigning a value of 1 to the ratio of densities obtained in control conditions. (For more details see Ref. 10.)

For immunoprecipitation, 200 µg of membrane protein were incubated for 20 seconds at 37°C in 0.5 ml of media with a composition similar to those used for Ca^{2+} uptake experiments. The reaction was stopped, let at 4°C for 20 minutes, and the primary antibodies against either PtdIns-4,5-P2 or NCX1 were added to a final dilution of 1/500; the tubes remained overnight at 4°C. Thereafter, protein A-agarose (Santa Cruz) was added at a ratio of 0.01 mg per milligram of total protein, continuing the incubation for 1 hour at 20–22°C. The washed inmunoprecipitates were subjected to electrophoresis and Western blotting. Negative controls consisted of mouse monoclonal antibody against SV40TAg (Santa Cruz) and nonimmune guinea pig serum.[10]

TLC separation of phospholipids from phosphorylated cardiac membrane vesicles was performed as described in Ref. 9. ^{32}P-labeled phospholipids were visualized in a phosphoimage instrument (Storm 840, Molecular Dynamics), while identification of phosphatidylinositol mono- and diphosphate was done by comigrating commercial standards and submitting the plates to an atmosphere of saturated iodine vapor. Quantifications were done with Image Quant software (Molecular Dynamics).

Solutions were made with deionized ultrapure water (18 MegaOhms). NaCl was Baker Ultrex; all other chemicals were analytical reagent grade. The sodium salt of ATP (Bohringer Mannheim) was transformed into Tris salts by passage through an Amberlite IR-120-P column. Tris, NMG, Mops, PtdIns- PLC from *Bacillus cereus*, PtdIns-4-P, and PtdIns-4,5-P2 were from Sigma Chemical Co., USA. SDS-PAGE reagents were from BioRad, and the molecular mass standards were Kaleidoscope prestained standards (BioRad). Free $[Mg^{2+}]$ was estimated using a dissociation constant of 0.091 mM for MgATP. Free $[Ca^{2+}]$ was estimated with the MaxChelator Program (http://www.stanford.edu/~cpatton/maxc.html).

RESULTS AND DISCUSSION

Preliminary Experiments

The experiment described in FIGURE 1A shows the existence of cross-immunoprecipitation of the exchanger with anti-PtdIns-4,5-P_2 and anti-NCX1 antibodies. Bands located in the same place (around 140 kDa) can be detected immunoprecipitating the membrane vesicles with the anti PtdIns-4,5-P2 antibody and blotting with the anti-NCX1 antibody or immunoprecipitating with the NCX1 exchange antibody and analyzing the blot with the anti-PtdIns-4,5-P_2 antibody. Therefore, this proves that the Na^+/Ca^{2+} exchanger in sarcolemmal membrane vesicles from bovine heart strongly binds PtdIns-4,5-P_2. This has also been observed with proteins such as ac-

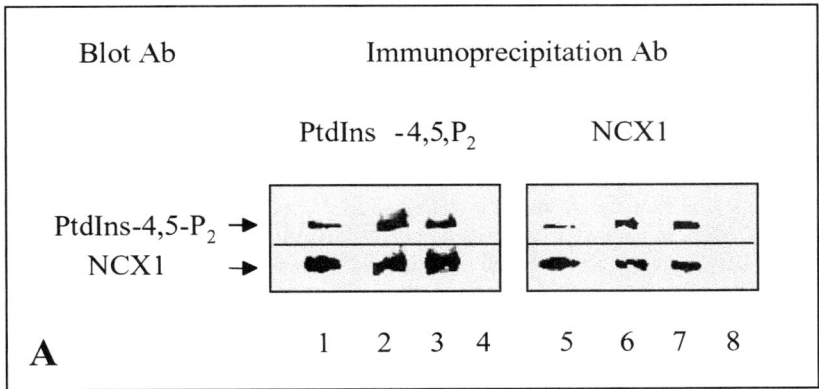

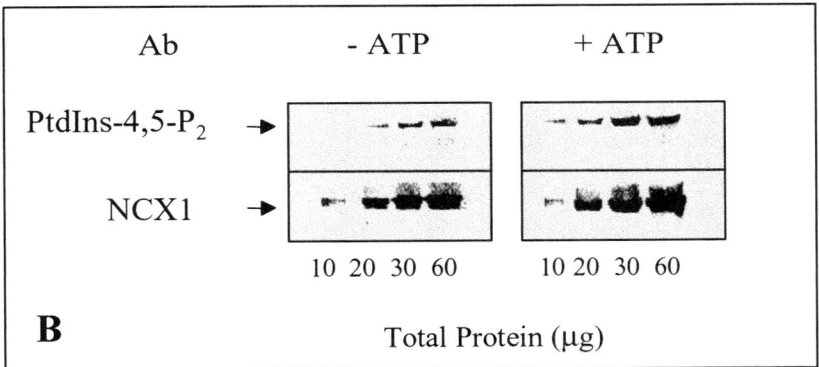

FIGURE 1. Detection of PtdIns-4,5-P_2 bound to bovine heart Na^+/Ca^{2+} exchanger by Western blot. (**A**) Cross-immunoprecipitation of PtdIns-4,5-P_2 and Na^+/Ca^{2+} exchanger protein. Bovine cardiac sarcolemmal membrane vesicles (200 µg) were incubated for 20 seconds at 37°C. The incubation solutions contained 1 µM Ca^{2+}, 3 mM Mg^{2+}, 25 mM NaCl, 135 mM NMG/HCl, 20 mM Mops/Tris (pH 7.4 at 37°C), and 0.4 mM vanadate with or without 1 mM ATP. Samples were then immunoprecipitated as indicated in **Methods** using anti-PtdIns-4,5-P2 antibody (*left panels*) or anti-NCX1 antibody (*right panels*). The inmunoprecipitates were subjected to SDS-PAGE (7.5% acrylamide) followed by immunoblot reacting with a monoclonal antibody anti-PtdIns-4,5-P2 (*top panels*) and developed using a chemiluminescence detection reagent. After stripping, the same blot was exposed to an antibody anti-Na^+/Ca^{2+} exchanger (*bottom panels*). Position of 140-kD molecular mass standard is indicated by *arrows*. Note: (i) *Lanes 1 and 5* are positive controls, without immunoprecipitation, using 40 µg total membrane protein. (ii) *Lanes 2 and 3* are duplicate inmunoprecipitates with anti-PtdIns-4,5-P2 antibody. (iii) *Lanes 6 and 7* are duplicate immunoprecipitates with anti-Na^+/Ca^{2+} exchanger antibody. (iv) *Lanes 4 and 8* are negative controls using a mouse monoclonal antibody against SV40TAg (Santa Cruz) (*lane 4*) and nonimmune guinea pig serum (*lane 8*). (**B**) As a function of total membrane protein concentrations. Cardiac sarcolemmal membrane vesicles (10–60 µg) were incubated for 20 seconds at 37°C in the same solutions as above and treated as indicated in **Methods**. Position of 140-kDa molecular mass standard is indicated by *arrows*. Note: (i) The amount of PtdIns-4,5-P2 and Na^+/Ca^{2+} protein increase in parallel with the increase in total protein. (ii) In the presence of ATP, the levels of PtdIns-4,5-P2 bound to Na^+/Ca^{2+} exchanger are markedly increased. (Reproduced with permission from Ref. 10.)

tinin and vinculin. These two proteins bind PtdIns-4,5-P_2 so tightly as to allow its detection by immunoprecipitation analysis.[13] FIGURE 1B (left and right panels) shows that with both anti-PtdIns-4,5-P_2 and anti-NCX1 antibodies, the intensity of the immunostained band varies linearly with the total amount of sample protein between 10 and 60 µg. In addition, comparing both panels indicates that when 1 mM ATP was added to the incubation solution (right panel), the intensity of the anti-PtdIns-4,5-P_2 antibody mark increased at all protein concentrations tested.

In these studies several crucial questions and points must be considered. On the one hand, there is no way to assess the actual amount of PtdIns-4,5-P_2 associated with the exchanger or to know how much of the PtdIns-4,5-P_2 is available for interaction with the specific antibody. In addition, we cannot determine *a priori* the amount of PtdIns-4,5-P_2 detected by its specific antibody that is relevant to exchanger regulation. As just mentioned, we must be aware that an increase in the antibody mark may reflect net synthesis of PtdIns-4,5-P_2, but it also changes in molecular packing. Conversely, a decrease in the antibody mark may be due to PtdIns-4,5-P_2 degradation as well as to molecule redistribution. We will discuss at least some of these points in this work.

Effects of Vanadate

We have shown[9] that in the absence of vanadate MgATP failed to stimulate Na^+/Ca^{2+} in these vesicles, and no clear explanation was found for this observation. This point was reexamined in relation to PtdIns-4,5-P_2 bound to the exchanger in the same batch of microsomal vesicles. The results are in FIGURE 2, where white columns correspond to vesicles incubated in the absence of vanadate, whereas those in black represent vanadate-containing solutions. FIGURE 2A confirms the original observation that MgATP does not stimulate the Na^+/Ca^{2+} exchanger in the absence of vanadate. In addition, it includes the new observation that in the absence of MgATP, the exchange fluxes are the same with or without vanadate (compare black and white columns). FIGURE 2B describes the PtdIns-4,5-P_2 bound to the exchanger. The results are very interesting. The PtdIns-4,5-P_2/NCX1 mark ratio obtained in the presence of vanadate and no ATP was use as the control and assigned a value of 1; that ratio was practically doubled with the addition of MgATP, a circumstance in which the nucleotide has stimulated the exchange fluxes. The white columns show that in the absence of vanadate MgATP is unable to promote an increase in the PtdIns-4,5-P_2 bound to the exchanger, that is, the lack of MgATP stimulation coincides with the lack of increase in the PtdIns-4,5-P_2 bound to the carrier. This result supports the notion that both processes are closely related. However, important additional information is also uncovered. First, the PtdIns-4,5-P_2/NCX1 ratios are much smaller in the absence than in the presence of vanadate, even when comparing vesicles not exposed and those exposed to MgATP. As the NCX1 marks were indistinguishable in the four cases (not shown here), those corresponding to PtdIns-4,5-P_2 were practically within the limits of detection. This happened 20 seconds after vesicles from a high Na^+ medium free of Mg^{2+} and Ca^{2+} were mixed with a solution containing low Na^+, 3 mM Mg^{2+}, and 1 µM Ca^{2+} with or without vanadate. Second, despite the marked difference in antibody-detected PtdIns-4,5-P_2, the exchange fluxes in the absence of MgATP are the same with and without vanadate in the media.

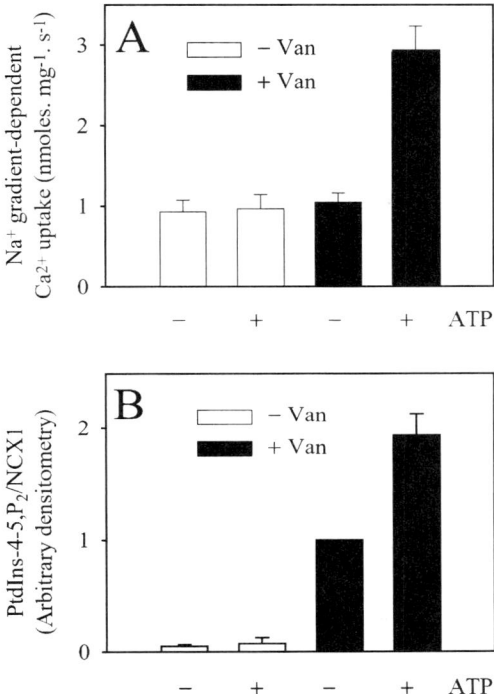

FIGURE 2. Effect of vanadate on Na^+/Ca^{2+} exchange fluxes (**A**) and on the amount of PtdIns-4,5-P_2 bound to bovine heart Na^+/Ca^{2+} exchanger (**B**) without and with 1 mM MgATP. Cardiac sarcolemmal membrane vesicles were incubated for 20 seconds at 37°C. The incubation solutions contained 1 μM Ca^{2+}, 3 mM Mg^{2+}, 25 mM NaCl, 135 mM NMG/HCl, and 20 mM Mops/Tris (pH 7.4 at 37°C) without (*white columns*) and with (*black columns*) 0.4 mM vanadate in the presence or absence of 1 mM ATP. The technique for flux measurements and estimation of PtdIns-4,5-P_2 is given in **Methods** and in the legend to FIGURE 1. All bars are the mean ± SEM of at least three determinations. In the case of immunoblots the PtdIns-4,5-P_2/NCX1 ratio in the presence of vanadate and with no ATP was arbitrarily given a value of 1.

In other experiments we followed the time course of PtdIns-4,5-P_2 combined with NCX1 in relation to ATP-stimulated exchange fluxes. These results are described in FIGURE 3. Note that following analysis of the results in FIGURE 2, we arbitrarily took as 1.0 the PtdIns-4,5-P_2/NCX1 band ratio at 20 seconds with vanadate and no ATP. In all cases, vesicles were allowed to reach the working temperature by keeping them for about 1 minute in their storing solution without Mg^{2+}, Ca^{2+}, vanadate, and ATP. After that, time aliquots of the vesicle suspension were mixed with the experimental solutions. Time zero corresponded to vesicles that were mixed directly with the Laemmli SDS-sample buffer. For other times, vesicles were incubated in Ca^{2+} uptake solutions (1 μM Ca^{2+}, 3 mM Mg^{2+}, 25 mM NaCl, 135 mM NMG/HCl, and 20 mM Mops/Tris (pH 7.4 at 37°C)) under the following conditions: absence of van-

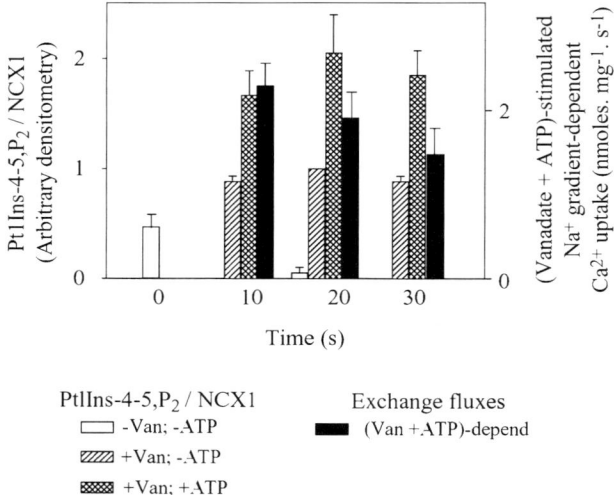

FIGURE 3. Effect of vanadate on the time dependence of the amount of PtdIns-4,5-P_2 bound to bovine heart Na^+/Ca^{2+} exchanger and on the MgATP-stimulated Na^+/Ca^{2+} exchange fluxes. In all cases, vesicles were allowed to reach working temperature by keeping them for ~1 minute in their storing solution without Mg^{2+}, Ca^{2+}, vanadate, or any nucleotide. After that time, aliquots of the vesicle suspension were mixed with the experimental solutions. Time zero correspond to vesicles that were mixed directly with the Laemmli SDS-sample buffer. For the other times the vesicles, treated as indicated in **Methods**, were incubated in Ca^{2+} uptake solutions (1 µM Ca^{2+}, 3 mM Mg^{2+}, 25 mM NaCl, 135 mM NMG/HCl, and 20 mM Mops/Tris [pH 7.4 at 37°C]) under the following conditions: (i) absence of vanadate and ATP (*white columns*); (ii) presence of 0.3 mM vanadate without ATP (*single oblique line pattern columns*); and (iii) presence of 0.3 mM vanadate and 1 mM ATP (*crossing oblique line pattern columns*). *Black columns* in the extreme right represent the levels of MgATP-stimulated Na^+/Ca^{2+} exchange fluxes in the presence of 0.3 mM vanadate. All bars are the mean ± SEM of at least 3 determinations. In the case of immunoblots the PtdIns-4,5-P_2/NCX1 ratio in the presence of vanadate and with no ATP was given arbitrarily a value of 1. The technique for flux measurements and estimation of PtdIns-4,5-P_2 is given in **Methods** and in the legend to FIGURE 1.

adate and ATP, 0.3 mM vanadate without ATP, and 0.3 mM vanadate and 1 mM ATP. Black columns at the extreme right represent the levels of MgATP-stimulated Na^+/Ca^{2+} exchange fluxes in the presence of 0.3 mM vanadate. In the absence of vanadate and ATP (white columns), the levels of phosphoinositide bound to NCX1 are much higher at $t = 0$ than those observed after 20 seconds of incubation in the Ca^{2+} uptake solution. On the other hand, in the presence of vanadate but not ATP (single oblique line pattern columns) there is twofold increase over the $t = 0$ values, already evident at 10 seconds and remaining constant over the 30-second period investigated. This difference is statistically significant at all times. As there is no ATP in the solutions, no net synthesis of PtdIns-4,5-P_2 could have occurred; plausible explanations, at least for the increase above the $t = 0$ values, are a repacking of phosphinositide molecules around NCX1, including some conformational change in the exchanger exposing more molecules to react with the antibody. In other experiments, not shown here, the extremely low levels of PtdIns-4,5-P_2 bound to NCX1 could partially be reversed by another 15-second incubation in the presence of vanadate (always without ATP). Therefore, vanadate, it seems, has two modes of increasing the PtdIns-4,5-P_2/NCX1 ratio in the absence of ATP: redistribution of PtdIns-4,5-P_2 bound to NCX1 and inhibition of PtdIns-4,5-P_2 hydrolysis. The second alternative, suggested by our results, is supported by the existence of a membrane-bound polyphosphoinositide phosphatase (3, 4, or 5 position in the inositol ring) that is inhibited by vanadate with a $K_{0.5}$ of 0.2 mM.[14] An additional point of note is that these changes in PtdIns-4,5-P_2 bound to NCX1 had no influence on exchange fluxes. The addition of ATP in the presence of vanadate (crossing oblique line pattern columns in FIG. 3) induces the already described increase in bound PtdIns-4,5-P_2; this increase, which has also been reported for *de novo* synthesis, reached its steady state value at 10 seconds and coincides with stimulation of the Na^+/Ca^{2+} exchange fluxes (black columns). The decline in ATP-stimulated fluxes as time progresses, particularly after 20 seconds, already described,[9] appears unrelated to the levels of bound PtdIns-4,5-P_2 and is likely due to the increase in both Na^+ and Ca^{2+} into the vesicles.

Effect of Pretreatment with a PtdIns Phospholipase C

Both in guinea pig cardiac myocytes under patch clamp[8] and in bovine cardiac sarcolemmal vesicles[9] pretreatment with a phospholipase that attacks phosphatidylinositides, including PtdIns, PtdIns-PLC leads to lack of ATP stimulation of the exchanger. In addition, PtdIns-PLC pretreated bovine heart vesicles showed no *de novo* synthesis of total membrane PtdIns-4,5-P_2 from [^{32}P]-γ-ATP.[9] This was reinvestigated comparing, in the same batch of vesicles, the effects of PtdIns-PLC on the exchange fluxes and the PtdIns-4,5-P_2 bound to NCX1. FIGURE 4A shows that pretreatment with PtdIns-PLC makes MgATP completely ineffective in stimulating the Na^+/Ca^{2+} exchange fluxes. FIGURE 4B shows that PtdIns-PLC reduced bound PtdIns-4,5-P_2 almost 5-fold. However, although under this condition no *de novo* synthesis of PtdIns-4,5-P_2 was detectable, when incubated with MgATP, these vesicles showed a sizable increase in the amount of PtdIns-4,5-P_2–bound NCX1. Nevertheless, that increase never surpassed the values observed in untreated vesicles not exposed to ATP. The discrepancy between the total *de novo* synthesis and the exchanger-bound PtdIns-4,5-P_2 could be reconciled if in PtdIns-PLC–treated vesicles small amounts of PtdIns-4,5-P_2 precursors, unable to influence the total membrane levels of *de novo* PtdIns-4,5-P2 synthesis, remain near the exchanger environment. A second discrepancy is that an increase in the exchanger-bound PtdIns-4,5-P_2 is not accompanied by an increase in the exchange fluxes. One way out of it is that there is a basal level of PtdIns-4,5-P_2 below which there is no effect on the Na^+/Ca^{2+} exchanger. (This could explain the results in FIG. 2A and B.) Another is that the newly synthesized PtdIns-4,5-P_2 is attached to an area on the exchanger with no relevance to its regulation.

Effects of XIP

The large intracellular loop of the NCX1 has a region that is related to Na^+ inactivation of the exchanger as well as to inhibition by XIP, a 20 amino acids peptide

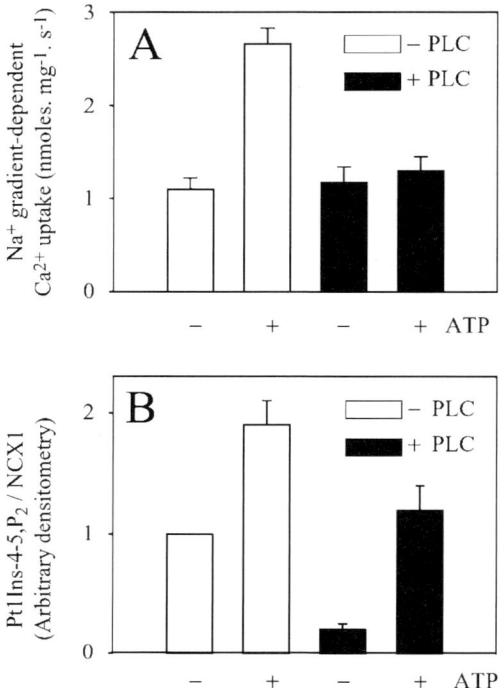

FIGURE 4. Effect of pretreatment with PtdIns-PLC on Na^+/Ca^{2+} exchange fluxes (**A**) and on the amount of PtdIns-4,5-P2 bound to bovine heart Na^+/Ca^{2+} exchanger (**B**) without and with 1 mM MgATP. Cardiac sarcolemmal membrane vesicles were incubated for 20 seconds at 37°C. The conditions were: untreated (*white columns*) and pretreated (*black columns*) with 4 U/ml PtdIns-PLC. The incubation solutions contained 1 μM Ca^{2+}, 3 mM Mg^{2+}, 25 mM NaCl, 135 mM NMG/HCl, 20 mM Mops/Tris (pH 7.4 at 37°C), and 0.4 mM vanadate without (−) and with (+) 1 mM ATP. All bars are the mean ± SEM of at least three determinations. In the case of immunoblots the PtdIns-4,5-P_2/NCX1 ratio in the absence of PtdIns-PLC and ATP was arbitrarily given a value of 1. Preincubation was for 4 minutes at room temperature in the absence of sodium gradient without or with 4 U/ml of PtdIns-PLC. The technique for flux measurements and estimation of PtdIns-4,5-P_2 is given in **Methods** and in the legend to FIGURE 1. (Reproduced with permission from Ref. 10.)

with the same sequence of that region.[15,16] The Na^+-dependent inactivation is counteracted by PtdIns-4,5-P_2.[8] In sarcolemmal vesicles of bovine heart, XIP also inhibits the Na^+/Ca^{2+} exchanger in both the absence and the presence of ATP.[9] In addition, recent evidence suggests that PtdIns-4,5-P_2 may exert its protection against Na^+ inactivation by actually binding onto the XIP region.[17]

FIGURE 5A confirms that in bovine heart membrane vesicles, 2 minutes of preincubation with 20 μM XIP inhibits both the basal and the ATP-stimulated Na^+/Ca^{2+} exchange fluxes. The fractional inhibition is slightly larger in the absence (85%) than in the presence (73%) of 1 mM MgATP, but the absolute levels of the inhibited fluxes are more than fourfold higher in the presence of the nucleotide. In addition,

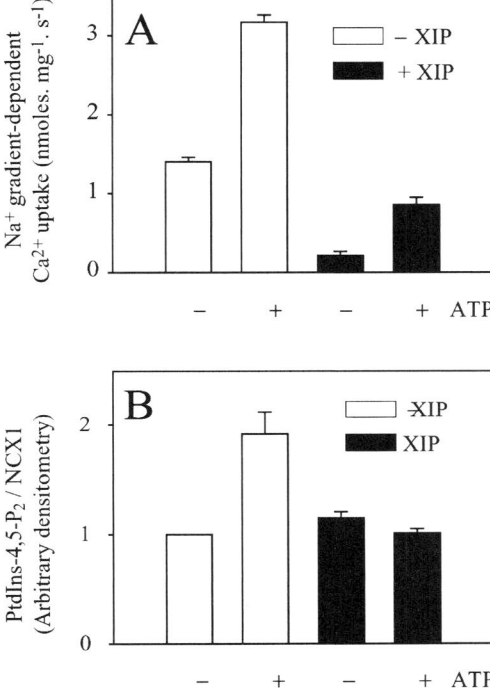

FIGURE 5. Effect of XIP on Na^+/Ca^{2+} exchange fluxes (**A**) and on the amount of $PtdIns$-$4,5$-P_2 bound to bovine heart Na^+/Ca^{2+} exchanger (**B**) without and with 1 mM MgATP. Cardiac sarcolemmal membrane vesicles were incubated for 20 seconds at 37°C. The vesicles were preincubated without (*white columns*) or with (*black columns*) 20 μM XIP. The incubation solutions contained 1 μM Ca^{2+}, 3 mM Mg^{2+}, 25 mM NaCl, 135 mM NMG/HCl, 20 mM Mops/Tris (pH 7.4 at 37°C), and 0.4 mM vanadate without (−) and with (+) 1 mM ATP. The techniques for flux measurements and estimation of $PtdIns$-$4,5$-P_2 are given in **Methods** and in the legend to FIGURE 1. All bars are the mean ± SEM of at least three determinations. In the case of immunoblots the $PtdIns$-$4,5$-P_2/NCX1 ratio in the absence of XIP and ATP was arbitrarily given a value of 1. Preincubation was for 2 minutes at 37°C in the absence of sodium gradient without or with XIP.

although in XIP-treated vesicles MgATP stimulation is fractionally relevant (4-fold), the absolute values of the stimulated fluxes are about one half those seen in control vesicles in the absence of ATP. The way this relates to the binding of $PtdIns$-$4,5$-P_2 to the exchanger is shown in FIGURE 5B. In control vesicles the $PtdIns$-$4,5$-P_2/NCX1 band density ratio is doubled by MgATP. Interestingly, XIP does not modify the ratio in the absence of MgATP, whereas it prevents its increase by the nucleotide. Summarizing the results in FIGURE 5A and B, in the absence of ATP, XIP inhibition of the fluxes does not correlate with the levels of $PtdIns$-$4,5$-P_2 bound to the exchange, which remain unchanged. On the other hand, in the presence of MgATP, XIP inhibition of the exchanger is accompanied by a reduction in bound $PtdIns$-$4,5$-P_2; however, the fractional reduction is more marked in fluxes that in the bound phosphoinositide. Two conclusions can be drawn from this section: (1) the PtdIns-4,5-P_2 relevant to regulation of the exchanger is mostly that newly synthesized; and (2) if XIP inhibition is wholly related to interactions with the binding of PtdIns-4,5-P_2 to the exchanger, some of these PtdIns-4,5-P_2 molecules are not accessible to antibody detection.

Summarizing, the results indicate (1) the existence of PtdIns-4,5-P_2 strongly bound to the Na^+/Ca^{2+} exchanger present in bovine heart sarcolemmal vesicles (NCX1), and (2) the amount of MgATP-stimulated binding of PtdIns-4,5-P_2 that parallels the MgATP stimulation of the Na^+/Ca^{2+} exchange fluxes.

ACKNOWLEDGMENTS

This work was supported by grants from the Agencia Nacional de Promoción Científica y Tecnológica (FONCYT PICT99 05-05158), Agencia Córdoba Ciencia (181/01), and Ministerio de Salud de la Nación (Beca Carrillo-Oñativia 2000) of Argentina. Two of the authors (G.B. and L.B.) are Established Investigators and one author (C.A.) is a Fellow of CONICET. The experiments presented here are part of the PhD thesis work of Carla Asteggiano.

REFERENCES

1. BLAUSTEIN, M.P. & W.J. LEDERER. 1999. Sodium/calcium exchange: its physiological implications. Physiol. Rev. **79:** 764–830.
2. DIPOLO, R. & L. BEAUGÉ. 1999. Metabolic pathways in the regulation of invertebrate and vertebrate Na^+/Ca^{2+} exchange. Biochem. Biophys. Acta **1422:** 57–71.
3. IWAMOTO, T., S. WAKABAYASHI & M. SHIGEKAWA. 1995. Growth factor-induced phosphorylation and activation of aortic smooth muscle Na^+-Ca^{2+} exchanger. J. Biol. Chem. **270:** 8996–9001.
4. IWAMOTO, T., Y. PAN, S. WAKABAYASHI, et al. 1996. Phosphorylation-dependent regulation of cardiac Na^+-Ca^{2+} via protein kinase C. J. Biol. Chem. **271:** 13609–13615.
5. BEAUGÉ, L., D. DELGADO, H. ROJAS, et al. 1996. A nerve cytosolic factor is required for MgATP stimulation of a Na^+ gradient-dependent Ca^{2+} uptake in plasma membrane vesicles from squid optic nerve. Ann. N.Y. Acad. Sci. **779:** 208–216.
6. DIPOLO, R., BERBERIÁN, G., DELGADO D., et al. 1997. A novel 13 kD cytoplasmic soluble protein is required for the nucleotide modulation of the Na/Ca exchange in squid nerve fibers. FEBS Lett. **401:** 6–10.
7. LUCIANI, S., M. ANTOLINI, S. BOVA, et al. 1995. Inhibition of cardiac sarcolemmal sodium-calcium exchanger by glycerophosphoinositol 4-phosphate and glycerophosphoinositol 4,5-bisphosphate. Biochim. Biophys. Res. Commun. **206:** 674–680.
8. HILGEMANN, D.W. & R. BALL. 1996. Regulation of cardiac Na^+, Ca^{2+} exchange and K_{ATP} potassium channels by PIP_2. Science **273:** 956–960.
9. BERBERIÁN, G., C. HIDALGO, R. DIPOLO & L. BEAUGÉ. 1998. ATP stimulation of Na^+/Ca^{2+} exchange in cardiac sarcolemmal vesicles. Am. J. Physiol. **274:** C724–C733.
10. ASTEGGIANO, C., G. BERBERIÁN & L. BEAUGÉ. 2001. Phosphatidylinositol-4,5-biphosphate bound to bovine cardiac Na/Ca exchanger displays a MgATP regulation similar to that of the exchange fluxes. Eur. J. Biochem. **268:** 437–442.
11. REEVES, J.P. 1985. The sarcolemmal sodium-calcium exchange system. Curr. Top. Membr. Transp. **25:** 77–119.
12. LAEMMLI, U.K. 1970. Cleavage of structural proteins during the assembly of the head of bacteriophage T4. Nature **227:** 680–685.

13. FUKAMI, K., T. ENDO, M. IMAMURA & T. TAKENAWA. 1994. Actinin and vinculin are PIP2-binding proteins involved in signaling by tyrosine kinase. J. Biol. Chem. **269:** 1518–1522.
14. RATH HOPE, H.M. & L.J. PIKE. 1994. Purification and characterization of a polyphosphoinositide phosphatase from rat brain. J. Biol. Chem. **269:** 23648–23654.
15. LI, Z.P., D.A. NICOLL, A. COLLINS, *et al.* 1991. Identification of a peptide inhibitor of the cardiac sarcolemmal Na^+/Ca^{2+} exchanger. J. Biol. Chem. **266:** 1014–1020.
16. MATSUAKA, S., D.A. NICOLL & K.D. PHILIPSON. 1997. Regulation of cardiac Na^+/Ca^{2+} exchanger by the endogenous XIP region. J. Gen. Physiol. **109:** 273–286.
17. HE, Z., S. FENG, Q. TONG, *et al.* 2000. Interaction of PIP_2 with the XIP region of the cardiac Na^+/Ca^{2+} exchanger. Am. J. Physiol. **278:** C661–C666.

Potassium-Dependent Sodium–Calcium Exchange through the Eye of the Fly

R. WEBEL, K. HAUG-COLLET, B. PEARSON, R.T. SZERENCSEI,[a]
R.J. WINKFEIN,[a] P.P.M. SCHNETKAMP,[a] AND N.J. COLLEY

Department of Ophthalmology & Visual Science, and Department of Genetics, University of Wisconsin, Madison, Wisconsin 53792, USA

[a]*Department of Physiology & Biophysics & MRC Group on Ion Channels/Transporters, University of Calgary, Faculty of Medicine, Calgary, Alberta T2N 4N1, Canada*

ABSTRACT: In this review, we describe the characterization of a *Drosophila* sodium/calcium-potassium exchanger, *Nckx30C*. Sodium/calcium (-potassium) exchangers (NCX and NCKX) are required for the rapid removal of calcium in excitable cells. The deduced protein topology for NCKX30C is similar to that of mammalian NCKX, with 5 hydrophobic domains in the amino terminus separated from 6 at the carboxy-terminal end by a large intracellular loop. NCKX30C functions as a potassium-dependent sodium–calcium exchanger and is expressed in adult neurons and during ventral nerve cord development in the embryo. *Nckx30C* is expressed in a dorsal/ventral pattern in the eye-antennal disc, suggesting that large fluxes of calcium may be occurring during imaginal disc development in the larvae. NCKX30C may play a critical role in modulating calcium during development as well as in the removal of calcium and maintenance of calcium homeostasis in adults.

KEYWORDS: calcium; development; *Drosophila*; photoreceptor; signal transduction; Na/Ca exchange

INTRODUCTION

Calcium is essential in phototransduction in both *Drosophila* and vertebrates. In both, phototransduction is initiated by the absorption of light by rhodopsin, and the light signal is transduced through G protein-coupled signaling cascades. However, distinct differences exist in the two phototransduction cascades (FIG. 1). In *Drosophila*, phototransduction occurs via a phospholipase C-mediated signaling cascade, and illumination results in the opening of the cation-selective channels and a large rise in intracellular calcium. The photoreceptors depolarize and calcium increases from about 100 nanomolar to as high as tens of micromolar.[1–7] The mechanism for calcium extrusion after illumination is thought to be sodium/calcium exchange. The measurement of sodium/calcium exchange in *Drosophila* and in other invertebrate

---

Address for correspondence: Dr. Nansi Jo Colley, Department of Ophthalmology & Visual Science, and Department of Genetics, K6/460CSC University of Wisconsin, Madison, WI 53792. Voice: 608-265-5398; fax: 608-265-6021.
njcolley@facstaff.wisc.edu

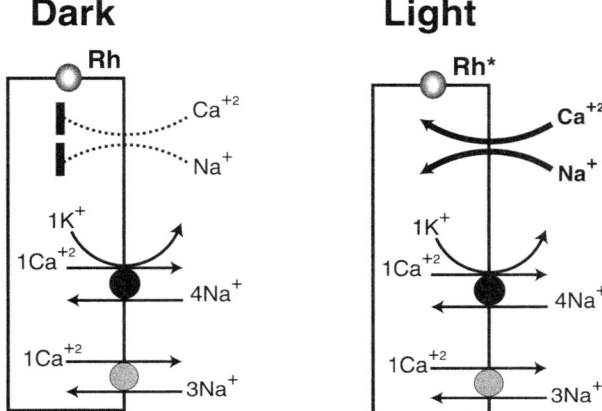

FIGURE 1. Phototransduction cascade in *Drosophila*. Absorption of a photon photoactivates rhodopsin, leading to the opening of the cation-selective channels (Light). Extracellular calcium (Ca^{+2}) and sodium (Na^+) enter the cell via the light-activated channels, causing a depolarization of the photoreceptor cells. Calcium levels rise about 100 nanomolar to greater than 10 micromolar upon light stimulation.[5] Calcium entering light-sensitive channels is thought to play a key role in deactivation of the light response and light adaptation. Rapid removal of calcium from the photoreceptor cells is key to the recovery from the light response. Removal of calcium following light stimulation may occur via NCKX (*dark circle*) and NCX (*shaded circle*). NCX uses a stoichiometry of $3Na^+/1Ca^{2+}$ to extrude calcium, and NCKX uses both the inward sodium gradient and the outward potassium gradient to extrude calcium at a stoichiometry of $4Na^+/1Ca^{2+}, 1K^+$. *Dashed arrows* indicate that the cation influx through the light-sensitive channel does not occur in the dark (*Dark*). Rh, rhodopsin molecule; Rh*, photoexcited rhodopsin molecule. This figure is an adaptation of one reproduced in Ref. 18. (Reproduced from J. Cell Biol. 1999. **147:** 659-669, by copyright permission of The Rockefeller University Press.)

photoreceptors has led to the proposal that exchange activity is critical in light adaptation.[2,4,8–12] Therefore, *Drosophila* photoreceptors may use an electrochemical exchanger that couples calcium removal to the inward sodium gradient, which has been demonstrated in vertebrate rod photoreceptors.[13–16] Unlike fly photoreceptor cells, illumination of vertebrate rod photoreceptors results in the closing of cGMP-gated cation channels and in hyperpolarization of the cell.[17,18] Cytosolic calcium falls from 500–600 nM in the dark to less than 50 nM in bright light.[19,20] This process plays a role in light adaptation in both retinal rods and cones.[21,22]

Calcium extrusion or sequestration to intracellular compartments is essential to all excitable cells, and prolonged increases in cytosolic calcium can be toxic, leading to cell death.[23,24] Two classes of plasma membrane proteins are responsible for calcium efflux. They are ATP-driven calcium pumps and sodium/calcium (-potassium) exchangers. The exchangers are essential in excitable cells that experience large fluxes of calcium across their plasma membrane, such as cardiac myocytes, skeletal and smooth muscle, photoreceptors, and other neurons (reviewed in Ref. 25).

Exchangers maintain low levels of intracellular calcium (100 nM or below) by using the transmembrane sodium gradient as an energy source. There are two well-known families of mammalian sodium/calcium exchangers. One family is NCX, and it uses an inward sodium gradient for extrusion of calcium. There are three NCX isoforms and they are expressed in a variety of tissues including heart, kidney, brain, as well as smooth and skeletal muscle.[26,27] The second family is NCKX, and it extrudes calcium using both an inward sodium gradient and an outward potassium gradient.[28,29] There are three NCKX-type exchangers, NCKX1, NCKX2, and NCKX3. Retinal rod NCKX1 exchangers have been cloned from a variety of mammalian species, and potassium-dependent sodium/calcium exchanger activity has been demonstrated.[30–36] NCKX2 has been isolated from human and chicken retinal cone photoreceptors as well as rat brain.[37,38] NCKX3 has been cloned from rat brain.[39] NCKX family members have been detected in both eukaryotic and prokaryotic genomes.[40,41] The NCKX- and NCX-type exchangers are thought to display a similar topology; however, there is little overall amino acid sequence identity. Despite this difference, they are thought to be evolutionarily related.[41,42]

Here we review the cloning and characterization of a *Drosophila* potassium-dependent sodium/calcium exchanger, *Nckx30C*.[43] We have shown that it functions as a potassium-dependent sodium/calcium exchanger and that it is distinct from *Calx*, an NCX-type exchanger in *Drosophila*.[41,44–47] Both *Nckx30C* and *Calx* are expressed in photoreceptor cells, in the adult brain and during embryogenesis and eye development. These exchangers are likely required for the removal of calcium generated during signaling processes in the fly.[43]

MATERIALS AND METHODS

Nckx30C was cloned and sequenced as previously described.[43] The BLAST search and the CLUSTAL W multiple sequence alignment programs were used to assess sequence similarity[48,49] (http://www.ncbi.nlm.nih.gov/BLAST) (http://pbil.ibcp.fr/NPSA/npsa_clustalw.html). The sequence data can be obtained in the DDBJ/EMBL/GenBank databases under accession number AF190455.

The full-length *Nckx30C* cDNA was expressed in High Five insect cells as previoiusly described.[43,50] High Five cells (BTI-TN-5B1-4) are derived from *Trichoplusia ni* egg cell homogenates and were purchased from Invitrogen. We do not detect endogenous exchanger activity in High Five cells. Potassium-dependent sodium/calcium exchange activity was measured as previously described.[43] The sodium-potassium ionophore monensin was used to load the cells with sodium in a medium containing 150 mM NaCl, 80 mM sucrose, 0.05 mM EDTA, and 20 mM Hepes (pH 7.4). This procedure was carried out according to the methods described for rod outer segments.[51] Monensin was removed, and the sodium-loaded cells were washed with and resuspended in 150 mM LiCl, 80 mM sucrose, 0.05 mM EDTA, and 20 mM Hepes (pH 7.4). The cells were resuspended in media containing 80 mM sucrose and 20 mM Hepes (pH 7.4) and either (1) 150 mM KCl, (2) 150 mM NaCl, or (3) 150 mM LiCl. The uptake of ^{45}Ca occurred by adding 35 µM $CaCl_2$ and 1 µCi ^{45}Ca to the media. ^{45}Ca uptake was measured as described previously.[52] Following ^{45}Ca uptake the cells were washed in an ice-cold medium containing 140 mM KCl, 80 mM sucrose, 5 mM $MgCl_2$, 1 mM EGTA, and 20 mM Hepes (pH 7.4).

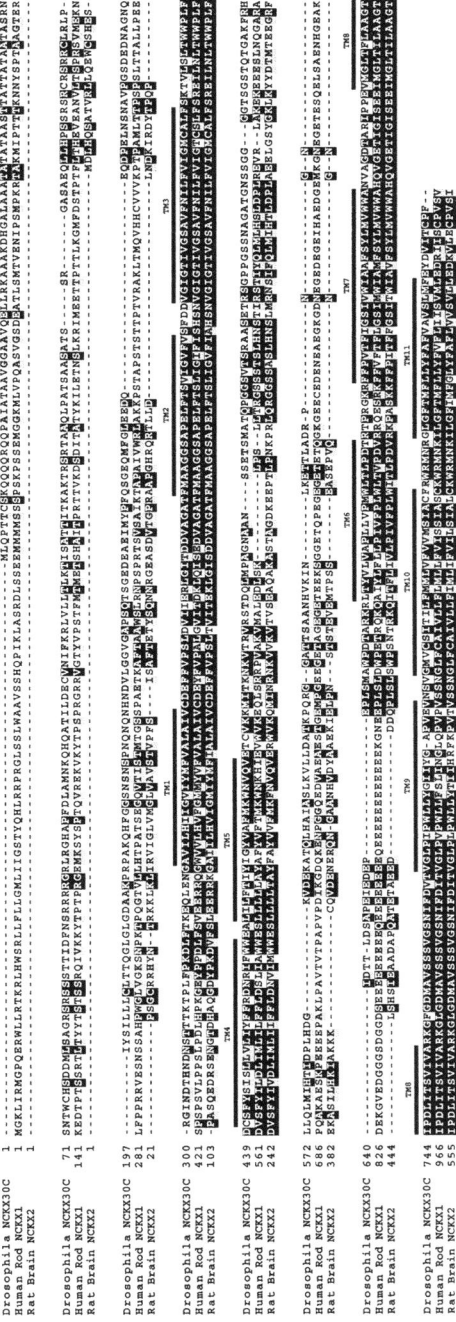

FIGURE 2. *Nckx30C* encodes a novel *Drosophila* NCKX. Clustal/W alignment between *Drosophila* NCKX30C, human rod NCKX1, and rat brain NCKX2.[31,32,56] *Dark shading* indicates identity; *light shading*, similarity. The 11 transmembrane domains of NCKX30C are indicated by lines above each row (Reproduced from Ref. 43, J. Cell. Biol. 1999. **147:** 659-669, by copyright permission of The Rockefeller University Press.)

Heads from wild-type flies (w^{1118}) were embedded in Tissue-Tek OCT Compound (Miles, Inc.), sectioned, and labeled as previously described.[43] Embryo and imaginal disc *in situ* hybridizations were carried out essentially as described by Panganiban *et al.*[53] Digoxigenin-labeled antisense and sense riboprobes were made by *in vitro* transcription, as recommended by the supplier (Boehringer Mannheim Corp., Indianapolis, IN). Five distinct *Nckx30C* cDNA probes, described previously,[43] were used. *Calx* cDNA was generously given to us by E. Schwarz.[41] *Chaoptin* cDNA was sent by D. Van Vactor and S. L. Zipursky.[54]

RESULTS AND DISCUSSION

Molecular and Functional Characteriztion of a Drosophila Potassium-Dependent Sodium–Calcium Exchanger

Potassium-dependent sodium/calcium exchangers (NCKX) are a group of calcium extrusion proteins that use an inward sodium gradient together with an outward potassium gradient to remove calcium from excitable cells. NCKX1 function was originally described in retinal rod photoreceptor cell outer segments (FIG. 1).[28,29] To isolate an NCKX from *Drosophila*, we constructed a cDNA probe from the bovine rod photoreceptor NCKX1 and used this probe to screen the *Drosophila* libraries. The *Drosophila* cDNA was obtained as described previously,[43] and the *Drosophila Nckx* was mapped to 2L at 30C5-7 cytologically by *in situ* hybridization to polytene chromosomes. Based on the chromosomal location we named the gene *Nckx30C*.[43] *Nckx30C* has a single open reading frame that encodes a protein of 856 amino acids. We compared the derived amino acid sequence of *Nckx30C* with the human NCKX1 and rat NCKX2 (FIG. 2). FIGURE 2 shows that there is 66% and 71% identity in two groups of amino acids that correspond to the predicted transmembrane domains of NCKX1 and NCKX2, respectively.

The hydropathy analysis predicts that *Drosophila* NCKX30C protein contains 11 hydrophobic regions that correspond to potential transmembrane (TM) helices predicted in NCXK1 and NCKX2 (FIG. 3A and B). NCKX30C displays a large cytoplasmic loop located between the hydrophobic amino acid clusters. This displays almost no amino acid identity with the intracellular cytoplasmic loops of NCKX1 or NCKX2 (FIG. 2). FIGURE 3B shows the domain structure of the exchangers. Hydropathy analysis reveals that NCKX30C may contain two additional membrane-spanning segments located in the N-terminus, not observed in either NCKX1 or NCKX2.

We examined the ability of NCKX30C to function as an exchanger in High Five cells.[43] Both NCX and NCKX can mediate both calcium efflux (forward exchange) and calcium influx (reverse exchange). The direction of calcium exchange is dictated by the direction of the sodium gradient across the membrane. *In vivo*, the inward sodium gradient removes calcium from the cell (forward exchange). However, *in vitro*, when external sodium is removed, the outward sodium gradient will drive calcium into the cell (reverse exchange). Taking advantage of reverse exchange of NCKX30C, we measured ^{45}Ca uptake in cells loaded with sodium (FIG. 4). We examined NCKX activity by using three different conditions of the cation gradient that are known to inhibit NCKX activity. In a medium containing high sodium, the cells

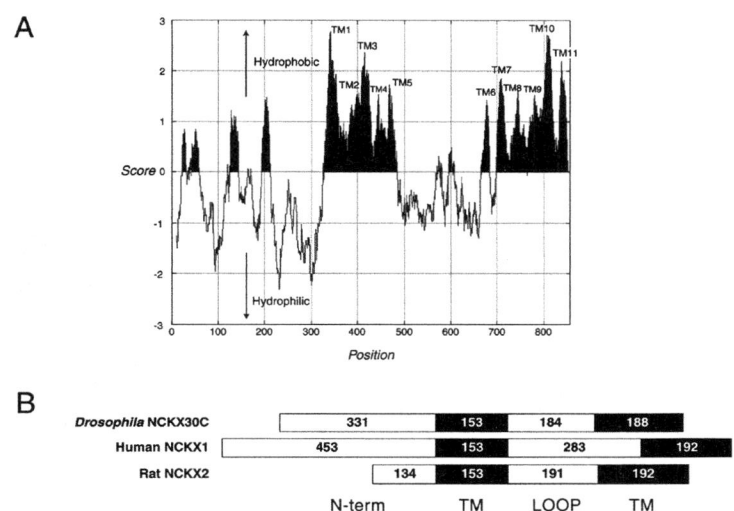

FIGURE 3. Hydropathy plot of the conceptual NCKX30C protein. (**A**) Hydropathy plot of the conceptual NCKX30C protein, analyzed by the Kyte-Doolittle algorithm.[74] Hydrophobic regions are designated in *black*. (**B**) Domain structure of *Drosophila* NCKX30C, human rod photoreceptor NCKX1, and rat brain NCKX2. *Shaded boxes* represent the two clusters of transmembrane segments (TM 1-5 and TM 6-11) that display high identity among the three sequences. Numbers represent the number of amino acids per segment. (Reproduced from Ref. 43, J. Cell Biol. 1999. **147:** 659-669, by copyright permission of The Rockefeller University Press.)

were filled with sodium using the sodium-potassium ionophore monensin.[51] After removal of the ionophore, the sodium-filled cells were washed with and resuspended in low sodium buffer, as described in Ref. 43. The cells were diluted into media containing 80 mM sucrose, 20 mM Hepes (pH 7.4), and either (1) 150 mM KCl, (2) 150 mM NaCl, or (3) 150 mM LiCl. The uptake of ^{45}Ca was initiated by the addition of 35 ∝M $CaCl_2$ and 1 µCi ^{45}Ca. Monensin causes the release of internal sodium and inhibits ^{45}Ca uptake by NCKX30C (FIG. 4B). These data demonstrate that intracellular sodium is required for calcium transport. In both NCK and NCKX, calcium uptake by reverse exchange is inhibited by high external sodium.[55] This property is thought to be due to competition of sodium and calcium for a common binding site.[55.] FIGURE 4C shows that ^{45}Ca uptake by NCKX30C is prevented by high external sodium. These data are consistent with its identity as a sodium/calcium exchanger.[43] NCKX and NCX can be distinguished by employing the property that calcium influx by NCKX requires the presence of potassium, and lithium cannot substitute for potassium.[56] NCKX30C requires potassium for ^{45}Ca uptake (FIG. 4D). demonstrating that NCKX30C functions as a potassium-dependent sodium/calcium exchanger.

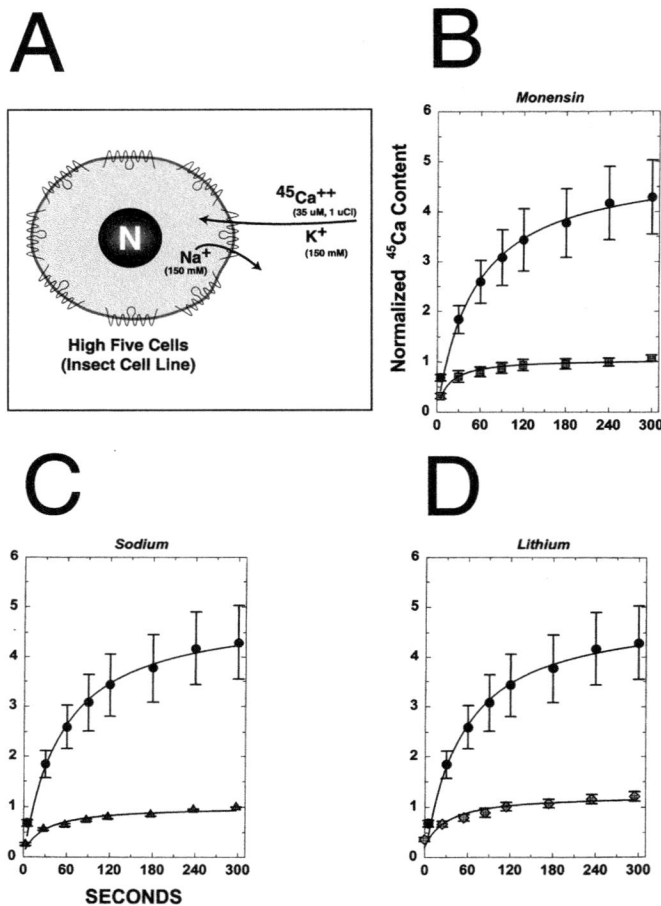

FIGURE 4. *NCKX30C* is a potassium-dependent sodium–calcium exchanger. (**A**) Reverse exchange was measured as Na_{inside} dependent ^{45}Ca uptake in High Five cells transformed with *Nckx30C* in the presence of external KCl (**B–D**, *filled circles*). Mean ± standard error of the mean (SEM). The same results are shown in **B–D** (*filled circles*). (**B**) Reverse sodium/calcium exchange requires intracellular sodium. ^{45}Ca uptake was measured in KCl medium with 20 μM monensin present (*shaded squares*) or without monensin (*filled circles*). Monensin was added 30 seconds before addition of ^{45}Ca and causes the release of intracellular sodium to the external medium. Average values ± SEM are shown for 13 experiments conducted in KCl medium and 4 experiments conducted in media containing KCl plus monensin. (**C**) Reverse sodium/calcium exchange is inhibited by extracellular sodium. ^{45}Ca uptake was measured in KCl medium (*filled circles*) or NaCl medium lacking potassium (*triangles*). Average values ± SEM are shown for 13 experiments conducted in KCl medium and 10 experiments conducted in media containing NaCl. (**D**) Reverse sodium/calcium exchange requires extracellular potassium. ^{45}Ca uptake was measured in KCl medium (*filled circles*) or LiCl medium lacking potassium (*diamonds*). Average values ± SEM are shown for 13 experiments conducted in KCl medium and 8 experiments conducted in media containing LiCl. (Reproduced from Ref. 43, J. Cell Biol. 1999. **147:** 659–669, by copyright permission of The Rockefeller University Press.)

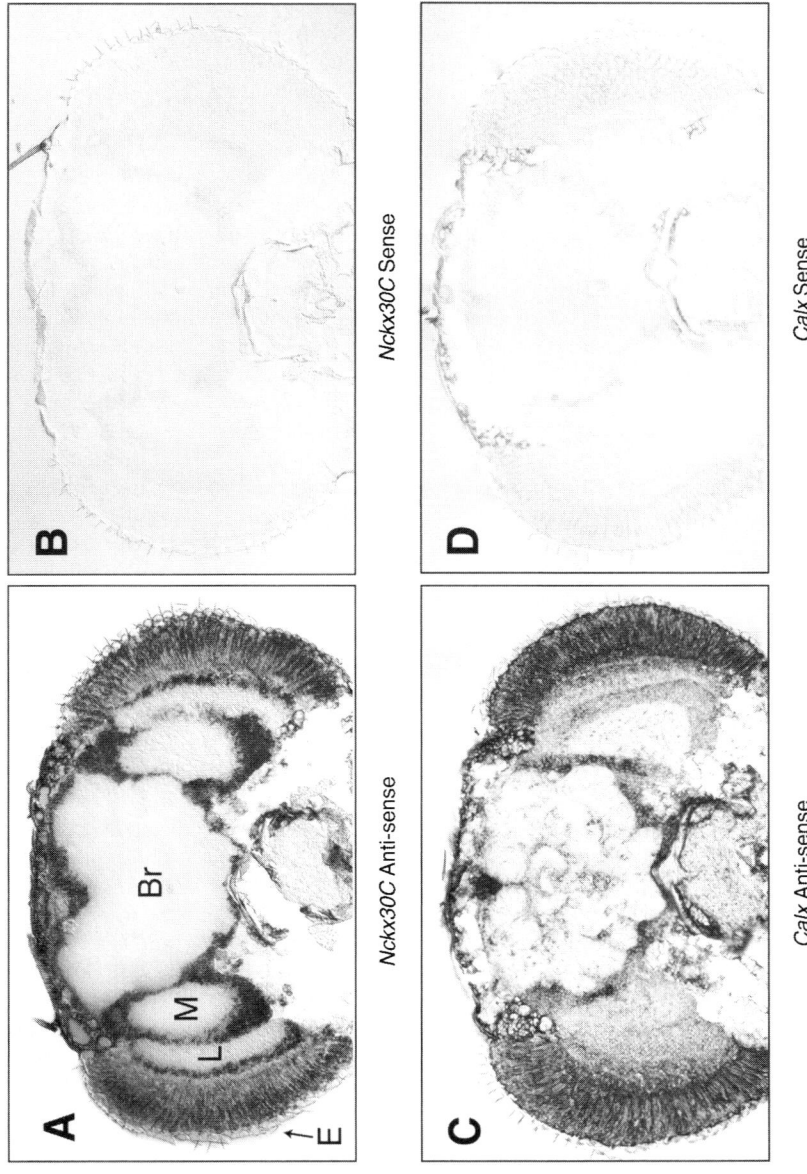

FIGURE 5. See Plate 6 in color section. *Nckx30C* and *Calx* are expressed in the adult eye and the brain of *Drosophila*. (**A**) Shown are *in situ* hybridizations to 14-μm cryostat sections of adult heads hybridized with digoxigenin-labeled riboprobes. (**A**) antisense riboprobe for *Nckx30C*; (**B**) sense probe for *Nckx30C*; (**C**) antisense riboprobe for *Calx*; (**D**) sense riboprobe for *Calx*; and (E) eye, (L) lamina, (M) medulla, and (Br) the brain. (Reproduced from J. Cell Biol. 1999. **147:** 659–669, by copyright permission of The Rockefeller University Press.)

Nckx30C and *Calx* Expression in Drosophila

A *Drosophila* sodium/calcium exchanger, *Calx*, was previously described and shown to be a member of the family of the NCX-type exchangers.[41,44,45] Expression patterns for *Calx* were reported in Refs. 41 and 43. We have shown that *Nckx30C* and *Calx* are expressed in the adult nervous system (FIG. 5). Both *Calx* and *Nckx30C* are expressed in the photoreceptor cells as well as in the lamina, medulla, and optic lobes of the brain (FIG. 5A and C). No signal was detected with the sense probes (FIG. 5B and D).

Both NCX and NCKX are responsible for extruding calcium in cells that are experiencing large calcium fluxes. *Drosophila* photoreceptors as well as other cells in the adult express both types of exchangers, indicating that there may be multiple mechanisms for calcium efflux in these cells. The NCKX exchangers have novel features that make them uniquely suited for calcium extrusion during phototransduction in photoreceptor cells. Light activation of *Drosophila* photoreceptors stimulates the opening of the cation-selective channels and a dramatic increase in intracellular calcium (FIG. 1). In addition, sodium also contributes to the inward current, leading to increased cytosolic sodium.[5,57–59] As the internal sodium concentration increases, the transmembrane sodium gradient is reduced and it possibly collapses during high light stimulation. NCX exchangers, such as *Calx*, likely reverse direction at much lower cytosolic sodium concentrations when compared with potassium-dependent NCKX exchangers.[60] Therefore, potassium-dependent exchangers are more effective for calcium removal during times of high sodium influx (high light intensities). The presence of both exchangers in photoreceptor cells could also be explained if the subcellular distributions of *NCKX30C* and *Calx* are very different, with each fulfilling differing functions. In addition to being present in the adult nervous system, *Nckx30C* and *Calx* are both expressed in the embryonic nervous system. *Calx* expression is present before activation of zygotic transcription, indicating that *Calx* transcripts are probably maternally inherited (FIG. 6B). *Calx* transcripts disappear in the cellular blastoderm (FIG. 6D) and are then detected again during embryonic stage 11 and 12 (FIG. 6F and H). Unlike *Calx*, *Nckx30C* transcripts were not detected in the preblastoderm or blastoderm stage (FIG. 6A and C). *Nckx30C* transcripts were first noted in cells at the ventral midline of the central nervous system during embryonic stage 13–14 (FIG. 6E). By stage 15, *Nckx30C* expression was detected in several neurons within the ventral nerve cord in the embryo (FIG. 6G). *Nckx30C* is expressed in many neurons within the ventral nerve cord and the embryonic brain in stage 16 (FIG. 6I). At similar times during embryogenesis, *Calx* expression is restricted to a smaller subset of neurons in the ventral nerve cord (FIG 6J). *Nckx30C* and *Calx* transcripts were observed in some cells outside the CNS, which may represent parts of the peripheral nervous system (FIG. 6I and J).

In addition to being expressed in the embryo, *Nckx30C* was also detected in larval imaginal discs. By contrast, *Calx* expression was not detected in any of the imaginal discs. *Drosophila* appendages develop from imaginal discs in a series of coordinated events. Photoreceptor differentiation is initiated at the posterior end of the eye disc and proceeds as a wave across the eye disc from posterior to anterior.[61–66] The morphogenetic furrow is a dorsoventral indentation in the eye disc, with the area anterior to the furrow being made up of actively dividing and unpatterned cells and the area posterior to the furrow containing differentiating photoreceptor cells.[66–68] FIGURE

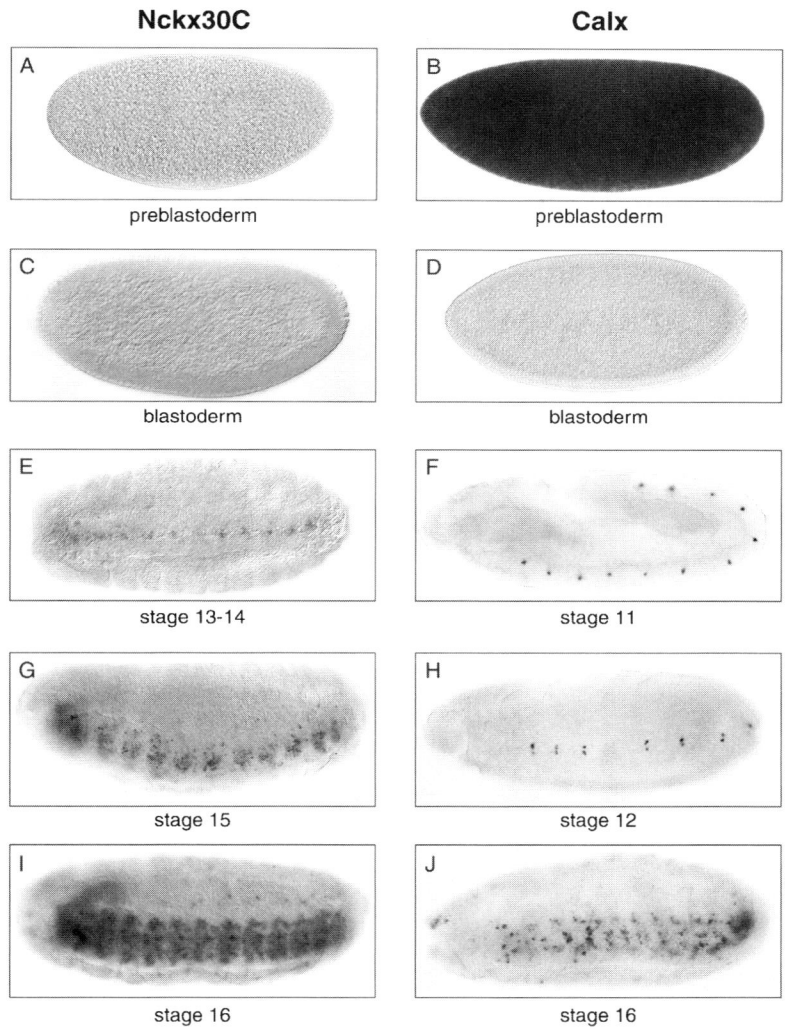

FIGURE 6. See Plate 7 in color section. *Nckx30C* and *Calx* are expressed in the ventral nerve cord of the *Drosophila* embryo. Shown are *in situ* hybridizations to whole-mount wild-type embryos (Canton S strain) hybridized with digoxigenin-labeled antisense riboprobes for *Nckx30C* and *Calx*. (**A**) Lateral view, preblastoderm, *Nckx30C*; (**B**) lateral view, preblastoderm, *Calx*; (**C**) lateral view, blastoderm, *Nckx30C*; (**D**) lateral view, blastoderm, *Calx*; (**E**) ventral view, stage 13-14, *Nckx30C*; (**F**) lateral view, stage 11, *Calx*; (**G**) ventrolateral view, stage 15, *Nckx30C*; (**H**) ventral view, stage 12, *Calx*; (**I**) ventrolateral view, stage 16, *Nckx30C*; and (**J**) ventrolateral view, stage 16, *Calx*. Note the labeling of the developing ventral nerve cord. Control hybridizations with sense probes did not produce signals (data not shown). All embryos are oriented anterior to the left. In lateral views, all embryos are oriented dorsal side up. (Reproduced from Ref. 43, J. Cell Biol. 1999. **147:** 659–669, by copyright permission of The Rockefeller University Press.)

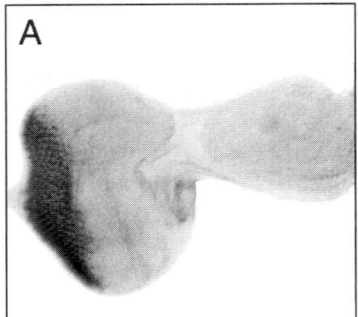
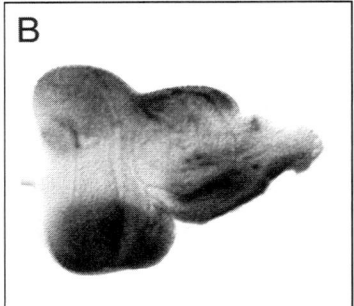
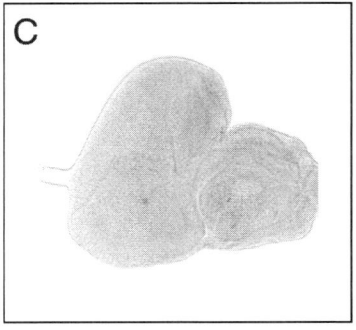

FIGURE 7. See Plate 8 in color section. *Nckx30C* is expressed in the third instar imaginal discs of larvae. Shown are *in situ* hybridizations to imaginal discs from wild-type larvae with digoxigenin-labeled riboprobes for *Nckx30C* and *chaoptin*. (**A**) Eye-antennal disc, antisense riboprobe for *chaoptin*; (**B**) eye-antennal disc, antisense riboprobe for *Nckx30C*; (**C**) eye-antennal disc, control sense riboprobe for *Nckx30C*. Note that control hybridizations with sense probes did not produce signals. The eye discs are oriented posterior to the left and dorsal up. (Reproduced from Ref. 43, J. Cell Biol. 1999. **147:** 659–669, by copyright permission of The Rockefeller University Press.)

7A shows expression of *chaoptin* in photoreceptor cells located posterior to the furrow. *Chaoptin* expression can be used as a marker for differentiated photoreceptor cells posterior to the morphogenetic furrow. Chaoptin is a photoreceptor cell surface glycoprotein.[54]. *Nckx30C* transcripts are present in a dorsal/ventral pattern, both anterior and posterior to the morphogenetic furrow, with no labeling in the midline (FIG. 7B and C). *Wingless* (*wg*) displays a similar expression pattern, and it is important for dorsal/ventral patterning in eye development.[61–63,69,70] *Nckx30C* expression is wider than that observed for *wingless*. It has been proposed that one of the vertebrate Wingless (Wnt) pathways may use a G-protein–mediated phosphatidylinositol signaling cascade that leads to an increase in intracellular calcium.[71–73] We suggest that NCKX30C may be playing a role in modulating calcium in this or in other patterning pathways.[43] Given what we know about the role of NCKX in extruding large amounts of calcium in rod photoreceptor cells, it is likely that cells that express *Nckx30C* may also be experiencing large and sustained rises in cytosolic of calcium.

Here, we review that both NCX and NCKX-type exchangers not only may function in the removal of calcium and maintenance of calcium homeostasis during signaling in the adult, but they also may play critical roles in signaling events during embryogenesis. In addition, NCKX may play a role in calcium modulation during cell differentiation and patterning in eye development.

ACKNOWLEDGMENTS

We thank The Rockefeller University Press for copyright permission for reproduction from the *Journal of Cell Biology* (1999. **147:** 659–669). We thank M. Jungwirth, E. Nielsen, and J. Loeffelholz for assistance with computer graphics. This work was supported by the Medical Research Council of Canada (to P.P.M.S) and National Institutes of Health Grant EY08768, Retina Research Foundation, HHMI, Foundation Fighting Blindness, Fight-For-Sight, and the Research to Prevent Blindness Foundation (to N.J.C.).

REFERENCES

1. PERETZ, A., C. SANDLER, K. KIRSCHFELD, *et al.* 1994. Genetic dissection of light induced Ca^{2+} influx into *Drosophila* photoreceptors. J. Gen. Physiol. **104:** 1057–1077.
2. RANGANATHAN, R., B.J. BACSKAI, R.Y. TSIEN & C.S. ZUKER. 1994. Cytosolic calcium transients: spatial localization and role in *Drosophila* photoreceptor cell function. Neuron **13**(4): 837–848.
3. RANGANATHAN, R., D.M. MALICKI & C.S. ZUKER. 1995. Signal transduction in *Drosophila* photoreceptors. Annu. Rev. Neurosci. **18:** 283–317.
4. HARDIE, R.C. 1995. Photolysis of caged Ca^{2+} facilitates and inactivates but does not directly excite light-sensitive channels in *Drosophila* photoreceptors. J. Neurosci. **15**(1 Pt 2): 889–902.
5. HARDIE, R.C. 1996. INDO-1 measurements of absolute resting and light-induced Ca^{2+} concentration in *Drosophila* photoreceptors. J. Neurosci. **16:** 2924–2933.
6. HARDIE, R.C. 1996. Calcium signalling: setting store by calcium channels. Curr. Biol. **6:** 1371–1373.
7. ZUKER, C.S. 1996. The biology of vision of *Drosophila*. Proc. Natl. Acad. Sci. USA **93**(2): 571–576.
8. LISMAN, J.E. & P.K. BROWN. 1972. The effects of intracellular iontophoretic injection of calcium and sodium ions in the light response of Limulus ventral photoreceptors. J. Gen. Physiol. **59:** 701–719.
9. MINKE, B. & E. ARMON. 1984. Activation of electrogenic Na-Ca exchange by light in fly photoreceptors. Vis. Res. **24**(2): 109–115.
10. O'DAY, P.M. & M.P. GRAY-KELLER. 1989. Evidence for electrogenic Na^+/Ca^{2+} exchange in Limulus ventral photoreceptors. J. Gen. Physiol. **93**(3): 473–494.
11. O'DAY, P.M., M.P. GRAY-KELLER & M. LONERGAN. 1991. Physiological roles of Na^+/Ca^{2+} exchange in Limulus ventral photoreceptors. J. Gen. Physiol. **97:** 369–391.
12. BAUER, P.J. *et al.* 1999. Direct evidence of Na^+/Ca^{2+} exchange in squid rhabdomeric membranes. Am. J. Physiol. **276**(3 Pt 1): C558–565.
13. YAU, K.-W. & K. NAKATANI. 1984. Electrogenic Na-Ca exchange in retinal rod outer segment. Nature **311:** 661–663.

14. SCHNETKAMP, P.P.M. 1986. Sodium-calcium exchange in the outer segments of bovine rod photoreceptors. Physiol. J. **373**: 25–45.
15. LAGNADO, L. & P.A. MCNAUGHTON. 1990. Electrogenic properties of the Na:Ca exchange. J. Membr. Biol. **113**(3): 177–191.
16. LAGNADO, L., L. CERVETTO & P.A. MCNAUGHTON. 1988. Ion transport by the Na-Ca exchange in isolated rod outer segments. Proc. Natl. Acad. Sci. USA **85**: 4548–4552.
17. STRYER, L. 1986. Cyclic GMP cascade of vision. Ann. Rev. Neurosci. **9**: 87–119.
18. YAU, K.-W. 1994. Phototransduction mechanism in retinal rods and cones. Invest. Ophthalmol. & Vis. Sci. **35**(1): 9–32.
19. GRAY-KELLER, M.P. & P.B. DETWILER. 1994. The calcium feedback signal in the phototransduction cascade of vertebrate rods. Neuron **13**: 849–861.
20. SAMPATH, A. *et al.* 1998. Bleached pigment produces a maintained decrease in outer segment Ca^{2+} in salamander rods. J. Gen. Physiol. **111**: 53–64.
21. MATTHEWS, H.R. *et al.* 1988. Photoreceptor light adaptation is mediated by cytoplasmic calcium concentration. Nature **334**(6177): 67–69.
22. NAKATANI, K. & K. YAU. 1988. Calcium and light adaptation in retinal rods and cones. Nature **334**: 69–71.
23. BERRIDGE, M.J. 1998. Neuronal calcium signaling. Neuron **21**(1): 13–26.
24. BERRIDGE, M.J., M.D. BOOTMAN & P. LIPP. 1998. Calcium: a life and death signal. Nature **395**: 645–648.
25. BLAUSTEIN, M.P. & J. LEDERER. 1999. Sodium/calcium exchange: its physiological implications. Physiol. Rev. **79**: 763–854.
26. PHILIPSON, K.D. *et al.* 1996. Molecular regulation of the Na^+-Ca^{2+} exchanger. Ann. N.Y. Acad. Sci. **779**: 20–28.
27. NICOLL, D.A. *et al.* 1996. Cloning of a third mammalian Na^+-Ca^{2+} exchanger, NCX3. J. Biol. Chem. **271**(40): 24914–24921.
28. SCHNETKAMP, P.P.M., D.K. BASU & R.T. SZERENCSEI. 1989. Na^+-Ca^{2+} exchange in bovine rod outer segments requires and transports K^+. Am. J. Physiol. **257**: C153–C157.
29. CERVETTO, L. *et al.* 1989. Extrusion of calcium from rod outer segments is driven by both sodium and potassium gradients. Nature **337**: 740–743.
30. REILÄNDER, H. *et al.* 1992. Primary structure and functional expression of the Na/Ca, K-exchanger from bovine rod photoreceptors. EMBO J. **11**: 1689–1695.
31. TUCKER, J.E. *et al.* 1998. cDNA cloning of the human retinal rod Na-Ca + K exchanger: comparison with a revised bovine sequence. Invest. Ophthalmol. & Vis. Sci. **39**(2): 435–440.
32. TUCKER, J. *et al.* 1998. Chromosomal localization and genomic organization of the human retinal rod Na^+-Ca+K exchanger. Human Genet. **103**(4): 411–414.
33. COOPER, C.B. *et al.* 1999. cDNA-cloning and functional expression of the dolphin retinal rod Na-Ca+K exchanger NCKX1: comparison with the functionally silent bovine NCKX1. Biochemistry **38**(19): 6276–6283.
34. COOPER, C., R. SZERENCSEI & P. SCHNETKAMP. 2000. Spectrofluorometric detection of Na-Ca+K exchange. Methods Enzymol. **315**: 847–864.
35. PRINSEN, C.F., R.T. SZERENCSCSEI & P.P.M. SCHNETKAMP. 2000. Molecular cloning and functional expression of the potassium-dependent sodium-calcium exchanger from human and chicken retinal cone photoreceptors. J. Neurosci. **20**(4): 1424–1434.
36. POON, S., S. LEACH, X.F. LI, *et al.* 2000. Alternatively spliced isoforms of the rat eye sodium/calcium+potassium exchanger NCKX1. Am. J. Physiol. Cell Physiol. **278**(4): C651–660.
37. CLEMENS, F. *et al.* 2000. Molecular cloning and functional expression of the potassium-dependent sodium-calcium exchangers from human and chicken retinal cone photoreceptors. J. Neurosci. **20**(4): 1424–1434.
38. TSOI, M, K.H. RHEE, D. BUNGARD, *et al.* 1998. Molecular cloning of a novel potassium-dependent sodium-calcium exchanger from rat brain. J. Biol. Chem. **273**(7): 4155–4162.
39. KRAEV, A., B.D. QUEDNAU, S. LEACH, *et al.* 2001. Molecular cloning of a third member of the potassium-dependent sodium-calcium exchanger gene family, NCKX3. J. Biol. Chem. **276**(25): 23161–23172.
40. WILSON, R. *et al.* 1994. 2.2 Mb of contiguous nucleotide sequence from chromosome III of C. elegans (see comments). Nature **368**(6466): 32–38.

41. SCHWARZ, E.M. & S. BENZER. 1997. Calx, a Na-Ca exchanger gene of *Drosophila melanogaster*. Proc. Natl. Acad. Sci. USA **94**: 10249–10254.
42. NICOLL, D. et al. 1996. Mutation of amino acid residues in the putative transmembrane segments of the cardiac sarcolemmal Na^+-Ca^{2+} exchanger. J. Biol. Chem. **271**(23): 13385–13391.
43. HAUG-COLLET, K. et al. 1999. Cloning and characterization of a potassium-dependent sodium/calcium exchanger in *Drosophila*. J. Cell Biol. **147**(3): 659–669.
44. HRYSHKO, L.V. et al. 1996. Anomalous regulation of the Drosophila Na^+-Ca^{2+} exchanger by Ca^{2+}. J. Gen. Physiol. **108**: 67–74.
45. RUKNUDIN, A. et al. 1997. Na^+/Ca^{2+} exchanger in *Drosophila*: cloning, expression, and transport differences. Am. J. Physiol. **273**(1 Pt 1): C257–265.
46. OMELCHENKO, A. et al. 1998. Functional differences in ionic regulation between alternatively spliced isoforms of the Na^+-Ca^{2+} exchanger from *Drosophila melanogaster*. J. Gen. Physiol. **111**(5): 691–702.
47. DYCK, C. et al. 1998. Structure-function analysis of CALX1.1, a Na^+-Ca^{2+} exchanger from Drosophila. Mutagenesis of ionic regulatory sites. J. Biol. Chem. **273**(21): 12981–12987.
48. ALTSCHUL, S.F. et al. 1990. Basic local alignment search tool. J. Molec. Biol. **215**(3): 403–410.
49. THOMPSON, J.D., D.G. HIGGINS & T.J. GIBSON. 1994. CLUSTAL W: improving the sensitivity of progressive multiple sequence alignment through sequence weighting, position-specific gap penalties and weight matrix choice. Nucleic Acids Res. **22**(22): 4673–4680.
50. FARRELL, P. et al. 1998. High-level expression of secreted glycoproteins in transformed lepidopteran insect cells using a novel expression vector. Biotechnol. Bioengineer. **60**: 656–663.
51. SCHNETKAMP, P.P.M., J.E. TUCKER & R.T. SZERENCSEI. 1995. Ca^{2+} influx into bovine retinal rod outer segments mediated by $Na^+/Ca^{2+}/K^+$ exchange. Am. J. Physiol. **269**: C1153–C1159.
52. SCHNETKAMP, P., R. SZERENCSEI & D. BASU. 1991. Unidirectional Na^+, Ca^{2+} and K^+ fluxes through the bovine rod outer segment Na-Ca-K exchanger. J. Biol. Chem. **266**: 198–206.
53. PANGANIBAN, G., L. NAGY & S.B. CARROLL. 1994. The role of the Distal-less gene in the development and evolution of insect limbs. Curr. Biol. **4**(8): 671–667.
54. REINKE, R. et al. 1988. Chaoptin, a cell surface glycoprotein required for Drosophila photoreceptor cell morphogenesis, contains a repeat motif found in yeast and human. Cell **52**: 291–301.
55. SCHNETKAMP, P. et al. 1991. Regulation of free cytosolic Ca^{2+} concentration in the outer segments of bovine retinal rods by Na-Ca-K exchange measured with Fluo-3. 1. Efficiency of transport and interactions between cations. J. Biol. Chem. **266**: 22975–22982.
56. TSOI, M. et al. Molecular cloning of a novel potassium-dependent sodium-calcium exchanger from rat brain. J. Biol. Chem. 1998. **273**(7): 4155–4162.
57. COLES, J.A. & R.K. ORKAND. 1985. Changes in sodium activity during light stimulation in photoreceptors. J. Physiol. **362**: 415–435.
58. NAKATANI, K. & K. YAU. 1988. Calcium and magnesium fluxes across the plasma membrane of the toad rod outer segment. J. Physiol. **395**: 675–730.
59. GERSTER, U. 1997. A quantitative estimate of flash-induced Ca^+-and Na^+-influx and Na^+/Ca^+-exchange in blowfly Callophora photoreceptors. Vis. Res. **37**: 2477–2485.
60. SCHNETKAMP, P.P.M., D.K. BASU & R.T. SZERENCSEI. 1991. The stoichiometry of Na–Ca+K exchange in rod outer segments isolated from bovine retinas. Ann. N.Y. Acad. Sci. **639**: 10–21.
61. TREISMAN, J.E. & G.M. RUBIN. 1995. Wingless inhibits morphogenetic furrow movement in the *Drosophila* eye disc. Development **121**(11): 3519–3527.
62. TREISMAN, J.E. & U. HEBERLEIN. 1998. Eye development in Drosophila: formation of the eye field and control of differentiation. Curr. Topics Dev. Biol. **39**: 119–158.
63. HEBERLEIN, U., E.R. BOROD & F.A. CHANUT. 1998. Dorsoventral patterning in the Drosophila retina by wingless. Development **125**(4): 567–577.

64. READY, D.F., T.E. HANSON & S. BENZER. 1976. Development of the Drosophila retina, a neurocrystalline lattice. Dev. Biol. **53:** 217–240.
65. TOMLINSON, A. & D.F. READY. 1987. Neuronal differentiation in the *Drosophila* ommatidium. Dev. Biol. **120:** 366–376.
66. TOMLINSON, A. 1988. Cellular interactions in the developing *Drosophila* eye. Development **104:** 183–193.
67. READY, D.F. 1989. A multifaceted approach to neural development. Trends Neurosci. **12:** 102–110.
68. BANERJEE, O. & S.L. ZIPURSKY. 1990. The role of cell-cell interaction during development of the Drosophila visual system. Neuron **4:** 177–187.
69. HEBERLEIN, U. & K. MOSES. 1995. Mechanisms of Drosophila retinal morphogenesis: the virtues of being progressive. Cell **81:** 987–990.
70. MA, C. & K. MOSES. 1995. Wingless and patched are negative regulators of the morphogenetic furrow and can affect tissue polarity in the developing *Drosophila* compound eye. Development **121:** 2279–2289.
71. SLUSARSKI, D.C. *et al.* 1997. Modulation of embryonic intracellular Ca^{2+} signaling by Wnt-5A. Dev. Biol. (Orlando) **182**(1): 114–120.
72. SLUSARSKI, D.C., V.G. CORCES & R.T. MOON. 1997. Interaction of Wnt and a Frizzled homologue triggers G-protein-linked phosphatidylinositol signalling. Nature **390**(6658): 410–413.
73. MOON, R.T., J.D. BROWN & M. TORRES. 1997. WNTs modulate cell fate and behavior during vertebrate development. Trends Genet. **13**(4): 157–162.
74. KYTE, J. & R.F. DOOLITTLE. 1982. A simple model for displaying the hydropathic character of a protein. J. Mol. Biol. **157:** 105–132.

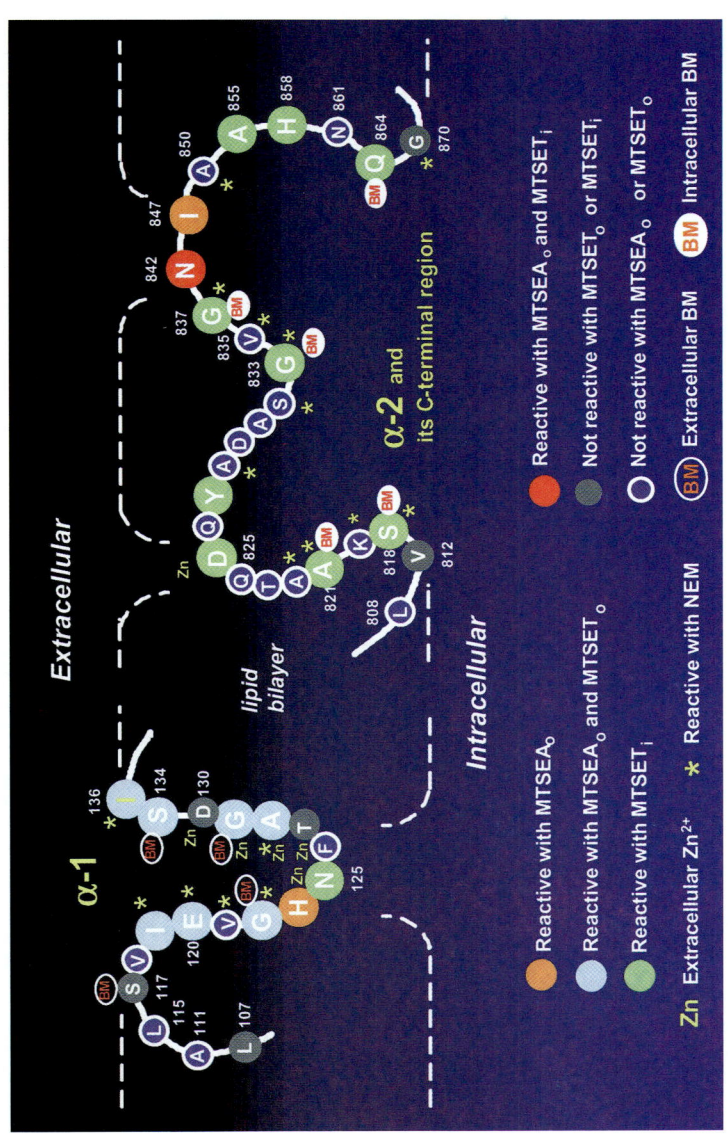

PLATE 1. See Shigekawa *et al.*, page 22.

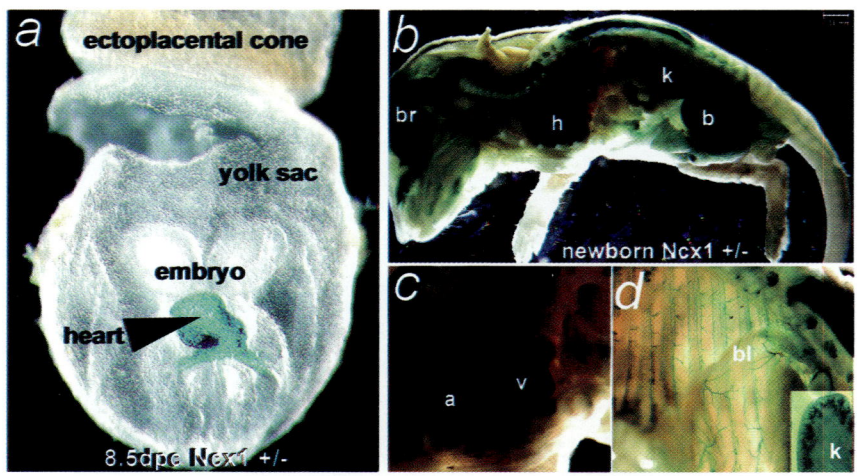

PLATE 2. See Conway *et al.*, page 271.

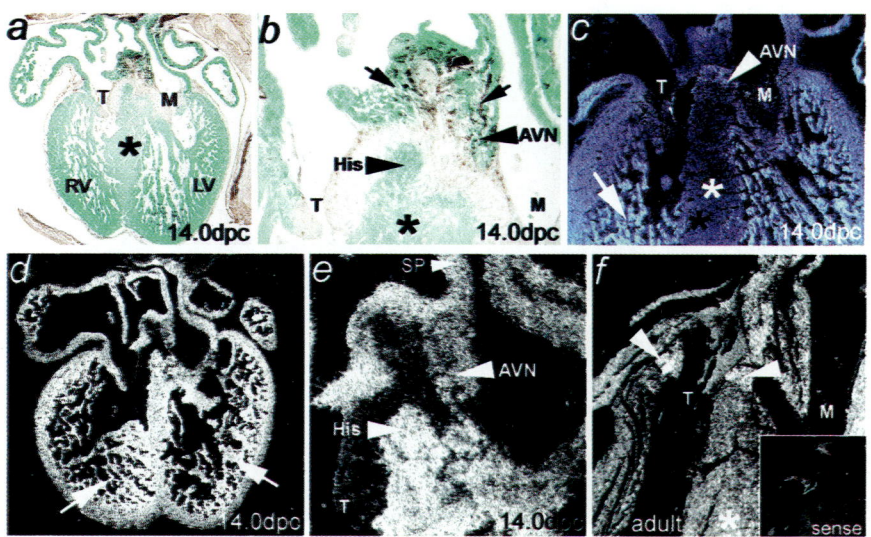

PLATE 3. See Conway *et al.*, page 274.

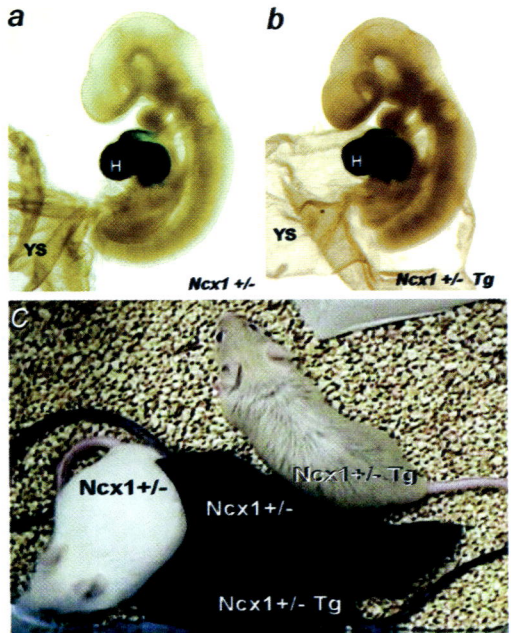

PLATE 4.
See Conway *et al.*, page 275.

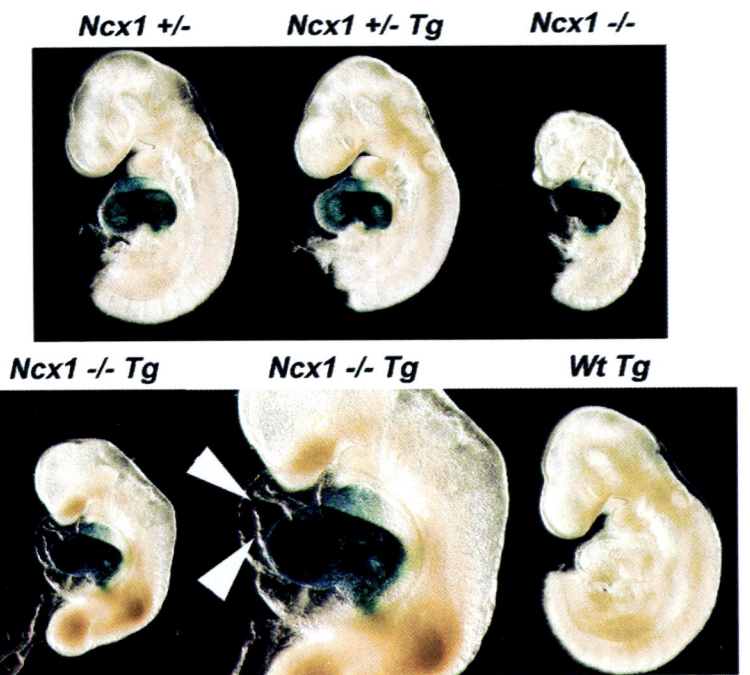

PLATE 5. See Conway *et al.*, page 276.

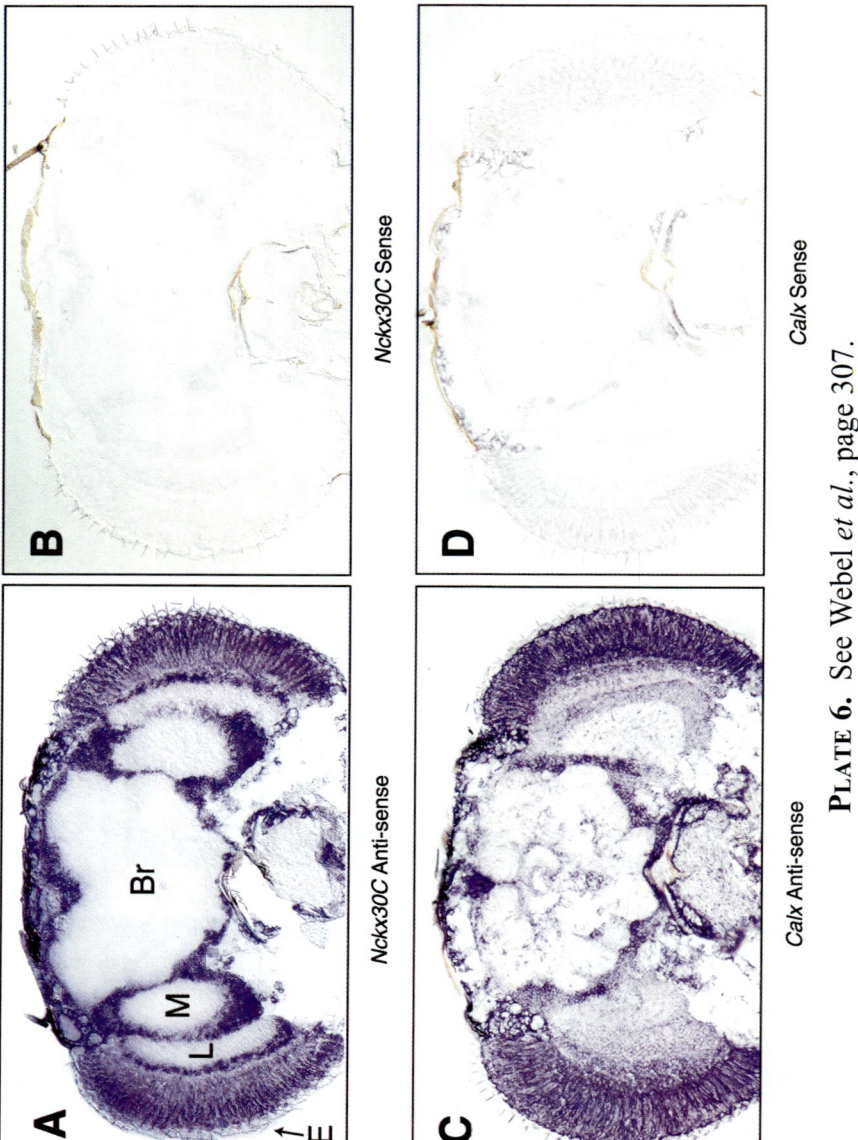

PLATE 6. See Webel et al., page 307.

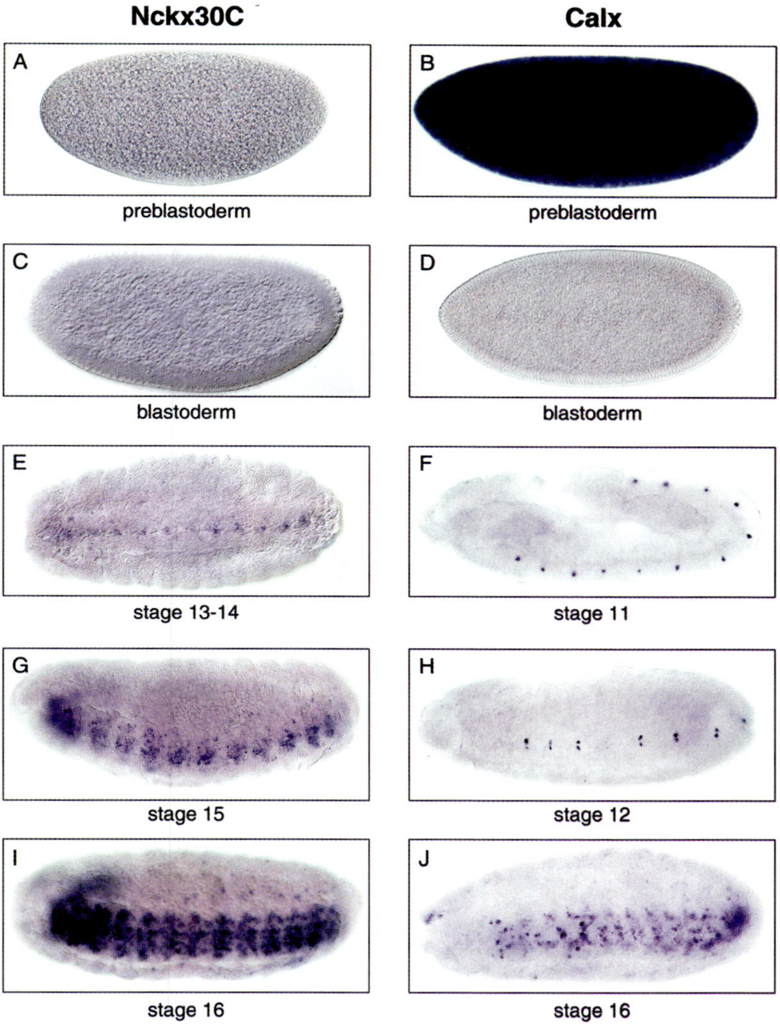

Nckx30C Calx

A preblastoderm **B** preblastoderm
C blastoderm **D** blastoderm
E stage 13-14 **F** stage 11
G stage 15 **H** stage 12
I stage 16 **J** stage 16

PLATE 7. See Webel *et al.*, page 309.

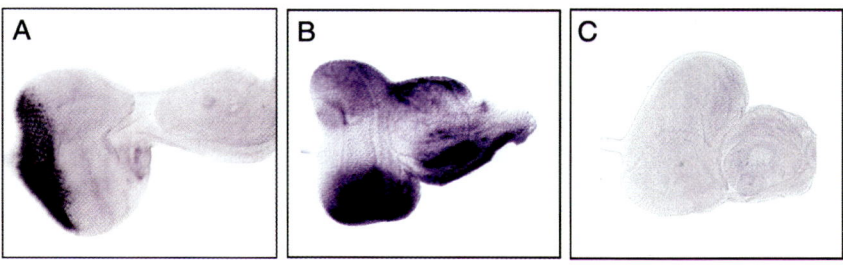

PLATE 8. See Webel *et al.*, page 310.

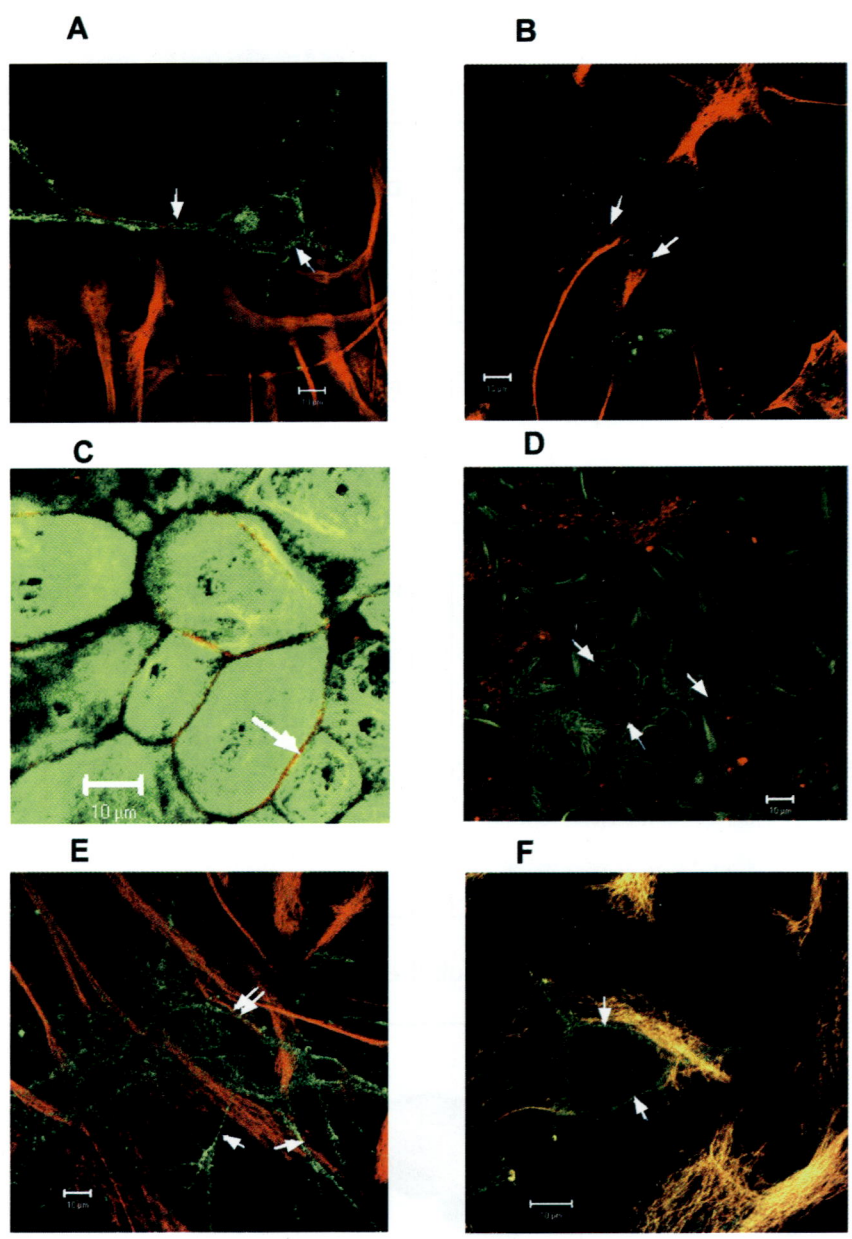

PLATE 9. See Thurneysen, *et al.*, page 372.

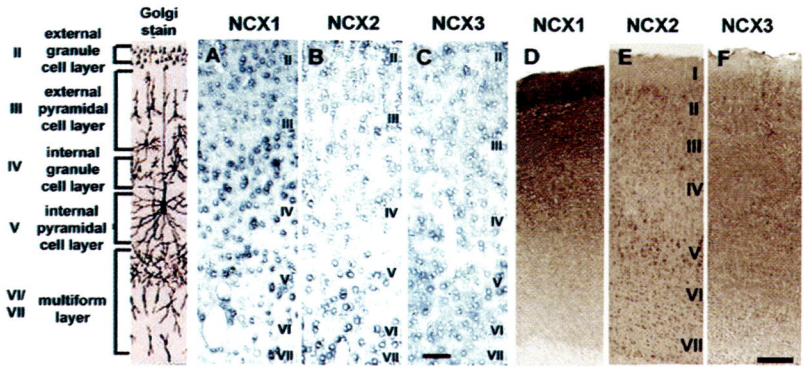

PLATE 10. See Canitano, *et al.*, page 397.

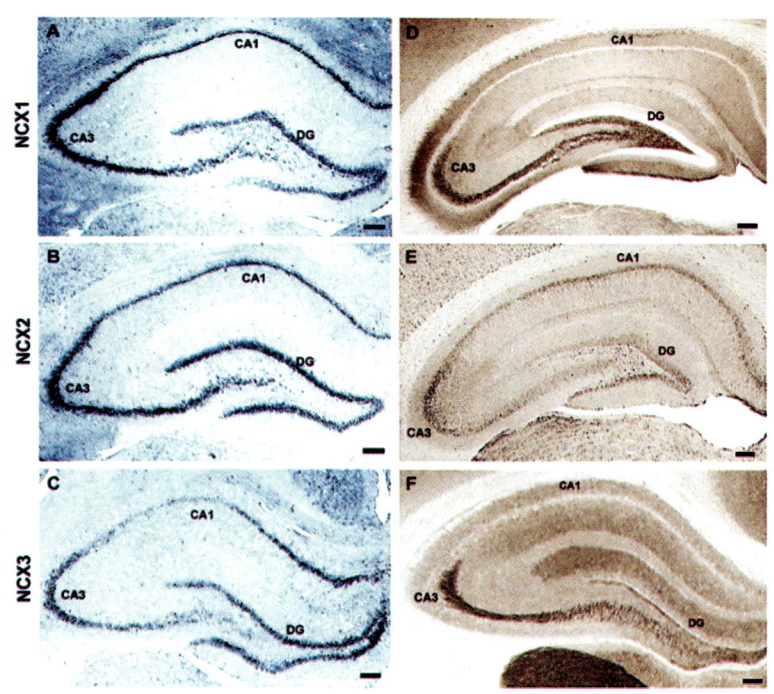

PLATE 11. See Canitano, *et al.*, page 398.

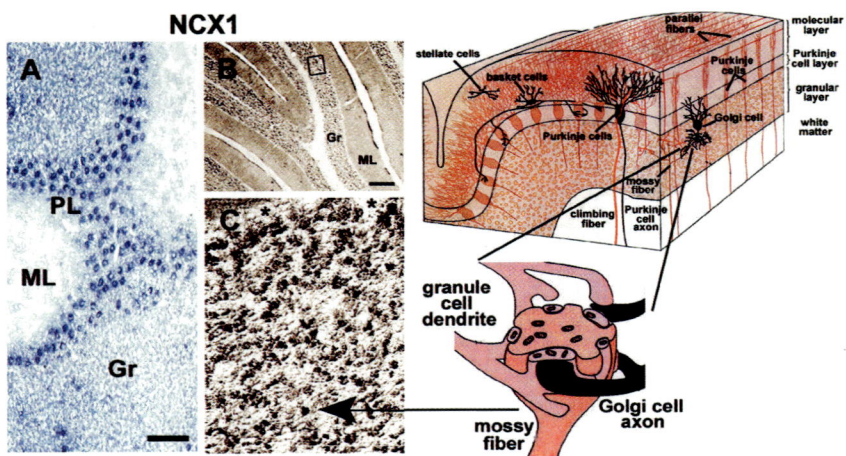

PLATE 12. See Canitano, *et al.*, page 400.

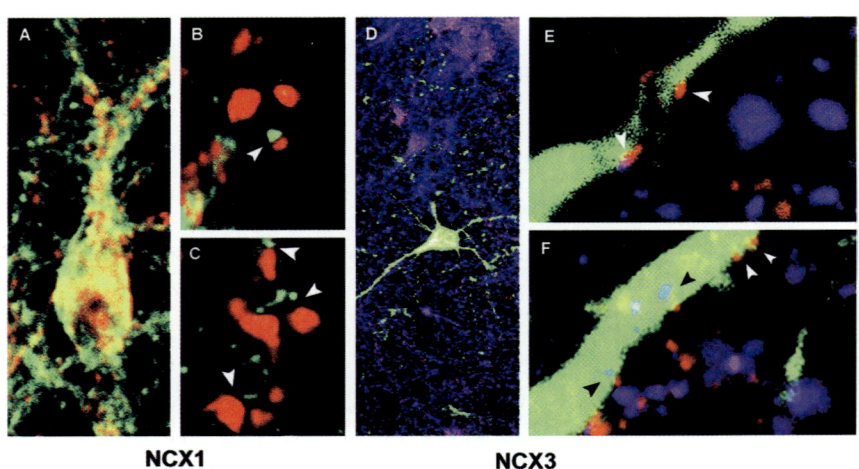

PLATE 13. See Canitano, *et al.*, page 401.

Na/Ca Exchange in Function, Growth, and Demise of β-Cells

ANDRÉ HERCHUELZ, OSCAR DIAZ-HORTA,[b] AND FRANÇOISE VAN EYLEN[a]

[a]*Laboratory of Pharmacology, Brussels University School of Medicine, B-1070, Brussels, Belgium*

[b]*National Institute of Endocrinology, Havana, Cuba*

ABSTRACT: Recent knowledge concerning the Na/Ca exchanger (NCX) in the pancreatic β-cell is reviewed. The β-cell expresses various NCX1 splice variants in a species-specific pattern (NCX1.3 and 1.7 in the rat; NCX1.2, 1.3, and 1.7 in the mouse) and in variable and different proportions. In the rat β-cell, the exchanger displays a high capacity, accounts for about 70% of Ca^{2+} extrusion, and participates in Ca^{2+} inflow during membrane depolarization. In the mouse, however, the contribution of the exchanger to Ca^{2+} extrusion is more modest, and to Ca^{2+} inflow, less evident. The exchanger has a stoichiometry of 3 Na^+ for 1 Ca^{2+}, is electrogenic, and displays a reversal potential at –20 mV. Although being of low magnitude, the current generated by the exchanger shapes glucose-induced β-cell electrical activity and intracellular Ca^{2+} oscillations. Intracellular Ca^{2+} may also trigger apoptosis. For instance, overexpression of the exchanger increases Ca^{2+}-dependent and Ca^{2+}-independent β-cell death by apoptosis, a phenomenon resulting from the depletion of ER Ca^{2+} stores with subsequent activation of caspase-12. Na/Ca exchange overexpression also reduces β-cell growth. Hence, the Na/Ca exchanger is a versatile system that appears to play an important role in the function, growth, and demise of the β-cell.

KEYWORDS: Na/Ca exchange; pancreatic β-cell; Ca^{2+} oscillations; apoptosis; caspase-12; endoplasmic reticulum stress

INTRODUCTION

The pancreatic β-cell secretes insulin in response to a rise in plasma glucose concentration, and insulin, in turn, stimulates glucose uptake in peripheral tissues such as skeletal muscles and liver. Any alteration in this loop may lead to glucose intolerance or diabetes mellitus, which may be of type 1 (insulin-dependent) or type 2 (insulin-independent). Type 1 diabetes results from the autoimmune destruction of the β-cell, whereas type 2 results in part from alterations in the process of insulin release,[1,2] the earliest manifestations of which are alterations of the initial peak and the oscillatory pattern in insulin release.[3]

Address for correspondence: A. Herchuelz, Laboratoire de Pharmacodynamie et de Thérapeutique, Université Libre de Bruxelles, Faculté de Médecine, Route de Lennik, 808-Bâtiment G.E., B-1070 Bruxelles, Belgium. Voice: (32) 2 555 62 01; fax: (32) 2 555 63 70.
herchu@ulb.ac.be

The pancreatic β-cell is unique in that glucose, the main physiologic stimulus of insulin release, not only constitutes a fuel but also is a cellular signal that triggers the process of insulin release. According to current knowledge, sugar stimulates insulin release as follows: after entry into the β-cell, glucose is metabolized and the resulting increase in ATP concentration or ATP/ADP ratio closes the ATP-sensitive K^+ channels, which depolarizes the plasma membrane and leads to the opening of voltage-sensitive Ca^{2+} channels. The ensuing entry of Ca^{2+} increases cytosolic-free Ca^{2+} concentration ($[Ca^{2+}]_i$), which triggers insulin release.[4]

In a previous review of the subject presented at the 2nd International Conference on Na/Ca exchange in Baltimore 10 years ago, we summarized the role played by Na/Ca exchange in β-cell Ca^{2+} homeostasis and the process of insulin release.[5] In the present review we summarize recent contributions to the role of Na/Ca exchange within the two aforementioned lines that may lead to diabetes mellitus.

Na/Ca EXCHANGER ISOFORMS AND SPLICE VARIANTS

The Na/Ca exchanger was cloned in various tissues about 10 years ago. Three genes coding for 3 different exchangers (NCX1, NCX2, and NCX3) were identified in mammals. Further variability results from alternative splicing of NCX1, and tissue-specific variants have been identified and called NCX.1.1, NCX1.2, etc. Although NCX1 appears to be widely distributed, NCX2 and NCX3 have been located only in brain and skeletal muscle. (For a review, see Ref. 6.)

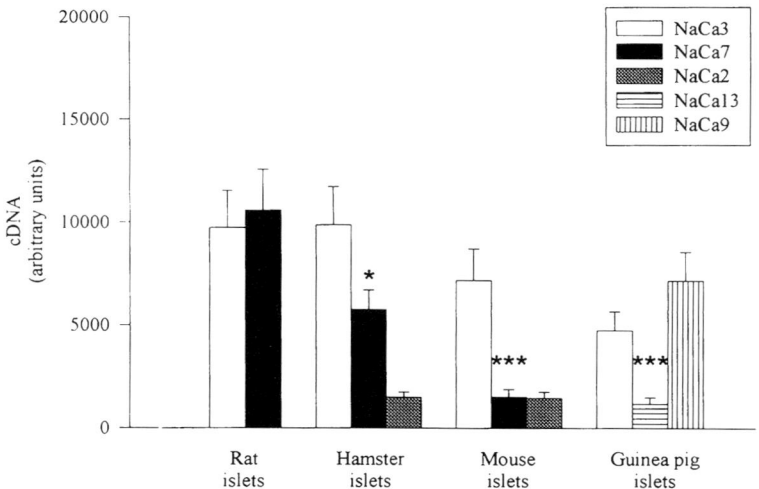

FIGURE 1. Expression and transcription pattern of the five NCX1 splice variants present in rat, hamster, mouse, and guinea pig islet cells (NCX1.3, NCX1.7, NCX1.2, NCX1.13, and NCX1.9). The cDNA values were determined from the fluorescence of PCR fragment by densitometry (arbitrary units). Mean ± SEM refers to 5 determinations. (*P <0.05; ***P <0.001). (Reprinted with permission from Van Eylen et al.[11])

To characterize the isoforms expressed in pancreatic β cells, reverse-transcribed polymerase chain reaction (RT-PCR) analysis was performed on mRNA from rat pancreatic islets, purified β cells (99% pure), and a rat insulinoma β-cell line (RINm5F cells).[7] PCR amplification did not yield a DNA fragment for NCX2, but it did yield 2 NCX1 bands in the 3 preparations, corresponding to NCX1.3 and NCX1.7. (The presence of NCX3 was not looked for.) Interestingly, while NCX1.3 and NCX1.7 were equally transcribed in pancreatic islets and purified β cells, a difference was seen in RINm5F cells, NCX1.7 being 3 times less transcribed than in β-cells.[7]

In view of this difference and the existence in the literature of apparently contradictory data on the role of Na/Ca exchange in Ca^{2+} homeostasis in rat and mouse β-cells,[8–10] we examined which NCX1 splice variants were expressed in 4 rodent species and especially in rat and mouse β-cells, which represent the 2 major experimental models for the study of β-cell function and insulin release. FIGURE 1 shows the expression and transcription pattern of NCX1 splice variants, as measured by quantitative PCR in these 4 rodent species. Although the rat β-cell indeed showed similar amounts of NCX1.3 and NCX1.7 transcripts, the mouse showed smaller amounts of these 2 transcripts and expressed, in addition, a third splice variant, NCX1.2.[11] Hamster and guinea pig β-cells showed distinct transcription patterns from rat and mouse β-cells. Interestingly, the level of NCX1.3 and NCX1.7 mRNA transcript correlated with Na/Ca exchange activity in these 4 species. For instance, Na/Ca exchange activity was twice as large in rat than in mouse β-cells.[11]

CONTRIBUTION OF Na/Ca EXCHANGE TO Ca^{2+} INFLOW AND OUTFLOW

Although previous work from this laboratory indicated that Na/Ca exchange participates in β-cell Ca^{2+} homeostasis and insulin release,[5,12,13] the exact contribution of the exchanger to Ca^{2+} inflow and outflow remained to be determined. Because there are no specific inhibitors of Na/Ca exchange, we used the antisense oligonucleotide (AS-oligos) strategy to knockout Na/Ca exchange in rat β-cells.[10]

Phosphorothioated AS-oligos were designed to target the region encompassing the AUG start codon of the rat NCX1 mRNA. Exposure of rat β-cells to 500 nmol/L of the AS-oligos inhibited Na/Ca exchange activity by ~77%. By contrast, control oligonucleotides (scrambled and mismatched) did not affect Na/Ca exchange activity as measured by the effect of extracellular Na^+ removal on $[Ca^{2+}]_i$. To ascertain that AS-oligos specifically acted by knocking down Na/Ca exchange, we studied the time course of the effect of AS-oligos on Na/Ca exchange activity and mRNA levels. AS-oligos induced a rapid and parallel decrease in both parameters, a maximal effect being observed after ~16 hours of exposure to AS-oligos. Moreover, the effect on both parameters was transient and reversible, indicating that it was not toxic or of a nonspecific nature. The contribution of Na/Ca exchange to Ca^{2+} inflow and outflow could then be quantified by measuring the effect of membrane depolarization on $[Ca^{2+}]_i$. FIGURE 2 shows that compared to control, treatment with AS-oligos induced 2 major effects on the K^+-induced $[Ca^{2+}]_i$ increase: first, it reduced both the rate of $[Ca^{2+}]_i$ rise and the maximal amplitude of the initial peak by about 30% and 50%, respectively; second, it reduced the rate of $[Ca^{2+}]_i$ decrease on membrane repolar-

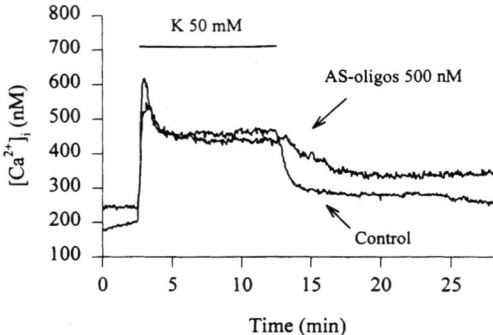

FIGURE 2. Effect of KCl (50 mM) on [Ca^{2+}]$_i$ in control and AS-oligos–treated pancreatic β-cells. Periods of exposure to KCl are indicated by the bar above the curves. The curves shown are the mean of more than 50 traces in each case. (Reprinted with permission from Van Eylen et al.[10])

ization by about 70%. A similar picture was observed when the hypoglycemic sulfonylurea tolbutamide was used to depolarize the plasma membrane. As a whole, the latter data indicate that in rat β-cells, Na/Ca exchange accounts for about 70% of Ca^{2+} outflow and that during the upstroke of the action potential, the Na/Ca exchanger may reverse and contribute to Ca^{2+} entry (about 25% of the initial peak).[10]

In parallel experiments, we examined the contribution of Na/Ca exchange to Ca^{2+} inflow and outflow in mouse β-cells by combining the use of the patch-clamp technique and [Ca^{2+}]$_i$ microfluorimetry.[14] The study showed that in mouse β-cells, the exchanger contributed to Ca^{2+} outflow to a lesser extent than it did in the rat (35% at [Ca^{2+}]$_i$ above 1 mM). In addition, we could not obtain any evidence that Na/Ca exchange could reverse during the mouse β-cell burst of electrical activity.[14] This confirms the existence of a major difference in Na/Ca exchange activity between rat and mouse β-cells. In the heart, similar species differences in Na/Ca exchange activity have been evidenced.[15] Nevertheless, in the mouse, we established that the stoichiometry of the exchanger was 3 Na$^+$ for 1 Ca^{2+}, that its reversal potential was at –20 mV, and that by generating an inward current, the exchanger contributed to the duration of the β-cell burst of electrical activity. Indeed, extracellular Na$^+$ removal induced a shortening of the burst duration (FIG. 3).[14] Mathematical simulations implementing 2 existing models for bursting in pancreatic β-cells confirmed that the presence of Na/Ca exchange activity could substantially increase the plateau fraction of the bursting of electrical activity[14] and hence suppress the oscillations of membrane potential and [Ca^{2+}]$_i$.

ROLE OF Na/Ca EXCHANGE ACTIVITY IN β-CELL FUNCTION

In view of such differences in Na/Ca exchanger transcription level and activity between the rat and mouse β-cell, and the view that the inward current generated by the exchanger working in its forward mode (Ca^{2+} efflux) may participate in the duration of the burst of electrical activity, we wondered if the difference in electrical

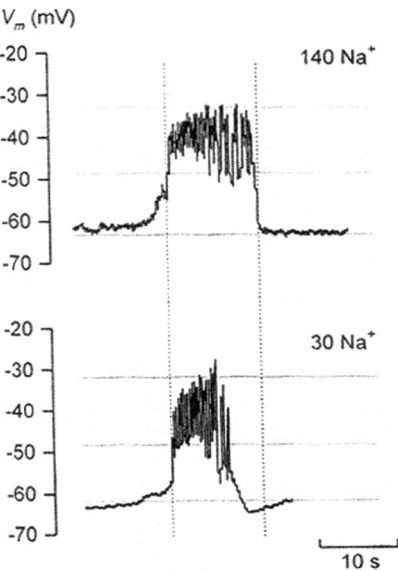

FIGURE 3. Effects of lowering extracellular Na$^+$ concentration on glucose-induced burst of electrical activity. (Reprinted with permission from Gall et al.[14])

activity and [Ca^{2+}]$_i$ behavior seen in the mouse and rat pancreatic β-cell may not result from a difference in Na/Ca exchange activity. Indeed, in response to an intermediate glucose concentration (11.1 mM), the mouse β-cell displays slow oscillations of the membrane potential and [Ca^{2+}]$_i$, whereas the rat β-cell displays a stepwise increase in both phenomena.[16]

To test such a hypothesis, NCX1.7 was cloned from a human insulinoma and stably transfected in the insulinoma cell line BRIN-BD11, produced by electrofusion of RINm5F cells with native rat pancreatic β-cells. Several clones showing high overexpression, as assessed at the mRNA and protein level, could be isolated and used for functional studies. The protein was appropriately targeted to the plasma membrane, as assessed by immunofluorescence and the fact that the various clones showed a 3- to 5-fold increase in activity compared to control cells.[17] Moreover, compared to control cells, overexpressing cells showed (1) a more rapid increase in [Ca^{2+}]$_i$ on membrane depolarization induced by 50 mM K$^+$, the rate of increase being doubled, and (2) a more rapid and earlier return of [Ca^{2+}]$_i$ to basal levels on membrane repolarization. These data are complementary to those obtained in AS-treated cells (see above) and confirm the prominent role played by the Na/Ca exchanger in rat β-cells.[10,17]

In response to an intermediate glucose concentration (11.1 mM), control cells showed a rapid increase in [Ca^{2+}]$_i$ that displayed a clear oscillatory pattern (FIG. 4A). By contrast, in overexpressing cells, glucose induced a stepwise increase in [Ca^{2+}]$_i$ without such oscillations (FIG. 4B). Because the cells where plated in high

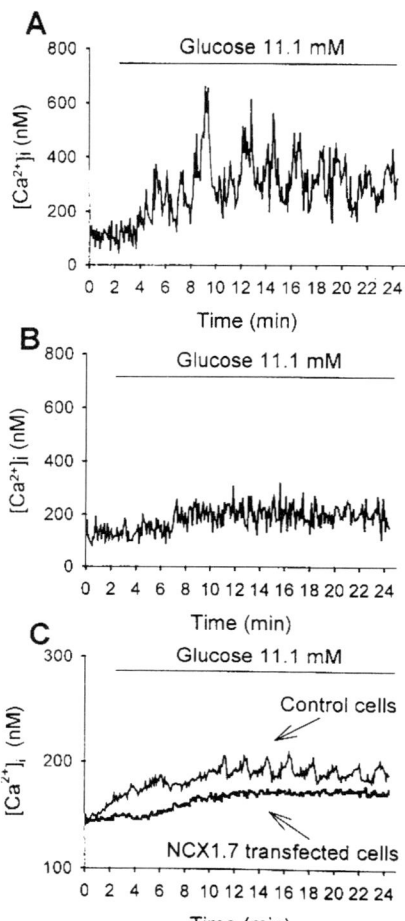

FIGURE 4. Effect of glucose 11.1 mM on $[Ca^{2+}]_i$ in control and NCX1.7-overexpressing BRIN-BD11 cells. **A** and **B** show typical individual traces observed in control and NCX1.7-transfected cells, respectively. The curves shown in **C** are the mean of 87 and 84 traces, respectively. Periods of exposure to glucose are indicated by the bars above the curves. (Reprinted with permission from Van Eylen et al.[17])

density on the coverslips, we considered the cells to be coupled and hence calculated the mean of all traces obtained in individual preparations. FIGURE 4C shows that although control cells displayed clear oscillations, overexpressing cells did not. Similar observations could be made using the hypoglycemic sulfonylurea tolbutamide instead of glucose. Therefore, these data confirm that the difference in electrical activity and $[Ca^{2+}]_i$ behavior seen in mouse and rat pancreatic β-cell indeed results from a difference in Na/Ca exchange activity in these species. Further confirmation was found in the observation that slight hyperpolarization of the plasma membrane by a low concentration of the plasma membrane K^+-channel opener diazoxide (2.5–10 μM) restored $[Ca^{2+}]_i$ oscillations in glucose-stimulated NCX1.7 overexpressing cells.[17]

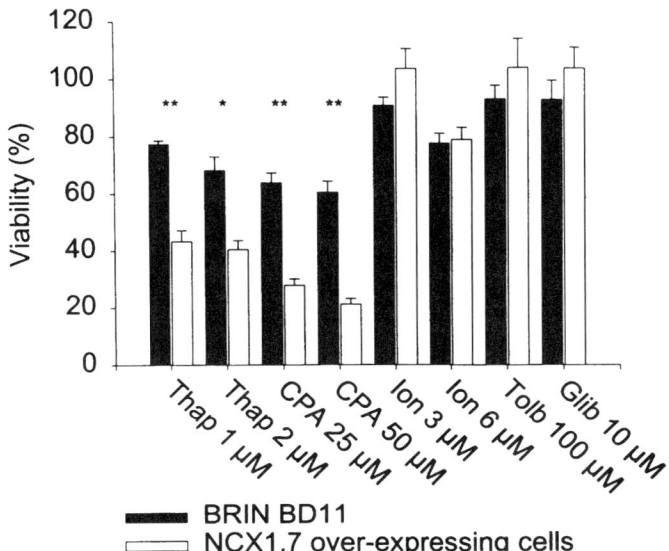

FIGURE 5. Effect of NCX1.7 overexpression on cell viability. Viability was tested in the presence of thapsigargin (Thap; 1 and 2 μM), cyclopiazonic acid (CPA; 25 and 50 μM), ionomycin (Ion; 3 and 6 μM), tolbutamide (Tolb; 100 μM), and glibenclamide (Glib; 10 μM). The MTT assay was used to measure cell viability (survival). Data are given as means ± SEM from at least 4 individual experiments comprising each of 4 replicates. Statistical significance: *P <0.005; **P <0.001. (Reprinted with permission from Diaz-Horta et al.[27])

ROLE OF Na/Ca EXCHANGE IN β-CELL APOPTOSIS

In addition of being of major importance as a cellular signaling agent, Ca^{2+} has been implicated as a proapoptotic agent involved in triggering apoptosis and regulating death-specific enzymes.[18] The family of cysteine proteases, the so-called caspases, are critical mediators of apoptosis[19] that can be triggered by 3 distinct pathways: oligomerization of death receptors located on the plasma membrane, mitochondrial damage, and the recently discovered endoplasmic reticulum (ER) pathway.[20,21] According to the latter triggering mechanism, "ER stress" as induced by any condition that may interfere with ER function, including disruption of ER Ca^{2+} homeostasis or accumulation of excess protein in the ER, causes cleavage of procaspase-12[22] with subsequent activation of caspase-12, which triggers the process of apoptosis.[21] Therefore, we wondered if Na/Ca exchanger overexpression would affect Ca^{2+}-induced apoptosis in insulin-secreting cells. This appeared to be of potential clinical significance, because in type 1 diabetes (insulin-dependent), the autoimmune destruction of pancreatic β-cells appears to be mediated by apoptosis,[1,23] a phenomenon that could involve Ca^{2+}.[24–26]

FIGURE 5 illustrates the effect of Na/Ca exchanger overexpression on a Ca^{2+}-dependent proapototic agent-induced decrease in cell viability, as measured by the MTT test. Whereas the sarco(endoplasmic) Ca^{2+}-ATPase (SERCA) inhibitor thapsigargin reduced viability by about 20% in control cells, the agent reduced viability by 60% in overexpressing cells.[27] Similar results were observed when the other SERCA inhibitor cyclopiazonic acid (CPA) was used. By contrast, NCX1 overexpression did not influence the effect of ionomycin and hypoglycemic sulfonylureas on cell viability. The reduction in cell viability resulted from apoptosis rather than necrosis, as shown by nuclear DNA staining using the fluorescent probe HOE 33342 and by DNA laddering.[27] The measurement of $[Ca^{2+}]_i$ showed that the SERCA inhibitors induced a rapid and transient increase in $[Ca^{2+}]_i$ in control cells, a phenomenon that was reduced by about 40–60% in Na/Ca exchanger-overexpressing cells. Ionomycin had a greater effect on $[Ca^{2+}]_i$ than SERCA inhibitors had on control cells, and its magnitude was less reduced in overexpressing cells, indicating that apoptosis resulted from a depletion of ER Ca^{2+} stores rather than from an excessive rise in $[Ca^{2+}]_i$. This depletion was directly objectified by measurement of the ER Ca^{2+} concentration using furaptra. In agreement with the view that apoptosis resulted from the depletion of ER Ca^{2+} stores, the increased rate of apoptosis was accompanied by the activation of caspase-12, a caspase that is specifically localized in the ER and that is activated by ER stress, including a decrease in ER Ca^{2+} stores, but not by membrane- or mitochondrial-targeted apoptotic signals.[21] Interestingly, activation of caspase-12 was increased in overexpressing cells even when untreated, namely, in a condition not associated with increased apoptosis. As a consequence, such caspase-12 activation sensitized the cells to apoptosis even when triggered by non-Ca^{2+}-dependent signaling pathways.[27]

On harvesting the cells, it was apparent that the growth of NCX1.7 overexpressing cells was less than that of control cells. Using the MTT test and cell counting to measure cell proliferation, a 40% decrease in the cell proliferation rate was shown in overexpressing cells. Thus, the latter data show that Na/Ca exchange overexpression, by depleting ER Ca^{2+} stores, triggers the activation of caspase-12 and increases apoptotic cell death by Ca^{2+}-dependent and Ca^{2+}-independent pathways. Overexpression of the exchanger led, in addition, to a decrease in cell proliferation. Because excessive cell proliferation and decreased ability to undergo apoptosis are two hallmarks of cancer and malignancies, overexpression of Na/Ca exchanger in cancer cells by gene therapy may represent a new potential approach in cancer therapy. On the other hand, our results open the way to the development of new strategies to control cellular Ca^{2+} homeostasis that could, on the contrary, prevent the process of apoptosis that mediates, in part, β-cell autoimmune destruction in type 1 diabetes. Indeed, if it is possible to increase apoptosis by overexpressing the Na/Ca exchanger, it should be possible to reduce it, as by transfecting a Na/Ca exchanger antisense oligonucleotide or by overexpressing a sarco(endoplasmic) reticulum Ca^{2+}-ATPase, which may lead to an increase in ER Ca^{2+}-stores.

In agreement with this view, nitric oxide was recently found to exert its proapoptotic action by inducing ER stress.[28] Furthermore, overexpression of calreticulin, a major Ca^{2+}-binding protein located in the lumen of the ER that regulates Ca^{2+} homeostasis in the ER lumen, increased the ER Ca^{2+} content and protected insulin-producing cells from NO-induced apoptosis.[28]

CONCLUSIONS

Na/Ca exchange is an important regulator of Ca^{2+} homeostasis in the β-cell and may, as such, regulate glucose-induced membrane potential and $[Ca^{2+}]_i$ oscillations as well as cell growth and apoptosis. Our results open the way to the development of new strategies to control cellular Ca^{2+} homeostasis that could either enhance or prevent the process of apoptosis and hence treat diseases characterized by too little or too much apoptosis, such as cancers and type 1 diabetes, respectively.

ACKNOWLEDGMENTS

We thank Christiane De Bruyne and Anne Van Praet for technical help. This work was supported by The Belgian Fund for Scientific Research (FRSM 3.4545.96, LN 9.4514.93, LN 9.4510.95, and 3.456.00), of which F.V.E. is a Postdoctoral Researcher, by the Concerted Action IREN in the Biomed 2 programme, and the ALFA programme IRELAN of the European Union of which O.D.-H is a grant holder coming from the National Institute of Endocrinology, Havana, Cuba.

REFERENCES

1. KURRER, M.O., V.P. SYAMASUNDAR, H.L. HANSON & J.D. KATZ. 1997. Beta cell apoptosis in T cell-mediated autoimmune diabetes. Proc. Natl. Acad. Sci. USA **94:** 213–218.
2. LEFEBVRE, P.J., G. PAOLISSO, A.J. SCHEEN & J.C. HENQUIN. 1987. Pulsatility of insulin and glucagon release: physiological significance and pharmacological implications. Diabetologia **30:** 443–452.
3. O'RAHILLY, S., R.C. TURNER & D.R. MATTHEWS. 1988. Impaired pulsatile secretion of insulin in relatives of patients with non-insulin-dependent diabetes. N. Engl. J. Med. **318:** 1225–1230.
4. FLATT, P. & S. LENZEN, Eds. 1994. Frontiers of Insulin Secretion and Pancreatic B-Cell Research. :1–621. Smith-Gordon. London.
5. HERCHUELZ, A. & P.-O. PLASMAN. 1991. Na/Ca exchange in the pancreatic B cell. Ann. N.Y. Acad. Sci. **639:** 642–656.
6. PHILIPSON, K.D. & D.A. NICOLL. 2000. Sodium-calcium exchange: a molecular perspective. Annu. Rev. Physiol. **62:** 111–133.
7. VAN EYLEN, F., M. SVOBODA & A. HERCHUELZ. 1997. Identification, expression pattern and potential activity of Na/Ca exchanger isoforms in rat pancreatic B-cells. Cell Calcium **21:** 185–193.
8. NADAL, A., M. VALDEOLMILLOS & B. SORIA. 1994. Metabolic regulation of intracellular calcium concentration in mouse pancreatic islets of Langerhans. Am. J. Physiol. **267:** E769–E774.
9. GARCIA-BARRADO, M.-J., P. GILON, Y. SATO, et al. 1996. No evidence for a role of reverse Na^+-Ca^{2+} exchange in insulin release from mouse pancreatic islets. Am. J. Physiol. **271:** E426–E433.
10. VAN EYLEN, F., C. LEBEAU, J. ALBUQUERQUE-SILVA & A. HERCHUELZ. 1998. Contribution of Na/Ca exchange to Ca^{2+} outflow and entry in the rat pancreatic β-cell. Diabetes **47:** 1873–1880.
11. VAN EYLEN, F., A. BOLLEN & A. HERCHUELZ. 2001. NCX1 Na/Ca exchanger splice variants in pancreatic islet cells. J. Endocrinol. **168:** 517–526.
12. HERCHUELZ, A. & P. LEBRUN. 1993. A role for Na/Ca exchange in the pancreatic B cell. Studies with thapsigargin and caffeine. Biochem. Pharmacol. **45:** 7–11.

13. VAN EYLEN, F., M.-H. ANTOINE, P. LEBRUN & A. HERCHUELZ. 1994. Inhibition of Na/Ca exchange stimulates insulin release from isolated rat pancreatic islets. Fundam. Clin. Pharmacol. **8:** 425–429.
14. GALL, D., J. GROMADA, I. SUSA, et al. 1999. Significance of Na/Ca exchange for Ca^{2+} buffering and electrical activity in mouse pancreatic β-cells. Biophys. J. **76:** 2018–2028.
15. BERS, D.M. 1991. Species differences and the role of sodium-calcium exchange in cardiac muscle relaxation. Ann. N.Y. Acad. Sci. **639:** 375–385.
16. ANTUNES, C.M., A.P. SALGADO, L.M. ROSARIO & R.M. SANTOS. 2000. Differential patterns of glucose-induced electrical activity and intracellular calcium responses in single mouse and rat pancreatic islets. Diabetes **49:** 2028–2038.
17. VAN EYLEN, F., O. DIAZ-HORTA, A. BAREZ, et al. 2002. Overexpression of the Na/Ca exchanger shapes stimulus-induced cytosolic Ca^{2+} oscillations in insulin producing BRIN-BD11 cells. Diabetes **51:** 366–375.
18. NICOTERA, P. & S. ORENIUS. 1998. The role of calcium in apoptosis. Cell Calcium **213:** 173–180.
19. CRYNS, V. & J. YUAN. 1998. Proteases to die for. Genes & Dev. **12:** 1551–1570.
20. MEHMET, H. 2000. Caspases find a new place to hide. Nature **403:** 29–30.
21. NAKAGAWA, T., H. ZHU, N. MORISHIMA, et al. 2000. Caspase-12 mediates endoplasmic-reticulum-specific apoptosis and cytotoxicity by amyloid-β. Nature **403:** 98–103.
22. VAN DE CRAEN, M., P. VANDENABEELE, W. DECLERCQ, et al. 1997. Characterization of seven murine caspase family members. FEBS Lett. **403:** 61–69.
23. LALLY, F.J., H. RATCLIFF & A.J. BONE. 2001. Apoptosis and disease progression in the spontaneouly diabetic BB/S rat. Diabetologia **44:** 320–324.
24. JUNTTI-BERGGREN, L., O. LARSSON, P. RORSMAN, et al. 1993. Increased activity of L-type Ca^{2+} channels exposed to serum from patients with type 1 diabetes. Science **261:** 86–90.
25. EFANOVA, J.B., S.V. ZAITSEV, B. ZHIVOTOVSKY, et al. 1998. Glucose and tolbutamide induce apoptosis in pancreatic β-cells. J. Biol. Chem. **273:** 33501–33507.
26. WANG, L., A. BHATTACHARJEE, Z. ZUO, et al. 1999. A low voltage-activated Ca^{2+} current mediates cytokine-induced pancreatic β-cell death. Endocrinology **140:** 1–5.
27. DIAZ-HORTA, O., A. HERCHUELZ & F. VAN EYLEN. 2002. Na/Ca exchanger overexpression induces endoplasmic reticulum-related apoptosis and caspase-12 activation in insulin-releasing BRIN-BD11 cells. Diabetes **51:** 1815–1824.
28. OYADOMARI, S., K. TAKEDA, M. TAGIKUCHI, et al. 2001. Nitric oxide-induced apoptosis in pancreatic β-cell is mediated by the endoplasmic reticulum stress pathway. Proc. Natl. Acad. Sci. USA **98:** 10845–10850.

Binding of the Retinal Rod Na^+/Ca^{2+}-K^+ Exchanger to the cGMP-Gated Channel Indicates Local Ca^{2+}-Signaling in Vertebrate Photoreceptors

PAUL J. BAUER

Institute for Biological Information Processing (IBI-1), Research Center Juelich, D-52425 Juelich, Germany

ABSTRACT: Ca^{2+} ions enter the outer segment of rod or cone photoreceptors exclusively through the cGMP-gated channel and are extruded by the Na^+/Ca^{2+}-K^+ exchanger. Recent evidence indicates that in the plasma membrane, the Na^+/Ca^{2+}-K^+ exchanger is associated with the cGMP-gated channel. In this contribution, the possible physiologic significance of this protein complex is considered. Based on recent experimental evidence, the possibility of a direct functional interaction between the cGMP-gated channel and the Na^+/Ca^{2+}-K^+ exchanger is discussed. Furthermore, a quantitative estimation of the cytoplasmic Ca^{2+} diffusion at the cGMP-gated channel indicates that Ca^{2+} diffusion is largely confined to the complex of the cGMP-gated channel and the associated Na^+/Ca^{2+}-K^+ exchanger molecules.

KEYWORDS: Na^+/Ca^{2+} exchange; regulation; cGMP-gated channel; local Ca^{2+} signaling; cross-linking

INTRODUCTION

The retinal rod Na^+/Ca^{2+}-K^+ exchanger (NCKX) in the outer segments (ROS) of rod photoreceptors is distinct from the more widespread Na^+/Ca^{2+} exchanger (NCX). Little sequence homology exists between the NCKX and the NCX.[1] Moreover, NCKX shows an absolute requirement of K^+ for Ca^{2+} transport.[2,3] FIGURE 1 demonstrates this characteristic feature of the NCKX: in contrast to the presence of K^+, there is virtually no Ca^{2+} transport when K^+ is isotonically replaced with Li^+.

K^+ does not simply act as a cofactor; it is cotransported along with Ca^{2+} ions.[2,3] Therefore, both the Na^+ inward and the K^+ outward gradient drive Ca^{2+} export. The utilization of 2 ion gradients for Ca^{2+} export implies theoretically a very low $[Ca^{2+}]_{in}$ equilibrium level of about 0.18 nM,[3] which is never reached physiologically.[4–6] If experimentally a K^+ inward gradient is applied, then both Ca^{2+} and K^+ have to be

---

Address for correspondence: Dr. Paul J. Bauer, Institute for Biological Information Processing (IBI-1), P.O. Box 1913, Research Center Juelich, D-52425 Juelich, Germany. Voice: (0049)-2461-614352; fax: (0049)-2461-614216.

p.j.bauer@fz-juelich.de

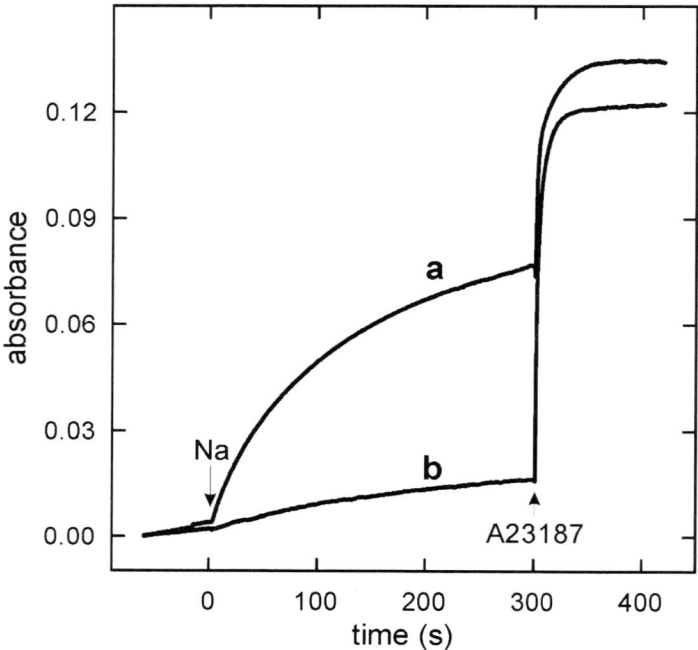

FIGURE 1. The NCKX transports Ca^{2+} only in the presence of K^+ ions. Fused ROS membrane vesicles, prepared by freeze-thawing and extrusion as described,[29] were loaded with 10 mM Ca^{2+} either in 100 mM KCl (*curve a*) or in 100 mM LiCl (*curve b*), buffered with 10 mM Hepes adjusted to pH 7.4 with Tris base. The external Ca^{2+} was removed with K^+–Chelex-100 beads (BioRad, mesh 200–400). The Ca^{2+} concentration was monitored with arsenazo III (50 µM) by recording the absorption difference between 650 nm and 700 nm with a dual wavelength photometer (Aminco DW2000). At time zero, Na^+ (95 mM) was added. The Ca^{2+}-ionophore A23187 (2.5 µM) was added after 300 seconds. The slope of the baseline is due to passive Ca^{2+} leakage.

transported against their concentration gradients, resulting in a reduced Ca^{2+} transport rate that scales with the inward K^+ gradient (FIG. 2).

This contribution briefly reviews recent biochemical evidence indicating that in rod membrane, NCKX and the cGMP-gated channel (CNG channel) form a complex that is part of a larger protein complex in the intact photoreceptor. Next, the possible physiologic role of the direct interaction between NCKX and CNG channel is discussed. Finally, the implications of the complex of NCKX and CNG channel is considered in terms of a novel mechanism of local Ca^{2+} signaling in photoreceptors.

ULTRASTRUCTURAL ORGANIZATION

The bovine retinal NCKX contains six cysteines, five of which are highly conserved between NCKX molecules from different isoforms and different species.[1,7–9] Thiol-specific cross-linking of ROS membranes efficiently yields dimers of the

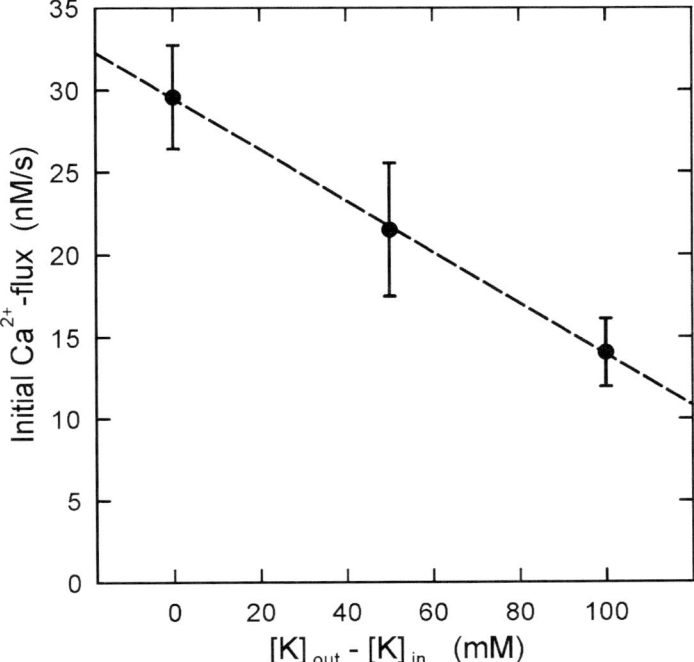

FIGURE 2. Ca^{2+} export is driven by both the Na^+ and the K^+ gradient. If the K^+ concentration is higher outside than inside the vesicles, the Na^+/Ca^{2+} exchange rate declines, because K^+ is cotransported with Ca^{2+} against the K^+ gradient. The NCKX has been purified by DEAE cellulose followed by AF Red Fractogel (Merck) chromatography.[30] The purified NCKX was reconstituted in soybean phosphatidyl choline vesicles in 50 mM KCl, 10 mM $CaCl_2$, and 10 mM Hepes/Tris, pH 7.4, by dialysis overnight at 4°C. External Ca^{2+} was removed, as described in FIGURE 1, and the K^+ concentration was increased by adding a small volume (1–2% v/v) of 3 M KCl. The initial Ca^{2+} flux upon the addition of 95 mM NaCl was determined as the difference between the initial slope after the addition of NaCl and the slope of the baseline due to passive leakage.

NCKX, which run at 490 kDa (FIG. 3A, lane c). Due to the paucity of cysteines in the NCKX, this cross-link is likely to be specific, indicating that the NCKX is dimerized in the ROS plasma membrane.[10] However, the NCKX dimer does not survive detergent solubilization. If the ROS membranes are solubilized in CHAPS, thiol-specific cross-linking yields no NCKX dimer.[10] Moreover, when the solubilized photoreceptor proteins are reconstituted into liposomes, only very little NCKX dimer can be detected by cross-linking.[10] The latter findings suggest that the efficient cross-linking of the NCKX to a dimer in the ROS membrane does not reflect an inherent propensity of the NCKX to dimerize; rather, it is mediated by the interaction of the NCKX with another protein.

In fact, thiol-specific cross-linking of ROS membranes in the presence of cGMP discloses additional cross-links of the NCKX (FIG. 3A, lane d).[11] If the CNG-gated channel is purified after cross-linking in the presence of 8Br-cGMP, these additional

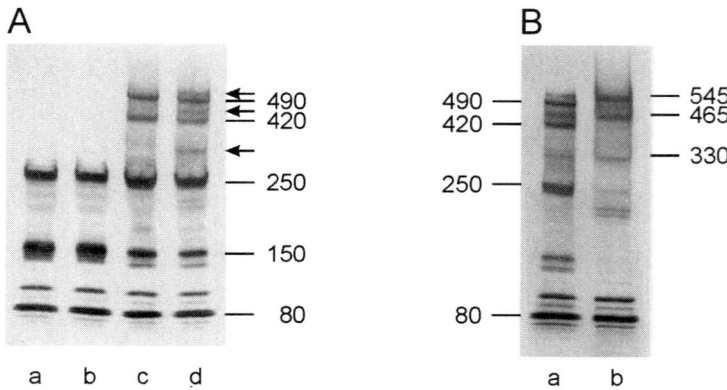

FIGURE 3. Thiol-specific cross-linking of bovine ROS membranes. **(A)** Hypotonically washed ROS membranes before cross-linking in the absence (*lane a*) and the presence of 50 µM 8Br-cGMP (*lane b*) and after cross-linking with BMDB in the absence (*lane c*) and the presence (*lane d*) of 50 µM 8Br-cGMP. The SDS-PAGE was blotted and labeled with the monoclonal antibody PMe 1B3 against the bovine NCKX. The NCKX monomer band has an apparent molecular mass of 250 kDa. The dimer band appears at 490 kDa. The bands below 250 kDa are due to degradation of the NCKX, which is commonly observed; however, the intensity of these bands is deceptive because the electro-transfer of small masses is much more efficient than that of greater molecular masses. The band at 420 kDa is due to a cross-link of the 150 kDa proteolytic fragment with a NCKX monomer. The *arrows* in *lane d* point to additional bands, which appear only upon cross-linking in the presence of 8Br-cGMP. This panel is taken from FIGURE 3 of Schwarzer *et al.*[11] **(B)** Purification of the CNG channel after cross-linking with BMDB in the presence of 50 µM 8Br-cGMP. The strong cross-link bands of the NCKX, which are apparent before purification (*lane a*), are absent after purification of the CNG channel; however, the additional bands are significantly enriched, indicating that the NCKX is cross-linked to the channel. This panel is taken from Figure 6 of Schwarzer *et al.*[11] Reproduced with permission from the American Society for Biochemistry and Molecular Biology.

cross-link products are copurified (FIG. 3B, lane b). This important finding indicates that the NCKX is associated with the CNG-channel, because the NCKX *per se* does not bind cGMP. The association of CNG channel and NCKX was initially inferred from Ca^{2+}-flux experiments.[12] In addition, we and others observed a significant affinity between the solubilized channel and NCKX molecules.[11,13] More specifically, we demonstrated that the NCKX binds directly to the α-subunit, but not the β-subunit, of the CNG channel.[11]

Together, these findings indicate that a complex of two NCKX molecules associated with one CNG channel constitutes a functional unit in photoreceptors. Generally, the channel is assumed to comprise 2 α-subunits and 2 β-subunits, but the subunit arrangement ($α_2β_2$ versus αβαβ) is still a matter of debate.[14,15] Our cross-linking data yield a distinct α-dimer, suggesting an $α_2β_2$ arrangement.[11] However, it should be stressed that the β-subunit by itself does not form a channel.[16] Therefore, whether this subunit belongs, indeed, to the pore-forming subunits of the CNG channel or simply constitutes an accessory regulatory protein of the channel remains unresolved. It should be noted that upon solubilization in 20 mM CHAPS, the β-subunit can readily be separated from the α-subunit at 750 mM NaCl by AF Red Fractogel (Merck) chromatography (P.J. Bauer, unpublished data).

Recently, Poetsch et al.[17] reported strong evidence that the β-subunit of the channel binds to peripherin-2, a disk rim protein that is dimerized. Peripherin-2 dimers either form higher homo-oligomers or associate with Rom-1 dimers, another disk membrane protein that is required for disk morphogenesis.[18] Thus, the channel-NCKX complex interacts directly with a protein complex of the disk rim.

SELF-INHIBITION OF THE DIMERIZED NCKX

The interaction of NCKX and CNG channel suggests that these proteins influence each other functionally. Unfortunately, this idea is hard to test experimentally. We asked whether modification of this interaction by thiol-specific cross-linking with reagents of different size affects the transport rate of the NCKX. This is, indeed, the case: surprisingly, thiol-specific cross-linking enhances the transport rate. Moreover, this enhancement of the transport rate depends on the size of the reagent.[19] This finding suggests that the contact site of the NCKX dimer is inhibitory.

An increase in the transport rate of the NCKX by a factor of almost 2 has also been observed on partial proteolysis of ROS membranes.[20,21] We examined whether cleavage of the inhibitory site is responsible for this phenomenon. We solubilized and reconstituted the ROS membrane proteins into liposomes, because this procedure restores only very little dimers of the NCKX (see above), thus precluding most of the inhibitory interaction between dimers. We observed no statistically significant difference in the transport rate between proteolytically treated and nontreated proteoliposomes. This finding is in agreement with the view that in the ROS membrane there is an inhibitory interaction between the two molecules of the NCKX dimer.[19]

Together, these experiments suggest a possible regulatory mechanism of the NCKX transport rate based on this intermolecular interaction. It is tempting to speculate that activation of the CNG channel reduces the inhibitory interaction between the two molecules of the NCKX dimer, thus resulting in an enhanced Ca^{2+} export rate. Although the increase in the transport rate by relief of the inhibitory contact is at most 2-fold, such an effect would be significant for the local Ca^{2+} concentration at the NCXK channel complex (see below).

SIGNIFICANCE OF A CHANNEL-EXCHANGER OMPLEX FOR LOCAL Ca^{2+}-SIGNALING

In the dark, a small depolarizing current of Na^+ and Ca^{2+} ions enters the outer segment of the photoreceptor. Most of the so-called "dark-current" is carried by Na^+; only 10–15% is carried by Ca^{2+} ions.[22] Upon illumination, the visual transduction cascade shuts down this ion current by hydrolysis of cGMP. We asked whether the association of the NCKX with the CNG channel results in restriction of Ca^{2+} diffusion to the close vicinity of the channel-NCKX complex.

Two conditions must be met to confine the Ca^{2+} diffusion locally to the channel-NCKX complex: (1) Ca^{2+} influx must equal Ca^{2+} efflux to establish a steady state; and (2) Ca^{2+} extrusion must be efficient enough not to be defeated by Ca^{2+} diffusion.

To examine the former condition, we have to recall that physiologically the CNG channels are only slightly activated (~1%).[23] Moreover, the cation influx is so strongly blocked by Ca^{2+} binding to glutamates in the pore region of the GMP-gated channel[24] that single channel recordings are only feasible in the absence of Ca^{2+}.[25] Based on published data, we estimated that under physiologic conditions the Ca^{2+} influx is only about 15 Ca^{2+}/second per CNG channel.[19] Because each NCKX molecule transports maximally at least 60 Ca^{2+}/second,[19] the NCKX molecules associated with the channel are readily capable of keeping abreast with a Ca^{2+} influx of 15 Ca^{2+}/second.

Can local Ca^{2+} extrusion compete with Ca^{2+} diffusion? To answer this question, the Ca^{2+} concentration due to Ca^{2+} influx through a single CNG channel will be compared with the Ca^{2+} concentration due to Ca^{2+} influx and simultaneous Ca^{2+} extrusion in a channel-NCKX complex. The CNG channel will be treated as a point-source and the NCKX as a point-sink for Ca^{2+}. For simplicity, the model has only 1 Ca^{2+} sink, although there are two NCKX molecules bound per CNG channel;[11] the 2 NCKX molecules will be considered as doubling the Ca^{2+} export rate. With these assumptions, a steady Ca^{2+} influx Φ_{in} generates near the channel at some intracellular point, P (FIG. 4), the Ca^{2+} concentration (in moles/l):

$$[Ca^{2+}]_{ch} = \frac{\Phi_{in}}{2\pi D_{Ca} N R_{ch}} \tag{1}$$

where D_{Ca} denotes the local Ca diffusion coefficient, N the Avogadro constant, and R_{ch} the distance between P and the channel pore.[26] If at some distance, d, Ca^{2+} is extruded with the rate Φ_{out}, then the resulting Ca^{2+} concentration results by superposition:

$$[Ca^{2+}]_{ch-ex} = [Ca^{2+}]_{ch} - \frac{\Phi_{out}}{2\pi D_{Ca} N R_{ex}} \tag{2}$$

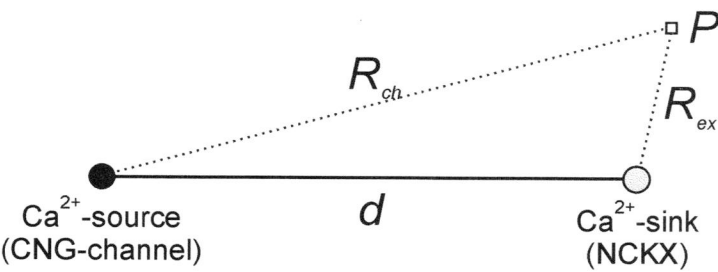

FIGURE 4. Schematic drawing of a channel-NCKX complex to introduce the geometric parameters used to calculate the local Ca^{2+} concentration.

where R_{ex} denotes the distance between the exchanger and point P (FIG. 4). The export rate Φ_{out} is given by the first-order Michaelis relation:[27]

$$\Phi_{out} = V_{max} \cdot \frac{[Ca^{2+}]_d}{[Ca^{2+}]_d + K_{Ca}} \qquad (3)$$

where V_{max} and K_{Ca} denote the maximal export rate and the Ca^{2+} dissociation constant, respectively, and $[Ca^{2+}]_d$ the Ca^{2+} concentration at the Ca^{2+}-binding site of the NCKX. This concentration is assumed to be due to the Ca^{2+} concentration gradient generated by the channel, that is, according to *Equation 1*:

$$[Ca^{2+}]_d = \frac{\Phi_{in}}{2\pi D_{Ca} N d} \qquad (4)$$

To estimate the fraction of local Ca^{2+} current, we calculate the cytoplasmic total Ca^{2+} content of a hemisphere centered at a single channel and centered at a channel-exchanger complex. The fraction of Ca^{2+} that does not leave the close vicinity of the channel-exchanger complex is given as:

$$\text{fraction} = \frac{\left(\int_{\text{hemisphere}} [Ca^{2+}]_{ch}\,d\tau - \int_{\text{hemisphere}} [Ca^{2+}]_{ch-ex}\,d\tau\right)}{\int_{\text{hemisphere}} [Ca^{2+}]_{ch}\,d\tau} \qquad (5)$$

To calculate this fraction, the cytoplasmic Ca^{2+} binding affinity K_{Ca} of the NCKX was taken to be 1.6 μM.[27]

The most critical parameter is the local Ca^{2+} diffusion constant D_{Ca} which is unknown. The fraction of locally extruded Ca^{2+} calculated with *Equation 5* for different assumed Ca^{2+} diffusion constants D_{Ca} is plotted in FIGURE 5. The local Ca^{2+} current decreases, as expected, with increasing Ca^{2+} diffusion. For free cytoplasmic Ca^{2+} diffusion, that is, for $D_{Ca} = 220$ μm^2/second,[28] only about 2% of the Ca^{2+} influx contributes to this local Ca^{2+} current. However, this condition is attained only at high cytoplasmic Ca^{2+} concentrations of about 1 mM.[28] Under physiologic conditions, the cytoplasmic Ca^{2+} diffusion constant is reduced to about 15 μm^2/second by Ca^{2+} buffering.[28] Then, *Equation 5* yields a fraction of about 30% of local Ca^{2+} current, assuming a distance of 5 nm between the channel pore and the NCKX. Moreover, both CNG channel and NCKX of bovine and human rod photoreceptors contain huge acidic protein domains: the β-subunit of the channel contains a so-called GARP (glutamic acid-rich protein) domain, and the NCKX has a highly acidic intracellular loop. These acidic domains are likely to reduce the local Ca^{2+} diffusion further. Finally, it should be noted that the transport rate of the NCKX was determined when the channel was closed. If, as hypothesized above, activation of the channel results in enhancement of the NCKX export rate by a factor of about 2, then such a mechanism would lead to a further increase in the fraction of local Ca^{2+} current (FIG. 5, dashed lines).

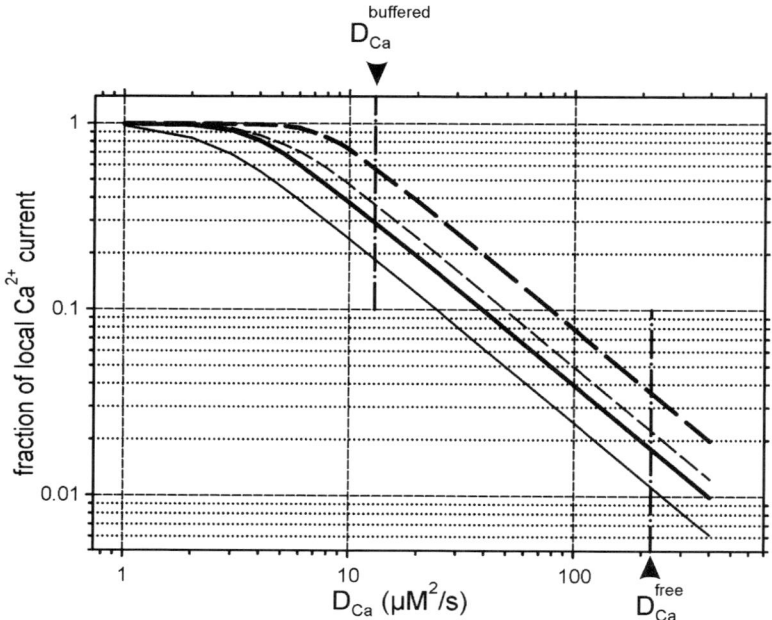

FIGURE 5. Fraction of local Ca^{2+} current, that is, the fraction of Ca^{2+} flux though the CNG channel, which is extruded by the associated NCKX molecules. This fraction is calculated by *Equation 5* for different Ca^{2+} diffusion constants. The parameters entering the calculation are: $K_{Ca} = 1.6\ \mu M$, $\Phi_{in} = 15\ Ca^{2+}/s$, and $V_{max} = 120\ Ca^{2+}/s$, $d = 5$ nm (*thick solid line*), $V_{max} = 240\ Ca^{2+}/s$, $d = 5$ nm (*thick dashed line*), $V_{max} = 120\ Ca^{2+}/s$, $d = 8$ nm (*thin solid line*), and $V_{max} = 240\ Ca^{2+}/s$, $d = 8$ nm (*thin dashed line*). The *dash-dotted vertical lines* indicate the Ca^{2+} diffusion constant, $D_{Cd}^{free} = 200\ \mu m^2/s$, for free cytoplasmic Ca^{2+} diffusion (i.e., at millimolar Ca^{2+} concentrations) and the Ca^{2+} diffusion constant, $D_{Ca}^{buff} = 13\ \mu m^2/s$ for buffered Ca^{2+} diffusion, which applies to Ca^{2+} diffusion at micro- and submicromolar Ca^{2+} concentrations, that is, the normal physiological situation.[28]

Together, I infer that a sizeable fraction of the Ca^{2+}-flux entering the outer segment of the photoreceptor through the CNG channel is directly extruded by the associated NCKX molecules without diffusing away from the channel-NCKX complex.

OUTLOOK

Although the association of channel and NCKX has only been demonstrated for bovine rod photoreceptor, it is probably a general feature of all rod photoreceptors. Moreover, a similar protein complex may also exist in cone photoreceptors. Chemical cross-linking is an appropriate technique to examine these problems. Physiologically, the coupling of Ca^{2+} source and Ca^{2+} sink to a functional unit is a novel mechanism for creating a spatially confined Ca^{2+} signal.

ACKNOWLEDGMENT

I thank Dr. Paul P.M. Schnetkamp for carefully reading the manuscript, Heike Schauf for excellent technical assistance, and Dr. R.S. Molday for generously providing the monoclonal antibody, PMe 1B3, against the bovine Na^+/Ca^{2+}-K^+ exchanger. This work was supported by a Grant Ba 721/1-3 from the Deutsche Forschungsgemeinschaft.

REFERENCES

1. REILÄNDER, H. et al. 1992. Primary structure and functional expression of the Na/Ca,K-exchanger from bovine rod photoreceptors. EMBO J. **11:** 1689–1695.
2. SCHNETKAMP, P.P.M., D.K. BASU & R.T. SZERENCSEI. 1989. Na^+-Ca^{2+} exchange in bovine rod outer segments requires and transports K^+. Am. J. Physiol. **257:** C153–C157.
3. CERVETTO L., L. LAGNADO, R.J. PERRY, et al. 1989. Extrusion of calcium from rod outer segments is driven by both sodium and potassium gradients. Nature (Lond.) **337:** 740–743.
4. RATTO, G.M., R. PAYNE, W.G. OWEN & R.W. TSIEN. 1988. The concentration of cytosolic free calcium in vertebrate rod outer segments measured with Fura-2. J. Neurosci. **8:** 3240–3246.
5. GRAY-KELLER, M.P. & P.B. DETWILER. 1994. The calcium feedback signal in the phototransduction cascade of vertebrate rods. Neuron **13:** 849–861.
6. MCCARTHY, S.T., J.P. YOUNGER & W.G. OWEN, 1994. Free calcium concentrations on bullfrog rods determined in the presence of multiple forms of Fura-2. Biophys. J. **67:** 2076–2089.
7. Tsoi, M. et al. 1998. Molecular cloning of a novel potassium-dependent sodium-calcium exchanger from rat brain. J. Biol. Chem. **273:** 4155–4162.
8. Szerencsei, R.T. et al. 2000. Minimal domain requirement for cation transport by the potassium-dependent Na/Ca-K exchanger. J. Biol. Chem. **275:** 669–676.
9. Prinsen, C.F.M., R.T. Szerencsei & P.P.M. Schnetkamp. 2000. Molecular cloning and functional expression of the potassium-dependent sodium-calcium exchanger from human and chicken retinal cone photoreceptors. J. Neurosci. **20:** 1424–1434 .
10. SCHWARZER, A., T.S.Y. KIM, V. HAGEN, et al. 1997. The Na/Ca-K exchanger of rod photoreceptor exists as dimer in the plasma membrane. Biochemistry **36:** 13667–13676.
11. SCHWARZER, A., H. SCHAUF & P.J. BAUER. 2000. Binding of the cGMP-gated channel to the Na/Ca-K exchanger in rod photoreceptors. J. Biol. Chem. **275:** 13448–13454.
12. BAUER, P.J. & M. DRECHSLER. 1992. Association of cyclic GMP-gated channels and Na^+-Ca^{2+}-K^+ exchangers in bovine retinal rod outer segment plasma membranes. J. Physiol. (Lond.). **451:** 109–131.
13. MOLDAY, R.S. & L.L. MOLDAY. 1998. Molecular properties of the cGMP-gated channel of rod photoreceptors. Vision Res. **38:** 1315–1323.
14. SHAMMAT, I.M. & S.E. GORDON. 1999. Stoichiometry and arrangement of subunits in rod cyclic nucleotide-gated channels. Neuron **23:** 809–819.
15. HE, Y., M.L. RUIZ & J.W. KARPEN. 2000. Constraining the subunit order of rod cyclic nucleotide-gated channels reveals a diagonal arrangement of like subunits. Proc. Natl. Acad. Sci. USA **97:** 895–900.
16. CHEN, T.-Y. et al. 1993. A new subunit of the cyclic nucleotide-gated cation channel in retinal rods. Nature (Lond.) **362:** 764–767.
17. POETSCH, A., L.L. MOLDAY & R.S. MOLDAY. 2001. The cGMP-gated channel and related glutamic acid rich proteins interact with peripherin-2 at the rim region of rod photoreceptor disc membranes. J. Biol. Chem. **276:** 48009–48016.
18. CLARKE, G. et al. 2000. Rom-1 is required for rod photoreceptor viability and the regulation of disk morphogenesis. Nat. Genet. **25:** 67–73.

19. BAUER, P.J. & H. SCHAUF. 2002. Mutual inhibition of the dimerized Na/Ca-K exchanger in rod photoreceptors. Biochim. Biophys. Acta **1559:** 121–134.
20. HUPPERTZ, B. & P.J. BAUER. 1994. Na^+-Ca^{2+},K^+ exchange in bovine retinal rod outer segments: quantitative characterization of normal and reversed mode. Biochim. Biophys. Acta **1189:** 119–126.
21. KIM, T.S.Y., D.M. REID & R.S. MOLDAY. 1998. Structure-function relationships and localization of the Na/Ca-K exchanger in rod photoreceptors. J. Biol. Chem. **273:** 16561–16567.
22. NAKATANI, K. & K.-W. YAU. 1988. Calcium and magnesium fluxes across the plasma membrane of the toad rod outer segment. J. Physiol. (Lond.) **395:** 695–729.
23. NAKATANI, K. & K.-W. YAU. 1988. Guanosine 3′,5q′-cyclic monophosphate-activated conductance studied in a truncated rod outer segment of the toad. J. Physiol. (Lond.) **395:** 731–753.
24. EISMANN, E., F. MÜLLER, S.H. HEINEMANN & U.B. KAUPP. 1994. A single negative charge within the pore region of a cGMP-gated channel controls rectification, Ca^{2+} blockage, and ionic selectivity. Proc. Natl. Acad. Sci. USA **91:** 1109–1113.
25. HAYNES, L.W., A.R. KAY & K.-W. YAU, 1986. Single cyclic GMP-activated channel activity in excised patches of rod outer segment membrane. Nature (Lond.) **321:** 66–70.
26. BAUER, P.J. 2001. The local Ca concentration profile in the vicinity of a Ca channel. Cell Biochem. Biophys. **35:** 49–61.
27. LAGNADO, L., L. CERVETTO & P.A. MCNAUGHTON. 1992. Calcium homeostasis in the outer segments of retinal rods from the tiger salamander. J. Physiol. (Lond.) **455:** 111–142.
28. ALLBRITTON, N.L., T. MEYER & L. STRYER. 1992. Range of messenger action of calcium ion and inositol 1,4,5-triphosphate. Science **258:** 1812–1815.
29. BAUER, P.J. 1996. Cyclic GMP-gated channels of bovine rod photoreceptors: affinity, density and stoichiometry of Ca^{2+}-calmodulin binding sites. J. Physiol. (Lond.) **494:** 675–685.
30. COOK, N.J. & U.B. KAUPP. 1988. Solubilization, purification, and reconstitution of the sodium-calcium exchanger from bovine retinal rod outer segments. J. Biol. Chem. **263:** 11382–11388.

Exercise Training Enhances Coronary Smooth Muscle Cell Sodium–Calcium Exchange Activity in Diabetic Dyslipidemic Yucatan Swine

E.A. MOKELKE, M. WANG, AND M. STUREK

Departments of Physiology, Internal Medicine, and Diabetes and Cardiovascular Biology Program, University of Missouri, Columbia, Missouri 65212, USA

KEYWORDS: exercise; Ca^{2+}; diabetes; dyslipidemia; fura-2; swine

Calcium (Ca^{2+}) regulation in vascular smooth muscle cells is tightly controlled at the membrane level by channels, pumps, and exchangers. Any perturbation in the function or number of these Ca^{2+} regulatory mechanisms could potentially alter the balance between influx and efflux, ultimately resulting in overload of Ca^{2+} stores or elevated cytosolic or nuclear Ca^{2+}. Both diabetes[1,2] and exercise[3] independently have been shown to alter components of vascular smooth muscle Ca^{2+} regulation. We tested the hypothesis that diabetic dyslipidemia impairs vascular smooth muscle Ca^{2+} regulatory proteins, specifically the Na^+/Ca^{2+} exchanger (NCX), in the Yucatan miniature swine model. We further hypothesized that a program of endurance exercise training would reverse or prevent this diabetes-induced impairment.

Male Yucatan swine were randomly assigned to 4 groups: control (C), fat (F), diabetic high fat (DF), or treadmill-trained DF (DFX). F, DF, and DFX animals were fed an atherogenic, hypercaloric diet (2% of calories from cholesterol), which increased the %kcal provided from fat from 8–46% and stimulated weight gain 3-fold above C animals. Chronically elevated blood glucose levels (300–400 mg/dL) were maintained in DF and DFX for 20 weeks. DFX were exercised 4 days/week, 30 minutes/day at 65–75% HR_{max} for 16 weeks. Smooth muscle cells were dispersed from the right coronary artery within 24 hours after sacrifice, and myoplasmic $[Ca^{2+}]$ (Ca_m) was measured by ratiometric (340/380) fura-2 digital imaging.[1] The protocol was designed to examine the contribution of the NCX, sarco/endoplasmic reticulum CaATPase (SERCA), and the plasmalemmal CaATPase (PMCA) in buffering depolarization-induced Ca^{2+} entry and subsequent changes in accumulated Ca_m. Cells were depolarized with 80 mM KCl (80K) to induce Ca^{2+} influx and increase Ca_m. Changes in Ca_m were determined using the area under the curve obtained in response

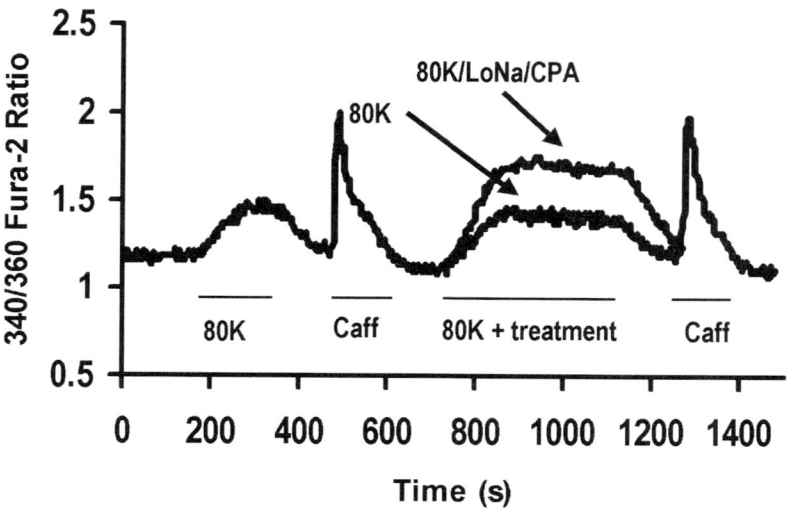

FIGURE 1. Representative fura-2 ratio values were obtained from freshly dispersed smooth muscle cells using two different protocols. *Horizontal bars* under traces indicate change in the superfusate from a physiological salt solution. For clarity, since the data of interest were obtained during the various pharmacological treatments ("80K + treatment"), two separate data traces are depicted for this portion only. As expected, superfusing cells in an 80K, low-Na (5 mM) solution containing CPA (10 µM) markedly increased the fura-2 ratio transient, compared to application of 80K alone.

to 340 and 380 nm excitation. This 80K protocol was repeated in a low (5 mM) NaCl condition, in a solution containing cyclopiazonic acid (CPA, 10 µM), and in a solution containing CPA and low Na. Caffeine-containing solution (5 mM, Caff) was applied before and after these 80K treatments to transiently deplete the sarcoplasmic reticulum of Ca^{2+} stores. A typical trace of the response to the 80K protocol with no inhibitors superimposed with a trace in response to the low Na/CPA condition is shown in FIGURE 1. In 80K alone, the baseline-corrected maximal Ca_m response (ΔCa_m) was lower in DFX versus C and F ($P < 0.05$), suggesting that exercise training resulted in either a reduction in Ca^{2+} influx or an increase in Ca^{2+} efflux during 80K-induced depolarization. Our laboratory has shown that L-type Ca^{2+} current is returned to control levels in vascular smooth muscle cells of DFX animals;[4] therefore, the reduction in Ca_m is more likely due to increased removal of Ca^{2+}. With SERCA inhibited, ΔCa_m was elevated in F versus DF ($P < 0.05$) animals, but no other differences were observed. This suggests that the activity of SERCA is depressed in DF animals, so that inhibition of SERCA does not result in a marked increase in ΔCa_m as seen in the other groups. With only NCX inhibited, ΔCa_m was greater in F versus C and DF animals; DFX was greater than DF ($P < 0.05$). These data suggest that an increase in NCX activity may be secondary to hyperlipidemia and exercise training and that diabetic dyslipidemia (DF) prevented the compensatory increase in NCX activity or protein expression. With both NCX and SERCA inhibited, ΔCa_m

increased in all groups as expected, but ΔCa_m was significantly (P <0.05) greater only in F versus DF; C, F, and DFX were not different from each other. One explanation for these data is that because diabetic dyslipidemia impairs both NCX and SERCA function, when both are inhibited, a blunted increase in ΔCa_m occurs. Finally, an increase occurred in the maximal rate of Ca^{2+} accumulation ($Ca_{m\ slope}$) of 41 and 57% (P <0.05) in the F and DFX than in the 80K alone, further suggesting that NCX activity is elevated in hyperlipidemia and exercise.

These data suggest that in diabetic dyslipidemia, both SERCA and NCX activity is blunted, but NCX activity is elevated in response to hyperlipidemia and endurance exercise training. The enhanced NCX activity could increase the ability of cells to buffer Ca^{2+}, possibly in an attempt to prevent cellular Ca^{2+} overload.

ACKNOWLEDGMENTS

This work was supported by Grants RR13223, HL62552, and HL10474-01 from the National Institutes of Health.

REFERENCES

1. HILL, B.J.F., J.L. DIXON & M. STUREK. 2001. Effect of atorvastatin on intracellular calcium uptake in coronary smooth muscle cells from diabetic pigs fed an atherogenic diet. Atherosclerosis **159:** 117–124.
2. WAMHOFF, B.R., J.L. DIXON & M. STUREK. 2002. Atorvastatin treatment prevents alterations in coronary smooth muscle nuclear Ca^{2+} signaling associated with diabetic dyslipidemia. J. Vasc. Res. **39:** 208–220.
3. HEAPS, C.L., D.K. BOWLES, M. STUREK, et al. 2000. Enhanced L-type Ca^{2+} channel current density in coronary smooth muscle of exercise-trained swine is compensated to limit myoplasmic free Ca^{2+} accumulation. J. Physiol.(Lond.) **528:** 435–445.
4. WAMHOFF, B.R., J.L. DIXON & M. STUREK. 2001. Exercise training prevents altered coronary smooth muscle L-type calcium channel function in diabetic dyslipidemia. Circulation **104:** II-157–I-157.

Dysregulation of $[Ca^{2+}]_i$ in OK-PTH Cells Expressing a Mesangial Cell Na^+/Ca^{2+} Exchanger Isoform from Dahl/Rapp Salt-Sensitive Rats

T. UNLAP, E. HWANG, G. KOVACS, J. PETI-PETERDI, B. SIROKY, I. WILLIAMS, AND P.D. BELL

Departments of Medicine and Physiology, University of Alabama at Birmingham, Birmingham, Alabama 35217, USA

KEYWORDS: dysregulation; $[Ca^{2+}]_i$; OK-PTH cells; mesangial cells; Na^+/Ca^{2+} exchanger isoform; Dahl/Rapp rats; salt sensitivity

INTRODUCTION

Our previous studies have shown that the Na^+/Ca^{2+} exchanger is differentially regulated in renal tissues of salt-sensitive and salt-resistant Dahl/Rapp rats.[1–3] In studying the regulation of the Na^+/Ca^{2+} exchanger (NCX) in renal contractile cells, we found that PKC upregulated the Na^+/Ca^{2+} exchanger in afferent arterioles and mesangial cells of salt-resistant but not salt-sensitive Dahl/Rapp rats.[1–3] We also showed that the upregulation of the Na^+/Ca^{2+} exchanger in salt-resistant mesangial cells occurred, in part, through translocation of the exchanger from the cytosol to the plasma membrane of mesangial cells of salt-resistant but not salt-sensitive Dahl/Rapp rats.[3] To assess the basis of the difference in regulation of the Na^+/Ca^{2+} exchanger in these two strains of rats, we cloned and sequenced the exchanger from primary cell culture of salt-sensitive and salt-resistant Dahl/Rapp rats using reverse transcription-polymerase chain reaction (RT-PCR).[2]

RESULTS AND DISCUSSION

The exchanger isoform from salt-sensitive (SNCX) and salt-resistant (RNCX) rats showed differences at the amino acid level. RNCX and SNCX differ at amino acid 218 where, through an A to T transversion, RNCX contains isoleucine, whereas SNCX contains phenylalanine.[2] The two isoforms also differ at the alternative splice site, where RNCX is encoded by exons B and D, whereas SNCX is encoded by exons

---

Address for correspondence: Dr. T. Unlap, Department of Medicine, University of Alabama at Birmingham, 7th Avenue South SC 865, Birmingham, AL 35217. Voice: 205-975-5761; fax: 205-934-1147.
unlap@uab.edu

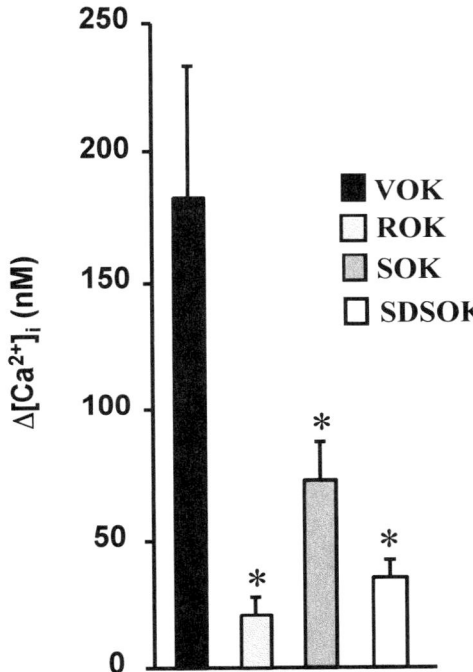

FIGURE 1. The abilities of ROK, SOK, and SDSOK to buffer ATP-induced $[Ca^{2+}]_i$ increases were assessed in VOK, ROK, SOK, and SDSOK after treatment with 100 µM ATP to elevate $[Ca^{2+}]_i$. *Bar graph* shows $\Delta[Ca^{2+}]_i$ in VOK, ROK, SOK, and SDSOK. ROK, SOK, and SDSOK had lower $\Delta[Ca^{2+}]_i$ or greater capacity to buffer $[Ca^{2+}]_i$ compared to VOK, with ROK and SDSOK having greater $[Ca^{2+}]_i$ buffering abilities. Data were analyzed for statistical significance using ANOVA. Means ± SEM ($n = 9$ for VOK and SDSOK; $n = 12$ for SOK and ROK). *P <0.05 compared with VOK. [See text for abbreviations.]

B, D, and F. Because the salt-sensitive and salt-resistant rats were derived from Sprague-Dawley rats, the Na^+/Ca^{2+} exchanger was also cloned and sequenced from mesangial cells of Sprague-Dawley rats. Sprague-Dawley mesangial cells express an isoform (SDRNCX) that is identical to the RNCX that is expressed in salt-resistant Dahl/Rapp rats. Mesangial cells of Sprague-Dawley rats also express an isoform (SDSNCX) that is identical to the SNCX that is expressed in mesangial cells of salt-sensitive Dahl/Rapp rats at every amino acid residue except at amino acid 218, where it is isoleucine in SDSNCX and not phenylalanine. To assess the abilities of these NCX isoforms to regulate $[Ca^{2+}]_i$, RNCX, SNCX, SDSNCX, or the vector alone was stably transfected into OK-PTH cells. Vector-transfected OK-PTH cells (VOK) and OK-PTH cells expressing RNCX (ROK), SNCX (SOK), or SDSNCX (SDSOK) were loaded with fura-2 to measure ATP (100 µM or 1 mM)-induced $[Ca^{2+}]_i$ increase. In response to low-level (100 µM) ATP, $[Ca^{2+}]_i$ was 185 ± 55 nM ($n = 9$) in VOK and was significantly lower in cells expressing the exchanger isoforms (21 ± 9 nM ($n = 12$) in ROK, 69 ± 18 nM ($n = 12$) in SOK, and 41 ± 7 nM ($n = 9$) in SDSOK cells (FIG. 1). NCX activity was assessed by examining the ability of these 3 NCX isoforms to reduce (1 mM) ATP-induced $[Ca^{2+}]_i$ increases back to baseline levels. ROK and SDSOK cells exhibited comparable Na^+/Ca^{2+} exchange activities that were significantly higher (over 200%) than those for SOK cells (FIG. 2a and b). Protein kinase C downregulation significantly attenuated the rates of return of $[Ca^{2+}]_i$ after ATP treatment in ROK and SDSOK cells, but it had no effect in SOK cells (FIG. 3).

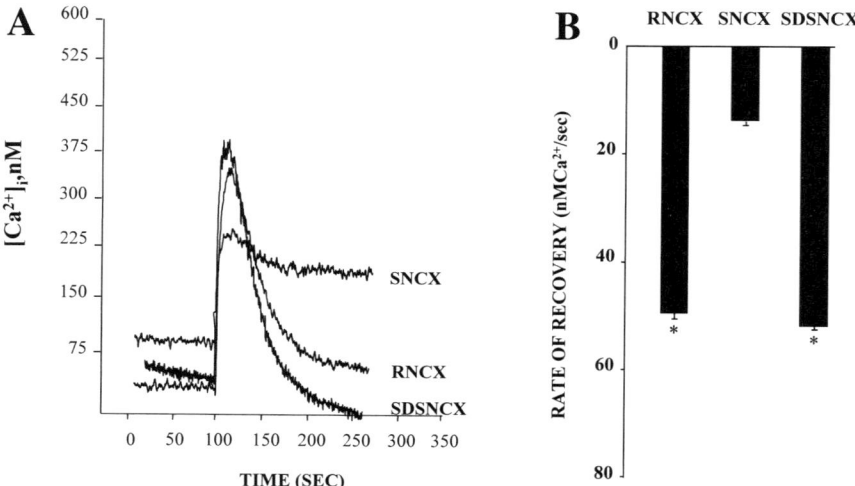

FIGURE 2. The abilities of ROK, SOK, and SDSOK to reduce ATP-induced $[Ca^{2+}]_i$ increases back towards baseline were assessed in fura-2–loaded cells after treatment with 1 mM ATP to elevate $[Ca^{2+}]_i$. Figure shows representative $[Ca^{2+}]_i$ tracings (**a**) and bar graph (**b**) showing the rates at which ROK, SOK, and SDSOK reduce ATP-induced $[Ca^{2+}]_i$ increases back towards baseline levels. The data were analyzed for statistical significance using ANOVA. Means ± SEM ($n = 12$). *$P < 0.05$ compared with VOK. [See text for abbreviations.]

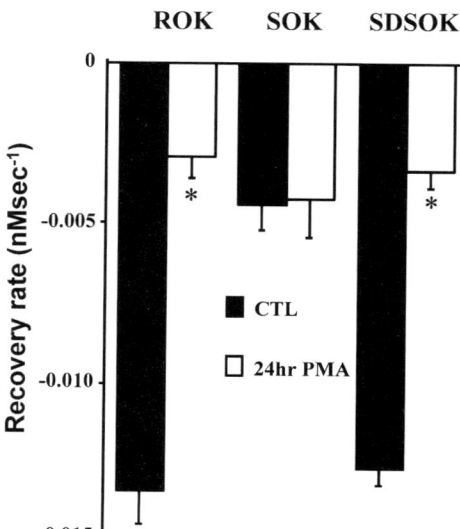

FIGURE 3. Effect of protein kinase C (PKC) downregulation on $[Ca^{2+}]_i$ recovery rate in response to 1 mM ATP in ROK, SOK, and SDSOK was examined in fura 2-loaded cells after ATP treatment with or without 24 hours of PMA (300 nM) pretreatment to downregulate PKC. *Bar graph* shows the effect of PKC downregulation on $[Ca^{2+}]_i$ recovery rate in ROK, SOK, and SDSOK. Data were analyzed for statistical significance using ANOVA. Means ± SEM ($n = 12$). *$P < 0.05$ compared with non-PMA–treated cells. [See text for abbreviations.]

CONCLUSION

The isoleucine to phenylalanine mutation that occurs at amino acid 218 in SNCX may therefore account, in part, for the reduced efficiency of SNCX to regulate $[Ca^{2+}]_i$. This mutation may result in defective regulation of $[Ca^{2+}]_i$, which could lead to increased preglomerular vascular resistance and mesangial cell contraction and therefore contribute to the decreased glomerular filtration rate, renal blood flow, and eventual renal failure, the hallmark of this model of salt-sensitive hypertension.

REFERENCES

1. NELSON, L.D., M.T. UNLAP, J. L. LEWIS & P.D. BELL. 1999. Renal arteriolar Na^+/Ca^{2+} exchange in salt sensitive hypertension. Am. J. Physiol. **276:** F567–573.
2. UNLAP, M.T., J. PETI-PETERDI & P.D. BELL. 2000. Cloning and functional expression of mesangial cell Na^{2+}/Ca^{2+} exchangers from Dahl/Rapp salt-sensitive/resistant rats. Am. J. Physiol. **279:** F177–184.
3. MASHBURN, N.A., M.T. UNLAP, J. RUNQUIST, et al. 1999. Altered protein kinase c activation of Na^+/Ca^{2+} exchange in mesangial cells from salt sensitive rats. Am. J. Physiol. **276:** F574–580.

Enhanced Susceptibility of a Na^+/Ca^{2+} Exchanger Isoform from Mesangial Cells of Salt-Sensitive Dahl/Rapp Rats to Oxidative Stress Inactivation

T. UNLAP, E.H. HWANG, B.J. SIROKY, J. PETI-PETERDI, G. KOVACS, I. WILLIAMS, AND P.D. BELL

Departments of Medicine and Physiology and Biophysics, University of Alabama at Birmingham, Birmingham, Alabama 35217, USA

KEYWORDS: susceptibility; Na^+/Ca^{2+} exchanger isoform; mesangial cells; salt sensitivity; Dahl/Rapp rats; oxidative stress

INTRODUCTION

Using reverse transcription-polymerase chain reaction (RT-PCR), we cloned and sequenced the Na^+/Ca^{2+} exchanger from mesangial cells of salt-sensitive (SNCX) and salt-resistant (RNCX) Dahl/Rapp rats.[1] These two isoforms differ by 1 amino acid residue in the fourth membrane-spanning domain at amino acid 218, where it is isoleucine in RNCX but is phenylalanine in SNCX. Our functional studies using OK-PTH cells stably expressing either RNCX (ROK) or SNCX (SOK) showed that RNCX was upregulated by protein kinase C (PKC), whereas SNCX was not.[1] Our studies also showed that SNCX demonstrated a diminished capacity to regulate agonist-induced increases in intracellular calcium ($[Ca^{2+}]_i$).[2] There is increasing evidence that oxidative stress, characterized by elevated levels of reactive oxygen species and dysregulation of $[Ca^{2+}]_i$, is increased under hypertensive conditions.[3] Evidence also indicates that oxidative stress attenuates the activity of the Na^+/Ca^{2+} exchanger.[4,5] Therefore, we sought to examine the effect of oxidative stress on the activity of RNCX and SNCX in ROK and SOK cells, respectively, using a Na-dependent $^{45}Ca^{2+}$ uptake assay.[1]

RESULTS AND DISCUSSION

Hydrogen peroxide (500 µM, 20 minutes) and peroxynitrite (50 µM, 20 minutes) treatments diminished SNCX activity by 31 and 40%, respectively, whereas the same treatments failed to affect the activity of RNCX (FIG. 1). Increasing the

Address for correspondence: Dr. T. Unlap, Department of Medicine, University of Alabama at Birmingham, 7th Avenue South SC 865, Birmingham, AL 35217. Voice: 205-975-5761; fax: 205-934-1147.
unlap@uab.edu

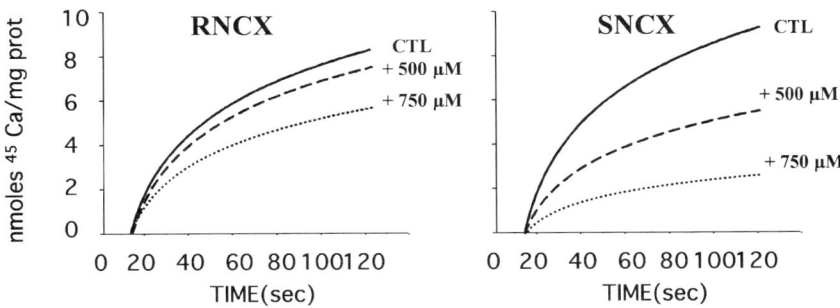

FIGURE 1. Effect of H_2O_2 on exchange activity was assessed in cells expressing either RNCX or SNCX by treating the cells with H_2O_2 (500 or 750 µM) for 20 minutes prior to $^{45}Ca^{2+}$ uptake assay.

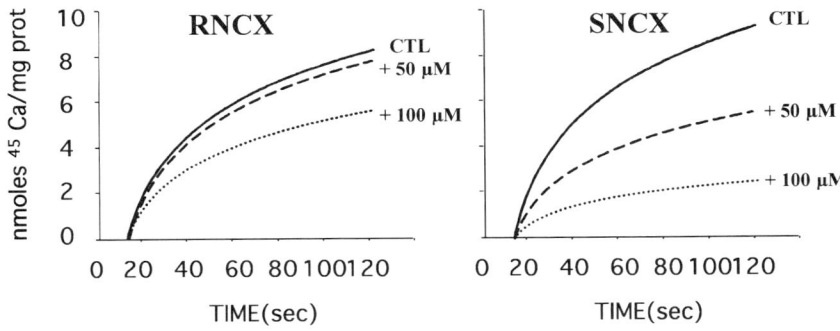

FIGURE 2. Effect of $ONOO^-$ on exchange activity was assessed in cells expressing either RNCX or SNCX by treating the cells with $ONOO^-$ (50 and 100 µM) for 20 minutes prior to $^{45}Ca^{2+}$ uptake assay.

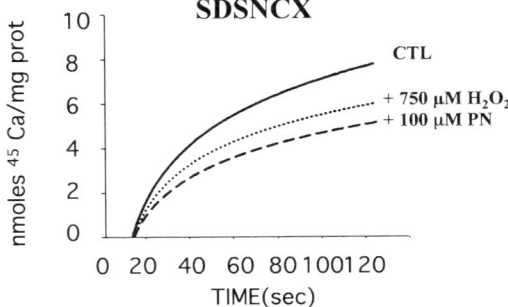

FIGURE 3. Effect of H_2O_2 (750 µM) and $ONOO^-$ (100 µM) on exchange activity was assessed in cells expressing the parental salt-sensitive exchanger isoform (SDSNCX) by treating the cells with H_2O_2 or $ONOO^-$ for 20 minutes prior to $^{45}Ca^{2+}$ uptake assay.

amounts of hydrogen peroxide and peroxynitrite to 750 and 100 µM, respectively, in the treatments nearly abolished SNCX activity while diminishing RNCX activity by 23 and 25%, respectively. Treatment of cells expressing a Na^+/Ca^{2+} exchanger isoform that is expressed in mesangial cells of the parental strain, Sprague-Dawley rats (SDSNCX), showed that the activity of SDSNCX was not affected by treatment with low concentrations of hydrogen peroxide and peroxynitrite, whereas treatment with higher concentrations (750 and 100 µM, 20 minutes) attenuated SDSNCX activity by 19 and 28%, respectively.

CONCLUSION

Because SNCX and SDNCX are identical at every amino acid residue except amino acid 218, where it is isoleucine in SDSNCX but is phenylalanine in SNCX, we conclude that the enhanced susceptibility of SNCX to oxidative stress may be mediated, in part, through the isoleucine to phenylalanine mutation. The susceptibility of SNCX to oxidative stress may lead to dysregulation of $[Ca^{2+}]_i$ in mesangial cells expressing SNCX and may account for the enhanced sensitivity of the salt-sensitive Dahl/Rapp rat to salt-induced hypertension.

REFERENCES

1. UNLAP, M.T., J. PETI-PETERDI & P.D. BELL. 2000. Cloning and functional expression of mesangial cell Na^{2+}/Ca^{2+} exchangers from Dahl/Rapp salt-sensitive/resistant rats. Am. J. Physiol. **279:** F177–184.
2. HWANG, E.F., G. KOVACS, J. PETI-PETERDI, et al. 2001. Impaired ability of the Na^+/Ca^{2+} exchanger from the Dahl/Rapp salt-sensitive rat to regulate cytosolic calcium. Submitted.
3. ABRAHAMSE, S.L., B.L. POOL-ZOBEL & G. RECHKEMMER. 1999. Potential short chain fatty acids to modulate the induction of DNA damage and changes in intracellular calcium concentration by oxidative stress in isolated rat distal colon cells. Carcinogenesis **20:** 629–634.
4. THOMPSON, L.E., G.J. RINALDI & D.F. BOHR. 1990. Decreased activity of the sodium-calcium exchanger in tail artery of stroke-prone spontaneously hypertensive rats. Blood Vessels **27**(2-5): 197-201.
5. RAMIREZ-GIL, J.F., P. TROUVE, N. MOUGENOT, et al. 1998. Modifications of myocardial Na^+,K^+-ATPase isoforms and N^+-Ca^{2+} exchanger in aldosterone/salt-induced hypertension in guinea pigs. Cardiovasc. Res. **38**(2): 451–462.

Reverse Mode Na$^+$/Ca^{2+} Exchange in the Collagen Activation of Human Platelets

DIANE E. ROBERTS AND RATNA BOSE

Department of Pharmacology and Therapeutics, University of Manitoba, Winnipeg, Manitoba R3E 0W3, Canada

KEYWORDS: Na$^+$/Ca^{2+} exchange; collagen activation; human platelets

INTRODUCTION

Calcium is an essential second messenger in the activation of platelets. Elevations in cytosolic calcium influence almost all of the platelet responses to stimulation including shape change, secretion, and thrombus formation.[1] Activation of platelets by collagen has been shown to cause an increase in cytosolic calcium. There are two components to the increase in platelet cytosolic calcium in response to activation: calcium entry across the plasma membrane and calcium release from intracellular stores.[1,2]

The purpose of this study is to quantify the source of the collagen-induced increase in cytosolic calcium in human platelets and to determine what role sodium plays in that increase.

METHODS

Blood was drawn from human volunteers into EDTA-containing vacutainer tubes. Platelets were isolated by centrifugation, resuspended in autologous platelet poor plasma, and incubated for 1 hour at 37°C with calcium-green-AM (20 µM) and fura-red-AM (40 µM) dyes to measure changes in cytosolic calcium ([Ca^{2+}]$_i$). Plasma and extracellular dye was removed by sepharose CL-2B gel filtration column with 1.9% sodium citrate and calcium-free HEPES buffer (140 mM NaCl, 4.9 mM KCl, 1.2 mM MgCl$_2$, 1.4 mM KH$_2$PO$_4$, 11 mM glucose, and 20 mM HEPES, pH 7.4). The platelet number was counted and adjusted so that each experiment was run with 2×10^7 platelets in 500 µL. All assays were preformed at 37°C. The Na$^+$/Ca^{2+} exchanger (NCX) blockers employed in this study were 5-(4-chlorobenzyl)-2',4'-dimethylbenzamil (CBDMB, obtained from E.J. Cragoe, Nacogdoches, TX) or 2-[2{4-4-nitrobenzyloxy)phenyl}ethyl]isothiourea (KB-R7943, purchased from

---

Address for correspondence: Diane E. Roberts, Department of Pharmacology and Therapeutics, University of Manitoba, A311, 753 Mc Dermot Avenue, Winnipeg, Manitoba R3E 0W3, Canada. Voice: 204-789-3562; fax: 204-789-3932.
diaroberts@hotmail.com

Tocris Cookson Ltd.). Platelets were activated with collagen (10 μg/ml; purchased from Nycomed Arzneimittel). The effects of sodium on the collagen-induced change in $[Ca^{2+}]_i$ were determined by resuspending platelet samples in calcium-free HEPES buffer in which sodium (Na^+) was substituted with either N-methyl glucamine, 140 mM potassium, or 280 mM sucrose. Fluorescence and aggregation were simultaneously measured in a Jasco 110 ion analyzer. The excitation wavelength for calcium measurement was 500 nm and the emission wavelengths were 540 m and 660 nm. Cytosolic calcium was calculated according to previously published formulas.[3,6]

RESULTS AND DISCUSSION

Effects of Extracellular Calcium on the Collagen-Induced Change in Cytosolic Calcium

The purpose of this experiment was to quantify the contribution of intracellular calcium release in response to stimulation with collagen. To achieve this platelet, samples were stimulated with collagen in the presence of 1 mM Ca^{2+} and in the absence of extracellular Ca^{2+}. Calcium-free medium was produced by chelation of extracellular Ca^{2+} with the addition of EGTA (5 mM) to the sample 1 minute prior to stimulation with collagen. FIGURE 1a demonstrates the typical response of a platelet sample in 1 mM Ca^{2+} and in calcium-free medium. There is a significant reduction in the collagen-induced change in $[Ca^{2+}]_i$ in the EGTA-treated sample. As shown in

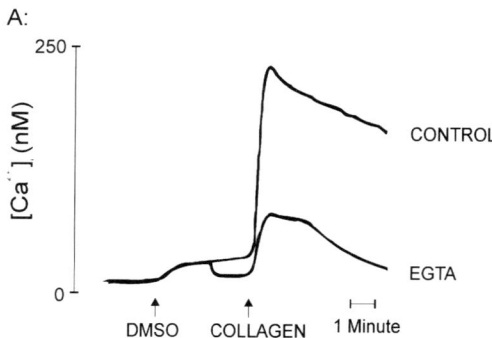

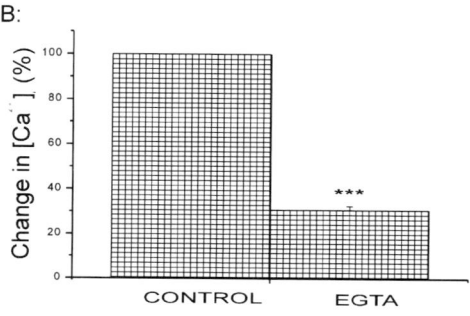

FIGURE 1. (a) Superimposed tracings for a typical sample of platelets suspended in 1 mM Ca^{2+} (*control*) and a sample of platelets in which Ca^{2+} has been chelated by EGTA (5 mM) 1 minute before stimulation with collagen. (b) Average collagen-induced change in cytosolic calcium measured 3 minutes after stimulation with collagen for platelet samples in 1 mM Ca^{2+} or where Ca^{2+} has been chelated with EGTA (5 mM), ***$P < 0.0005$, $n = 5$.

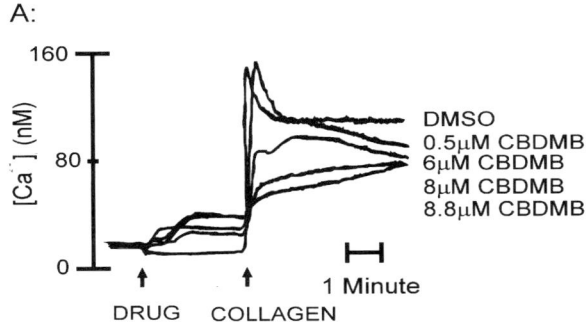

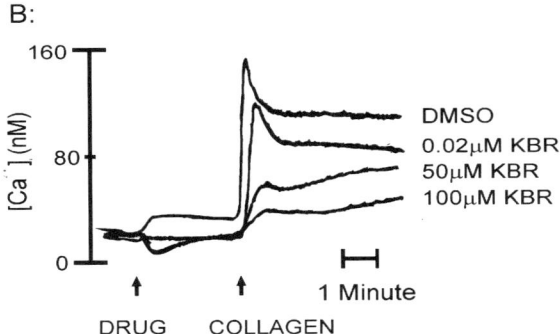

FIGURE 2. Superimposed tracings for a characteristic sample of platelets treated with increasing doses of the NCX blockers (**a**) CBDMB and (**b**) KB-R7943.

FIGURE 1b, 3 minutes after activation with collagen, under conditions of zero calcium, there is a 31 ± 2% increase in $[Ca^{2+}]_i$. Therefore, 31 ± 2% of the collagen-induced increase in $[Ca^{2+}]_i$ is due to the release from intracellular stores and 69 ± 2% of the collagen-induced increase in $[Ca^{2+}]_i$ is due to the influx of Ca^{2+} from the extracellular milieu. The total collagen-induced increase in $[Ca^{2+}]_i$, in 1 mM Ca^{2+}-containing medium, was taken as 100%.

Effects of Na^+/Ca^{2+} Blockers on the Collagen-Induced Change in Cytosolic Calcium

Three minutes before stimulation of platelet samples with collagen, the NCX was blocked with increasing doses of CBDMB or KB-R7943 to determine the role of the NCX in the collagen-induced increase in $[Ca^{2+}]_i$. FIGURE 2 demonstrates a typical tracing of a sample of platelets treated to increasing doses of (a) CBDMB or (b) KB-R7943. In each case the collagen-induced increase in $[Ca^{2+}]_i$ was reduced with increasing doses of CBDMB or KB-R7943. Work performed by Takano et al.[4] has demonstrated the ability of KB-R7943 to inhibit platelet aggregation induced by adrenaline and 5-HT.

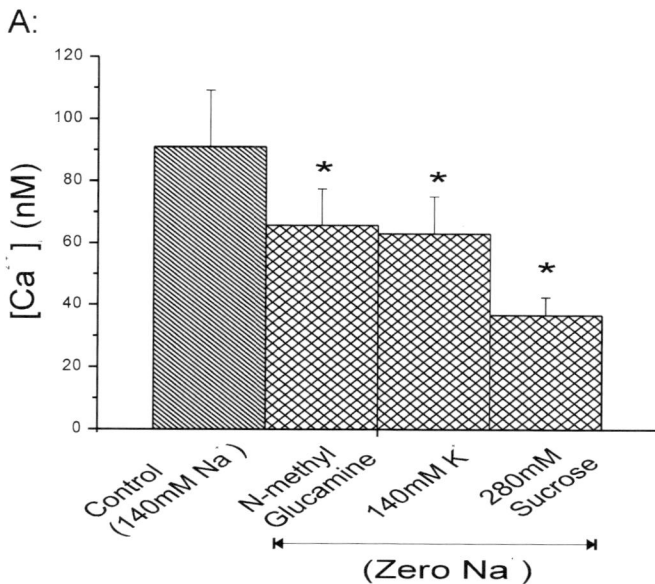

FIGURE 3. Average collagen-induced change in cytosolic calcium measured 3 minutes after stimulation with collagen for platelet samples suspended in 1 mM Ca^{2+} HEPES buffer containing 140 mM Na^+ (*control*) or zero Na^+ substituted with N-methyl glucamine, 140 mM K^+, or 280 mM sucrose, $*P <0.05$, $n = 3$.

Effects of Extracellular Sodium on the Collagen-Induced Change in Cytosolic Calcium

Our experiments suggest that NCX played an important role in the increase in $[Ca^{2+}]_i$ in response to platelet activation with collagen. To function in this capacity, the reverse mode activity of importing Ca^{2+} into the platelet cytosol in exchange for Na^+ is necessary. The NCX mode of action is determined by the Na^+ gradient, the Ca^{2+} gradient, and the membrane potential. Previous research has shown that collagen has no effect on the platelet membrane potential;[5] therefore, the membrane potential is not responsible for the platelet NCX reversal in response to stimulation with collagen. The role of Na^+ can be elucidated by maintaining the concentration of Ca^{2+} (1 mM) and performing the experiment in the absence of extracellular Na^+. FIGURE 3a shows the average change in $[Ca^{2+}]_i$ measured 3 minutes after activation with collagen for platelets suspended in either 140 mM Na^+ (*control*) or zero sodium substituted with N-methyl glucamine, 140 mM potassium, or 280 mM sucrose. All 3 zero sodium treatment groups show a significant reduction in their collagen-induced change in $[Ca^{2+}]_i$ when compared to control (140 mM Na^+). Not all agonists for platelet activation use the reverse mode NCX. As shown elsewhere (Bose *et al.*, this volume), thrombin activation is modulated by a forward mode activity of the NCX.[6]

CONCLUSION

A significant portion of the increase in cytosolic calcium, in response to platelet activation with collagen, is due to the influx of calcium through the NCX into the platelet cytosol. NCX function is determined by the sodium gradient, and platelets suspended in sodium-free medium show a reduction in their collagen-induced change in cytosolic calcium.

ACKNOWLEDGMENTS

This work was supported by grants from the Heart and Stroke Foundation and Manitoba Medical Services Foundation.

REFERENCES

1. SARGEANT, P. & S.O. SAGE. 1994. Calcium signalling in platelets and other nonexcitable cells. Pharmacol. Ther. **64:** 395–443.
2. SHIRAISHI, M., M. IKEDA, T. FUJISHIRO, et al. 2000. Characteristics of collagen-induced Ca^{2+} mobilization in bovine platelets. Cell Calcium **27:** 53–60.
3. GRYNKIEWICZ, G., M. POENIE & R. TSIEN. 1985. A new generation of Ca^{2+} indicators with greatly improved fluorescence properties. J. Biol. Chem. **260:** 3440–3450.
4. TAKANO, S., J. KIMURA & O. TOMOYUKI. 2001. Inhibition of aggregation of rabbit and human platelets induced by adrenaline and 5-hydroxytryptamine by KB-R7943, a Na^+/Ca^{2+} exchange inhibitor. Br. J. Pharmacol. **132:** 1383–1388.
5. MACINTYRE, D.E. & T.J. RINK. 1982. The role of platelet membrane potential in the initiation of platelet aggregation. Thromb. Haemost. **47:** 22–26.
6. LI, Y., V. WOO & R. BOSE. 2001. Platelet hyperactivity and abnormal Ca^{2+} homeostasis in diabetes mellitus. Am. J. Physiol. Heart Circ. Physiol. **280:** H1480–H1489.

Na^+/Ca^{2+} Exchange in Activated and Nonactivated Human Platelets

RATNA BOSE, YUN LI, AND DIANE ROBERTS

Departments of Pharmacology and Therapeutics, University of Manitoba, Winnipeg, R3E 0W3, Canada

KEYWORDS: Na^+/Ca^{2+} exchange; human platelets

INTRODUCTION

Na^+/Ca^{2+} exchange activity has been demonstrated in human platelet plasma membrane vesicles;[1] however, the results in intact platelets are inconsistent. Some studies found that changing the Na^+ gradient and/or pretreating platelets with ouabain had some effect on cytosolic calcium ($[Ca^{2+}]_i$).[2–4] Other studies, however, could not demonstrate such an effect.[5,6] Recently, human platelets were shown to display Na^+/Ca^{2+} K^+ exchange activity;[7] however, the physiologic function of the Na^+/Ca^{2+} exchanger in platelets is not clear. Therefore, the objective of this study was to evaluate the existence of a Na^+/Ca^{2+} exchanger in intact platelets and to further characterize it.

METHODS

Measurement of Cytosolic Ca^{2+}

Platelets loaded with Ca^{2+}-sensitive fluorescent dyes were separated from plasma and extracellular dyes by gel filtration.[8] Fluorescence was measured simultaneously with aggregation in a Jasco 110 ion analyzer. A combination of long wavelength dyes (excitation wavelength 500 nm) was used so that Na^+/Ca^{2+} exchanger (NCX) blockers could be employed. Ca green-AM and Fura red-AM were used in dual emission mode for ratiometric measurements for $[Ca^{2+}]$. NCX blockers 5-(4-chlorobenzyl)-2,4-dimethylbenzamil (CBDMB) and KB-R7943 (KBR) were dissolved in dimethyl sulfoxide (DMSO) and their effects were compared with those of DMSO alone. Iso-osmolar substitution for sodium was done either with *n*-methyl glucamine (0NanMGluc) or potassium chloride (0NaK140). Platelets were stored in Ca^{2+}-free medium, and 1 mM Ca^{2+} was added during the 3–5 minutes of preincubation at 37°C. All assays were performed at 37°C. Platelet count was adjusted to 2 ×

Address for correspondence: Ratna Bose, Ph.D., Professor, Department of Pharmacology & Therapeutics, University of Manitoba, A311–753 McDermot Ave., Winnipeg, Manitoba, R3E 0W3, Canada. Voice: 204-789-3562; fax: 204-789-3932.
    rbose@ms.umanitoba.ca

10^8 platelets/ml. HEPES-buffered assay medium contained (in mM) 140 NaCl, 4.9 KCl, 1.2 $MgCl_2$, 1.4 KH_2PO_4, 11 glucose, and 20 HEPES (6-N-2-hydroxyethyl-piperazine-N'-2-ethanesulfolic acid) at pH 7.4.

RESULTS AND DISCUSSION

Na^+-Dependent Ca^{2+} Efflux

The forward mode of the Na^+/Ca^{2+} exchanger can be estimated by measuring the $[Ca^{2+}]_i$ decrease in response to extracellular Na^+ in intact platelets. First, the loaded platelets were suspended in 0 Na^+, 1 mM calcium medium for 15 minutes. When the cytosolic $[Ca^{2+}]_i$ reaches a steady state, required concentrations of NaCl were added to the medium. This extracellular addition of NaCl produced a decrease in $[Ca^{2+}]_i$. This Na^+-dependent $[Ca^{2+}]_i$ decline was taken as a measure of the forward mode of Na^+/Ca^{2+} exchanger (Na^+-dependent Ca^{2+} efflux). The observation that a decrease in $[Ca^{2+}]_i$ was indeed due to efflux of Ca^{2+} has been documented by measuring the increase in Ca^{2+} in the extracellular medium.[9] In the same reference, the Na^+-dependent Ca^{2+} efflux in intact platelets was inhibited by exchange inhibitory peptide (XIP) and CBDMB. As shown in FIGURE 1, 140 K^+-substituted sodium-free medium did not prevent the decrease in $[Ca^{2+}]_i$ upon readdition of Na^+ to a Na-free medium. In addition, XIP and KBR also inhibited the Na^+-induced decrease in $[Ca^{2+}]_i$. NCKX is not expected to be inhibited by these agents.[10]

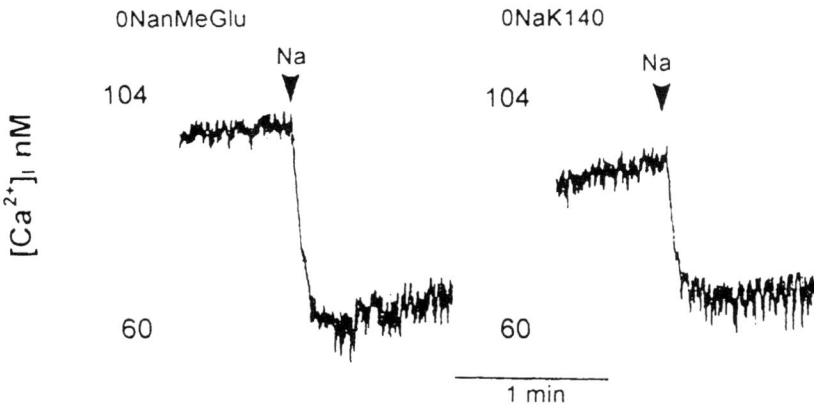

FIGURE 1. Sodium (120 mM)-dependent decline in cytosolic Ca^{2+} in the same batch of dye-loaded platelets in two different Na^+-free media. Tracing on the *left* is from a medium with n-methyl glucamine as a substitute for Na^+ and on the *right* KCl was substituted for NaCl.

TABLE 1. Cytosolic Ca^{2+} [a]

	% Change	SE	N	P
Basal				
CBDMB	150	27.5	9	0.05
KBR	189.1	54.3	5	0.04
28 mM Na	153.7	15.6	15	0.004
Thrombin				
CBDMB	128.4	9.3	11	0.006
KBR	178.3	21.9	4	0.019
28mM Na	149.1	16.3	5	0.02
Collagen				
CBDMB	41.6	15.5	5	0.01
KBR	60.6	5.7	5	0.001
28mM Na	78.1	18.1	4	0.16

[a]$[Ca^{2+}]_i$ in DMSO control was considered 100% to compare the effects of different treatments in each batch of platelets. Concentration of CBDMB was 4 µM, KBR 25 µM, collagen 10 µg/ml, and thrombin 0.5 U/ml. Paired Student's t test was used to compare the difference from DMSO controls.

Effect of NCX Blockers and Reduced Transmembrane Na^+ Gradient on Cytosolic Calcium in Nonactivated Platelets

The effect of partial replacement of external Na^+ with isomolar N-methyl-D-glucamine on Ca^{2+} homeostasis in intact human platelets was studied. This protocol was designed to determine whether decreasing extracellular Na^+ concentration by partially substituting Na^+ with nMGluc medium could alter Ca^{2+} homeostasis. The low Na^+ buffer (28 mM) was selected because in 28 mM extracellular Na^+, platelet $[Na^+]_i$ did not decline.[2] A part of the platelet suspension in standard buffer and that in Na^+-free buffer were mixed to adjust the extracellular Na^+ concentration to 28 mM (1 volume of standard buffer and 4 volumes of nMGluc buffer were used). This eliminated the wash or centrifugation step in sodium-free medium. In nonactivated platelets, NCX functions in the forward mode, because inhibiting it by pharmacologic tools or decreasing the Na^+ gradient causes the basal $[Ca^{2+}]_i$ to increase significantly (TABLE 1).

Effect of NCX Blockers and Reduced Transmembrane Na^+ Gradient on $[Ca^{2+}]_i$ in Platelets Activated by Thrombin or Collagen

On activation with thrombin, NCX continues to operate in the forward mode. This could be demonstrated by decreasing Na^+ gradient or by adding NCX blockers. These treatments significantly increased the thrombin-induced $[Ca^{2+}]_i$ response (second phase 1.5–3 minutes after thrombin) (TABLE 1). There was no effect on peak Ca^{2+} release from intracellular stores at 0.5 minutes;[8] only a reduced rate of removal of $[Ca^{2+}]_i$ was observed when NCX blockers were present or the Na^+ gradient was decreased. In contrast to activation with thrombin, collagen induces a reverse mode NCX activity in platelets. This inference was based on the observation that NCX blockers or a decrease of the Na^+ gradient attenuated collagen-induced increases in $[Ca^{2+}]_i$ (TABLE 1 and Roberts and Bose, this volume).

CONCLUSION

In nonactivated platelets, a Na^+-dependent decrease in $[Ca^{2+}]_i$ in nMGluc or 140 K^+-substituted, Na-free medium was observed on the addition of Na^+ to a Na-free medium. If the exchanger was K dependent, then a disruption in the K^+ gradient should have attenuated the decrease in $[Ca^{2+}]_i$ in response to extracellular Na^+. In addition, XIP and KBR also inhibited the Na^+-induced decrease in $[Ca^{2+}]_i$. NCKX is not expected to be inhibited by these agents.[10] In normal physiologic medium, NCX inhibitors affected the agonist-induced increase in $[Ca^{2+}]_i$ in opposite ways for collagen and thrombin, implicating different roles for the NCX during activation of platelets.

ACKNOWLEDGMENTS

This work was supported by funds from the Heart and Stroke Foundation (R.B.).

REFERENCES

1. RENGASAMY, A., S. SOURA & H. FEINBERG. 1987. Platelet Ca^{2+} homeostasis: Na^+-Ca^{2+} exchanger in plasma membrane vesicles. Thromb. Homeostasis **57**: 337–340.
2. OSHIMA, T., T. ISHIDA, H. MATSUURA, et al. 1994. Lack of effect of ouabain on calcium homeostasis in rat platelets: comparative study with human platelets. Am. J. Physiol. **266**: R651–R657.
3. SCHAEFFER, J. & M.P. BLAUSTEIN. 1989. Platelet free calcium concentrations measured with Fura-2 are influenced by transmembrane sodium gradient. Cell Calcium **10**: 101–113.
4. LEES, A.D., J. WILSON, C.H. ORCHARD & M.A. ORCHARD. 1989. Ouabain enhances basal and stimulus-induced cytoplasmic concentrations in platelets. Thromb. Haemostasis **62**: 1000–1005.
5. BRASS, L.F. 1984. The effect of Na^+ on Ca^{2+} homeostasis in unstimulated platelets. J. Biol. Chem. **259**: 12571–12575.
6. DOYLE, V.M. & V.T. RUEGG. 1985. Lack of evidence for voltage dependent calcium channels on platelets. Biochem. Biophys. Res. Commun. **127**: 161–167.
7. KIMURA, M., E.M. JEANCLOS, R.J DONELLY, et al. 1999. Physiological and molecular characterization of the Na^+/Ca^{2+} exchanger in human platelets. Am. J. Physiol. **277**: H911–H917.
8. LI, Y., V. WOO & R. BOSE. 2001. Platelet hyperactivity and abnormal Ca^{2+} homeostasis in diabetes mellitus. Am. J. Physiol. **280**: H1480–H1489.
9. BOSE, R., Y. LI & V WOO. 2001. Sodium-calcium exchange in platelets of diabetics. Proc. West Pharmacol. Soc. **44**: 183-184.
10. IWAMOTO, T., S. KITA, A. UEHERA, et al. 2001. Structural domains influencing sensitivity to isothiourea derivative KB-R7943 in cardiac Na^+/Ca^{2+} exchanger. Mol. Pharmacol. **59**: 524–531.

Effect of Glucose on the Expression Level of the Plasma Membrane Ca^{2+}-ATPase and the Na^+/Ca^{2+} Exchanger in Pancreatic Islet Cells

HELENA MARIA XIMENES,[a,b] ADAMA KAMAGATE,[a] FRANÇOISE VAN EYLEN,[a] AND ANDRÉ HERCHUELZ[a]

[a]*Laboratory of Pharmacology, Brussels University School of Medicine, Brussels, Belgium*

[b]*Department of Human Physiology and Biophysics / Institute of Biomedical Science, University of São Paulo, São Paulo, Brazil*

KEYWORDS: glucose; plasma membrane Ca^{2+}-ATPase; PMCA; Na^+/Ca^{2+} exchanger; NCX; pancreatic islet cells

INTRODUCTION AND AIMS

When stimulated by glucose, pancreatic β-cells display large oscillations of the intracellular free Ca^{2+} concentration ($[Ca^{2+}]_i$).[1,2] These increases in $[Ca^{2+}]_i$ activate several Ca^{2+}-dependent enzymes mediating insulin secretion. A rise in $[Ca^{2+}]_i$ also triggers compensatory mechanisms with the aim of reducing $[Ca^{2+}]_i$ back to resting levels. This removal of Ca^{2+} from the cell can be accomplished by two main processes: the plasma membrane Ca^{2+}-ATPase (PMCA) and the Na^+/Ca^{2+} exchanger (NCX).[3,4] The expression pattern of these two proteins has been evidenced in pancreatic β-cells,[5–8] but the level of expression, the activity, and the relative contributions of these two mechanisms to Ca^{2+} homeostasis remain unclear. The present work was designed to investigate the level of expression of PMCA and NCX in pancreatic islets exposed to various glucose concentrations.

METHODS

Pancreatic islets were cultured in the presence of different concentrations of glucose during different periods of time. Quantitative reverse transcription-polymerase chain reaction (RT-PCR) was performed to evaluate the pattern of mRNA transcription and immunoblotting to quantify the protein expression.

Address for correspondence: André Herchuelz, Laboratory of Pharmacology, Brussels University School of Medicine, Route de Lennik 808, Brussels, Belgium. Voice: 32-2-5556201; fax: 32-2-5556370.

herchu@ulb.ac.be

RESULTS AND CONCLUSIONS

Results obtained from quantitative RT-PCR show that glucose can regulate the transcription of PMCA and NCX in a time- and concentration-dependent manner. Although the mRNA of PMCA decreases after 24 hours of culture with high glucose concentration, the NCX mRNA increases. These results agree with the functional differences existing between the two proteins, PMCA being a high-affinity but low-capacity Ca^{2+} removal system and NCX being a low-affinity but high-capacity transport system. This is consistent with the view that in response to a high glucose concentration, the β-cell needs a more efficient system to remove Ca^{2+} from the cell and to maintain Ca^{2+} homeostasis.

ACKNOWLEDGMENTS

This work was supported by the ALFA programme of the E.U of which H.M.X. is a grant holder and by the Belgian Fund for Scientific Research (FRSM 3.4562.00) of which F.V.E. is a Postdoctoral Researcher.

REFERENCES

1. HERCHUELZ, A., R. POCHET, C. PASTIELS & A. VAN PRAET. 1991. Cell Calcium **12**: 577–586.
2. LENZEN, S., M. LERCH, T. PECKMANN & M. TIEDGE. 2000. Biochem. Biophys. Acta **1523**: 65–72.
3. CARAFOLI, E. 1988. Method. Enzymol. **157**: 3–11.
4. BLAUSTEIN, M.P., W.F. GOLDMAN, G. FONTANA, et al. 1991. Ann. N.Y. Acad. Sci. **639**: 254–274.
5. VARADI, A., E. MOLNAR & S.J. ASHCROFT. 1996. Biochem. J. **314**: 663–669.
6. KAMAGATE, A., A. HERCHUELZ, A. BOLLEN & F. VAN EYLEN. 2000. Cell Calcium **27**: 231–246.
7. VAN EYLEN, F., A. BOLLEN & A. HERCHUELZ. 2001. J. Endocrinol. **168**: 517–526.
8. VAN EYLEN, F., C. LEBEAU, J. ALBUQUERQUE-SILVA & A. HERCHUELZ. 1998. Diabetes **47**: 1873–1880.

Na/Ca Exchanger and PMCA Localization in Neurons and Astrocytes

Functional Implications

M.P. BLAUSTEIN, M. JUHASZOVA,[a] V.A. GOLOVINA, P.J. CHURCH,[b] AND E.F. STANLEY[c]

Department of Physiology, University of Maryland School of Medical School, Baltimore, Maryland 21210, USA

[a]*Gerontology Research Center, NIA, Baltimore, Maryland 21224, USA*

[b]*Department of Physiology and Biophysics, Mt. Sinai School of Medicine, New York, New York 10029, USA*

[c]*Cellular and Molecular Biology Section, Toronto Western Research Institute, Toronto, Ontario MST 258, Canada*

ABSTRACT: Immunocytochemistry reveals that the Na/Ca exchanger (NCX) in neuronal somata and astrocytes is confined to plasma membrane (PM) microdomains that overlie sub-PM (junctional) endoplasmic reticulum (jER). By contrast, the PM Ca^{2+} pump (PMCA) is more uniformly distributed in the PM. At presynaptic nerve terminals, the NCX distribution is consistent with that observed in the neuronal somata, but the PMCA is clustered at the active zones. Thus, the PMCA, with high affinity for Ca^{2+} ($K_d \approx 100$ nM), may keep active zone Ca^{2+} very low and thereby "reprime" the vesicular release mechanism following activity. NCX, with lower affinity for Ca^{2+} ($K_d \approx 1{,}000$ nM), on the other hand, may extrude Ca^{2+} that has diffused away from the active zones and been temporarily sequestered in the endoplasmic reticulum. The PL microdomains that contain the NCX also contain Na^+ pump high ouabain affinity α2 (astrocytes) or α3 (neurons) subunit isoforms ($IC_{50} \approx 5$–50 nM ouabain). In contrast, the α1 isoform (low ouabain affinity in rodents; $IC_{50} > 10{,}000$ nM), like the PMCA, is more uniformly distributed in these cells. The sub-PM endoplasmic reticulum in neurons (and probably glia and other cell types as well) and the adjacent PM form junctions that resemble cardiac muscle dyads. We suggest that the PM microdomains containing NCX and α2/α3 Na^+ pumps, the underlying jER, and the intervening tiny volume of cytosol ($<10^{-18}$ l) form functional units (PLasmERosomes); diffusion of Na^+ and Ca^{2+} between these cytosolic compartments and "bulk" cytosol may be markedly restricted. The activity of the Na^+ pumps with α2/α3 subunits may thus regulate NCX activity and jER Ca^{2+} content. This view is supported by studies in mice with genetically reduced (by ≈50%) α2 Na^+ pumps: evoked Ca^{2+} transients were augmented in these cells despite normal cytosolic Na^+ and resting Ca^{2+} concentrations ($[Na^+]_{CYT}$ and $[Ca^{2+}]_{CYT}$). We conclude that α2/α3 Na^+ pumps control

Address for correspondence: Mordecai P. Blaustein, M.D., Department of Physiology, University of Maryland School of Medicine, 655 W. Baltimore Street, Baltimore, MD 21201. Voice: 410-706-3345; fax: 410-706-8341.

mblauste@umaryland.edu

PLasmERosome (*local*) [Na$^+$]$_{CYT}$. This, in turn, via NCX, modulates *local* [Ca^{2+}]$_{CYT}$, jER Ca^{2+} storage, Ca^{2+} signaling, and cell responses.

KEYWORDS: Na/Ca exchanger; PMCA localization; neurons; astrocytes

DIFFERENT LOCATIONS OF TWO PLASMA MEMBRANE Ca^{2+} EXTRUSION MECHANISMS

Two major plasma membrane (PM) mechanisms are involved in extrusion of Ca^{2+} from cells. One is the Na/Ca exchanger (NCX) and the other is the ATP-driven PM Ca^{2+} pump (PMCA). PMCA has a high affinity for Ca^{2+} ($K_d \approx 100$ nM) but a low turnover number (30–250 s^{-1}), whereas NCX has a lower affinity (K$_d \approx 1000$ nM) but a much higher turnover number (2,000–5,000 s^{-1}). Also, NCX is activated by non-transported intracellular Ca^{2+}.[1] Moreover, the NCX can move Ca^{2+} either out of or into cells, depending on the prevailing Na$^+$ electrochemical gradient and the difference between the NCX reversal potential and the membrane potential.[1]

Functional studies first established that the NCX and PMCA are simultaneously expressed in most cells.[2,3] This has now been confirmed by immunoblotting and immunocytochemistry.[4–7] For example, both types of transporters are simultaneously

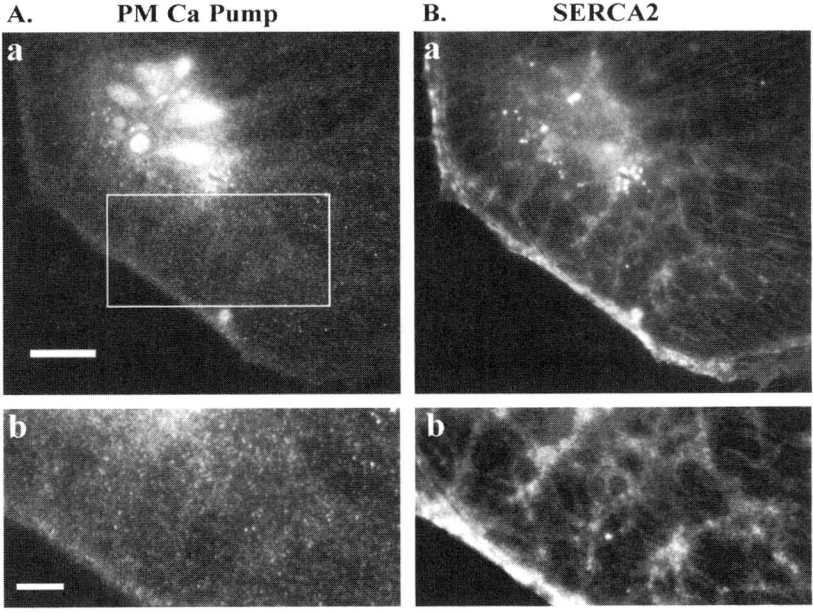

FIGURE 1. Immunochemical localization of the PMCA (**A**) and SERCA-2b (**B**) in a primary cultured mouse brain astrocyte. The cell was immunolabeled with polyclonal anti-PMCA antibodies and then with monoclonal anti-SERCA-2. The PMCA antibodies label the cell surface relatively uniformly and differently from the pattern of the underlying ER labeled with anti-SERCA antibodies. Unpublished experiment of V.A. Golovina.

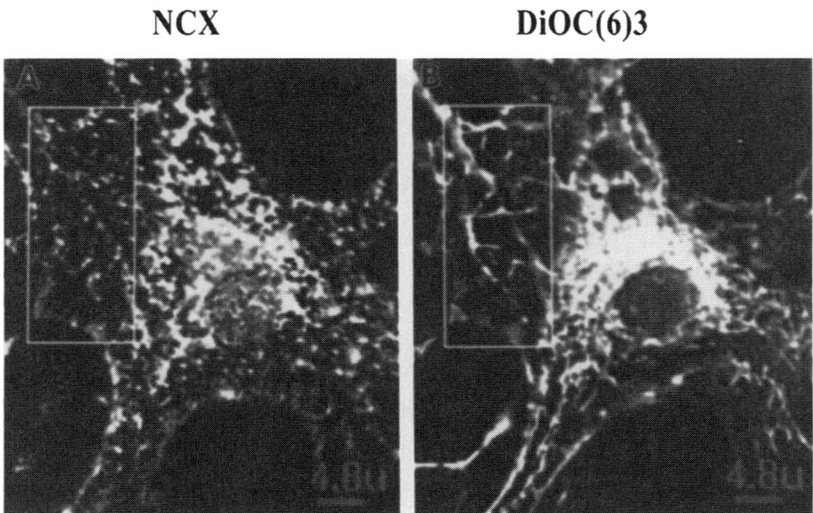

FIGURE 2. Immunochemical localization of the NCX (**A**) and subsequent DiOC(6)3 staining of ER (**B**)[8] in a primary cultured rat brain astrocyte. The cell was immunolabeled with affinity-purified polyclonal anti-NCX rabbit antiserum and then stained with 500 ng/ml DiOC(6)3. Both dyes exhibit a lacy, reticular staining pattern, especially in the cell periphery. The overlap between the two stains is clearly evident in the areas outlined by the white rectangles in **A** and **B**. Here, 40% of the areas within the box that stained for NCX also stained with DiOC(6)3; conversely, 25% of the DiOC(6)3 labeled regions (ER) also stained for NCX. Both values are highly significant ($P < 0.005$). (Reprinted from Juhaszova et al.[5] with permission.)

expressed in both neurons and astrocytes.[2–5] This implies that the two transporters, with different kinetic properties, must have different functions in the cell. Therefore, a critical question is: What are the specific roles of the NCX and PMCA?

Clues to the answer come from studies of the localization of the PMCA and NCX in the PM of neurons and astrocytes. Immunocytochemical data reveal that the PMCA in primary cultured astrocytes (FIG. 1), as in other types of cells, is distributed relatively uniformly over the cell surface[5] (but see Ref. 7 for a contrasting view). FIGURE 1A illustrates the "ground glass" appearance of the fluorescent label corresponding to a uniform distribution of the PMCA in a mouse astrocyte. For comparison, FIGURE 1B shows the location of the endoplasmic reticulum (ER) in this cell, as indicated by the lace-like distribution pattern of the anti-SERCA-2 (sarcoendoplasmic reticulum Ca^{2+} pump type 2) antibodies.

In contrast to PMCA, the NCX in astrocytes (FIG. 2A) and neurons[4,5] is distributed in the PM in a lace-like reticular pattern (FIG. 2A) that closely parallels the distribution of the underlying ER. The ER can be visualized by staining with the lipophilic cationic dye DiOC(6)3 [8] (FIG. 2B) or anti-SERCA antisera.[5] This relation is noteworthy because the sub-PM ER in neurons forms specialized junctions with the adjacent PM.[9,10] Structurally, these PM-ER junctions closely resemble the PM-SR (sarcoplasmic reticulum) junctions observed in skeletal and cardiac muscles[11,12] and smooth muscle.[13]

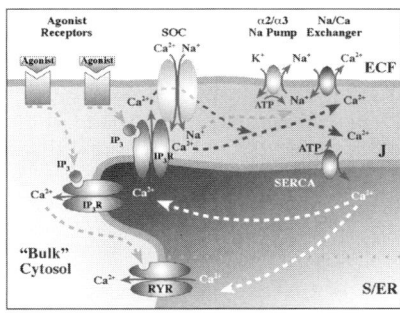 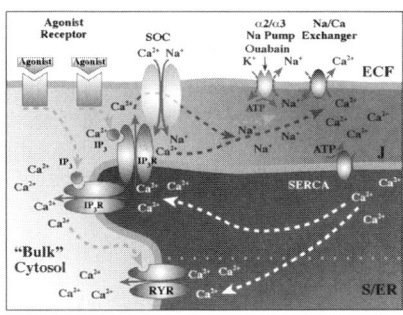

FIGURE 3. Model of the PM-jER region showing key transport proteins involved in local control of jER Ca^{2+} stores and modulation of Ca^{2+} signaling. The PM region shows agonist receptors, a nearby PM microdomain (containing store-operated channels or SOCs, $\alpha2/\alpha3$ Na^+ pumps, and Na/Ca exchanger), and adjacent jER (with SERCA and IP_3R and RYR), and intervening "diffusion-restricted" junctional cytosolic space (J). (**Left**) Normal conditions; (**right**) after inhibition of $\alpha2/\alpha3$ Na^+ pumps by low-dose ouabain. Shading indicates relative concentrations of Na^+ and/or Ca^{2+}. ECF, extracellular fluid. $\alpha1$ Na^+ pumps are widely distributed in the PM, but they may be excluded from these microdomains. The model shows a physical association between SOCs and IP_3R that some authors have suggested (see text). (Modified from Ref. 25 and reproduced with permission.)

These findings suggest that the PMCA plays a major role in regulating the low resting Ca^{2+} concentration in bulk cytosol ($[Ca^{2+}]_{CYT}$). The NCX, however, because of its restriction to the PM microdomains that overlie junctional ER (jER), apparently controls primarily the Ca^{2+} concentration in the tiny junctional space (J) between the PM and the jER, that is, $[Ca^{2+}]_J$ (FIG. 3). Consequently, the NCX also must help regulate the Ca^{2+} concentration in the jER ($[Ca^{2+}]_{jER}$). Indeed, these distributional and functional differences between the PMCA and the NCX indicate how the two Ca^{2+} transport mechanisms, with different kinetic properties, can function in parallel in the same membrane: they operate on different cytosolic Ca^{2+} compartments.

DIFFERENT LOCATIONS OF PM Na^+ PUMP CATALYTIC SUBUNIT ISOFORMS

Ca^{2+} transport mediated by the NCX is governed largely by the activity of the Na^+ pump. It is likely, therefore, that studies on the distribution of Na^+ pump catalytic (α) subunit isoforms in neurons and astrocytes will provide additional clues to test the foregoing hypothesis. Virtually all cells express the Na^+ pump $\alpha1$ isoform, which has a very low affinity for ouabain in rodents ($IC_{50} > 10$ μM) and a high affinity for Na^+ ($EC_{50} \approx 10$ mM). In addition, most cells also express one of the higher ouabain affinity ($IC_{50} < 0.05$ μM), lower Na^+ affinity ($EC_{50} = 15-25$ mM) alpha isoforms ($\alpha2$ or $\alpha3$). For example, rat cortical astrocytes express both $\alpha1$ and $\alpha2$ isoforms,[14,15] and many brain neurons express both $\alpha1$ and $\alpha3$ isoforms.[5,14,15] The $\alpha1$ isoform is uniformly distributed in the PM of both primary cultured rat astrocytes and neurons.[5,15] By contrast, the $\alpha2$ and $\alpha3$ isoforms are expressed in small clusters that distribute in reticular patterns in the PM of neurons and astrocytes, respectively.[5,15] Double-labeling immunocytochemical studies demonstrate that this distribution of the α2 and α3 isoforms in the PM is very similar to this distribution of the NCX as well as the underlying ER.[5,15] Thus, there appear to be small patches (microdomains or lipid rafts) of PM that contain both the NCX and one of the high ouabain affinity isoforms of the Na$^+$ pump catalytic subunit (α2 in astrocytes; α3 in neurons). These PM patches also contain store-operated channels (SOCs) that open in response to ER unloading and participate in ER Ca^{2+} store replenishment.[16–18]

ORGANIZATION AND FUNCTION OF THE PM-jER (PLasmERosome) REGION

FIGURE 3 illustrates the hypothesized organization of the PM-jER region and shows some of the key transporters located there. We have named this region the PLasmERosome[17] because the PM and jER, with intervening junctional space, act as a functional unit. Diffusion of Na$^+$ and Ca^{2+} between the junctional space and bulk cytosol appears to be markedly restricted. This is inferred from observations on the effects of low-dose ouabain and a Na$^+$ pump α2 subunit null mutation on Ca^{2+} signaling in astrocytes.

FIGURE 4A shows that 10–100 nM ouabain, which should inhibit Na$^+$ pumps with α2 subunits selectively, did not significantly elevate bulk [Na$^+$]$_{CYT}$ (measured with SBFI; Refs. 19 and 20) in primary cultured mouse cortical astrocytes. Nevertheless, these low doses of ouabain significantly increased Ca^{2+} transients (measured with Fura-2) evoked by cyclopiazonic acid (CPA, a SERCA inhibitor). Comparable findings, namely, normal [Na$^+$]$_{CYT}$ and augmented CPA-evoked [Ca^{2+}]$_{CYT}$ transients, were obtained in astrocytes[21] from mice with a single Na$^+$ pump α2 allele null mutation.[22] Astrocytes[21] and cardiac myocytes[22] from these Na$^+$ pump heterozygotes express the normal complement of α1 Na$^+$ pumps, but only about half the normal complement of α2 Na$^+$ pumps. Moreover, [Ca^{2+}]$_{CYT}$ transients in rat astrocytes, evoked by CPA (FIG. 4B) or the purinergic receptor agonist ATP (FIG. 4C), were augmented reversibly by 100 nM ouabain (FIG. 4B, C, and D). Clearly, partial reduction of Na$^+$ pump α2 isoform activity alone is sufficient to augment the Ca^{2+} transients.

Low-dose ouabain-induced augmentation of the Ca^{2+} transients appears to depend on NCX-mediated, enhanced Ca^{2+} entry/inhibited Ca^{2+} extrusion, and augmented Ca^{2+} storage in the jER. This role of NCX in ouabain-augmented responses has been verified with NCX knockdown by an antisense oligodeoxynucleotide in vascular myocytes[23] and in cardiac myocytes from embryonic mice with null mutations in both NCX-1 genes.[24] We infer that these particular Na$^+$ pumps control the Na$^+$ concentration in the tiny junctional space (J; FIG. 3) that is not measurable with SBFI because the Na$^+$ pumps with α2 subunits are confined to PM microdomains that are apparently coupled to underlying jER in astrocytes. Therefore, the PM transporters located in the PM microdomains within the PLasmERosomes regulate not only [Na$^+$]$_J$ and [Ca^{2+}]$_J$, but also [Ca^{2+}]$_{jER}$ and Ca^{2+} signaling as well. Augmentation of Ca^{2+} signaling may occur by two mechanisms. The augmented CPA-evoked Ca^{2+} transients indicate that reduced activity of Na$^+$ pumps with α2 subunits is associated with increased Ca^{2+} storage in the jER. In addition, if some inositol tri-

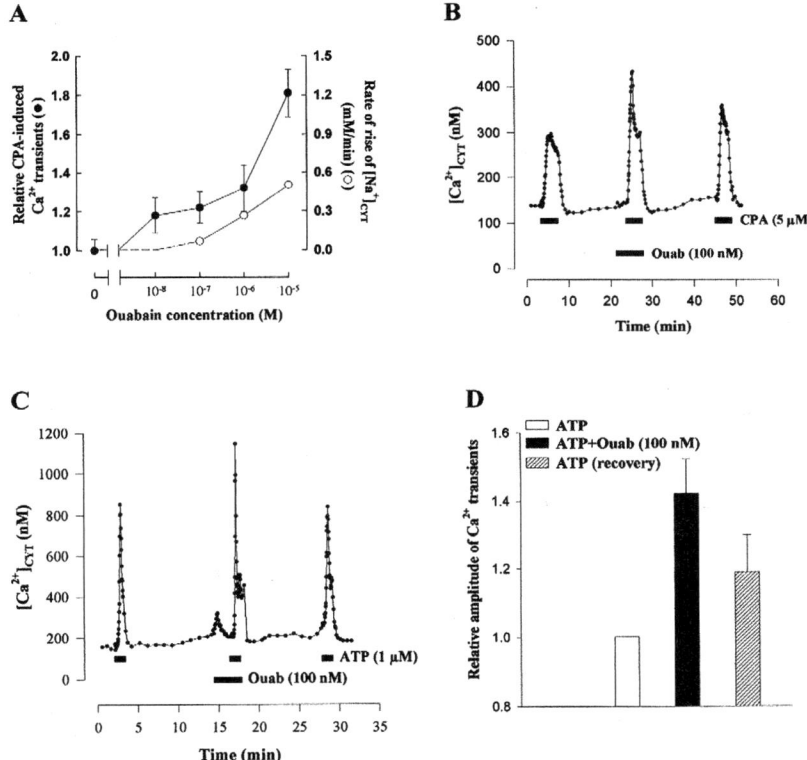

FIGURE 4. Effects of low-dose ouabain on $[Na^+]_{CYT}$ (measured with SBFI) and $[Ca^{2+}]_{CYT}$ (measured with Fura-2) in primary cultured mouse (**A**) and rat (**B–D**) cortical astrocytes. (**A**) Ouabain dose-response curve shows that 10^{-7} M ouabain had negligible effect on $[Na^+]_{CYT}$, whereas 10^{-8} M ouabain significantly increased the cyclopiazonic acid (CPA, an inhibitor of SERCA) evoked $[Ca^{2+}]_{CYT}$ transient. Ouabain (100 nM) significantly and reversibly augmented the transient $[Ca^{2+}]_{CYT}$ increases evoked by CPA (**B**) and ATP (an agonist; **C,D**). The bar graph in **D** summarizes data from 28 cells comparable to the one illustrated in **C**. Unpublished data of V.A. Golovina and M.P Blaustein.

phosphate receptors (IP_3R) — perhaps one specific isoform — are located in the jER within the PLasmERosome, the increased $[Ca^{2+}]_J$ may sensitize these IP_3R to IP_3.[25]

As just noted and illustrated in FIGURE 3, the PLasmERosome PM microdomains also contain store-operated channels (SOCs).[18] These channels mediate Na^+ as well as Ca^{2+} entry.[25] They therefore contribute in a major way to regulating the Na^+ and Ca^{2+} electrochemical gradients across the PLasmERosome PM microdomains, thereby influencing NCX activity.

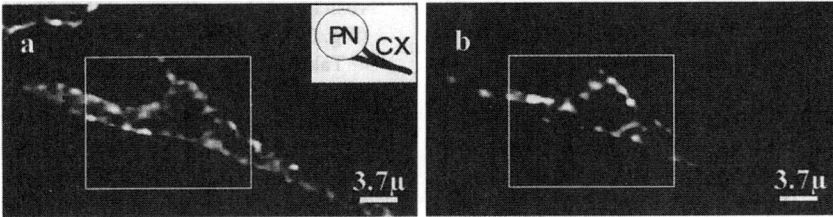

FIGURE 5. Localization of PMCA in the chick ciliary ganglion. A partially dissociated calyx was immunolabeled with antibodies raised against the PMCA (**a**) and synaptotagmin (**b**); a polyclonal antibody was used to label PMCA. A single focal plane image of this calyx is shown. *Inset* in **a** is a diagram of the relation between the presynaptic terminal (calyx, CX; black) and the postsynaptic neuron (PN), a tiny portion of which is seen at the upper left corner of panel **a**. The *boxed areas* of **a** and **b** show the presynaptic terminal region (defined by the staining with antibodies to synaptotagmin). Note the extensive colocalization of the two proteins; in this experiment, 33.7% of synaptotagmin-labeled pixels overlap with PMCA-labeled pixels. SC, Schwann cell. Reprinted from Juhaszova et al.[6] with permission.

DIFFERENT LOCATIONS OF PMCA AND NCX AT PRESYNAPTIC NERVE TERMINALS

It has been suggested that the PMCA is expressed in synaptic vesicle membranes at presynaptic nerve terminals.[26] We explored this possibility in chick ciliary ganglia. Ganglia were homogenized and synaptic vesicles were separated from PM vesicles in a partially purified vesicle fraction by size-exclusion chromatography on a Sephacryl-1000 column. Western blots of the eluate from consecutive column fractions revealed that the PM markers, NCX, the Na^+ pump, and PMCA, could be separated completely from the synaptic vesicle marker SV-2.[6] Another synaptic vesicle marker, synaptotagmin, was found in both the PM and the synaptic vesicles, presumably because some of this marker is bound to the PM. These data demonstrate that the PMCA and the NCX are both expressed in the PM and not in synaptic vesicle membranes.

Functional evidence indicates that presynaptic nerve terminals exhibit relatively high NCX activity.[27–29] We used immunocytochemistry to localize these transporters as a step in determining their role in regulating nerve terminal $[Ca^{2+}]_{CYT}$ and neurotransmitter release. We explored transporter localization at the calyx nerve terminal of the chick ciliary ganglion. This terminal is unusually large and contains numerous neurotransmitter release sites (active zones) that can be localized by staining for the secretory clusters with antisynaptotagmin antisera.

As shown in FIGURE 5, both PMCA (panel a) and synaptotagmin (panel b) are abundantly expressed in the presynpatic glycocalyx of the chick ciliary ganglion (regions within the white boxes). Moreover, in this partially dissociated glycocalyx (inset in panel a shows the presynaptic calyx, CX, and the postsynaptic terminal, PN), the two labels show substantial, albeit incomplete, overlap. On the average (in 4 experiments), 36–41% of the voxels (imaged volume elements) that stained for 1 label also stained for the other. Complete overlap would not be expected because not

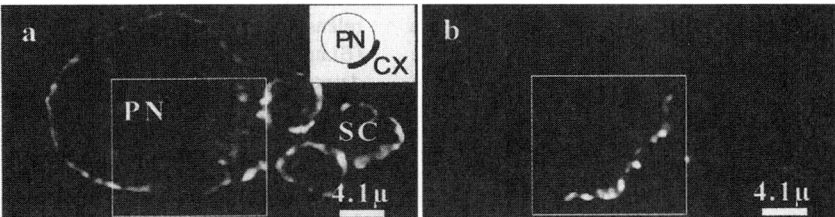

FIGURE 6. Localization of NCX in the chick ciliary ganglion. The figure compares the distribution of NCX (**a**) and synaptotagmin (**b**) in the single plane image of an attached calyx; a polyclonal antibody was used to label NCX. *Inset* in **a** is a diagram of the relation between the presynaptic terminal (calyx, CX; black) and the postsynaptic neuron (PN). The *boxed area* of the calyx (**a** and **b**) shows the presynaptic terminal region (defined by the staining with antibodies to synaptotagmin). Note the relative paucity of colocalization of the two proteins; only 1.1% of synaptotagmin-labeled pixels overlap with NCX-labeled pixels.[6]

all synaptic vesicles are located close to, or are fused with, the PM. Nevertheless, these data indicate that the some of the PMCA is located very close to the neurotransmitter release sites (active zones) in the presynaptic terminals.

In contrast to the foregoing, the overlap between the NCX and synaptotagmin (panels a and b, FIG. 6) was negligible even though NCX also is abundantly expressed at the terminals (FIG. 6, panel a).[6] This was confirmed by quantitative analysis, which revealed insignificant (<4%) overlap between the NCX and the synaptotagmin-labeled voxels. Indeed, these data indicate that the NCX may even be excluded from the neurotransmitter release sites (active zones) of the presynaptic terminal PM.

Ca^{2+} HOMEOSTASIS AT PRESYNAPTIC TERMINALS: DIFFERENT ROLES OF NCX AND PMCA

These findings on the chick ciliary ganglion appear to provide new clues to the respective roles of the PMCA and NCX at presynaptic nerve terminals. When an action potential invades the nerve terminal, the terminal depolarizes, thereby opening voltage-gated Ca^{2+} channels (VGCC) which are located close to the active zones (FIG. 7). The resulting Ca^{2+} entry elevates the local (sub-PM) $[Ca^{2+}]_i$. This triggers fusion of synaptic vesicles (SV) with the PM (#1, FIG. 7) and promotes neurotransmitter exocytosis. Upon repolarization of the terminal and closing of the Ca^{2+} channels, the Ca^{2+} diffuses away from the active zones; some of the Ca^{2+} binds reversibly to Ca^{2+} binding proteins (Ca^{2+}BPr; #2, FIG. 7). Residual Ca^{2+} at the active zones is then rapidly extruded by the nearby PMCA (#3, FIG. 7; FIG. 5) in order to "reprime" the release sites in preparation for the next action potential. Much of the Ca^{2+} that diffuses away from the active zones is taken up into the ER via ATP-driven Ca^{2+} pumps (SERCA; #4, FIG. 7). Some Ca^{2+} is probably also accumulated by mitochondria,[28] at the expense of ATP hydrolysis or the energy from oxidative phosphorylation (#5, FIG. 7); this Ca^{2+} may then be slowly transferred, via the cytosol, to the nearby ER (not shown in FIG. 7). The net Ca^{2+} gained by the terminal and stored in the ER is then slowly released into a cytosolic compartment located between the ER

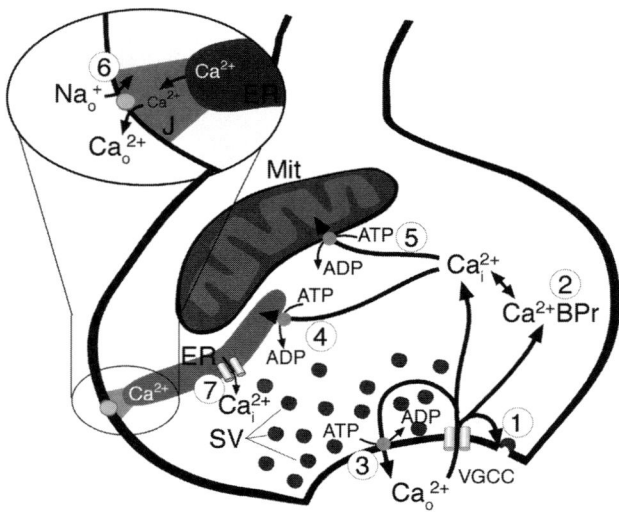

FIGURE 7. Diagram illustrating many of the elements (carriers, pumps, and channels) involved in Ca^{2+} homeostasis and neurotransmitter release at a presynaptic nerve terminal. Shown here are: voltage-gated Ca^{2+} channels (VGCC), fusion of a synaptic vesicle (SV) with the PM at an active zone (1), Ca^{2+}binding protein (Ca^{2+}BPr, #2), the PMCA (3), SERCA (4), mitochondrial Ca^{2+}-ATPase (5), NCX (6), and ER Ca^{2+} channels (both caffeine/ryanodine-sensitive and inositol trisphosphate-sensitive; #7). The tiny junctional cytosolic compartment (J) between the jER and the PM is indicated by shading; cation diffusion between this compartment and bulk cytosol appears to be markedly restricted (also see *inset* and FIG. 3).

and the adjacent PM that exhibits restricted cation exchange with bulk cytosol (J, FIG. 7; also FIG. 3). Ca^{2+} within this compartment is then extruded across the PM via the NCX that is confined to the adjacent PM microdomains (#6, FIG. 7). This is the final step in restoring normal Ca^{2+} balance after nerve terminal activation.

Although not directly addressed here, the "active" release of Ca^{2+} from the ER, via the channels associated with ryanodine receptors and/or inositol trisphosphate receptors, or both (#7, FIG. 7), likely helps to modulate neurotransmitter release. This may be especially important for such phenomena as facilitation and post-tetanic potentiation. Moreover, local Na^+ pump α3 subunit activity may be particularly influential in regulating the Ca^{2+} content of the jER and thus modulating nerve terminal Ca^{2+} signaling.

ACKNOWLEDGMENTS

This work was supported by National Institutes of Health Grant NS-16106 (to M.P.B.) and by a Grant-in-Aid from the American Heart Association Mid-Atlantic Affiliate (to V.A.G.).

REFERENCES

1. BLAUSTEIN, M.P. & W.J. LEDERER. 1999. Sodium/calcium exchange: its physiological implications. Physiol. Rev. **79:** 763–854.
2. DIPOLO, R. & L. BEAUGE. 1983. The calcium pump and sodium-calcium exchange in squid axons. Annu. Rev. Physiol. **45:** 313–324.
3. DIPOLO, R. & L. BEAUGE. 1988. Ca^{2+} transport in nerve fibers. Biochim. Biophys. Acta **947:** 549–569.
4. JUHASZOVA, M., H. SHIMIZU, M. BORIN, et al. 1996. Localization of the Na^+-Ca^{2+} exchanger in vascular smooth muscle, and in neurons and astrocytes. Ann. N.Y. Acad. Sci. **779:** 318–335.
5. JUHASZOVA, M. & M.P. BLAUSTEIN. 1997. Distinct distribution of different Na^+ pump alpha subunit isoforms in plasmalemma. Physiological implications. Ann. N.Y. Acad. Sci. **834:** 524–536.
6. JUHASZOVA, M., P. CHURCH, M.P. BLAUSTEIN & E.F. STANLEY. 2000. Location of calcium transporters at presynaptic terminals. Eur. J. Neurosci. **12:** 839–886.
7. SACCHETTO, R., A. MARGRETH, M. PELOSI & E. CARAFOLI. 1996. Colocalization of the dihydropyridine receptor, the plasma-membrane calcium ATPase isoform 1 and the sodium/calcium exchanger to the junctional-membrane domain of transverse tubules of rabbit skeletal muscle. Eur. J. Biochem. **237:** 483–488.
8. TERASKI, M. 1989. Fluorescent labeling of endoplasmic reticulum. Methods Cell Biol. **29:** 125–135.
9. HENKART, M., D.M. LANDIS & T.S. REESE. 1976. Similarity of junctions between plasma membranes and endoplasmic reticulum in muscle and neurons. J. Cell Biol. **70:** 338–347.
10. WATANABE, H. & G. BURNSTOCK. 1976. Junctional subsurface organs in frog sympathetic ganglion cells. J. Neurocytol. **5:** 125–136.
11. FRANZINI-ARMSTRONG, C. 1996. Functional significance of membrane architecture in skeletal and cardiac muscle. Soc. Gen. Physiol. Ser. **51:** 3–18.
12. FRANZINI-ARMSTRONG, C., F. PROTASI & V. RAMESH. 1998 Comparative ultrastructure of Ca^{2+} release units in skeletal and cardiac muscle. Ann. N.Y. Acad. Sci. **853:** 20–30.
13. SOMLYO, A.V. & C. FRANZINI-ARMSTRONG. 1985. New views of smooth muscle structure using freezing, deep-etching and rotary shadowing. Experientia **41:** 841–856.
14. MCGRAIL, K.M., J.M. PHILLIPS & K.J. SWEADNER. 1991. Immunofluorescent localization of three Na,K-ATPase isozymes in the rat central nervous system: both neurons and glia can express more than one Na,K-ATPase. J. Neurosci. **11:** 381–391.
15. JUHASZOVA, M. & M.P. BLAUSTEIN. 1997. Na^+ pump low and high ouabain affinity alpha subunit isoforms are differently distributed in cells. Proc. Natl. Acad. Sci. USA **94:** 1800–1805.
16. GOLOVINA, V.A. & M.P. Blaustein. 2000. Unloading and refilling of two classes of spatially resolved endoplasmic reticulum Ca^{2+} stores in astrocytes. Glia **31:** 15–28.
17. BLAUSTEIN, M.P. & V.A. GOLOVINA. 2001. Structural complexity and functional diversity of endoplasmic reticulum Ca^{2+} stores. Trends Neurosci. **24:** 602–608.
18. GOLOVINA, V. 2002. Store-operated Ca^{2+} influx and plasmalemmal TRP1 proteins colocalize to the endoplasmic reticulum. Biophys. J. **82** (No. 1, Part 2): 61a.
19. BORIN, M.L., W.F. GOLDMAN & M.P. BLAUSTEIN. 1993 Intracellular free Na^+ in resting and activated A7r5 vascular smooth muscle cells. Am. J. Physiol. **264:** C1513–C1524.
20. MONTEITH, G.R. & M.P. BLAUSTEIN. 1998 Different effects of low and high dose cardiotonic steroids on cytosolic calcium in spontaneously active hippocampal neurons and in co-cultured glia. Brain Res. **795:** 325–340.
21. GOLOVINA, V., H. SONG, P. JAMES, et al. 2002. Na^+ pump α2 subunit expression modulates Ca^{2+} signaling. Am. J. Physiol. Cell Physiol. In press.
22. JAMES, P.F, I.L. GRUPP, G. GRUPP, et al. 1999. Identification of a specific role for the Na,K-ATPase alpha 2 isoform as a regulator of calcium in the heart. Molec. Cell **3:** 555–563.
23. SLODZINSKI, M.K. & M.P. BLAUSTEIN. 1998. Physiological effects of Na^+/Ca^{2+} exchanger knockdown by antisense oligodeoxynucleotides in arterial myocytes. Am. J. Physiol. **275:** C251–C259.

24. REUTER, H., S.A. HENDERSON, T. HAN, et al. 2002. The Na^+-Ca^{2+} exchanger is essential for the action of cardiac glycosides. Circ. Res. **90:** 305–308.
25. ARNON, A., J.M. HAMLYN & M.P. BLAUSTEIN. 2000. Na^+ entry via store-operated channels modulates Ca^{2+} signaling in arterial myocytes. Am. J. Physiol Cell Physiol. **278:** C163–C173.
26. FUJII, J.T., F.T. SU, D.J. WOODBURY, et al. 1996. Plasma membrane calcium ATPase in synaptic terminals of chick Edinger-Westphal neurons. Brain Res. **734:** 193–202.
27. FONTANA, G., R.S. ROGOWSKI & M.P. BLAUSTEIN. 1995. Kinetic properties of the sodium-calcium exchanger in rat brain synaptosomes. J. Physiol. **485:** 349–364.
28. REUTER, H. & H. PORZIG. 1995. Localization and functional significance of the Na^+/Ca^{2+} exchanger in presynaptic boutons of hippocampal cells in culture. Neuron **15:** 1077–1084.
29. SCOTTI, A.L., J.Y. CHATTON & H. REUTER. 1999. Roles of Na^+-Ca^{2+} exchange and of mitochondria in the regulation of presynaptic Ca^{2+} and spontaneous glutamate release. Philos. Trans. R. Soc. Lond. B Biol. Sci. **354:** 357–364.

Immunohistochemical Detection of the Sodium–Calcium Exchanger in Rat Hippocampus Cultures Using Subtype-Specific Antibodies

T. THURNEYSEN, D.A. NICOLL, K.D. PHILIPSON, AND H. PORZIG

[a]*Pharmacological Institute, University of Bern, CH 3010 Bern, Switzerland*

[b]*Cardiovascular Research Laboratories, UCLA School of Medicine, Los Angeles, California 90095-1760, USA*

ABSTRACT: All of the known Na^+/Ca^{2+} exchanger subtypes, NCX1–3, are expressed in the brain, albeit with marked regional differences. On the mRNA level, overall expression seems most prominent for NCX2, intermediate for NCX1, and, except for a few regions, low for NCX3. Using three subtype-specific antibodies, we have now studied the cellular expression of the NCX subtypes in rat hippocampus cultures by immunohistochemical techniques. Our results provide evidence for a highly cell-specific expression pattern of NCX subtypes and show surprisingly little colocalization. NCX1 and NCX3 are both primarily expressed in neuronal cells. While NCX1 is found in the large majority of neurons, NCX3 expression was restricted to a small minority of cells. By contrast, NCX2 was almost exclusively present in glial cells. The NCX2 antibody, a IgM, stained glial cell membranes as well as an intermediate fibrillar system. In spite of extensive screening, the nature of this fiber system has not yet been identified.

KEYWORDS: sodium–calcium exchanger subtypes; subtype-specific antibodies; immunohistochemistry; confocal microscopy; rat hippocampus cultures

The Na^+/Ca^{2+} exchanger, first cloned by Nicoll and Philipson[1] represents one of the major membrane transport systems involved in regulating free Ca^{2+} concentrations in excitable, and also some nonexcitable, tissues (for review see Blaustein and Lederer[2] and Philipson and Nicoll[3]). In mammals, three subtypes of the exchanger protein (NCX1, NCX2, NCX3) have been identified as being products of different genes. (Gene products such as NCX1–3 are called "subtypes," whereas splice variants are generally identified as "isoforms.") Moreover, several splice variants of NCX1 and NCX3 were shown to be expressed in a partially tissue-specific manner.[4]

Address for correspondence: Dr. Hartmut Porzig, Pharmakologisches Institut, der Universität Bern, Friedbuehlstrasse 49, CH 3010 Bern, Switzerland. Voice: x41-31-632-2524; fax: x41-31-632-4992.

hartmut.porzig@pki.unibe.ch

However, only the NCX1 subtype and its splice variants appear to be widely distributed in different organs and cell types, whereas NCX2 and NCX3 seem to be exclusively expressed in brain and skeletal muscle. The highly subtype- and splice variant-specific distribution of exchanger protein or mRNA expression has prompted a search for functional differences that might allow us to rationalize this intriguing pattern. A number of recent reports have indeed succeeded in defining significant though subtle functional characteristics that distinguish individual subtypes or even splice variants.[5,6]

In brain, the expression of all three subtypes of the exchanger protein has been studied mainly on the mRNA level and found to vary significantly between major regions.[7] For NCX1 splice variants, a cell-specific distribution pattern in rat neurons and astrocytes has been established. However, for lack of suitable antibodies, little information is available for cell-specific expression on the protein level. We have now generated antibodies for each of the three NCX subtypes that were then used for immunohistochemical labeling studies in cultures of rat hippocampus cells.

METHODOLOGICAL CONSIDERATIONS

Characterization of Antibodies

Monoclonal antibodies directed against NCX1 (R3F1) or NCX2 (W1C3) were prepared as described by Porzig et al.[8] The polyclonal NCX3 antibody (aNCX3) was generated in rabbits. We used as antigens either the recombinant protein (NCX1) expressed in Sf9 insect cells or His-tagged fusion proteins containing the large hydrophilic intracellular loop connecting transmembrane domains 5 and 6 of either the

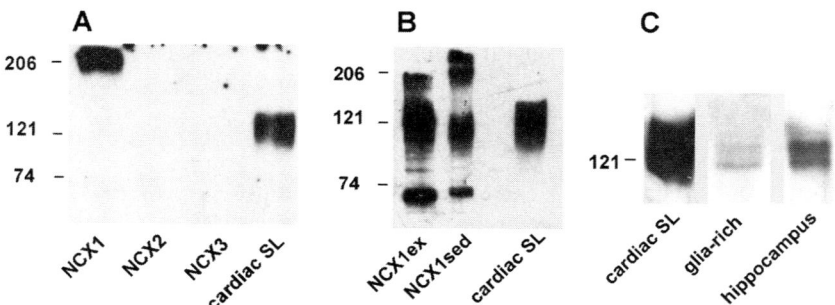

FIGURE 1. Immunoblots showing the reaction of the monoclonal NCX1-specific antibody R3F1 with proteins from recombinant BHK cells (**A, B**) or hippocampus cell cultures (**C**). In (**A**), cell homogenates were used as starting material, whereas in (**B**) proteins from NCX1 cells were extracted with Triton-X100 before PAGE (NCX1ex, Triton-solubilized; NCX1sed, proteins remaining in sediment). NCX1, NCX2, NCX3 indicate proteins from BHK cells expressing the respective exchanger subtypes. Cardiac SL indicates proteins from equine cardiac sarcolemma vesicles that were run for comparison. In (**C**), "glia rich" refers to cell homogenates from cultures containing mostly glial cells and a few neurons, whereas "hippocampus" refers to homogenates from baby rat hippocampus (the starting material for our cell cultures). Generally 4–5 µg protein were applied/lane.

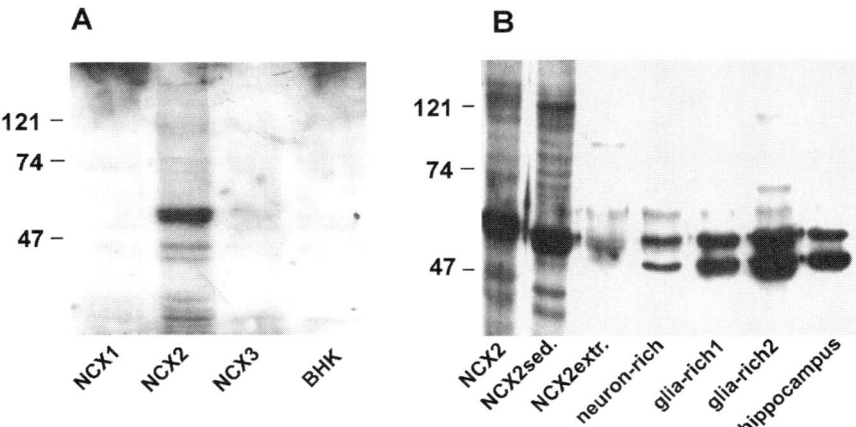

FIGURE 2. Immunoblots showing the reaction of the monoclonal NCX2-specific antibody W1C3 with proteins from recombinant BHK cells (**A**) or hippocampus cell cultures (**B**). Panel **A** documents the reaction of the antibody with PAGE-separated and blotted Triton-X100 extracts of protein homogenates from native BHK cells or from BHK cells expressing either NCX1 or NCX2 or NCX3. Note that an immune reaction is limited to proteins from the NCX2-expressing cells. In (**B**) the immune reaction with NCX2-BHK cell proteins is contrasted with the reaction in hippocampus cell proteins. "NCX2sed" and "NCX2extr" refer to proteins from sediment and supernatant of Triton X-100 extracts. "Glia-rich" refers to cultures that contain mostly glial cells and a few neurons, "neuron-rich" refers to cultures in which glial cell growth was inhibited with cytosine β-D-arabinofuranoside (3 μM), whereas "hippocampus" refers to homogenates from native baby rat hippocampus. Generally 4–5 μg protein were applied/lane. Note that of the two immunoreactive bands in hippocampal tissue, only one corresponds to the dominant immunoreactive band in BHK cells.

NCX2 or the NCX3 exchanger. Immunolabeling was facilitated by the fact that the three antibodies belonged to three different immunoglobulin types: mouse IgG (R3F1), mouse IgM (W1C3), and rabbit IgG (aNCX3). Therefore, simultaneous labeling with different non-cross-reacting secondary antibodies appeared possible. The subtype specificity of the antibodies was tested using three recombinant baby hamster kidney (BHK) cell lines, each expressing one of the exchanger subtypes[9] (compare FIGS. 1A, 2A, 3A). With the exception of a faint cross-reaction of mAb R3F1 with NCX2, the antibodies labeled protein bands only in immunoblots of proteins from the corresponding BHK cell line. For NCX1 and NCX3, membrane proteins were separated by PAGE after extraction from the cell homogenate using Triton X-100 in the presence of a cocktail of protease inhibitors. NCX2 could not be extracted by this method but was recovered from the Triton-insoluble sediment. R3F1 and aNCX3 both detected a major band close to a molecular mass of 120 kDa, presumably corresponding to the full-length exchanger (FIGS. 1B and 3A). By contrast the only major immunoreactive band detected by W1C3 in NCX2-expressing BHK cells had a molecular mass of about 60 kDa, a rather faint band appeared close to 120 kDa (FIG. 2A, B). Because NCX2 appears to be particularly prone to proteolytic cleavage,[9] we suggest that this band corresponds to a fragment of the intracellular loop that contains the epitope for W1C3.

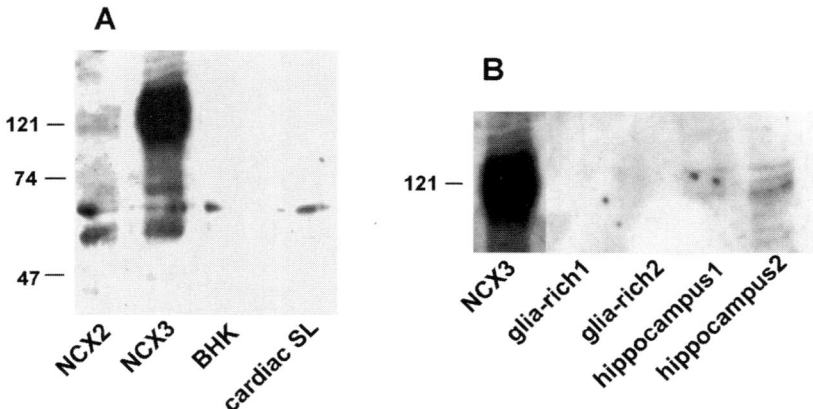

FIGURE 3. Immunoblots showing the reaction of the polyclonal NCX3-specific antibody with proteins from recombinant BHK cells (**A**) or hippocampus cell cultures (**B**) Panel **A** shows the reaction of the antibody with blots from BHK cells expressing NCX2 or NCX3. Moreover, the reaction with wild-type BHK cells and with cardiac sarcolemma is documented. The reaction with NCX1-expressing BHK cells has been tested in a different blot (not shown) and was also negative. Panel **B** shows the reaction of this antibody with proteins either from glia-rich hippocampus culture homogenate or from native baby rat hippocampus homogenate. Designations "1" and "2" refer to two different protein concentrations (7 and 14 μg/lane, respectively, for glia and 2.5 and 5 μg/lane, respectively, for hippocampus). Note, that no immune reaction was detected in the glia-rich cultures.

RESULTS AND DISCUSSION

In immunoblots using protein extracts from hippocampus cultures as well as from native hippocampus tissue of newborn rats, only the NCX1- and NCX2-reactive mAbs detected major bands, while the labeling with aNCX3 remained very weak (compare FIGS. 1C, 2B, 3B). The banding pattern corresponded to the one in BHK cells, except for s strong additional band at 54 kDa that was consistently labeled by mAb W1C3 in hippocampus tissue but not in NCX2-expressing BHK cells. Since hippocampus cultures always contain glial cells together with neurons, it was not clear from these data whether the two exchanger subtypes were expressed to the same extent in both tissues or predominantly in either neuronal or glial cells. Therefore, we prepared immunoblots from cultures that either contained more than 80% glial cells ("glia-rich") or in which the growth of glia was inhibited by adding cytosine-β-D-arabinofuranoside (3 μM).[10] Although most of the protein extracted from these latter cultures originated from neurons ("neuron-rich"), glial cells could still be discovered upon microscopic examination. In fact, a certain proportion of glial cells appeared to be required in order to maintain the viability of neuronal cells during the 3-week culture period. Expression of the W1C3-reactive protein was significantly stronger in immunoblots from glia-rich cultures (FIG. 2B), while the R3F1-reactive protein dominated in neuron-rich tissue (FIG. 1C).

Immunolocalization of NCX1 and NCX2

In a first set of immunohistochemical studies, we studied the distribution of NCX1 and NCX2 immunoreactivity in a mixed neuron–glia coverslip culture after fixation and permeabilization of the cells with ethanol. After simultaneous labeling with mAbs R3F1 and W1C3, the antibodies were detected with FITC-conjugated goat-anti-mouse (GaM) IgG or rhodamine-conjugated donkey-anti-mouse (DaM) IgM, respectively. Second antibodies were selected that did not show any IgG–IgM cross-reactivity and that did not produce any background staining of hippocampus cells at the standard settings for fluorescence measurements. Fluorescence was detected using a Zeiss LSM 510 laser scanning confocal microscope equipped with argon and HeNe lasers and driven by the Windows NT 4.0-based software of the manufacturer.

As shown in FIGURE 4A, mAb R3F1 (green) exclusively labeled neuronal cells including the whole dendritic network. By contrast, mAb W1C3 (red) predominantly stained cytoskeletal glial cell structures. Surprisingly, these doubly labeled cultures failed to show any obvious colocalization of the two antibodies. Structural elements of glial or nerve cells (e.g., cell membranes) labeled by both antibodies would appear yellow under these conditions. However, careful examination of pictures scanned at variable depth of focus revealed that most, if not all, yellow staining could be explained by overlay of glial tissue on neurons rather than by colocalization within a defined structure. Previous studies by our laboratory had already shown preferential labeling of neuronal membranes and, in particular, synaptic boutons by mAb R3F1.[10] On the other hand, functional studies had clearly demonstrated Na^+/Ca^{2+} exchange activity in glial cells.[11,12] Yet, neither mAb R3F1 nor W1C3 seemed to label glial cell membranes.

One characteristic feature of mixed neuron/glia cultures is the highly polymorphous structure of glial cells that are woven in an intricate pattern around neuronal cells or show a multitude of differently shaped extensions. Because of the staining of glial cytoskeletal fibers by mAb W1C3, these structures are clearly visualized in the confocal images as in FIGURE 4A. Moreover, glial cells are extremely thin. Confocal stacks across glial cells attached to the growth surface rarely extend beyond 1.5 μm. We concluded from these morphologic observations that the fluorescent signal emanating from the glial cell membrane would not be separable from the cytoskeletal labeling in case NCX2 is the dominant exchanger subtype expressed in glial cell membranes. By contrast, green fluorescence from mAb R3F1 or yellow staining resulting from NCX1/NCX2 colocalization might have been detected.

To confirm this assumption, we used three different approaches:

(i) In neuron-rich, glia-depleted cultures, individual glial cells contact neuronal cells via long processes that often end in growth cone–like structures. We searched for individual glial cell processes that might allow us to discriminate membrane from cytoskeletal immunofluorescence. In several instances, we detected such glial "growth cones" where mAb W1C3, in addition to staining cytoskeletal fibers, also labeled the membrane at the tip of the glial cell extension (FIG. 4B). By contrast, mAb R3F1 failed to label any of the glial cell processes.

(ii) We used glia-rich cultures, depleted of neurons, that show a different cell growth pattern. In these cultures glial cells grow in a regular, cob-

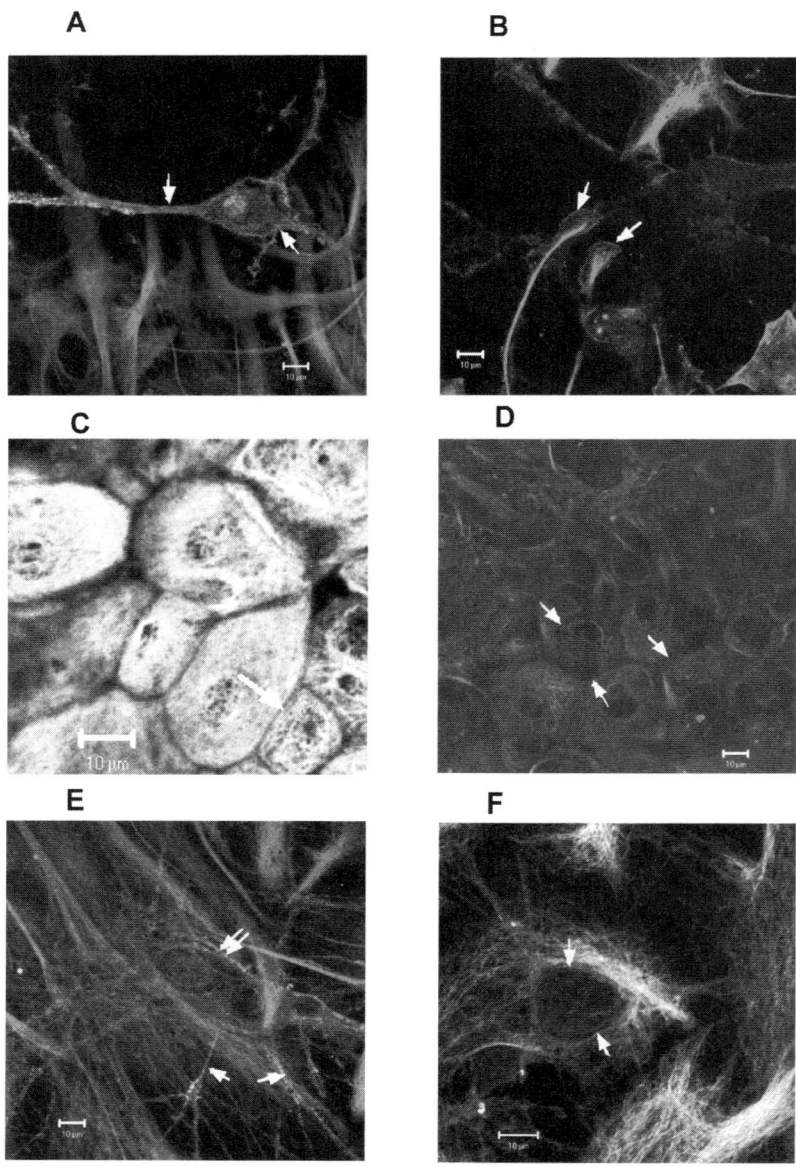

FIGURE 4. See Plate 9 in color section. Confocal images of ethanol-fixed hippocampus cultures after immunocytochemical staining with NCX subtype-specific antibodies. **(A)** Staining with R3F1 (anti-NCX1, FITC-conjugated 2nd Ab), and W1C3 (anti-NCX2, rhodamine-conjugated 2nd Ab). Absence of the fusion color yellow indicates lack of colocalization. The *arrows* point to the neuronal cell and its dendrites. **(B)** Similar stain. The tips of the glial growth processes (*arrows*) show the presence of W1C3 immunoreactivity in the

blestone-like fashion. The borders of individual cells are better defined and more easily detected in transmission microscopy. Again, mAb W1C3 clearly stained the membranes between the glial cells that formed a single homogeneous cell layer on the growth surface (FIG. 4C). In parallel experiments on the same type of cultures, no comparable labeling could be observed with mAb R3F1.

(iii) Finally, we used a red fluorescent membrane tracker dye (CM-DIL) to stain membranes by a method not dependent on immune complex formation. Glial cells were also labeled by mAb W1C3 and visualized with FITC-conjugated DaM IgM. In mixed glia/neuron cultures CM-DIL seemed to stain only the membranes of neurons including the dendritic network. In addition, there was patchy red staining interspersed with glial cells. However, it was impossible to decide whether these stained elements corresponded to tips of glial cell processes or to dendritic structures. Therefore, we used the same homogeneous glial cell cultures described above *(ii)* to localize CM-DIL-stained membranes. CM-DIL labeled the same pattern of intercellular membranes as did mAb W1C3 (FIG. 4D).

The results of all three approaches were consistent with the assumption that glial cell membranes could be distinctly labeled by mAb W1C3, whereas only very little, if any, staining by mAb R3F1 could be detected.

In view of earlier observations that NCX1 splice variant mRNA is expressed in cortical astrocytes,[11] we did not expect such strict cell-specific expression of NCX1 and NCX2. We considered two possible explanations: (1) The pattern of mRNA isoform expression differs in neurons and glia.[13] In families of splice variants containing the mutually exclusive exon A (neurons) or B (astrocytes), a sequence that forms the mAb R3F1 epitope is partially altered. Hence, a lack of glia labeling could result if this change would cause a significant decrease in the affinity of mAb R3F1. However, this explanation was judged less likely since the antibody recognizes so well the kidney isoforms of NCX1[14] that are known to belong to the B family. (2) Alternatively, the expression of NCX1 in glial cells might depend on culture conditions or on the developmental stage of the cultivated cells. In most experiments we used hippocampus cultures 14 to 21 days of age. To assess a possible dependency of NCX1 expression on culture age, we studied a few astrocyte cultures that were main-

---

cell membrane. **(C)** Glia-rich culture stained with W1C3, detected with a rhodamine-conjugated 2nd Ab and with an anti-vimentin monoclonal antibody, detected with a FITC-labeled 2nd Ab. The *arrow* points to the glial cell membrane that interacts exclusively with mAb W1C3. **(D)** Similar culture as in **C**, except that cell borders are stained with the "membrane tracker dye" CM-DIL rather than with W1C3. This technique visualizes glial cell membranes without requiring immune reactions. **(E)** Co-staining with the anti-NCX3 antibody (detected with a FITC-labeled 2nd Ab) and mAb W1C3 (detected with a rhodamine-labeled 2nd AB). Note the exclusive localization of NCX3 immunoreactivity in neuronal cells. The arrows point to the cell body and associated dendrite structures of a neuron. **(F)** Co-staining with mAb R3F1 and mAb W1C3 as in **A**, but using as 2nd antibodies a specific anti-mouse IgM (coupled to rhodamine) together with an anti-IgG Ab that cross-reacted with IgM (coupled to FITC). Note homogeneous colocalization of the two 2nd antibodies on glial fibers but lack of co-localization on the neuronal cell marked by arrows.

tained for more than 28 days. In these aged cultures, some glial cell membranes, indeed, showed NCX1 immunoreactivity, even though NCX2 labeling remained dominant. Nevertheless, no significant colocalization of NCX1 and NCX2 immunoreactivity could be detected.

We conclude from these observations that NCX1 represents the predominant exchanger subtype in hippocampus neurons, while NCX2 expression is mostly confined to glial cells.

Immunolocalization of NCX3

Immunoblots from native hippocampus homogenates as well as from hippocampus cell cultures had shown only weak immunoreactivity with the specific polyclonal antibody (FIG. 3B). Consequently, the cell population expressing NCX3 was expected to be small. Careful analysis of mixed neuron/glia cultures revealed a small minority of neurons that showed strong NCX3 immunoreactivity in the membranes of the cell body as well as in the corresponding dendritic network (FIG. 4E) but lacks NCX1. Because no significant NCX1/NCX3 colocalization was detected, the expression of NCX3 appears strictly cell-specific. However, from these immunohistochemical studies, it was not possible to associate NCX3-expressing cells with one of the specific, functionally identified subgroups of hippocampus neurons.

Association of NCX2 with a Glial Cytoskeletal Fiber Network

A prominent feature of mAb W1C3-labeling in hippocampus cultures was its extensive reaction with an intracellular fiber system essentially confined to glial cells. As already mentioned in the context of FIGURE 2, this mAb reacted with a protein band in hippocampus culture extracts that did not show up in NCX2-expressing BHK cells. Therefore, we assumed that this second strongly immunoreactive band at 54 kDa might represent a cross-reacting epitope associated with one of the intermediary fiber proteins in glial cells. These proteins migrate in PAGE separations with apparent molecular masses of around 50 kDa. If this assumption were true, we should expect to observe perfect colocalization of mAb W1C3 with antibodies directed against either vimentin or glial fibrillary acidic protein (GFAP), the most prominent intermediary fibrous proteins in glial cells. Monoclonal antibodies directed against vimentin or GFAP indeed labeled cellular fiber networks that phenotypically resembled the one detected with mAb W1C3. However, colocalization studies in each case failed to identify the W1C3-reactive protein as being tightly associated with either vimentin or GFAP. In both cases, a variable mosaic of green-, yellow-, and red-stained fibers was visible in different focal plains compatible with areas of overlay among separate fiber systems rather than with homogeneous colocalization. Similarly, antibodies raised against tubulin or actin did not colocalize with mAb W1C3.

As a positive control for colocalization, we established a staining procedure that utilized the capacity of many anti-IgG second antibodies to cross-react strongly with IgM. Hippocampus cultures were first doubly labeled with mAbs R3F1 and W1C3 and then stained with rhodamine-conjugated µ-chain-specific DaM IgM as well as with FITC-conjugated GaM (whole molecule) IgG. Under these conditions binding of mAb W1C3 would create binding sites for both secondary antibodies. Hence, any structure carrying the mAb W1C3 epitope would show yellow fluorescence, the red-green fusion color, while structures exclusively carrying the mAb R3F1 epitope would appear green. The result of this staining procedure is given in FIGURE 4F. The glial fiber system has assumed a yellow-orange color that appears homogeneous throughout all focal plains within the glial cell layer. The neuronal cell within this field of view has retained its green fluorescence. This result clearly confirmed our conclusion that (1) proteins reacting with mAb W1C3 are indeed mainly confined to glial cells, whereas mAb R3F1 immunoreactivity resides in neurons and that (2) there is no true colocalization of W1C3 immunolabeling with major glial fiber proteins. The nature of the fibrous protein remains to be determined.

ACKNOWLEDGMENT

The authors would like to thank Dr. A. Scotti from our Institute for introducing us in the techniques of confocal microscopy.

REFERENCES

1. NICOLL, D.A., S. LONGONI & K.D. PHILIPSON. 1990. Molecular cloning and functional expression of the cardiac sarcolemmal Na^+-Ca^{2+} exchanger. Science **250**: 562–565.
2. BLAUSTEIN, M.P. & W.J. LEDERER. 1999. Sodium/calcium exchange: its physiological implications. Physiol. Rev. **79**: 763–854.
3. PHILIPSON, K.D. & D.A. NICOLL. 2000. Sodium–calcium exchange: a molecular perspective. Annu. Rev. Physiol. **62**: 111–133.
4. QUEDNAU, B.D., D.A. NICOLL & K.D. PHILIPSON. 1997. Tissue specificity and alternative splicing of the Na^+/Ca^{2+} exchanger isoforms NCX1, NCX2, and NCX3 in rat. Am. J. Physiol. **272**: C1250–C1261.
5. DYCK, C., A. OMELCHENKO, C.L. ELIAS, et al. 1999. Ionic regulatory properties of brain and kidney splice variants of the NCX1 Na^+–Ca^{2+} exchanger. J. Gen. Physiol. **114**: 701–711.
6. RUKNUDIN, A., S. HE, W.J. LEDERER & D.H. SCHULZE. 2000. Functional differences between cardiac and renal isoforms of the rat Na^+–Ca^{2+} exchanger NCX1 expressed in *Xenopus* oocytes. J. Physiol. **529**(Pt. 3): 599–610.
7. YU, L. & R.A. COLVIN. 1997. Regional differences in expression of transcripts for Na^+/Ca^{2+} exchanger isoforms in rat brain. Brain Res. Mol. Brain Res. **50**: 285–292.
8. PORZIG, H., Z. LI, D.A. NICOLL & K.D. PHILIPSON. 1993. Mapping of the cardiac sodium–calcium exchanger with monoclonal antibodies. Am. J. Physiol. (Cell Physiol.) **265**: C748–C756.
9. LINCK, B., Z. QIU, Z. HE, et al. 1998. Functional comparison of the three isoforms of the Na^+/Ca^{2+} exchanger (NCX1, NCX2, NCX3). Am. J. Physiol. **274**: C415–423.
10. REUTER, H. & H. PORZIG. 1995. Localization and functional significance of the Na^+/Ca^{2+} exchanger in presynaptic boutons of hippocampal cells in culture Neuron **15**: 1077–1084.
11. GOLDMAN, W.F., P.J. YAROWSKI, M. JUHASZOVA, et al. 1994. Sodium/calcium exchange in rat cortical astrocytes J. Neurosci. **14**: 5834–5843.
12. TAKUMA, K., T. MATSUDA, H. HASHIMOTO, et al. 1994. Cultured rat astrocytes possess Na^+-Ca^{2+} exchanger. Glia **12**: 336–342
13. HE, S., A. RUKNUDIN, L.L. BAMBRICK, et al. 1998. Isoform-specific regulation of the Na^+/Ca^{2+} exchanger in rat astrocytes and neurons by PKA. J. Neurosci. **18**: 4833–4841.
14. SMITH, L., H. PORZIG, H.W. LEE & J.B. SMITH. 1995. Phorbol esters downregulate expression of the sodium/calcium exchanger in renal epithelial cells Am. J. Physiol. Cell Physiol. **38**: C457–463.

A Study of the Activity of the Plasma Membrane Na/Ca Exchanger in the Cellular Environment

MARISA BRINI, SABRINA MANNI, AND ERNESTO CARAFOLI

Department of Biochemistry and Venetian Institute of Molecular Medicine (VIMM), University of Padova, Padova, Italy

ABSTRACT: The Na/Ca exchanger is a large-capacity system that is supposed to be predominant in the ejection of calcium in excitable tissues. Its activity is normally evaluated in electro-physiological experiments or using Na-loaded cells or plasma membrane vesicles in experiments in which the activity is estimated by measuring the uptake of isotopic calcium. Here the activity of the exchanger has been evaluated in model cells that coexpress the exchanger together with the (protein) Ca indicator aequorin targeted to the different subcellular compartments. Exchanger isoform 1 (NCX1) has been used for experiments in which the effect of its overexpression has been evaluated on the Ca homeostasis in the cytoplasm at large, in the endoplasmic reticulum, in mitochondria, and in the layer of cytoplasm underneath the plasma membrane. The experiments have shown that, despite its very low calcium affinity, the exchanger was active in cells at rest in which calcium is around 100 nM. They have also shown that expressed NCX1 significantly reduced the cytosolic and mitochondrial calcium transients produced by the emptying of the ER calcium stores with $InsP_3$-linked agonists. The effect of the expressed NCX1 on the calcium homeostasis in the domain underneath the plasma membrane was more complex: The exchanger apparently catalyzed a reverse operation, leading to increased penetration of calcium.

KEYWORDS: Na/Ca exchanger; plasma membrane; cytosolic calcium transients; mitochondrial calcium transients

INTRODUCTION

In animal tissues the Na/Ca exchanger (NCX) is the product of a multigene family. NCX1 is the isoform expressed in heart, whereas other tissues also express the other isoforms. Some cells, for example, cerebellar granule neurons, express all three exchangers: They have been used to evaluate the respective contribution of NCX2 and NCX1 to the overall exchanger activity of the cells.[1] The work on cerebellar granule neurons is one of the very few cases in which the exchanger activity has been evaluated in the native cellular environment. Routinely, NCX activity has

---

Address for correspondence: Ernesto Carafoli, Department of Biochemistry, University of Padova, 35121 Padova, Italy. Voice: +39 049 8276137; fax: +39 049 8276125.
ernesto.carafoli@unipd.it

been estimated electrophysiologically on giant patches of the plasma membrane or in plasma membrane vesicles artificially loaded with Na^+ to establish a gradient that is then discharged to promote the uptake of radioactive calcium. One acknowledged problem with the exchanger is its very low affinity for Ca^{2+}, which should in principle negate its activity in resting cells, whose calcium concentration is in the vicinity of 100 nM. Another problem is the portion of the total calcium ejection activity of cells that can be traced back to the exchanger compared to the other plasma membrane calcium ejection system, the calcium pump.

Calcium homeostasis in cells is a complex spatio-temporal operation performed by the concerted operation of the transporting systems of the organelles and of the plasma membrane. The main cellular Ca^{2+} store, the endoplasmic reticulum, releases Ca^{2+} through agonist-activated channels that create circumscribed domains of high Ca^{2+} concentration which activate the low Ca^{2+} affinity of the uptake system of neighboring mitochondria. In the absence of these vicinal Ca^{2+} hotspots, mitochondria would be unable to take up Ca^{2+} at a significant rate. Therefore, the reversible uptake of Ca^{2+} by mitochondria *in vivo* is an indirect means to asses the amount of releasable Ca^{2+} in the endoplasmic reticulum. The general problem of the activity of the exchanger in cells at rest could be circumvented if the Ca^{2+} concentration in the cytosol were inhomogeneous and much higher in the layer immediately beneath the plasma membrane. If this should be so, that is, if the Ca^{2+} concentration there should reach the micromolar level, the exchanger could be active in unstimulated cells despite its very poor affinity for Ca^{2+}. Some indirect evidence that this occurs has indeed been provided.

In this work the activity of the exchanger has been evaluated in CHO cells expressing the recombinant NCX1 together with the luminescent probe aequorin directed to the various cell compartments. We have shown very evident effects of the exchanger on all cellular Ca^{2+} pools and have documented that the system is indeed active in unstimulated cells. The work will also describe experiments on the respective contributions of NCX1, NCX3, and NCX2 to the homeostasis of Ca in cerebellar granule neurons.

RESULTS AND DISCUSSION

Activity of NCX2 in Cultured Cerebellar Neurons

Granular cells isolated from the cerebellum of newborn rats and cultured *in vitro* mature to fully competent neurons in 3 to 5 days. However, if maintained in the usual culturing medium containing about 5 mM KCl, they undergo apoptosis in less than a week. To promote their prolonged survival, they must be cultured in media containing higher amounts of KCl (25 mM). Under these conditions, cytosolic calcium increases threefold, to about 150 nM, due to the partial depolarization of the plasma membrane that induces the opening of plasma membrane L-type calcium channels.[2] The molecular mechanisms by which the modest calcium increase switches off the apoptotic program are not known, but models have been provided (e.g., Yano *et al*.[3]). The maintenance of cytosolic Ca^{2+} at about 150 nM is made possible by the dramatic rearrangement of the pattern of expression of the plasma membrane Ca^{2+}-ejecting systems, the PMCA pumps and the NCXs.[1,2,4,5] One of the exchanger isoforms,

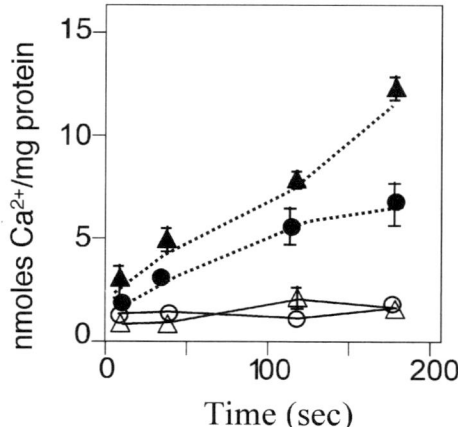

FIGURE 1. Exchange activity mediated by NCX2 in cultured cerebellar granules. Cells were plated on poly-lysine-coated plates and cultured for 7 days in a medium containing 25 mM KCl and 300 nM FK506. Before the measurements, cells were washed twice in the presence of 140 mM NaCl, 2 mM $MgCl_2$, 1 mM ouabain, 25 µM nystatin, and 20 mM MOPS-Tris pH 7.4 and incubated for 15 min in the same medium at 27°C. After two washes in the same medium, Ca^{2+} uptake was initiated by overlaying the cells with a medium containing 140 mM KCl, 50 µM $CaCl_2$ (labeled with ^{45}Ca), 1 mM ouabain, and 20 mMOPS-Tris pH 7.4. In the control experiments, KCl in the uptake buffer was replaced with NaCl. The reaction was stopped at the times indicated in a buffer containing 10 mM LaCl3, 100 mM $MgCl_2$, and 20 mM MOPS-Tris pH 7.4. The amount of Ca^{2+} taken up was determined after lysing the cells in 10 mM Tris-HCl, pH 8.0/SDS 2%. *Filled-in triangles*, 140 mM KCl, 300 nM FK506; *filled-in circles*, 140 mM KCl; *open triangles*, 140 mM NaCl, 300 nM FK506; *open circles*, 140 mM NaCl.[2]

NCX2, disappears in less than 1 hour, whereas NCX3 becomes significantly upregulated. Instead, NCX1 remains nearly unchanged. The discovery that the disappearance of NCX2 is mediated by the activity of the protein phosphatase calcineurin[1] has made it possible to evaluate the percent of total exchanger activity in the cells that is due to NCX2 as opposed to NCX1 and NCX3. This has been determined experimentally by treating cells with the immunosuppressive drug FK506, which inhibits calcineurin, and then loading them with NaCl to establish an outwardly directed Na^+ gradient (FIG. 1). Dilution of the Na^+-loaded cells into an external medium containing KCl instead of NaCl and radioactive Ca^{2+} promoted the uptake of Ca^{2+} into the cells. As FIGURE 1 shows, the exchanger activity was about 40% higher in cells treated with FK506, in which no downregulation of NCX2 had occurred. Therefore, in the ambient of cultured cerebellar granules, the activity of NCX2 corresponded to about 40% of the total exchanger activity.

Activity of NCX1 Expressed in CHO Cells Together with Targeted Aequorin

The luminescent Ca^{2+}-sensitive protein probe aequorin can be targeted to cellular organelles or to specific cell domains by adding to its construct suitable signal sequences.[6] Unmodified constructs express instead aequorin in the cytosol. The signal sequences used have directed aequorin to the endoplasmic reticulum, to the mitochondrial matrix, and to the rim of cytoplasm immediately beneath the plasma membrane: they have been described in preceding publications.[7] CHO cells do not normally express Na^+/Ca^{2+} exchange activity; therefore, they are a good experimental tool to evaluate the activity of the recombinant exchanger. After transfection of a DNA coding for NCX1 high levels of expression of the protein correctly targeted to the plasma membrane were obtained in 20 to 40% of the cell population.

The experiments have yielded a number of interesting results. The first was the clear demonstration of exchanger activity in cells at rest. To evaluate endoplasmic reticulum $[Ca^{2+}]$, aequorin was reconstituted with the prostetic group coelenterazine in a Ca^{2+}-free medium containing EGTA and ionomycin to empty the intracellular stores. After removal of ionomycin by extensive washes, Ca^{2+} was readmitted to the medium to reinitiate the replenishment of the stores. Under these conditions, the expression of NCX1 clearly reduced the steady-state level of Ca^{2+} in the endoplasmic reticulum, showing that the NCX had evidently managed to keep the concentration of Ca^{2+} in the cytosol to a lower level. Therefore, despite the low concentration of Ca^{2+} in these cells at rest (well below the μM range), the exchanger was still able to operate with reasonable efficiency.

A second interesting result was generated by experiments in which the aequorin probe had been directed to the inner mitochondrial space, to the cytosol at large, or to the cortical rim of the cell by adding to it one of the components of the exocytosis pathway (SNAP25).[8] In the experiments on mitochondria and the cytosol, Ca^{2+} was discharged from the endoplasmic reticulum by applying to the cell an agonist able to promote the formation of the intracellular messenger $InsP_3$. Ca^{2+} transients were produced in the cytoplasm and in the inner space of mitochondria. They were significantly lower in cells expressing the NCX, showing that the exchanger was considerably active under these conditions (not shown).

Previous work from our laboratory on CHO cells overexpressing the PMCA pump had shown that the (increased) pump activity in cells stimulated with $InsP_3$-producing agonists affected mainly the rate of the declining leg of the cytosolic Ca^{2+} transient induced by the agonist (corresponding to the capacitative influx phase) rather than the agonist-induced peak amplitude.[9] It was thus important to establish whether the NCX activity mainly influenced the Ca^{2+} mobilization from the ER or the capacitative Ca^{2+} influx into the cytosol from the external spaces. To dissect the two components, advantage was taken of the fact that agonists such as ATP are coupled to both Ca^{2+} mobilization from intracellular stores (via $InsP_3$ formation) and Ca^{2+} influx through plasma membrane channels. Cells were first incubated in a Ca^{2+}-free medium supplemented with EGTA to prevent the influx of Ca^{2+}, and then stimulated with ATP. A cytosolic Ca^{2+} transient was promptly generated as expected. After the transient, the addition of Ca^{2+} to the medium caused the second transient (linked to the capacitative Ca^{2+} influx triggered by the emptying of the ER store) shown in FIGURE 2. The figure shows that the first and second peak in control cells reached similar heights.

The peaks produced by the influx of Ca^{2+} and, particularly, that due to Ca^{2+} release from the ER in cells expressing the NCX were lower than those in the controls. The kinetics of the $[Ca^{2+}]$ transients is worth a comment, however, in particular because the rate of the Ca^{2+} influx phase was markedly faster in NCX-expressing cells than in the controls. The finding is an indication that NCX could have operated in

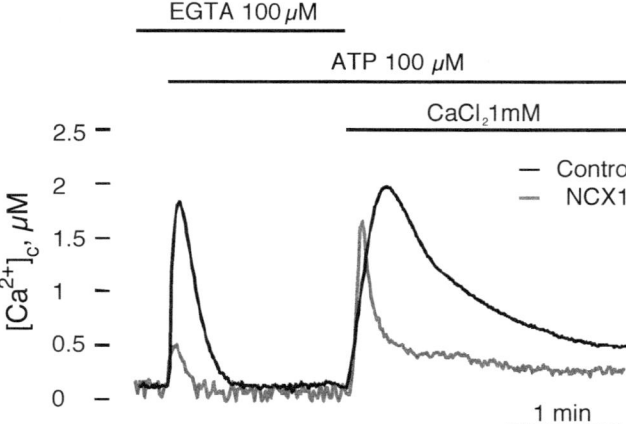

FIGURE 2. Effects of ATP-induced internal Ca^{2+} mobilization and of Ca^{2+} influx on $[Ca^{2+}]_c$ in control (*black trace*), and NCX1-expressing (*gray trace*) CHO cells. Where indicated, the medium was supplemented with 100 μM EGTA, 100 μM ATP, and 1 mM $CaCl_2$. The first and the second peak reflect the contribution of $InsP_3$-induced Ca^{2+} mobilization and of capacitative Ca^{2+} influx from the external medium, respectively. CHO cells were grown in Ham's F12 medium, supplemented with 10% fetal calf serum (FCS), in 75 cm^2 Falcon flasks. Before transfection, they were seeded onto 13 glass cover slips and allowed to grow to 50% confluence. At this stage, transfection with 3 μg of plasmid DNA (or 1.5:1.5 μg in the case of cotransfection) was performed using a Ca-phosphate procedure. Aequorin measurements were performed 36 h later. Transfected cytAEQ was reconstituted by incubating the cells for 1–3 h with the aequorin prostetic group, 5 μM coelenterazine in Dulbecco's modified Eagle's medium supplemented with 1% FCS, at 37°C in 5% CO_2 atmosphere. After this incubation, the cells were washed with Krebs-Ringer buffer (KRB) and transferred to the perfused, thermostated chamber of a purpose-built luminometer. The chamber was placed in close proximity to a low-noise photomultiplier, with a built-in amplifier discriminator. The output of the discriminator was captured by a Thorn-EMI photon counting board and stored in an IBM-compatible computer for further analyses. The additions to the KRB medium were made as specified in the figure. The experiments were terminated by lysing the cells with 100 μM digitonin in a hypotonic Ca^{2+}-rich solution (10 mM $CaCl_2$ in distilled H_2O), to discharge the remaining aequorin pool. The aequorin luminescence data were calibrated off-line into $[Ca^{2+}]$ values, using a computer algorithm based on the Ca^{2+} response curve of wt aequorins as previously described.[10]

the reverse mode, contributing to the entry of Ca^{2+} from the extracellular medium. The declining phase of $[Ca^{2+}]$ transient was instead markedly faster in NCX expressing cells, indicating that NCX was particularly efficient in restoring Ca^{2+} basal values by dissipating microdomains of high Ca^{2+} generated by the opening of the plasma membrane Ca^{2+} influx channels.

The sub-plasma membrane Ca^{2+} transient generated by the opening of capacitative plasma membrane Ca^{2+} channels triggered by the emptying of the ER Ca^{2+} store tended to be lower in NCX-expressing cells (not shown). Unexpectedly, however, the post-transient Ca^{2+} in the sub-plasma membrane domain of the cells remained at a value that was about threefold higher in NCX-expressing cells. This was a further indication that the NCX could have contributed to the influx of Ca^{2+} into the cortical cell rim.

ACKNOWLEDGMENTS

The work described here has been made possible by the financial contribution of the Italian National Research Council (Target Project Biotechnology), the Italian Ministry of Education and Research (PRIN 2001 and 2002), the Harvard Armenise Foundation, the Human Frontier Science Organization, Telethon-Italy (Grants No. 963 and GPO193Y01), and the National Research Council of Italy (Agency 2000).

REFERENCES

1. LI, L., D. GUERINI & E. CARAFOLI. 2000. Calcineurin controls the transcription of Na^+/Ca^{2+} exchanger isoforms in developing cerebellar neurons. J. Biol. Chem. **275**: 20903–20910.
2. GENAZZANI, A.A., D. GUERINI & E. CARAFOLI. 1999. Calcium controls the transcription of its own transporters and channels indeveloping neurons. Biochem. Biophys. Res. Commun. **266**: 624–632.
3. YANO, S., H. TOKUMITSU & T.R. SODERLING. 1998. Calcium promotes cell survival through CaM-K kinase activation of the protein-kinase-B pathway. Nature **396**: 584–587.
4. GUERINI, D., E. GARCIA-MARTIN, A. GERBER, et al. 1999. The expression of plasma membrane Ca^{2+} pump isoforms in cerebellar granule neurons is modulated by Ca^{2+}. J. Biol. Chem. **274**: 1667–1676.
5. GUERINI, D., X. WANG, L. LI & A. GENAZZANI. 2000. Calcineurin controls the expression of isoform 4CII of the plasma membrane Ca^{2+} pump in neurons. J. Biol. Chem. **275**: 3706–3712.
6. RIZZUTO, R., M. BRINI, C. BASTIANUTTO, et al. 1995. Photoprotein mediated measurement of $[Ca^{2+}]$ in mitochondria of living cells. Meth. Enzymol. **260**: 417–428.
7. BRINI, M., P. PINTON, T. POZZAN & R. RIZZUTO. 1999. Targeted recombinant aequorins: tools for monitoring Ca^{2+} in the various compartments of a living cell. Microsc. Res. Tec. **46**: 380–389.
8. MARSAULT, R., M. MURGIA, T. POZZAN & R. RIZZUTO. 1997. Domains of high Ca^{2+} beneath the plasma membrane of living A7r5 cells. EMBO J. **16**: 1575–1581.
9. BRINI, M., D. BANO, S. MANNI, et al. 2000. Effects of PMCa and SERCA pump overexpression on the kinetics of cell Ca^{2+} signalling. EMBO J. **19**: 4926–4935.
10. BRINI, M., R. MARSAULT, C. BASTIANUTTO, et al. 1995. Transfected aequorin in the measurement of cytosolic Ca^{2+} concentration ($[Ca^{2+}]_c$): a critical evaluation. J. Biol. Chem. **270**: 9896–9903.

K^+-Dependent Na^+/Ca^{2+} Exchangers in the Brain

JONATHAN LYTTON,[a] XIAO-FANG LI,[a] HUI DONG,[a] AND
ALEXANDER KRAEV[b]

[a]*Cardiovascular Research Group, Departments of Biochemistry & Molecular Biology and Physiology & Biophysics, University of Calgary, Calgary, AB, Canada T2N 4N1*

[b]*Samuel Lunenfeld Research Institute, Mt. Sinai Hospital, Toronto, ON, Canada M5G 1X5*

ABSTRACT: Sodium–calcium exchange was first characterized in heart myocytes and squid axon more than 3 decades ago. Since then, it has been appreciated that functioning of the Na/Ca exchanger molecule plays a critical role in calcium homeostasis in neurons. Genome analysis indicates that Na/Ca exchangers are a superfamily encoded by 7 different genes divided into 2 groups: the Na/Ca exchangers (NCX; *SLC8*) and the Na/Ca+K exchangers (NCKX; *SLC24*). Two different NCX genes, NCX1 and NCX2, are highly expressed in brain. We recently described the widespread expression of 2 NCKX-type exchangers in brain, NCKX2 and NCKX3, and uncovered evidence for expression of another, NCKX4. The unique role that each different exchanger plays in neuronal calcium homeostasis, however, awaits further investigation. To begin exploring this central question, we examined both the expression pattern and the functional properties of the K-dependent Na/Ca exchanger isoforms expressed in brain and compared and contrasted these with NCX-type exchangers. Distinct patterns of transcript abundance, regional distribution, and developmental expression were noted for each isoform. Functional properties, including stoichiometry and the kinetic characteristics of ion binding, were determined for NCKX2 and are discussed in the context of cellular Ca^{2+} signaling.

KEYWORDS: Northern blotting; *in situ* hybridization; recombinant expression; stoichiometry

INTRODUCTION

Ca^{2+} homeostasis in the brain is essential for a variety of functions at different locations, including signal transmission, neuronal excitability, signal reception and integration, and synaptic plasticity underlying learning and memory.[1] Unfortunately, dysregulation of Ca^{2+} homeostasis is responsible for excitotoxic and neurodegenerative states. It seems likely that temporal and spatial control over Ca^{2+} concentration

---

Address for correspondence: Jonathan Lytton, Ph.D., Department of Biochemistry & Molecular Biology, University of Calgary Health Sciences Centre, Room 2518, 3330 Hospital Drive NW, Calgary, AB, Canada T2N 4N1. Voice: 403-220-2893; fax: 403-283-4841.
jlytton@ucalgary.ca

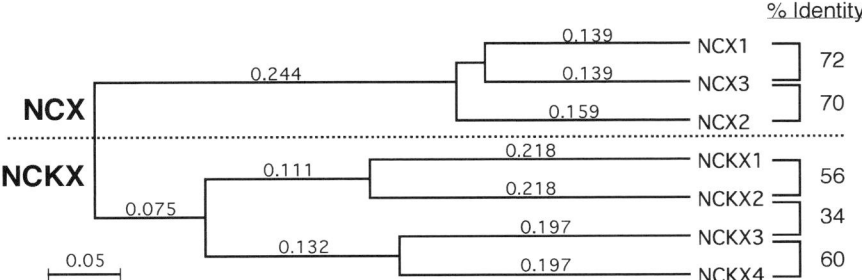

FIGURE 1. Phylogenetic tree of the Na^+/Ca^{2+} exchanger superfamily. The 2 main branches, NCX and NCKX, of the Na^+/Ca^{2+} exchanger gene family are clearly defined. There is only about 20% sequence identity between these branches. The approximate % amino acid identity between individual family members is given on the *right*. The numbers above each branch of the tree correspond to the *P* values, which are a measure of proportional differences between sequences. Calculation was performed using CustalW multiple sequence alignment with MacVector software and confirmed by bootstrapping.

is at the root of the versatility observed in Ca^{2+} signaling that leads to diverse cellular responses. Part of the explanation for such control comes about as a consequence of unique and varied expression patterns for multiple Ca^{2+} transporters.[2]

Na^+/Ca^{2+} exchange has long been recognized as a prominent contributor to Ca^{2+} homeostasis in neurons and other cells.[3] Evidence gathered during the past 5 years indicates that Na^+/Ca^{2+} exchangers are encoded by a superfamily of 7 different genes divided into 2 groups (FIG. 1): those that are independent of K^+ (NCX; SLC8) and those that depend on and transport K^+ in addition to Na^+ and Ca^{2+} (NCKX; SLC24).[3,4] Two NCX genes, NCX1 and NCX2, and three NCKX genes, NCKX2, NCKX3, and NCKX4, are all highly expressed in brain.[5-10] The unique role that each different exchanger plays in neuronal Ca^{2+} homeostasis is the subject of current investigation.

As a first approach towards understanding the physiologic role of the exchangers, we used Northern blot analysis to compare the expression of each exchanger gene product among different tissues (FIG. 2). NCX1 has an obvious broad pattern of expression, with a major band of about 7 kb that is particularly abundant in heart, brain, and kidney. (Lower intensity bands, especially in brain, can be seen at both larger and smaller sizes.) In contrast, the NCKX1 transcript of 6 kb is expressed only in eye, although the possibility that it is also expressed at lower levels in tissues we have not tested cannot be excluded, as suggested by cloning experiments from hematopoietic cells.[11] NCKX2 expression is largely restricted to an 11-kb product present at high levels in brain, although expression in peripheral neurons outside the central nervous system is suggested by the appearance of a band in adrenal and intestinal RNA and bands of lower intensity and smaller size in some other tissues, such as heart and aorta. NCKX3 transcripts of 5 kb are broadly expressed, but they are particularly abundant in different brain regions, aorta, lung, and intestine. A variety of minor NCKX3 species of larger size are also observed. This pattern suggests expression in smooth muscle. NCKX4 is also present in a wide selection of tissues as a combination of 11-kb and 4-kb bands, and it is also particularly abundant

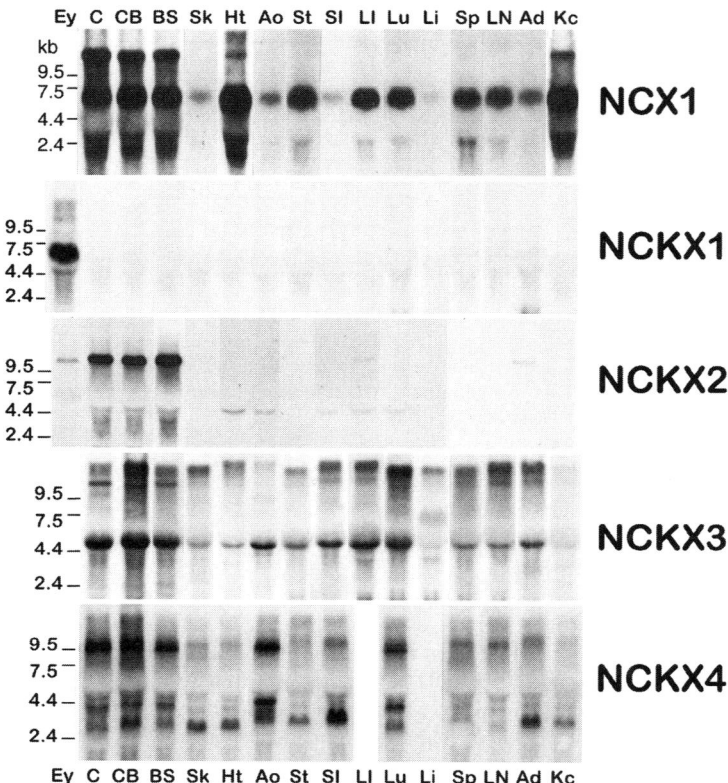

FIGURE 2. Northern analysis of Na^+/Ca^{2+} exchanger family gene expression. Ten micrograms of total RNA isolated from the indicated rat tissues was analyzed by Northern blotting using probes and hybridization conditions specific for the Na^+/Ca^{2+} exchanger gene products indicated on the *right* of the figure. These data have been collected from a variety of blots and combined for presentation purposes. Note that in the NCKX1 panel, the eye lane has been exposed only $1/10^{th}$ as long as the remainder of the sample lanes; in the NCKX2 blot, the lanes corresponding to heart, aorta, small and large intestine, and lung have been exposed 10 times longer than the remainder of the lanes. In the NCKX4 panel, there is no sample corresponding to large intestine. Ey, eye; C, cerebral cortex; CB, cerebellum; BS, brainstem; Sk, skeletal muscle; Ht, heart; Ao, aorta; St, stomach; SI, small intestine; LI, large intestine; Lu, lung; Li, liver; Sp, spleen; LN, lymph node; Ad, adrenal gland; Kc, kidney cortex. Modified, in part, from Tsoi *et al.*,[7] Poon *et al.*,[10] and Kraev *et al.*,[8] with permission.

in brain and aorta. Overall, however, NCKX4 expression is not quite so ubiquitous as that of NCKX3.

As NCKX2, NCKX3, and NCKX4 were all expressed at high levels in brain, we examined the time course of expression during development. NCKX2 turned on only at birth, and reached maximal and fairly stable levels of expression by 4 weeks of age. NCKX3 was expressed even in the embryo and seemed to be maximally ex-

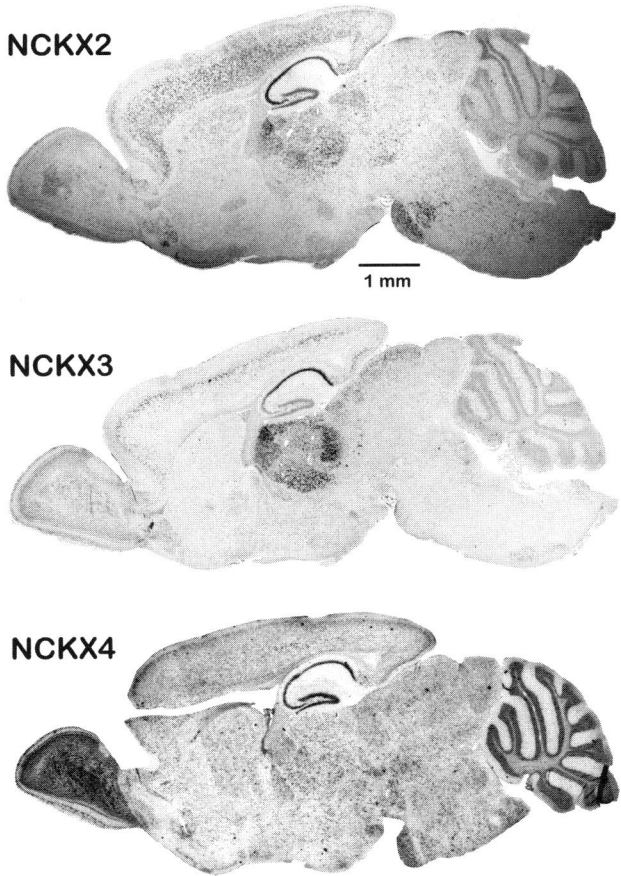

FIGURE 3. Analysis of NCKX transcript distribution in mouse brain. *In situ* hybridization using digoxigenin-labeled antisense probes specific for the indicated Na^+/Ca^{2+} exchanger gene product was performed on parasagittal sections of mouse brain. Modified, in part, from Kraev *et al.*[8] with permission.

pressed already by about day 2. NCKX4, in contrast, was much more abundantly expressed in embryonic brain, decreasing slightly during the first few weeks of life, then increasing again following weaning, and finally decreasing in older animals. Quantitative comparison of expression in adult whole brain indicated comparable levels of transcripts for each of NCKX2, NCKX3, and NCKX4.

Next, we turned to *in situ* hybridization to examine more precisely the pattern of NCKX expression in the brain (FIG. 3). These results indicated that NCKX2 was present in neurons throughout the brain, but it was particularly evident in the deeper levels of the cortex, in the CA3 pyramidal neurons of the hippocampus, in selected thalamic locations, in the pontine nuclei and the reticulotegmental nuclei of pons,

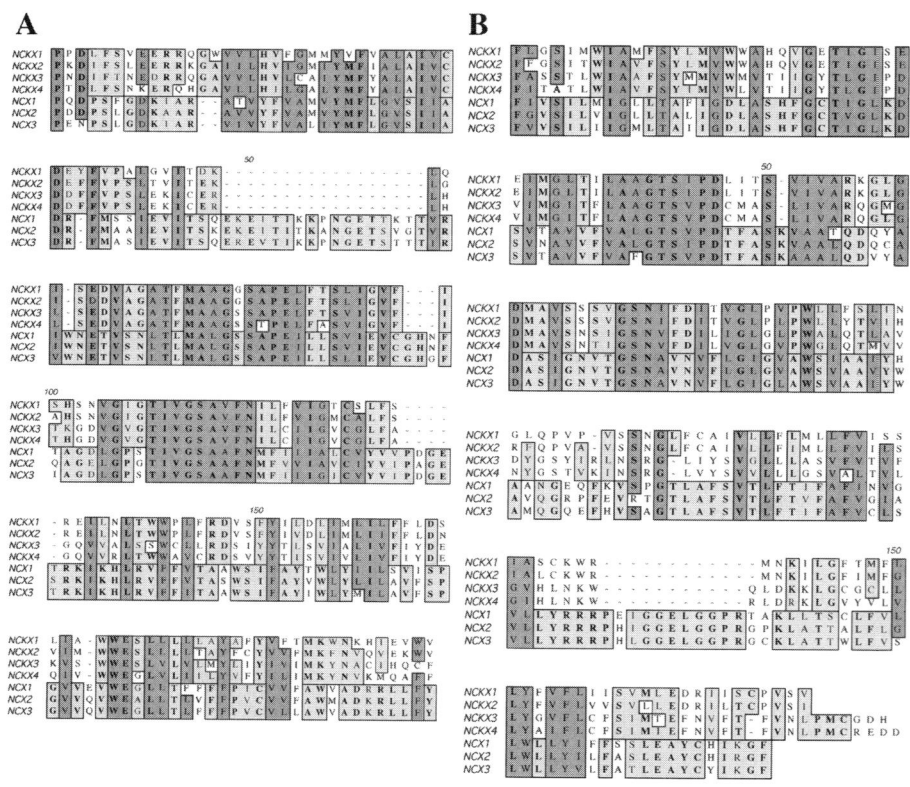

FIGURE 4. Sequence alignment of the transmembrane regions of all Na^+/Ca^{2+} exchanger paralogs. The amino acid sequences from the first (panel **A**) and second (panel **B**) cluster of hydrophobic transmembrane spans of all known human Na^+/Ca^{2+} exchanger gene products are compared using CustalW multiple sequence alignment calculated with the MacVector software package. The regions compared correspond to amino acids 441-610 and 941-1099 of NCKX1; 121-290 and 503-661 of NCKX2; 96-265 and 480-644 of NCKX3; 69-238 and 440-605 of NCKX4; 66-259 and 804-973 of NCX1; 60-253 and 752-921 of NCX2; and 65-258 and 758-927 of NCX3. A single-letter amino acid code is used, with *dashes* indicating gaps introduced to maximize the alignment. Conserved amino acids, i.e., those within groups of shared chemical properties, are enclosed within *shaded boxes*; identical amino acids are also *highlighted in bold*. *Dark grey boxes* correspond to homology among all 7 gene products at the 80% level or higher. *Light grey boxes* correspond to homology within each separate subfamily (NCX or NCKX) at the 70% level or higher.

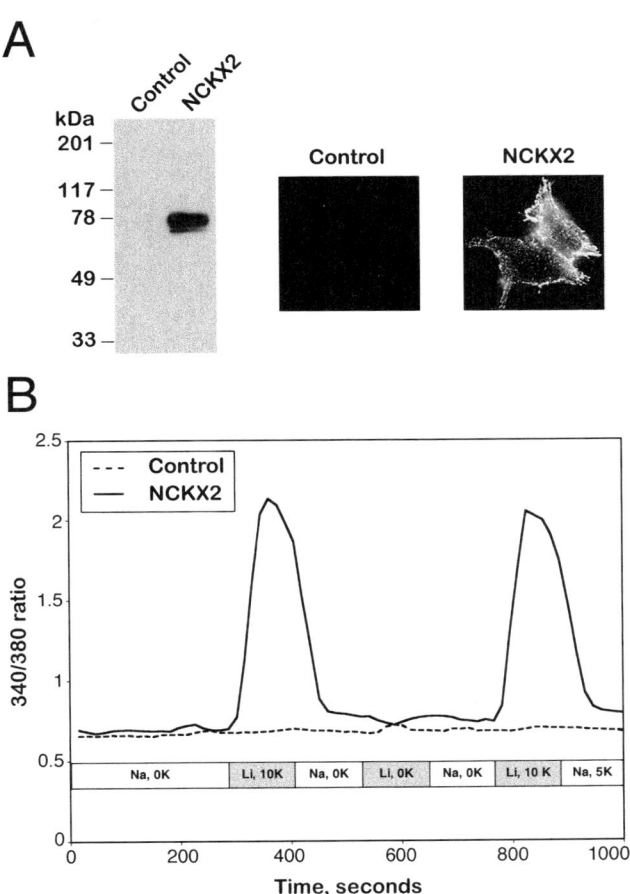

FIGURE 5. Detection of FLAG-tagged NCKX2 expressed in HEK-293 cells. Rat NCKX2, tagged with the FLAG epitope in the first extracellular loop, was expressed by transient transfection in HEK-293 cells. (**A**) Anti-FLAG antibody M2 was used to detect the expressed protein either by immunoblot (*left panel*) or by immunofluorescence (*right panels*). The control corresponds to cells transfected with vector only. (**B**) Transfected cells were loaded with the Ca^{2+}-sensitive dye, fura-2, and analyzed by digital imaging. Bathing perfusate contained 10 mM Tris-HEPES, pH 7.4, 10 mM D-glucose, 0.1 mM $CaCl_2$, and either: 145 mM NaCl only (Na, 0K); 135 mM LiCl and 10 mM KCl (Li, 10K); 145 mM LiCl only (Li, 0K); or 140 mM NaCl and 5 mM KCl (Na, 5K). An increase in cytoplasmic $[Ca^{2+}]$ is recorded as an increase in the ratio of fura-2 fluorescence excited at 340 over 380 nm. Modified, in part, from Tsoi *et al.*[7] with permission.

and within the molecular layer of the cerebellum. NCKX3 had a strikingly laminar distribution within the cortex, selectively expressed in neurons of layer IV, as well as strong expression in the hippocampal CA1 pyramidal cells and particularly strong expression in several distinct thalamic nuclei and in cells of the cerebellar molecular layer. NCKX4 had the most uniform pattern of expression of all, present in neurons in virtually all parts of the brain, although expression is particularly striking in hippocampal CA1, CA3, and dendate gyrus neurons, in the granular layer of the cerebellum, and in the olfactory bulb.

Early transport studies revealed a considerable amount of mechanistic similarity between the NCX and NCKX branches of the Na^+/Ca^{2+} exchangers.[3,12] It was therefore of some surprise that when both of these molecules were cloned, the deduced amino acid sequences displayed very limited identity, restricted exclusively to the highly conserved α-repeat regions thought to be critically involved in lining the ion binding and translocation pathway.[13–15] Within each subfamily, however, there is considerable identity that varies from about 35% to more than 70% (FIG. 1). Indeed, if one's attention is focused on the hydrophobic, presumably membrane-embedded, parts of the protein, the level of identity rises to 100% over many stretches of sequence (FIG. 4). Although this level of identity suggests a considerable degree of functional similarity among the K^+-dependent Na^+/Ca^{2+} exchangers, a detailed understanding of their individual functional properties is critical towards defining the roles of these molecules in cellular Ca^{2+} homeostasis.

We began this analysis by expressing the FLAG-epitope tagged NCKX2 protein in HEK-293 cells.[7] Abundant, surface expression of a 75-kDa glycosylated protein was observed, as anticipated from the cDNA sequence (FIG. 5A). We were able to characterize the function of NCKX2 in this system using fura-2–loaded cells to monitor Ca^{2+} entry when Na^+ in the bathing medium was replaced with Li^+. Such experiments clearly demonstrated that NCKX2 was a K^+-dependent Na^+/Ca^{2+} exchanger and that HEK-293 cells themselves displayed no appreciable background Na^+/Ca^{2+} exchange activity (FIG. 5B).

To characterize further the interactions of the transported ions with their binding sites, we used whole cell patch-clamp analysis of HEK-293 cells transfected with rat NCKX2 cDNA.[16] Transfected cells, identified by coexpression of the green fluorescent protein as a marker, were dialyzed with a pipette solution containing 122 mM NaCl and bathed in a solution containing 145 mM LiCl and 0.5 mM EGTA. A perfusion switch to a bath solution containing both 1 mM $CaCl_2$ and 40 mM KCl, but not either alone, elicited large outward currents in the cells that had been transfected with NCKX2 cDNA, but not in the control vector-transfected cells (FIG. 6). These currents arise from the unequal movement of charge during the turnover cycle of the exchanger, positive charge moving in the same direction as Na^+ ions. It is important to note that a combination of factors makes it possible to measure currents purely due to the activity of NCKX2, even though the magnitude of these currents is small. First, the membranes of HEK-293 cells are electrically quiet under the ionic conditions used to measure NCKX2 function. Second, the small cell size (about 33 pF on average) affords excellent and precise control over intracellular ionic environment. Third, the NCKX2 protein can be expressed at high levels in HEK-293 plasma membranes.

The so-called "reverse-mode" of NCKX2 operation, in which Na^+ moves out of the cell in exchange for the simultaneous entry of Ca^{2+} and K^+, allowed us to characterize the interaction of these ions at the external face of the exchanger. Current

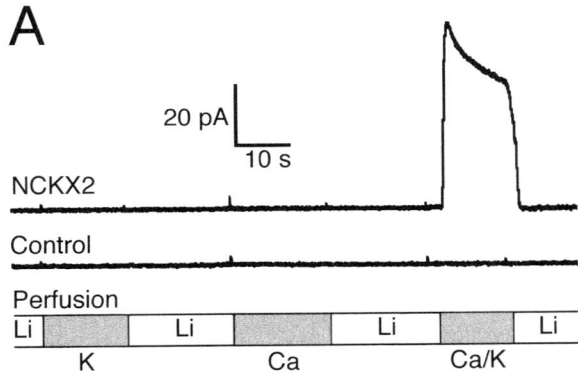

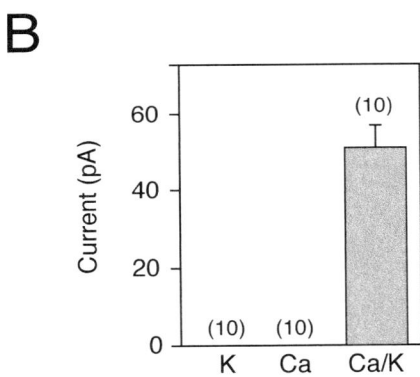

FIGURE 6. Outward currents mediated by NCKX2. HEK-293 cells transfected with either vector alone (Control) or with FLAG-tagged rat NCKX2 were analyzed by whole-cell patch clamp using a holding potential of 0 mV and a pipette solution containing 120 mM NaCl, 5 mM KCl, 2 mM $MgCl_2$, 20 mM TEA-Cl, 1 mM Na_2ATP, 8 mM D-glucose, 10 mM HEPES/TMA, 5 mM EGTA, 4.28 mM $CaCl_2$, ($[Ca^{2+}]_{free}$ = 1 µM), pH 7.2. Bath perfusion solutions contained 1 mM $MgCl_2$, 10 mM D-glucose, 10 mM HEPES/TMA, pH 7.4, and either 145 mM LiCl and 0.5 mM EGTA (Li); 105 mM LiCl, 40 mM KCl, and 0.5 mM EGTA (K); 145 mM LiCl and 1 mM $CaCl_2$ (Ca); or 105 mM LiCl, 40 mM KCl, and 1 mM $CaCl_2$ (Ca/K). (**A**) Traces from representative experiments, illustrating currents elicited by bath perfusion switches for cells transfected with the indicated cDNA. (**B**) Summary data from the indicated number of experiments. Averages ± SEM are shown. Modified from Dong et al.[16] with permission.

magnitude increased with increasing extracellular $[Ca^{2+}]$ in a manner consistent with a single Ca^{2+} binding site with an apparent affinity of 1.4 µM. The presence of 1 mM $MgCl_2$ in the bath had a competitive effect on Ca^{2+} binding, shifting the apparent affinity to 101 µM without affecting the maximal current amplitude. Similarly, we were able to analyze the interaction of K^+ with the exchanger, using either choline or lithium as the substituting cation. These experiments revealed that K^+ activated NCKX2-mediated current with an apparent affinity of about 12 mM in choline chloride and 36 mM in LiCl, but these values were not determined with high precision, because true saturation could not be obtained under conditions compatible with the whole cell patch.

These data indicate that under typical physiologic extracellular conditions of about 1 mM Ca^{2+} and 5 mM K^+, the Ca^{2+} binding site would be ~90% saturated and the K^+ site about 10–25% saturated. Such a situation might favor the inward movement of Ca^{2+} by NCKX2. However, previous work on the retinal rod NCKX1 protein documented a competitive interaction between Na^+ and both Ca^{2+} and K^+.[17,18] Therefore, we tested the effect of increasing extracellular [Na^+] on outward NCKX2 currents. At 120 mM, Na^+ inhibited NCKX2-mediated current by more than 80%. Thus, it seems likely that under close to physiologic extracellular ionic conditions, occupancy of the Ca^{2+} and K^+ binding sites will be low and therefore operation of NCKX2 in the "reverse" Ca^{2+} entry mode would not be favored kinetically.

As well as inhibiting the interactions of Ca^{2+} and K^+ with NCKX2, Na^+ will bind to drive the so-called "forward-mode" operation of the exchanger to extrude Ca^{2+} from the cell. We examined this interaction using a pipette dialysis solution containing 5 μM Ca^{2+} and 120 mM KCl. Under these conditions, Na^+ activated inward NCKX2 currents with an apparent affinity of 30 mM and a Hill coefficient of 2.8, suggesting at least 3 Na^+ binding sites. Thus, under physiologic conditions we would expect that the Na^+ binding sites would be fully saturated, kinetically favoring Ca^{2+} extrusion. It has been reported that [K^+] can vary considerably at the extracellular surface of actively firing neurons,[19] and so we also investigated whether raising [K^+] would have an influence on forward-mode NCKX2-mediated inward currents. Although K^+ inhibited NCKX2 current magnitude modestly, the effect was already maximal at 1 mM [K^+] and therefore is probably not related to the interaction of K^+ with binding sites on the exchanger itself.

The direction of operation of the NCKX2 exchanger (either Ca^{2+} influx or efflux) will be determined not only by the kinetic considerations above, but also by the thermodynamic driving forces of the ion gradients and the stoichiometry with which these ions bind to the exchanger. To examine stoichiometry, we chose a thermodynamic equilibrium approach, in which the ion gradients are varied until they balance and no net flux occurs. Under these conditions, the membrane potential is a sum of the individual Nernstian potentials (E) for each transported ion, weighted according to their stoichiometry (n):

$$E_{NCKX} = \frac{1}{(n_{Na} - 2n_{Ca} - n_K)}(n_{Na}E_{Na} - 2n_{Ca}E_{Ca} - n_K E_K).$$

Substituting for the individual Nernstian potentials and under conditions in which the concentration of transported ions on both sides of the membrane is held fixed, except for a single ion that is varied only on the extracellular side of the membrane ($[X]_o$), gives the following equation:

$$E_{NCKX} = \left(\frac{n_X}{(n_{Na} - 2n_{Ca} - n_K)}\right)(2.303RT/F)\log[X]_o + \sum C_i.$$

Therefore, a plot of NCKX2 reversal potential (E_{NCKX}) versus log of the varying extracellular ion concentration yields a slope that is a measure of the stoichiometry

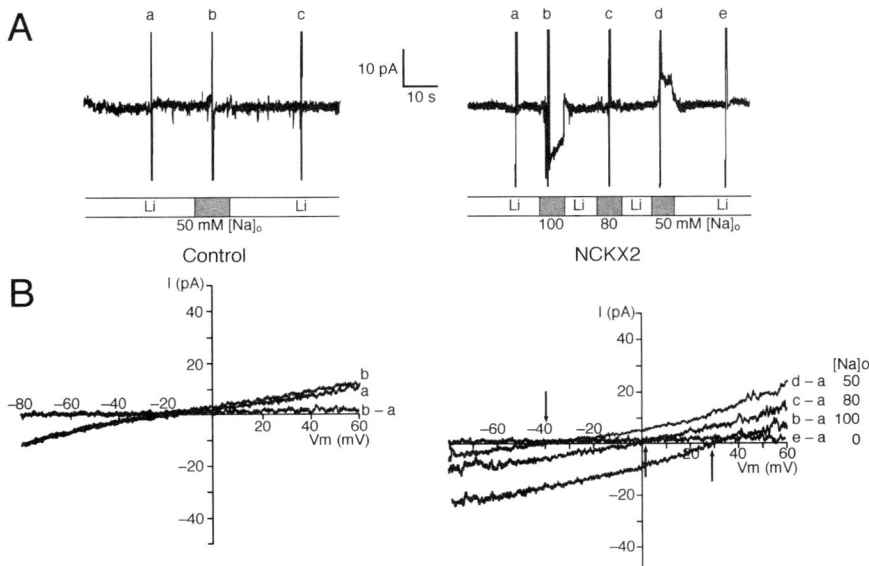

FIGURE 7. Reversal potential determinations for NCKX2.HEK-293 cells transfected with either vector alone (Control) or FLAG-tagged rat NCKX2 were analyzed by whole-cell patch clamp using a holding potential of 0 mV and a pipette solution containing 18 mM NaCl, 100 mM K-gluconate, 20 mM TEA-Cl, 1 mM Na$_2$ATP, 10 mM EGTA, 6.40 mM CaCl$_2$ ([Ca^{2+}]$_{free}$ = 0.3 μM), 10 mM D-glucose, and 10 mM HEPES/TMA, pH 7.2. Bath perfusion solutions contained 1 mM MgCl$_2$, 10 mM D-glucose, 20 mM TEA-Cl, 10 mM HEPES/TMA, pH 7.4, and either 125 mM LiCl, 0.5 mM EGTA (Li), or 0.5 mM CaCl$_2$ and various concentrations of Na$^+$ (replacing Li$^+$), as indicated. Voltage ramps from −100 to +80 mV were applied during each perfusion and in Li at the start and the end of the experiment. (A) Representative current traces for Control- or NCKX2-transfected cells subjected to perfusion switches of varying [Na$^+$]. (B) I-V curves obtained from the indicated ramps during the perfusions illustrated in A, with the I-V curve obtained in Li at the start of the experiment (marked a) digitally subtracted in some cases, as shown. *Arrows* in the right panel of B indicate the reversal potentials recorded for NCKX2 under the 3 different [Na$^+$] perfusions. Reproduced from Dong *et al.*[16] with permission.

for that ion. Note that the denominator term in this equation, $n_{Na}-2n_{Ca}-n_K$, corresponds to the net number of charges transported per exchange cycle. FIGURE 7 illustrates the ramp protocols employed to assess reversal potential at different [Na$^+$]$_o$. Currents recorded in LiCl/EGTA perfusion solution (conditions under which NCXK2 is inactive) were digitally subtracted from the traces to reveal NCKX2 ramp currents. Note that in control transfected cells there is no NCKX2 current and that in NCKX2-transfected cells, the reversal potential for the NCKX2 current shifts according to the [Na$^+$]$_o$, as anticipated. Conceptually identical experiments were also performed in which either [Ca^{2+}]$_o$ or [K$^+$]$_o$ were varied.

The averaged data from several different experiments examining stoichiometry are summarized in FIGURE 8. The slope of the relation between reversal potential and log of the extracellular ion concentration was used to determine an NCKX2 stoichi-

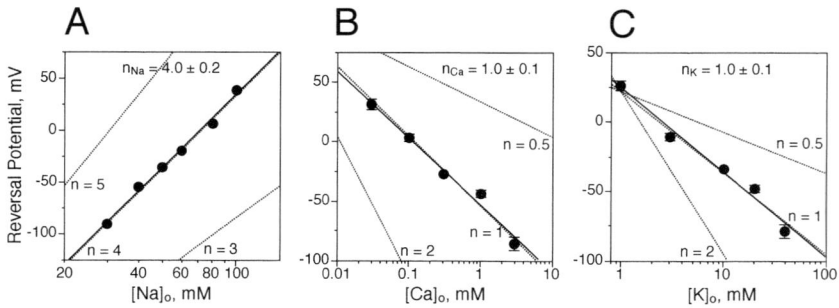

FIGURE 8. Stoichiometry determinations for NCKX2. Plots of reversal potential against log of the ion concentration for **A**, Na$^+$; **B**, Ca^{2+}; or **C**, K$^+$, determined as illustrated in FIGURE 7. Averages ± SEM for between 5 and 8 determinations at each data point are shown; in some cases the excursion of the error is smaller than the symbol size. *Solid lines* show the fit of this data to the equation described in the text. The stoichiometry (n_X, the number of ions X moved per net charge) extracted from the slope of these fits using a value of 58.5 for the term 2.303(RT/F) is shown in each panel. For comparison, *dashed lines* show the theoretical relation described by the equation in the text with the indicated number (n) of ions moved per net charge. Modified from Dong et al.[16] with permission.

ometry per exchange cycle of 4 Na$^+$ ions transported in exchange for 1 Ca^{2+} and 1 K$^+$, and the net movement of 1 positive charge in the same direction as the Na$^+$ ions. These data are consistent with previous data obtained on the retinal rod photoreceptor Na$^+$/Ca^{2+}+K$^+$ exchanger[20,21] as well with recent quantitative flux data on the retinal cone Na$^+$/Ca^{2+}+K$^+$ exchanger.[22]

Given this 4:1:1 Na$^+$:Ca^{2+}:K$^+$ stoichiometry, NCKX2 would be expected to operate in the Ca^{2+} extrusion mode under all but the most extreme conditions. If the normal physiologic extracellular ion concentrations for Na$^+$, K$^+$, and Ca^{2+} are 140 mM, 5 mM, and 1 mM, whereas the intracellular concentrations are 10 mM, 140 mM, and 0.1 μM, respectively, then NCKX2 would only reverse direction to allow Ca^{2+} influx at a membrane potential more positive than +125 mV. During an action potential or just after a train of repeated potentials, these ionic concentrations may change substantially. For example, if extracellular concentrations of Na$^+$, K$^+$, and Ca^{2+} were 120 mM, 60 mM, and 1 mM, whereas the intracellular concentrations were 25 mM, 120 mM, and 1 μM, respectively, then NCKX2 would reverse function at only +2 mV. Even so, the kinetic analysis just discussed would suggest that Ca^{2+} flux into the cell should be very low. From such considerations, it seems likely that NCKX2 is designed to operate largely in the Ca^{2+} extrusion mode, possibly to assist in the control of cytosolic Ca^{2+} in those neurons that undergo frequent and large ionic fluctuations across their plasma membranes, where other transport pathways would be inadequate.

REFERENCES

1. BERRIDGE, M.J. 1998. Neuronal calcium signaling. Neuron **21:** 13–26.
2. BERRIDGE, M.J., M.D. BOOTMAN & P. LIPP. 1998. Calcium: a life and death signal. Nature **395:** 645–648.

3. BLAUSTEIN, M.P. & W.J. LEDERER. 1999. Sodium/calcium exchange: its physiological implications. Physiol. Rev. **79:** 763–854.
4. PHILIPSON, K.D. & D.A. NICOLL. 2000. Sodium-calcium exchange: a molecular perspective. Annu. Rev. Physiol. **62:** 111–133.
5. LEE, S.-L., A.S.L. YU & J. LYTTON. 1994. Tissue-specific expression of Na/Ca exchanger isoforms. J. Biol. Chem. **269:** 14849–14852.
6. LI, Z., S. MATSUOKA, L.V. HRYSHKO, et al. 1994. Cloning of the NCX2 isoform of the plasma membrane Na-Ca exchanger. J. Biol. Chem. **269:** 17434–17439.
7. TSOI, M., K.-H. RHEE, D. BUNGARD, et al. 1998. Molecular cloning of a novel potassium-dependent sodium-calcium exchanger from rat brain. J. Biol. Chem. **273:** 4115–4162.
8. KRAEV, A., B.D. QUEDNAU, S. LEACH, et al. 2001. Molecular cloning of a third member of the potassium-dependent sodium-calcium exchanger gene family, NCKX3. J. Biol. Chem. **276:** 23161–23172.
9. LI, X.F., A. KRAEV & J. LYTTON. 2002. Molecular cloning of a fourth member of the potassium-dependent sodium–calcium exchanger gene family, NCKX4. J. Biol. Chem. In press.
10. POON, S., S. LEACH, X.F. LI, et al. 2000. Alternatively spliced isoforms of the rat eye sodium/calcium+potassium exchanger NCKX1. Am. J. Physiol. Cell Physiol. **278:** C651–660.
11. KIMURA, M., E.M. JEANCLOS, R.J. DONNELLY, et al. 1999. Physiological and molecular characterization of the Na+/Ca2+ exchanger in human platelets. Am. J. Physiol. Heart Physiol. **277:** H911–H917.
12. SCHNETKAMP, P.P.M. 1995. Calcium homeostasis in vertebrate retinal rod outer segments. Cell Calcium **18:** 322–330.
13. NICOLL, D.A., S. LONGONI & K.D. PHILIPSON. 1990. Molecular cloning and functional expression of the cardiac Na,Ca-exchanger. Science **250:** 562–565.
14. REILÄNDER, H., A. ACHILLES, U. FRIEDEL, et al. 1992. Primary structure and functional expression of the Na/Ca,K-exchanger from bovine rod photoreceptors. EMBO J. **11:** 1689–1695.
15. SCHWARZ, E.M. & S. BENZER. 1997. Calx, a sodium-calcium exchanger gene of *Drosophila melanogaster*. Proc. Natl. Acad. Sci. USA **94:** 10249–10254.
16. DONG, H., P.E. LIGHT, R.J. FRENCH & J. LYTTON. 2001. Electrophysiological characterization and ionic stoichiometry of the rat brain K^+-dependent Na^+/Ca^{2+} exchanger, NCKX2. J. Biol. Chem. **276:** 25919–25928.
17. SCHNETKAMP, P.P.M., J.E. TUCKER & R.T. SZERENCSEI. 1995. Ca2+ influx into bovine retinal rod outer segments mediated by Na+/Ca2+/K+ exchange. Am. J. Physiol. Cell Physiol. **269:** C1153–1159.
18. SCHNETKAMP, P.P.M., X.B. LI, D.K. BASU & R.T. SZERENCSEI. 1991. Regulation of free cytosolic Ca2+ concentration in the outer segments of bovine retinal rods by Na-Ca-K exchange measured with fluo-3. I. Efficiency of transport and interactions between cations. J. Biol. Chem. **266:** 22975–22982.
19. RANSOM, C.B., B.R. RANSOM & H. SONTHEIMER. 2000. Activity-dependent extracellular K+ accumulation in rat optic nerve: the role of glial and axonal Na+ pumps. J. Physiol. **522:** 427–442.
20. SCHNETKAMP, P.P.M., D.K. BASU & R.T. SZERENCSEI. 1989. Na^+-Ca^{2+} exchange in bovine rod outer segments requires and transports K^+. Am. J. Physiol. Cell Physiol. **257:** C153–C157.
21. CERVETTO, L., L. LAGNADO, R.J. PERRY, et al. 1989. Extrusion of calcium from rod outer segments is driven by both sodium and potassium gradients. Nature **337:** 740–743.
22. SZERENCSEI, R.T., C.F. PRINSEN & P.P. SCHNETKAMP. 2001. Stoichiometry of the retinal cone Na/Ca-K exchanger heterologously expressed in insect cells: comparison with the bovine heart Na/Ca exchanger. Biochemistry **40:** 6009–6015.

Brain Distribution of the Na^+/Ca^{2+} Exchanger-Encoding Genes NCX1, NCX2, and NCX3 and Their Related Proteins in the Central Nervous System

ADRIANA CANITANO,[a] MICHELE PAPA,[b] FRANCESCA BOSCIA,[a] PASQUALINA CASTALDO,[a] STEFANIA SELLITTI,[b] MAURIZIO TAGLIALATELA,[a] AND LUCIO ANNUNZIATO[a]

[a]*Division of Pharmacology, Department of Neuroscience, School of Medicine, University of Naples Federico II, 80131 Naples, Italy*

[b]*Institute of Anatomy, 2nd University of Naples, 80131 Naples, Italy*

ABSTRACT: In the central nervous system, the Na^+/Ca^{2+} exchanger plays a fundamental role in controlling changes in the intracellular concentrations of Na^+ and Ca^{2+} ions that occur in physiologic conditions such as neurotransmitter release, cell migration and differentiation, gene expression, as well as neurodegenerative processes. Three genes, NCX1, NCX2, and NCX3, encoding for Na^+/Ca^{2+} exchanger isoforms have been cloned. In this review, by using nonradioactive *in situ* hybridization and light immunohistochemistry with NCX isoform-specific riboprobes and antibodies, respectively, a systematic brain mapping for both transcripts and proteins encoded by all three NCX genes is described. Intense expression of NCX transcripts and proteins was detected in the cerebral cortex, hippocampus, thalamus, metathalamus, hypothalamus, brainstem, spinal cord, and cerebellum. In these areas, NCX transcripts and proteins were often found with an overlapping distribution pattern, although specific brain areas displaying a peculiar expression of each exchanger isoform were also found. Furthermore, immunoelectron and confocal microscopy revealed the expression of the NCX1 isoform of the exchanger at both pre- and postsynaptic sites as well as in association with membranes of the endoplasmic reticulum. Collectively, these data suggest that the different isoforms of the Na^+/Ca^{2+} exchanger appear to be selectively expressed in several CNS regions where they might underlie different functional roles.

KEYWORDS: brain; calcium; sodium; sodium/calcium exchanger; *in situ* hybridization; immunohistochemistry; neurodegeneration

Address for correspondence: Lucio Annunziato, M.D., Division of Pharmacology, Department of Neuroscience, School of Medicine, University of Naples Federico II, Ed. 19, Via Pansini 5, 80131 Naples, Italy. Voice: 0039-081-7463318/3325; fax: 0039-081-7463323.
lannunzi@unina.it

INTRODUCTION

In neuronal cells, the Na^+/Ca^{2+} exchanger plays a fundamental role in controlling Na^+ and Ca^{2+} homeostasis.[1] In fact, increases in the intracellular concentrations of Na^+ and Ca^{2+} ions are known to occur in physiologic conditions, such as neurotransmitter and hormone release,[2–4] and in pathophysiologic states, such as neurodegenerative diseases and hypoxic–anoxic states.[5–7] Interestingly, in the central nervous system (CNS) three genes encoding for Na^+/Ca^{2+} exchanger isoforms were cloned.[8–10] Although the NCX1 gene is expressed in several tissues including heart, transcripts encoded by the NCX2 and NCX3 genes were found exclusively in neural and skeletal muscle tissues.[9,10] Both NCX1 and NCX3, but not NCX2, give rise to several splicing variants that appear to be selectively expressed in different tissues and, within the brain, in different brain regions and cellular populations.[11,12]

Although two previous studies[13,14] provided insights into the regional distribution of mainly the NCX1 isoform in the brain, a systematic study of CNS distribution of both transcripts and proteins encoded by the three NCX isoforms was not yet available. In this review, detailed brain mapping of NCX1, NCX2, and NCX3 isoforms is provided (TABLE 1) with the help of *in situ* hybridization and immunohistochemistry with NCX isoform-specific riboprobes and antibodies, respectively. The subcellular distribution of NCX1 and NCX3 isoforms is also shown by electron and confocal microscopy.

These data show that the three exchanger isoforms appear to be differentially expressed in several CNS regions, suggesting that each exchanger may play a different functional role in distinct CNS regions.

TABLE 1. Distribution of NCX1, NCX2, and NCX3 mRNA in the CNS

	NCX1	NCX2	NCX3
Olfactory system			
Olfactory bulb	+	++	++
Anterior olfactory nucleus	++	++	+
Cerebral cortex			
Cingulate cortex	++	++	++
Motor cortex: layers III/V	+++	+	+
Motor cortex: layers V/VI	++	++	+
Sensory cortex: layers I/III	+	+++	+
Sensory cortex: layers V/VII	+	+++	+/−
Piriform cortex	+++	+++	++
Hippocampus			
Pyramidal cell layer, CA1	++	++	+
Pyramidal cell layer, CA2	+++	+++	+
Pyramidal cell layer, CA3	+++	+++	+
Granule cell layer, dentate gyrus	++	+++	+++
Polymorph layer, dentate gyrus	++	+	−
Amygdala			
Lateral nucleus	++	+	++

TABLE 1. Distribution of NCX1, NCX2, and NCX3 mRNA in the CNS (*Continued*)

	NCX1	NCX2	NCX3
Basolateral/basomedial nuclei	++	+/−	++/+++
Central nucleus	++	+	++
Basal ganglia			
Nucleus accumbens/core	+	+++	+
Nucleus accumbens/shell	−	+++	+
Caudate putamen	+/−	+++	+/−
Globus pallidus	+/−	+++	+/−
Thalamus			
Habenula	++	++	+/++
Anterodorsal thalamic nuclei	++	++	++
Ventromedial nucleus	++/+++	++	++
Ventrolateral	++	++	+/++
Ventroposterior (VP)	+	++/+++	+/++
Dorsolateral thalamic nuclei	++	++/+++	++/+++
Reticular nucleus	+	++	++
Geniculate nucleus	+	++	−
Hypothalamus			
Magnocellular preoptic nucleus	+++	−	+/−
Supraoptic nucleus	+++	+	+
Lateral hypothalamic nuclei	++	+	+
Mammillary nuclei	+/−	++	+
Brainstem			
Substantia nigra, pars compacta, VTA	++/+++	+/−	+/−
Reticular formation	++	++	++
Raphe nuclei	++	++	++/+++
Cerebellum			
Granular cell layer	+	+	++
Purkinje cells	+++	+	++
Molecular cell layer	−	−	−
Cerebellar nuclei	++	++	−
Medulla and spinal cord	+	++	++

DISTRIBUTION OF NCX1, NCX2, AND NCX3 TRANSCRIPTS AND PROTEINS IN THE RAT CNS

Cerebral Cortex

In the cerebral cortex, *in situ* hybridization experiments revealed expression of the mRNAs encoding for all three exchanger isoforms. In particular, the upper neurons of the motor system and the terminal neurons of the sensory system preferen-

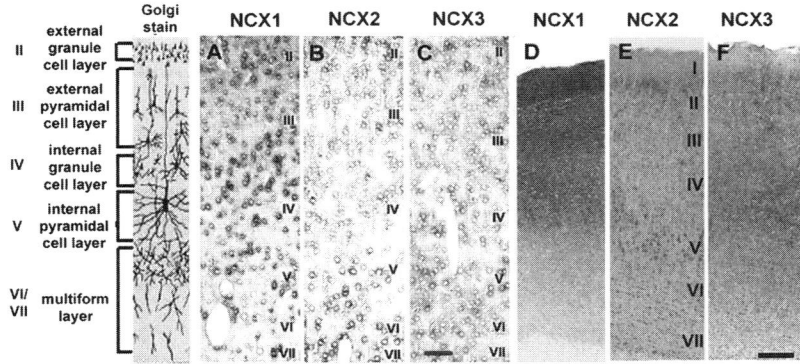

FIGURE 1. See Plate 10 in color section. Expression of NCX1, NCX2, and NCX3 transcripts (**A, B,** and **C**) and proteins (**D, E,** and **F**) in the different layers of the motor cortex. Roman numbers refer to the cortical layers, schematically represented by the Golgi stain on the left. Scale bars, 100 μm.

tially express distinct NCX isoforms (FIG. 1); in fact, pyramidal neurons of layers III and V in the motor cortex intensively expressed mainly NCX1 transcripts, whereas neurons within the somatosensory cortical area appeared to be most intensively stained with the NCX2-specific riboprobe. More superficially, within the molecular layer of the motor cortex containing the terminal dendritic field of the pyramidal cells, immunoreactivity for NCX1 was much more intense than that for NCX2, a result consistent with the preferential expression of NCX1 transcripts in neurons having their cell bodies in layers III and V of this cortical area. NCX2 isoform was characteristically present in neurons of layers V and VI of the sensory cortex, where corticothalamic projections controlling afferent thalamic inputs originate.

Hippocampus

Within the hippocampus, riboprobes for all three NCX isoforms showed intense labeling of most neuronal populations, whereas immunocytochemistry experiments revealed differential expression of selective NCX isoforms in the hippocampal circuitry components (FIG. 2). NCX1 protein expression was particularly intense in the granule cell layer and hilus of the dentate gyrus, representing the terminal field of the perforant pathway, the major excitatory input to the hippocampus, originating from the entorhinal cortex. In accordance with this result, neurons within this cortical area showed intense positivity for NCX1 transcripts. Both NCX1 and NCX3 antibodies strongly labeled the mossy fiber projections to the CA3 region, whereas NCX3 protein was mainly found in the CA1 field. These results suggest that distinct Na^+/Ca^{2+} exchanger isoforms may play a crucial role in controlling intracellular Na^+ and Ca^{2+} homeostasis of the major afferent, intrinsic, and efferent hippocampal projections. These circuitries (FIG. 3) are crucial for synaptic plasticity phenomena, such as long-term potentiation and depression;[15] as a matter of fact, changes in Ca^{2+} homeostasis have been linked to LTP and LTD occurrences in the hippocampus.[16] Current results on NCX distribution in the hippocampus may provide the anatomic

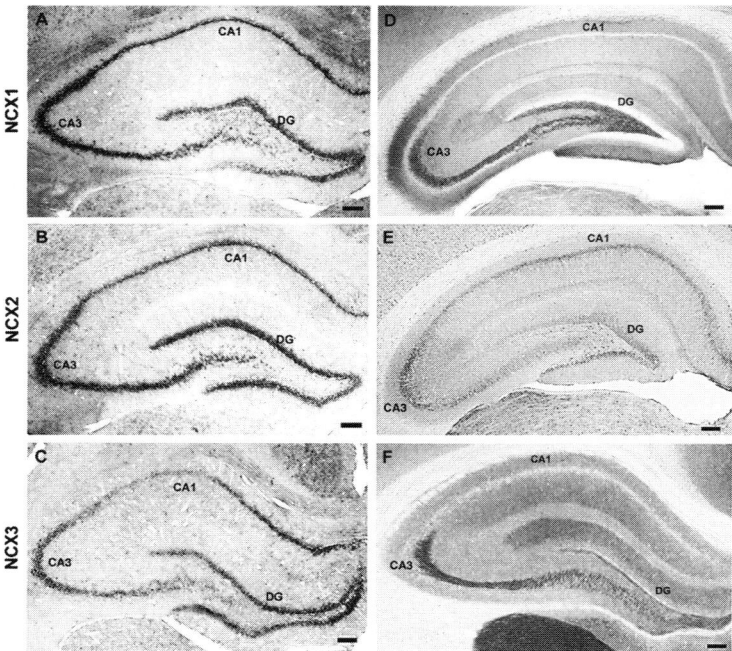

FIGURE 2. See Plate 11 in color section. Expression of NCX1, NCX2, and NCX3 transcripts (panels **A**, **B**, and **C**) and proteins (panels **D**, **E**, and **F**) in the different fields of the hippocampus of the adult rat brain. CA1, CA3: subfields in the Cornus Ammonis (CA); DG: dentate gyrus. Scale bars, 200 μm.

basis for the suggested pathophysiologic role played by the antiporter in pathophysiologic conditions within this brain re-gion, such as anoxic/glucopenic conditions[7,17] and excitatory neurotransmitter release.[18,19]

Basal Ganglia and Amygdala

NCX isoforms were all expressed in crucial areas for extrapyramidal control of motor coordination. In particular, both *in situ* and immunohistochemical experiments showed NCX2 labeling in the caudate putamen and more so in the globus pallidus, two regions mainly endowed with intrinsic GABAergic neuronal elements. By contrast, *in situ* hybridization experiments revealed that NCX1 transcripts were heavily expressed in the substantia nigra pars compacta, where dopaminergic cell bodies are located. Consistent with this result, NCX1 antibodies detected a diffuse pattern of immunoreactivity in the striatum, the terminal projection field of dopaminergic nigrostriatal neurons. Interestingly, although in the basal ganglia, as in other brain regions, NCX3 was expressed less than the other NCX isoforms, both the mRNA and protein encoded by this gene were abundantly expressed in the nucleus accumbens, a brain region involved in the motivational control of motor coordination. Noteworthy is the recent report that in striatal synaptosomes, NCX inhibitors potentiate metamphetamine-induced neurotoxicity,[20] arguing in favor of a relevant

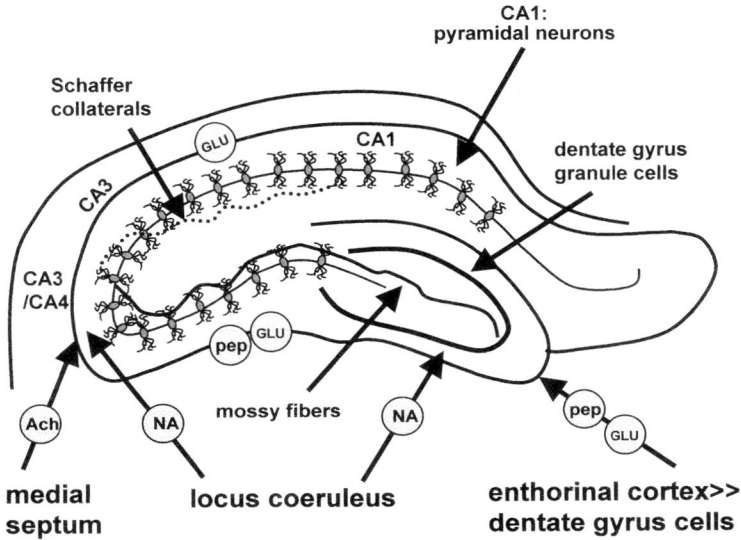

FIGURE 3. In this scheme of the hippocampus are represented the major synaptic circuits with some of the neurotrasmitter involved. (Modified from *The Biochemical Basis of Neuropharmacology.* J.R Cooper et al., Eds.: 470. Oxford University Press. New York.)

protective role played by NCX isoforms during the course of action of this dopaminergic neurotoxin. In this regard, NCX1 and NCX3 isoforms on striatal nerve terminals appear to be major candidates as molecular targets for this neuroprotective action.

As for the amygdala complex, it has been proposed that activation of glutamate receptors in the basolateral amygdala triggers a Na^+/Ca^{2+} exchanger-mediated inward current;[21] indeed, this brain region displayed high levels of transcripts encoding for the three exchanger isoforms.

Diencephalon

Within several thalamic nuclei, both RNAs and proteins of all three NCX isoforms were detected. Particularly for NCX1 and NCX2, the reticular nucleus displayed many intensely labeled neurons, whereas at the protein level, intense staining was evident for NCX1 in the reticular nucleus, for NCX2 in the midline nuclei such as the paraventricular nucleus, and for NCX3 in the laterodorsal nuclei.

In the hypothalamus, several neurons in the arcuate nucleus and mammillary nuclei showed intense NCX2 mRNA labeling. It is noteworthy that excitation of histaminergic tuberomammillary neurons induced by serotonin and orexin involves activation of the Na^+/Ca^{2+} exchanger.[22,23]

Receptor-dependent regulation of NCX activity in the hypothalamus is not restricted to serotonin. It has been suggested that histamine-induced depolarization of vasopressin neurons in the supraoptic nucleus depends on Na^+/Ca^{2+} exchanger activation.[24] Interestingly, large amounts of transcripts encoding for the NCX1 isoform were found in this area. Furthermore, within the ventromedial nucleus, a region that displayed strong labeling with NCX1- and NCX2-specific riboprobes, it was suggested that activation of metabotropic group I glutamate receptors depolarizes and increases the firing frequency of neurons, an effect possibly mediated by G-protein–dependent activation of the Na^+/Ca^{2+} exchanger.[25]

Finally, the median eminence, a neuroendocrine region devoid of cell bodies and containing the nerve terminals of hypothalamic tuberoinfundibular neurons, showed intense immunostaining for all three exchanger isoforms. These results provide morphologic support for the functional evidence that pharmacologic manipulation of the Na^+/Ca^{2+} exchanger may influence endogenous dopamine release from tuberoinfundibular neurons[2,3] and for the presynaptic localization of the three exchanger isoforms in this area.

Cerebellum

Among the brain regions examined, the cerebellum showed the greatest density of cells that stained positively for the mRNAs encoding for all NCX isoforms. In the cerebellum, a different neuronal population showed expression of transcripts encoding for selective NCX isoforms. NCX1 immunoreactivity was associated with the glomerular structure formed by mossy fiber terminals, granule cell dendrites, and Golgi cell axons (FIG. 4), suggesting that this exchanger isoform plays a fundamental role in controlling the main input to the cerebellar cortex. Interestingly, functional studies on cultured cerebellar granule cells suggest that the reverse mode of the exchanger mediates a glutamate-induced intracellular Ca^{2+} increase,[26] an effect that, in view of the current results, could be mediated by NCX1. NCX1 and NCX3 tran-

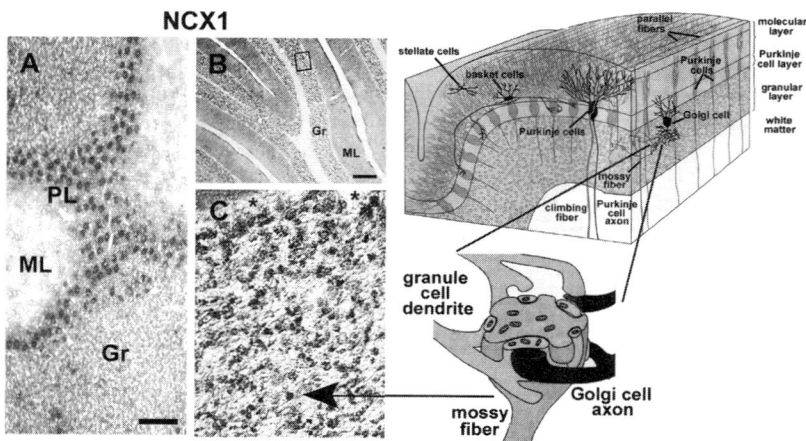

FIGURE 4. See Plate 12 in color section. Expression of NCX1 transcript (**A**) and protein (**B** and **C**) in the cerebellum. The box in panel **B** shows the region magnified in panel **C**. The large arrow head labels the cerebellar glomerulus schematically presented on the right; *asterisks* indicate the location of Purkinje cell bodies. PL, Purkinje layer; ML, molecular layer; Gr, granule cell layer; Scale bars: 100 μm (**A**, **B**, and **C**); 200 μm (**D**, **E**, and **F**); and 20 μm (**G**, **H**, and **I**).

scripts were the most represented isoforms in the Purkinje cell layer; consistent with this pattern of expression, the molecular layer of the cerebellar cortex, where parallel fibers of the granule cells synapse on the dendritic harborization of the Purkinje cells, was labeled with particular intensity with NCX1- and NCX3-specific antibodies. These findings suggest possible postsynaptic localization of these exchanger proteins in the superficial layers of the cerebellar cortex. NCX3 immunoreactivity was characteristically associated with axonal harborization of the basket cells around the Purkinje cell bodies and axon hillock, suggesting that this exchanger isoform participates in inhibitory feedback control of Purkinje cells representing the main cerebellar cortical output.

SUBCELLULAR LOCALIZATION OF Na^+/Ca^{2+} EXCHANGER ISOFORMS

Confocal Laser Microscopy

Double or triple immunofluorescence labeling of hippocampal neurons with antibodies raised against NCX1 and NCX3 showed a cluster distribution of immuno-

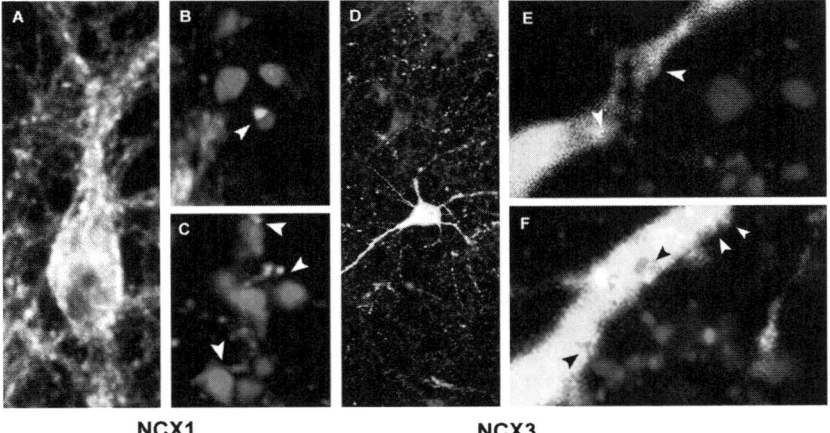

FIGURE 5. See Plate 13 in color section. Confocal visualization of NCX1 and NCX3 isoforms in 14–21 DIV cultured hippocampal neurons. In panels **A**, **B**, and **C**, hippocampal neurons were double-labeled with the NCX1-specific antibody (visualized in red) and with an anti-synapsin antibody (visualized in green). In panels **B** and **C**, *arrowheads* point towards regions where NCX1 and the presynaptic marker, although adjacent, are not colocalized. The low density of yellow fluorescence in panel **B**, indicating regions of possible colocalization of the two antibodies, suggests that the two proteins were mostly expressed in distinct subcellular regions. Panels **D**, **E**, and **F** show the results of a triple-labeling experiments with anti NCX3-specific antibody (visualized in red) and with anti-synaptophysin antibody (visualized in blue), in a single hippocampal neuron which was injected with Alexa-568 (visualized in green). Panels **E** and **F**, representing a magnification of panel **D**, show that in several regions of the field, NCX3 appeared not to colocalize with the presynaptic marker synaptophysin.

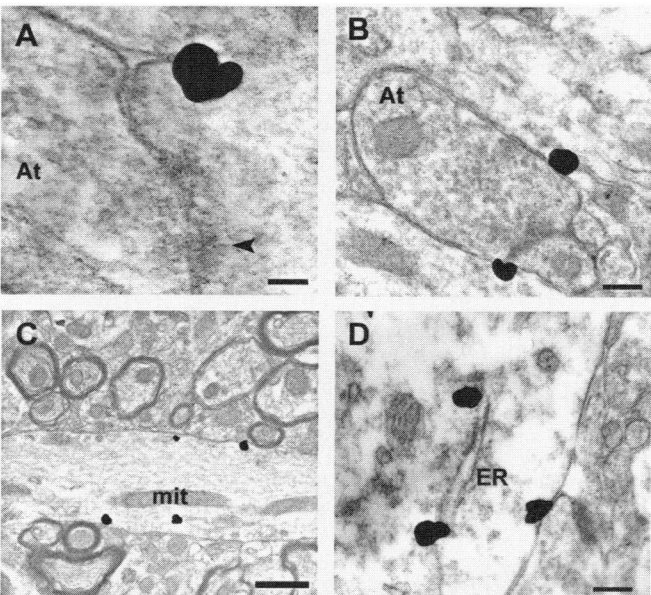

FIGURE 6. Immunoelectron microscopy of NCX1 expression in adult rat brain sections. In panel **A** and **B**, the gold particle labeling NCX1 is localized both on the postsynaptic (**A**) and the presynaptic (**B**) sites of an excitatory synapse, although at a distance from the synaptic cleft identified by the postsynaptic density. NCX1 was also localized along the axonal plasmamembrane; (**C**). The possible association of the exchanger with the subplasmalemmal endoplasmic reticulum, as recently suggested,[28] is shown in panels **C** and **D**. Scale bars: 4 μm (**A**); 20 μm (**B**); 40 μm (**C**); 15 μm (**D**).

reactivity that was distinct from synapsin- or synaptophysin-identified presynaptic sites, possibly indicating a preferential location for these exchanger isoforms outside the release sites (FIG. 5), as previously suggested for NCX1.[27] Nevertheless, some regions in which NCX1 and NCX3 immunofluorescence overlapped with that of presynaptic markers could also be visualized.

Immunogold Electron Microscopy

Immunogold silver-enhanced electron microscopy enabled us to detect NCX1 particles on both post- and presynaptic sites of excitatory synapses (FIG. 6A and B, respectively), but in regions located outside the release sites. NCX1 was associated with membranes of the endoplasmic reticulum (FIG. 6D), as recently suggested,[28] as well as with axonal (FIG. 6C) and dendritic (FIG. 6D) plasma membranes.

CONCLUSIONS

The described mapping of the distribution of the three NCX isoforms in the mammalian brain may help to unravel the physiologic and possibly the pathophysiologic role of the different NCX isoforms in neuronal function. Therefore, despite the lack of knowledge of the functional characteristics of each NCX isoform with respect to kinetics, ligand affinity, regulation,[29,30] and pharmacology,[31] it is tempting to speculate that the significant heterogeneity of NCX isoforms generated by the three genes might serve selective functional roles in distinct regions of the CNS. Further studies on the localization of each NCX isoform in chemically defined neuronal systems may enhance current knowledge of the pathophysiologic role of the Na^+/Ca^{2+} exchanger in the CNS.

ACKNOWLEDGMENTS

We are deeply indebted to Prof. Kenneth Philipson (Department of Physiology, UCLA, Los Angeles, USA) for kindly providing NCX1, NCX2, and NCX3 cDNAs, and NCX1- and NCX3-specific antibodies; to Prof. Hartmut Porzig (Universität Bern, Bern, Switzerland) for the NCX2-specific antibody; and to F. Iammuno (Stazione Zoologica A. Dohrn, Naples, Italy) for assistance in the electron microscopy experiments.

This study was supported by the following grants: National Research Council (CNR) 97.5184ST74, 98.03371ST74, and 00.D132-001 (to L.A.); MURST COFIN 2000, 2001 (to L.A.); MURST-COFIN 2000 (to M.P.); MURST-CNR Biotechnology Program L.95/95, 99.000192.PF31, and 01.00169.PF31 (to L.A.); and Regione Campania Legge 41/94 (to L.A. and M.P.) and Ist. Sup. Sanità 2002 (to L.A.).

REFERENCES

1. BLAUSTEIN, M.P. & W.J. LEDERER. 1999. Sodium/calcium exchange: its physiological implications. Physiol. Rev. **79:** 763–854.
2. TAGLIALATELA, M., S. AMOROSO, L.M. CANZONIERO, et al. 1988. Membrane events and ionic processes involved in dopamine release from tuberoinfundibular neurons. II. Effect of the inhibition of the Na^+–Ca^{2+} exchange by amiloride. J. Pharmacol. Exp. Ther. **246:** 689–694.
3. TAGLIALATELA, M., L.M. CANZONIERO, E.J. CRAGOE, JR., et al. 1990. Na^+–Ca^{2+} exchange activity in central nerve endings. II. Relationship between pharmacological blockade by amiloride analogues and dopamine release from tuberoinfundibular hypothalamic neurons. Mol. Pharmacol. **38:** 393–400.
4. DI RENZO, G., S. AMOROSO, A. BASSI, et al. 1995. Role of the Na^+–Ca^{2+} and Na^+–H^+ antiporters in prolactin release from anterior pituitary cells in primary culture. Eur. J. Pharmacol. **294:** 11–15.
5. LIPTON, P. 1999. Ischemic cell death in brain neurons. Physiol. Rev. **79:** 1431–1568.
6. CANZONIERO, L.A., A. ROSSI, M. TAGLIALATELA, et al. 1992. The Na^+–Ca^{2+} exchanger activity in cerebrocortical nerve endings is reduced in old compared to young and mature rats when it operates as a Ca^{2+} influx or efflux pathway. Biochim. Biophys. Acta **1107:** 175–178.
7. AMOROSO, S., A. TORTIGLIONE, A. SECONDO, et al. 2000. Sodium nitroprusside prevents chemical hypoxia-induced cell death through iron ions stimulating the activity of the Na^+–Ca^{2+} exchanger in C6 glioma cells. J. Neurochem. **74:** 1505–1513.
8. NICOLL, D.A., S. LONGONI & K.D. PHILIPSON. 1990. Molecular cloning and functional expression of the cardiac sarcolemmal Na^+–Ca^{2+} exchanger. Science **250:** 562–565.
9. LI, Z., S. MATSUOKA, L.V. HRYSHKO, et al. 1994. Cloning of the NCX2 isoform of the plasma membrane Na^+–Ca^{2+} exchanger. J. Biol. Chem. **269:** 17434–17439.

10. NICOLL, D.A., B.D. QUEDNAU, Z. QUI, et al. 1996. Cloning of a third mammalian Na^+–Ca^{2+} exchanger, NCX3. J. Biol. Chem. **271:** 24914–24921.
11. LEE, S.L., A.S. YU & J. LYTTON. 1994. Tissue-specific expression of Na^+–Ca^{2+} exchanger isoforms. J. Biol. Chem. **269:** 14849–14852.
12. QUEDNAU, B.D., D.A. NICOLL & K.D. PHILIPSON. 1997. Tissue specificity and alternative splicing of the Na^+/Ca^{2+} exchanger isoforms NCX1, NCX2, and NCX3 in rat. Am. J. Physiol. **272:** C1250–C1261.
13. MARLIER, L.N., T. ZHENG, J. TANG & D.R. GRAYSON. 1993. Regional distribution in the rat central nervous system of a mRNA encoding a portion of the cardiac sodium/calcium exchanger isolated from cerebellar granule neurons. Brain Res. Mol. Brain Res. **20:** 21–39.
14. YU, L. & R.A. COLVIN. 1997. Regional differences in expression of transcripts for Na^+/Ca^{2+} exchanger isoforms in rat brain. Brain Res. Mol. Brain Res. **50:** 285–292.
15. MADISON, D.V., R.C. MALENKA & R.A. NICOLL. 1991. Mechanisms underlying long-term potentiation of synaptic transmission. Annu. Rev. Neurosci. **14:** 379–397.
16. GHOSH, A. & M.E. GREENBERG. 1995. Calcium signaling in neurons: molecular mechanisms and cellular consequences. Science **268:** 239–247.
17. SCHRODER, U.H., J. BREDER, C.F. SABELHAUS & K.G. REYMANN. 1999. The novel Na^+/Ca^{2+} exchange inhibitor KB-R7943 protects CA1 neurons in rat hippocampal slices against hypoxic/hypoglycemic injury. Neuropharmacology **38:** 319–321.
18. AMOROSO, S., S. SENSI, G. DI RENZO & L. ANNUNZIATO. 1993. Inhibition of the Na^+–Ca^{2+} exchanger enhances anoxia and glucopenia-induced [^{3}H]aspartate release in hippocampal slices. J. Pharmacol. Exp. Ther. **264:** 515–520.
19. TRUDEAU, L.E., V. PARPURA & P.G. HAYDON. 1999. Activation of neurotransmitter release in hippocampal nerve terminals during recovery from intracellular acidification. J. Neurophysiol. **81:** 2627–2635.
20. CALLAHAN, B.T., B.J. CORD, J. YUAN, et al. 2001. Inhibitors of Na^+/H^+ and Na^+/Ca^{2+} exchange potentiate methamphetamine-induced dopamine neurotoxicity: possible role of ionic dysregulation in methamphetamine neurotoxicity. J. Neurochem. **77:** 1348–1362.
21. KEELE, N.B., V.L. ARVANOV & P. SHINNICK-GALLAGHER. 1997. Quisqualate-preferring metabotropic glutamate receptor activates Na^+–Ca^{2+} exchange in rat basolateral amygdala neurones. J. Physiol. **499** (Pt 1): 87–104.
22. ERIKSSON, K.S., D.R. STEVENS & H.L. HAAS. 2001. Serotonin excites tuberomammillary neurons by activation of Na^+/Ca^{2+} exchange. Neuropharmacology **40:** 345–351.
23. ERIKSSON, K.S., O. SERGEEVA, R.E. BROWN & H.L. HAAS. 2001. Orexin/hypocretin excites the histaminergic neurons of the tuberomammillary nucleus. J. Neurosci. **21:** 9273–9279
24. SMITH, B.N. & W.E. ARMSTRONG. 1996. The ionic dependence of the histamine-induced depolarization of vasopressin neurones in the rat supraoptic nucleus. J. Physiol. **495** (Pt 2): 465–478.
25. LEE, K. & P.R. BODEN. 1997. Characterization of the inward current induced by metabotropic glutamate receptor stimulation in rat ventromedial hypothalamic neurones. J. Physiol. **504** (Pt 3): 649–663.
26. KIEDROWSKI L., G. BROOKER, E. COSTA & J.T. WROBLEWSKI. 1994. Glutamate impairs neuronal calcium extrusion while reducing sodium gradient. Neuron **12:** 295–300.
27. JUHASZOVA M., P. CHURCH, M.P. BLAUSTEIN & E.F. STANLEY. 2000. Location of calcium transporters at presynaptic terminals. Eur. J. Neurosci. **12:** 839–846.
28. BLAUSTEIN M.P. & V.A. GOLOVINA. 2001. Structural complexity and functional diversity of endoplasmic reticulum Ca^{2+} stores. Trends Neurosci. **24:** 602–608
29. LINCK B., Z. QIU, Z. HE, et al. 1998. Functional comparison of the three isoforms of the Na^+/Ca^{2+} exchanger (NCX1, NCX2, NCX3). Am. J. Physiol. **274:** C415–C423.
30. LI L., D. GUERINI & E. CARAFOLI. 2000. Calcineurin controls the transcription of Na^+/Ca^{2+} exchanger isoforms in developing cerebellar neurons. J. Biol. Chem. **275:** 20903–20910.
31. IWAMOTO, T. & M. SHIGEKAWA. 1998. Differential inhibition of Na^+/Ca^{2+} exchanger isoforms by divalent cations and isothiourea derivative. Am. J. Physiol. **275:** C423–C430.

Neurotransmitter-Induced Activation of Sodium–Calcium Exchange Causes Neuronal Excitation

KRISTER S. ERIKSSON, OLGA A. SERGEEVA, DAVID R. STEVENS, AND HELMUT L. HAAS

Department of Neurophysiology, Heinrich-Heine-University, D-40225 Dusseldorf, Germany

KEYWORDS: NCX; excitation; brain; tuberomammillary nucleus

Neurons of the posterior hypothalamic tuberomammillary (TM) nucleus innervate most parts of the brain and release histamine (HA), which acts as a neurotransmitter, and the HAergic system is implicated in the regulation of waking and feeding.[1] Both serotonin (5-HT), released by the neurons in the raphe nucleus, and the orexin (ORX) peptides, which are produced by neurons in the perifornical hypothalamus, are involved in the regulation of waking and feeding.[2]

To investigate the relations between these neurotransmitter systems, we studied the actions of ORX and 5-HT on the TM nucleus, using sharp electrode intracellular recordings from rat hypothalamic slices. Both ORX and 5-HT strongly depolarized the TM neurons dose-dependently and increased their firing rate via an action on postsynaptic receptors (FIG. 1A and B).[3,4] Single-cell polymerase chain reaction (PCR) studies showed that TM neurons express both ORX receptors (1 and 2), whereas pharmacologic experiments showed that the 5-HT effect was mediated by the 5-HT$_{2C}$ receptor. The depolarization by ORX was unaffected by increasing $[K^+]_o$ and attenuated by replacement of $[Na^+]_o$ with N-methyl-D-glucamine (NMDG). It was suppressed by KB-R7943 and abolished by 3 mM Ni^{2+}. Depolarization by 5-HT was dependent on temperature (33°C vs. room temperature) and was suppressed by replacement of $[Na^+]_o$ with Li^+ or NMDG. Pretreatment with Ni^{2+}, 2',4'-dichlorobenzamil or KB-R7943 also attenuated the effect (FIG. 1C). These features indicate that excitation of TM neurons by 5-HT and ORX is the result of activation of a Na^+/Ca^{2+}-exchanger (NCX), which leads to a net inward current that depolarizes the neurons. Only a few other reports attribute neuronal excitation to the current produced by an NCX operating in its forward mode. Other investigators have shown that depolarization of neurons in the amygdala and ventromedial hypothalamus via class

Address for correspondence: Dr. Krister S. Eriksson, Department of Neurophysiology, Heinrich-Heine-University, Moorenstr. 5, D-40225 Dusseldorf, Germany. Voice: +492118112688; fax: +492118114231.
krister@uni-duesseldorf.de

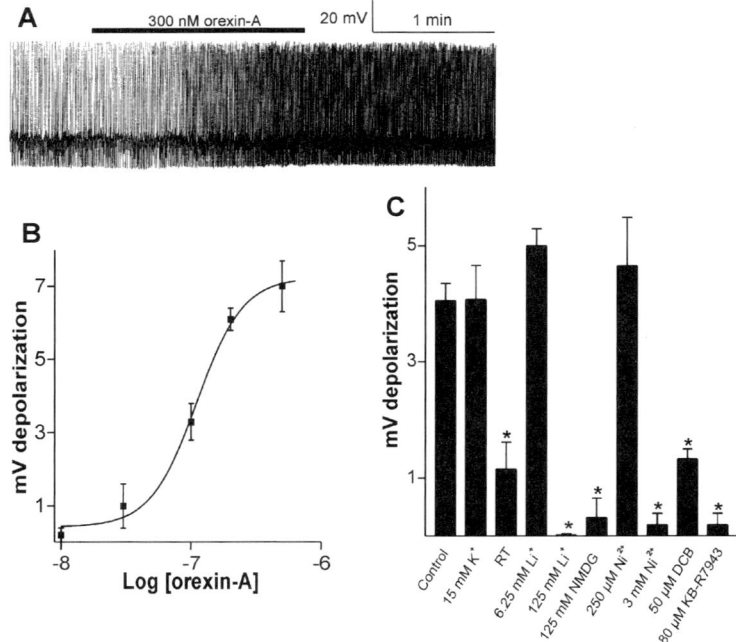

FIGURE 1. Intracellular recordings from tuberomammillary (TM) neurons. (**A**) Treatment with orexin causes a marked increase in the spontaneous firing rate of the neuron. (**B**) Depolarization of TM neurons by orexin is dose dependent. Data were obtained under tetrodotoxin, which inhibits firing and causes synaptic isolation. Thus, the effect is mediated by postsynaptic receptors. (**C**) Overview of 5-HT–induced depolarization of TM neurons under different conditions. Depolarization is attenuated by treatments that are known to inhibit NCX activity. See text for further explanations (mean ± SEM; * $P < 0.005$.)

I metabotropic glutamate receptors as well as H_1 receptor-mediated depolarization of supraoptic nucleus neurons occurs via activation of an NCX.[5-7] Because the involved receptors are all coupled to IP_3 production, activation of NCX is most likely an effect secondary to a surge in $[Ca^{2+}]_i$. This Ca^{2+} is probably mainly released from intracellular stores, because no major Ca^{2+}-channel component or change in membrane conductance has been associated with NCX activation.

These results support the view that t NCX can play a prominent role in determining the excitability of neurons. The results also provide a functional connection between three transmitter systems, histaminergic, orexinergic, and serotonergic, which modulate many physiologic functions, such as sleep and feeding, in the brain.

REFERENCES

1. BROWN. R.E., D.R. STEVENS & H.L. HAAS. 2001. The physiology of brain histamine. Prog. Neurobiol. **63:** 637–672.

2. WILLIE, J.T., et al. 2001. To eat or to sleep? Orexin in regulation of feeding and wakefulness. Annu. Rev. Neurosci. **24:** 429–458.
3. ERIKSSON, K.S., D.R. STEVENS & H.L. HAAS. 2001. Serotonin excites tuberomammillary neurons by activation of Na^+/Ca^{2+}-exchange. Neuropharmacology **40:** 345–351.
4. ERIKSSON, K.S., et al. 2001. Orexin/hypocretin excites the histaminergic neurons of the tuberomammillary nucleus. J. Neurosci. In press.
5. KEELE, N.B., et al. 2000. Epileptogenesis up-regulates metabotropic glutamate receptor activation of sodium-calcium exchange current in the amygdala. J. Neurophysiol. **83:** 2458–2462.
6. SMITH, B.N. & W.E. ARMSTRONG. 1996. The ionic dependence of the histamine-induced depolarization of vasopressin neurones in the rat supraoptic nucleus. J. Physiol. **495:** 465–478.
7. LEE. K. & P.R. BODEN. 1997. Characterization of the inward current induced by metabotropic glutamate receptor stimulation in rat ventromedial hypothalamic neurones. J. Physiol. **504:** 649–663.

Na^+/Ca^{2+} Exchanger in Na^+ Efflux–Ca^{2+} Influx Mode of Operation Exerts a Neuroprotective Role in Cellular Models of *in Vitro* Anoxia and *in Viv*o Cerebral Ischemia

A. TORTIGLIONE,[a] G. PIGNATARO,[a] M. MINALE,[a] A. SECONDO,[a]
A. SCORZIELLO,[a] G.F. DI RENZO,[a] S. AMOROSO,[a] G. CALIENDO,[b]
V. SANTAGADA,[b] AND L. ANNUNZIATO[a]

[a]*Division of Pharmacology, Department of Neuroscience, School of Medicine, "Federico II" University of Naples, 80131 Naples, Italy*

[b]*School of Pharmacy, "Federico II" University of Naples, Naples, Italy*

KEYWORDS: Na^+/Ca^{2+}; NCX; anoxia; cerebral ischemia

INTRODUCTION

The Na^+/Ca^{2+} exchanger (NCX) plays a fundamental role in controlling intracellular levels of Na^+ and Ca^{2+} in neuronal cells.[1] Because under anoxic/ischemic conditions the intracellular homeostasis of these cations is altered, the role played by NCX in conditions of cellular anoxia or *in vivo* cerebral ischemia was evaluated. Primary culture of cortical neurons and C6 glioma cells, which can express the same Na^+/Ca^{2+} exchanger isoform (NCX1) of astrocytes, were exposed to chemical hypoxia and intracerebroventricularly cannulated rats were exposed to permanent middle cerebral artery occlusion (pMCAO.). It was recently reported that sodium nitroprusside (SNP), a nitric oxide (NO)-generating compound, possesses the ability to activate the Na^+/Ca^{2+} exchanger in cultured rat astrocytes.[3] On the other hand, the SNP molecule contains iron besides the NO group. This metal ion, mediating the transfer of electrons between the cellular redox compounds and appropriate disulfide-thiol groups of the NCX molecule, can stimulate antiporter activity.[4] Therefore, we investigated whether: (a) SNP could prevent chemical hypoxia-induced C6 glioma cell death, (b) this neuroprotective action was due to a stimulation of the Na^+/Ca^{2+} exchanger activity, and (c) these effects were due to the NO donor property or the iron-releasing effect of SNP.

Address for correspondence: Lucio Annunziato, M.D., Division of Pharmacology, Department of Neuroscience, School of Medicine, "Federico II" University of Naples, Via Pansini 5, 80131 Naples, Italy. Voice: +39-081-7463318/3325; fax: +39-081-7463323.
lannunzi@unina.it

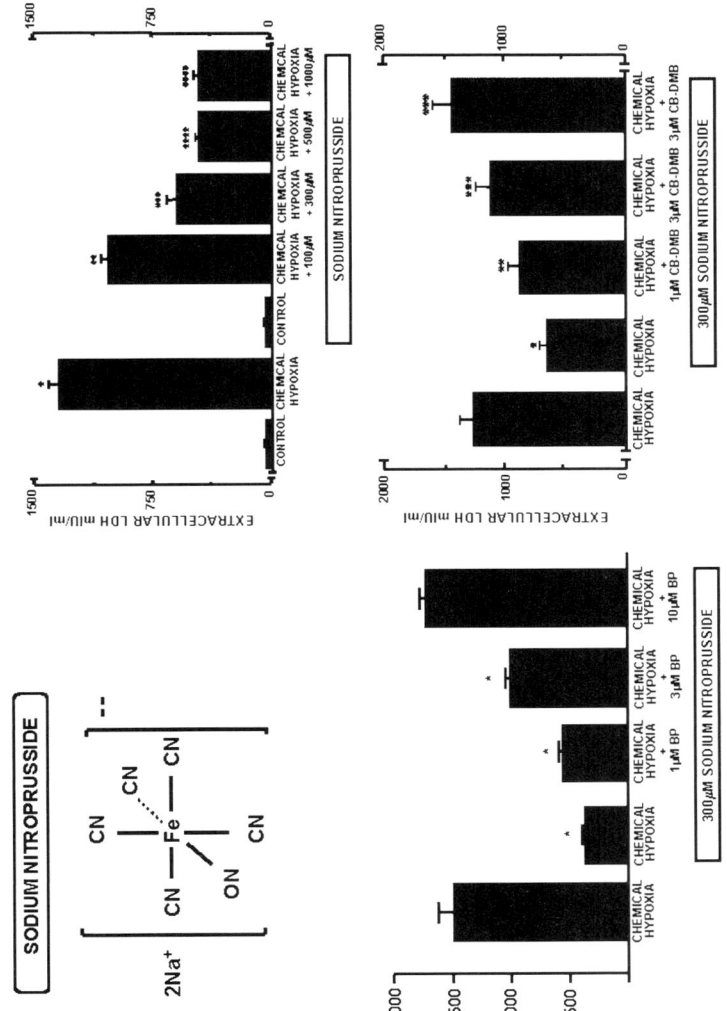

FIGURE 1. Effect of SNP on chemical hypoxia-induced LDH release and its reversal by Bepridil and CB-DMB in C6 glioma cells. (Reprinted with permission from Amoroso et al.[8])

To extrapolate these *in vitro* observations on the role played by the Na^+/Ca^{2+} exchanger during anoxia to *in vivo* models of cerebral ischemia, pMCAO was performed in male rats.

RESULTS AND DISCUSSION

C6 glioma cells were exposed to chemical hypoxia obtained by adding oligomycin (5 μg/ml) plus 2-deoxy-D-glucose (2-DG, 2 mM) in glucose-free medium for 6 hours. This combination produces a fall in cell ATP content, because oligomycin inhibits oxidative phosphorylation, whereas 2-DG causes inhibition at the first step of the glycolytic pathway, that is, at the reaction catalyzed by hexokinase. This experimental condition induced cell death, as revealed by propidium iodide/fluorescein diacetate (PI/FDA) intravital staining, and an increase of lactate dehydrogenase (LDH) release.[5] SNP is a molecule that, besides its ability to donate NO, can release iron ions. In this way, it could modify the conformation of the Na^+/Ca^{2+} exchanger from a less active (reduced) to a more active (oxidized) form, and it could increase the activity of the antiporter as a Na^+ efflux-Ca^{2+} influx pathway. Indeed, treatment of C6 glioma cells exposed to chemical hypoxia with SNP prevented, in a dose-dependent manner, cell injury as shown by the reduction of LDH release and propidium iodide (PI)-positive cells. Conversely, when C6 cells were treated with bepridil and CB-DMB, two specific inhibitors of the Na^+/Ca^{2+} exchanger, the SNP protective effect was abolished (FIG. 1). To evaluate whether the SNP effects were dependent on iron ions present in its molecule, the well known iron ion chelator deferoxamine (DFX, 1 mM) was used. This iron chelator completely counteracted the reduction of cell injury produced by SNP. These findings further support the hypothesis that the protective effect played by SNP during chemical hypoxia is due to a direct increase of Na^+/Ca^{2+} exchanger activity.[6–8] The role of the Na^+/Ca^{2+} exchanger was also investigated on cortical neurons in which the two isoforms, NCX1 and NCX2, were expressed. Experiments were performed by using microfluorimetric analysis with Fura 2-AM, a specific marker of $[Ca^{2+}]_i$. Preliminary results showed that bepridil and CB-DMB were able to avoid the $[Ca^{2+}]_i$ increase elicited by Na^+ removal. Altogether, these observations suggested that activation of NCX, when it is operating as Na^+ efflux-Ca^{2+} influx, could prevent anoxia-induced neuronal death in cell cultures.

The role played by the Na^+/Ca^{2+} exchanger was also evaluated in a model of focal cerebral ischemia *in vivo*. Permanent middle cerebral artery occlusion (pMCAO) was performed in male rats. The infarctual volume was evaluated 24 hours after pMCAO by the TTC method in cerebral slices ipsilateral to the pMCAO. Inhibition of the NCX by bepridil,[9] administered icv at a dose of 40 μg/kg 3, 6, and 22 hours after pMCAO, produced an increase in the infarctual volume (75% vs control animals). The same findings were observed with the glycosylated form of the exchange inhibitory peptide (GLU-XIP).[10] GLU-XIP is a synthetic 20 amino acid peptide, with a sequence the same as that of the exchange inhibitory peptide (XIP) region of the NCX protein, in which glycosylation was performed to facilitate peptide entrance into the cells via the membrane glucose-transporter. Glycosylated XIP was perfused icv with an osmotic pump (1 μl/hr) for 24 hours from the beginning of pMCAO at a dose of 40 μg/kg. GLU-XIP induced a 40% increase in infarctual volume compared with saline-treated ischemic rats (FIG. 2).

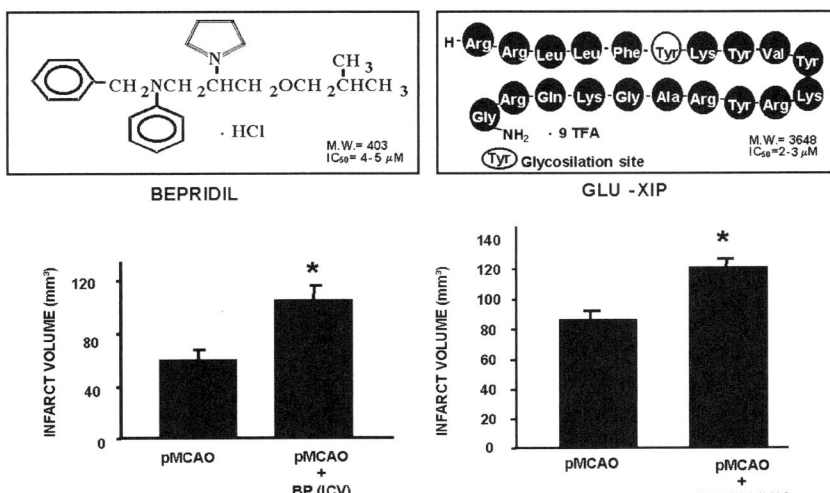

FIGURE 2. Effect of Bepridil and GLU-XIP on cerebral ischemia induced by pMCAO in the rat.

Collectively, these *in vivo* and *in vitro* studies suggest that during anoxic/ischemic conditions, increased activation of the NCX as a Na^+ efflux-Ca^{2+} influx pathway may exert a protective effect, whereas its pharmacologic blockade can exacerbate brain injury.

ACKNOWLEDGMENTS

We are deeply indebted to Prof. Kenneth Philipson (Department of Physiology, UCLA, Los Angeles, USA) for kindly providing NCX1, NCX2, and NCX3 cDNAs, and NCX1- and NCX3-specific antibodies; to Prof. Hartmut Porzig (Universität Bern, Bern, Switzerland) for the NCX2-specific antibody; and to F. Iammuno (Stazione Zoologica A. Dohrn, Naples, Italy) for assistance in the electron microscopy experiments.

This study was supported by the following grants: National Research Council (CNR) 97.5184ST74, 98.03371ST74, and 00.D132-001 (to L.A.); MURST COFIN 2000, 2001 (to L.A.); MURST-COFIN 2000 (to M.P.); MURST-CNR Biotechnology Program L.95/95, 99.000192.PF31, and 01.00169.PF31 (to L.A.); and Regione Campania Legge 41/94 (to L.A. and M.P.) and Ist. Sup. Sanità 2002 (to L.A.).

REFERENCES

1. SANCHEZ-ARMASS, S. & M.P. BLAUSTEIN. 1987. Role of sodium-calcium exchange activity in regulation of intracellular calcium in nerve terminals. Am. J. Physiol. **252:** C595–C603.

2. TAMURA, A., D.I. GRAHAM, J. MCCULLOCH & G.M. TEASDALE. 1981. Focal cerebral ischemia in the rat. I. Description of technique and early neuropathological consequences following middle cerebral artery occlusion. J. Cerebr. Blood Flow Metab. **1:** 53–60.
3. ASANO, S., T. MATSUDA, K. TAKUMA, *et al.* 1995. Nitroprusside and cyclic GMP stimulate Na^+-Ca^{2+} exchange activity in neuronal preparation and cultured rat astrocytes. J. Neurochem. **64:** 2437–2441.
4. REEVES, J.P., C.A. BAILEY & C.C. HALE. 1986. Redox modification of the sodium-calcium exchange activity in cardiac sarcolemmal vesicles. J. Biol. Chem. **261:** 4948–4955.
5. CHOI, D.W., J. KOH & S. PETERS. 1988. Pharmacology of glutamate neurotoxicity in cortical cell culture: attenuation by NMDA antagonists. J. Neurosci. **8:** 185–196.
6. AMOROSO, S., S. SENSI, G.F. DI RENZO & L. ANNUNZIATO. 1993. Inhibition of the Na^+-Ca^{2+} exchanger enhances anoxia and glucopenia-induced ^{3}H-aspartate release in hippocampal slices. J. Pharmacol. Exp. Ther. **264:** 515–520.
7. AMOROSO, S., M. DE MAIO, G.M. RUSSO, *et al.* 1997. Pharmacological evidence that the activation of the Na^+-Ca^{2+} exchanger protects C6 glioma cells during chemical hypoxia. Br. J. Pharmacol. **121:** 303–309.
8. AMOROSO, S., A. TORTIGLIONE, A. SECONDO, *et al.* 2000. Sodium nitroprusside prevents chemical hypoxia-induced cell death through iron ions stimulating the activity of the Na^+-Ca^{2+} exchanger in C6 glioma cells. J. Neurochem. **74:** 1505–1513.
9. GILL, A., S.F. FLAIM, BP. DAMIANO, *et al.* 1992. Pharmacology of bepridil. Am. J. Pharmacol. **69** (suppl. D): 11D–16D.
10. MATSUOKA, S., D.A. NICOLL, Z. HE & K.D. PHILPSON. 1997. Regulation of the cardiac Na^+-Ca^{2+} exchanger by the endogenous XIP region. J. Gen. Physiol. **109:** 273–286.

Mitochondria Buffer Sodium-Dependent Ca^{2+} Influx in Cultured Cerebellar Granule Cells

LECH KIEDROWSKI[a,b] AND ANETA CZYŻ[a,c]

[a]*Department of Psychiatry, The Psychiatric Institute, University of Illinois, Chicago, Illinois 60612, USA*

[b]*Department of Pharmacology, The Psychiatric Institute, University of Illinois, Chicago, Illinois 60612, USA*

KEYWORDS: sodium–calcium exchange; mitochondria; oligomycin; rotenone; gramicidin; neurons; excitotoxicity

INTRODUCTION

Depolarization of the neuronal plasma membrane occurs early during brain ischemia and is associated with the elevation of cytosolic [Na$^+$].[1] Under such conditions, plasmalemmal Na/Ca exchange may reverse and mediate potentially toxic Ca^{2+} influx. In depolarized cultured cerebellar granule cells (CGCs), this type of Ca^{2+} influx contributes to NMDA-mediated excitotoxicity.[2] It is well established that excessive Ca^{2+} accumulation in mitochondria leads to neuronal death.[3,4] Therefore, in the current work, we tested whether Ca^{2+} entering CGCs via reversed plasmalemmal Na/Ca exchange accumulates in mitochondria. We reversed Na/Ca exchange by imposing Na$^+$ influx through gramicidin-formed pores, which permeate alkali cations and protons but exclude polyvalent cations.[5] We then studied how mitochondrial inhibitors affect gramicidin-elicited cytosolic [Ca^{2+}] ([Ca^{2+}]$_c$) transients and ^{45}Ca^{2+} accumulation. Negative controls were performed in the presence of Li$^+$ or Cs$^+$, cations that do not substitute for Na$^+$ in plasmalemmal Na/Ca exchange.[6] Because Na/Ca exchange is pH sensitive,[7] we also monitored cytosolic pH (pH$_c$).

MATERIALS AND METHODS

CGCs prepared from 8-day-old rats and plated on poly-D-lysine–treated glass coverslips were cultured at 25 mM KCl and were used for experiments at 8–10 days *in vitro*.[8] [Ca^{2+}]$_c$ and pH$_c$ were simultaneously monitored using fura-2 FF and BCECF fluorescence,[8] respectively. (K_d of 6 μM for fura-2FF Ca^{2+} affinity[9] was used to calculate [Ca^{2+}]$_c$.) Background fluorescence was measured in cell-free areas

---

Address for correspondence: Lech Kiedrowski, Ph.D., The Psychiatric Institute, 1601 W. Taylor Street, Chicago, IL 60612. Voice: 312-413-4559; fax: 312-413-4544

lkiedr@psych.uic.edu

[c]Aneta Czyż's permanent address: Instytut Biologii Doswiadczalnej im. M. Nenckiego, Pasteura 3, 02-093 Warszawa, Poland.

and was subtracted from total fluorescence prior to calculating $[Ca^{2+}]_c$ and pH_c. Cytosolic $[Na^+]$ ($[Na^+]_c$) was measured in CGCs loaded with SBFI,[10] and $^{45}Ca^{2+}$ accumulation was measured using scintillation spectroscopy.[10] Preliminary experiments showed that application of gramicidin to CGCs incubated in Locke's buffer causes cell swelling that often leads to rupture of the plasma membrane. To reduce this swelling, we decreased $[Na^+]$ or $[Li^+]$ or $[Cs^+]$ in the medium to 100 mM. This medium contained (in mM) 100 NaCl or LiCl or CsCl, 2 KCl, 3.6 KHCO$_3$, 57.6 N-methyl-D-gluconium-HCl (NMG), 1.3 CaCl$_2$, 1 MgCl$_2$, 10 HEPES, pH 7.4, adjusted with Tris and was supplemented with 1 mM ouabain to inhibit Na,K-ATPase. Moreover, to minimize Ca^{2+} influx by pathways other than reverse Na/Ca exchange, L-type voltage channels, NMDA, and AMPA/kainate receptors were blocked by 10 μM nifedipine, 10 μM (+)5-methyl-10-11-dihydro-5H-dibenzocyclohepten-5,10-imine (MK-801) and 10 μM 2,3-dioxo-6-nitro-1,2,3,4-tetrahydrobenzo[f]quinoxaline-7-sulfonamide (NBQX), respectively.

RESULTS AND DISCUSSION

Application of 5 μM gramicidin promptly elevated $[Na^+]_c$ to ambient levels (FIG. 1A) and led to a transient acidification of cytosol (arrowhead in FIG. 1B), which was followed by a slow equilibration of intracellular and extracellular pH (FIG. 1B). Interestingly, although exposed to a drastic increase in $[Na^+]_c$ caused by gramicidin, CGCs were able to maintain $[Ca^{2+}]_c$ homeostasis for over 10 min despite inhibition of mitochondrial and glycolytic ATP production by oligomycin and iodoacetate (FIG. 1B). Depolarization of mitochondria by rotenone,[11] however, caused a prompt increase in $[Ca^{2+}]_c$ to fura-2FF–saturating levels (FIG. 1B). This indicates that before the addition of rotenone, Ca^{2+} entering the cells was buffered by mitochondria, resembling what happens in CGCs after activation of NMDA receptors.[11–13] The initial gramicidin-induced decrease in pH_c likely indicates H^+ influx down its electrochemical gradient via gramicidin-formed pores; the subsequent increase in pH_c is a consequence of gramicidin-induced plasma membrane depolarization.

Although oligomycin (without rotenone) failed to destabilize $[Ca^{2+}]_c$ homeostasis in gramicidin-treated CGCs (FIG. 1B), it robustly enhanced gramicidin-induced $^{45}Ca^{2+}$ accumulation (FIG. 1C). Na^+ influx was mandatory for this enhancement, because oligomycin failed to affect $^{45}Ca^{2+}$ accumulation when Na^+ was substituted by Li^+ or Cs^+ (FIG. 1C). The data are consistent with the hypothesis that gramicidin-induced Na^+ influx and depolarization of the plasma membrane reverse plasmalemmal Na/Ca exchange and that Ca^{2+} entering via this pathway accumulates in mitochondria. Oligomycin hyperpolarizes the inner mitochondrial membrane[14] and therefore enhances Ca^{2+} accumulation in the matrix. Furthermore, oligomycin promotes ATP depletion (data not shown), which compromises ATP-dependent Ca^{2+} extrusion from the cells. Taken together, the data show that reversal of plasmalemmal Na/Ca exchange leads to Ca^{2+} accumulation in the mitochondria and may compromise neuronal viability.

Neurons express several types of plasmalemmal Na/Ca exchange,[6] and currently it is unclear how various isoforms of Na/Ca exchange operate during brain ischemia. Interestingly, pyramidal neurons in the CA1 region of the hippocampus, particularly susceptible to ischemic damage,[15] abundantly express a K-dependent Na/Ca ex-

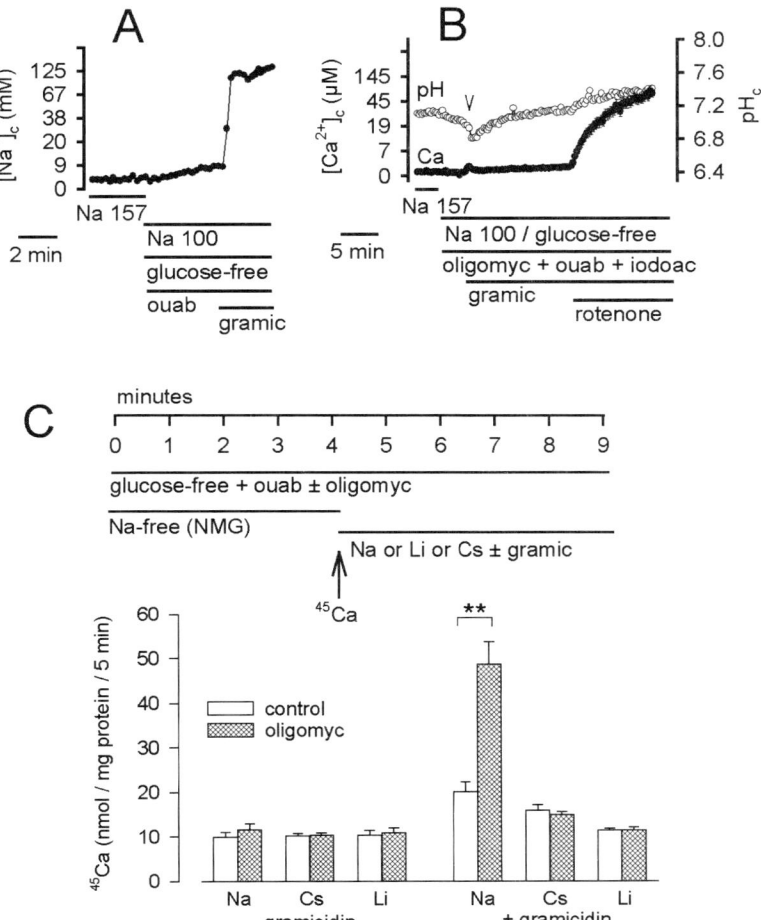

FIGURE 1. Effects of gramicidin on $[Na^+]_c$ (A), $[Ca^{2+}]_c$ and pH_c (B), and $^{45}Ca^{2+}$ accumulation (C) in CGCs. **(A)** Gramicidin (gramic) (5 µM) causes a prompt elevation in $[Na^+]_c$ to ambient levels. CGCs initially were incubated in Locke's buffer containing 157.6 mM Na^+ and 5 mM glucose (Na 157). This was replaced by a glucose-free experimental buffer containing 100 mM Na^+, 1 mM ouabain (ouab), 10 µM nifedipine, 10 µM MK-801, and 10 µM NBQX (Na 100) followed by an application of 5 µM gramicidin (gramic). The data are means ± SEM from 50 cells and were repeated with similar results in three such experiments. The error bars are smaller than the symbols indicating data points. **(B)** $[Ca^{2+}]_c$ (*filled circles*) and pH_c (*open circles*) in CGCs treated with gramicidin as described in (A). The experimental buffer was supplemented with 3 µg/ml oligomycin (oligomyc) and 1 mM iodoacetate (iodoac). When indicated, 2 µM rotenone was added to depolarize mitochondria. The data are means ± SEM from 47 cells and were repeated with similar results in three such experiments. **(C)** In the presence of Na^+, but not if Na^+ is replaced by Cs^+ or Li^+, oligomycin enhances the gramicidin-induced $^{45}Ca^{2+}$ accumulation. The cells were treated for 4 min with Na-free Lockes's buffer (Na^+ substituted by NMG^+) and then were exposed for

changer, NCKX3, which is relatively resistant to reversal.[16] CGCs cultured at 25 mM K$^+$, as in our study, express NCX3 but not NCX2,[17] and we currently are exploring the possibility that CGCs also may express a K-dependent Na/Ca exchange. Physiological consequences of differential expression of various NCXs and NCKXs in the central nervous system need to be addressed by further investigations.

ACKNOWLEDGMENTS

This work was supported by National Institutes of Health Grant NS 37390. We are grateful to Gabriela Martin for help in preparing the manuscript.

REFERENCES

1. ERECIŃSKA, M. & I.A. SILVER. 1994. Ions and energy in mammalian brain. Prog. Neurobiol. **43:** 37–71.
2. CZYZ, A., G. BARANAUSKAS & L. KIEDROWSKI. 2002. Instrumental role of Na$^+$ in NMDA excitotoxicity in glucose-deprived and depolarized cerebellar granule cells. J. Neurochem. **81:** 379–389.
3. REYNOLDS, I.J. 1999. Mitochondrial membrane potential and the permeability transition in excitotoxicity. Ann. N.Y. Acad. Sci. **893:** 33–41.
4. NICHOLLS, D.G. & S.L. BUDD. 2000. Mitochondria and neuronal survival. Physiol. Rev. **80:** 315–360.
5. HLADKY, S.B. & D.A. HAYDON. 1972. Ion transfer across lipid membranes in the presence of gramicidin A. I. Studies of the unit conductance channel. Biochim. Biophys. Acta **274:** 294–312.
6. BLAUSTEIN, M.P. & W.J. LEDERER. 1999. Sodium/calcium exchange: its physiological implications. Physiol. Rev. **79:** 763–854.
7. DOERING, A.E., D.A. EISNER & W.J. LEDERER. 1996. Cardiac Na-Ca exchange and pH. Ann. N.Y. Acad. Sci. **779:** 182–198.
8. KIEDROWSKI, L. 1999. N-methyl-D-aspartate excitotoxicity: relationships among plasma membrane potential, Na$^+$/Ca^{2+} exchange, mitochondrial Ca^{2+} overload, and cytoplasmic concentrations of Ca^{2+}, H$^+$, and K$^+$. Mol. Pharmacol. **56:** 619–632.
9. HYRC, K.L., J.M. BOWNIK & M.P. GOLDBERG. 2000. Ionic selectivity of low-affinity ratiometric calcium indicators: mag-Fura-2, Fura-2FF and BTC. Cell Calcium **27:** 75–86.
10. KIEDROWSKI, L. 1999. Elevated extracellular K$^+$ concentrations inhibit N-methyl-D-aspartate-induced Ca^{2+} influx and excitotoxicity. Mol. Pharmacol. **56:** 737–743.
11. BUDD, S.L. & D.G. NICHOLLS. 1996. A reevaluation of the role of mitochondria in neuronal Ca^{2+} homeostasis. J. Neurochem. **66:** 403–411.
12. KIEDROWSKI, L. & E. COSTA. 1995. Glutamate-induced destabilization of intracellular calcium concentration homeostasis in cultured cerebellar granule cells: role of mitochondria in calcium buffering. Mol. Pharmacol. **47:** 140–147.
13. KHODOROV, B., V. PINELIS, O. VERGUN, et al. 1996. Mitochondrial deenergization underlies neuronal calcium overload following a prolonged glutamate challenge. FEBS Lett. **397:** 230–234.

---

5 min to gramicidin and ^{45}Ca^{2+} in the presence of 100 mM Na$^+$ or Li$^+$ or Cs$^+$ as illustrated. Note that oligomycin robustly enhances gramicidin-induced ^{45}Ca^{2+} accumulation but only in the presence of Na$^+$. The data are means ± SEM from four independent experiments. **$P < 0.01$, one-way ANOVA followed by Student Newman–Keuls test.

14. ICHAS, F., L.S. JOUAVILLE & J.P. MAZAT. 1997. Mitochondria are excitable organelles capable of generating and conveying electrical and calcium signals. Cell **89:** 1145–1153.
15. PULSINELLI, W.A. 1985. Selective neuronal vulnerability: morphological and molecular characteristics. Prog. Brain Res. **63:** 29–37.
16. KRAEV, A., B.D. QUEDNAU, S. LEACH, et al. 2001. Molecular cloning of a third member of the potassium-dependent sodium-calcium exchanger gene family, NCKX3. J. Biol. Chem. **276:** 23161–23172.
17. LI, L., D. GUERINI & E. CARAFOLI. 2000. Calcineurin controls the transcription of Na^+/Ca^{2+} exchanger isoforms in developing cerebellar neurons. J. Biol. Chem. **275:** 20903–20910.

ATP Stimulation of Na^+/Ca^{2+} Exchanger in Bovine Brain Membrane Vesicles Is Similar to That of the Heart and Independent of Ionic Strength of Assay or Preparation

GRACIELA BERBERIÁN, CARLA ASTEGGIANO, AND CUONG PHAM

Laboratorio de Biofísica, Instituto de Investigación Médica Mercedes y Martín Ferreyra, INIMEC-CONICET, Córdoba, Argentina

KEYWORDS: Na^+/Ca^{2+} exchange; ATP; bovine brain; phosphoinositides

There is enough evidence for the upregulation of Na^+/Ca^{2+} exchange activity by MgATP in several tissues and species,[1–3] but the actual mechanism varies between species and tissues of the same species.[1–5] In squid nerves, MgATP stimulation of the Na^+/Ca^{2+} exchanger requires the presence of a low-molecular-weight, soluble cytosolic protein,[6] and this effect does not involve phosphatidylinositides.[7] In the mammalian heart Na^+/Ca^{2+} exchanger, MgATP stimulation does not need a soluble cytosolic protein,[8,9] and it is indeed related to phosphatidylinositides.[8–10] The differences observed between squid nerve cells and mammalian heart could have more than one explanation: they could represent tissue rather than species differences, or they could obey the different ionic strength and/or osmolarity used in preparing the vesicles and measuring the exchange fluxes. We therefore decided to explore these possibilities using, as the source of Na^+/Ca^{2+} exchanger, bovine brain membrane vesicles.

Brain membrane vesicles were prepared by differential gradient centrifugation from bovine brain obtained immediately after killing the animal. When the preparation was done at normal ionic strength (approximately 160 mM), the method was as indicated for cardiac sarcolemmal vesicles[9,11]; when the vesicles were made at high ionic strength (approximately 300 mM), we followed the technique applied for squid brain.[6] The preparation consisted of a mixed population of vesicles of approximately 30% inside out, 40% right side out, and 30% leaky; this was established by the Na^+, K^+–ATPase activity.[9] Na^+/Ca^{2+} exchange activity was measured as Na^+ gradient–dependent Ca^{2+} uptake using $^{45}Ca^{2+}[CaCl_2]$ under conditions previously described.[9] Briefly, 5 μl of vesicles were diluted in 100 μl of the uptake solutions.

Address for correspondence: Graciela Berberián, Laboratorio de Biofísica, Instituto de Investigación Médica Mercedes y Martín Ferreyra, Casilla de Correo 389, 5000 Córdoba, Argentina. Voice: (54-351) 468-1465; fax: (54-351) 469-5163.
 gelso@immf.uncor.edu

TABLE 1. Effect of ionic strength on the ATP stimulation of Na$^+$ gradient-dependent Ca^{2+} uptake

Ionic strength (vesicles–medium)	Na$^+$ gradient–dependent ^{45}Ca^{2+} uptake (nmol · mg^{-1} · 15 s^{-1})		
	Basal	ATP	ATP-stimulated
High–high	0.373 ± 0.046	1.072 ± 0.060	0.699 ± 0.075
High–low	0.327 ± 0.022	0.825 ± 0.104	0.498 ± 0.121
Low–low	0.374 ± 0.034	1.038 ± 0.075	0.664 ± 0.082
Low–high	0.130 ± 0.029	0.321 ± 0.058	0.191 ± 0.065

NOTE: (i) Aliquots of both groups of vesicles (prepared at low and high ionic strength as indicated in the text) then were loaded, overnight at 4°C, in solutions containing 160 mM NaCl, 20 mM MOPS-Tris (pH 7.4 at 37°C) and 0.1 mM EDTA (normal ionic strength for assay) or 300 mM NaCl, 20 mM MOPS-Tris (pH 7.4 at 37°C), and 0.1 mM EDTA (high ionic strength for assay). (ii) The final [ATP] was 1 mM. (iii) The entries are mean ± SE of triplicate determinations. (iv) There is no appreciable difference in the fractional ATP stimulation of the Na$^+$/Ca^{2+} exchanger under all conditions investigated. (Taken from Berberián et al.[12] Used with permission.)

TABLE 2. Effect of 1 mM ATP on Na$^+$ gradient–dependent Ca^{2+} uptake of bovine brain plasma membrane vesicles with or without phosphoinositide incorporation

Condition	Na$^+$ gradient–dependent ^{45}Ca^{2+} uptake (nmol · mg^{-1} · 15 s^{-1})	
	Total	Stimulated
Basal	0.32 ± 0.05	—
ATP	1.45 ± 0.15	1.13 ± 0.16
PtdIns-4P	0.80 ± 0.16	0.47 ± 0.17
PtdIns-4P + ATP	1.60 ± 0.21	1.28 ± 0.22
PtdIns-4,5P$_2$	1.75 ± 0.27	1.43 ± 0.27
PtdIns-4,5P$_2$ + ATP	1.70 ± 0.31	1.38 ± 0.31

NOTE: (i) In the absence of ATP, the incorporation of PtdIns-4P leads to a stimulation of the which amounts to approximately one-half of that of MgATP, and the incorporation of PtdIns-4,5P$_2$ produces a stimulation of the exchanger similar to that seen with MgATP and (ii) the addition of MgATP to membrane vesicles with had been exposed to PtdIns-4,5P$_2$ does not increase the rate of Na$^+$/Ca^{2+}. (Taken from Berberián et al.[12] Used with permission.)

These solutions had a final [Na$^+$] of 10 mM or 30 mM (low-Na$^+$ medium) and 160 mM or 300 mM (high-Na$^+$ medium), for normal and high ionic strength, respectively. In low-Na$^+$ medium, the osmolarity was compensated with N-methylglucamine-Cl. In addition, the usual extravesicular solutions contained 1.0 μM of [Ca^{2+}] and 3 mM of [Mg^{2+}]. These vesicles have an MgATP-dependent Ca^{2+} pump, which was inhibited by adding vanadate (0.3 mM) to the incubation solutions. The uptake periods lasted 15 s or 20 s at 37°C and pH 7.4. Incorporation of phosphoinositides into membrane vesicles was done by incubating concentrate vesicles for 20 min at 0°C and then for 1 min at 37°C with liposomes made of PtdIns, PtdIns-4P, or PtdIns-4,5P$_2$ (0.2 mg/mg total protein).[7]

The results show that MgATP activates the bovine brain Na^+/Ca^{2+} exchanger with a $K_{0.5}$ of 336 μM. As a difference compared with the squid nerve, this effect has no need of any cytosolic component. Also, stimulation is the same in vesicles prepared and/or assayed at mammal (160 mM) or marine animal (300 mM) ionic strength (see TABLE 1). Another difference between squid and bovine nerve is that in bovine brain Na^+/Ca^{2+} exchanger stimulation by MgATP is related to the production of polyphosphatidylinositides (see TABLE 2). In this regard, bovine heart and brain Na^+/Ca^{2+} exchangers behave similarly. These results indicate that metabolic regulation of the squid and mammalian nerve Na^+/Ca^{2+} exchangers are not alike and represent differences between species.

ACKNOWLEDGMENTS

This work was supported by grants from the Agencia Nacional de Promoción Científica y Tecnológica (FONCYT PICT99 05-05158), Agencia Córdoba Ciencia (181/01) of Argentina, and Ministerio de Salud de la Nación (Beca "Carrillo-Oñativia," 2001). G.B. is Established Investigator and C.A. is a Fellow of CONICET. C.P. was a Passant Student from UCLA Medical School.

REFERENCES

1. HILGEMANN, D.W. 1997. Cytoplasmic ATP-dependent regulation of ion transporters and channels: mechanisms and messengers. Annu. Rev. Physiol. **59:** 193–220.
2. BLAUSTEIN, M.P. & W.J. LEDERER. 1999. Sodium/calcium exchange: its physiological implications. Physiol. Rev. **79:** 763–854.
3. DIPOLO, R. & L. BEAUGÉ. 1999. Metabolic pathways in the regulation of invertebrate and vertebrate $Na^+/Ca2^+$ exchange. Biochim. Biophys. Acta **1422:** 57–71.
4. IWAMOTO, T., P.Y. WAKABAYASHI, T. IMAGAWA, et al. 1996. Phosphorylation-dependent regulation of cardiac Na^+-Ca^{2+} via protein kinase C. J. Biol. Chem. **271:** 13609–13615.
5. HE, S., A. RUKNUDIN, L. BAMBRIK, et al. 1998. Isoform-specific regulation of Na^+/Ca^{2+} exchanger in rat astrocytes and neuron by PKA. J. Neurosci. **18:** 4833–4841.
6. DIPOLO, R., G. BERBERIÁN, D. DELGADO, et al. 1997. A novel 13 kDa cytoplasmic soluble protein is required for the nucleotide (MgATP) modulation of the Na/Ca exchange in squid nerve fibers. FEBS Lett. **401:** 6–10.
7. DIPOLO, R., G. BERBERIÁN & L. BEAUGÉ. 2000. In squqid nerves intracellular Mg promotes deactivation of the ATP-unregulated Na/Ca exchanger. Am. J. Physiol. **279:** C1631–C1639.
8. HILGEMANN, D.W. & R. BALL. 1996. Regulation of cardiac Na^+, Ca^{2+} exchange and K_{ATP} potassium channels by PIP_2. Science **273:** 956–959.
9. BERBERIÁN, G., C. HIDALGO, R. DIPOLO & L. BEAUGÉ. 1998. ATP stimulation of Na^+/Ca^{2+} exchange in cardiac sarcolemmal vesicles. Am. J. Physiol. **274:** C724–C733.
10. ASTEGGIANO, C., G. BERBERIÁN & L. BEAUGÉ. 2001. Phosphatidylinositol-4,5-biphosphate bound to bovine cardiac Na/Ca exchange displays an MgATP regulation similar to that of the exchange fluxes. Eur. J. Biochem. **268:** 437–442.
11. REEVES, J.P. 1985. The sarcolemmal sodium–calcium exchange system. Curr. Top. Membr. Transp. **25:** 77–119.
12. BERBERIÁN, G., C. ASTEGGIANO, C. PHAM, et al. 2002. MgATP and phosphoinositides activate Na^+/Ca^{2+} exchange in bovine brain vesicles. Comparison with other Na^+/Ca^{2+} exchangers. Pflugers Arch. Eur. J. Physiol. **444:** 677–684.

Is Na/Ca Exchange during Ischemia and Reperfusion Beneficial or Detrimental?

ELIZABETH MURPHY,[a] HEATHER R. CROSS,[b] AND CHARLES STEENBERGEN[b]

[a]*Laboratory of Signal Transduction, NIEHS, Research Triangle Park, North Carolina 27709, USA*

[b]*Department of Pathology, Duke University, Durham, North Carolina, USA*

ABSTRACT: Cytosolic calcium increases to approximately 3 μM after 15 min of global ischemia. Manipulations that attenuate this increase in cytosolic Ca^{2+} reduce myocyte death and dysfunction. The increase in cytosolic Ca^{2+} during ischemia is dependent on an increase in intracellular Na^+, suggesting a role for Na/Ca exchange. Typical ischemic values for ionized intra- and extracellular Na^+, Ca^{2+}, and membrane potential are consistent with the Na/Ca exchanger operating near equilibrium during ischemia. Studies were undertaken using hearts from mice that overexpress the Na/Ca exchanger to determine if Na/Ca exchanger overexpression enhances or reduces ischemic injury. These studies suggest that overexpression of the Na/Ca exchanger enhances injury in males, but females are protected by a gender-related mechanism.

KEYWORDS: cytosolic calcium; ischemia; Na/Ca exchanger

A substantial increase in intracellular free Ca^{2+} (Ca_i) has been shown to occur before lethal ischemic injury, and inhibition or delay of this increase in Ca_i is protective.[1-3] Furthermore, an increase in Ca_i is involved in proteolysis of contractile proteins, providing a mechanism by which an increase in Ca_i causes injury.[4-6] An understanding of the mechanism responsible for the increase in Ca_i during ischemia would be important for the development of pharmacological tools to inhibit this increase in Ca_i and thus lead to cardioprotection. There are several possible candidate mechanisms for the increase in Ca_i during ischemia, including entry via Ca^{2+} channels, altered Na/Ca exchange, and release from the sarcoplasmic reticulum (SR). There are substantial data suggesting an important role for Na/Ca exchange in the increase in Ca_i during ischemia.[7-11]

Many studies report an increase in intracellular Na^+ (Na_i) during ischemia.[9-22] The mechanism responsible for this increase in Na_i is debated. There are data supporting a role for Na^+ entry via the Na/H exchanger[7-11,23] as well as data supporting Na^+ entry via noninactivating Na^+ channels.[21] In support of a role for the Na/H exchanger, there are several studies reporting that inhibitors of the Na/H exchanger reduce the increase in Na_i and Ca_i during ischemia and reduce injury on

Address for correspondence: Elizabeth Murphy, 111 Alexander Drive, Building 101, NIEHS, Research Triangle Park, NC 27709. Voice: 919-541-3873; fax: 919-541-3385.
murphy1@niehs.nih.gov

reperfusion.[9–11,24] FIGURE 1 presents typical changes in intracellular pH (pH_i), Na_i, and Ca_i during ischemia. These ion measurements were made in Langendorff-perfused rat hearts by using nuclear magnetic resonance methods described previously.[9] The increase in Na_i during ischemia occurs with similar kinetics to the increase in Ca_i. We find that after 20 minutes of ischemia pH_i decreases to approximately pH 6.0, Na_i increases to approximately 28 mM, and Ca_i increases to approximately 3 μM. FIGURE 1 also shows that pretreating hearts with the Na/H exchanger inhibitor, amiloride, or the more specific analogue dimethylamiloride has no effect on the pH_i reached during ischemia. We and others have interpreted this finding as showing that there are multiple pathways for H^+ efflux, and if one pathway is blocked, H^+ can exit via alternative pathways. However, as expected for inhibition of Na/H exchanger, we did find that amiloride largely blocks the increase in Na_i during ischemia and also attenuates the increase in Ca_i, consistent with altered Na/Ca exchange. We also find that the reduction in the increase in Na_i and Ca_i occurs with a concomitant reduction in postischemic contractile dysfunction, as measured via a balloon in the left ventricle. Normally after 30 minutes of ischemia, hearts recover approximately 40% of preischemic left ventricular developed pressure. Hearts pretreated with amiloride recover approximately 80% of preischemic function. Similar results are obtained with a more specific inhibitor of the Na/H exchanger, dimethylamiloride, which inhibits the increase in Ca_i (FIG. 1). These data are consistent with an increase in Na/H exchange resulting in an elevated Na_i, leading to altered Na/Ca exchange, which leads to an increase in Ca_i. The observation that inhibiting the increase in Na_i blocked the increase in Ca_i during ischemia was somewhat surprising because low pH, as would occur during ischemia, inhibits Na/Ca exchange. Philipson et al.[25] reported that H^+ competes with Na^+ for binding to the Na/Ca exchanger, and therefore low intracellular pH reduces the activity of the Na/Ca exchanger; however, they further showed that elevated Na^+ stimulated Na/Ca exchange, and thus the increase in Na_i may serve to offset the inhibition because of the low pH. Taken together, the data clearly show a role for Na^+-dependent Ca^{2+} entry during ischemia.

However, it has been shown that many of the amiloride-derived Na/H exchanger inhibitors can also block Na^+ channels.[21,22] Therefore, the reduction in Na_i during ischemia observed with some inhibitors of Na/H exchanger is likely to be an overestimate of the Na^+ entry resulting from the Na/H exchanger alone. Furthermore, in myocytes, lidocaine and other inhibitors of Na^+ channels completely block the increase in Na_i during ischemia, suggesting that Na^+ entry is primarily mediated by entry through noninactivating Na^+ channels.[21,22] However, Malloy et al. showed in perfused heart that lidocaine attenuates but does not totally block the increase in Na_i.[14] The difference in workload between myocytes and perfused heart could contribute to this discrepancy. Perfused hearts have a higher workload and therefore an increased metabolic production of H^+. Consistent with this concept, the increase in Na_i during ischemia is dependent on pacing rate, which affects metabolic rate.[19] Taken together, the data suggest that the proportion of Na^+ entry via Na/H exchanger may be influenced by metabolism. In a metabolically active perfused heart, it appears that both Na/H exchange and Na^+ channels contribute. This is confirmed by studies with new Na/H exchanger inhibitors (HOE 694, HOE 642), which do not appear to inhibit Na^+ channels.[22] In perfused heart, HOE 694 is reported to partially reduce the increase in Na_i and Ca_i during ischemia,[26] consistent with a role for both Na/H exchange and Na^+ channels in the increase in Na_i during ischemia.

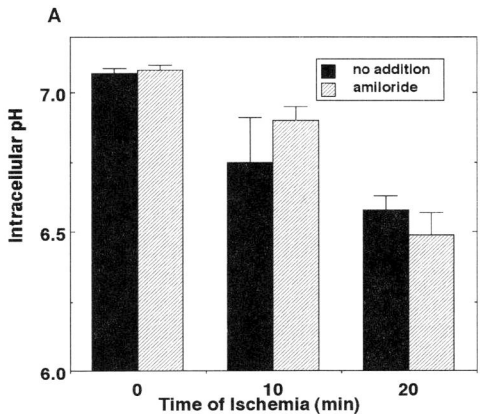
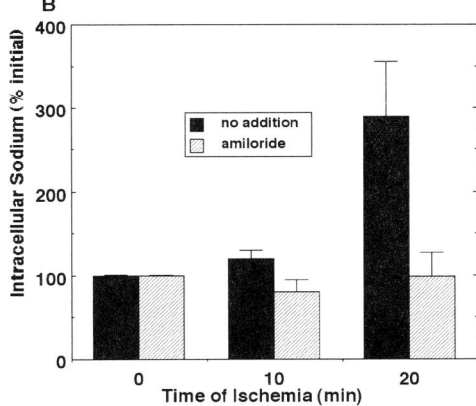
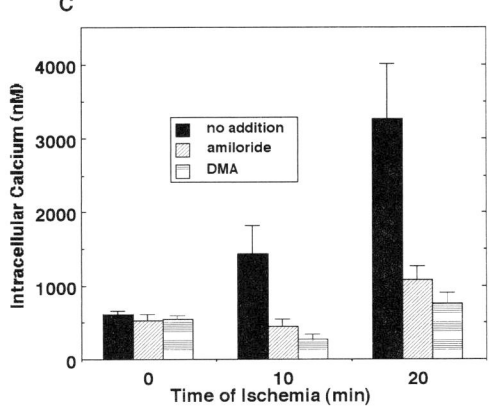

FIGURE 1. Changes in intracellular ion concentrations during ischemia with and without inhibition of the Na/H exchanger. For details see Murphy et al.[9] DMA, dimethylamiloride.

Further support for a role for Na/H exchange in the increase in Na_i during ischemia comes from studies showing that reducing H^+ generation during ischemia also reduces the increase in Na_i.[27–29] Preconditioning is a good illustration of this concept. Brief periods of intermittent ischemia and reperfusion, termed "preconditioning," reduce injury during a subsequent sustained period of ischemia. Preconditioning reduces anaerobic glycolysis and generation of lactate, and this leads to less H^+ generation,[27] which in turn leads to less Na/H exchange, although not because of inhibition of the Na/H exchanger but because of reduced substrate (H^+) supply. The reduction in Na_i results in less Na/Ca exchange and less of an increase in Ca_i.[11,29] These data support a role for Na/H exchange in the increase in Na_i during ischemia.

Another issue is the importance of Na^+ entry during ischemia versus reperfusion. Na^+ clearly enters myocytes during ischemia (see FIG. 1), and inhibiting the increase in Na_i reduces the increase in Ca_i during ischemia. Thus, Na^+ entry during ischemia is important for the increase in Ca_i during ischemia. How much does Na/H exchange activity, during reperfusion, contribute to the increase in Ca_i? Data in the literature are consistent with enhanced Na/H exchanger activity at the start of reflow that does not cause a measurable increase in Na_i because of activity of the Na-K ATPase.[10,16,24,30] Although enhanced Na/H exchange activity at the start of reflow may not increase steady state Na_i, it might provide an additional source of Na_i to exchange with extracellular Ca^{2+}. Data show that on reperfusion, pH_i quickly returns to normal (pH ~7.1).[10,28] Thus, the first few seconds of reflow are the crucial period for Na/H exchange activity. It is difficult, if not impossible, to deliver a Na/H exchange inhibitor to the myocyte during this crucial time if it is added only during reperfusion. This is the likely explanation as to why in a large clinical trial, the Na/H exchanger inhibitor, cariporide, was only protective when added before the ischemic event such as during coronary artery bypass surgery.[31]

Another important issue is whether the increase in Ca_i during ischemia occurs because there is decreased Ca^{2+} efflux via the Na/Ca exchanger, because of decreased driving force for Na^+ entry, or whether the decrease in the Na^+ gradient, coupled with the depolarization of the membrane potential, leads to reverse Na/Ca exchange, in which Ca^{2+} enters in exchange for Na^+.[32] Under normoxic conditions when Na_i is low and the resting membrane potential is approximately –80 mV, it is thought that the Na/Ca exchanger operates to pump Ca^{2+} from the cell. During total ischemia, Na_i increases to 28 mM, and extracellular Na^+ correspondingly declines to approximately 120 mM, because extracellular Na^+ is the only source for the increase in Na_i. In addition, the membrane potential depolarizes to –40 or –60 mV during ischemia. Assuming these values and a stoichiometry of 3 Na^+ to 1 Ca^{2+}, the equilibrium equation ($Ca_i = (Ca_o)(Na_i)^3/(Na_o)^3 e^{-Em(F/RT)}$) suggests that the Na/Ca exchanger operates near equilibrium. TABLE 1 presents the calculated values for Ca_i, assuming different values for intra- and extracellular Na^+ and membrane potential. However, these data do not distinguish whether the Na/Ca exchanger operates in Ca^{2+} entry or Ca^{2+} efflux mode. It is unclear whether the increase in Ca_i during ischemia occurs because of reverse Na/Ca exchange or because of reduced Ca^{2+} efflux (see FIG. 2). If Ca_i during ischemia is maintained significantly below that calculated from the Na/Ca exchanger equilibrium (see TABLE 1), it would be strong evidence that the Na/Ca exchanger does not run in reverse during ischemia. However, during ischemia Ca_i reaches approximately 3 µM, a level consistent with Na/Ca exchanger equilibrium (see TABLE 1); thus, it is likely that the Na/Ca exchanger operates in the Ca^{2+} influx

TABLE 1. Calculated value for cytosolic free Ca^{2+} (Ca_i)

Calculated Ca_i (μM)	Ca_o (mM)	Na_i (mM)	Na_o (mM)	Em (mV)
0.097	1.75	15	145	−80
0.055	1.00	15	145	−80
0.016	1.00	10	145	−80
0.021	1.25	10	145	−80
2.909	1.25	28	100	−60
2.798	1.25	28	130	−40
3.557	1.25	28	120	−40
0.035	1.25	12	145	−80
0.370	1.25	16	145	−40
0.630	1.25	19	145	−40

NOTE: Calculated value for cytosolic free Ca^{2+} (Ca_i) in μM using the following equation and assuming the indicated values for extracellular Ca^{2+} (Ca_o), intra- and extracellular Na^+ (Na_i and Na_o) and membrane potential (Em). Equation: $Ca_i = (Ca_o)(Na_i)^3/(Na_o)^3 e^{-Em(F/RT)}$.

mode at least some of the time. However, if the stoichiometry is 4 Na^+ to 1 Ca^{2+}, the Na/Ca exchanger would not reverse even with a membrane potential of −40 mV and intra- and extracellular Na^+ of 28 and 120 mM, respectively (see TABLE 1).

The issue of whether the Na/Ca exchanger runs in reverse during ischemia has important implications for the use of Na/Ca exchanger inhibitors to limit ischemic injury. To further examine the issue of whether the increased Ca_i during ischemia and development of injury is caused by the Na/Ca exchanger operating in the reverse mode, we performed studies using transgenic mice, developed by Philipson et al.,[33,34] that overexpress the Na/Ca exchanger. As illustrated in FIGURE 2, if there were increased Na/Ca exchanger, and if the Na/Ca exchanger operates in the reverse mode during ischemia, this might be predicted to enhance injury (model C), whereas if the increase in Ca_i during ischemia was caused by reduced Ca^{2+} efflux, then increased Na/Ca exchanger might be predicted to be protective (model A). In the study with hearts from Na/Ca exchanger overexpressor mice, we found that after 20 minutes of ischemia wild-type (WT) mice recover approximately 35% of function, but Na/Ca exchanger overexpressor mice recover significantly less (~8%).[35] These data support model C (FIG. 2), suggesting that the increase in the Na^+ gradient leads to reverse Na/Ca exchange. Also consistent with increased injury in the Na/Ca exchanger overexpressor hearts, postischemic recovery of creatine phosphate is lower in Na/Ca exchanger overexpressor hearts than in the WT. We also find that on reperfusion the Na/Ca exchanger overexpressor hearts develop pressure alternans. We find that all (5/5) of the transgenic hearts develop alternans whereas alternans occur in none (0/6) of the WT littermates.[35] However, hearts from female Na/Ca exchanger overexpressors did not exhibit enhanced injury. There was no difference in recovery of function between female WT and female Na/Ca exchanger overexpressor hearts. Furthermore, when we ovariectomized the females approximately 2 weeks before experimentation, they showed enhanced injury similar to the males. Thus, the data

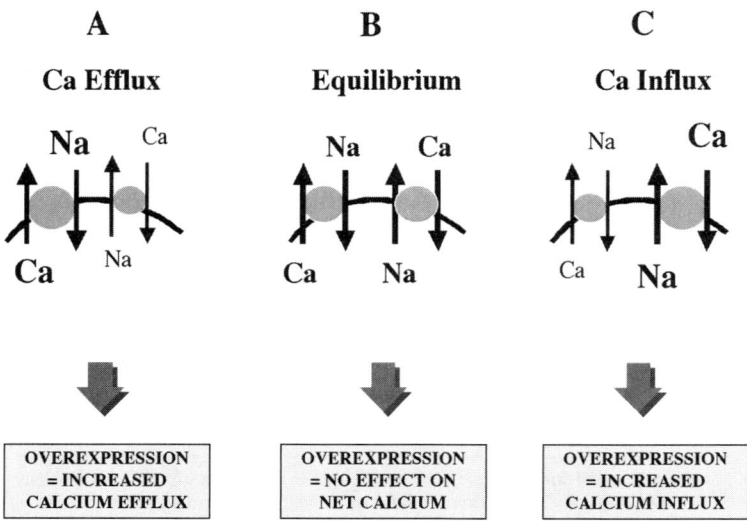

FIGURE 2. Possible effects of overexpression of NCX on the increase in cytosolic Ca^{2+} during global ischemia. **(A)** Increased Na/Ca exchanger leads to increased Ca^{2+} efflux (increased forward-mode Na/Ca exchanger) and decreased injury; **(B)** increased Na/Ca exchanger does not alter net $[Ca^{2+}]_i$ levels (only the rate at which they are attained); and **(C)** increased Na/Ca exchanger leads to increased Ca^{2+} influx (reverse-mode Na/Ca exchanger) and increased injury.

show that male, but not female, Na/Ca exchanger overexpressor hearts have increased ischemia-reperfusion injury.[35] The data from the males are consistent with an increase in Na/Ca exchanger leading to an increase in Ca_i during ischemia and increased injury; these data support the concept that the Na/Ca exchanger is active during ischemia and operates in reverse. In contrast, females that overexpress the Na/Ca exchanger do not show increased injury. In no case does overexpression of the Na/Ca exchanger lead to reduced injury, as might be expected if Na/Ca exchanger overexpression enhanced Ca^{2+} efflux during ischemia (FIG. 2A).

What are the possible reasons for the differential response in males versus females? One obvious possibility is that female hearts have less Na/Ca exchanger overexpression than male hearts. However, transgenic male and female hearts have similar levels of Na/Ca exchanger expression.[35] Why do females have reduced injury despite similar levels of Na/Ca exchanger? One possibility is that the increase in Ca_i during ischemia is similar in males and females, but perhaps females are protected because of estrogen-stimulated mechanisms (such as increased PI3 kinase), leading to reduced cell death.[36] However, Sugishita et al. studied myocytes isolated from the Na/Ca exchanger overexpressor mice and reported a significantly higher Ca_i during metabolic inhibition in males versus females.[37] As male and female Na/Ca exchanger overexpressor hearts have similar levels of Na/Ca exchanger expression, one would expect an increased Ca_i during ischemia in females as well as males. FIGURE 2 presents the possibilities: (1) increased Na/Ca exchanger leads to increased

Ca^{2+} efflux (increased forward mode Na/Ca exchanger) and decreased injury; (2) increased Na/Ca exchanger does not alter net Ca_i levels (only the rate at which they are attained); and (3) increased Na/Ca exchanger leads to increased Ca^{2+} influx (reverse-mode Na/Ca exchanger) and increased injury. Because neither male nor female Na/Ca exchanger overexpressor hearts are protected, there is no support for protection mediated by increased forward mode Na/Ca exchanger (model A). Overexpression of the Na/Ca exchanger would not be expected to alter Na/Ca exchanger equilibrium; therefore, overexpression of the Na/Ca exchanger might not increase Ca^{2+} entry during ischemia, only the rate at which Ca^{2+} is transported (FIG. 2B). If this is the case, then neither male nor female Na/Ca exchanger overexpressor hearts should have increased Ca_i during ischemia, and the increase in Ca_i observed in males is caused by other factors secondary to Na/Ca exchanger overexpression. Consistent with this hypothesis, it has been reported that Na/Ca exchanger overexpressor hearts have increased basal SR Ca levels.[38] It is suggested that the increase in SR Ca^{2+} is caused by enhanced reverse mode Na/Ca exchange during rest and the latter part of the Ca^{2+} transient leading to increased SR Ca^{2+} load. If there were a gender difference in SR Ca^{2+} loading or differences in SR Ca^{2+} handling, this could account for the increased Ca_i during ischemia in males, but not females. Consistent with this hypothesis, in other models with increased SR Ca^{2+}, Cross et al. report increased ischemia-reperfusion injury in males versus females.[39]

Another possibility is that increased expression of the Na/Ca exchanger may lead to enhanced Na_i transport from the cell via the Na/Ca exchanger versus other Na^+ efflux pathways thereby increasing the level of Ca_i during ischemia (FIG. 2C). In this case, both male and female Na/Ca exchanger overexpressor hearts should have increased Ca_i during ischemia. Because females do not show increased Ca_i during ischemia, they would have additional compensatory mechanisms to reduce Ca_i. However, there must be an explanation why these compensatory Ca_i-reducing mechanisms are not observed in WT littermates. It is possible that overexpression of Na/Ca exchanger potentiates some gender-dependent difference. Sugishita et al. report a gender difference in the increase in Na_i during ischemia.[37] Because of the stoichiometry of 3 Na^+ to 1 Ca^{2+}, a small male–female difference in Na_i would lead to a larger difference in Ca_i. Sugishita et al. report that during metabolic inhibition Na_i increases to 16 mM in females versus 19 mM in males.[37] As shown in TABLE 1, using these values for Na_i, 145 mM for extracellular Na^+ (in the myocyte model extracellular Na^+ is infinite, and it will not be decreased by Na^+ uptake into the cell), 1.25 mM for extracellular Ca^{2+}, and −40 mV for the membrane potential, the Na/Ca exchanger would be in equilibrium, with Ca_i values of 370 and 630 nM for females and males, respectively. These values are very similar to the values of Ca_i measured by Sugishita et al. (426 and 688 nM). Because of the 3 to 1 stoichiometry of Na/Ca exchanger, any gender difference in Na_i would lead to amplified differences in Ca_i. The mechanism responsible for the gender difference in Na_i during ischemia is unclear. The two main candidates for a reduced increase in Na_i during ischemia are a reduction in substrate (i.e., a reduction in generation of H^+) or an altered set point for Na/H exchanger. The gender difference in Na_i during ischemia does not appear to be caused by a reduced generation of H^+ in females and therefore less Na/H exchange, as Cross et al. did not observe a male–female difference in pH_i during ischemia.[35] It is also possible that during ischemia females have less Na^+ entry via noninactivating Na^+ channels, or that females have enhanced Na^+ efflux via Na-K ATPase.

Thus, from the male Na/Ca exchanger overexpressor data alone, it appears that the Na/Ca exchanger reverses during ischemia and leads to increased Ca_i and injury. However, the lack of increased Ca_i and injury in female Na/Ca exchanger overexpressors requires a refinement of that conclusion. Na/Ca exchanger overexpression may cause an increase in Ca_i during ischemia in males via an alternative mechanism, such as a gender-dependent increase in SR Ca^{2+} which indeed may be a result of increased Na/Ca exchanger, but which occurs at baseline, before ischemia. Alternatively, Ca^{2+} entry during ischemia may occur primarily via the Na/Ca exchanger, but if females have less of an increase in Na_i during ischemia this would result in less of a subsequent increase in Ca_i in female Na/Ca exchanger overexpressors. Future studies will address these issues.

REFERENCES

1. STEENBERGEN, C., E. MURPHY, L. LEVY & R. LONDON. 1987. Elevation in cytosolic free calcium concentration early in myocardial ischemia in perfused rat heart. Circ. Res. **60:** 700–707.
2. MARBAN, E., Y. KORESTUNE, M. CORRETTI, et al. 1989. Calcium and its role in myocardial cell injury during ischemia and reperfusion. Circulation. **80**(Suppl. IV): IV17–IV22.
3. STEENBERGEN, C., E. MURPHY, J. WATTS & R. LONDON. 1990. Correlation between cytosolic free calcium, contracture, ATP, and irreversible injury in perfused rat heart. Circ. Res. **66:** 135–146.
4. GAO, W., D. ATAR, Y. LIU, et al. 1997. Role of troponin I proteolysis in the pathogenesis of stunned myocardium. Circ. Res. **80:** 393–399.
5. WESTFALL, M. & R. SOLARO. 1992. Alterations in myofibrillar function and protein profiles after complete global ischemia in rat hearts. Circ. Res. **70:** 302–313.
6. MATSUMURA, Y., E. SAEKI, M. INOUE, et al. 1996. Inhomogeneous disappearance of myofiliment-related cytoskeletal proteins in stunned myocardium of guinea pig. Circ. Res. **79:** 447–454.
7. TANI, M. & J. NEELY. 1989. Role of intracellular Na^+ in Ca^{2+} overload and depressed recovery of ventricular function of reperfused ischemic rat hearts: possible involvement of H^+-Na^+ and Na^+-Ca^{2+} exchange. Circ. Res. **65:** 1045–1056.
8. KARMAZYN, M. 1988. Amiloride enhances postischemic ventricular recovery: possible role of Na-H exchange. Am. J. Physiol. **255:** H608–H615.
9. MURPHY, E., M. PERLMAN, R.E. LONDON & C. STEENBERGEN. 1991. Amiloride delays the ischemia-induced rise in cytosolic free calcium. Circ. Res. **68:** 1250–1258.
10. PIKE, M.M., C.S. LUO, M.D. CLARK, et al. 1993. NMR measurements of Na^+ and cellular energy in ischemic rat heart: role of $Na(^+)$-H^+ exchange. Am. J. Physiol. **265:** H2017–H2026.
11. ANDERSON, S., E. MURPHY, C. STEENBERGEN, et al. 1990. Na-H exchange in myocardium: effects of hypoxia and acidification on Na and Ca. Am. J. Physiol. **259:** C940–C948.
12. MALLOY, C., D. BUSTER, M. CASTRO, et al. 1990. Influence of global ischemia on intracellular sodium in the perfused rat heart. Magn. Reson. Med. **15:** 33–44.
13. PIKE, M., M. KITAKAZE & E. MARBAN. 1990. 23Na-NMR measurements of intracellular sodium in intact perfused ferret hearts during ischemia and reperfusion. Am. J. Physiol. **259:** H1767–H1773.
14. BUTWELL, N., R. RAMASSAMY, I. LAZAR, et al. 1993. Effect of lidocaine on contracture, intracellular sodium and pH in ischemic rat hearts. Am. J. Physiol. **264:** H1884–H1889.
15. NAVON, G., J.G. WERRMANN, R. MARON & S.M. COHEN. 1994. 31P NMR and triple quantum filtered 23Na NMR studies of the effects of inhibition of Na^+/H^+ exchange on intracellular sodium and pH in working and ischemic hearts. Magn. Reson. Med. **32:** 556–564.

16. CROSS, H., G. RADDA & K. CLARKE. 1995. The role of Na^+/K^+ ATPase activity during low flow ischemia in preventing myocardial injury: a 31P, 23Na and 87Rb NMR spectroscopic study. Magn. Reson. Med. **34:** 673–685.
17. VAN EMOUS, J., M. NEDERHOFF, T. RUIGROK & C. VAN ECHTELD. 1997. The role of the Na^+ channel in the accumulation of intracellular Na^+ during myocardial ischemic: consequences for post-ischemic recovery. J. Mol. Cell. Cardiol. **29:** 85–96.
18. CHOY, I., V. SCHEPKIN, T. BUDINGER, et al. 1997. Effects of specific sodium/hydrogen exchange inhibitor during cardioplegic arrest. Ann. Thorac. Surg. **64:** 94–99.
19. DIZON, J., D. BURKHOFF, J. TAUSKELA, et al. 1998. Metabolic inhibition in the perfused rat heart: evidence for glycolytic requirement for normal sodium homeostasis. Am. J. Physiol. **274:** H1082–H1089.
20. SATOH, H., H. HAYASHI, H. KATOH, et al. 1995. Na^+/H^+ and Na^+/Ca^{2+} exchange in regulation of $[Na^+]i$ and $[Ca^{2+}]i$ during metabolic inhibition. Am. J. Physiol. **268:** H1239–H1248.
21. HAIGNEY, M., E. LAKATTA, M. STERN & H. SILVERMAN. 1994. Sodium channel blockade reduces hypoxic sodium loading and sodium-dependent calcium loading. Circulation **90:** 391–399.
22. RUSS, U., C. BALSER, W. SCHOLZ, et al. 1996. Effects of the Na^+/H^+-exchange inhibitor Hoe 642 on intracellular pH, calcium and sodium in isolated rat ventricular myocytes. Pflugers Arch. **433:** 26–34.
23. KARMAZYN, M., M. RAY & J.V. HAIST. 1993. Comparative effects of Na^+/H^+ exchange inhibitors against cardiac injury produced by ischemia/reperfusion, hypoxia/reoxygenation, and the calcium paradox. J. Cardiovasc. Pharmacol. **21:** 172–178.
24. LIU, H., P.M. CALA & S.E. ANDERSON. 1997. Ethylisopropylamiloride diminishes changes in intracellular Na, Ca and pH in ischemic newborn myocardium. J. Mol. Cell. Cardiol. **29:** 2077–2086.
25. PHILIPSON, K., M. BERSOHN & A. NISHIMOTO. 1982. Effect of pH on Na^+-Ca^{2+} exchange in canine cardiac sarcolemmal vesicles. Circ. Res. **50:** 287–293.
26. HENDRIKX, M., K. MUBAGWA, F. VERDONCK, et al. 1994. New $Na(^+)$-H^+ exchange inhibitor HOE 694 improves postischemic function and high-energy phosphate resynthesis and reduces Ca^{2+} overload in isolated perfused rabbit heart. Circulation **89:** 2787–2798.
27. MURRY, C., V. RICHARD, K. REIMER & R. JENNINGS. 1990. Ischemic preconditioning slows energy metabolism and delays ultrastructural damage during as sustained ischemic episode. Circ. Res. **66:** 913–931.
28. LIU, H., P. CALA & S. ANDERSON. 1998. Ischemic preconditioning: effects on pH, Na and Ca in newborn rabbit hearts during ischemia/reperfusion. J. Mol. Cell. Cardiol. **30:** 685–697.
29. STEENBERGEN, C., M. PERLMAN, R. LONDON & E. MURPHY. 1993. Mechanism of preconditioning: Ionic alterations. Circ. Res. **72:** 112–125.
30. VAN EMOUS, J.G., J.H. SCHREUR, T.J. RUIGROK & C.J. VAN ECHTELD. 1998. Both Na^+-K^+ ATPase and Na^+-H^+ exchanger are immediately active upon post-ischemic reperfusion in isolated rat hearts. J. Mol. Cell. Cardiol. **30:** 337–348.
31. THEROUX, P., B. CHAITMAN, N. DANCHIN, et al. 2000. Inhibition of the sodium-hydrogen exchanger with cariporide to prevent myocardial infarction in high-risk ischemic situations. Main results of the GUARDIAN trial. Guard during ischemia against necrosis (GUARDIAN) Investigators. Circulation **102:** 3032–3038.
32. CH'EN, F., R. VAUGHAN-JONES, K. CLARKE & D. NOBLE. 1998. Modelling myocardial ischemia and reperfusion. Prog. Biophys. Mol. Biol. **69:** 515–538.
33. LI, Z., R.-Y. WU, D. NICOLL & K. PHILIPSON. 1994. Expression of the canine Na^+-Ca^{2+} exchanger in transgenic mouse hearts. Biophys. J. **66:** A331.
34. YAO, A., Z. SU, A. NONAKA, et al. 1998. Effects of overexpression of the Na-Ca exchanger on $[Ca^{2+}]i$ transients inmurine ventricular myocytes. Circ. Res. **82:** 657–665.
35. CROSS, H., L. LU, C. STEENBERGEN, et al. 1998. Overexpression of the cardiac Na^+/Ca^{2+} exchanger increases susceptibility to ischemia/reperfusion injury in male, but not female, transgenic mice. Circ. Res. **83:** 1215–1223.
36. CAMPER-KIRBY, D., S. WELCH, A. WALKER, et al. 2001. Myocardial Akt activation and gender: increased nuclear activity in females versus males. Circ. Res. **88:** 1020–1027.

37. SUGISHITA, K., Z. SU, F. LI, et al. 2001. Gender influences [Ca^{2+}]i during metabolic inhibition in myocytes overexpressing the Na^+–Ca^{2+} exchanger. Circulation **104:** 2101–2106.
38. TERRACCIANO, C., A. SOUZA, K. PHILIPSON & K. MACLEOD. 1998. Na^+–Ca^{2+} exchange and sarcoplasmic reticular Ca^{2+} regulation in ventricular myocytes from transgenic mice overexpressing the Na^+–Ca^{2+} exchanger. J. Physiol. **512:** 651–667.
39. CROSS, H.R., C. STEENBERGEN & E. MURPHY. 1999. Phospholamban knock-out mice exhibit increased susceptibility to myocardial ischemic injury: a gender-specific effect. Circulation **100:** 493–493.

Simulation of Na/Ca Exchange Activity during Ischemia

DENIS NOBLE

University Laboratory of Physiology, Oxford University, Oxford OX1 3PT, UK

> ABSTRACT: Simulation of sodium–calcium exchange activity during the rise of intracellular sodium that occurs during ischemia suggests that the exchanger may not reverse direction except transiently during calcium oscillations. This conclusion depends on the presence of a small resting leak of calcium into the cell, consistent with radioactive calcium flux measurements. The conditions for intracellular calcium to rise to around 3 µM were explored. A combination of extracellular potassium accumulation and extracellular sodium depletion is sufficient to explain this result. The computations also show a counterintuitive result concerning the role of the exchanger in the mechanism of calcium oscillations. Reducing its activity would be expected to enhance these oscillations, whereas increasing it can reduce or suppress oscillations. If such oscillations play a role in acute ischemic arrhythmias, then block of Na/Ca exchange may not be therapeutic.
>
> KEYWORDS: Na/Ca exchange; ischemia; computational modeling

The interactions between metabolites, ion concentrations, and ionic currents that occur during the development of a disease state such as ischemia are highly complex. Intuition alone frequently fails us when faced with such biological complexity. This is precisely the kind of situation in which computational modeling becomes essential if we are to begin to understand what is going on.[1] My contribution to this debate is to show what is required to reconstruct some of the known experimental data in ischemic hearts. I will also demonstrate an important counterintuitive result.

However, first, let us be clear about the role of modeling in biological work. When we construct a model, our aim, of course, is to succeed in reproducing the phenomena that we are seeking to explain. At this stage, modeling is a method of asking what are the expected theoretical consequences of whatever assumptions were built into the model. This will show whether our assumptions are adequate. A second stage is when the model is used to determine what we would expect to happen under conditions that were not used to construct the model. Here, we will learn something, no matter whether the model succeeds or whether it fails. In fact, we often learn most when models *fail* (see, e.g., Noble and Rudy[2]), particularly if they do so in a way that enables us to see the reason for failure. A third stage is when a complex model is stripped down to the minimal number of interactions required to explain the phenomena.

Address for correspondence: Denis Noble, Department of Physiology, Oxford University, Parks Road, Oxford OX1 3PT, United Kingdom.
denis.noble@physiol.ox.ic.uk

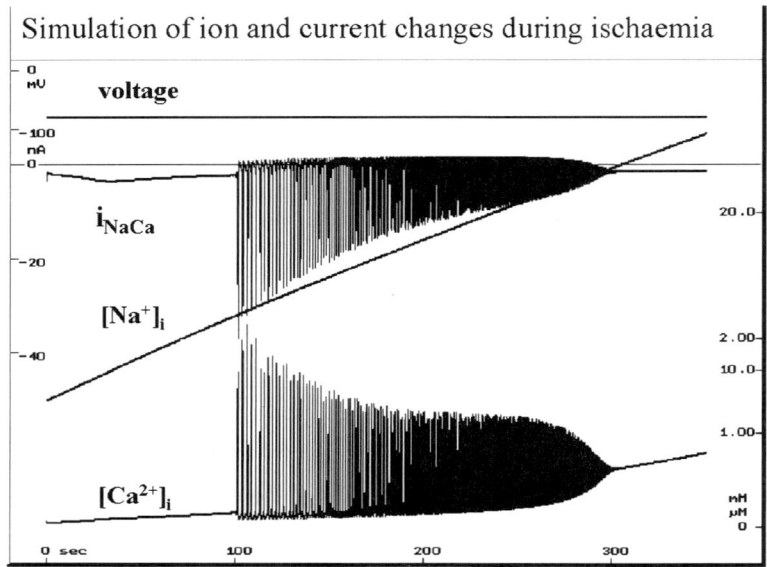

FIGURE 1. This computation uses a partial model of ischemia.[3] It isolates that part of the process that is secondary to the increase in [Na]$_i$. The model is based on a multicellular Purkinje fiber model,[4] which is why the exchanger current is plotted in nA rather than pA, as would be appropriate in a single cell. The zero baseline indicates the direction of the exchange current. All the computations shown here used OXSOFT HEART version 4.8.

The partial model of ischemia that we developed in our laboratory[3] has been used in all three of these stages. Here, I will present results using a stripped-down model in which the driver of the changes is the increase of [Na]$_i$ that occurs in ischemia as a result of inhibition of the Na-K pump and of sodium transport to counter acidity.

FIGURE 1 shows the main result of such a computation. This was done using the multicellular Purkinje fiber model of DiFrancesco and Noble.[4] The voltage was set constant at the resting potential, −80 mV, to simplify the calculation. It took 6 min for [Na]$_i$ to increase from its normal value (8 mM) to approximately 26 mM. As the increase in [Na]$_i$ develops, [Ca]$_i$ also increases as a result of the smaller sodium gradient available for the forward mode of the Na/Ca exchanger. At first, this increase is monotonic. At approximately 100 seconds (this time depends critically on the precise conditions of the computation), the increase in [Ca]$_i$ is sufficient to induce calcium oscillations through the mechanism of calcium-induced calcium release. These oscillations drive oscillatory variations in Na/Ca exchange activity, creating large inward currents that are known to underlie one of the forms of arrhythmia. Eventually, the increase in [Ca]$_i$ becomes monotonic again.

Notice two important facts about this calculation. First, the level of [Ca]$_i$ that is reached is much less than that observed experimentally. [Ca]$_i$ is still less than 1 μM instead of the value of 3.5 μM observed experimentally.[5] The second important feature is that, except transiently during each oscillation, the exchanger does not reverse. Here, we have a mixture of success and failure. The model succeeds beautifully in reproducing the role of Na/Ca exchange in mediating the increase in $[Ca]_i$ (this is the so-called "sodium-lag" mechanism and, incidentally, the model does also reproduce the correct sequence of inotropic changes consequent on this mechanism), and it succeeds in reproducing one of the mechanisms of arrhythmia. However, it fails to reconstruct the magnitude of the observed increase in calcium. Something, therefore, is missing (FIG. 1).

One way of explaining the larger-than-expected increase in $[Ca]_i$ would be to suppose that, in reality, the exchanger *does* reverse direction and starts to generate a net influx of calcium, instead of doing its best to pump calcium out. The sodium–calcium exchanger does not consume ATP (being ATP *dependent* is not the same thing at all!). Therefore, all it has to pump calcium is the energy of the sodium gradient. Except transiently, it cannot indefinitely maintain a net influx of calcium against the energy gradient available to it. This is a necessary thermodynamic fact. This point can also been seen by a simple mechanistic approach. Suppose that, for whatever reason, $[Ca]_i$ does increase beyond the level expected (and there could be many such reasons in a complicated situation such as ischemia). What will the exchanger then do? It then must pump calcium out even harder ($[Ca]_i$ is the main activator of the forward mode), and so the current that it generates will become even more inward. Far from reversing, it will be driven even harder in forward mode.

The only way therefore to modify the model in this way would be to build in a Maxwell's demon (a device that escapes the laws of thermodynamics). Most of us theoreticians will go to the stake before doing that! My conclusion therefore is that there is something missing from the data used in the model. During the debates at the meeting, we discovered what that may be. First, Elizabeth Murphy correctly observed that the potential will not stay at −80 mV. Potassium accumulation during ischemia depolarizes the cells by approximately 20 mV. Inserting $E = -60$ mV into the computation helps ($[Ca]_i$ gets closer to 1 µM), but it does not solve the problem.

This idea did, however, trigger a further train of thought regarding ion accumulation and depletion. Where does the sodium come from for the increase in $[Na]_i$? It must, of course, come from the extracellular space, *which is not being perfused*. A decrease in $[Na]_o$ therefore is obligatory. The model can be used to answer the question, How large a decrease in $[Na]_o$ would be sufficient? The answer is shown in FIGURE 2. Setting external sodium to 100 mM (instead of 140) is sufficient to ensure that, at the time when $[Na]_i$ has increased to 25 mM, $[Ca]_i$ exceeds 3 µM.

Does this actually happen?

Several groups have used nuclear magnetic resonance techniques to measure ion concentrations in whole hearts undergoing ischemia. One of these is in Oxford, directed by Kieran Clarke. She has kindly permitted me to quote the following unpublished data as a personal communication:

Preischemia: $[Na]_i = 9$ mM, $[Na]_o = 144$ mM, ratio 0.06 (rounded)

Ischemia: $[Na]_i = 9 \times 3.5 = 31.5$, ratio 0.3, so $[Na]_o = 105$ mM

which is *very* close indeed to the change in external sodium required in the computer model.

There are two further observations to make about the computation shown in FIGURE 2. First, the changes in membrane potential and in intracellular sodium do

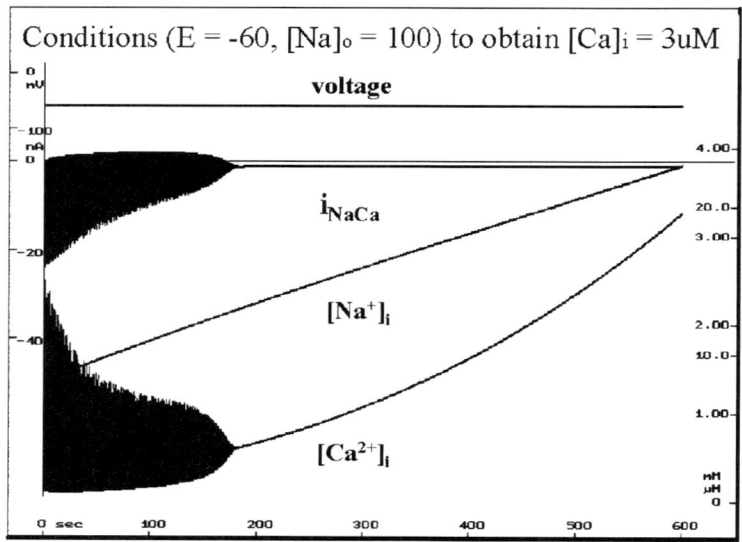

FIGURE 2. The computation shown in FIGURE 1 was rerun with E set to −60 mV and $[Na]_o$ set to 100 mM.

not occur immediately. They also have a time course. I therefore have repeated the computation with the changes in E and $[Na]_o$ phased over time. This makes a difference to the time of onset of calcium oscillations (they occur later), but it does not affect the end point. Nor should it, because this is determined primarily by thermodynamic considerations.

Second, apart from transiently during oscillations, the exchanger is always operating in the forward mode (inward current). I therefore conclude that, although its virtue is compromised during ischemia by the lowering of the sodium gradient that feeds its ability to move calcium, the exchanger is still doing what it can to promote calcium efflux. In this respect, it is one of the "good guys." This conclusion depends on the value of the resting calcium influx (or leak). In the model used here, this corresponds to approximately 12 pA/cell. There is no direct electrophysiological evidence for such a small resting current, but it would be surprising if the resting membrane were to be completely impermeable to calcium ions. The best estimate of resting calcium influx is that of Neidergerke,[6] who measured a Ca^{45} flux of 0.009 $pmol/cm^2/s$, which, for a cell with a capacitance of 100 pF, translates to approximately 20 pA.

The respect in which the exchanger is a bad guy is in the generation of arrhythmias via the mechanism of calcium oscillations. Can we use the model to say anything about this process? This is the point at which I introduce the counterintuitive result I mentioned to earlier.

A standard therapeutic approach to cardiac arrhythmias has been to seek to block the transporter (channel or exchanger) that carries the current immediately responsible for the abnormal excitations underlying the arrhythmia. FIGURE 3 shows that

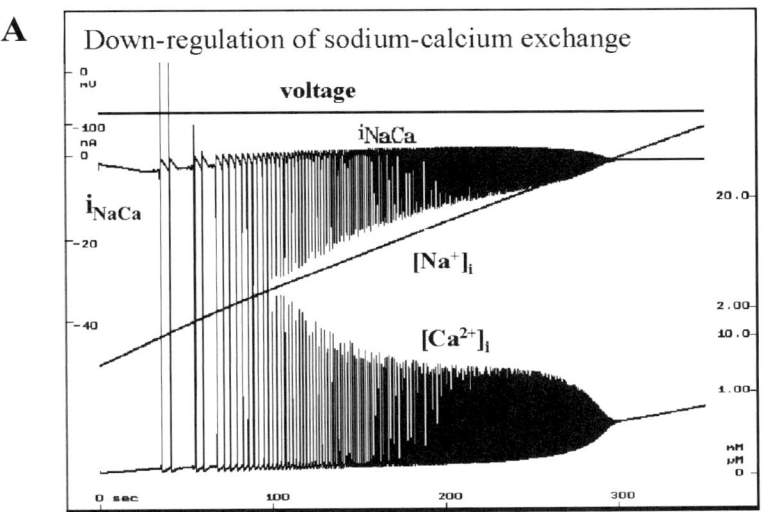

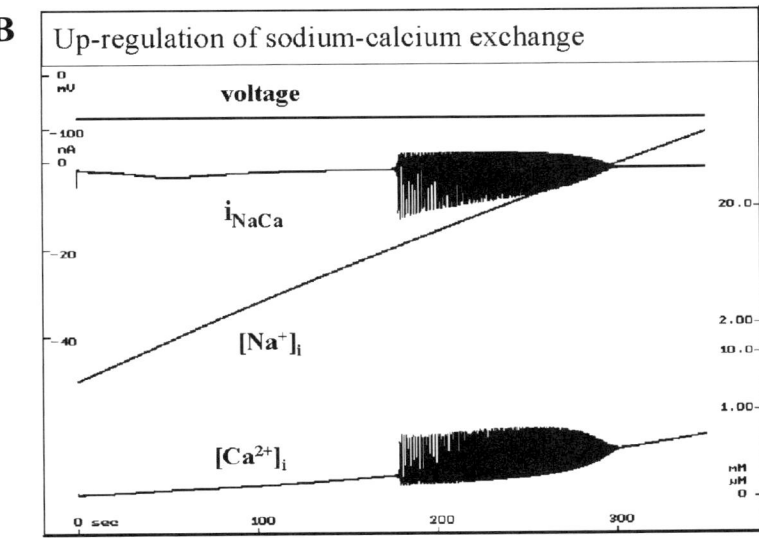

FIGURE 3. The computation shown in FIGURE 1 is rerun here with 10% block of the exchanger (*left*) and a 50% upregulation (*right*).

this approach may be incorrect. The left panel in the figure shows the effect of a 10% reduction in the activity of the exchanger. Far from reducing the oscillations, a comparison with FIGURE 1 shows that the situation is made worse. The oscillations occur earlier and are stronger. Of course, much larger inhibition of the exchanger (in particular a complete block) would succeed, but that would not be therapeutic for other, very obvious reasons. The right panel shows the effect of upregulating the exchanger by 50%. The oscillations now occur later and are much smaller. A 100% increase in activity removes the oscillations altogether (see Ch'en et al.[3]). Rob Hinch (personal communication) has shown that these results are necessary mathematically. They are not a quirk of the particular computer model being used (FIG. 3).

At the least, this result should make us pause for thought about antiarrhythmia strategies. A channel or transporter may play more than one role in a complex set of biological interactions. The sodium–calcium exchanger certainly does this. Its role as guardian of the cell's calcium incidentally makes it a carrier of potentially lethal ionic current. However, weakening its ability to counter calcium overload actually may compromise the situation more than any possible benefit from reducing its capacity to carry depolarizing current.

It is difficult, though, to see further than this generalization. The jury is still out on whether blockers of sodium–calcium exchange may be therapeutic. At least one study[7] shows that KB-R9032 is not effective against mortality in ischemic cardiac conditions in dogs. One of our difficulties in putting the modeling to experimental test is that none of the agents that affect the sodium–calcium exchange are highly specific. On the contrary, agents that target the sodium–calcium exchanger also target most of the ionic channels.[8] The results shown here might provide one reason why many of them are not very effective as therapeutic agents. In seeking therapeutic agents, we need to pay much more attention to biological complexity. There may well be subtle combinations of drug actions that would be therapeutic, but we have to work out what these may be. Otherwise we are shooting in the dark.

ACKNOWLEDGMENTS

This work was supported by the British Heart Foundation, the Medical Research Council, and The Wellcome Trust.

REFERENCES

1. NOBLE, D. 2002. Biological computation. *In* Encyclopedia of the Life Sciences. Nature Publishing Group. London. http://www.els.net
2. NOBLE, D. & Y. RUDY. 2001. Models of cardiac ventricular action potentials: iterative interaction between experiment and simulation. Phil. Trans. R. Soc. Ser. A **359**: 1127–1142.
3. CH'EN, F.T., R.D. VAUGHAN-JONES, K. CLARKE & D. NOBLE. 1998. Modelling myocardial ischaemia and reperfusion. Prog. Biophys. Mol. Biol. **69**: 515–537.
4. DIFRANCESCO, D. & D. NOBLE. 1985. A model of cardiac electrical activity incorporating ionic pumps and concentration changes. Philos. Trans. R. Soc. B **307**: 353–398.
5. MURPHY, E., H.R. CROSS & C. STEENBERGEN. 2002. Is Na/Ca exchange during ischemia and reperfusion beneficial or detrimental? Ann. N. Y. Acad. Sci. This volume.

6. NIEDERGERKE, R. 1967. Movements of Ca in frog heart ventricles at rest and during contractures. J. Physiol. **167:** 515–550
7. HASHIMOTO, K., A. SUGIYAMA, S. MIYAMOTO, *et al.* 2000. Block of Na exchangers on ischemia induced arrhythmias. Jpn. J. Electrocardiol. **20**(Suppl 3): 101–104.
8. KIMURA, J. 2002. Pharmacology of Na^+/Ca^{2+} exchanger. Ann. N.Y. Acad. Sci. This volume.

Role of the Na/Ca Exchanger in Arrhythmias in Compensated Hypertrophy

KARIN R. SIPIDO,[a] PAUL G. A. VOLDERS,[b] MARIEKE SCHOENMAKERS,[b]
S. H. MARIEKE DE GROOT,[b] FONS VERDONCK,[c] AND MARC A. VOS[b]

[a]*Laboratory of Experimental Cardiology, University of Leuven, Leuven, Belgium*

[b]*Department of Cardiology, University Hospital Maastricht, Maastricht, Belgium*

[c]*Interdisciplinary Research Center, University of Leuven, Kortrijk, Belgium*

ABSTRACT: Sudden, presumably arrhythmic, death is common in heart failure patients. Although total mortality is highest in end-stage failure, the fraction of sudden death in total mortality is higher in the early stages. In each of these stages various, not necessarily identical, ionic mechanisms may contribute to arrhythmogenesis. Dogs with chronic complete atrioventricular block (6–8 weeks) have an increased risk for arrhythmias and sudden death and have compensated biventricular hypertrophy. In this animal model, Ca^{2+} release from the sarcoplasmic reticulum (SR) is not reduced. For low frequencies of stimulation, the SR Ca^{2+} content is increased, related to a higher activity of the Na/Ca exchanger. Spontaneous Ca^{2+} release induces inward Na/Ca exchange current, which can lead to delayed afterdepolarizations (DADs) triggering a new action potential. Such arrhythmogenic DADs and ectopic beats also can be observed *in vivo* during monophasic action potential recording. They appear after pacing protocols, and/or administration of ouabain, which result in contractile potentiation, suggestive of a enhanced sarcoplasmic reticulum Ca^{2+} content. Other arrhythmogenic mechanisms related to increased dispersion of repolarization also can be identified *in vivo*. Downregulation of delayed K^+ currents is an important factor in prolongation of action potentials. In conclusion, in this animal model of compensated hypertrophy, Ca^{2+} handling is different from end-stage heart failure. It is possible that arrhythmogenic mechanisms related to a higher Ca^{2+} load contribute to the high incidence of sudden death in stages of compensated hypertrophy before overt heart failure. However, more than one ionic remodeling process is likely to be present, and different cellular mechanisms of arrhythmias can coexist.

KEYWORDS: Na/Ca exchanger; arrhythmias; Ca^{2+} content

INTRODUCTION

Sudden cardiac death, presumed as arrhythmic in origin, is a major cause of death in patients with heart failure. Sudden cardiac death is, however, not restricted to end-stage heart failure patients but is actually more common among patients with a less

Address for correspondence: Karin R. Sipido, M.D., Ph.D., Laboratory of Experimental Cardiology, KUL, Campus Gasthuisberg O/N 7th floor, Herestraat 49, B-3000 Leuven, Belgium. Voice: 32-16-347153; fax: 32-16-345844.
karin.sipido@med.kuleuven.ac.be

severe degree of chronic heart failure (NYHA class II, III), in which it may account for greater than 50% of total mortality.[1,2] In addition, population studies show that cardiac hypertrophy is an independent risk factor for sudden death (reviewed in Zipes and Wellens[1]). It therefore is important to examine the ionic changes contributing to the increased risk for arrhythmias in all stages of heart failure, and some mechanisms may be of particular importance in compensated hypertrophy. Valuable information on mechanisms of electrical remodeling in human heart failure has come from the study of human cells obtained from cardiac tissue at the time of heart transplantation.[3–7] Although there are reports on cells isolated from biopsies obtained during cardiac surgery,[8,9] it is much more difficult to obtain tissue samples and good quality cells from hearts with compensated hypertrophy. Animal models therefore currently offer an alternative. Electrical remodeling has been demonstrated in many animal models of heart failure, with similarities to the observations in humans (reviewed in Tomaselli and Marban[10]). For compensated hypertrophy, fewer models are available, but it has been pointed out that major differences with overt heart failure may exist, and most particularly for Ca^{2+} handling (e.g., Shorofsky et al.[11]). In a longitudinal study on mice after aortic banding, Ito et al.[12] could also document time-dependent changes in myocyte function and protein expression during the evolution from hypertrophy to failure, including changes in sarcoplasmic reticulum Ca^{2+}-ATPase (SERCA) and in Na/Ca exchange (NCX).

In this paper we review our current knowledge on the changes in Ca^{2+} handling and the role of NCX that may be particular to compensated hypertrophy and to the arrhythmias in this condition, based on the data we obtained in a dog model.

COMPENSATED HYPERTROPHY AND RISK OF ARRHYTHMIAS IN A DOG MODEL

After destruction of the atrioventricular (AV) node in the dog, the ventricular rate decreases from approximately 100 beats/min to approximately 40 beats/min (see Vos et al.[13] for methodological details). At the time of creation of AV block, this sudden decrease in ventricular rate is associated with a decrease in cardiac output, elevation of left-ventricular end-diastolic pressure (LVEDP) and maintained LV pressure development, +LV dP/dt.[14,15] In the following weeks these values normalize, with even an increase in the +LV dP/dt when measured at 6–8 weeks of chronic AVB (CAVB). At this time, the dogs have a pronounced biventricular eccentric hypertrophy, and there are no signs of neurohormonal changes as associated with heart failure.[14] Therefore, at this time point they can be defined as being in compensated hypertrophy. Sudden death rate in these dogs is elevated and related to the occurrence of polymorphic ventricular tachyarrhythmias.[16] Electrical stimulation as well as pharmacological interventions are more likely to induce arrhythmias at this time of compensated hypertrophy[13] than at the time of creation of AVB.[14] Electrical remodeling is present, as supported by a prolongation of the QT time and of the LV and RV monophasic action potential (MAP) *in vivo*. This remodeling is intrinsic to the myocardium, as proved by studies on isolated cells.[17] The remodeling is not homogeneous throughout both ventricles, and LV myocytes have a more pronounced increase in action potential duration (APD) than RV myocytes. Downregulation of delayed rectifier K^+ currents, but not of I_{K1} or I_{to}, are an important factor in this electrical remodeling[18] and the increased susceptibility to drug-induced polymorphic ventricular tachycardias.

CHANGES IN $[Ca^{2+}]_i$ HANDLING CONTRIBUTE TO THE CONTRACTILE ADAPTATION

Cells isolated from dogs with CAVB have no impairment of contraction, and the extent of shortening is even enhanced at low frequencies of stimulation.[19] The difference in contractility between CAVB and control, as evaluated by unloaded cell shortening, is, however, less pronounced than the differences observed *in vivo*, using +LV dP/dt as a parameter for contractility. This suggests that *in vivo* additional, probably load-dependent, factors are important. Measurements of $[Ca^{2+}]_i$ transients parallel the changes in contraction.[19] The larger initial increase in $[Ca^{2+}]_i$ in cells from CAVB stimulated at low frequency is consistent with an increase in Ca^{2+} release from the sarcoplasmic reticulum (SR). The voltage dependence of SR release remained bell shaped, indicating that I_{CaL} was the major trigger for SR Ca^{2+} release,[20] and peak I_{CaL} was unchanged. Using fast caffeine applications to empty the SR after conditioning steps at different stimulation frequencies, we found that the SR Ca^{2+} content was significantly larger in cells from CAVB, particularly at low frequencies (FIG. 1A). This is consistent with the larger contractions and SR Ca^{2+} release at this low frequency.

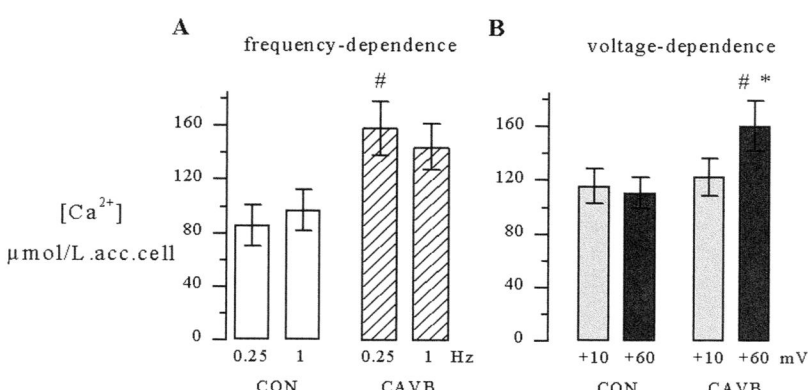

FIGURE 1. SR Ca^{2+} content is larger in hypertrophied myocytes from dogs with CAVB. SR Ca^{2+} content was measured by integrating the Na/Ca exchange inward current transient during Ca^{2+} release induced by rapid application of caffeine (10 mM, during 5 s), and expressed as concentration, assuming a surface/volume ratio of 5 pF/pl and an accessible cell volume fraction of 0.65.[19,36] In panel **A**, we compare the SR Ca^{2+} content after a conditioning train of 300-ms depolarizing steps from −70 to +10 mV at the indicated frequency (n_{cells} = 18 for CAVB, n_{cells} = 13 for control, CON). In panel **B**, we compare the effect of increasing the depolarization from +10 to +60 during a 1-Hz conditioning train (n_{cells} = 12 for CAVB, n_{cells} = 8 for control, CON). *, $P < 0.05$ for CAVB versus CON; #, $P < 0.05$ for steps to +10 versus to +60 mV.

INCREASE IN SR LOAD IS RELATED TO ALTERATIONS IN NCX ACTIVITY

The increased SR Ca^{2+} load could not be attributed to changes in Ca^{2+} current density. Peak I_{CaL} was unchanged,[19] and recent experiments indicate that T-type Ca^{2+} currents are not present (K.R. Sipido, Antoons, P.G.A. Volders, M.A.Vos, unpublished). NCX current density, however, was increased.[19] Additional evidence for an increased contribution of NCX in loading of the SR is shown in FIGURE 1B. With conditioning steps from −70 to +10 mV at 1 Hz, the difference between control and CAVB is not significant. Increasing the depolarization to +60 mV, to enhance Ca^{2+} influx via the exchanger, significantly increased the SR content in CAVB, but not in control, confirming the larger role of NCX in CAVB. The mechanisms of increased SR loading via NCX are not fully understood. Increased expression or activity *per se* could be sufficient as in one study of mice with overexpression of NCX in which a higher SR content was found.[21] The preferential increase of SR content at low frequencies is reminiscent of conditions of increased $[Na^+]_i$.[22,23] Indeed, with higher $[Na^+]_i$, gain of Ca^{2+} occurs in diastole.[24,25] We currently have preliminary data that show that indeed $[Na^+]_i$ is higher,[26] and this would increase reverse mode NCX. The increase in inward current cannot be explained by an increase in $[Na^+]_i$ but may be related to altered regulation.[19]

ARRHYTHMIAS RELATED TO ALTERED $[Ca^{2+}]_i$ AND Na/Ca EXCHANGE ACTIVITY

The finding of an increased SR content is in contrast with the findings in heart failure.[27,28] One type of arrhythmia that is commonly associated with increased SR content is the ventricular tachycardia triggered by ectopic beats as seen during digitalis intoxication and consequent $[Na^+]_i$ overload.[29] *In vitro*, spontaneous Ca^{2+} release induces a delayed afterdepolarization (DAD) which can trigger an action potential if of sufficient amplitude.[30] The major current underlying these afterdepolarizations is the Na/Ca exchanger, although other Ca-dependent currents may contribute (review in Volders *et al.*[31]). An increased tendency for spontaneous Ca^{2+} release in cells from CAVB could be demonstrated by applying long depolarizing pulses.[19] De Groot *et al.* examined whether *in vivo* arrhythmias triggered by DADs could be documented.[15] After a pacing train (300-ms interval, 5–13 pulses) DADs and ectopic beats could be evoked with CAVB for 6–8 weeks (FIG. 2, top panel), but not at baseline. A direct relation could be demonstrated between the increase in LV dP/dt postpacing and the incidence of DADs (observed in monophasic action potential recordings) and ectopic beats. *In vivo* administration of ryanodine could suppress these DADs and ectopic activity (FIG. 2, bottom panel). This is consistent with the idea that a critical increase in SR content had occurred, inducing spontaneous release and DADs, which could be achieved only in CAVB. The sensitivity for ouabain-induced ventricular tachycardia was also more pronounced in CAVB than in control,[15] again suggesting that Ca^{2+} overload was induced more easily in CAVB.

The contribution of NCX to arrhythmias is not limited to the increase in SR content. Larger currents at the time of spontaneous release are likewise important. In

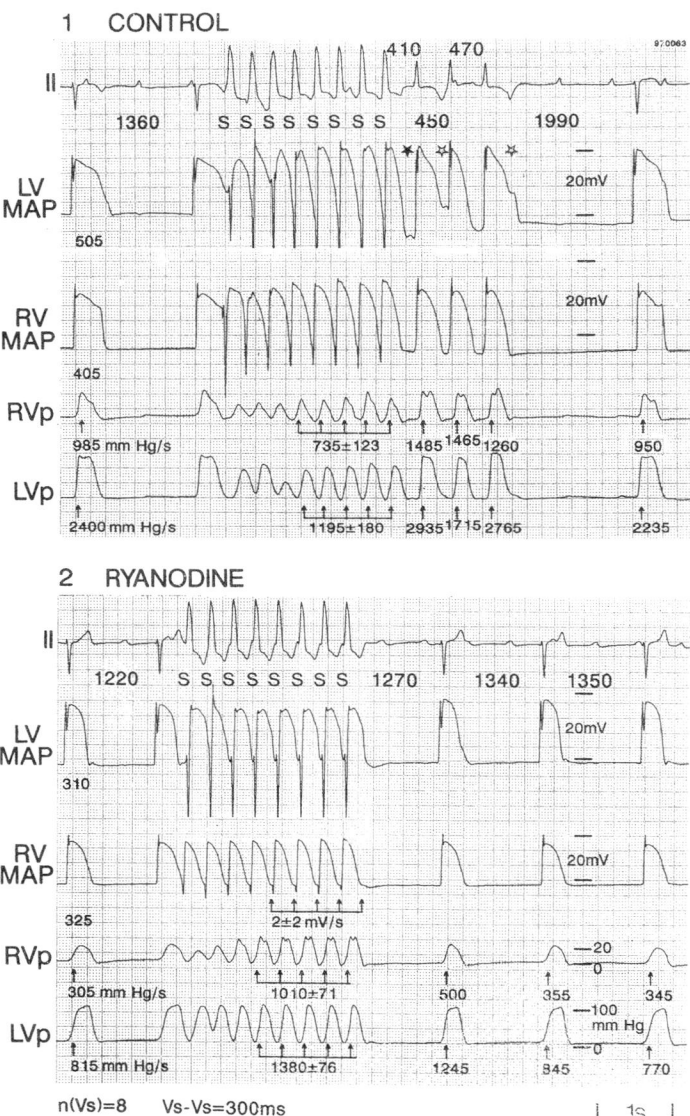

FIGURE 2. Suppression of triggered ectopic beats in CAVB dog by ryanodine. One ECG lead (II), two monophasic action potential recordings (LV MAP, RV MAP), and the intraventricular pressure (RVp, LVp) were registered at a paper speed of 25 mm/s in a dog with CVAB before (control, *top*) and after administration of 10 µg/kg ryanodine over 10 min. (*bottom*). After a pacing train (300-ms interval, $n = 8$) DADs (*filled stars*) and ectopic beats are observed, with some EADs (*unfilled stars*), and the contraction is potentiated (see increase in LV dP/dt). Ryanodine suppresses the afterdepolarizations and ectopic activity.

other models, such as the rabbit with heart failure in which SR content is not increased, this will be the major mechanism.[28] In this model, less release is required to induce a DAD because of the increased NCX current at the time of release and because of a reduced I_{K1}.

ARRHYTHMOGENESIS IN CARDIAC REMODELING: NO "SINGLE CULPRIT," NO SINGLE PHENOTYPE

The DAD-dependent arrhythmias are not the only type of ventricular arrhythmias in this animal model. More common, dogs with CAVB have an increased propensity to develop Torsade-de-Pointes (TdP), which can occur spontaneously or be induced by drugs and/or pacing.[13] This type of arrhythmia occurs at low heart rates when APD is highest, and dispersion most pronounced. TdP often is preceded by early afterdepolarizations (EADs) which can be seen *in vivo* on the MAP. Both EADs and a large dispersion have been implicated in the initiation of the TdP by triggering ectopic beats and/or by local re-entry.[16,32,33] Loss of repolarizing K^+ currents is a major factor in the APD increase, whereas enhanced NCX currents during the action potential plateau also may contribute.[18] Therefore, at least in this model it is clear that several changes in ionic currents contribute to the electrical remodeling and the enhanced risk of (several types of) arrhythmias. For the rabbit model of heart failure, an interplay between enhanced NCX current and decreased I_{K1} leads to the increased occurrence of DADs and triggered arrhythmias, despite a lower SR Ca^{2+} content.[28] In dogs with tachycardia-induced heart failure, downregulation of the transient outward current and slow inactivation of I_{CaL} due to decreased SR Ca^{2+} release, but not the upregulation of NCX, appear to be the major mechanisms of the AP prolongation.[34] These examples illustrate that there will be different phenotypes depending on the stimulus for hypertrophy/failure, and on associated changes in ion channels and Ca^{2+} handling. Furthermore, review of the literature on expression of NCX in cardiac hypertrophy and heart failure[35] clearly shows that upregulation of NCX is not a general feature of cardiac remodeling but is dependent on functional state as well as dependent on the stimulus for hypertrophy/failure.

CONCLUSIONS

In dogs with CAVB and compensated hypertrophy, several mechanisms may contribute to increased susceptibility for arrhythmias. These include increased NCX activity and increased SR Ca^{2+} load, which are at the same time contributing to the preserved contractile function. These mechanisms are different from the ones described so far in heart failure. It is tempting to speculate that in humans differences in arrhythmia incidence in compensated versus decompensated heart failure may be related to altered Ca^{2+} handling, and a different role for NCX. This issue merits further investigation through longitudinal studies, examining the role of the initial stimulus for hypertrophy/failure and of the functional stage on the characteristics of remodeling.

REFERENCES

1. ZIPES, D.P. & H.J.J. WELLENS. 1998. Sudden cardiac death. Circulation **98:** 2334–2351.
2. MERIT-HF STUDY GROUP. 2001. Effect of metoprolol CR/XL in chronic heart failure: Metoprolol CR/XL Randomised Intervention Trial in Congestive Heart Failure (MERIT-HF). Lancet **353:** 2001–2007.
3. BEUCKELMANN, D.J., M. NABAUER & E. ERDMANN. 1992. Intracellular calcium handling in isolated ventricular myocytes from patients with terminal heart failure. Circulation **85:** 1046–1055.
4. WETTWER, E., G.J. AMOS, H. POSIVAL & U. RAVENS. 1994. Transient outward current in human ventricular myocytes of subepicardial and subendocardial origin. Circ. Res. **75:** 473–482.
5. LI, G.R., J. FENG, L. YUE, et al. 1996. Evidence for two components of delayed rectifier K^+ current in human ventricular myocytes. Circ. Res. **78:** 689–696.
6. SIPIDO, K.R., T. STANKOVICOVA, W. FLAMENG, et al. 1998. Frequency dependence of Ca^{2+} release from the sarcoplasmic reticulum in human ventricular myocytes from end-stage heart failure. Cardiovasc. Res. **37:** 478–488.
7. MATIELLO, J.A., K.B. MARGULIES, V. JEEVANANDAM & S.R. HOUSER. 1998. Contribution of reverse mode sodium-calcium exchange to contractions in failing human left ventricular myocytes. Cardiovasc. Res. **37:** 424–431.
8. BENITAH, J.P., P. BAILLY, M.C. D'AGROSA, et al. 1992. Slow inward current in single cells isolated from adult human ventricles. Pflügers Arch. **421:** 176–187.
9. KONARZEWSKA, H., G.A. PEETERS & M.C. SANGUINETTI. 1995. Repolarizing K^+ currents in nonfailing human hearts. Similarities between right septal subendocardial and left subepicardial ventricular myocytes. Circulation **92:** 1179–1187.
10. TOMASELLI, G.F. & E. MARBAN. 1999. Electrical remodeling in hypertrophy and heart failure. Cardiovasc. Res. **42:** 270–284.
11. SHOROFSKY, S.R., R. AGGARWAL, M. CORRETTI, et al. 1999. Cellular mechanisms of altered contractility in the hypertrophied heart. Circ. Res. **84:** 424–434.
12. ITO, K., X. YAN, M. TAJIMA, et al. 2000. Contractile reserve and intracellular calcium regulation in mouse myocytes from normal and hypertrophied failing hearts. Circ. Res. **87:** 588–595.
13. VOS, M.A., S.C. VERDUYN, A.P. GORGELS, et al. 1995. Reproducible induction of early afterdepolarizations and torsade de pointes arrhythmias by D-sotalol and pacing in dogs with chronic atrioventricular block. Circulation **91:** 864–872.
14. VOS, M.A., S.H. DE GROOT, S.C. VERDUYN, et al. 1998. Enhanced susceptibility for acquired torsade de pointes arrhythmias in the dog with chronic, complete AV block is related to cardiac hypertrophy and electrical remodeling. Circulation **98:** 1125–1135.
15. DE GROOT, S.H., M. SCHOENMAKERS, M.M. MOLENSCHOT, et al. 2000. Contractile adaptations preserving cardiac output predispose the hypertrophied canine heart to delayed afterdepolarization-dependent ventricular arrhythmias. Circulation **102:** 2145–2151.
16. VAN OPSTAL, J.M., S.C. VERDUYN, H.D. LEUNISSEN, et al. 2001. Electrophysiological parameters indicative of sudden cardiac death in the dog with chronic complete AV-block. Cardiovasc. Res. **50:** 354–361.
17. VOLDERS, P.G.A., K.R. SIPIDO, M.A. VOS, et al. 1998. Cellular basis of biventricular hypertrophy and arrhythmogenesis in dogs with chronic complete atrioventricular block and acquired torsade de pointes. Circulation **98:** 1136–1147.
18. VOLDERS, P.G.A., K.R. SIPIDO, M.A. VOS, et al. 1999. Downregulation of delayed rectifier K^+ currents in dogs with chronic complete atrioventricular block and acquired torsades de pointes. Circulation **100:** 2455–2461.
19. SIPIDO, K.R., P.G.A. VOLDERS, S.H. DE GROOT, et al. 2000. Enhanced Ca^{2+} release and Na/Ca exchange activity in hypertrophied canine ventricular myocytes: a potential link between contractile adaptation and arrhythmogenesis. Circulation **102:** 2137–2144.
20. SIPIDO, K.R., M.M. MAES & F. VAN DE WERF. 1997. Low efficiency of Ca^{2+} entry through the Na/Ca exchanger as trigger for Ca^{2+} release from the sarcoplasmic reticulum. Circ. Res. **81:** 1034–1044.

21. TERRACCIANO, C.M., A.I. DE SOUZA, K.D. PHILIPSON & K.T. MACLEOD. 1998. Na^+-Ca^{2+} exchange and sarcoplasmic reticular Ca^{2+} regulation in ventricular myocytes from transgenic mice overexpressing the Na^+-Ca^{2+} exchanger. J. Physiol. **512:** 651–667.
22. DIAZ, M.E., S.J. COOK, J.P. CHAMUNORWA, et al. 1996. Variability of spontaneous Ca^{2+} release between different rat ventricular myocytes is correlated with Na^+-Ca^{2+} exchange and $[Na^+]_i$. Circ. Res. **78:** 857–862.
23. MUBAGWA, K., W. LIN, K.R. SIPIDO, et al. 1997. Monensin-induced reversal of positive force-frequency relationship in cardiac muscle: role of intracellular sodium in rest-dependent potentiation of contraction. J. Mol. Cell. Cardiol. **29:** 977–989.
24. BENNETT, D.L., S.C. O'NEILL & D.A. EISNER. 1999. Strophanthidin-induced gain of Ca^{2+} occurs during diastole and not systole in guinea-pig ventricular myocytes. Pflügers Arch. **437:** 731–736.
25. MEME, W., S. O'NEILL & D.A. EISNER. 2001. Low sodium inotropy is accompanied by diastolic Ca^{2+} gain and systolic loss in isolated guinea-pig ventricular myocytes. J. Physiol. (London) **530:** 487–495.
26. VERDONCK, F., P.G.A. VOLDERS, M.A. VOS & K.R. SIPIDO. 2001. Cardiac hypertrophy is associated with an increase in subsarcolemmal Na^+. Biophys. J. **80:** 589A
27. LINDNER, M., E. ERDMANN & D.J. BEUCKELMANN. 1998. Calcium content of the sarcoplasmic reticulum in isolated ventricular myocytes from patients with terminal heart failure. J. Mol. Cell. Cardiol. **30:** 743–749.
28. POGWIZD, S.M., K. SCHLOTTHAUER, L. LI, et al. 2001. Arrhythmogenesis and contractile dysfunction in heart failure: roles of sodium–calcium exchange, inward rectifier potassium current, and residual β-adrenergic responsiveness. Circ. Res. **88:** 1159–1167.
29. DE GROOT, S.H., M.A. VOS, A.P. GORGELS, et al. 1995. Combining monophasic action potential recordings with pacing to demonstrate delayed afterdepolarizations and triggered arrhythmias in the intact heart. Value of diastolic slope. Circulation **92:** 2697–2704.
30. WIER, W.G. & P. HESS. 1984. Excitation-contraction coupling in cardiac Purkinje fibers. Effects of cardiotonic steroids in the intracellular $[Ca^{2+}]$ transient, membrane potential and contraction. J. Gen. Physiol. **83:** 395–415.
31. VOLDERS, P.G.A., A. KULCSAR, M.A. VOS, et al. 1997. Similarities between early and delayed afterdepolarizations induced by isoproterenol in canine ventricular myocytes. Cardiovasc. Res. **34:** 348–359.
32. EL-SHERIF, N., E.B. CAREF, H. YIN & M. RESTIVO. 1996. The electrophysiological mechanism of ventricular arrhythmias in the long QT syndrome. Tridimensional mapping of activation and recovery patterns. Circ. Res. **79:** 474–492.
33. ANTZELEVITCH, C., Z.Q. SUN, Z.Q. ZHANG & G.X. YAN. 1996. Cellular and ionic mechanisms underlying erythromycin-induced long QT intervals and torsade de pointes. J. Am. Coll. Cardiol. **28:** 1836–1848.
34. WINSLOW, R.L., J.J. RICE, S. JAFRI, et al. 1999. Mechanisms of altered excitation-contraction coupling in canine tachycardia-induced heart failure, II: model studies. Circ. Res. **84:** 571–586.
35. SIPIDO, K.R., P.G.A. VOLDERS, M.A. VOS & F. VERDONCK. 2002. Altered Na/Ca exchange activity in cardiac hypertrophy and failure: a new target for therapy? Cardiovasc. Res. **53:** 782–805.
36. VARRO, A., N. NEGRETTI, S.B. HESTER & D.A. EISNER. 1993. An estimate of the calcium content of the sarcoplasmic reticulum in rat ventricular myocytes. Pflügers Arch. **423:** 158–160.

Enhanced Sodium–Calcium Exchange in the Infarcted Heart

Effects on Sarcoplasmic Reticulum Content and Cellular Contractility

SHELDON E. LITWIN AND DONGFANG ZHANG

Division of Cardiology, Salt Lake City Veterans Affairs Medical Center, and the University of Utah, Salt Lake City, Utah 84148, USA

ABSTRACT: Arrhythmias and contractile dysfunction both contribute to the high morbidity and mortality in patients with congestive heart failure. Contractile dysfunction is generally believed to reflect a decrease in the amplitude of intracellular Ca^{2+} transients, whereas tachyarrythmias are often initiated in the setting of cellular Ca^{2+} overload. In a rabbit model of left ventricular dysfunction due to myocardial infarction, we found evidence that myocyte sarcoplasmic reticulum Ca^{2+} content may be normal or even increased at slow stimulation rates. This may occur because prolonged action potential duration promotes Ca^{2+} influx via the Na^+/Ca^{2+} exchanger. Despite preserved SR Ca^{2+} content, intracellular Ca^{2+} transients and contractions may be reduced in amplitude because of impaired synchronization of Ca^{2+} release events throughout the myocyte.

KEYWORDS: Na/Ca exchange; infarcted heart; sarcoplasmic reticulum; cellular contractility

INTRODUCTION

Severe left ventricular (LV) dysfunction is associated with increased mortality due to both malignant arrhythmias and progressive heart failure. Ventricular arrhythmias may initiate when increased sarcoplasmic reticulum (SR) Ca^{2+} content causes spontaneous SR Ca^{2+} release and afterdepolarizations.[1] In contrast, contractile failure generally is thought to result from reduced SR Ca^{2+} release.[2] Triggered arrhythmias and progressive contractile dysfunction often occur simultaneously in the same patient, and the risk of sudden cardiac death appears to increase as the LV ejection fraction decreases.[3] Based on current notions of how intraluminal SR Ca^{2+} content modulates arrhythmogenesis and contractility, it is difficult to reconcile how these seemingly different circumstances can exist concurrently in the same heart.

Address for correspondence: Sheldon E. Litwin, M.D., University of Utah, Cardiovascular Division, 50 N. Medical Drive, Salt Lake City, UT 84148. Voice: 801-581-7715; fax: 801-581-7735.
   sheldon.litwin@hsc.utah.edu

The sarcolemmal Na^+/Ca^{2+} exchanger can produce Ca^{2+} entry or extrusion in myocytes. The direction of Ca^{2+} movement depends on the driving forces for Na^+ and Ca^{2+} (i.e., the electrochemical gradients across the sarcolemma for each ion). Because of its ability to move Ca^{2+} in both directions across the sarcolemmal membrane and to compete with the SR Ca^{2+} ATPase, the exchanger may be a key regulator of SR Ca^{2+} content. Expression and/or activity of the Na^+/Ca^{2+} exchanger is increased in many forms of cardiac hypertrophy and failure.[4] Thus, the issue of how enhanced Na^+/Ca^{2+} exchange activity affects SR content in myocytes from diseased hearts is a crucial question.

Herein, we review evidence suggesting that Ca^{2+} influx by the Na^+/Ca^{2+} exchanger may increase SR Ca^{2+} content in a model of postinfarction LV dysfunction. We believe that enhanced SR content could explain a predisposition to triggered arrhythmias, whereas poor coordination of subcellular Ca^{2+} release events may cause impaired contractility, even with normal or increased SR Ca^{2+} content.

METHODS

Myocardial infarctions were induced in male New Zealand white rabbits by ligation of the circumflex coronary artery.[5,6] Rabbits were killed 3–8 weeks after myocardial infarction (MI), and LV myocytes from the periinfarct borderzone were selectively isolated. Myocytes were studied during perfusion with a HEPES-buffered modified Tyrode's solution. Cell motion was monitored with a video edge detection system (Crescent Electronics). Solutions superfusing cells were rapidly changed using a two-barrel glass solution switcher. Cellular function was studied using the following techniques: (1) field stimulation, (2) current clamp, and (3) whole-cell, ruptured patch voltage clamp. Dynamic confocal line scan images were obtained in fluo-3 acetoxy methylester (AM)-loaded myocytes during field stimulation. Individual protocols are described in RESULTS.

RESULTS

Total occlusion of the circumflex artery resulted in transmural infarctions occupying approximately 20% of the total LV surface. MI rabbits had significant LV enlargement and reduced LV systolic function Myocytes from infarcted hearts were increased in length by approximately 20% with minimal change in width compared with those from control hearts.[5]

Maximal Na^+/Ca^{2+} exchange current (I_{NaCa}) density was recorded in voltage-clamped myocytes by rapidly changing the superfusing solution to one in which LiCl was substituted for NaCl.[5] The sudden decrease in extracellular Na^+ causes rapid extrusion of Na^+ by the Na^+/Ca^{2+} exchanger which produces an outward current. Using this method, we found a 32% increase in maximal I_{NaCa} density in MI myocytes versus controls (FIG. 1).

The increase in maximal I_{NaCa} is functionally significant in the MI myocytes. First, we show that Ca^{2+} extrusion by the exchanger is enhanced in MI myocytes compared with controls. This is demonstrated by accelerated rest decay in the MI myocytes (FIG. 2). Myocytes were field-stimulated for 2 min to achieve stable SR

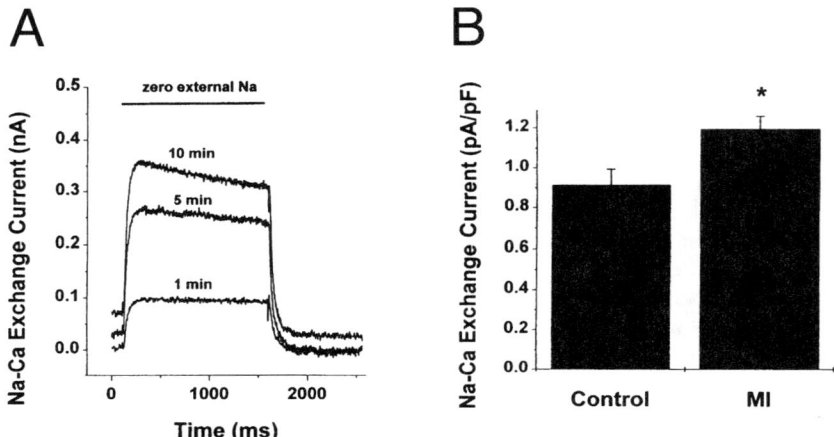

FIGURE 1. Maximal Na^+/Ca^{2+} exchange currents in control and MI myocytes. Cell were held at –40 mV, and the superfusing solution was rapidly changed from the normal Tyrode's solution to one in which NaCl was replaced by LiCl. Rapid solution switches were performed every 2–3 minutes until current amplitude reached steady state. The largest value of the exchange current was recorded for each cell and expressed relative to membrane capacitance. **(A)** Example of currents recorded using this protocol. **(B)** Summary data for control ($n = 10$) and MI ($n = 13$) myocytes. Maximal I_{NaCa} expressed relative to membrane capacitance was increased by approximately 30% in MI myocytes compared with controls. From Litwin and Bridge[5]; used with permission. * $P < 0.05$ vs. control.

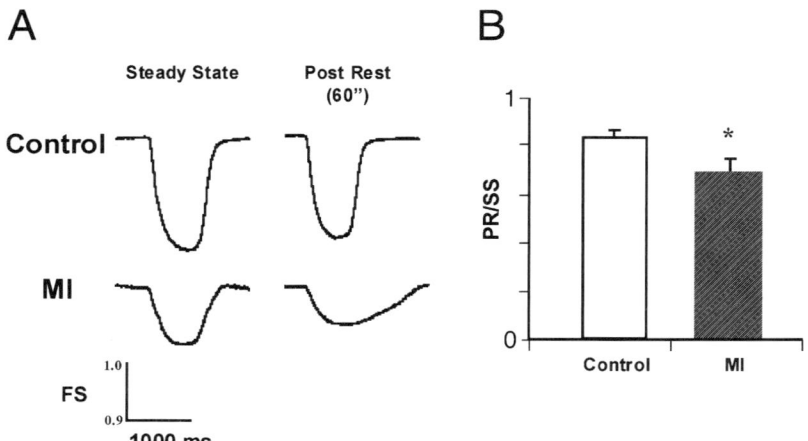

FIGURE 2. Rest decay in control and MI myocytes. Cells were field stimulated to allow steady-state (SS) conditions to develop. Stimulation then was halted for 60 seconds and resumed thereafter. The final steady-state contraction and the first postrest (PR) contraction for a control and an MI myocyte are shown in **A**. Summary data for control and MI myocytes ($n = 29$ and 42, respectively) are shown in **B** (data expressed as relative amplitude of PR to SS contraction). Rest decay is greater in the MI myocytes, suggesting enhanced Ca^{2+} extrusion via forward Na^+/Ca^{2+} exchange. * $P < 0.05$ vs. control.

loading. A period of 60 seconds of rest was interposed, and stimulation was resumed. During the period of rest, Ca^{2+} "leaks" from the SR and is extruded from the cell by the Na^+/Ca^{2+} exchanger. Increased rest decay is believed to reflect enhanced activity of the forward mode of Na^+/Ca^{2+} exchange (Ca^{2+} extrusion).

"Reverse" mode Na^+/Ca^{2+} exchange (Ca^{2+} entry) also is enhanced in MI myocytes.[5] In these experiments, we held cells at –40 mV and applied five conditioning pulses to +10 mV to achieve steady-state SR content. We then applied a 400-ms voltage clamp step to a potential between –70 and +90 mV ("loading pulses"). Three seconds later, another clamp step to +10 mV was applied ("test pulse"). We assumed that the cellular contraction during the test pulse would reflect any changes in SR content that occurred during the loading pulse. Because I_{Ca} has a bell-shaped relationship with voltage and Na^+/Ca^{2+} exchange has an exponential I-V relationship, pulses to positive potentials should produce a small I_{Ca} and a relatively larger I_{NaCa}. Thus, after loading pulses to positive potentials, changes in SR content should predominantly reflect the effects of reverse Na/Ca exchange. We expected that loading pulses to positive potentials might cause more augmentation of the subsequent contraction (test pulse) in the MI than the control myocytes. The peak cellular shorten-

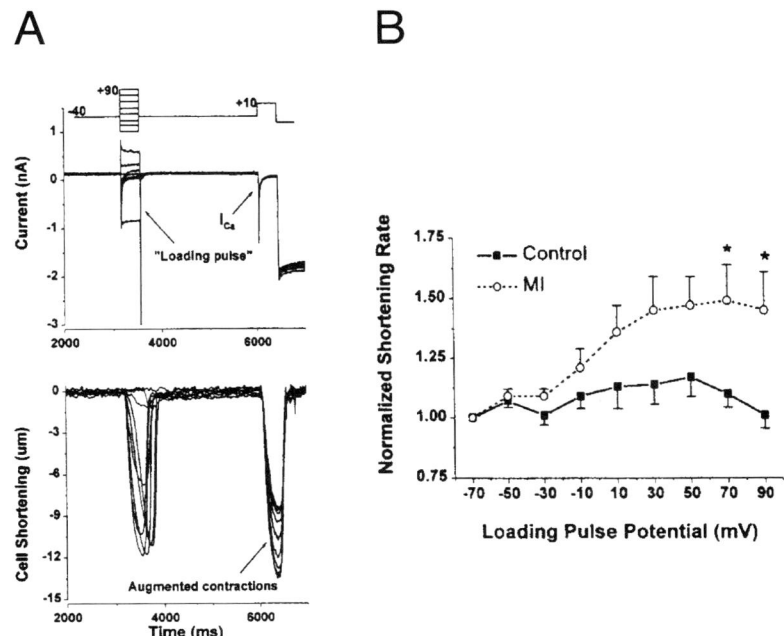

FIGURE 3. Effects of Na^+/Ca^{2+} exchange on SR loading. **(A)** Example of loading protocol (see text for details). **(B)** Depolarization to more positive potentials causes greater augmentation of the subsequent contraction in MI ($n = 20$) than in control ($n = 10$) myocytes. This suggests that Ca^{2+} entry via the Na^+/Ca^{2+} exchanger is more effective at loading the SR in MI compared with control myocytes. Modified from Litwin and Bridge[5] with permission. * $P < 0.05$ vs. control.

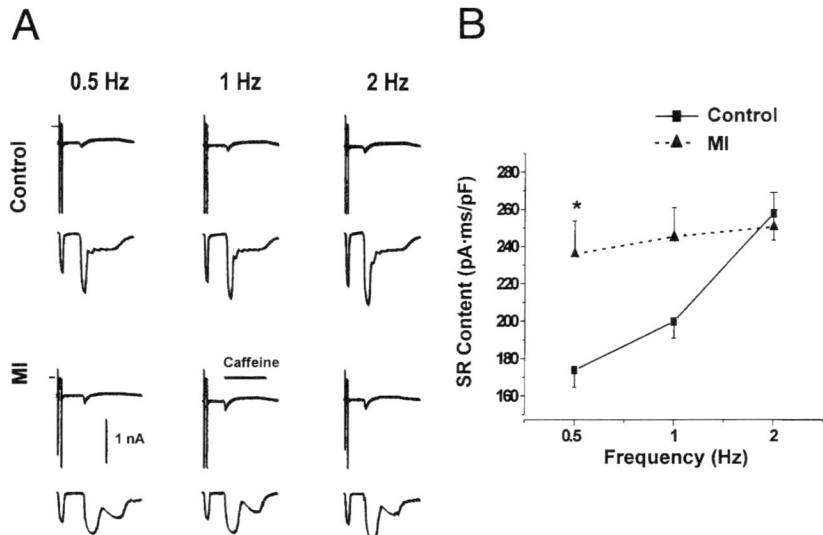

FIGURE 4. (A) Examples of SR Ca^{2+} content measurement using caffeine application in control and MI myocytes. Methods are described in the text. (B) Summary data for SR content measured after conditioning trains at different stimulation frequencies ($n = 28$ in each group). SR content is higher in MI myocytes at low stimulation rates but does not increase at faster stimulation rates as is seen in control cells. Figure modified from Litwin et al.[6] with permission. * $P < 0.05$ vs. control.

ing rate of each contraction was expressed relative to that of the first postloading contraction for the same cell. To prove that augmentation of contractions was caused by changes in SR content, we performed the protocol in cells ($n = 3$) that were pretreated with ryanodine (1 µM) and thapsigargin (1 µM). This treatment abolished the augmentation seen with loading pulses to positive potentials. These experiments showed that a step to strongly positive potentials (> +50 mV) produced more augmentation of the subsequent contraction in MI myocytes than in control myocytes when SR function was intact (FIG. 3).

SR Ca^{2+} content was estimated after a train of six conditioning pulses (400-ms clamp steps at 2-s intervals), which was used to establish steady-state SR loading. After the train, the cell was rapidly superfused with Tyrode's solution containing 20 mM caffeine. The application of caffeine induced a large contraction accompanied by an inward current (FIG. 4A). The inward current induced by caffeine is caused by Na^+/Ca^{2+} exchange, which extrudes the Ca^{2+} released from the SR.[7] Integration of the inward current gives an estimate of the total amount of Ca^{2+} released from the SR. Because cell size differs between control and infarcted hearts, we expressed the caffeine current integral relative to cell capacitance. We found that SR Ca^{2+} content was greater in MI than in control myocytes when the train of conditioning pulses was applied at a slow rate. However, SR content increased with faster stimulation in control, but not MI myocytes (FIG. 4B).

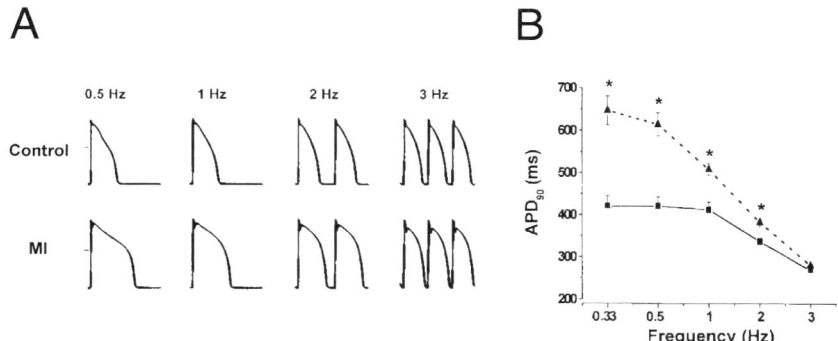

FIGURE 5. Representative examples of action potential duration in control and MI myocytes recorded at different stimulation frequencies (**A**). APD versus stimulation rate relationship is steeper in MI myocytes ($n = 15$) compared with controls ($n = 16$) (**B**). AP prolongation at slow stimulation rates may promote Ca^{2+} influx via the Na^+/Ca^{2+} exchanger and thus promote loading of the SR. Modified from Litwin et al.[6] with permission. $P < 0.05$ vs. control.

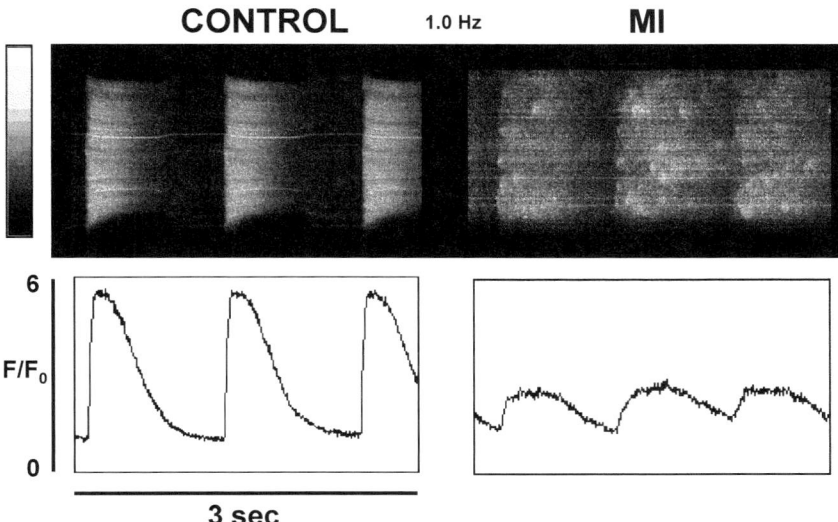

FIGURE 6. Dyssynchronous Ca^{2+} sparks in MI myocytes. Stacked line scan images in a control myocyte (*left*) and a periinfarct myocyte (*right*). Control myocyte shows coordinated Ca^{2+} release immediately after electrical stimulation. Ca^{2+} decline occurs homogeneously along the length of the cell. In contrast, the MI myocyte shows a scalloped appearance of the leading edge of the Ca^{2+} transient. Discrete release events are noted very late after the electrical stimulus, even as global Ca^{2+} appears to be declining. Integrated Ca^{2+} transients are shown (*bottom panels*). This altered coordination of SR Ca^{2+} release may account for slowed upstroke and decline of whole cell transients, even with relatively replete SR content. From Litwin et al.[9]; used with permission.

Depolarization to potentials positive to E_{NaCa} result in Ca^{2+} influx. Depending on several factors, including $[Na]_i$, the peak and plateau of the cardiac action potential may be positive to E_{NaCa}. Thus, longer action potentials may enhance Ca^{2+} influx via the exchanger. We found that action potential duration (APD) was significantly prolonged in MI myocytes at slow stimulation rates (FIG. 5A). However, the slope of the relationship between APD and stimulation frequency was steeper in MI myocytes than controls (FIG. 5B). This finding suggests that increased Ca^{2+} influx via the Na^+/Ca^{2+} exchanger at slow stimulation rates may support contractility that otherwise would be reduced. This beneficial effect may be lost at faster stimulation rates when APD is shorter, and thus may contribute to blunting of the force–frequency relationship.

If SR Ca^{2+} content is maintained at normal or near normal levels in the myocytes from damaged hearts, why would myocyte contractility be reduced? One possibility is a reduction in the "gain" of Ca^{2+}-induced Ca^{2+} release.[8] Our initial studies of macroscopic gain suggest that this is not the main mechanism accounting for decreased contractility in the periinfarct rabbit myocytes. However, studies performed using dynamic confocal microscopy suggest that dyssynchronous subcellular Ca^{2+} release events might contribute to a diminution of the peak Ca^{2+} transient, even with adequate intraluminal Ca^{2+} content (FIG. 6).[9]

CONCLUSIONS

These data support the general notion that Ca^{2+} influx via the sarcolemmal Na^+/Ca^{2+} exchanger can enhance SR Ca^{2+} content in cardiac myocytes. This effect appears to be amplified in myocytes from infarcted hearts in which there is a modest increase in Na^+/Ca^{2+} exchange activity. Whether this phenomenon occurs in other models or stages of LV dysfunction is not certain. However, the presence and magnitude of Ca^{2+} influx by the Na^+/Ca^{2+} exchanger will depend to a large extent on the activity of the SR Ca^{2+}-ATPase and $[Na]_i$.

ACKNOWLEDGMENTS

Supported by grants from the Department of Veterans Affairs, and the National Heart Lung and Blood Institute (HL 52338-06).

REFERENCES

1. ARONSON, R.S. & Z. MING. 1993. Cellular mechanisms of arrhythmias in hypertrophied and failing myocardium. Circulation **87**(Suppl. VII): VII-76–VII-83.
2. HOUSER, S.R. & E.G. LAKATTA. 1999. Function of the cardiac myocyte in the conundrum of end-stage, dilated human heart failure. Circulation **99**: 600–604.
3. BIGGER, J.T., JR., J.L. FLEISS, R. KLEIGER, et al. 1984. The relationships among ventricular arrhythmias, left ventricular dysfunction, and mortality in the 2 years after myocardial infarction. Circulation **69**: 250–258.
4. BARRY, W.H. 2000. Na(+)-Ca(2+) exchange in failing myocardium: friend or foe? Circ. Res. **87**: 529–531.

5. Litwin, S.E. & J.H.B. Bridge. 1997. Enhanced sodium-calcium exchange in the infarcted heart: implications for excitation–contraction coupling. Circ. Res. **81:** 1083–1093.
6. LITWIN, S.E., D. ZHANG, P. ROBERGE & G.D. PENNOCK. 2000. DITPA prevents the blunted contraction-frequency relationship in myocytes from infarcted hearts. Am. J. Physiol. **278:** H862–H870.
7. VARRO, A., N. NEGRETTI, S.B. HESTER & D.A. EISNER. 1993. An estimate of the calcium content of the sarcoplasmic reticulum in rat ventricular myocytes. Pflugers Arch. **423:** 158–160.
8. GOMEZ, A.M., H.H. VALDIVIA, H. CHENG, et al. 1997. Defective excitation-contraction coupling in experimental cardiac hypertrophy and heart failure. Science **276:** 800–806.
9. LITWIN, S.E., D. ZHANG & J.H. BRIDGE. 2000. Dyssynchronous Ca(2+) sparks in myocytes from infarcted hearts. Circ. Res. **87:** 1040–1047.

Na/Ca Exchange in Heart Failure

Contractile Dysfunction and Arrhythmogenesis

STEVEN M. POGWIZD[a] AND DONALD M. BERS[b]

[a]*Department of Medicine, University of Illinois at Chicago, Chicago, Illinois 60612, USA*
[b]*Department of Physiology, Loyola University Chicago, Maywood, Illinois 60153, USA*

> ABSTRACT: Congestive heart failure (HF) is characterized by contractile dysfunction and a high incidence of sudden death from nonreentrant ventricular arrhythmias, both of which involve altered intracellular calcium handling. The focus of this article is on the critical role of the Na/Ca exchanger. We demonstrate that upregulation of Na/Ca exchanger unloads the sarcoplasmic reticulum (SR), leading to contractile dysfunction. At the same time, Na/Ca exchanger underlies the arrhythmogenic transient inward current (I_{ti}) in HF. Preserved β-adrenergic responsiveness in HF plays a crucial role in increasing SR Ca load, leading to SR Ca release and activation of I_{ti}. In addition, decreased I_{K1} (inward rectifier) current in HF destabilizes resting membrane potential (E_m) and further enhances arrhythmogenesis mediated by the upregulated Na/Ca exchanger. We thus propose a new paradigm in which upregulated Na/Ca exchanger in HF plays a dual role in underlying both the contractile dysfunction and arrhythmogenesis in the failing heart. Therapeutic approaches to the treatment of HF will need to balance increasing SR Ca load with the arrhythmogenic effects of SR Ca overload that involve activation of I_{ti} carried by Na/Ca exchanger.
>
> KEYWORDS: Na/Ca exchanger; heart failure; cardiomyopathy; arrhythmia; tachycardia; contractile dysfunction

INTRODUCTION: CONTRACTILE DYSFUNCTION AND ARRHYTHMOGENESIS IN HEART FAILURE

Heart failure (HF) is characterized by marked contractile dysfunction that often leads to death from pump failure.[1] However, nearly one-half of HF patients die suddenly, primarily from ventricular arrhythmias including ventricular tachycardia and ventricular fibrillation.[1] Thus, the development of effective therapeutic strategies will require an understanding of the biochemical and cellular mechanisms involving both contractile dysfunction and arrhythmogenesis.

There is considerable evidence that altered intracellular Ca handling contributes to altered contraction in the failing heart.[2–6] Failing myocardium is characterized by decreased Ca transients along with altered expression and function of a number of

---

Address for correspondence: Steven M. Pogwizd, M.D., Department of Medicine, University of Illinois at Chicago, 840 South Wood Street, M/C 715, Chicago, IL 60612. Voice: 312-996-9342; fax: 312-413-1616.
spogwizd@uic.edu

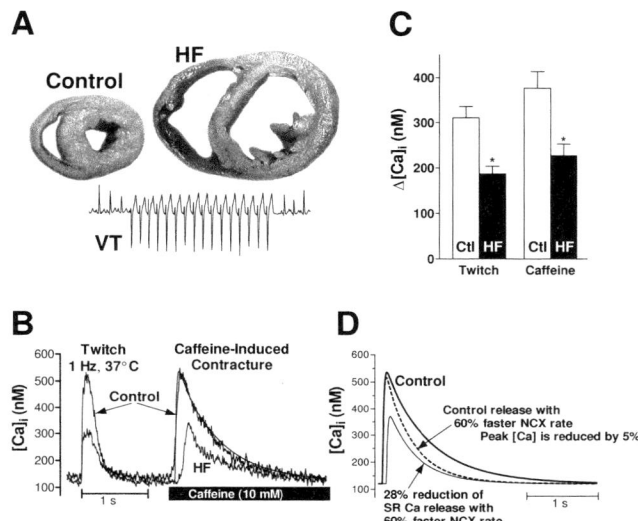

FIGURE 1. Arrhythmogenic rabbit HF model. **(A)** Cross-sections of a control and HF rabbit heart and an example of nonsustained ventricular tachycardia (VT) in HF rabbit Holter recording. **(B, C)** Contractile function and SR Ca load are reduced in HF. Steady-state twitches $\Delta[Ca]_i$ during AP recording (1 Hz, 37°C) with caffeine-induced $\Delta[Ca]_i$ to assess SR Ca load ($n = 18,17$; $P < 0.001$ for twitch and <0.01 for caffeine). **(D)** Control caffeine-induced Ca transient simulation (*thick line*, also shown in FIG. 1B). The *dashed curve* shows the same control SR Ca release with Na/Ca exchanger function increased by 60% (resulting in 5% reduction in peak $[Ca]_i$). *Thin curve* shows a 40% reduction in peak $[Ca]_i$, which required a 28% decrease in SR Ca release along with the 60% increase in Na/Ca exchanger function (see text). Data are from Pogwizd et al.[12]

Ca-handling proteins including the Na/Ca exchanger, the sarcoplasmic reticulum Ca-ATPase (SERCA), and the Ca release channel (or ryanodine receptor, RyR).[5–10] The focus of this article is the central role of Na/Ca exchanger in the failing heart.

ARRHYTHMOGENIC MODEL OF HEART FAILURE

Defining biochemical mechanisms for contractile dysfunction and arrhythmogenesis requires an experimental animal model of heart failure that exhibits both of these features. We have performed studies in an arrhythmogenic rabbit model of heart failure induced by combined aortic insufficiency and aortic constriction.[6,11,12] Rabbits with combined volume and pressure overload demonstrate marked ventricular dilatation and severely depressed LV function (FIG. 1A). Left ventricular end-diastolic dimension is increased 37%, and LV fractional shortening is decreased 38%. The hearts are hypertrophied (75% increase in heart weight to body weight, 89% increase in cell size) and exhibit interstitial fibrosis.[6,11,12]

Along with the contractile dysfunction, there is arrhythmogenesis. Holter monitoring has demonstrated nonsustained VT in 90% of HF rabbits (compared with none in controls; FIG. 1A), and approximately 10% of rabbits die suddenly.[6,12] In three-dimensional cardiac-mapping studies,[11] we have shown that spontaneous VT initiates by a nonreentrant mechanism such as triggered activity from delayed afterdepolarizations (DADs) or early afterdepolarizations (EADs).[13] Three-dimensional mapping studies of VT in hearts from patients with end-stage nonischemic heart failure (idiopathic dilated cardiomyopathy) at the time of cardiac transplantation show a similar focal nonreentrant mechanism.[14] Cellular studies in HF rabbits have demonstrated DADs, aftercontractions, and activation of an arrhythmogenic transient inward current (I_{ti}) by catecholamines,[12,15] suggesting that I_{ti} and DADs underlie nonreentrant initiation of VT in HF.

Thus, this heart failure model exhibits the functional, pathological, and electrophysiological alterations seen in the human nonischemic HF and allows us an opportunity to define alterations in Ca handling that are relevant to human HF.

DECREASED SR Ca LOAD AS THE BASIS FOR CONTRACTILE DYSFUNCTION IN HF RABBITS

Twitch amplitude and SR Ca load were assessed in field-stimulated LV myocytes from control and HF rabbits. We found that the 40% decrease in twitch amplitude was matched by a 40% decrease in SR Ca load (assessed by rapid caffeine administration, FIG. 1B and C).[12] It could be argued that the enhanced Na/Ca exchanger expression (see below) could contribute to the smaller caffeine-induced Ca transient in HF by curtailing the peak $[Ca]_i$.[16] We addressed this using quantitative flux analysis in the following manner.

A control caffeine-induced Ca transient was simulated using the measured Na/Ca exchanger transport flux dependence on $[Ca]_i$: $J_{NCX} = (242 \, \mu M/s)/(1 + \{316 \, nM/[Ca]_i\}^{3.7}$ (from Bassani et al.,[16] with V_{max} increased ~3-fold for 37 vs. 23°C), along with cytosolic Ca buffering: $[Ca]_{Tot} = (244 \, \mu M)/(1 + 673 \, nM/[Ca]_i)$ (from Bers[17]), and a simple SR Ca release flux that is the product of an increasing and decreasing exponential, with amplitude and time constants adjusted to match the caffeine-induced Ca transient in FIGURE 1B. The superimposed smooth curve in FIGURE 1B is the simulation (shown again as the control in FIG. 1D). If we simply increase the rate of Na/Ca exchanger by 60% (as we measured in this HF model), the rate of $[Ca]_i$ decline is faster, but the peak $[Ca]_i$ is reduced by only 5%. Thus, this curtailment effect is minimal. Finally, to assess the SR Ca release required to cause a 40% decrease in caffeine-induced Ca transient in the face of this curtailment, we progressively reduced the simulated SR Ca release to match the Ca transient amplitude observed experimentally (bottom trace in FIG. 1D). This required a 28% reduction of SR Ca release. We conclude that SR Ca content is reduced by at least 28% in this HF model and accounts for the 40% decrease in twitch amplitude.

Our findings of decreased SR Ca load in HF myocytes are consistent with those of Hobai and O'Rourke in the canine pacing model[18] and by Lindner et al. in the failing human heart.[19] Thus, the decreased SR Ca load explains the contractile dysfunction without the need to invoke other alterations in excitation–contraction (EC) coupling such as altered Ca gain.[4] Although more detailed study would be needed to rule out alterations in intrinsic EC-coupling gain, we saw no differences in the ratio of twitch $\Delta[Ca]_i$: caffeine $\Delta[Ca]_i$ ratio in HF,[12] consistent with a comparable SR Ca release.

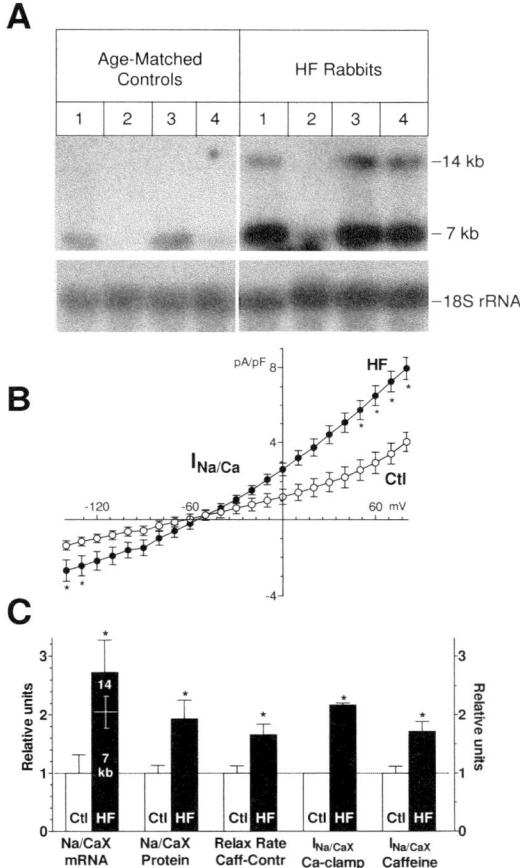

FIGURE 2. Na/Ca exchanger is upregulated in HF. **(A)** Northern blot analysis of Na/Ca exchanger mRNA in myocardial tissue from control and HF rabbits. In the four left lanes are samples from age-matched controls and in the four right lanes are samples from HF rabbits. The uppermost gel was hybridized with a cDNA probe specific for cardiac Na/Ca exchanger. All rabbits demonstrate expression of a 7-kb transcript (although one of the controls expressed low levels that were more apparent on prolonged exposures). However, HF rabbits 1, 3, and 4 also demonstrated an approximately 14-kb transcript that was evident in only one of the six control rabbits. At bottom is shown the same gel which has been reprobed with cDNA specific for 18S rRNA to reflect comparable degree of loading. Quantitation of Na/Ca exchanger mRNA was normalized to the levels of the 18S rRNA. **(B)** Current–voltage relationship for $I_{Na/Ca}$ recorded in HF and control rabbit myocytes in which $[Ca]_i$ was clamped to 100 nM. **(C)** Na/Ca exchanger expression, activity and current. Values are shown for Na/Ca exchanger mRNA (7- and 14-kb transcripts) from Northern blotting, Na/Ca exchanger protein from immunoblotting, pseudorate constant of relaxation of the caffeine contraction ($\lambda_{Na/Ca} = \ln 2/t_{1/2}$), and $I_{Na/Ca}$ (inward and outward, excluding values near the reversal potential, −70 to −50 mV). Results are normalized to values from control myocytes. Asterisk (*) denotes statistical significance with $P < 0.05$ versus controls. Data are from Pogwizd et al.[6,12]

UPREGULATION OF Na/Ca EXCHANGER IN HF RABBITS

Because SR Ca load is affected by the expression and function of Na/Ca exchanger and SERCA (which can change in HF[7–9]), we assessed these parameters in HF rabbit tissue and myocytes. Northern blot analysis showed upregulation of Na/Ca exchanger mRNA in HF with a twofold increase in the 7-kb transcript. In addition, there was a high molecular weight (~14 kb) transcript present in the HF rabbits (FIG. 2A) that has been reported in rabbit heart, kidney, and brain that also was present in very low levels in one of the controls.[6] With the inclusion of this of this 14-kb transcript, Na/Ca exchanger mRNA expression was increased 2.7-fold compared with controls. There was also a 93% increase in Na/Ca exchanger protein expression measured by immunoblotting that correlated with the mRNA upregulation.[6]

Na/Ca exchanger function was assessed in indo-loaded myocytes. The rate constant ($\lambda = \ln 2/t_{1/2}$) of the decline of the caffeine-induced Ca transient (when net SR uptake is inhibited) was increased greater than 60% in HF, consistent with enhanced Na/Ca exchanger activity. Inward and outward Na/Ca exchanger current (measured by voltage clamp) were increased twofold, whether $[Ca]_i$ was clamped to 100 nM (FIG. 2B) or allowed to change in a dynamic fashion (during caffeine-induced $I_{Na/Ca}$).[12] Overall, $I_{Na/Ca}$ for any given $[Ca]_i$ was increased whether measured as the $I_{Na/Ca}$ current density at $[Ca]_i = 500$ nM or as the slope of the plot of $I_{Na/Ca}$ versus $[Ca]_i$. Thus, as shown in FIGURE 2C, there was a consistent twofold upregulation of Na/Ca exchanger on the level of mRNA, protein, Ca decline, and $I_{Na/Ca}$.[6]

We assessed SERCA expression and function. Although there were no significant differences in SERCA mRNA and protein or in the rate constant of twitch decline (which is primarily caused by SERCA), on further analysis we found evidence of a 24% decrease in SERCA function.[6] Our findings of upregulated Na/Ca exchanger in the presence of relatively preserved SERCA (which are consistent with those reported in failing human hearts by Hasenfuss et al.[10]) suggest that the upregulated Na/Ca exchanger competes better with the SR during relaxation. The result is unloading of the SR and contractile dysfunction.

UPREGULATED Na/Ca EXCHANGER AND ARRHYTHMOGENESIS

So how does the upregulation of Na/Ca exchanger relate to arrhythmogenesis? DADs arise from activation of I_{ti}, a Ca-dependent inward current that arises when SR Ca overload leads to spontaneous SR Ca release.[13,20] I_{ti} could be caused by forward-mode Na/Ca exchanger (three monovalent Na ions in and one divalent Ca ion out, resulting in net inward current), a nonspecific cationic current ($I_{NS(Ca)}$),[21] or a Ca-activated chloride current ($I_{Cl(Ca)}$).[22] To determine what currents underlie DADs, we studied DADs induced by rapid administration of caffeine (cDADs) in current clamped rabbit HF myocytes that were loaded with indo-1 AM.[12,23] Progressive SR Ca loading induced by increasing stimulation frequency led to cDADs that increased in amplitude until they triggered an action potential (AP; FIG. 3A). Caffeine administration in the presence of 0 $[Na]_o$ and 0 $[Ca]_o$ (Li-substituted, where $I_{Cl(Ca)}$ and $I_{NS(Ca)}$ can still flow) nearly abolished cDADs. In contrast, niflumic acid, which blocks $I_{Cl(Ca)}$ had no significant effect on cDADs.[12] Thus, the upregulated Na/Ca exchanger in HF underlies I_{ti} that can lead to arrhythmogenesis.

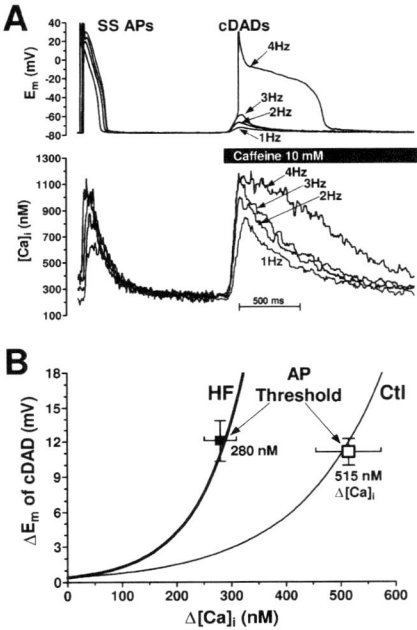

FIGURE 3. (**A**) Caffeine-induced Ca transients and delayed afterdepolarizations (cDADs). Rabbit ventricular myocytes were studied at 37°C under current clamp. Steady state (SS) APs were induced at different frequencies, resulting in altered twitch and caffeine-induced D[Ca]$_i$. Larger Δ[Ca]$_i$ caused larger cDADs and at some point trigger APs. (**B**) Mean subthreshold data are fit by $0.4\exp(k\Delta[Ca]_i)$ such that ΔE_m doubles for each Δ[Ca]$_i$ of $\ln(2)/k$. The ΔE_m doubles with 44% less Δ[Ca]$_i$ in HF. *Large squares* indicate threshold for triggering APs. Data at top are redrawn from Schlotthauer and Bers,[23] and data below are drawn from Pogwizd *et al.*[12]

However, does the upregulated Na/Ca exchanger actually contribute to HF myocytes being more prone to triggered arrhythmias? To address this, we assessed the level of threshold [Ca]$_i$ needed for a cDAD to trigger an AP.[12] We found that a given Δ[Ca]$_i$ led to a greater change in membrane potential (ΔE_m) in HF and that the threshold Δ[Ca]$_i$ to trigger an AP was approximately 50% lower in HF (FIG. 3B). These findings are consistent with enhanced arrhythmogenesis caused by an upregulated Na/Ca exchanger (although other currents could contribute, see below).

PRESERVED β-ADRENERGIC RESPONSIVENESS ALLOWS SR Ca LOAD TO INCREASE IN HF

So how can the upregulated Na/Ca exchanger play a dual role in which it unloads the SR of Ca (to underlie contractile dysfunction) yet mediates the arrhythmogenic current that arises with SR Ca overload? The resolution of this paradox lies with preserved β-adrenergic responsiveness that can substantially increase SR Ca load. We found that I_{Ca} and Ca transients of HF myocytes retained β-adrenergic responsiveness.[12] We then measured SR Ca load (as the integral of the caffeine-induced $I_{Na/Ca}$) in voltage-clamped myocytes after a series of conditioning pulses plus an additional depolarizing pulse (1–3 seconds, +50 mV) designed to load the SR via Na/Ca exchanger. With the addition of isoproterenol, SR load significantly increased and I_{ti} was more easily induced in HF myocytes (although the threshold level of SR Ca load required to induce I_{ti} was unchanged compared to controls).[12] This finding is consistent with our other studies in which isoproterenol led to a greater incidence of aftercontractions in isolated HF myocytes, and to PVCs and ventricular tachycardia in intact HF rabbits.[12] Thus, preserved β-adrenergic responsiveness plays an important role in mediating the arrhythmogenesis arising from upregulated Na/Ca exchanger. These findings may explain why arrhythmic deaths are common in patients with compensated HF (when β-adrenergic responsiveness is better preserved), whereas pump failure deaths are more common in end-stage HF (when β-adrenergic responsiveness is minimal).[24–26] This may also explain why treatment with β-adrenergic blockers might be efficacious in decreasing the incidence of sudden death in HF patients.[27]

As mentioned above, an upregulated Na/Ca exchanger could explain a lower threshold $\Delta[Ca]_i$ for an AP, but other currents could contribute. For instance, a decrease in the inward rectifying current I_{K1} could destabilize the resting membrane potential and lead to a greater depolarization for any given inward current. We measured I_{K1} in HF myocytes and found that it was decreased by 49% (compared with controls) across a wide range of E_m (FIG. 4A).[12] We then did current injection (using current waveforms similar to measured I_{ti} pseduo-I_{ti}) to determine whether there was

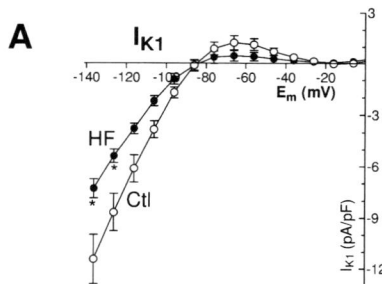

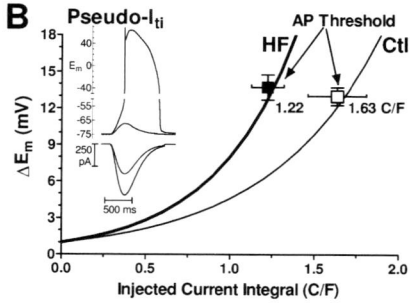

FIGURE 4. (A) Current–voltage relationship for I_{K1} recorded in HF and control rabbit myocytes (*$P < 0.05$). (B) Pseudo-I_{ti}–induced depolarization. Ca-independent current injection (pseudo-I_{ti}, with time course shown in *inset*) causes depolarization fit to an exponential

$$[\Delta E_m = \exp\left(k\int I_{ti} dt\right)].$$

The ΔE_m doubles with 27% less charge in HF. Threshold current integral was reduced 25% in HF versus control (*large squares*, $P < 0.05$). Data are from Pogwizd et al.[12]

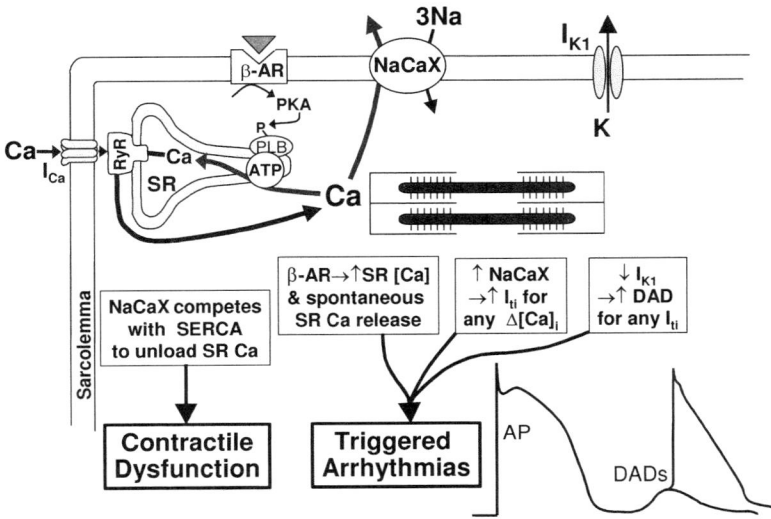

FIGURE 5. Schema of Na/Ca exchanger, I_{K1} and β-adrenergic receptors (βAR) in contractile dysfunction and arrhythmogenesis in HF (redrawn from Pogwizd et al.[12]).

a lower threshold current to trigger an AP in HF. Indeed, we found that for any given I_{ti} (with no associated change in $\Delta[Ca]_i$) there was a greater ΔE_m (FIG. 4B).[12] To determine whether this was caused by a decreased I_{K1}, we repeated these studies in control myocytes using a concentration of barium (3 µM) that we found to inhibit I_{K1} by approximately 50%. This Ba block of I_{K1} mimicked the pseudo-I_{ti} shift in HF,[12] indicating that the reduced I_{K1} quantitatively explained the shift and therefore contributes to the enhanced arrhythmogenicity of HF myocytes.

Thus, we propose a new paradigm for the central role of Na/Ca exchanger in HF (FIG. 5).[12] Upregulated Na/Ca exchanger plays a dual role. First it unloads the SR leading to contractile dysfunction. Second, it underlies I_{ti} leading to arrhythmogenesis. Preserved β-adrenergic responsiveness plays a critical role in raising SR Ca load to the point of spontaneous release and activation of I_{ti}. In addition, decreased I_{K1} further contributes to arrhythmogenesis by destabilizing resting E_m and producing a greater depolarization for any given I_{ti}. Recent computer modeling studies[28] suggest that the upregulated Na/Ca exchanger and the decreased I_{K1} contribute equally to enhanced arrhythmogenesis in HF.

IMPLICATIONS

The implication of this paradigm is that whereas Na/Ca exchanger plays a critical role in the contractile dysfunction and arrhythmogenesis in HF, it is not a straightforward target for therapeutic intervention. Enhancing Na/Ca exchanger to offset unloading of the SR could directly enhance the arrhythmogenic I_{ti} (especially in the presence of β-adrenergic stimulation). Inhibition of Na/Ca exchanger would attenuate I_{ti}, but this could come at the expense of impaired relaxation (diastolic dysfunction) and enhancement of SR Ca load that could still lead to spontaneous Ca release and arrhythmias (albeit with reduced I_{ti}). Even attempts to enhance inotropy by increasing SR Ca load, either with overexpression of SERCA or inhibition of phospholamban,[29,30] have the potential for arrhythmogenesis if SR Ca overload occurs. Perhaps there would be a greater therapeutic window in which to modulate SR Ca load and/or I_{ti} in the setting of β-adrenergic blockade. Enhancing I_{K1} could be beneficial but runs the risk of decreasing excitability that could alter conduction and be arrhythmogenic. Overall, therapeutic approaches to the failing heart will require balancing the need for increased SR Ca load with the need to avoid Ca overload and arrhythmogenesis. This point is best supported by the lesson learned from the use of inotropes such as phosphodiesterase inhibitors in HF, which improve contractile performance but are associated with increased arrhythmic deaths.[31]

Other Ca-handling proteins also could play a role in modulating the effects of Na/Ca exchanger in HF. For instance, RyR hyperphosphorylation, described by Marx et al. in human HF[5] and recently demonstrated by us in our HF rabbits,[32] could dissociate FKBP12.6 from the RyR and the lead to a subconductance state that increases SR Ca leak (and unloads the SR). Greater downregulation of SERCA also would enhance SR Ca unloading. Although these and other Ca-handling proteins are important in HF, Na/Ca exchanger clearly plays a critical role.

UPREGULATED Na/Ca EXCHANGER IN HF—GOOD OR BAD?

Therefore, is upregulation of Na/Ca exchanger in HF good or bad? In our arrhythmogenic heart failure preparation, an upregulated Na/Ca exchanger would unload the SR and lead to contractile dysfunction. The extent of this SR Ca unloading might be limited by increased $[Na]_i$ in HF[33] which would shift Na/Ca exchanger more in favor of Ca entry. However, the primary finding is that the SR is unloaded, so the overall effect of upregulated Na/Ca exchanger on LV function appears to be "bad." In fact, if $[Na]_i$ were not elevated in HF, there would likely be even greater SR unloading by the upregulated Na/Ca exchanger. From an arrhythmogenic standpoint, enhanced Na/Ca exchanger means enhanced I_{ti}, and the overall effect of this is also bad. Thus, in the setting of severe (but not end-stage HF) where there is preserved β-adrenergic responsiveness an upregulated Na/Ca exchanger appears to be detrimental.

This may not necessarily be the case in end-stage heart failure, in which there is marked SERCA downregulation, limited β-adrenergic responsiveness, and a high incidence of deaths from pump failure (with fewer arrhythmic deaths).[24,25] There, the combination of AP prolongation, lower peak systolic $[Ca]_i$, increased $[Na]_i$, and increased Na/Ca exchanger all could contribute to increased Ca entry during the AP by reverse Na/Ca exchanger.[34] This effect might be beneficial from a functional standpoint and, with limited β-adrenergic responsiveness, could be neutral from an arrhythmia standpoint.

Overall the role of Na/Ca exchanger can vary drastically, depending on the cellular conditions. In fact, overexpression of Na/Ca exchanger produces different SR Ca loads depending, at least in part, on the level of $[Na]_i$. That is, overexpression of Na/Ca exchanger in transgenic mice (in which $[Na]_i$ is high) can result in increased SR

Ca load,[35] whereas Na/Ca exchanger overexpression in rabbit myocytes where $[Na]_i$ is low results in SR Ca depletion.[36]

CONCLUSIONS

In conclusion, upregulated Na/Ca exchanger plays a dual role in modulating contractile function and arrhythmogenesis in the failing heart. Na/Ca exchanger is a difficult therapeutic target because of the delicate balance of SR load (too low a load leading to contractile dysfunction, too high leading to activation of I_{ti} carried by Na/Ca exchanger and arrhythmias). Therapeutic approaches to treat patients with HF will need to be developed in such a way to achieve this balance so as to enhance inotropy without enhancing arrhythmogenesis.

ACKNOWLEDGMENTS

This work was supported by National Institutes of Health Grants HL-46929 (to S.M.P.) and HL-30077 and HL-64724 (to D.M.B.).

REFERENCES

1. PACKER, M. 1985. Sudden unexpected death in patients with congestive heart failure: a second frontier. Circulation **72**: 681–685.
2. PIESKE, B., B. KRETSCHMANN, M. MEYER, et al. 1995. Alterations in intracellular calcium handling associated with the inverse force-frequency relation in human dilated cardiomyopathy. Circulation **92**: 1169–1178.
3. O'ROURKE, B., D.A. KASS, G.F. TOMASELLI, et al. 1999. Mechanisms of altered excitation-contraction coupling in canine tachycardia-induced heart failure. I. Experimental studies. Circ. Res. **84**: 562–570.
4. GÓMEZ, A.M., H.H. VALDIVIA, H. CHENG, et al. 1997. Defective excitation-contraction coupling in experimental cardiac hypertrophy and heart failure. Science **276**: 800–806.
5. MARX, S.O., S. REIKEN, Y. HISAMATSU, et al. 2000. PKA phosphorylation dissociates FKBP12.6 from the calcium release channel (ryanodine receptor): defective regulation in failing hearts. Cell **101**: 365–376.
6. POGWIZD, S.M., M. QI, W. YUAN, et al. 1999. Upregulation of Na^+/Ca^{2+} exchanger expression and function in an arrhythmogenic rabbit model of heart failure. Circ. Res. **85**: 1009–1019.
7. MERCADIER, J.J., A.M. LOMPRE, P. DUC, et al. 1990. Altered sarcoplasmic reticulum Ca^{2+}-ATPase gene expression in the human ventricle during end-stage heart failure. J. Clin. Invest. **85**: 305–309.
8. ARAI, M., N.R. ALPERT, D.H. MACLENNAN, et al. 1993. Alterations in sarcoplasmic reticulum gene expression in human heart failure. A possible mechanism for alterations in systolic and diastolic properties of the failing myocardium. Circ. Res. **72**: 463–469.
9. STUDER, R., H. REINECKE, J. BILGER, et al. 1994. Gene expression of the cardiac Na^+-Ca^{2+} exchanger in end-stage human heart failure. Circ. Res. **75**: 443–453.
10. HASENFUSS, G., W. SCHILLINGER, M. PREUSS, et al. 1999. Relationship between Na^+-Ca^{2+}-exchanger protein levels and diastolic function of failing human myocardium. Circulation **99**: 641–648.
11. POGWIZD, S.M. 1995. Nonreentrant mechanism underlying spontaneous ventricular arrhythmias in a model of nonischemic heart failure in rabbits. Circulation **92**: 1034–1048.

12. POGWIZD, S.M., K. SCHLOTTHAUER, L. LI, et al. 2001. Arrhythmogenesis and contractile dysfunction in heart failure. Roles of sodium-calcium exchange, inward rectifier potassium current, and residual β-adrenergic responsiveness. Circ. Res. **88:** 1159–1167.
13. WIT, A.L. & M.R. ROSEN. 1992. Afterdepolarizations and triggered activity: distinction from automaticity as an arrhythmogenic mechanism. In: Heart and Cardiovascular System: Scientific Foundations. H.A. Fozzard, E. Habert, R.B. Jennings, A.M. Katz & H.E. Morgan, Eds.: 2113–2163. 2nd ed. Raven Press. New York.
14. POGWIZD, S.M., J.P. MCKENZIE & M.E. CAIN. 1998. Mechanisms underlying spontaneous and induced ventricular arrhythmias in patients with idiopathic dilated cardiomyopathy. Circulation **98:** 2404–2414.
15. VERMEULEN, J.T., M.A. MCGUIRE, T. OPTHOF, et al. 1994. Triggered activity and automaticity in ventricular trabeculae of failing human and rabbit hearts. Cardiovasc. Res. **28:** 1547–1554.
16. BASSANI, J.W.M., R.A. BASSANI & D.M. BERS. 1994. Relaxation in rabbit and rat cardiac cells: species-dependent differences in cellular mechanisms. J. Physiol. **476:** 279–293.
17. BERS, D.M. 2001. Excitation-Contraction Coupling and Cardiac Contractile Force. 2nd ed. Kluwer Academic Press. Dordrecht, The Netherlands.
18. HOBAI, I.A. & B. O'ROURKE. 2001. Decreased sarcoplasmic reticulum calcium content is responsible for defective excitation-contraction coupling in canine heart failure. Circulation **103:** 1577–1584.
19. LINDNER, M., E. ERDMANN & D.J. BEUCKELMANN. 1998. Calcium content of the sarcoplasmic reticulum in isolated ventricular myocytes from patients with terminal heart failure. J. Mol. Cell. Cardiol. **30:** 743–749.
20. JANUARY, C.T. & J.M. RIDDLE. 1989. Early afterdepolarizations: mechanism of induction and block. A role for L-type Ca^{2+} current. Circ. Res. **64:** 977–990.
21. COLQUHOUN, D., E. NEHER, H. REUTER & C.F. STEVENS. 1981. Inward current channels activated by intracellular Ca in cultured cardiac cells. Nature **294:** 752–754.
22. ZYGMUNT, A.C., R.J. GOODROW & C.M. WEIGEL. 1998. $I_{Na/Ca}$ and $I_{Cl(Ca)}$ contribute to isoproterenol-induced delayed afterdepolarizations in midmycardial cells. Am. J. Physiol. **275:** H1979–H1992.
23. SCHLOTTHAUER, K. & D.M. BERS. 2000. Sarcoplasmic reticulum Ca^{2+} release causes myocyte depolarization—underlying mechanism and threshold for triggered action potentials. Circ. Res. **87:** 774–780.
24. FOWLER, M.B., J.A. LASER, G.L. HOPKINS, et al. 1986. Assessment of β-adrenergic receptor pathway in the intact failing human heart: progressive receptor down-regulation and subsensitivity to agonist response. Circulation **74:** 1290–1302.
25. KJEKSHUS, J. 1990. Arrhythmias and mortality in congestive heart failure. Am. J. Cardiol. **65:** 42I–48I.
26. BRISTOW, M.R., R. GINSBURG, W. MINOBE, et al. 1982. Decreased catecholamine sensitivity and beta-adrenergic-receptor density in failing human hearts. N. Engl. J. Med. **307:** 205–211.
27. MERIT-HF STUDY GROUP. 1999. Effect of metoprolol CR/XL in chronic heart failure: metoprolol CR/XL randomised intervention trial in congestive heart failure (MERIT-HF). Lancet **353:** 2001–2007.
28. PUGLISI, J.L. & D.M. BERS. 2001. LabHEART: an interactive computer model of rabbit ventricular myocyte ion channels and Ca transport. Am. J. Physiol. **281:** C2049–C2060.
29. MINAMISAWA, S., M. HOSHIJIMA, G. CHU, et al. 1999. Chronic phospholamban-sarcoplasmic reticulum calcium ATPase interaction is the critical calcium cycling defect in dilated cardiomyopathy. Cell **99:** 313–322.
30. MIYAMOTO, M.I., F. DEL MONTE, U. SCHMIDT, et al. 2000. Adenoviral gene transfer of SERCA2a improves left-ventricular function in aortic-banded rats in transition to heart failure. Proc. Natl. Acad. Sci. USA **97:** 793–798.
31. PACKER, M., J.R. CARVER, R.J. RODEHEFFER, et al. 1991. Effect of oral Milrinone on mortality in severe chronic heart failure. N. Engl. J. Med. **325:** 1468–1475.

32. AI, X., S.R. REIKEN, A.R. MARKS, et al. 2001. Altered calcium release channel expression and RyR2 hyperphosphorylation in an arrhythmogenic rabbit model of heart failure. Circulation **104:** II-131.
33. DESPA, S., M.A. ISLAM, S.M. POGWIZD & D.M. BERS. 2002. Intracellular [Na] is elevated in heart failure, but sodium-pump function is not the primary cause. Circulation **105:** 2543–2548.
34. DIPLA, K., J.A. MATTIELLO, K.B. MARGULIES, et al. 1999. The sarcoplasmic reticulum and the Na^+/Ca^{2+} exchanger both contribute to the Ca^{2+} transient of failing human ventricular myocytes. Circ. Res. **84:** 435–444.
35. TERRACCIANO, C.M.N., A.I. DE SOUZA, K.D. PHILIPSON & K.T. MACLEOD. 1998. Na^+-Ca^{2+} exchange and sarcoplasmic reticular Ca^{2+} regulation in ventricular myocytes from transgenic mice overexpressing the Na^+-Ca^{2+} exchanger. J. Physiol. **512:** 651–667.
36. SCHILLINGER, W., P.M.L. JANSSEN, S. EMAMI, et al. 2000. Impaired contractile performance of cultured rabbit ventricular myocytes after adenoviral gene transfer of Na^+-Ca^{2+} exchanger. Circ. Res. **87:** 581–587.

Modulation of Contractility in Failing Human Myocytes by Reverse-Mode Na/Ca Exchange

VALENTINO PIACENTINO III,[a] CHRISTOPHER R. WEBER,[b]
JOHN P. GAUGHAN,[a] KENNETH B. MARGULIES,[a]
DONALD M. BERS,[b] AND STEVEN R. HOUSER[c]

[a]*Department of Physiology, Temple University, Philadelphia, Pennsylvania 19140, USA*

[b]*Department of Physiology, Loyola University Chicago, Maywood, Illinois 60153, USA*

[c]*Department of Physiology, Temple University, Philadelphia, Pennsylvania 19140, USA*

ABSTRACT: A decrease in the peak systolic $[Ca]_i$ and slow decay of the Ca_i transient are common features of the end-stage failing human ventricular myocyte and may underlie the contractile abnormalities observed in congestive heart failure. The role of the Na/Ca exchanger has been a great area of interest given the changes observed at the molecular level. Results from these experiments have been inconsistent, however, and therefore cellular-based experiments may be required to characterize the role of the Na/Ca exchanger in failing human myocardium. We review recent data that suggest an increased ability of the Na/Ca exchanger to transport Ca into the cytoplasm in failing human myocytes. We hypothesize that this increased Ca influx can explain the slowed decay and impaired relaxation of failing human ventricular myocytes.

KEYWORDS: heart failure; cardiac Na/Ca exchange; sarcoplasmic reticulum; contractility

REVIEW

Congestive heart failure (CHF) is characterized by an inability of the heart muscle to generate sufficient force to maintain an adequate cardiac output, initially during exercise and eventually at rest. Abnormalities of force development and relaxation at the whole organ level and in isolated muscle and myocytes underlie much of this pump dyfunction.[1] CHF can result from several different diseases, but, at the level of the myocyte, there appears to be several common features. The most notable of these are prolongation of the action potential (AP) duration, decreased contraction magnitude, and slow relaxation.[2,3] These contractile alterations are largely responsible for the systolic and diastolic defects of the failing heart.

Numerous studies have attempted to define the molecular basis for this dysfunction; however, inconsistencies in these reports have led to a need for more integrative

Address for correspondence: Steven R. Houser, Ph.D., Molecular and Cellular Cardiology Laboratories, Cardiovascular Research Group, Department of Physiology, Temple University School of Medicine, 3400 North Broad Street, Philadelphia, PA 19140. Voice: 215-707-3278; fax: 215-707-4003.

srhouser@unix.temple.edu

and cellular-based investigations. In particular, molecular studies have focused on the abundance of key Ca regulatory proteins such as the sarcoplasmic reticulum (SR) Ca ATPase (SERCA) and the Na/Ca exchanger (NCX). Western blot analysis has shown an increase, a decrease, or no change in the protein or mRNA abundance of these molecules.[4–7] Therefore, studies designed to determine the relative contributions of SERCA and NCX in the intact cell and how they may function in the normal and failing human heart are the topic of this report.

The NCX is an electrogenic countertransporter protein on the sarcolemmal membrane that can translocate three Na ions into or out of the cell in exchange for a Ca ion.[8] When Ca inside the cell is elevated and the membrane is either repolarizing or near the resting membrane potential, the thermodynamics that dictate the direction of NCX promote Ca efflux from the myoplasm. When the Ca is low and the membrane is depolarized (plateau of the AP), the thermodynamic factors dictate Ca influx (reverse-mode NCX). Therefore, the direction of Ca transport by the NCX is determined by the electrochemical gradients for Na and Ca and the shape and size of the AP.

Most cellular experiments on the NCX have been performed in smaller mammals and suggest that the main role of NCX is to transport Ca out of the cell via forward-mode NCX.[8] It is important to remember that smaller animals have shorter action potentials than larger mammals, including cat and human, such that the membrane

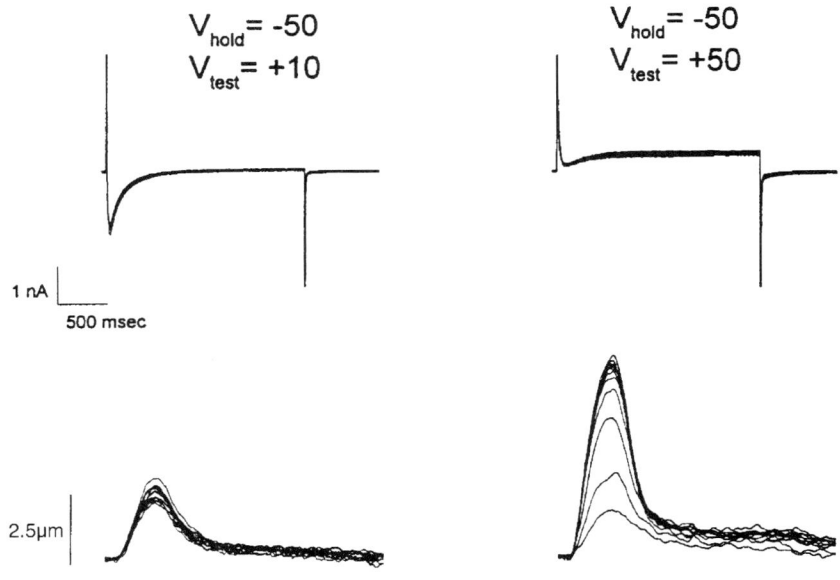

FIGURE 1. Voltage-clamp experiments on a failing ventricular myocyte at 37°C in Tyrode solution with 1 mM Ca. Multiple test steps from a holding potential (V_{hold}) of –50 mV to a test potential (V_{test}) are shown. Steps to +10 mV activated $I_{Ca,L}$ and small contractions with very little contractile staircase (*left*). Steps to +50 mV activated reverse-mode exchange, small $I_{Ca,L}$, and progressively larger contractions (*right*). Reprinted from Gaughan *et al.*[11]

potential is repolarizing during most of the Ca_i transient. These factors promote forward-mode NCX. However, during the plateau phase of the AP in large mammals (with a V_m trajectory through +40 to +20 mV) the NCX may work in reverse-mode and bring Ca into the cell. This Ca could directly activate the myofilaments or be a source of Ca to load the SR. Data from our laboratory suggest that this can occur in feline ventricular myocytes.[9] Our goal was to extend these studies to nonfailing and failing human myocytes.

An early observation from our laboratory was a profound inotropic effect of strong voltage clamp depolarizations in isolated failing human myocytes.[10,11] We sought to determine whether this positive inotropic effect was caused by Ca entry via the L-type Ca current or NCX. To separate the two processes, we depolarized the resting myocyte to +10 mV (the peak of the $I_{Ca,L}-V_m$ relationship) or +50 mV (a V_m where $I_{Ca,L}$ would be smaller and the electrochemistry for NCX is increasing). When the voltage clamp protocol depolarized the myocyte to +10 mV, a small staircase in contractility was observed from rest (FIG. 1) A different behavior was observed when the same myocyte was stepped to +50 mV. A large staircase was observed, and contractions were larger then those observed during depolarization to +10 mV. These data suggest that NCX-mediated Ca influx can increase SR Ca load and has a greater influence on SR Ca content than the L-type Ca current.

We next performed experiments in nonfailing and failing human myocytes to further define the respective contributions of Ca influx via the L-type Ca channel and NCX to SR Ca loading.[12] A voltage protocol was designed to first elicit SR Ca release similar to that during the early phases of the AP and then to change the V_m during the plateau to promote Ca influx via the L-type Ca channel and/or NCX. A third step was incorporated to assess the effect of these fluxes on SR Ca content (see FIG. 2, *inset*). In nonfailing myocytes, increasing the magnitude of the depolarization during the plateau had little effect on $[Ca]_i$ and had a small effect on the size of the Ca_i transient during the SR load assessment. When the same protocol was performed in failing myocytes, increasing the magnitude of the depolarization caused $[Ca]_i$ to increase until repolarization. This increase in $[Ca]_i$ was accompanied by a positive inotropic effect on the SR load (FIG. 2) These data suggest that Ca influx via NCX is increased in failing myocardium and can be a source of Ca to load the SR.

The above experiments addressed whether Ca entry during the AP plateau was caused by the L-type Ca current or NCX, and the data support a greater role for NCX. One explanation for this behavior is that because of depressed SR activity and smaller Ca_i transients the NCX is biased toward reverse-mode during the AP plateau, and this can bring Ca into the cell to support SR loading.[13] We tested this hypothesis by inducing SR dysfunction with cyclopiazonic acid in myocytes with strong SR function (nonfailing or LVAD-supported myocytes). With application of CPA, the Ca_i transient during the early V_m step was depressed and exhibited a prolonged decay. During the increase in V_m during the second step, $[Ca]_i$ increased in a similar fashion to failing myocytes. These data suggest that SR dysfunction "allows" the NCX to bring Ca into the cell during the AP plateau.

If SR function modulates how the NCX behaves during the AP, then increasing SERCA activity in a failing myocyte should alter the effect of NCX on $[Ca]_i$ during depolarized voltage steps. Experiments were designed to enhance SERCA activity by increasing the protein kinase A phosphorylation state of the myocyte. The multistep protocol described above was performed in failing myocytes under control con-

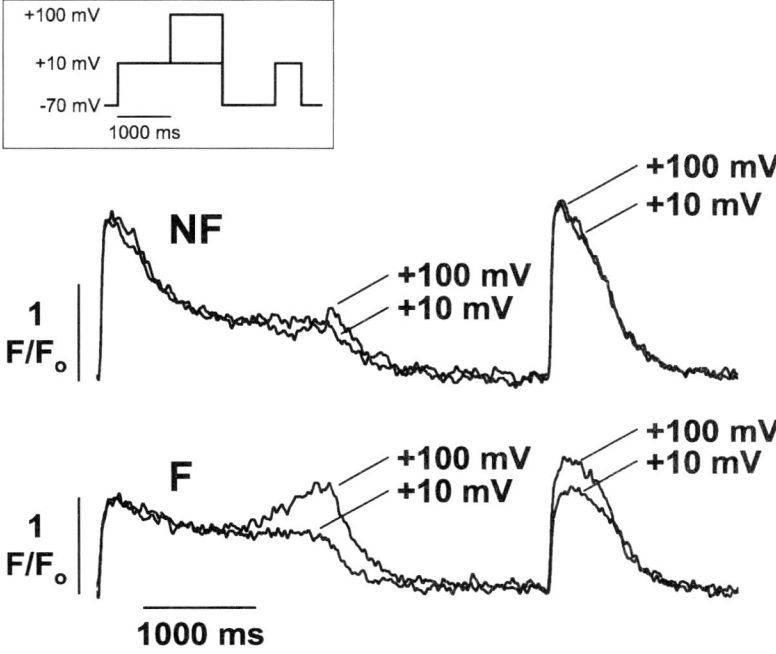

FIGURE 2. $[Ca]_i$ measured in a nonfailing and failing human myocyte during the multistep protocol (*inset*, 37°C in Tyrode solution with 1 mM Ca). In nonfailing myocytes (*top*), the second step was unable to increase $[Ca]_i$. SR Ca load was unaffected by the preceding step, possibly because of the ability of the SERCA to replete the SR. Note the "tail" contraction upon repolarization of the second step. In the failing myocyte (*bottom*), $[Ca]_i$ increased as the second step became more positive. This step influenced the SR load, with more positive potentials increasing SR Ca load. Reprinted from Pieske et al.[12]

ditions. Forskolin then was applied to failing myocytes, because this agent bypasses the receptor pathway to activate PKA. Application of forskolin caused an increase in the initial SR Ca release and accelerated the Ca_i transient decay. During the increase in V_m, Ca did not increase, similar to the behavior seen in nonfailing myocytes. The SR load was not strongly potentiated, as in nonfailing myocytes, but was large and exhibited a fast decay. These findings suggest that increasing SERCA activity maintains low $[Ca]_i$ and the SR sequesters Ca brought into the cell via the NCX.[14] Together, these results suggest that the behavior of the NCX during the AP is largely determined by SERCA activity. Thus, depressed SERCA activity causes decreased SR content which leads to reduced SR Ca release and promotes Ca entry via the NCX (see Piacentino et al. [extended abstract] in this volume for more detailed information).

The experiments described above used V_m to modify the Na and Ca electrochemistry and measured Ca to assess whether the NCX was bringing Ca into or out of the cell. Because the NCX is electrogenic, it is possible to record the NCX current to determine the direction and magnitude of Ca transport. One limitation of the multistep experiment is that it does not directly measure the NCX current because of issues related to separation of interdependent membrane currents. With these limitations in mind, we set out to design a set of experiments to determine the NCX current during the AP and test whether reverse-mode NCX current is present during the plateau of the AP in failing human myocytes.

Our experimental design was a refinement of the approach described by Egan et al.[15] Using AP voltage clamp protocols, we measured forward-mode I_{NCX} at various time points during the AP by interrupting the waveform and repolarizing to −70 mV in nonfailing and failing human myocytes. With the I_{NCX} at each time point and a steady-state relationship between I_{NCX} and $[Ca]_i$, we could predict the submembrane Ca (Ca_{sm}) during the AP. Next, using the same steady-state relationship NCX equation, we predicted the I_{NCX} during the AP using the known ions (Na_o, Ca_o, Na_i), V_m and $[Ca]_{sm}$.[16] Using this technique, we predict that outward NCX is favored in failing human myocytes throughout most of the AP plateau (see Weber et al. [extended abstract] in this journal for more detailed information). These data support earlier experiments that showed greater NCX-mediated Ca influx in failing human myocytes. Note that the prediction for I_{NCX} holds true using the same $[Na]_i$ for both nonfailing and failing human myocytes. Therefore, the difference in NCX behavior may be even greater, given the recent data from our laboratory and others that $[Na]_i$ is elevated in heart failure.[12]

In summary, there is strong evidence that the behavior of the NCX is different between nonfailing and failing human myocytes. The longer AP, lower systolic $[Ca]_I$, and higher $[Na]_i$ in failing myocytes all favor more reverse-mode NCX. These changes in NCX activity are likely to contribute to slow $[Ca]_i$ decline and impaired relaxation in the failing heart. Also, these changes may translate into diastolic dysfunction at higher heart rates and to the abnormal force–frequency relationship in human heart failure.

REFERENCES

1. HOUSER, S.R., V. PIACENTINO III & J. WEISSER. 2000. Abnormalities of calcium cycling in the hypertrophied and failing heart. J. Mol. Cell. Cardiol. **32:** 1595–1607.
2. KÄÄB, S., H.B. NUSS, N. CHIAMVIMONVAT, et al. 1996. Ionic mechanism of action potential prolongation in ventricular myocytes from dogs with pacing-induced heart failure. Circ. Res. **78:** 262–273.
3. DAVIES, C.H., K. DAVIA, J.G. BENNETT, et al. 1995. Reduced contraction and altered frequency response of isolated ventricular myocytes from patients with heart failure. Circulation **92:** 2540–2549.
4. MEYER, M., W. SCHILLINGER, B. PIESKE, et al. 1995. Alterations of sarcoplasmic reticulum proteins in failing human dilated cardiomyopathy. Circulation **92:** 778–784.
5. SCHWINGER, R.H., M. BOHM, U. SCHMIDT, et al. 1995. Unchanged protein levels of SERCA II and phospholamban but reduced Ca^{2+} uptake and Ca^{2+}-ATPase activity of cardiac sarcoplasmic reticulum from dilated cardiomyopathy patients compared with patients with nonfailing hearts. Circulation **92:** 3220–3228.
6. FLESCH, M., R.H. SCHWINGER, F. SCHIFFER, et al. 1996. Evidence for functional relevance of an enhanced expression of the Na^+–Ca^{2+} exchanger in failing human myocardium. Circulation **94:** 992–1002.
7. SCHWINGER, R.H., J. WANG, K. FRANK, et al. 1999. Reduced sodium pump alpha$_1$, alpha$_3$, and beta$_1$-isoform protein levels and $Na^+$$K^+$-ATPase activity but unchanged Na^+–Ca^{2+} exchanger protein levels in human heart failure. Circulation **99:** 2105–2112.

8. BERS, D.M. 2001. Excitation-contraction coupling and cardiac contractile force. Kluwer Academic Publishers. Dordrecht/Boston.
9. NUSS, H.B. & S.R. HOUSER. 1992. Sodium-calcium exchange-mediated contractions in feline ventricular myocytes. Am. J. Physiol. **263:** H1161–H1169.
10. MATTIELLO, J.A., K.B. MARGULIES, V. JEEVANANDAM & S.R. HOUSER. 1998. Contribution of reverse-mode sodium-calcium exchange to contractions in failing human left ventricular myocytes. Cardiovasc. Res. **37:** 424–431.
11. GAUGHAN, J.P., S. FURUKAWA, V. JEEVANANDAM, *et al.* 1999. Sodium/calcium exchange contributes to contraction and relaxation in failed human ventricular myocytes. Am. J. Physiol. **277:** H714–H724.
12. PIESKE, B., L.S. MAIER, V. PIACENTINO III, *et al.* 2002. Altered $[Na^+]_i$ homeostasis is related to disturbed Ca^{2+} handling and contractile dysfunction in failing human myocardium. Circulation **106:** 447–453.
13. NUSS, H.B. & S.R. HOUSER. 1994. Effect of duration of depolarization on contraction of normal and hypertrophied feline ventricular myocytes. Cardiovasc. Res. **28:** 1482–1489.
14. DIPLA, K., J.A. MATTIELLO, K.B. MARGULIES, *et al.* 1999. The sarcoplasmic reticulum and the Na^+/Ca^{2+} exchanger both contribute to the Ca^{2+} transient of failing human ventricular myocytes. Circ. Res. **84:** 435–444.
15. EGAN, T.M., D. NOBLE, S.J. NOBLE, *et al.* 1989. Sodium-calcium exchange during the action potential in guinea-pig ventricular cells. J. Physiol. **411:** 639–661.
16. WEBER, C.R., V. PIACENTINO III, K.S. GINSBURG, *et al.* 2002. Na^+–Ca^{2+} exchange current and submembrane $[Ca^{2+}]$ during the cardiac action potential. Circ. Res. **90:** 182–189.

Decreased β-Adrenergic Responsiveness of Na/Ca Exchange Current in Failing Pig Myocytes

SHAO-KUI WEI, STEPHEN U. HANLON, AND MARK C.P. HAIGNEY

Department of Medicine, Uniformed Services University of the Health Sciences, Bethesda, Maryland 20184, USA

KEYWORDS: β-adrenergic responsiveness; failing pig myocytes; Na/Ca exchange current

BACKGROUND

Na/Ca exchanger (NCX) protein is significantly upregulated in heart failure, and increased NCX current (I_{NCX}) has been reported in most but not all preparations.[1,2] Increased inward I_{NCX} during β-adrenergic receptor (β-AR) stimulation-induced calcium loading may lead to triggered arrhythmias.[3] We asked whether I_{NCX} is augmented during activation of the β-AR-associated signaling complex in control and failing swine myocytes.

METHODS

Ventricular myocytes were isolated from control and ventricularly paced failing Yorkshire swine hearts (200 bpm until LV dysfunction determined by 2-D echo). I_{NCX} was measured during whole-cell patch clamp after other currents were blocked. Currents were elicited by a descending voltage ramp (from +80 mV to −120 mV at 100 mV/sec). Internal pipette solution contained (mM): CsCl 70, NaCl 20, Hepes 10, TEA 20, EGTA 21, $CaCl_2$ 6, Na_2ATP 5, $MgCl_2$ 4.0, ryanodine 0.5 μM, pH 7.2 (CsOH). External solution contained (mM): NaCl 143, $MgCl_2$ 1, Hepes 5, glucose 10, $CaCl_2$ 2, CsCl 5, ouabain 0.02, nifedipine 0.01. Ni^{2+} (5 mmol/l) is added to define the fraction of current that derives from NCX (total current remaining after subtraction of post-Ni^{2+} trace). All experiments were performed at 37°C in the presence or absence of either isoproterenol (2 μmol/l), forskolin (5 μmol/l), 8-Br-cAMP (1 mmol/l), or okadaic acid (1 μmol/l).

---

[a]DISCLAIMER: The views expressed in this paper reflect the opinions of the authors only and not the official policy of the Uniformed Services University or the Department of Defense.

Address for correspondence: Mark C.P. Haigney, M.D., A-3060, Division of Cardiology, Uniformed Services University of the Health Sciences, 4301 Jones Bridge Road, Bethesda, MD 20814. Voice: 301-295-3826; fax: 301-295-3557.

MCPH@aol.com

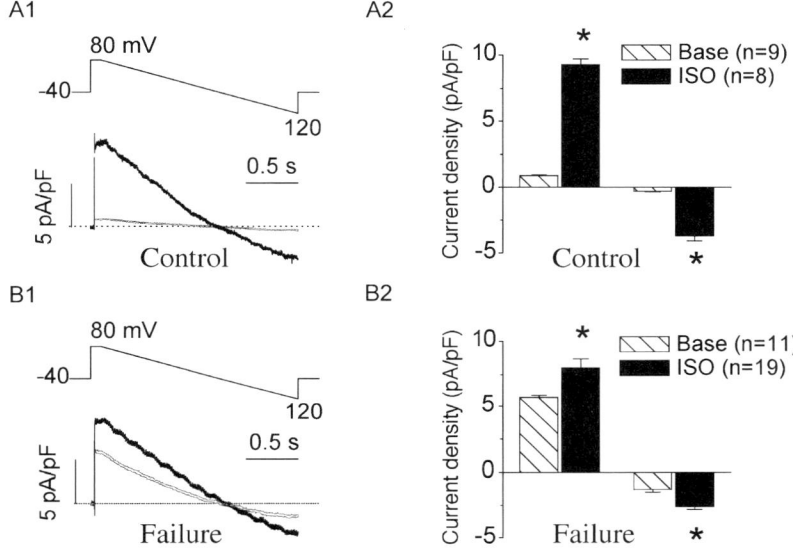

FIGURE 1. Effect of isoproterenol on I_{NCX} in control and failing myocytes. **(A1)** The representative I_{NCX} trace from control myocytes in presence (*upper*) or absence (*lower*) of ISO (2 μmol/l). **(A2)** Mean data of peak outward and inward (at +80/–120 mV) I_{NCX} from myocytes in **(A1)**. *$P < 0.05$ ISO vs. base. **(B)** I_{NCX} from failing myocytes in presence or absence of ISO. Format analogous to that in panel A.

RESULTS

We confirmed a significant increase in NCX protein in failing cells (26–69%) by Western blot (data not shown). At baseline, outward and inward I_{NCX} was significantly increased in failing myocytes compared with control myocytes (outward/inward: $5.67 \pm 0.20/-1.28 \pm 0.22$ pA/pF for failure, $n = 11$, verse $0.88 \pm 0.06/-0.32 \pm 0.04$ pA/pF for control, $n = 9$, $p < 0.01$). Isoproterenol (ISO) increased I_{NCX} by nearly one order of magnitude in control myocytes ($p < 0.01$, see FIG. 1). However, in failing myocytes, ISO only increased I_{NCX} by 50%, such that the ISO-stimulated I_{NCX} in failing cells was less than the ISO-stimulated current in control cells. To test whether this decrease in recruitable current was due to reduced β-AR number or G-protein uncoupling, we exposed failing cells to either forskolin ($n = 7$), or 8-Br-cAMP ($n = 9$); whereas both inward and outward I_{NCX} increased compared with baseline, there was no difference compared to ISO. Finally, to test whether the reduced response to β-AS was due to largely phosphorylated NCX channel at baseline owing to a decline in phosphatase activity, both control and failing myocytes were exposed to the protein phosphatase inhibitor, okadaic acid (OKA).[4] In control myocytes, OKA significantly increased NCX current (10 times), but in failing myocytes, the increase in NCX current was again blunted (see FIG. 2).

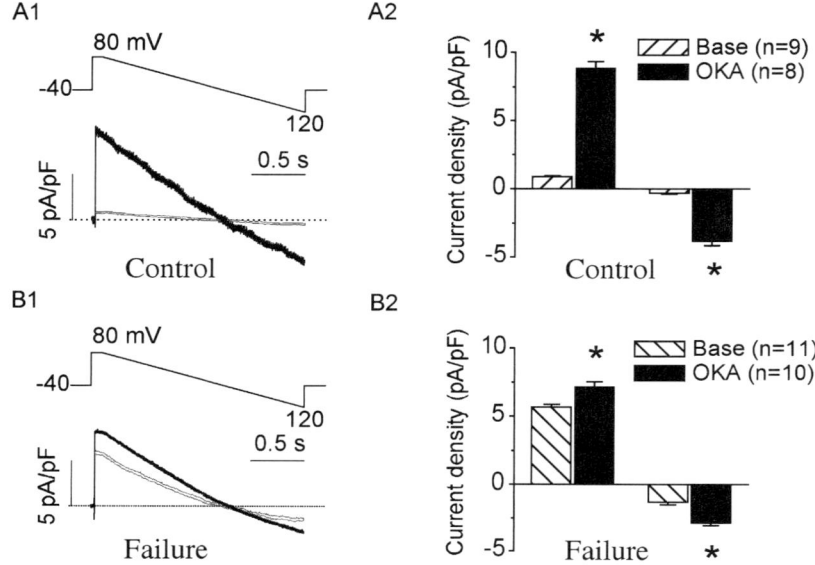

FIGURE 2. Effects of phosphatase inhibitor-okadaic acid on I_{NCX} in control and failing myocytes. **(A1)** The representative I_{NCX} trace from control myocytes in presence (*upper*) or absence (*lower*) of okadaic acid (1 µmol/l). **(A2)** Mean data of peak outward and inward (at +80/−120 mV) I_{NCX} from myocytes in A1. *$P < 0.05$ OKA vs. base. **(B)** I_{NCX} from failing myocytes in presence or absence of OKA. Format analogous to that in panel A.

DISCUSSION

Consistent with both human and experimental data, heart failure resulted in a significant increase in NCX protein in our model. At baseline, failing cells manifested significantly increased I_{NCX}, but β-AR stimulation in control cells resulted in a greater relative and absolute increase in I_{NCX}. This defect was not overcome by directly increasing cAMP. Similarly, inhibition of the phosphatase by okadaic acid increased I_{NCX} in failure and control cells to similar degrees as isoproterenol. This argues that failing cells retain some residual phosphatase activity, but that decreased phosphatase activity at baseline might be partially responsible for the augmented I_{NCX} seen in failing cells. The failure to respond to β-AR stimulation is consistent with either a defect in signaling distal to cAMP that might limit protein phosphorylation, alteration of the protein phosphorylation site, a change in NCX stoichiometry, or the presence of increased levels of nonfunctional NCX protein.

REFERENCES

1. STUDER, R., H. REINECKE, J. BILGER, *et al.* 1994. Gene expression of the cardiac Na^+–Ca^{2+} exchanger in end-stage human heart failure. Circ. Res. **75:** 443–453.

2. O'ROURKE, B., D.A. KASS, G.F. TOMASELLI, *et al.* 1999. Mechanisms of altered excitation–contraction coupling in canine tachycardia-induced heart failure, I: experimental studies. Circ. Res. **84:** 562–570.
3. POGWIZD, S.M., K. SCHLOTTHAUER, L. LI, *et al.* 2001. Arrhythmogenesis and contractile dysfunction in heart failure: roles of sodium-calcium exchange, inward rectifier potassium current, and residual beta-adrenergic responsiveness. Circ. Res. **88**(11): 1159–1167.
4. SCHRODER, F., R. HANDROCK, D. BEUCKELMANN, *et al.* 1998. Increased availability and open probability of single L-type calcium channels from failing compared with nonfailing human ventricle. Circulation **98:** 969–976.

Ca Influx via the Na/Ca Exchanger Maintains Sarcoplasmic Reticulum Ca Content in Failing Human Myocytes

VALENTINO PIACENTINO III, KENNETH B. MARGULIES, AND
STEVEN R. HOUSER

Department of Physiology, Temple University School of Medicine, Philadelphia, Pennsylvania 19140, USA

KEYWORDS: heart failure; cardiac Na/Ca exchange; sarcoplamasic reticulum; contractility

BACKGROUND

The L-type Ca current ($I_{Ca,L}$) is both the major trigger of sarcoplasmic reticulum (SR) Ca release and source of Ca to change the SR Ca content in cardiac myocytes.[1] It has been suggested that the efficiency of the $I_{Ca,L}$ to load the SR is reduced in diseased myocytes and that the Na/Ca exchanger (NCX) may play a more prominent role.[2] The goal of this present study was to determine the relative contributions of $I_{Ca,L}$ and NCX to SR Ca loading in nonfailing (NF) and failing (F) human ventricular myocytes.

METHODS

Myocytes were isolated from failing human hearts ($n = 5$) at the time of cardiac transplantation and nonfailing human hearts ($n = 3$) deemed unsuitable for donation. Whole-cell voltage clamp (VC) was used to control V_m, and Flou-3 was used to assess $[Ca]_i$ (37°C, normal $Na_{i,o}$ and $K_{i,o}$, 1 mM Ca_o). A three-step VC protocol was designed to determine the V_m dependency of $[Ca]_i$ and SR Ca loading. (see Piacentino *et al.*, this volume, for more information). SR Ca ATPase activity was reduced or enhanced with cyclopiazonic acid (CPA, 75 mM) or forskolin (FK, 1 mM), respectively.

Address for correspondence: Steven R. Houser, Ph.D., Molecular and Cellular Cardiology Laboratories, Cardiovascular Research Group, Department of Physiology, Temple University School of Medicine, 3400 North Broad Street, Philadelphia, PA 19140. Voice: 215-707-3278; fax: 215-707-4003.
  srhouser@unix.temple.edu

RESULTS

$[Ca]_i$ was higher with depolarizations to positive V_m and followed the energetics of the NCX to a greater extent in F ($n = 7$) compared to NF ($n = 7$) myocytes. The SR Ca load assessed on the next beat was influenced by NCX activity in F myocytes, where the V_m–SR Ca load relationship was more bell-shaped in NF myocytes. CPA application in NF myocytes caused $[Ca]_i$ to rise with depolarization, but did not increase SR Ca content. FK application in F myocytes increased the magnitude of $[Ca]_i$ and accelerated the rate of decay. An increase in Ca during the second V_m step was not observed in the presence of FK.

CONCLUSIONS

Ca influx via $I_{Ca,L}$ is a major contributor to SR Ca loading in nonfailing human ventricular myocytes. The switch in failing human ventricular myocytes to a greater reliance on the NCX may explain the dependency of contractility of the failing heart on the action potential morphology and duration.

ACKNOWLEDGMENTS

This work was supported by the National Institues of Health (HL-61495 to SRH) and American Heart Association (0010103U to VPIII).

REFERENCES

1. BERS, D.M. 2001. Excitation–contraction coupling and cardiac contractile force. Kluwer Academic Publishers. Dordrecht/Boston.
2. GAUGHAN, J.P., S. FURUKAWA, V. JEEVANANDAM, *et al.* 1999. Sodium/calcium exchange contributes to contraction and relaxation in failed human ventricular myocytes. Am. J. Physiol. **277:** H714–724.

Calcium Influx via I_{NCX} Is Favored in Failing Human Ventricular Myocytes

CHRISTOPHER R. WEBER,[a] VALENTINO PIACENTINO III,[b]
KENNETH B. MARGULIES,[b] DONALD M. BERS,[a] AND STEVEN R. HOUSER[b]

[a]*Department of Physiology, Loyola University Chicago, Maywood, Illinois 60153, USA*

[b]*Department of Physiology, Temple University School of Medicine, Philadelphia, Pennsylvania 19140, USA*

KEYWORDS: cardiac Na/Ca exchange; submembrane [Ca]; heart failure

In failing (F) human myocytes, average peak $[Ca]_i$ is reduced, time to 50% $[Ca]_i$ decay is prolonged, and sarcoplasmic reticulum (SR) Ca content is reduced. There is heterogeneity in the F myocyte population, but a fair proportion of F myocytes display a low-amplitude $[Ca]_i$ transient with a rapid initial increase, but then a much slower increase over greater than 300 ms (until action potential repolarization). This phenotype was rare in nonfailing (NF) myocytes, in which $[Ca]_i$ peaked at approximately 100 ms and declined during the action potential (AP). We hypothesize that the F versus NF phenotype can be explained by predominance versus absence of outward I_{NCX} during the plateau of the AP.

After SR Ca release, submembrane $[Ca]_i$ near the NCX molecules ($[Ca]_{sm}$) increases more quickly and to a higher value than bulk $[Ca]_i$ measured with fluorescent indicators. I_{Ca} prevents the direct measurement I_{NCX} during an AP, and blockade of I_{Ca} would directly and indirectly (via SR Ca release) alter $[Ca]_{sm}$, making direct measurement of I_{NCX} during an AP impractical. Model predictions of I_{NCX} during the ventricular AP, which are based solely on $[Ca]_i$, yield substantial outward I_{NCX} at physiological $[Na]_i$, but when one considers severalfold higher $[Ca]_{sm}$, I_{NCX} is inward throughout most of an AP under normal conditions.

We use a modeling approach to predict I_{NCX} based on measured $[Ca]_{sm}$ in myocytes from F and NF human hearts. A typical failing human AP, recorded under current-clamp at 37°C, was used as a template for AP-clamp experiments. We interrupted AP clamps at various times, allowing us to record initial I_{NCX} tails at −70 mV without contaminating currents. I_{NCX} at the time of AP interruption was used to determine $[Ca]_{sm}$ using the steady-state relationship between I_{NCX} and $[Ca]_{sm}$, deter-

---

Address for correspondence: Donald M. Bers, Ph.D., Department of Physiology, Loyola University Chicago, Stritch School of Medicine, 2160 South First Avenue, Maywood, IL 60153. Voice: 708-216-1018; fax: 708-216-6308.
dbers@luc.edu

mined on the same cell at -70 mV (Eq. 1). Furthermore, using the same equation, we predicted I_{NCX} during an AP using $[Ca]_{sm}$, and E_m during an AP.

K currents and Ca-activated Cl channels were blocked, P/8 protocols were used to remove capacitative tail currents, and fast I_{Ca} tail currents (t ≈ 1 ms) were analytically removed when present. $[Na]_o$, $[Na]_i$, and $[Ca]_o$ were 150, 12.5, and 1 mM, respectively (pH$_i$ 7.2, pH$_o$ 7.4). The steady-state I_{NCX} relationship was:

$$I_{Na/Ca} = \frac{V_{max}([Na]_i^3[Ca]_o e^{\eta Vk} - [Na]_o^3[Ca]_i e^{(\eta-1)Vk})/(1 + \{K_{mCaAct}/[Ca]_i\}^2)}{\begin{bmatrix} K_{mCao}[Na]_i^3 + K_{mNao}^3[Ca]_i + K_{mNai}^3[Ca]_o(1 + [Ca]_i/K_{mCai}) + \\ K_{mCai}[Na]_i^3(1 + \{[Na]_i/K_{mNai}\}^3) + [Na]_i^3[Ca]_o + [Na]_o^3[Ca]_i \end{bmatrix} \begin{bmatrix} 1 + k_{sat} e^{(\eta-1)Vk} \end{bmatrix}} \tag{1}$$

k = F/RT, K_{mCaAct} = 125 nmol/L, K_{mNao} = 87.5 mmol/L, K_{mNai} = 12.3 mmol/L, K_{mCai} = 3.6 μmol/L, K_{mCao} = 1.30 mmol/L, k_{sat} = 0.32, and η = 0.27. V_{max} (in A/F) and a small linear leak component was determined separately on every cell.

Average peak $[Ca]_{sm}$ was lower and later in the F phenotype (0.92 μM at 205 ms; SEM = 0.2 μM, n = 4) compared with the NF phenotype (3.4 μM at 15 ms; SEM = 1.7 μM, n = 5). $[Ca]_{sm}$ in NF decayed throughout the AP, but in F, $[Ca]_{sm}$ remained fairly constant until repolarization. During the AP plateau, predicted I_{NCX} was substantially outward in F, and close to zero in NF using 12.5 mM $[Na]_i$ (pipette value). These differences in predicted I_{NCX} between NF and F are further increased if one considers that $[Na]_i$ is approximately 3 mM higher in F compared with NF, and that the AP is prolonged in F. Thus, reduced Ca efflux during the AP via inward I_{NCX} may account for the slow increase and decay in $[Ca]_i$ that often is observed in the F myocyte.

REFERENCES

1. TRAFFORD, A.W., M.E. DÍAZ, S.C. O'NEILL & D.A. EISNER. 1995. Comparison of subsarcolemmal and bulk calcium concentration during spontaneous calcium release in rat ventricular myocytes. J. Physiol. **488**: 577–586.
2. EGAN, T.M., D. NOBLE, S.J. NOBLE, et al. 1989. Sodium–calcium exchange during the action potential in guinea-pig ventricular cells. J. Physiol. **411**: 639–661.
3. WEBER, C.R., V. PIACENTINO III, K.S. GINSBURG, et al. 2002. Na/Ca exchange current and submembrane [Ca] during the cardiac action potential. Circ. Res. **90**: 182–189.
4. WEBER, C.R., K.S. GINSBURG, K.D. PHILIPSON, et al. 2001. Allosteric regulation of Na/Ca exchange current by cytosolic Ca in intact cardiac myocytes. J. Gen. Physiol. **117**: 119–131.
5. DESPA, S., M.A. ISLAM, C.R. WEBER, et al. 2002. Intracellular Na$^+$ concentration is elevated in heart failure, but Na/K pump function is unchanged. Circ. Res. **28**: 2543–2548.

Overexpression of Na^+/Ca^{2+} Exchanger Attenuates Postinfarcted Myocardial Dysfunction

JIANG-YONG MIN,[a] MATTHEW F. SULLIVAN,[a] VICTOR CHU,[a]
JU-FENG WANG,[a] IVO AMENDE,[a] JAMES P. MORGAN,[a]
KENNETH D. PHILIPSON,[b] AND THOMAS G. HAMPTON[a]

[a]*Cardiovascular Division, Beth Israel Deaconess Medical Center and Harvard Medical School, Boston, Massachusetts 02146, USA*

[b]*Department of Physiology and Medicine, University of California School of Medicine, Los Angeles, California 90095, USA*

KEYWORDS: Na/Ca exchanger; postinfarcted myocardial dysfunction; myocardial ischemia

We have[1] demonstrated that cardiac function and intracellular Ca^{2+} during myocardial ischemia are better preserved in hearts with increased expression of the Na^+/Ca^{2+} exchanger (NCX). However, the functional significance of enhanced expression of the NCX gene in postinfarcted heart remains speculative.

METHODS AND RESULTS

Myocardial infarction (MI) was performed in age- and gender-matched wild-type (WT) mice and transgenic mice with overexpression of the NCX gene (TG), a strain well characterized by Philipson's group.[2] The animal model of MI was induced by ligation of the left anterior descending coronary artery under anesthesia. The study was composed of infarcted transgenic mice with overexpression of the NCX gene (TG, $n = 11$) and infarcted WT mice ($n = 13$). Left ventricular pressure (LVSP), left ventricular end-diastolic pressure (LVEDP), and the maximum rising rate of pressure (+dP/dtmax) were monitored for functional evaluation. Isometric contractility was assessed subsequently in isolated papillary muscles. Infarct size was measured in subsets of mice (five of each). Five weeks after MI, myocardial function was less impaired in TG than in WT mice. (See FIG. 1. LVSP, 41 ± 2 mmHg in WT vs. 58 ± 3 mmHg in TG, $P < 0.05$; LVEDP, 7 ± 1 mmHg in WT vs. 4 ± 0.5 mmHg in TG,

Address for correspondence: Thomas G. Hampton, Department of Medicine, Beth Israel Deaconess Medical Center and Harvard Medical School, 330 Brookline Avenue, Boston, MA 02215. thampton@caregroup.harvard.edu

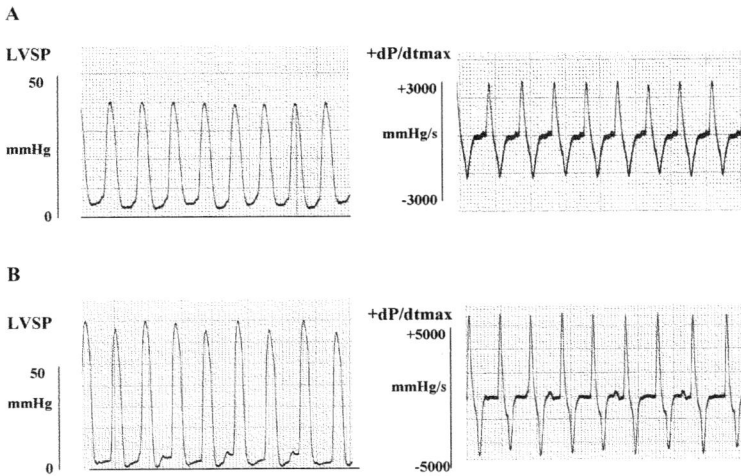

FIGURE 1. (A) Hemodynamics measured from a WT mouse. **(B)** Hemodynamics measured from a TG mouse.

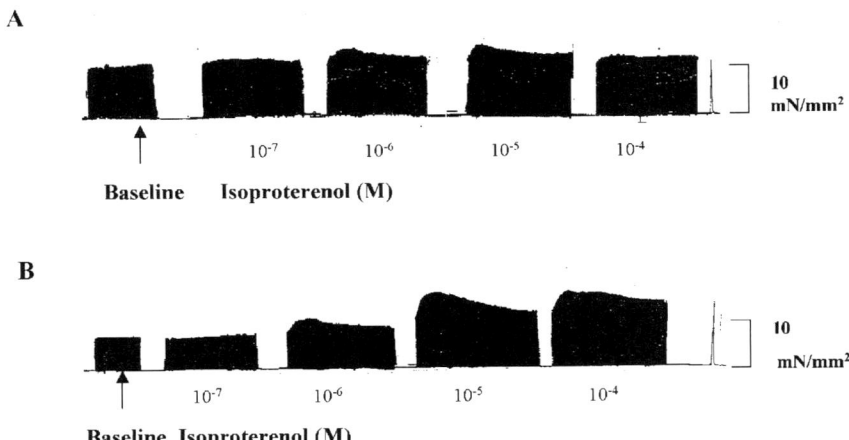

FIGURE 2. (A) Isometric contractility to isoproterenol stimulation in a papillary muscle isolated from a WT mouse. **(B)** Isometric contractility to isoproterenol stimulation in a papillary muscle isolated from a TG mouse.

$P < 0.05$; +dP/dt, 3750 ± 346 mmHg/s in WT vs. 5075 ± 334 mmHg/s in TG, $P < 0.05$.) The isometric contractile response to β-adrenergic stimulation was more enhanced in cardiac muscles from TG than in WT mice (see FIG. 2. 13.2 ± 1.4 mN/mm^2 in WT vs. 16.3 ± 2.2 mN/mm^2 in TG with 10^{-4} M of isoproterenol stimulation, $P < 0.05$), although the infarct size was similar in both groups ($34 \pm 3\%$ in WT vs. $36 \pm 3\%$ in TG). Furthermore, there was a trend for reduced mortality after infarction in mice with overexpression of the NCX gene than WT mice during the 5 weeks of follow-up by the log-rank test (18% in TG vs. 31% in WT; $\chi^2 = 3.9$; $P < 0.05$).

DISCUSSION

Our findings point to the functional relevance of upregulation of the NCX gene in a pathological setting in which sarcoplasmic reticulum Ca^{2+} regulation is impaired. Cardiac function and inotropic response to β-adrenergic stimulation are better preserved in TG mice with overexpression of the NCX gene at 5 weeks after MI. Our previous study[1] demonstrated that enhanced gene expression of the NCX preserves intracellular Ca^{2+} homeostasis during ischemia. The underlying mechanism of improved cardiac function in postinfarcted myocardium with overexpression of the NCX gene might be partially related to modification of intracellular Ca^{2+} handling to compensate the impaired sarcoplasmic reticulum and improve myocardial performance. Further experiments are required to investigate intracellular Ca^{2+} handling in post-MI myocardium with overexpression of the NCX gene.

REFERENCES

1. HAMPTON, T.G., J.F. WANG, J. DEANGELIS, et al. 2000. Enhanced gene expression of Na^+/Ca^{2+} exchanger attenuates ischemic and hypoxic contractile dysfunction. Am. J. Physiol. **279**: H2846–H2854.
2. ADACHI-AKAHANE, S., L. LU, Z. LI, et al. 1997. Calcium signaling in transgenic mice overexpressing cardiac Na^+–Ca^{2+} exchanger. J. Gen. Physiol. **109**: 717–729.

Ischemic Tolerance of Homozygous Transgenic Mouse Hearts Overexpressing the Sodium–Calcium Exchanger

MALCOLM M. BERSOHN,[a,b] CECIL R. CARMACK,[a] AND KENNETH D. PHILIPSON[b]

[a]*V.A. Greater Los Angeles Healthcare System and* [b]*University of California at Los Angeles, Los Angeles, California 90073, USA*

KEYWORDS: sodium–calcium exchange; ischemia; reperfusion; heart; transgenic; mouse

INTRODUCTION

The sarcolemmal sodium–calcium exchanger NCX1 is the major route for Ca^{2+} efflux from cardiac myocytes. Because ischemic injury has been linked to cellular Ca^{2+} overload, we hypothesized that the overexpression of Na^+/Ca^{2+} exchange might protect hearts from the deleterious effects of ischemia. In previous experiments, we were surprised to find no effect on ischemic tolerance of hearts from mice that were heterozygous for a transgene causing cardiac-specific expression of the Na^+/Ca^{2+} exchanger.[1] We have now repeated similar experiments using hearts from mice that were homozygous for the transgene to provide for greater Na^+/Ca^{2+} exchange overexpression.

METHODS

A transgenic mouse line that exhibits cardiac-specific overexpression of the canine Na^+/Ca^{2+} exchanger was used as a model of NCX1 overexpression. Hearts from male and female mice homozygous for the transgene were compared with nontransgenic age-matched controls. Animals were studied at 261 ± 18 and 266 ± 17 days of age for controls and transgenics, respectively. Hearts were removed from anesthetized mice, and the aortas were cannulated with a 20-gauge cannula for perfusion in a Langendorff apparatus with 80 cm liquid perfusion pressure. We used a modified Krebs-Henseleit solution with Na 142 mM, Cl 125.5 mM, HCO_3 25 mM, K 4.7 mM, Mg 1.2 mM, PO_4 1.2 mM, SO_4 1.2 mM, Ca 1.5 mM, glucose 11.1 mM, pH 7.4 at 37°C when equilibrated with 95% O_2, 5% CO_2. The left ventricle was vented with a PE50 catheter inserted through the apex, and a latex rubber balloon (Radnoti, Inc.)

Address for correspondence: Malcolm M. Bersohn, M.D., Ph.D., Cardiology (111E), VAGLAHS, 11301 Wilshire Boulevard, Los Angeles, CA 90073.
mbersohn@ucla.edu

Ann. N.Y. Acad. Sci. 976: 483–486 (2002). © 2002 New York Academy of Sciences.

was inserted through the mitral valve into the ventricle to control preload and measure isovolumic pressure development. Hearts were paced to maintain a rate of 390 beats per minute throughout aerobic and ischemic periods.

Hearts were perfused for 30 min with the intraventricular balloon volume adjusted to give an end-diastolic pressure of approximately 10 mm Hg. The final left ventricular end-diastolic pressure and developed pressure measured after 30 min of aerobic perfusion were used as a baseline for comparison with postischemic recovery. No adjustments were made to the left ventricular volume following the stabilization period. Hearts then were made globally ischemic for 40 min followed by 30 min of reperfusion.

At the end of the experiment, hearts were immediately cooled to 4°C. The membranes were prepared by homogenizing the hearts in sucrose 250 mM, Tris maleate 20 mM, pH 7.6, extracting contractile proteins with the addition of KCl 300 mM and Na-pyrophosphate 25 mM, and sedimentation at 240,000 × g max for 40 min. The pellets were resuspended in 140 mM NaCl, 10 mM MOPS (pH 7.4 at 37°C). Purification of the whole cell membrane preparation was assessed by the ratio of membrane to initial homogenate K^+-dependent p-nitrophenylphosphatase activity. To confirm that the transgenic hearts did indeed exhibit increased Na^+/Ca^{2+} exchange activity after ischemia and reperfusion, we measured the initial velocity of Na^+/Ca^{2+} exchange as Na^+_i-dependent $^{45}Ca^{2+}$ uptake 1 s after dilution of 50-μg aliquots of membranes into 140 mM KCl, 10 mM MOPS, 0.4 μM valinomycin, pH 7.4, at 37°C. The same measurements also were made using whole cell membranes from hearts removed from mice and not subjected to any *in vitro* perfusion. All data are presented as mean ± SE.

RESULTS

The left ventricular pressures recorded before and after ischemia and reperfusion are shown in TABLE 1. There were no significant differences between control and transgenic hearts in any of the pressures before or after ischemia and reperfusion. Recovery of developed pressure for control and transgenic hearts was 60 ± 13% and 63 ± 7%, respectively. Results of membrane purification and kinetic parameters for Na^+/Ca^{2+} exchange are shown in TABLE 2. There were no differences in Ca^{2+} affinity between controls and transgenics. The calculated V_{max} for Na^+/Ca^{2+} exchange was more than twofold greater for the transgenics both before and after ischemia. There were no differences in any of the parameters reported in either table between males and females (data not shown).

DISCUSSION

The baseline left ventricular performance of hearts from transgenic mice homozygous for a transgene causing cardiac overexpression of NCX1 was the same as that of age-matched controls. After 40 min of ischemia and 30 min of reperfusion, there was approximately 60% recovery of left ventricular developed pressure in both transgenics and controls, with an equal degree of contracture in both. The activity of the Na^+/Ca^{2+} exchanger remained approximately 2.5-fold greater in whole-cell mem-

TABLE 1. Left ventricular pressure

	Controls ($n = 9$)	Transgenics ($n = 12$)
Aerobic perfusion		
Systolic (mmHg)	38 ± 3	45 ± 3
Diastolic (mmHg)	9 ± 1	10 ± 1
Developed pressure (mmHg)	29 ± 3	34 ± 3
After ischemia/ reperfusion		
Systolic (mmHg)	29 ± 6	36 ± 5
Diastolic (mmHg)	12 ± 5	14 ± 3
Developed pressure (mmHg)	17 ± 4	22 ± 4
Recovery of developed pressure (%)	60 ± 13	63 ± 7

TABLE 2. Sodium–calcium exchange

	Controls	Transgenics
Number of unperfused hearts	8	5
Membrane purification factor	2.7 ± 0.4	3.0 ± 0.3
K_m for Ca^{2+}	19 ± 4	18 ± 2
V_{max} (nmol·mg^{-1}·sec^{-1})	0.31 ± 0.02	0.69 ± 0.02*
Number after ischemia/reperfusion	9	12
Membrane purification factor	2.1 ± 0.5	2.1 ± 0.2
K_m for Ca^{2+} (μM)	15 ± 4	17 ± 2
V_{max} (nmol·mg^{-1}·sec^{-1})	0.24 ± 0.03	0.60 ± 0.03*

*$P < 0.05$ for control vs. transgenic.

brane preparations from transgenic hearts compared with controls after ischemia and reperfusion. In this model of myocardial ischemia and reperfusion, there is no effect of NCX1 overexpression on recovery from ischemia, and we saw no effects of gender in these experiments. Our results in the present experiments differ from those of Cross et al.[2] who found less recovery from ischemia in male than in female mice that were heterozygotic for the NCX1 transgene; as a result, there was a detrimental effect of NCX1 overexpression apparent in males only. NCX1 overexpression would be expected to improve recovery from ischemia if the sodium–calcium exchanger remains more important for Ca^{2+} efflux than influx during ischemia and reperfusion, but overexpression would be detrimental if reverse sodium–calcium exchange predominates during ischemia and reperfusion. The absence of any effect suggests that either sodium–calcium exchange is relatively inhibited and therefore unimportant in Ca^{2+} regulation during ischemia or that increases in both efflux and influx modes of sodium–calcium exchange are roughly balanced in the transgenics.

REFERENCES

1. BERSOHN, M.M., C.R. CARMACK, A.E. SHALABY & K.D. PHILIPSON. 1999. Ischemic tolerance of hearts from transgenic mice overexpressing the sodium–calcium exchanger [abstract]. Biophys. J. **76:** A253.
2. CROSS, H.R., L. LU, C. STEENBERGEN, *et al.* 1998. Overexpression of the cardiac Na^+/Ca^{2+} exchanger increases susceptibility to ischemia/reperfusion injury in male, but not female, transgenic mice Circ. Res. **83:** 1215–1223.

Na/Ca Exchanger Overexpression Induces Endoplasmic Reticulum–Related Apoptosis and Caspase-12 Release

O. DIAZ-HORTA,[a,b] A. HERCHUELZ,[a] AND F. VAN EYLEN[a]

[a]*Laboratory of Pharmacology, Free University of Brussels, Brussels, Belgium*
[b]*National Institute of Endocrinology, Havana, Cuba*

> KEYWORDS: Na/Ca exchanger; endoplasmic reticulum; apoptosis; caspase-12; ER stress

Ca^{2+} may trigger programmed cell death (apoptosis) and regulate death-specific enzymes. Therefore, the development of strategies to control Ca^{2+} homeostasis may represent a potential approach to prevent or enhance cell apoptosis. To test this hypothesis, we overexpressed the plasma membrane Na/Ca exchanger (NCX1.7 isoform) stably in insulin-secreting tumoral cells. NCX1.7 overexpression increased apoptosis induced by endoplasmic reticulum (ER) Ca^{2+}-ATPase inhibitors but not by agents increasing cytosolic Ca^{2+} concentration ($[Ca^{2+}]_i$) through opening of plasma membrane Ca^{2+} channels. NCX1.7 overexpression reduced the increase in $[Ca^{2+}]_i$ induced by all agents, depleted ER Ca^{2+} stores, sensitized the cells to Ca^{2+}-independent proapoptotic signaling pathways and reduced cell proliferation by approximately 40%. ER Ca^{2+} stores depletion was attended by activation of the ER-specific caspase (caspase-12), the activation being enhanced by ER Ca^{2+}-ATPase inhibitors. Hence, Na/Ca exchanger overexpression, by depleting ER Ca^{2+} stores, induces ER stress with resulting activation of caspase-12 and increase in apoptotic cell death. By increasing apoptosis and decreasing cell proliferation, overexpression of Na/Ca exchanger may represent a new potential approach in cancer gene therapy.

ACKNOWLEDGMENTS

This work was supported by the Belgian Fund for Scientific Research (FRSM 3.4562.00) of which F.V.E. is a Senior Research Assistant. O.D.-H. is a grant holder of the ALFA programme of the European Union.

Address for correspondence: André Herchuelz, Laboratory of Pharmacology, Brussels University School of Medicine, Route de Lennik 808, Brussels, Belgium. Voice: 32-2-5556201; fax: 32-2-5556370.
 herchu@ulb.ac.be

The Molecular Architecture of Calcium Microdomains in Rat Cardiomyocytes

DAVID R. L. SCRIVEN, AGNIESZKA KLIMEK, KELLY L. LEE, AND EDWIN D. W. MOORE

Department of Physiology, University of British Columbia, Vancouver, British Columbia, V6T 1Z3 Canada

ABSTRACT: We have used standard indirect immunofluorescence techniques in combination with wide-field microscopy and image deconvolution to assess the distribution of proteins implicated in excitation–contraction coupling and Ca^{2+} homeostasis in adult rat cardiomyocytes. We begin by discussing our earlier results and summarizing what is known about the molecular architecture of this species to provide a rationale for the work presented here. The previous results showed that the dyads contain Ca^{2+} channels and ryanodine receptors, but few Na^+ channels or Na^+/Ca^{2+} exchangers. The latter proteins were not colocalized elsewhere on the membrane, and we have now found that they appear to be minimally associated with caveolin-3. None of the molecules examined are distributed uniformly in the membranes in which they are located but are organized into discrete clusters attached to the underlying cytoskeleton, an arrangement that, at the level of light microscopy, does not appear to be affected by the enzymatic dissociation used to study single cells. Analysis of how the clusters are organized and distributed throughout the volume of the cell suggests that there may be differences in excitation–contraction coupling between the cell surface and the interior.

KEYWORDS: fluorescence microscopy; deconvolution; NCX; ryanodine receptors; Na^+ channels; Ca^{2+} channels; calsequestrin; caveolin-3; cytoskeleton, α-actinin; vinculin; excitation–contraction coupling; calcium; collagenase; digital image processing

Contraction of an adult ventricular myocyte begins with an influx of Ca^{2+} from the extracellular space. The amount of Ca^{2+} entering the cell is too small to produce a significant contraction, but it provokes the release of a much larger amount of Ca^{2+} from intracellular stores, the sarcoplasmic reticulum (SR).[1] This mechanism, Ca^{2+}-induced calcium release (CICR), displays a high gain because of the juxtaposition of L-type Ca^{2+} channels in the membrane of the sarcolemma to the Ca^{2+}-sensitive SR release channels, ryanodine receptors, in the membrane of the adjacent SR.[2,3] This

---

Address for correspondence: Dr. Edwin D. W. Moore, Department of Physiology, University of British Columbia, 2146 Health Sciences Mall, Vancouver, BC, Canada V6T 1Z3. Voice: 604-822-3423; fax: 604-822-6048.
  edmoore@interchange.ubc.ca

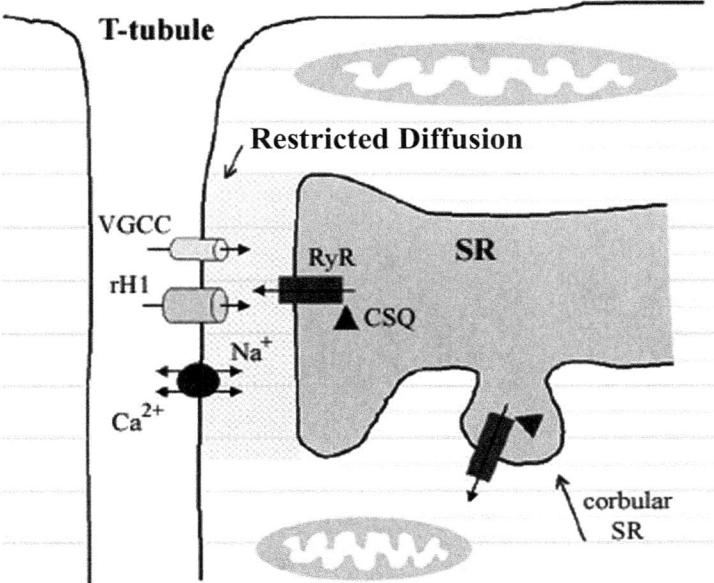

FIGURE 1. Proposed molecular architecture of the cardiomyocyte dyad. VGCC, L-type voltage-gated Ca^{2+} channel; rH1, Na^+ channel; RyR, ryanodine receptor; CSQ, calsequestrin; SR, sarcoplasmic reticulum.

is the most commonly accepted mechanism for excitation–contraction coupling in cardiac muscle, but there may be other ways to produce phasic SR Ca^{2+} release in a healthy myocyte.

Leblanc and Hume[4] have suggested that reverse-mode operation of the Na^+/Ca^{2+} exchanger (NCX) can also trigger CICR. The proposed mechanism requires a localized increase in $[Na^+]_i$ during the rising phase of the action potential reversing the exchanger and increasing the local $[Ca^{2+}]_i$ enough to trigger release from the SR. The changes in intracellular ion concentrations must be highly localized, because if the incoming Na^+ disperses throughout the myoplasm the change in $[Na^+]_i$ would be insufficient to reverse the exchanger. As Lederer[5] has discussed, this constraint requires that the cell have a space where diffusion is restricted (the so-called "fuzzy space"), to which NCX, the Na^+ channels, and the ryanodine receptors must all have access (FIG. 1). The dimensions and location of this space have not been defined, but the most likely candidate is the coupling region between the t-tubule and SR membranes, that is, the dyads. There is no consensus as to whether reversal of the exchanger contributes significantly to contraction, because electrophysiological and functional experiments have produced conflicting results.[6–11]

To test this hypothesis directly, we studied the molecular architecture of the adult rat ventricular cardiomyocyte by using standard indirect immunofluorescence techniques.[12] Cells were dissociated using established protocols using collagenase to disrupt the cell–extracellular matrix bonds[13] and then fixed, permeabilized, and attached to coverslips using poly-L-lysine as an adhesive. To locate the position of specific molecules, we used antibodies against the ryanodine receptor (monoclonal, Dr. Gerhard Meissner), NCX (polyclonal, Dr. K. Philipson), calsequestrin (polyclonal, Dr. K. Campbell), the α_{1C}-subunit of the L-type Ca^{2+} channel (polyclonal, Dr. W. Catterall) and rH1, the predominant isoform of the Na^+ channel in rat myocardium (polyclonal, Dr. S. Cohen). Secondary antibodies fluorescently tagged with fluorescein isothiocyanate (FITC) or Texas Red were used because of their wide spectral separation. Three-dimensional image stacks then were acquired using a wide-field microscope equipped with a thermoelectrically cooled CCD camera (SITe S1502AB chip; peak Q.E. 80%) and deconvolved to reassign out-of-focus light. This combination of image acquisition and processing has proved more sensitive than other optical microscopy techniques[14,15] and gives a comparable to superior resolution.[16] Voxel sizes were $122 \times 122 \times 250$ nm, which satisfied the Nyquist theorem for sampling. Small multispectral fluorescent beads were used to align the data sets. The strategy was to label two proteins, one with FITC and the other with Texas Red, and then calculate the percentage of colocalized voxels. The results can be viewed at http://www.interchg.ubc.ca/edmoore/

We observed the maximum possible colocalization between voxels containing the α_{1C} subunit of the Ca^{2+} channel with voxels containing ryanodine receptors. This was a positive control because these proteins are known to be associated *in vivo*. The colocalization between voxels containing ryanodine receptors and calsequestrin, proteins that are also functionally and physically coupled,[17] were similarly at the maximum we could record. There were significantly more voxels containing ryanodine receptors than Ca^{2+} channels, indicating that not all of the ryanodine receptors were associated with Ca^{2+} channels. We concluded that these uncoupled ryanodine receptors were most likely located within corbular SR.[18] Less than 10% of voxels identified as containing either NCX or rH1 colocalized with the ryanodine receptors; therefore, both are largely excluded from the dyad. In addition, they did not colocalize with each other. The impact of these results on theories of excitation–contraction coupling depends on the presumed structure and location of the fuzzy space.

If this space is limited to the area encompassed by the dyads, then our results indicate that the Na^+ channel and NCX communicate with the bulk myoplasm. With this architecture, the $[Na^+]_i$ cannot increase sufficiently during the rising phase of the action potential to reverse the exchanger. Yet, there is compelling evidence for an area of restricted diffusion: NCX equilibrates with a discrete subsarcolemmal compartment,[19] the $[Ca^{2+}]$ sensed by NCX is different from that in the bulk myoplasm,[20] and there are discrete subsarcolemmal microdomains of elevated $[Ca^{2+}]$.[21] There is also evidence of multiple subsarcolemmal domains where Na^+ diffusion is restricted.[22] These data raise the possibility of multiple restricted spaces, either disjointed or continuous. With this architecture of the fuzzy space, reverse-mode operation of the exchanger might possibly trigger CICR. Until the dimensions of the space have been clarified, and the amount of Ca^{2+} entering the cell via reverse-mode operation of NCX quantified, a definitive answer is not possible. Our data indicate that with either model, Ca^{2+} would have to diffuse from NCX at least 122 nm to reach either junctional or corbular SR.

Because rH1 and NCX were not within the dyad, or adjacent to each other, a possible location was within caveolae. These small membranous invaginations, formed

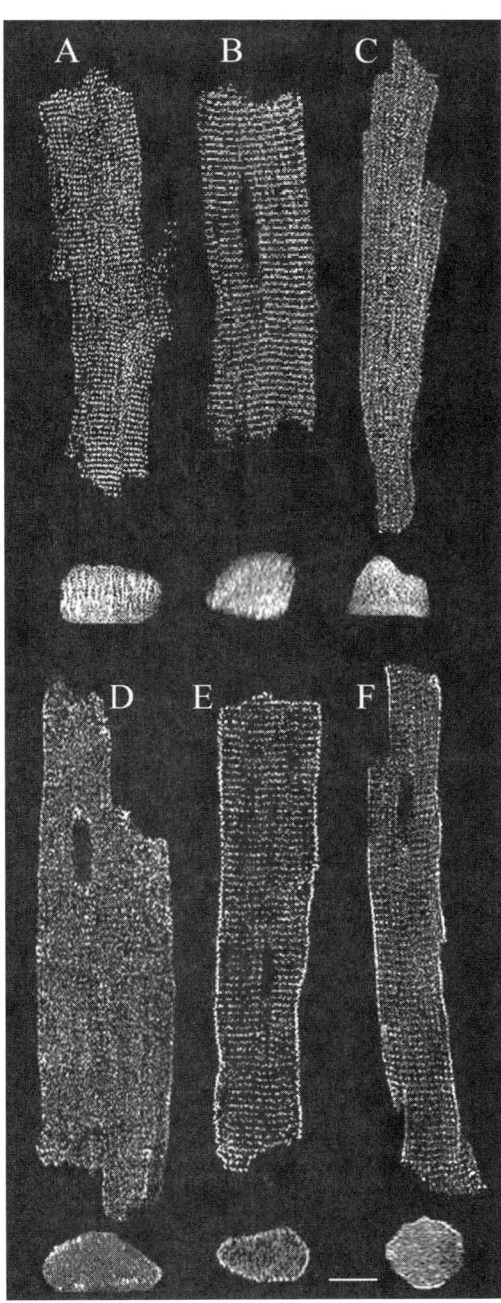

FIGURE 2. Images of single cells acquired at low magnification. Voxels are 400 × 400 × 400 nm. For clarity only an 800-nm-thick section is displayed from the center of each cell along with a representative cross-section, 4 μm thick, from the three-dimensional image. Scale bar = 10 μm. **(A)** α_{1C} subunit of the Ca^{2+} channel. **(B)** Ryanodine receptor. **(C)** NCX. **(D)** rH1 isoform of the Na^{+} channel. **(E)** Caveolin-3. **(F)** Calsequestrin.

by homomultimers of caveolin-3, are located both at the cell surface and within t-tubules.[23,24] They are thought to anchor a variety of signaling molecules, cytoskeletal proteins, and ion pumps, channels, and exchangers,[25] including NCX.[26] Some of the molecules located within caveolae regulate, and are regulated by, $[Na^+]_i$ and $[Ca^{2+}]_i$, yet the distribution of caveolae relative to most ion-transporting proteins has been largely unexplored.[27,28] We therefore labeled cells with an antibody specific for caveolin-3 (monoclonal, C38320; BD Transduction Laboratories, Lexington, KY) in combination with an antibody for either rH1, NCX, or the α_{1C} subunit of the Ca^{2+} channel. Our preliminary analyses indicate that, as expected, the L-type Ca^{2+} channels are not associated with caveolin-3, but neither are NCX or rH1. The observed structural organization, with the ion-transporting proteins separate from caveolae, may insulate the enzymes within the caveolae from large beat-to-beat fluctuations in ion concentration.

To further clarify the observed staining patterns, we collected images of labeled cells at low power so that the entire cell could be viewed (FIG. 2). Significant differences in the distribution of the molecules are apparent from these images. The L-type Ca^{2+} channels (FIG. 2A) were distributed only along the membranes of transverse t-tubules and therefore are located solely at the Z-lines. The ryanodine receptors (FIG. 2B) are located in the adjacent junctional SR and have a similar distribution, but the lack of specific labeling in the middle of the sarcomeres indicates that uncoupled ryanodine receptors in corbular SR are also near the Z-lines. This distribution is consistent with their hypothesized role as secondary sites of Ca^{2+} release. The Na^+/Ca^{2+} exchangers (FIG. 2C) are located in both transverse and longitudinal elements of the t-tubules, the latter visible as short segments running parallel to the long axis of the cell. The rH1 isoform of the Na^+ channel (FIG. 2D) also was distributed in transverse and longitudinal t-tubules, but it must be in areas separate from NCX because there was little colocalization between the two proteins. Caveolin-3 (FIG. 2E) was distributed all along the surface of the cell but was only on the transverse t-tubules in the cell interior; none was visible in its longitudinal elements. Calsequestrin (FIG. 2F) was largely along the Z-lines. The cross-section of the caveolin-3 image displays a bright edge, and that of ryanodine a dim edge, indicating possible changes in fluorescent intensity/voxel or a change in the density of illuminated voxels through the cell. Such variations in the distribution of proteins that either directly or indirectly regulate $[Ca^{2+}]_i$ could have functional consequences. We therefore quantified the number of illuminated voxels, and the fluorescence intensity per illuminated voxel, as a function of distance from the cell surface in all of these images.

In FIGURE 3A, we present an image of the distribution of the ryanodine receptor in an entire cell. The coordinates of the surface voxels were determined by positioning a cylinder around the edges of, but not touching, the cell (FIG. 3B). The image of the cell then was convolved with a small gaussian point-spread function which had a FWHM of eight voxels in all dimensions, and the coordinates of the cylinder were then allowed to deform along the intensity gradient created by the convolution. The voxel with maximum intensity was identified as a surface voxel, and to prevent distortion between these coordinates the algorithm penalized sharp curvature so as to create as smooth a surface as possible.[29] The resultant image of the surface is presented as a wire mesh surrounding the initial image (FIG. 3C), the wire mesh alone (FIG. 3D), and as a solid, in which the surface contours are more readily apparent

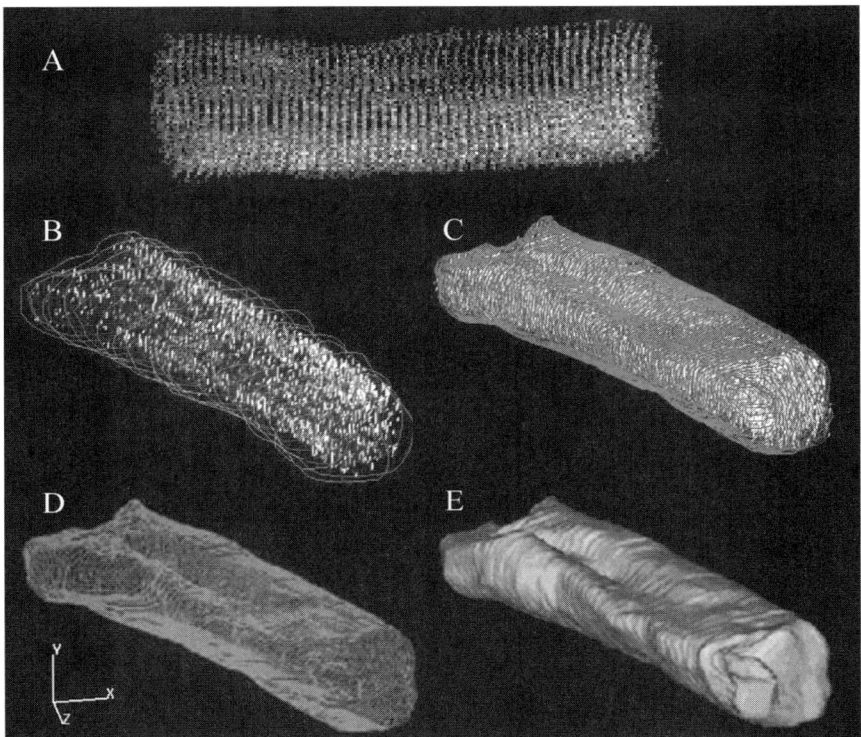

FIGURE 3. Finding the cell surface. (**A**) Maximum intensity projection of the distribution of the ryanodine receptor in a single cardiomyocyte. Scale bar = 10 μm in each dimension. (**B**) Manually drawn cylinder surrounding the cell in **A**. (**C**) Convergence of a wire mesh defining the coordinates of the cell surface with the image of the cell. (**D**) The wire mesh. (**E**) The cell surface depicted as a solid object.

(FIG. 3E). This allowed us to calculate the average cell volume, 37.6 ± 8.3 pl ($n = 15$), which is comparable to that previously measured in this species.[30] To determine the relative intensity as a function of distance from the cell surface, we cut a large section of the cell, excluding the nucleus, from the image. Total intensity was calculated for a series of irregular cylinders with a wall thickness of one voxel. The outermost cylinder was the cell surface, and each successive cylinder was one voxel farther in from the surface. Three values were determined for each cylinder: the total number of voxels, the number of lit voxels, and the summed intensity. The total intensity for each cylinder was divided by the number of lit voxels present to compensate for the fact that as the cylinder shrinks and approaches the center of the cell there are fewer voxels and therefore less total fluorescence. Intensity values were normalized to that of the surface to allow comparison of cells labeled and recorded at different times. This procedure was repeated for cells individually labeled with the other antibodies; three cells were analyzed for each antibody. These results are presented as graphs plotting

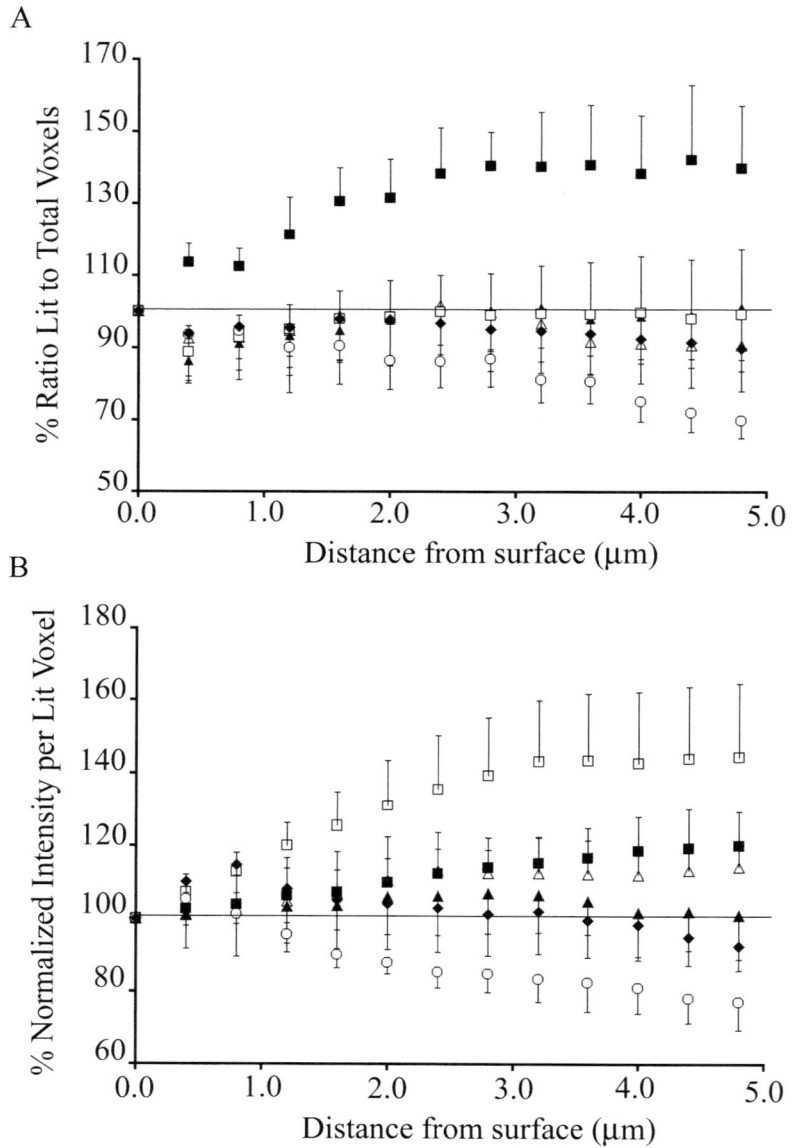

FIGURE 4. (**A**) Cluster density, represented by percentage of voxels lit, as a function of distance from the cell surface. (**B**) Fluorescent intensity/lit voxel as a function of distance from the cell surface. ▲ – rH1 isoform of the Na^+ channel; □ – ryanodine receptor; △ – NCX; ○ – caveolin-3; ■ – α_{1C} subunit of the Ca^{2+} channel; ◆ – calsequestrin.

(a) the ratio of lit to total voxels, and (b) the fluorescence intensity per lit voxel versus distance through the cell for each of the antibodies (FIG. 4).

The graphs indicate that both the percentage of voxels lit and the voxel intensity are almost constant in cells labeled for the Na^+ channel, the exchanger, and calsequestrin. This suggests that the density of the clusters in which these molecules are distributed is uniform throughout the cell. Cells labeled with antibody against the ryanodine receptor display a significant increase in the amount of fluorescence/voxel toward the cell interior, but no difference in the percentage of lit voxels/volume. The most likely explanation of these data is that the number of ryanodine receptors/voxel increases toward the center of the cell. The voltage-gated calcium channels displayed the opposite: an increasing percentage of lit voxels in the cell's interior, but no change in fluorescent intensity. Our interpretation of these data is that the deeper one goes within the cell, the greater the area involved in Ca^{2+} influx across the sarcolemma. Taken together, these observations may suggest that in the interior of the cell there is more Ca^{2+} available to trigger release and an increase in the rate at which the stores can be emptied, but there is a constant amount of Ca^{2+} stored/volume of cell (the calsequestrin curves were flat). Caveolin-3 decreases both in the amount/volume of cell and in the intensity/voxel as the cell is penetrated, possibly indicating that the concentration of caveolin and caveolae in the cell's interior is lower than at the surface. We speculate that this may represent differences between the caveolae in these two regions.

Although the low-magnification images revealed the distribution density, the high-magnification images indicated that none of the molecules were distributed continuously throughout the compartment in which they were located; all appeared as discrete clusters. This pattern likely reflects binding of the intramembranous proteins to the underlying cytoskeleton, which in addition to providing a scaffold for organizing membrane distribution[31,32] can regulate their function.[33] An intact cytoskeleton would seem to be necessary to anchor these proteins, and because our experiments are performed in cells isolated by collagenase we were concerned by reports that even brief exposure to collagen fragments can induce gross cytoskeletal degradation and rearrangement.[34]

Experiments in cultured smooth muscle cells indicate that exposure to collagen fragments for as little as 30 seconds produces an integrin-mediated activation of calpain-I with subsequent cleavage of focal adhesion kinase, paxillin, and talin.[34] This resulted in significant reorganization of cytoskeletal proteins, including vinculin, actin, and α-actinin, with corresponding changes in cell shape. To determine whether this was occurring in our cells, we compared the distribution of ryanodine receptors (FIG. 5A,B), α-actinin (FIG. 5C,D), and vinculin (FIG. 5E–G) in enzymatically dissociated cells with that in fresh-frozen tissue sections. Because the ryanodine receptor is located in the SR membrane, we anticipated that its distribution would not be affected by collagenase. In both images, the receptor is distributed in discrete clusters along the Z-lines, but not on the surface between Z-lines. This was a positive control and demonstrated that the cryosections and enzymatically dissociated cells should produce indistinguishable results at the level of light microscopy. In these preliminary experiments, there were no obvious rearrangements in the distribution of the cytoskeletal proteins examined. α-Actinin cross-links actin filaments at the Z-line, and the distance between the bands in both the cryosections and the isolated cells corresponds to the expected distance between Z-lines. Vinculin is lo-

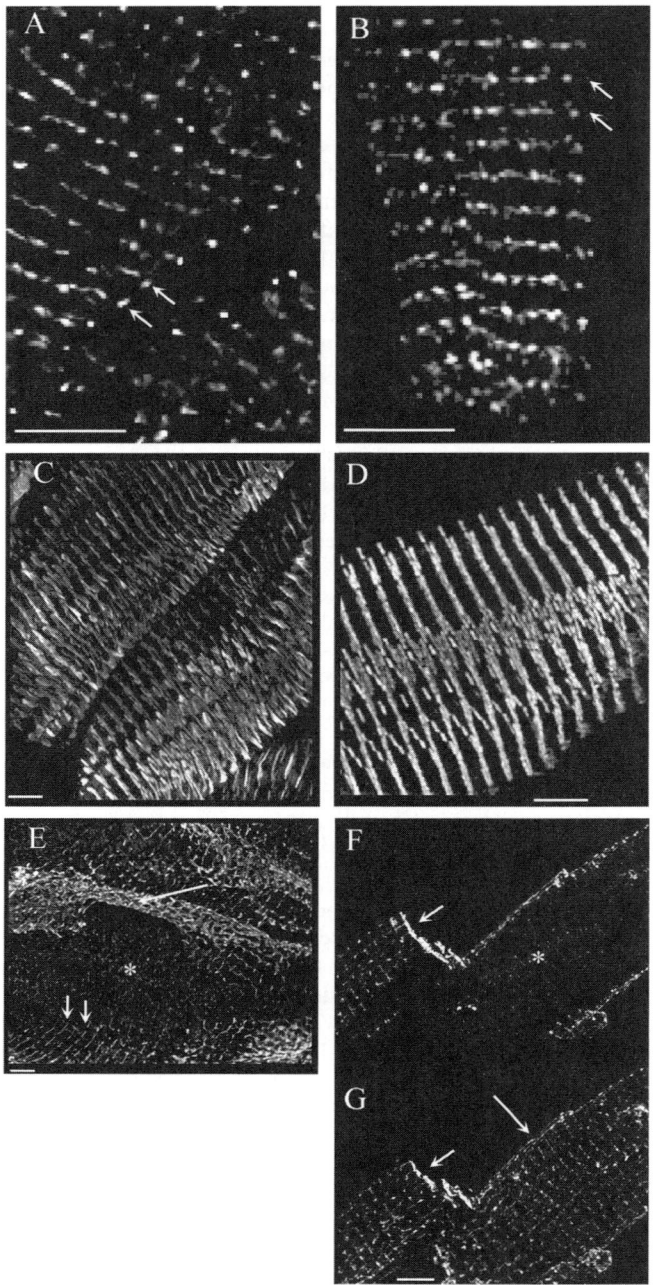

FIGURE 5. *See following page for legend.*

cated at intercalated discs and in the costameres where it anchors the Z lines to the sarcolemma.[35] This was the observed pattern of distribution in both the cryosections and the dissociated cells, and in both preparations there was little vinculin within the interior of the cell. These results demonstrate that the organizational patterns that we have observed likely reflect those *in vivo*.

In conclusion, we have extended our previous colocalization analysis and have begun to examine the possible locations of NCX and the rH1 isoform of the Na^+ channel. Our preliminary results indicate that neither of these molecules is associated with caveolin-3. All of the molecules examined are distributed in discrete clusters which likely reflect their attachment to the underlying cytoskeleton, which appears to be unaffected by the enzymatic dissociation. The distribution of clusters of molecules throughout the cell volume suggests that there may be changes in excitation–contraction coupling through the cell volume. These results apply to the rat myocardium; other species may be organized differently. In either case, the structure and location of restricted diffusion spaces requires continued investigation.

ACKNOWLEDGMENTS

The authors gratefully acknowledge support from the Canadian Institutes of Health Research and the Heart and Stroke Foundation of British Columbia and the Yukon.

REFERENCES

1. FABIATO, A. 1985. Time and calcium dependence of activation and inactivation of calcium-induced release of calcium from the sarcoplasmic reticulum of a skinned canine cardiac Purkinje cell. J. Gen. Physiol. **85:** 247–289.
2. CARL, S.L., K. FELIX, A.H. CASWELL, *et al.* 1995. Immunolocalization of sarcolemmal dihydropyridone receptor and sarcoplasmic reticular triadin and ryanodine receptor in rabbit ventricle and atrium. J. Cell Biol. **129:** 673–682.
3. SUN, X.H., F. PROTASI, M. TAKAHASHI, *et al.* 1995. Molecular architecture of membranes involved in excitation-contraction coupling of cardiac muscle. J. Cell Biol. **129:** 659–671.
4. LEBLANC, N. & J.R. HUME. 1990. Sodium current-induced release of calcium from cardiac sarcoplasmic reticulum. Science **248:** 372–376.
5. LEDERER, W.J., E. NIGGLI & R.W. HADLEY. 1990. Sodium-calcium exchange in excitable cells: fuzzy space. Science **248:** 283.
6. BOUCHARD, R.A., R.B. CLARK & W.R. GILES. 1993. Role of sodium–calcium exchange in activation of contraction rat ventricle. J. Physiol. **472:** 391–413.

FIGURE 5. A comparison of protein distribution in cryosections (**A, C,** and **E**) and enzymatically dissociated cells (**B, D, F,** and **G**). All cryosections are 20 µm thick. Scale bars are 5 µm. (**A** and **B**) Distribution of antiryanodine receptor. *Arrows* point to the cell surface and to two Z lines. (**C** and **D**) Distribution of anti–α-actinin. (**E–G**) Distribution of antivinculin. (**E**) The *small arrows* point to costameres. The *large arrow* to the cell surface curving out of the page. The asterisk labels the cell interior. (**F**) The image is 2 µm deep and was acquired from the middle of the cell. The arrow points to an intercalated disc. The asterisk marks the cell interior. (**G**) The image is 2 µm deep and was acquired from the cell surface. The *small arrow* points to an intercalated disc, the large arrow to a costamere.

7. LEVI, A.J., P. BROOKSBY & J.C. HANCOX. 1993. A role for depolarisation induced calcium entry on the Na-Ca exchange in triggering intracellular calcium release and contraction in rat ventricular myocytes. Cardiovasc. Res. **27:** 1677–1690.
8. WASSERSTROM, J.A. & A.M. VITES. 1996. The role of Na^+-Ca^{2+} exchange in activation of excitation-contraction coupling in rat ventricular myocytes. J. Physiol. **493:** 529–542.
9. SHAM, J.S.K., L. CLEEMANN & M. MORAD. 1992. Gating of the cardiac Ca^{2+} release channel: the role of Na^+ current and Na^+-Ca^{2+} exchange. Science **255:** 850–853.
10. Satoh, H., K.S. Ginsburg, K. Qing, et al. 2000. KB-R7943 block of Ca^{2+} influx via Na^+/Ca^{2+} exchange does not alter twitches or glycoside inotropy but prevents Ca^{2+} overload in rat ventricular myocytes. Circulation **101:** 1441–1446.
11. HAN, C., P. TAVI & M. WECKSTROM. 2002. Role of the Na^+-Ca^{2+} exchanger as an alternative trigger of CICR in mammalian cardiac myocytes. Biophys. J. **82:** 1483–1496.
12. SCRIVEN, D.R.L., P. DAN & E.D.W. MOORE. 2000. Distribution of proteins implicated in excitation-contraction coupling in rat ventricular myocytes. Biophys. J. **79:** 2682–2691.
13. RODRIGUES, B. & D.L. SEVERSON. 1997. Preparation of cardiomyocytes. In Biochemical Techniques in the Heart. J.H. McNeil, Ed.: 101–115. CRC Press. Boca Raton, FL.
14. SWEDLOW, J.R., K. HU, P.D. ANDREWS, et al. 2002. Measuring tubulin content in *Toxoplasma gondii*: a comparison of laser-scanning confocal and wide-field fluorescence microscopy. Proc. Natl. Acad. Sci. USA **99:** 2014–2019.
15. CARRINGTON, W.E., K.E. FOGARTY & F.S. FAY. 1990. 3D fluorescence imaging of single cells using image restoration. In Noninvasive Techniques in Cell Biology. J.K. Foskett, S. Grinstein, Eds.: 53–72. Wiley-Liss. New York.
16. CARRINGTON, W.A., R.M. LYNCH, E.D.W. MOORE, et al. 1995. Superresolution three-dimensional images of fluorescence in cells with minimal light exposure. Science **268:** 1483–1487.
17. GUO, W. & K.P. CAMPBELL. 1995. Association of triadin with the ryanodine receptor and calsequestrin in the lumen of the sarcoplasmic reticulum. J. Biol. Chem. **270:** 9027–9030.
18. JORGENSEN, A.O., A.C. SHEN & K.P. CAMPBELL. 1985. Ultrastructural localization of calsequestrin in adult rat atrial and ventricular muscle cells. J. Cell Biol. **101:** 257–268.
19. LANGER, G.A. & T.L. RICH. 1992. A discrete Na-Ca exchange-dependent Ca compartment in rat ventricular cells: exchange and localization. Am. J. Physiol. **262:** C1149–C1153.
20. TRAFFORD, A.W., M.E. DIAZ, S.C. O'NEILL & D.A. EISNER. 1995. Comparison of subsarcolemmal and bulk calcium concentration during spontaneous calcium release in rat ventricular myocytes. J. Physiol. **488:** 577–586.
21. GALLITELLI, M.F., M. SCHULTZ, G. ISENBERG & F. RUDOLF. 1999. Twitch-potentiation increases calcium in peripheral more than in central mitochondria of guinea-pig ventricular myocytes. J. Physiol. **518:** 433–447.
22. WENDT-GALLITELLI, M.F., T. VOIGHT & G. ISENBERG. 1993. Microheterogeneity of subsarcolemmal sodium gradient. Electron probe microanalysis in guinea pig ventricular myocytes. J. Physiol. **472:** 33–44.
23. GABELLA, G. 1978. Inpocketings of the cell membrane (caveolae) in the rat myocardium. J. Ultrastruct. Res. **65:** 135–147.
24. PAGE, E. 1978. Quantitative ultrastructural analysis in cardiac membrane physiology. Am. J. Physiol. **235:** C147–C158.
25. RAZANI, B. & M.P. LISANTI. 2001. Caveolins and caveolae: molecular and functional relationships. Exp. Cell Res. **271:** 259–263.
26. BOSSUYT, J., B.E. TAYLOR, M. JAMES-KRACKE & C.C. HALE. 2002. Evidence for cardiac sodium–calcium exchanger association with caveolin-3. FEBS Lett. **511:** 113–117.
27. LI, S., T. OKAMOTO, M. CHUN, et al. 1995. Evidence for a regulated interaction between heterotrimeric G proteins and caveolin. J. Biol. Chem. **270:** 15693–15701.
28. RYBIN, V.O., X. XU & S.F. STEINBERG. 1999. Activated protein kinase C isoforms target to cardiomyocyte caveolae. Stimulation of local protein phosphorylation. Circ. Res. **84:** 980–988.

29. LIFSHITZ, L.M., J.A. COLLINS, E.D. MOORE & J. GAUCH. 1994. Computer vision and graphics in fluorescence microscopy. In IEEE Workshop on Biomedical Analysis. 166–175. IEEE Computer Society Press. Los Alamitos. California.
30. SATOH, H., L.M.D. DELBRIDGE, L.A. BLATTER & D.M. BERS. 1996. Surface:volume relationship in cardiac myocytes studied with confocal microscopy and membrane capacitance measurements: species-dependence and developmental effects. Biophys. J. **70:** 1494–1504.
31. LI, Z., E.P. BURKE, J.S. FRANK, et al. 1993. The cardiac Na^+–Ca^{2+} exchanger binds to the cytoskeletal protein ankyrin. J. Biol. Chem. **268:** 11489–11491.
32. GEE, S., R. MADHAVEN, S. LEVINSON, et al. 1998. Interaction of muscle and brain sodium channels with multiple members of the syntrophin family of dystrophin-associated proteins. J. Neurosci. **18:** 128–137.
33. MARUOKA, N.D., D.F. STEELE, B.P. AU, et al. 2000. α-Actinin-2 couples to cardiac Kv1.5 channels, regulating current density and channel localization in HEK cells. FEBS Lett. **473:** 188–194.
34. CARRAGHER, N.O., B. LEVKAU, R. ROSS & E.W. RAINES. 1999. Degraded collagen fragments promote rapid disassembly of smooth muscle focal adhesions that correlates with cleavage of $pp125^{FAK}$, paxillin, and talin. J. Cell Biol. **147:** 619–629.
35. KOSTIN, S., D. SCHOLZ, T. SHIMADA, et al. 1998. The internal and external protein scaffold of the T-tubular system in cardiomyocytes. Cell Tissue Res. **294:** 449–460.

Na/Ca Exchange Function in Intact Ventricular Myocytes

DONALD M. BERS AND CHRISTOPHER R. WEBER

Department of Physiology, Loyola University Chicago, Maywood, Illinois 60153, USA

ABSTRACT: Here, we address three issues in intact ventricular myocytes that specifically relate to the role of Na/Ca exchange (NCX) current under physiological conditions. First, we revisit the issue of NCX stoichiometry in light of some recent findings that the stoichiometry of the NCX may not be fixed at 3Na:1Ca. We discuss some data that strongly favor the 3:1 stoichiometry, at least under physiological conditions. Second, we address the controversy over the role of allosteric Ca regulation in intact myocytes. We show that outward and inward I_{NCX} can be activated dynamically by changing $[Ca]_i$ over the physiological range and that outward I_{NCX} can be activated quite rapidly with sarcoplasmic reticulum Ca release. These data are well described using an instantaneous equation for NCX current that includes an allosteric activation factor with K_{mCaAct} = 125 nM. Finally, we consider the effect on NCX current of submembrane elevations in $[Ca]_i$ (that are far greater than are measured in the bulk cytoplasm). Taken together with a NCX stoichiometry of 3, these findings have allowed us to make some predictions of the role of I_{NCX} during an AP. Our simulations suggest that NCX current is outward for less than approximately 10 ms at the beginning of the action potential.

KEYWORDS: Na/Ca exchange; cardiac muscle; excitation–contraction coupling; calcium transport; sarcoplasmic reticulum

INTRODUCTION

It is now clear that Na/Ca exchange (NCX) is the major mechanism by which Ca is extruded from cardiac myocytes.[1] Indeed, the rate and amount of Ca extrusion via NCX is greater than 20 times that carried by the sarcolemmal Ca-ATPase.[2] This means that during a steady-state heartbeat the amount of Ca influx via Ca current (I_{Ca}) plus NCX plus leak must be matched by Ca extrusion via NCX (plus the minor contribution via the sarcolemmal Ca-ATPase). Of course, Ca also is released and taken up by the sarcoplasmic reticulum (SR) with each beat, but again the amount released and taken back up during each steady-state beat must be equal.[3] There are major quantitative differences in the relative contributions of sarcolemmal versus SR Ca fluxes among cardiac myocytes from different species, different regions of the heart, different developmental stages, and during pathophysiological changes. For

---

Address for correspondence: Donald M. Bers, Ph.D., Department of Physiology, Loyola University Chicago, 2160 South First Avenue, Maywood, IL 60153. Voice: 708-216-1018; fax: 708-216-6308.

dbers@luc.edu

example, in adult rat and mouse ventricle, the SR is responsible for approximately 92% of the Ca involved in contractile activation, whereas only approximately 7% is supplied by I_{Ca} and removed by NCX (I_{Ca}/NCX). In rabbit, the SR is responsible for 75% of the activating Ca, whereas 20–25% enters via I_{Ca} and is extruded by NCX. Most mammalian ventricular myocytes behave more like the rabbit than the rat (including human, guinea pig, cat, ferret, and dog). In neonatal ventricle, the SR contribution is typically less and the I_{Ca}/NCX contribution is greater. During heart failure, there is usually an upregulation of NCX expression and reduced SR Ca content, and this may shift the balance in favor of I_{Ca}/NCX.[4–6] Indeed, in human and rabbit heart failure the I_{Ca}/NCX contribution may nearly equal the SR contribution.

NCX also carries an ionic current (I_{NCX}) which can be either an inward current (extruding Ca) or an outward current (bringing Ca in). The widely accepted stoichiometry of NCX is 3Na:1Ca, which means that one net charge is transported in the direction of Na transport (but see below).[1,7] Clearly, the direction of I_{NCX} depends on the transsarcolemmal [Na] and [Ca] gradients as well as membrane potential (E_m). The magnitude of I_{NCX} also will be influenced by the actual intracellular and extracellular concentrations of Na and Ca ([Na]$_i$, [Na]$_o$, [Ca]$_i$, and [Ca]$_o$) and the effect of E_m on NCX. NCX activity also is regulated allosterically by [Na]$_i$ and [Ca]$_i$. There is [Na]$_i$-dependent inactivation that is prominent only at unphysiological levels of [Na]$_i$ (>30 mM), so we usually ignore this effect in our experiments under more physiological conditions. NCX also is activated by [Ca]$_i$, but the apparent K_m and speed of this effect are controversial. This [Ca]$_i$-dependent regulation may be functionally important in activating NCX more fully when [Ca]$_i$ is high (to enhance Ca extrusion) and in turning off NCX as [Ca]$_i$ declines (preventing [Ca]$_i$ from decreasing too low). The [Na]$_i$-dependent inactivation also could be protective in the event that [Na]$_i$ increases to very high levels under pathophysiological conditions (where NCX would tend to bring Ca into the cell). Inactivating NCX under these conditions could protect the cell from more serious Ca overload.

I_{NCX} can contribute prominently to the cardiac action potential (AP) waveform and also can be an important arrhythmogenic current, underlying transient inward currents (I_{ti}) and delayed afterdepolarizations (DADs).[8] Thus, it is very important to understand exactly how I_{NCX} functions during the AP in myocytes. Isolation of I_{NCX} during a real AP and Ca transient, as typically done for other ion currents, is impossible for two major reasons. First, if one blocks I_{NCX} (even if we had a highly selective inhibitor), it would perturb the Ca transient that modulates other currents (e.g., I_{Ca}). Second, if you block all other contaminating currents (including Na and Ca channels), then the Ca transient that is critical in driving I_{NCX} is altered.

Three main issues will be discussed further here: (1) NCX stoichiometry, which recently has become more controversial, (2) allosteric regulation of NCX during dynamic Ca transients in intact cells, and (3) the influence of submembrane [Ca]$_i$ on real I_{NCX} during the normal cardiac AP and Ca transient.

NCX STOICHIOMETRY

The classic paper by Reeves and Hale clearly established the stoichiometry of cardiac NCX as 3Na:1Ca (based on reversal of sarcolemmal vesicle Ca fluxes).[7] A tremendous amount of data before then and in the subsequent 16 years have been entirely consistent with and often have confirmed that stoichiometry.[1,9,10] Indeed, this stoichiometry seemed as solidly established as the 3Na:2K transported by the Na/K-ATPase. Two provocative new studies in 2000 challenged the notion of a fixed 3:1 NCX stoichiometry. Fujioka *et al.* carefully measured I_{NCX} reversal potential in excised membrane patches from guinea pig ventricular myocytes over a wide range of ionic conditions and found that the apparent stoichiometry was closer to 4Na:1Ca versus 3:1.[11] Egger and Niggli measured Ca removal via NCX during Ca transients by measuring I_{NCX} and $[Ca]_i$ decline simultaneously (with SR Ca uptake disabled).[12] They found that at low extracellular pH (5.0–6.0) NCX still mediated robust $[Ca]_i$ decline but failed to produce much I_{NCX}, consistent with a stoichiometry closer to 2 under highly acidic conditions. The lower apparent stoichiometry with severe extracellular acidosis is intriguing, but it is unclear exactly how this comes about. The report of stoichiometry approaching 4 under relatively physiological conditions is disconcerting for several reasons.

First, for a 4:1 stoichiometry, I_{NCX} would not reverse until E_m above +20 mV (vs. −35 mV for $n = 3$). However, one can readily drive Ca entry and increase $[Ca]_i$ via I_{NCX} at E_m less than zero. This would be much more consistent with $n = 3$. Second, a 4:1 NCX ought to be capable of driving resting $[Ca]_i$ down to 130 pM (vs. 38 nM for a 3:1 NCX). Such low $[Ca]_i$ is not observed. $[Ca]_i$ rarely can be driven very much below 100 nM, perhaps approaching the limiting 38 nM value for 3:1 NCX. Third, Bridge *et al.* showed that the integrated I_{Ca} (bringing Ca into the cell) was exactly twice as large as the integrated I_{NCX} which extruded the same amount of Ca.[13] This is completely consistent with $n = 3$, but the I_{NCX} integral should have been twice as large if n was 4. Fourth, we can release all of the Ca in the SR by application of caffeine (FIG. 1) and measure the amount of charge moved by I_{NCX} and change in $[Ca]_i$. We independently measured the cytosolic Ca buffering, allowing transformation of $[Ca]_i$ to total $[Ca]$ ($[Ca]_{Tot}$).[14] This provides one measure of total SR Ca (118 μmol/l cytosol). Integrated I_{NCX}, multiplied by the measured surface:volume ratio (6.44 pF/pl cytosol)[15] and divided by 0.93 to correct for non-NCX-mediated Ca removal[2] yields another measure of SR content. Assuming 3:1 NCX stoichiometry, this value is 106 μmol/l cytosol, in remarkably good agreement with the NCX-independent measure. If it were 4:1, we would have expected twice as much charge moved by I_{NCX}. Finally, we can use I_{NCX} (assuming 3:1 stoichiometry) to measure Ca buffering (i.e., how much Ca is removed for a given change in $[Ca]_i$, FIG. 1B).[16] The Ca buffering measured this way was almost the same as that measured by more direct biochemical methods.

Thus, based on Ca flux and I_{NCX} measurements, it seems clear that the stoichiometry is close to 3:1 (rather than 4:1). We cannot explain why Fujioka *et al.*[11] obtained an estimate of 4:1. However, one possible complication with their measurements is that they did not measure $[Ca]_i$ or $[Na]_i$ during their measurements of reversal potential. When we do measure both $[Ca]_i$ and $[Na]_i$ during I_{NCX} measurements, $[Ca]_i$ and $[Na]_i$ both change via NCX at the holding potential (even with 6 mM BAPTA + Br$_2$BAPTA in the pipette).[17] Thus, we think it is premature to dismiss the extensive body of data indicative of 3:1 NCX stoichiometry, but there is also reason to revisit this important fundamental issue.

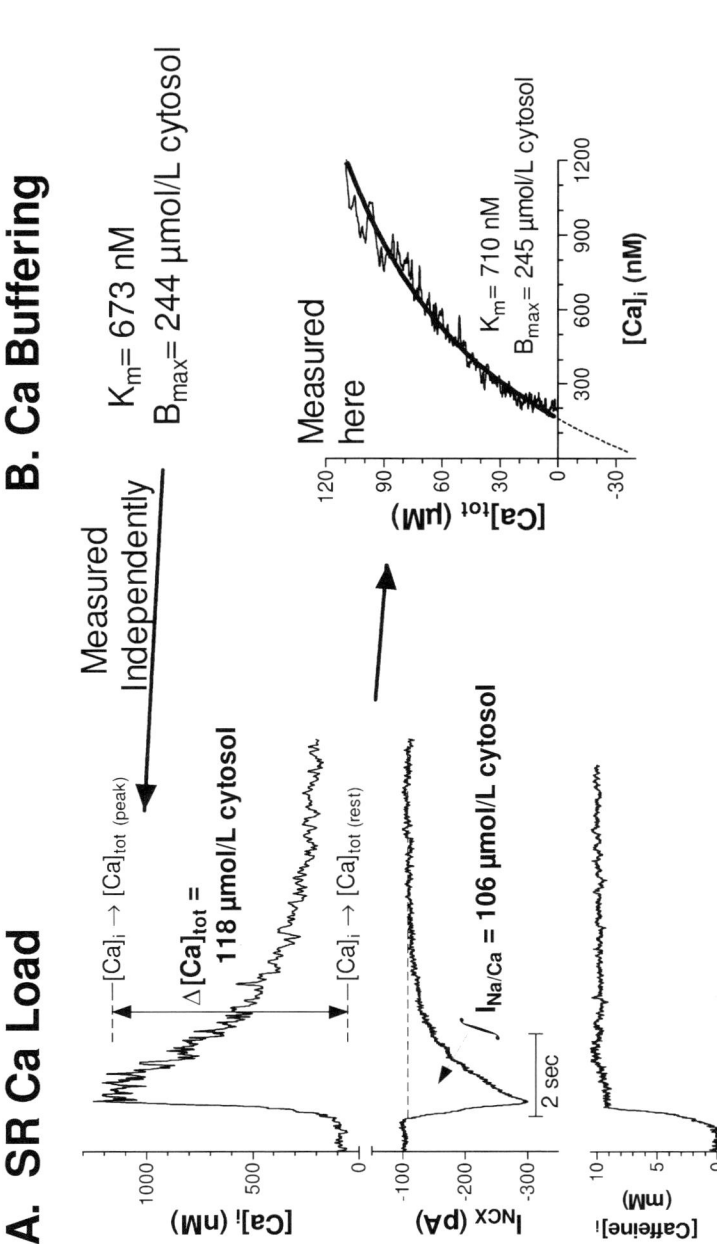

FIGURE 1. Independent measurements of SR load verify NCX stoichiometry. **(A)** SR Ca Load is measured from (1) peak [Ca]$_i$ converted [Ca]$_{Tot}$,[14] and from (2) I$_{NCX}$/Volume$_{cyt}$,[15] after application of 10 mM caffeine. These two measurements are in close agreement, and the latter depends on a NCX stoichiometry of 3. **(B)** These data also provide an independent measurement of Ca buffering,[16] by plotting cumulative I$_{NCX}$ versus [Ca]$_i$. This technique, utilized here, yields buffering parameters that are similar to Hove-Madsen and Bers,[14] which provide an additional consistency check for a NCX stoichiometry of 3. Adapted from a figure in Bers.[1]

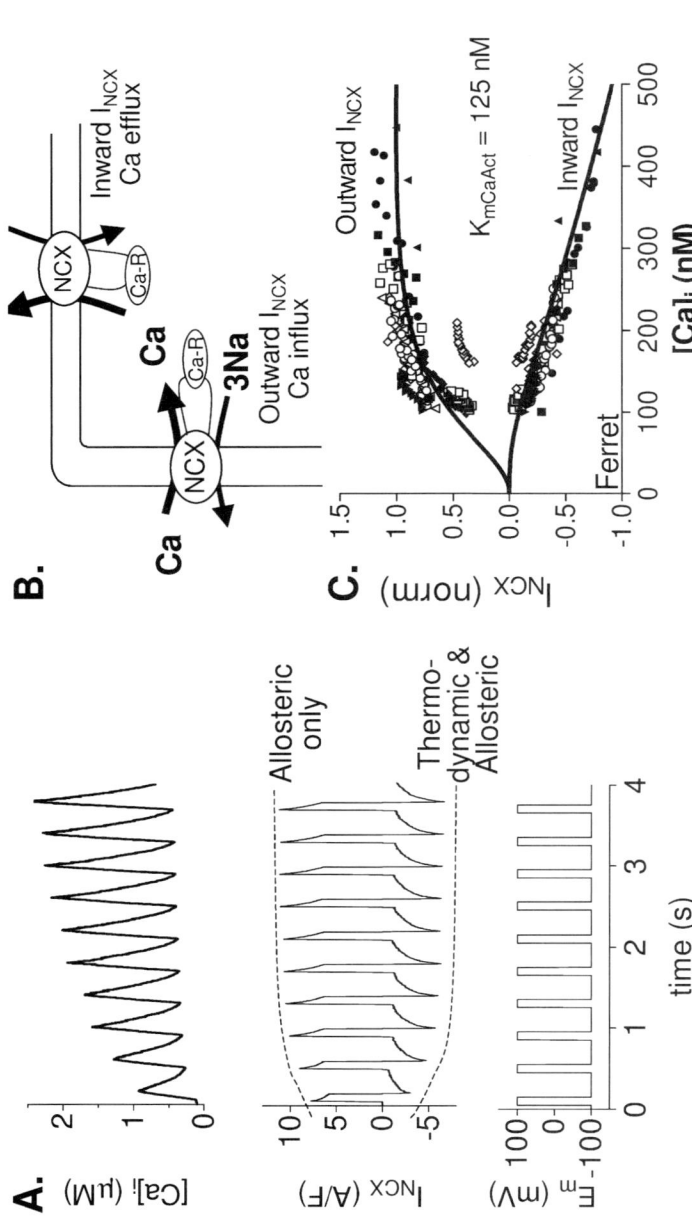

FIGURE 2. Protocol to measure effect of regulatory $[Ca]_i$ on I_{NCX} in intact myocytes. Currents other than I_{NCX} were blocked, and myocytes were pretreated with thapsigargin + caffeine to eliminate the contribution of the SR to $[Ca]_i$ oscillations. (**A**) Data from mouse myocytes overexpressing canine NCX1 show outward I_{NCX} at +100 mV (Ca entry) and inward I_{NCX} at −100 mV (Ca extrusion). The increasing outward envelope of I_{NCX} is explained only by allosteric Ca activation of I_{NCX}. (**B**) Outward I_{NCX} occurs at +100 mV, and inward I_{NCX} occurs at −100 mV. (**C**) Pooled data from 14 ferret myocytes. Each symbol represents averaged data during a single pulse to +100 mV or −100 mV, as in A. Data (from Weber *et al.*[20]) were fit using Equation 1 with average K_{mCaAct} = 125 nM.

ALLOSTERIC REGULATION BY INTRACELLULAR Ca

Allosteric regulation of NCX by $[Ca]_i$ has been reported based on both giant excised patch recordings and in cardiac myocytes dialyzed with Ca-buffered solutions.[18,19] The apparent $K_m(Ca)$ reported (K_{mCaAct}) has varied from approximately 22 nM to 6 µM, and the regulation appeared to be relatively slow in giant excised patches. These results have left it unclear whether this regulation is functionally relevant during dynamic Ca transients in ventricular myocytes. We measured allosteric Ca activation in intact voltage clamped ventricular myocytes during dynamic variations of $[Ca]_i$ in the physiological range.[20] FIGURE 2A shows the protocol used. E_m was switched between −100 mV (where Ca efflux is favored) and +100 mV (where Ca influx is favored). I_{Ca}, K currents, Cl currents, Na/K-ATPase, and SR Ca transport were blocked to better isolate I_{NCX}. In this context, Ni-sensitive currents were taken as I_{NCX}, and changes in $[Ca]_i$ are caused by I_{NCX}. The percentage of time spent at +100 mV was varied (analogous to varying stimulation frequency) to gradually drive $[Ca]_i$ up.

As $[Ca]_i$ increases as shown in FIGURE 2A,B, the amplitude of both inward and outward I_{NCX} get progressively larger. The inward I_{NCX} gets larger for two reasons: (1) higher $[Ca]_i$ increases the substrate concentration and thermodynamic driving force for inward I_{NCX}, and (2) there is allosteric activation by intracellular Ca binding to a site other than the transport site. However, the increase in $[Ca]_i$ should inhibit outward I_{NCX}, based on both substrate concentration and thermodynamic reasons. The only reason for the increase in outward I_{NCX} is the allosteric activation by Ca. Thus, we focus especially on the outward I_{NCX} to characterize allosteric regulation by $[Ca]_i$. Note also that the offsetting thermodynamic/substrate effects will mean that the growth in outward I_{NCX} is an underestimate of the extent of allosteric Ca regulation. FIGURE 2C shows the $[Ca]_i$ dependence of inward and outward I_{NCX} from 14 ferret ventricular myocyte experiments of this type. Data from each cell were fit with the following equation,[20,21] which includes a factor for allosteric $[Ca]_i$ regulation (first factor in numerator):

$$I_{NCX} = \frac{(V_{max}/(1 + \{K_{mCaAct}/[Ca]_i\}^2))([Na]_i^3[Ca]_o e^{\eta Vk} - [Na]_o^3[Ca]_i e^{(\eta-1)Vk})}{\left[\begin{array}{l} K_{mCao}[Na]_i^3 + K_{mNao}^3[Ca]_i + K_{mNai}^3[Ca]_o(1 + [Ca]_i/K_{mCai}) + \\ K_{mCai}[Na]_o^3(1 + \{[Na]_i/K_{mNai}\}^3) + [Na]_i^3[Ca]_o + [Na]_o^3[Ca]_i \end{array}\right][1 + k_{sat}e^{(\eta-1)Vk}]} \quad (1)$$

The K_m values are the Na and Ca dissociation constants for intracellular (i) and extracellular (o) Na and Ca. η is the position of the energy barrier of NCX in the membrane electric field, and k_{sat} is a factor that controls the saturation of I_{NCX} at negative E_m. Fixed parameters were: $K_{mNao} = 87.5$ mM, $K_{mNai} = 12.3$ mM, $K_{mCai} = 3.6$ (µM, $K_{mCao} = 1.30$ mM, $k_{sat} = 0.27$, $\eta = 0.32$, $k = F/RT$, and $T = 23°C$. V_{max} (in A/F) was determined separately on every cell to account for cell-to-cell variations in NCX expression. The mean K_{mCaAct} for ferret ventricular myocytes was 125 ± 16 nM. This would be consistent with dynamic allosteric regulation over the physiological range of $[Ca]_i$. Similar experiments in wild-type (WT) mice showed no evidence of allosteric regulation over the physiological range of $[Ca]_i$ studied (i.e., outward I_{NCX} only decreased as $[Ca]_i$ increased). In transgenic mice that overexpress WT canine NCX in heart (approximately threefold), I_{NCX} showed allosteric regulation quite similar to that seen in ferret myocytes. However, with transgenic mice overexpressing a mutant canine NCX which lacks a crucial 6-amino acid stretch in the large intracellular loop ((680-685), there is again no evidence of allosteric activation as $[Ca]_i$ increases. This confirms that this portion of the canine NCX (680–685) is crucial in allosteric Ca regulation.[20,22] Interestingly, this sequence (IIEESY) is different in the mouse NCX molecule (IIQESY). Although it is possible that WT mouse lacks allosteric Ca regulation, Maxwell et al. found some activation between 0 and 1 µM Ca in excised giant patch recording.[22] Thus, we think it likely that the mouse does have allosteric regulation, but that the K_{mCaAct} might be lower than in the ferret and canine NCX such that it is nearly fully activated at the physiological range of $[Ca]_i$ that we have explored.

FIGURE 3 shows that simulated data (using Eq. 1 and the driving E_m) to predict both I_{NCX} and $[Ca]_i$ matched experimental measurements rather well. With allosteric

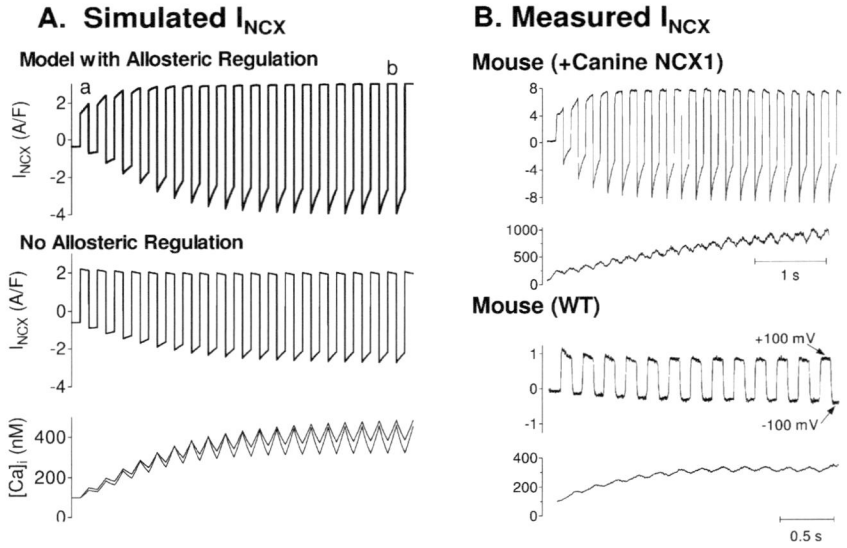

FIGURE 3. I_{NCX} data at +100 and −100 mV: simulations versus measurements. Simulations consider cytoplasmic buffering,[14] cytoplasmic volumes,[15] and I_{NCX}.[20] **(A)** The allosteric regulatory factor from Equation 1 determines if the envelope outward I_{NCX} increases (*top*) or not (*middle*), but it does not influence the overall shape of $[Ca]_i$ traces. As $[Ca]_i$ increases from rest, the shapes of the early pulses **(a)** are governed by allosteric Ca activation, but in the later pulses **(b)**, the allosteric effect is offset by the effects of $[Ca]_i$ at the transport sites as well as a reduced electrochemical driving force for outward I_{NCX}. **(B)** Data from mouse myocytes overexpressing canine NCX1 are similar to simulations that include the allosteric regulatory factor (*top*), and data from WT mouse myocytes are better represented by excluding the allosteric regulatory factor (*bottom*). Data taken from Weber et al.[2]

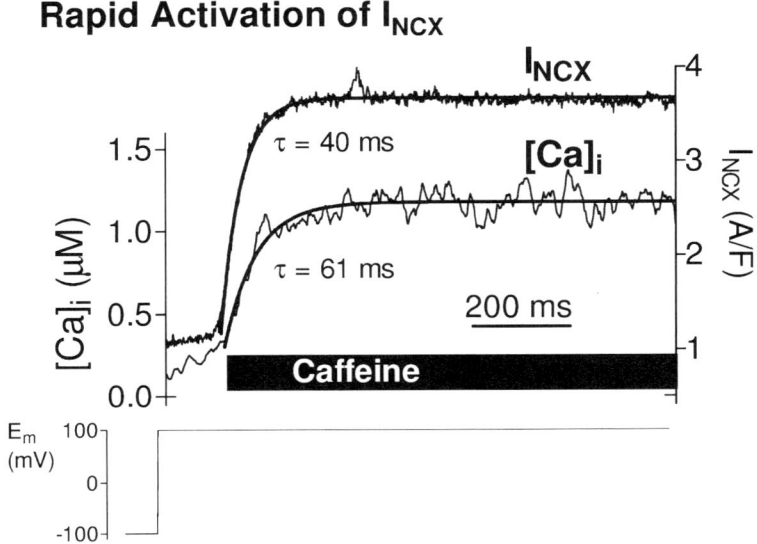

FIGURE 4. I_{NCX} activates rapidly with application of caffeine in ferret myocytes. SR function was not pharmacologically inhibited here. Caffeine was applied at +100 mV to cause rapid increase in $[Ca]_i$ and I_{NCX} increases on a similar time course. This justifies the use of an instantaneous equation (Eq. 1) to describe experiments in FIGURES 2 and 3, and it shows that I_{NCX} may respond rapidly to physiological increases in $[Ca]_i$. Modified from Weber et al.[20]

regulation functioning, the time course of I_{NCX} (inward and outward) was very much like that in mouse with WT canine NCX overexpression (or in normal ferret). This includes the allosteric activation with increasing outward I_{NCX} during the first few individual pulses (FIG. 3Aa) as well as the flat or slightly declining outward I_{NCX} at later pulses (FIG. 3Ab) where allosteric activation is nearly complete. Without allosteric activation (and in WT mouse) outward I_{NCX} declines slightly as $[Ca]_i$ increases, and this can even be seen during individual pulses at $E_m = +100$ mV.

ALLOSTERIC REGULATION IS RAPID

Although allosteric regulation clearly occurs at physiological $[Ca]_i$, the way I_{NCX} functions also will be importantly affected by how rapidly this activation occurs. We explored this in experiments in which SR Ca transport was not inhibited, allowing us to create rapid SR Ca release by application of caffeine (FIG. 4). Depolarizing to $E_m = +100$ mV activates outward I_{NCX}, which brings in some Ca (increasing $[Ca]_i$), and this causes a gradual increase in outward I_{NCX}. When caffeine is rapidly applied, there is a rapid SR Ca release and $[Ca]_i$ increases abruptly to approximately 1.2 μM with a time constant (τ) of 61 ms. Outward I_{NCX} increased with a similar τ (40 ms), suggest-

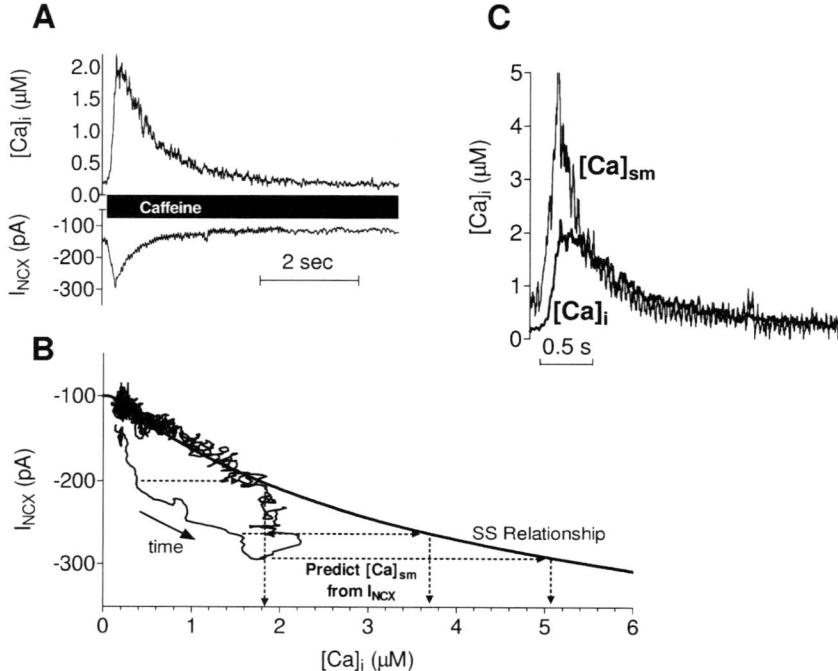

FIGURE 5. I_{NCX} can be used to predict $[Ca]_{sm}$. **(A)** Caffeine is applied to rapidly release SR Ca stores, resulting in a transient increase in $[Ca]_i$, which is extruded from the cell via inward I_{NCX} as in FIGURE 1. **(B)** when I_{NCX} is plotted versus $[Ca]_i$, it becomes evident that I_{NCX} increases rapidly and then later descends to the steady-state (SS) I_{NCX} relationship described by Equation 1. **(C)** This behavior can be explained by rapid increases in $[Ca]_{sm}$ in the vicinity of the NCX that are not measured by fluorescent indicators in the bulk cytoplasm.[28] Extrapolations from Equation 1, with known I_{NCX} and E_m, are used to predict the time course of $[Ca]_{sm}$.

ing that the allosteric regulation of I_{NCX} is very rapid. This largely justifies our implicit simplifying assumption in Equation 1 that allosteric regulation is instantaneous.

USING I_{NCX} TO MEASURE SUBMEMBRANE $[Ca]_i$

When caffeine is applied very rapidly, as shown in FIGURES 1 and 4, inward (and outward) I_{NCX} increases faster than the global $[Ca]_i$ signal sensed by the fluorescent indicator. This is undoubtedly because the NCX senses the higher local $[Ca]_i$ near the membrane ($[Ca]_{sm}$) during Ca influx and SR Ca release versus the bulk cytosolic $[Ca]_i$. We can take advantage of this temporal difference, using I_{NCX} as a bioassay for the $[Ca]_{sm}$ sensed by the NCX molecules. FIGURE 5A shows caffeine-induced Ca transient and I_{NCX}, and FIGURE 5B plots I_{NCX} as a function of $[Ca]_i$. The thick curve shows the steady-state I_{NCX} versus $[Ca]_i$ relationship (and the shape is dictated by

Eq. 1, as verified experimentally by Weber et al.[21]). Early on, I_{NCX} is higher than expected based on $[Ca]_i$, but one can extrapolate the $[Ca]_{sm}$, which must have occurred to produce this I_{NCX} (dotted lines). Using these values, the $[Ca]_{sm}$ sensed by NCX during the caffeine-induced Ca transient can be superimposed on the global $[Ca]_i$ transient (FIG. 5C). Clearly, $[Ca]_{sm}$ increases faster and to a higher peak than $[Ca]_i$. Note that if there is a delay between the increase in $[Ca]_i$ and activation of I_{NCX}, the peak $[Ca]_{sm}$ might be higher and earlier still.

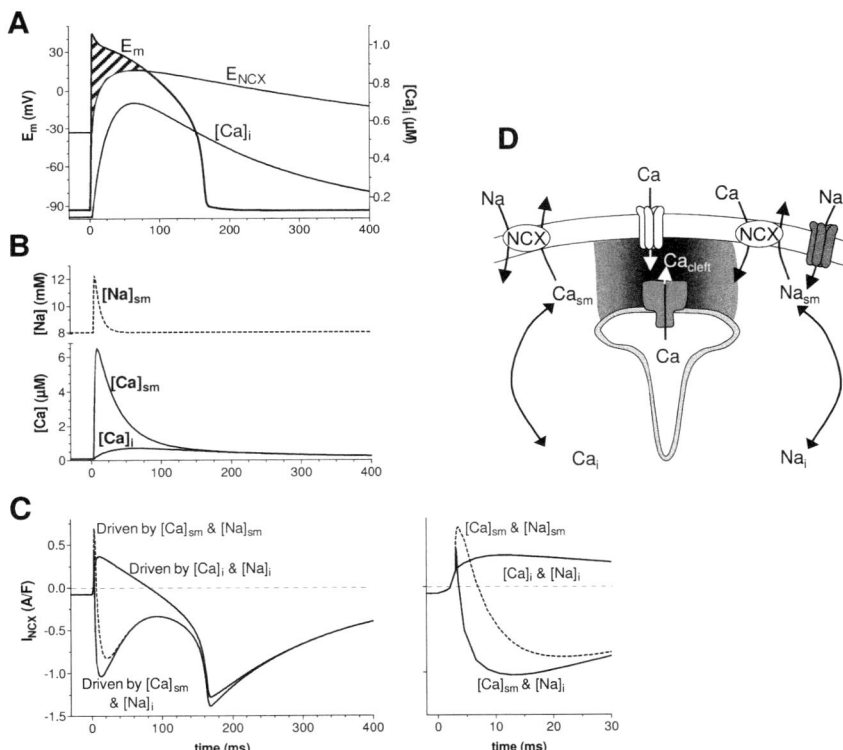

FIGURE 6. I_{NCX} during an AP. **(A)** The reversal potential for the NCX (E_{NCX}) can be used to predict the direction of I_{NCX} during an AP. When E_m is positive to E_{NCX}, outward I_{NCX} should occur. **(B)** The approximation in A does not consider that $[Na]_{sm}$ and $[Ca]_{sm}$ may be transiently higher than the values measured in the bulk cytoplasm. **(C)** When [Ca]sm is considered, I_{NCX} (defined by Eq. 1) becomes inward for most of the AP. When $[Na]_{sm}$ is also considered, there is slightly more outward I_{NCX} in the beginning of the AP. This is more obvious on an expanded timescale (*right*), but I_{NCX} will still become inward before 10 ms. **(D)** $[Ca]_{cleft}$ will be higher than $[Ca]_{sm}$, which will limit the role of outward I_{NCX} in triggering SR Ca release.

I_{NCX} DURING THE VENTRICULAR ACTION POTENTIAL

This same temporal and amplitude discrepancy for $[Ca]_{sm}$ versus bulk $[Ca]_i$ also occurs during the AP, where physiological Ca influx and SR Ca release are sensed at the membrane before they are sensed by the myofilaments.[21] FIGURE 6 shows the impact that both $[Ca]_{sm}$ and submembrane $[Na]_i$ ($[Na]_{sm}$) are expected to have on I_{NCX} during the AP.

I_{NCX} exhibits a reversal potential that is analogous to those of ion channels ($E_{NCX} = 3E_{Na} - 2E_{Ca}$, where E_{Na} and E_{Ca} are equilibrium potentials for Na and Ca). FIGURE 6A shows a rabbit ventricular AP, Ca transient, and E_{NCX}. At rest, $E_m < E_{NCX}$, so Ca extrusion is favored (initial inward I_{NCX} in FIG. 6C). Early in the AP, the E_m exceeds E_{NCX}, which would drive Ca entry via outward I_{NCX}, until $E_m = E_{NCX}$ during repolarization. Note that E_{NCX} changes because $[Ca]_i$ (and hence E_{Ca}) changes. Upon AP repolarization, the negative E_m and high $[Ca]_i$ drive a large inward I_{NCX}, and this reflects Ca extrusion from the cell. This simple expectation is based on global $[Ca]_i$ and $[Na]_i$, but possible elevations in $[Ca]_{sm}$ and $[Na]_{sm}$ (FIG. 6B) complicate this picture.[23–25] Although $[Ca]_{sm}$ may not get as high as cleft $[Ca]_i$ during I_{Ca} and SR Ca release, high $[Ca]_{sm}$ causes I_{NCX} to become inward very early in the AP duration, such that very little Ca enters via I_{NCX}. An I_{Na}-induced increase in $[Na]_{sm}$ during the AP might delay the reversal of I_{NCX} to inward (FIG. 6C), but outward I_{NCX} would still last for only less than 4 ms (with little integrated Ca entry). Thus, under physiological conditions NCX exchange works mainly in the Ca extrusion mode, driven mainly by the Ca transient. However, positive E_m during the AP plateau can limit Ca extrusion.

Although Na/Ca exchange normally may work mainly in the Ca efflux mode, the amount of Ca influx via I_{NCX} can be increased greatly if $[Na]_i$ is elevated. This can occur if the Na/K-ATPase is blocked by digitalis glycosides or if Na influx is increased. Ca influx via NCX is also more strongly favored if SR Ca release or I_{Ca} are inhibited, or if AP duration is prolonged.[26]

CONCLUSIONS

The stoichiometry of the NCX has historically been determined to be 3, but recent and intriguing new data suggest that the stoichiometry may be 4 in excised patches[11] or even 2 with extracellular acidificaiton.[12] However, as discussed above, most experiments under physiological conditions unequivocally support a stoichiometry very close to 3 (allowing us to make some mathematical descriptions of I_{NCX}). We have shown that allosteric regulation by $[Ca]_i$ occurs rapidly with a $K_{mCaAct} = 125$ nM in intact cardiac myocytes, making this allosteric regulation relevant during the cardiac cycle. This regulation also allows I_{NCX} to optimally control diastolic $[Ca]_i$ (e.g., prevent $[Ca]_i$ from going too low or too high). This keeps diastolic $[Ca]_i$ poised near the threshold of contractile activation but prevents diastolic stiffness.

Our third main point concerns $[Ca]_{sm}$, which differs from bulk $[Ca]_i$ and has a more direct role in governing the direction of I_{NCX} during the AP. Normally, I_{NCX} extrudes Ca for most of the cardiac AP (FIG. 6). However, under certain pathophysiological conditions (e.g., heart failure) in which $[Na]_i$ is elevated, SR Ca release is reduced, or AP duration is prolonged, there will be greater Ca influx via NCX.[29]

NCX molecules that are located further from the diadic cleft will sense lower $[Ca]_{sm}$ (allowing more local Ca entry via I_{NCX}), and those closer to the diadic cleft will sense higher $[Ca]_{sm}$ than average (and be carrying more exclusively inward I_{NCX}). Although I_{NCX}, on its own, is a less efficient release trigger than I_{Ca},[27] outward I_{NCX} might still contribute to EC coupling gain by (1) slowing diffusion of Ca from the cleft or (2) activating release if no local I_{Ca} flows. Thus, the predominant role of I_{NCX} under physiological conditions is in the inward direction to extrude Ca from the cytosol, contributing to relaxation.

[*Note added in proof*: After this manuscript was revised, a new very careful study of NCX stoichiometry in a heterologous expression system (HEK-293 cells) was published.[30] Those data provide additional strong evidence in favor of the 4:1 stoichiometry for the expressed cardiac NCX. While we still consider this issue incompletely resolved, this emphasizes the importance of further tests of the NCX stoichiometry, especially with respect to function in cardiac myocytes.]

ACKNOWLEDGMENTS

This work was supported by National Institutes of Health Grants HL-30077 and HL-64724 (D.M.B.) and American Heart Association Predoctoral Fellowship 0010180Z (C.R.W.). The authors thank Drs. Kenneth S. Ginsburg, Kenneth D. Philipson, and Thomas R. Shannon for contributions to this work.

REFERENCES

1. BERS, D.M. 2001. Excitation–Contraction Coupling and Cardiac Contractile Force. Kluwer Academic Publishers. Dordrecht/Boston.
2. BASSANI, J.W., R.A. BASSANI & D.M. BERS. 1994. Relaxation in rabbit and rat cardiac cells: species-dependent differences in cellular mechanisms. J. Physiol. **476:** 279–293.
3. EISNER, D.A., H.S. CHOI, M.E. DIAZ, et al. 2000. Integrative analysis of calcium cycling in cardiac muscle. Circ. Res. **87:** 1087–1094.
4. HASENFUSS, G., W. SCHILLINGER, S.E. LEHNART, et al. 1999. Relationship between Na^+-Ca^{2+}-exchanger protein levels and diastolic function of failing human myocardium. Circulation **99:** 641–648.
5. POGWIZD, S.M., K. SCHLOTTHAUER, L. LI, et al. 2001. Arrhythmogenesis and contractile dysfunction in heart failure: roles of sodium–calcium exchange, inward rectifier potassium current, and residual beta-adrenergic responsiveness. Circ. Res. **88:** 1159–1167.
6. POGWIZD, S.M., M. QI, W. YUAN, et al. 1999. Upregulation of Na^+/Ca^{2+} exchanger expression and function in an arrhythmogenic rabbit model of heart failure. Circ. Res. **85:** 1009–1019.
7. REEVES, J.P. & C.C. HALE. 1984. The stoichiometry of the cardiac sodium–calcium exchange system. J. Biol. Chem. **259:** 7733–7739.
8. SCHLOTTHAUER, K. & D.M. BERS. 2000. Sarcoplasmic reticulum Ca^{2+} release causes myocyte depolarization. Underlying mechanism and threshold for triggered action potentials. Circ. Res. **87:** 774–780.
9. PHILIPSON, K.D. & D.A. NICOLL. 2000. Sodium–calcium exchange: a molecular perspective. Annu. Rev. Physiol. **62:** 111–133.
10. BLAUSTEIN, M.P. & W.J. LEDERER. 1999. Sodium/calcium exchange: its physiological implications. Physiol. Rev. **79:** 763–854.

11. FUJIOKA, Y., M. KOMEDA & S. MATSUOKA. 2000. Stoichiometry of Na^+-Ca^{2+} exchange in inside-out patches excised from guinea-pig ventricular myocytes. J. Physiol. **523:** 339–351.
12. EGGER, M. & E. NIGGLI. 2000. Paradoxical block of the Na^+-Ca^{2+} exchanger by extracellular protons in guinea-pig ventricular myocytes. J. Physiol. **523**(2): 353–366.
13. BRIDGE, J.H., J.R. SMOLLEY & K.W. SPITZER. 1990. The relationship between charge movements associated with I_{Ca} and I_{Na-Ca} in cardiac myocytes. Science **248:** 376–378.
14. HOVE-MADSEN, L. & D.M. BERS. 1993. Passive Ca buffering and SR Ca uptake in permeabilized rabbit ventricular myocytes. Am. J. Physiol. **264:** C677–C686.
15. SATOH, H., L.M. DELBRIDGE, LA. BLATTER & D.M. BERS. 1996. Surface:volume relationship in cardiac myocytes studied with confocal microscopy and membrane capacitance measurements: species-dependence and developmental effects. Biophys. J. **70:** 1494–1504.
16. TRAFFORD, A.W., M.E. DIAZ & D.A. EISNER. 1999. A novel, rapid and reversible method to measure Ca buffering and time-course of total sarcoplasmic reticulum Ca content in cardiac ventricular myocytes. Pflugers Arch. Eur. J. Physiol. **437:** 501–503.
17. GINSBURG, K.S., C.R. WEBER, S. DESPA & D.M. BERS. 2002. Simultaneous measurement of $[Na]_i$, $[Ca]_i$, and I_{NCX} in intact cardiac myocytes. Ann. N. Y. Acad. Sci. This volume.
18. MIURA, Y. & J. KIMURA. 1989. Sodium–calcium exchange current. Dependence on internal Ca and Na and competitive binding of external Na and Ca. J. Gen. Physiol. **93:** 1129–1145.
19. HILGEMANN, D.W., A. COLLINS & S. MATSUOKA. 1992. Steady-state and dynamic properties of cardiac sodium-calcium exchange. Secondary modulation by cytoplasmic calcium and ATP. J. Gen. Physiol. **100:** 933–961.
20. WEBER, C.R., K.S. GINSBURG, K.D. PHILIPSON, et al. 2001. Allosteric regulation of Na/Ca exchange current by cytosolic Ca in intact cardiac myocytes. J. Gen. Physiol. **117:** 119–131.
21. WEBER, C.R., V. PIACENTINO III, K.S. GINSBURG, et al. 2002. Na/Ca exchange current and submembrane [Ca] during the cardiac action potential. Circ. Res. **90:** 182–189.
22. MAXWELL, K., J. SCOTT, A. OMELCHENKO, et al. 1999. Functional role of ionic regulation of Na^+/Ca^{2+} exchange assessed in transgenic mouse hearts. Am. J. Physiol. **277:** H2212–H2221.
23. LEBLANC, N. & J.R. HUME. 1990. Sodium current-induced release of calcium from cardiac sarcoplasmic reticulum. Science **248:** 372–376.
24. LEDERER, W.J., E. NIGGLI & R.W. HADLEY. 1990. Sodium–calcium exchange in excitable cells: fuzzy space. Science **248:** 283.
25. LIPP, P. & E. NIGGLI. 1994. Sodium current-induced calcium signals in isolated guinea-pig ventricular myocytes. J. Physiol. **474:** 439–446.
26. DIPLA, K., J.A. MATTIELLO, K.B. MARGULIES, et al. 1999. The sarcoplasmic reticulum and the Na^+/Ca^{2+} exchanger both contribute to the Ca^{2+} transient of failing human ventricular myocytes. Circ. Res. **84:** 435–444.
27. SIPIDO, K.R., M. MAES & F. VAN DE WERF. 1997. Low efficiency of Ca^{2+} entry through the Na^+-Ca^{2+} exchanger as trigger for Ca^{2+} release from the sarcoplasmic reticulum. A comparison between L-type $Ca2^+$ current and reverse-mode Na^+-Ca^{2+} exchange. Circ. Res. **81:** 1034–1044.
28. TRAFFORD, A.W., M.E. DÍAZ, S.C. O'NEILL & D.A. EISNER. 1995. Comparison of subsarcolemmal and bulk calcium concentration during spontaneous calcium release in rat ventricular myocytes. J. Physiol. **488:** 577–586.
29. DESPA, S., M.A. ISLAM, S.M. POGWIZD & D.M. BERS. 2002. Intracellular Na^+ concentration is elevated in heart failure, but Na/K-pump function is unchanged. Circulation **105:** 2543–2548.
30. DONG, H., J. DUNN & J. LYTTON. 2002. Stoichiometry of the cardiac Na^+/Ca^{2+} exchanger NCS1.1 measured in transfected HEK cells. Biophys. J. **82:** 1943–1952.

Pharmacology of Na^+/Ca^{2+} Exchanger

JUNKO KIMURA,[a] YASUHIDE WATANABE,[b] LIBING LI,[a] AND TOMOKAZU WATANO[a]

[a]*Department of Pharmacology, School of Medicine, Fukushima Medical University, Fukushima, Japan*

[b]*Department of Ecology and Clinical Therapeutics, School of Nursing, Fukushima Medical University, Fukushima, Japan*

ABSTRACT: KB-R7943 inhibits the Na^+/Ca^{2+} exchanger in an independent manner or in a manner dependent on the direction of the current. This effect may be due to the experimental protocols bawed on the competition between the drug and external substrate ions. Some antiarrhythmic drugs inhibit NCX. A new column of NCX was added in Sicilian Gambit.

KEYWORDS: Na^+/Ca^{2+} exchanger; NCX pharmacology; KBR7943; antiarrhythmic agents; amiodarone

INTRODUCTION

In the previous international conference on Na^+/Ca^{2+} exchange (NCX), there were very few presentations on the pharmacology of the Na^+/Ca^{2+} exchanger.[1] However, a variety of drugs and compounds have been reported to inhibit NCX since then. Using the whole-cell voltage clamp technique, we have reported that KB-R7943,[2-4] 2,3-butanedione monoxime,[5] and some antiarrhythmic drugs[6-9] block NCX current (I_{NCX}) of guinea pig cardiac ventricular cells. Here, we review KB-R7943 and some antiarrhythmic agents that inhibit I_{NCX}.

KB-R7943

In 1996, we reported that KB-R7943 (2-[2-[4-(4-nitrobenzyloxy)phenyl]ethyl]isothiourea methanesulfonate) inhibits I_{NCX} in a direction-dependent manner in guinea pig ventricular cells.[2] In this experiment, a unidirectional outward I_{NCX} was induced by raising external Ca^{2+} from 0 to 1 mM in the presence of 20 mM Na^+ in the pipette solution. A unidirectional inward I_{NCX} was induced by raising external Na^+ from 0 (replaced by 140 mM Li^+) to 140 mM in the presence of 433 nM free Ca^{2+}

Address for correspondence: Junko Kimura, Department of Pharmacology, Fukushima Medical University, 1 Hikariga-Oka, Fukushima 960-1295, Japan. Voice: 81-24548-2111, ext. 2150; fax: 81-24548-0575.
jkimura@fmu.ac.jp
Yasuhide Watanabe's current address: Department of Pathophysiology, Basic Nursing, Hamamatsu University School of Medicine, Hamamatsu 431-3192, Japan.

in the pipette solution. BAPTA 20 mM was present in the pipette solution to avoid current oscillation due to intracellular Ca^{2+} fluctuations. KB-R7943 inhibited the outward I_{NCX} more potently than inward I_{NCX} with IC_{50} of 0.3 and 17 µM, respectively.

The bidirectional I_{NCX} was recorded as a steady current over the voltage range between 30 and −150 mV with a reversal potential at around the holding potential of −60 mV.[4] The solutions contained Na^+ and Ca^{2+} in both the external and pipette so-

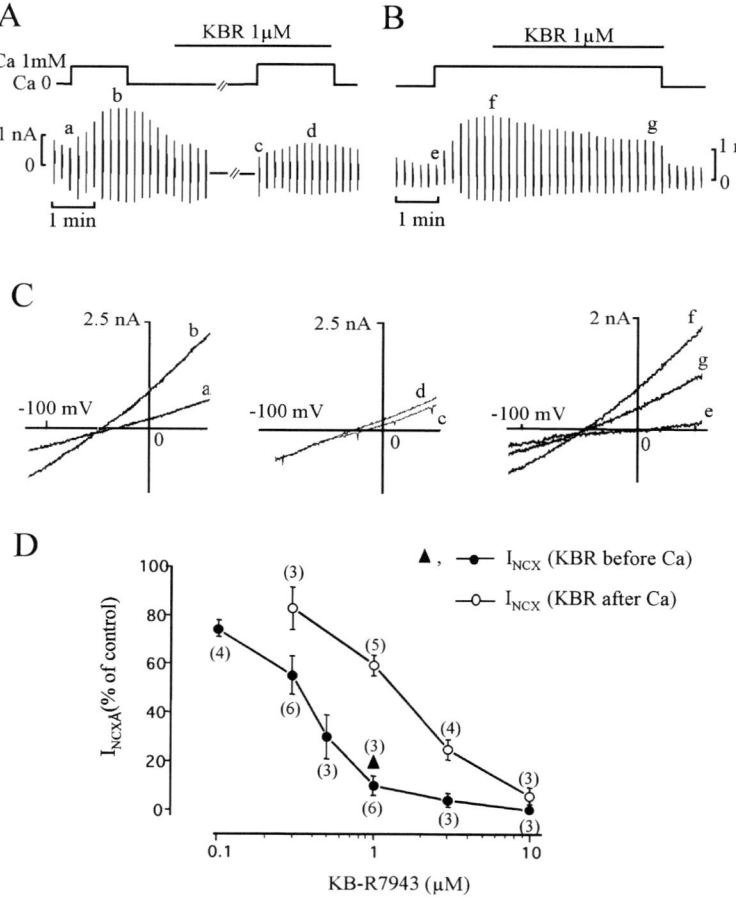

FIGURE 1. Effect of the order of application between KB-R7943 (KBR) and external Ca^{2+}. **(A)** Chart recordings of the current. External Ca^{2+} was increased to 1 mM to induce I_{NCX}. KBR was added before application of second Ca^{2+}. **(B)** KBR was added in the presence of 1 mM Ca^{2+}. **(C)** I–V curves of the corresponding labels in **(A)** and **(B)**. **(D)** Concentration-inhibition curves of KBR on bidirectional I_{NCX} with KBR applied before (*triangles*) and after (*open circles*) application of external Ca^{2+}. I_{NCX} was measured at 50 mV. A concentration-inhibition curve (*filled circles*) was of the unidirectional outward I_{NCX} from Figure 3 in Ref. 2. The inhibitory effect of 1 µM KBR is similar between unidirectional and bidirectional I_{NCX}.

lutions throughout the experiment. KB-R7943 inhibited both directions of bidirectional I_{NCX} equally with IC_{50} of 1 μM.[4]

Why does KB-R7943 inhibit unidirectional I_{NCX} in a direction-dependent manner and bidirectional I_{NCX} in a direction-independent manner? We previously showed that KB-R7943 is a competitive inhibitor for external Ca^{2+}.[2] Therefore, we speculated that the direction-dependent block of the drug might be related to external Ca^{2+} concentration. To test this possibility, we compared two different protocols. First, when 1 μM KB-R7943 was applied at 0 Ca^{2+}, it inhibited I_{NCX} by approximately 90%, which was induced by subsequent application of 1 mM external Ca^{2+} (FIG. 1A and C). In contrast, when KB-R7943 was applied after application of 1 mM Ca^{2+}, 1 μM KB-R7943 inhibited I_{NCX} by only approximately 50% (FIG. 1B and C). The concentration–response curves are shown in FIGURE 1D. The IC_{50} value shifted from 0.3 to 1.5 μM by changing the order of application of Ca^{2+} and KB-R7943. When KB-R7943 was applied before Ca^{2+}, the IC_{50} value was 0.3 μM. When KB-R7943 was applied in the presence of Ca^{2+}, the curve shifted to the right with the IC_{50} of 1.5 μM. This is consistent with our previous result using 1.8 mM external Ca^{2+}.[3] This suggests that the direction-dependent inhibition by KB-R7943 may be caused by the experimental conditions.

FIGURE 2 shows a schematic model of NCX. An outward I_{NCX} was induced by raising external Ca^{2+} from 0 to 1 mM. KB-R7943 is a competitive inhibitor for external Ca^{2+}.[10] The affinity of the exchanger for Na^+ is low. In the absence of external Ca^{2+}, the exchangers are available for KB-R7943 binding. Therefore, subsequent application of Ca^{2+} could induce less current and show apparently potent inhibition by KB-R7943. In contrast, an inward I_{NCX} was induced by raising external Na^+ from 0 to 140 mM. In the absence of external Na^+, the exchanger is available for Ca^{2+} to bind, because external Ca^{2+} and Na^+ are competitive.[10] Recently, we found that KB-R7943 is competitive with external Na^+ (unpublished observation). Therefore, in the absence of external Na^+, the affinity of Ca^{2+} is high, and thus KB-R7943 is less potent for inhibiting inward I_{NCX}. For bidirectional NCX in the presence of both Na^+ and Ca^{2+} in the external solution, the effect of KB-R7943 is similar to that when it is applied after Ca^{2+}. Thus, our results in FIGURE 1 and in Li and Kimura[11] in this

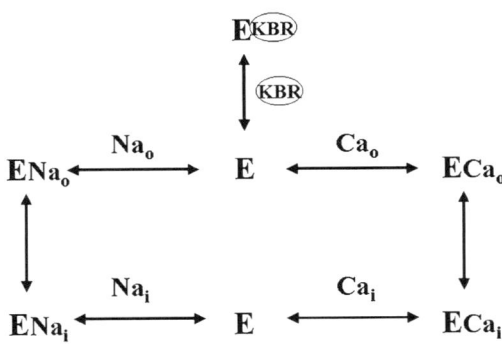

FIGURE 2. A consecutive model of NCX with a competitive inhibitor of KB-R7943.

issue suggest that the direction-dependent inhibition of I_{NCX} by externally applied KB-R7943 is caused by experimental protocols.

Recently, using inside-out giant patch membrane, Elias et al.[12] reported that KB-R7943 inhibits I_{NCX} in a direction-dependent manner and outward I_{NCX} more potently than inward I_{NCX}. Therefore, the competition with the external ions cannot explain the direction dependency. Elias et al.[12] claimed KB-R7943 as a noncompetitive inhibitor. If so, KB-R7943 could bind either ENa^+ or ECa^{2+} form of the exchanger. Because $[Ca^{2+}]_i$ could induce inward I_{NCX} in the presence of KB-R7943 in their preparation, we speculate that the drug might bind $ENa^+{}_i$ form of the exchanger. Iwamoto et al.[13] suggested that Gly833 in the α-2 reentrant loop is a residue most responsible for KB-R7943 binding, and that this residue may be accessible from both external and cytoplasmic sides. Therefore, we speculate that externally applied KB-R7943 may be competitive with the external substrate ions. Cytoplasmic KB-R7943 may bind the exchanger bound with $Na^+{}_i$. KB-R7943 cannot bind to NCX which already was bound with $Ca^{2+}{}_i$. Further experiments are necessary to elucidate the mechanism of the effect of KB-R7943.

SICILIAN GAMBIT AND NCX

The Sicilian Gambit[14] is a table of antiarrhythmic drugs showing the effects of each drug on different channels, receptors, and pumps. There are also columns showing clinical and electrocardiogram effects of drugs. The column of Na^+ channel is divided into fast, medium, and slow kinetics of the drug. There is a column of pumps that contains only the Na-K pump, and there is no column for NCX. Recently, we found that several antiarrhythmic drugs inhibit NCX.[6–9] Therefore, we were challenged to add a column for NCX.

Among the antiarrhythmic agents, amiodarone[6] was the first tested because it inhibits multiple channels and receptors and is the least selective drug on the Sicilian Gambit. Another reason is that amiodarone currently is regarded as the most promising drug for treating severe life-threatening arrhythmia.[15] We found that amiodarone inhibited NCX current of guinea pig cardiac ventricular cells with an IC_{50} value of 3.3 μM and a Hill coefficient of 1.[6]

We explored the binding site of amiodarone on the exchanger with the whole-cell voltage clamp technique. We first tested trypsin in the pipette solution because intracellular trypsin treatment abolishes various modulation of NCX such as Na^+-dependent inactivation,[16,17] Ca^{2+}-dependent activation,[17] and intracellular pH sensitivity.[18] The sites of trypsin proteolysis are in the large internal domain of the exchanger.[19] Trypsin treatment decreased the blocking effect of amiodarone on the NCX. We further explored this by using cells expressing NCX1 and several of its mutations.

We compared the effects of amiodarone on NCX current in guinea pig myocytes and CCL39 cells expressing NCX3, wild-type NCX1, deletion mutants of the first and second internal loops, deletion of the large internal domain including the XIP region or excluding XIP, and deletion of the COOH terminus. Only the mutants involving deletions in the large internal domain showed significantly different sensitivity to amiodarone. This result indicates that the large internal domain is involved in the blocking effect of amiodarone. However, the decrease in the blocking effect was not so robust. Therefore, other parts of the molecule also may be involved.

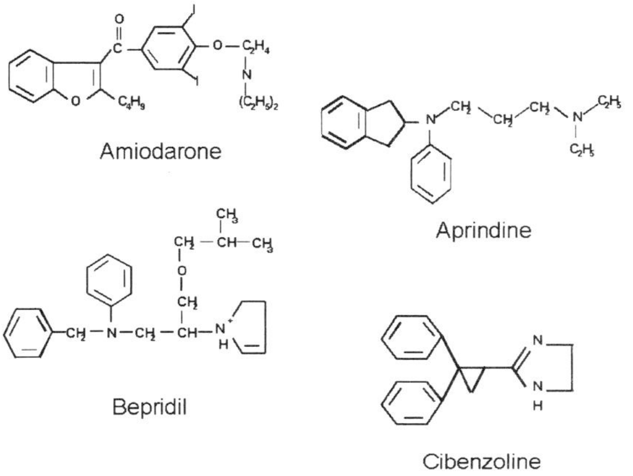

FIGURE 3. Chemical structures of antiarrhythmic drugs that inhibit the Na$^+$/Ca^{2+} exchanger.

Sicilian Gambit

DRUGS	CHANNELS			RECEPTORS			PUMPS	
	Na	Ca	K	α	β	M$_2$	Na/K	Na/Ca
lidocaine	○							x
procaineamide	○		○					x
disopyramide	○		○			○		x
quinidine	○		○	○		○		x
propafenone	○				○			x
aprindine	●	●	●					(●)
cibenzoline	●	●	●			●		(●)
pirmenol	○		○			○		x
flecainide	○		○					x
bepridil	●	●	●					(●)
verapamil	○	○		○				x
diltiazem		○						x
sotalol			○		○			x
amiodarone	●	●	●	●	●			●

FIGURE 4. Sicilian Gambit with a column of Na$^+$/Ca^{2+} exchanger. *Open circles* indicate inhibitory effects. *Closed circles* indicate effects of drugs that inhibit NCX; (x) those that do not inhibit NCX at 100 μM. Closed circles in parentheses indicate that the concentration range for NCX inhibition was higher than the therapeutic concentration range. Columns of subdivision of Na$^+$ channel and those of clinical and electrocardiogram effects are omitted.

In addition to amiodarone, we found that three other antiarrhythmic drugs, that is, bepridil,[7] aprindine,[8] and cibenzoline,[20] also inhibited I_{NCX} with IC_{50} values 8, 49, and 95 µM. Chemical structures of the four drugs are shown in FIGURE 3. Amiodarone was the only drug that inhibited I_{NCX} within a therapeutic concentration range of 0.1–10 µM. The therapeutic concentrations of bepridil, aprindine, and cibenzoline were in the ranges of 0.5–5, 2.5–7, and 1–4 µM, respectively. The Hill coefficient for all of the four drugs was 1. The drugs that showed trypsin sensitivity were amiodarone and bepridil. Blocking effects of aprindine and cibenzoline were not affected by intracellular treatment of trypsin, indicating that amiodarone and bepridil affect NCX from the intracellular side of the membrane, but that aprindine and cibenzoline affect it from the external side or from inside the membrane.

FIGURE 4 shows Sicilian Gambit with the addition of a new column for the Na/Ca exchanger together with that for the Na-K pump under the name of "pump." A common feature of the antiarrhythmic drugs that inhibit NCX is that they inhibit all three Na^+, Ca^{2+}, and K^+ channels. We tested all the other drugs in FIGURE 4, but they did not affect I_{NCX} at a concentration of 100 µM.

Drugs that are not antiarrhythmic drugs but inhibit NCX include KB-R7943, Ni^{2+}, and butanedione monoxime (BDM).[5] BDM is also trypsin sensitive, but KB-R7943 and Ni^{2+} are trypsin insensitive. More recently, two other compounds, SEA0400[21] and SN-6,[22] are reported to be selective NCX inhibitors. Many more inhibitors will be found in the near future before the next international NCX meeting.

ACKNOWLEDGMENTS

This work was supported by Grants-in-Aid from Japan Foundation for Promotion of Sciences (11670096, 11357020). We thank Dr. T. Ono and Ms. Sanae Sato for valuable technical assistance. We also thank Ryohei Otani, fourth-year medical student, for joining the experiment.

REFERENCES

1. NOBLE, D. 1996. Commentary. Ann. N.Y. Acad. Sci. **779:** 1–6.
2. WATANO, T., J. KIMURA, T. MORITA & H. NAKANISHI. 1996. A novel antagonist, No. 7943, of the Na^+/Ca^{2+} exchange current in guinea-pig cardiac ventricular cells. Br. J. Pharmacol. **119:** 555–563.
3. WATANO, T. & J. KIMURA. 1998. Calcium-dependent inhibition of the sodium-calcium exchange current by KB-R7943. Can. J. Cardiol. **14:** 259–262.
4. KIMURA, J., T. WATANO, M. KAWAHARA, et al. 1999. Direction-independent block of bi-directional Na^+/Ca^{2+} exchange current by KB-R7943 in guinea-pig cardiac myocytes. Br. J. Pharmacol. **128:** 969–974.
5. WATANABE, Y., T. IWAMOTO, I. MATSUOKA, et al. 2001. Inhibitory effect of 2,3-butanedione monoxime (BDM) on Na^+/Ca^{2+} exchange current in guinea-pig cardiac ventricular myocytes. Br. J. Pharmacol. **132:** 1317–1325.
6. WATANABE, Y. & J. KIMURA. 2000. Inhibitory effect of amiodarone on Na^+/Ca^{2+} exchange current in guinea-pig cardiac myocytes. Br. J. Pharmacol. **131:** 80–84.
7. WATANABE, Y. & J. KIMURA. 2001. Blocking effect of bepridil on Na^+/Ca^{2+} exchange current in guinea-pig cardiac ventricular myocytes. Jpn. J. Pharmacol. **85:** 370–375.
8. WATANABE, Y., T. IWAMOTO, M. SHIGEKAWA & J. KIMURA. 2002. Inhibitory effect of aprindine on Na^+/Ca^{2+} exchange current in guinea-pig cardiac ventricular myocytes. Br. J. Pharmacol. **136:** 361–366.

9. WATANABE, Y., R. OTANI, & J. KIMURA. 2002. Na^+/Ca^{2+} exchange current inhibitory drugs in Sicilian Gambit. Jpn. J. Pharmacol. **88:** Suppl. I. 273P.
10. MIURA, Y. & J. KIMURA. 1989. Sodium–calcium exchange current. Dependence on internal Ca and Na and competitive binding of external Na and Ca. J. Gen. Physiol. **93:** 1129–1145.
11. LI, L. & J. KIMURA. 2002. Effect of KB-R7943 on oscillatory Na^+/Ca^{2+} exchange current in guinea-pig ventricular myocytes. Ann. N.Y. Acad. Sci. This volume.
12. ELIAS, C.L., A. LUKAS, S. SHURRAW, et al. 2001. Inhibition of Na^+/Ca^{2+} exchange by KB-R7943: transport mode selectivity and antiarrhythmic consequences. Am. J. Physiol. **281:** H1334–H1345.
13. IWAMOTO, T., S. KITA, A. UEHARA, et al. 2001. Structural domains influencing sensitivity to isothiourea derivative inhibitor KB-R7943 in cardiac Na^+/Ca^{2+} exchanger. Mol. Pharmacol. **59:** 524–531.
14. TASK FORCE OF THE WORKING GROUP ON ARRHYTHMIAS OF THE EUROPEAN SOCIETY OF CARDIOLOGY. 1991. The Sicilian Gambit. A new approach to the classification of antiarrhythmic drugs based on their actions on arrhythmogenic mechanisms. Circulation **84:** 1831–1851.
15. KODAMA, I., K. KAMIYA & J. TOYAMA. 1997. Cellular electropharmacology of amiodarone. Cardiovasc. Res. **35:** 13–29.
16. HILGEMANN, D.W. 1990. Regulation and deregulation of cardiac Na^+-Ca^{2+} exchange in giant excised sarcolemmal membrane patches. Nature **344:** 242–245.
17. MATUSOKA, S. & D.W. HILGEMANN. 1994. Inactivation of outward Na-Ca exchange current in guinea-pig ventricular myocytes. J. Physiol. **476:** 443–458.
18. DOERING, A.E. & W.J. LEDERER. 1993. The mechanism by cytoplasmic protons inhibit the sodium–calcium exchanger in guinea-pig heart cells. J. Physiol. **476:** 443–458.
19. CHEN, M., Z. ZHANG, M. TAWIAH-BOATENG & P.M.D. HARDWICKE. 2000. A Ca^{2+}-dependent tryptic cleavage site and a protein kinase A phosphorylation site are present in the Ca^{2+} regulatory domain of scallop muscle Na^+-Ca^{2+} exchanger. J. Biol. Chem. **275:** 22961–22968.
20. WATANABE, Y. & J. KIMURA. 2002. Effects of cibenzoline on Na^+/Ca^{2+} exchange current in guinea-pig cardiac ventricular cells. Folia Pharmacol. Jpn. **119:** 2P (abstract in Japanese).
21. MATSUDA, T., N. ARAKAWA, K. TAKUMA, et al. 2001. SEA0400, a novel and selective inhibitor of the Na^+-Ca^{2+} exchanger, attenuates reperfusion injury in the in vitro and in vivo cerebral ischemic models. J. Pharm. Exp. Ther. **298:** 249–256.
22. HOTTA, Y., X. LU, M. YAJIMA, et al. 2002. Protective effect of SN-6, a selective Na^+-Ca^{2+} exchange inhibitor, on ischemia-reperfusion-injured hearts. Jpn. J. Pharmacol. **88:** Suppl. I, 257P.

Functional Consequences of Na/Ca Exchanger Overexpression in Cardiac Myocytes

CESARE TERRACCIANO

Imperial College of Science, Technology and Medicine, Faculty of Medicine, National Heart and Lung Institute, London, UK

ABSTRACT: Several factors are important in the relationship between Na/Ca exchanger overexpression and Ca cycling in physiological and pathophysiological situations. First, there are species differences. Transgenic mouse cardiac myocytes overexpressing Na/Ca exchanger showed a faster Ca transient associated with increased sarcoplasmic reticulum (SR) Ca content compared with wild-type myocytes. Cultured rabbit cardiac myocytes overexpressing Na/Ca exchanger showed reduced amplitude of the Ca transient and reduced SR Ca content. Second, the activity of other Ca regulatory proteins have to be considered. When Ca uptake via SR Ca ATPase (SERCA) was reduced by thapsigargin in transgenic mouse myocytes overexpressing Na/Ca exchanger, the time course of the Ca transient was slowed, and the SR Ca content was reduced to wild-type mouse myocyte levels, suggesting a potential compensatory role of Na/Ca exchanger overexpression when SERCA function is reduced. Finally, there are confounding factors related to the pathophysiological conditions. Our results suggest that Na/Ca exchanger overexpression can compensate for defects in SR Ca uptake in mouse myocytes. The consequences of Na/Ca exchanger overexpression in other species and conditions is unpredictable and require further investigation.

KEYWORDS: Na/Ca exchanger; sarcoplasmic reticulum; Ca cycling; cardiac myocytes

INTRODUCTION

In cardiac hypertrophy and failure, there are defects in Ca handling resulting in smaller and slower Ca transients.[1–3] Concomitant changes in Ca regulatory protein expression occur, noticeably a reduction in the sarcoplasmic reticulum (SR) Ca ATPase (SERCA) and an overexpression of the Na/Ca exchanger.[4,5] Reduction in SERCA expression, if associated with reduced function, can explain systolic and diastolic dysfunction. Restoring normal SERCA expression has be shown to be beneficial. When SERCA expression was increased in failing cardiac myocytes by adenoviral transfection, both contraction and relaxation were faster.[6]

Address for correspondence: Cesare Terracciano, National Heart & Lung Institute, Imperial College School of Medicine, Cardiac Medicine, Dovehouse Street, London SW3 6LY, U.K. Voice: +44-207-3518142; fax +44-207-351-8145.
c.terracciano@ic.ac.uk

What is then the functional role of overexpression of the Na/Ca exchanger in heart failure? Is this a factor contributing to the failing events, or a compensatory mechanism to balance dysfunction of other mechanisms? This point has been addressed by several investigators with contrasting results.[7–9]

The reasons behind this confusing picture are related to the complexity of action of the Na/Ca exchanger, to the ability of the exchanger to reverse, to competition with other Ca regulatory mechanisms, and to other factors (such as β-adrenergic stimulation) that directly or indirectly influence membrane potential and transmembrane gradients of [Na] and [Ca]. The stage of the disease also can be important. Ultimately, the effects of overexpression will depend on the contribution of Na/Ca exchanger to E-C coupling in that particular situation. This contribution varies, for example, in different species.[10]

The function of the Na/Ca exchanger in intact cells in normal, physiological conditions is poorly understood for several reasons. For example, there is a lack of specific pharmacological agents for the exchanger. There is also a complex interaction between the Na/Ca exchanger and other mechanisms, such as SR Ca uptake. For these reasons, the overexpression studies are valuable. Increasing the number of Na/Ca exchangers in the cell produces a direction of change in functional parameters that indicate, at least in qualitative terms, the direction of the exchanger contribution to function. But there are several important assumptions to make: First, the change in expression is functional; that is, the overexpressed exchangers have the same functional features of the native exchangers. Second, the spatial relationship between Na/Ca exchangers and other cellular mechanisms are not modified. Finally, regulating factors for Na/Ca exchanger function are not different. Note that these points are unresolved.

It is clear that the study of the functional consequences of Na/Ca exchanger overexpression needs to be done by performing experiments in several specific conditions and comparing the results. We have studied Na/Ca exchanger overexpression in isolated cardiac myocytes obtained in two different species and when SERCA function was altered. More studies on specific pathophysiological conditions are required.

METHODS

In all the experiments described below, we used two different technical approaches to achieve a minimal perturbation of intracellular conditions. This is an essential requirement when the Na/Ca exchanger contribution to ion regulation is studied.

Cytoplasmic [Ca] was determined using the acetoxymethylester form of the Ca-sensitive fluorescent indicator indo-1 (AM; Molecular Probes, Eugene, OR) with a technique previously described.[11,12] Myocytes were loaded for 10 minutes in Dulbecco MEM (Gibco, Grand Island, NY) solution containing 10 µM indo-1 AM. The supernatant was removed, and the cells were resuspended in fresh Dulbecco MEM. Calcium transients also were measured in field-stimulated cells by using a pair of platinum electrodes placed in the superfusing bath.

Electrophysiological experiments were performed in switch-clamp mode using an Axoclamp-2B system (Axon Instruments, Burlingame, CA). High-resistance microelectrodes (20–30 MΩ) to minimize intracellular dialysis were used. The microelectrode filling solution contained KCl, 2M; EGTA, 0.1 mM; HEPES, 5 mM; pH 7.2. The switching rate was 3–5 kHz. Gain was increased up to 0.8 nA/mV to obtain a maximally square voltage trace with no oscillations. Current clamp and voltage clamp protocols were controlled with pClamp 7 software (Axon Instruments).

Normal Tyrode (NT) solution contained NaCl, 140 mM; KCl, 6 mM; $MgCl_2$, 1 mM; $CaCl_2$, 1 mM; glucose, 10 mM; HEPES, 10 mM; pH to 7.4 with 2 M NaOH. Chemicals were purchased from BDH (Poole, U.K.). Caffeine (20 mM) was added as solid to the final solution.

The temperature of the superfusing solution was approximately 37°C. The rate of superfusion was 2–3 ml·min^{-1} except during fast application of caffeine when it was 12–15 ml·min^{-1}. Miniature solenoid valves (The Lee Company, Essex, UK) were used to produce fast changes in the superfusate.

A MOUSE MODEL OF Na/Ca EXCHANGER OVEREXPRESSION

The first important example of Na/Ca exchanger overexpression was made available by Philipson et al.[13] These investigators produced a transgenic mouse line in which the transgenic construct consisted of the open reading frame of the canine Na/Ca exchanger under the control of the α-myosin heavy chain promoter.

Note that the first studies were performed on cardiac myocytes from heterozygous mice.[13,14] Nontransgenic littermates were used as control. More recently, homozygous Na/Ca exchanger overexpressing mice were made available, and they have shown functional differences with the heterozygous mice (unpublished observations). More information on the consequences of this genotype on Ca regulation is needed.

We performed experiments on cardiac myocytes from the heterozygous mouse model to assess the question on whether the Na/Ca exchanger depletes the cell of Ca, and, in particular, the relation of the Na/Ca exchanger with SR Ca regulation.[12]

The first surprising result was that, when we studied the Na/Ca exchanger during Ca removal by blocking its function in a Na- and Ca-free solution in control mouse myocytes (i.e., littermates), we could not see a slower Ca decay but we detected an acceleration of Ca decline. This differed substantially from what has been found previously by us and others in other species (e.g., rabbit and guinea pig[10,15]) in which, in Na- and Ca-free solution, myocytes relax more slowly. Because other Ca extrusion mechanisms, such as mitochondrial uptake and sarcolemmal Ca ATPase, did not have a significant role in relaxation in mouse myocytes,[12] we speculated that the Na/Ca exchanger may, under normal conditions, bring Ca into the cell during the latter part of the Ca transient. This is important for the functional consequences of overexpression of the exchanger in mouse cardiac cells.

SR Ca CONTENT IS INCREASED IN TRANSGENIC MICE OVEREXPRESSING THE Na/Ca EXCHANGER

Using rapid application of 20 mM caffeine, we recorded a transient inward current ascribed to the Na/Ca exchanger,[16] which is generated as a consequence of the removal of the Ca released from the SR, and observed that the time course of this

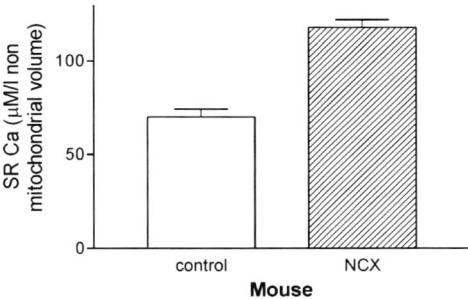

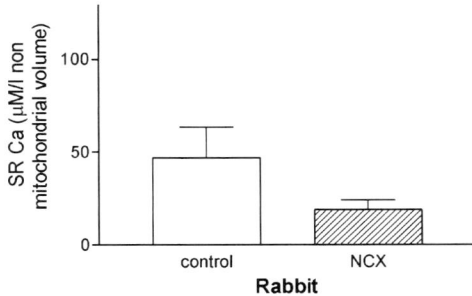

FIGURE 1. Overexpression of Na/Ca exchanger produces opposite results on the SR Ca content from rabbit and mouse ventricular myocytes. In mouse myocytes, SR Ca content was increased when Na/Ca exchanger was overexpressed ($n > 18$). Cultured rabbit cells already had a reduced SR Ca content compared with control mouse cells. However, when the Na/Ca exchanger was overexpressed in rabbit myocytes, SR Ca was even smaller than in control. Most cells did not show any transient inward current upon application of caffeine. The graph only includes results from the responding cells and refers to eight cells per group after 24 hours of culture. NCX, Na/Ca exchanger overexpressing myocytes.

current was faster and the amplitude was larger in transgenic mouse myocytes, and this was consistent with the overexpression of the Na/Ca exchanger. However, the integral of the current was increased. The integral depends on charge movement and ultimately on the amount of Ca released from the SR, suggesting that the SR contained more Ca in transgenic mouse myocytes (FIG. 1). This was despite an unchanged level of expression of SERCA and phospholamban, and an unchanged ability of the SR to uptake Ca. We concluded that, when the Na/Ca exchanger is overexpressed in mouse myocardium, there is an increased SR Ca load, which is consistent with a predominant role of Ca entry for the Na/Ca exchanger in this species.[12]

As mentioned above, the contribution of the Na/Ca exchanger to Ca regulation varies in different species, and in rabbit myocardium the Na/Ca exchanger has been shown to play a significant role in Ca extrusion.[10] The overexpression of the exchanger therefore could compete more efficiently with SERCA for Ca during relaxation and bring about SR Ca depletion. To test this hypothesis, we studied a model of overexpression of the Na/Ca exchanger in adult rabbit ventricular myocytes achieved with adenoviral technology. These experiments were presented at the Biophysical Society Meeting in 2000.[17,18]

OVEREXPRESSION OF THE Na/Ca EXCHANGER IN RABBIT VENTRICULAR MYOCYTES DEPLETES THE SR OF Ca

Cardiac myocytes from adult male New Zealand White rabbits were enzymatically dissociated as previously described.[19] The cells were cultured for up to 48 hours.[20,21] Adenoviral vectors were produced using an overall strategy developed by He et al.[22] Briefly, the canine Na/Ca exchanger cDNA was cloned into the shuttle vector pAdTrack-CMV. The construct was cotransformed with a supercoiled backbone adenoviral vector pAdEasy-1 into *Escherichia coli* strain BJ5183. GFP reporter also was incorporated into the viral backbone. The resultant virus (Ad.NCX.GFP) had a final yield at 7×10^{11} particles ml^{-1}. A reporter virus producing GFP alone under the control of a CMV promoter also was generated (Ad.GFP), at 1.2×10^{12} particles·ml^{-1}.

To 10^4 rod-shaped myocytes in each well of a 12-well plate was added Ad.NCX.GFP at an MOI of 10–1000 or Ad.GFP (MOI 1000). After culture at 37°C for 24 or 48 hours, myocytes were viewed using fluorescence microscopy with an excitation wavelength of 485 nm and emission wavelength of 520 nm to positively identify infected cells.

After 48-hour culture, myocytes infected with Ad.NCX.GFP had a reduced contractility and a smaller Ca transient. Rapid application of caffeine elicited a transient inward current in only approximately 16% of these cells. When the current was elicited, its integral was greatly reduced compared with noninfected myocytes[17,18] (FIG. 1). This suggested that the SR was depleted with Ca and that the Na/Ca exchanger predominantly removed Ca from the cells in rabbit hearts. Similar results also were reported by Schillinger et al.[23] Na/Ca exchanger overexpression depletes the SR of Ca, in clear contrast with the results obtained in the mouse myocytes.

What then is the difference between mouse and rabbit myocytes to produce this discrepancy? Although several factors can be considered, a particularly relevant parameter for Na/Ca exchanger function is intracellular [Na].

Surprising results showed that bulk intracellular [Na] in the mouse heart is approximately 15 mM, two to three times higher than that assumed to be normal in rabbit and guinea pig hearts.[14] Interestingly, in heart failure, intracellular [Na] increases. If this parameter governs the behavior of the exchanger, it is possible that, depending on intracellular [Na], the failing myocardium can either load or deplete the SR with Ca, and both behaviors can be explained by Na/Ca exchanger overexpression.

TOWARD THE ROLE OF THE Na/Ca EXCHANGER OVEREXPRESSION IN THE FAILING HEART: SERCA2a MODULATION

As mentioned above, however, the Na/Ca exchanger is only one of the many factors to be affected by the pathophysiological events in heart failure. Ca regulation in

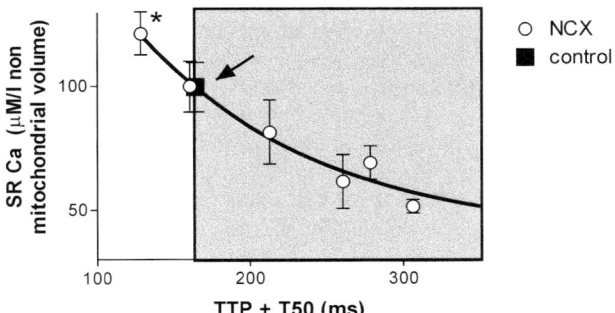

FIGURE 2. SR Ca content was increased in transgenic myocytes overexpressing Na/Ca exchanger (*asterisk*) compared with nontransgenic littermates (*arrow*). This was associated with a faster Ca transient (time to peak + time to 50% relaxation, TTP + T50). When 200 nM thapsigargin was applied on transgenic myocytes, it produced a reduction of SR Ca content and prolongation of Ca transients. The overexpression of the exchanger was able to produce a compensation for up to 28% inhibition of SERCA. Below this value, still in the presence of exchanger overexpression, SR Ca content was reduced and Ca transients prolonged (*gray box*). This experiment suggests that overexpressing the Na/Ca exchanger can be associated with different, even opposite consequences for Ca regulation, depending on the activity of other Ca regulatory mechanisms. Redrawn from Terracciano et al.[24]

the heart depends on several mechanisms working together. In myocytes, the SR Ca uptake and the Na/Ca exchanger are predominantly responsible for Ca extrusion on a beat to beat basis. It is clear that, if one of the two is altered, this would have an effect on the function of the other. A reduction in SERCA expression is associated with the overexpression of the Na/Ca exchanger in heart failure, and this relationship may be crucial for assessing the consequence of such overexpression. In the following experiments, we studied whether the exchanger is able to compensate for a reduction of SERCA function as this could be the case in heart failure.[24]

We slowly inhibited SERCA using 200 nM thapsigargin and studied Ca fluxes via the Na/Ca exchanger and SR Ca uptake (FIG. 2). We used the time course of the Ca transient (parameter affected in heart failure) as an indication of function. We found that myocytes overexpressing the Na/Ca exchanger (TR) displayed faster Ca transients and twitches compared with littermate myocytes (NON). Superfusion with thapsigargin prolonged the time course of Ca transients of TR myocytes until these were equal to the ones measured in NON myocytes. The amount of SERCA inhibition needed to obtain such compensation on the duration of Ca transients was calculated as a function of V_{max} for the Ca flux via SR CaATPase and found to be 28%. In TR myocytes, V_{max} for the Ca flux via Na/Ca exchange was 240% of NON myocytes. When Ca transients in TR myocytes were slowed by thapsigargin to similar values to the ones recorded in NON myocytes, SR Ca content also was correspondingly reduced. The results suggest that in pathophysiological conditions, in which there is a reduction in SERCA function, overexpression of Na/Ca exchanger can compensate and allow the time course of the Ca transient to be maintained. This is the proof of principle that in some conditions (mouse myocytes with overexpressing

Na/Ca exchanger and normal SERCA function) the SR Ca content is increased and the time course of the Ca transient is reduced. In other conditions with the same level of Na/Ca exchanger overexpression (but with reduction in SERCA function), the SR Ca content can be reduced, and the time course of the Ca transient is prolonged. This could lead to the speculation that, in the progression of heart failure, the relationship between overexpression of the Na/Ca exchanger and SERCA dysfunction can determine the transition between compensated and decompensated disease. However, heart failure is more likely to be the result of a very complex interaction between several other changes in myocyte regulation, and this hypothesis, although intriguing, could be too simplistic.

In conclusion, there are no straightforward answers to the questions about what are the functional consequences of the overexpression of the Na/Ca exchanger. From our experiments, these appear to be dependent on species and on the function of other mechanisms involved in cell regulation. Intracellular [Na] may have a crucial role.

ACKNOWLEDGMENTS

I thank Ken MacLeod, Sian Harding, Ken Philipson, Roger Hajjar, Ayesha de Souza, Hardeep Ranu, and Kerry Davia for contributing to the work described in this article. I also thank the Wellcome Trust and the British Heart Foundation for financial support.

REFERENCES

1. BEUCKELMANN, D.J., M. NABAUER & E. ERDMANN. 1992. Intracellular calcium handling in isolated ventricular myocytes from patients with terminal heart failure. Circulation **85:** 1046–1055.
2. O'ROURKE, B., D.A. KASS, G.F. TOMASELLI, et al. 1999. Mechanisms of altered excitation-contraction coupling in canine tachycardia-induced heart failure. I. Experimental studies. Circ. Res. **84:** 562–570.
3. POGWIZD, S.M., K. SCHLOTTHAUER, L. LI, et al. 2001. Arrhythmogenesis and contractile dysfunction in heart failure: roles of sodium–calcium exchange, inward rectifier potassium current, and residual beta-adrenergic responsiveness. Circ. Res. **88:** 1159–1167.
4. STUDER, R., H. REINECKE, J. BILGER, et al. 1994. Gene expression of the cardiac Na^+-Ca^{2+} exchanger in end-stage human heart failure. Circ. Res. **75:** 443–453.
5. HASENFUSS, G., M. MEYER, W. SCHILLINGER, et al. 1997. Calcium handling proteins in the failing human heart. Basic Res. Cardiol. **92**(Suppl. 1): 87–93.
6. DEL MONTE, F., S.E. HARDING, U. SCHMIDT, et al. 1999. Restoration of contractile function in isolated cardiomyocytes from failing human hearts by gene transfer of SERCA2a. Circulation **100:** 2308–2311.
7. LITWIN, S.E. & J.H. BRIDGE. 1997. Enhanced Na^+–Ca^{2+} exchange in the infarcted heart. Implications for excitation-contraction coupling. Circ. Res. **81:** 1083–1093.
8. POGWIZD, S.M., K. SCHLOTTHAUER, L. LI, et al. 2001. Arrhythmogenesis and contractile dysfunction in heart failure: roles of sodium-calcium exchange, inward rectifier potassium current, and residual beta-adrenergic responsiveness. Circ. Res. **88:** 1159–1167.
9. SIPIDO, K.R., P.G. VOLDERS, S.H. DE GROOT, et al. 2000. Enhanced Ca^{2+} release and Na/Ca exchange activity in hypertrophied canine ventricular myocytes: potential link between contractile adaptation and arrhythmogenesis. Circulation **102:** 2137–2144.
10. BASSANI, J.W.M., R.A. BASSANI & D.M. BERS. 1994. Relaxation in rabbit and rat cardiac cells: species-dependent differences in cellular mechanisms. J. Physiol. **476:** 279–293.

11. TERRACCIANO, C.M.N. & K.T. MACLEOD. 1997. Effects of lactate on the relative contribution of Ca^{2+} extrusion mechanisms to relaxation in guinea-pig ventricular myocytes. J. Physiol. **500:** 557–570.
12. TERRACCIANO, C.M.N., A. DE SOUZA, K.D. PHILIPSON & K.T. MACLEOD. 1998. Na^+/Ca^{2+} exchange and sarcoplasmic reticular Ca^{2+} regulation in ventricular myocytes from transgenic mice overexpressing the Na^+/Ca^{2+} exchanger. J. Physiol. **512–3:** 651–667.
13. ADACHI AKAHANE, S., L. LU, Z. LI, et al. 1997. Calcium signaling in transgenic mice overexpressing cardiac Na^+–Ca^{2+} exchanger. J. Gen. Physiol. **109:** 717–729.
14. YAO, A., Z. SU, A. NONAKA, et al. 1998. The effects of overexpression of the Na/Ca exchanger on $[Ca^{2+}]_i$ transients in murine ventricular myocytes. Circ. Res. **82:** 657–665.
15. TERRACCIANO, C.M.N. & K.T. MACLEOD. 1994. The effect of acidosis on Na^+/Ca^{2+} exchange and consequences for relaxation in isolated cardiac myocytes from guinea-pig. Am. J. Physiol. **267:** H477–H487.
16. VARRO, A., N. NEGRETTI, S.B. HESTER & D.A. EISNER. 1993. An estimate of the calcium content of the sarcoplasmic reticulum in rat ventricular myocytes. Pflugers Arch. **423:** 158–160.
17. TERRACCIANO, C.M.N., E. BERNOBICH, K. DAVIA, et al. 2000. Adenovirus-mediated Na/Ca exchanger overexpression reduces sarcoplasmic reticulum (SR) Ca content in adult rabbit cardiomyocytes. Biophys. J. **78:** 2199.
18. BERNOBICH, E., C.M.N. TERRACCIANO, K. DAVIA, et al. 2000. Adenovirus-mediated Na/Ca exchanger overexpression depresses contractile function in adult rabbit cardiomyocytes. Biophys. J. **78:** 2199.
19. HARDING, S.E., G. VESCOVO, M. KIRBY, et al. 1988. Contractile responses of isolated adult rat and rabbit cardiac myocytes to isoproterenol and calcium. J. Mol. Cell. Cardiol. **20:** 635–647.
20. DAVIA, K., R.J. HAJJAR, C.M. TERRACCIANO, et al. 1999. Functional alterations in adult rat myocytes after overexpression of phospholamban with use of adenovirus. Physiol. Genomics **1:** 41–50.
21. DAVIA, K., E. BERNOBICH, H.K. RANU, et al. 2001. SERCA2A overexpression decreases the incidence of aftercontractions in adult rabbit ventricular myocytes. J. Mol. Cell. Cardiol. **33:** 1005–1015.
22. HE, T.C., S. ZHOU, L.T. DA COSTA, et al. 1998. A simplified system for generating recombinant adenoviruses. Proc. Natl. Acad. Sci. USA **95:** 2509–2514.
23. SCHILLINGER, W., P.M. JANSSEN, S. EMAMI, et al. 2000. Impaired contractile performance of cultured rabbit ventricular myocytes after adenoviral gene transfer of Na^+–Ca^{2+} exchanger. Circ. Res. **87:** 581–587.
24. TERRACCIANO, C.M.N., K.D. PHILIPSON & K.T. MACLEOD. 2001. Overexpression of the Na^+/Ca^{2+} exchanger and inhibition of the sarcoplasmic reticulum Ca ATPase in ventricular myocytes from transgenic mice. Cardiovasc. Res. **49:** 38–47.

Na^+/Ca^{2+} Exchanger Is Functional in Both Ca^{2+} Influx and Efflux Modes in Rat Myocytes

XUE-QIAN ZHANG,[a] JIANLIANG SONG,[a] GEORGE M. TADROS,[a] LAWRENCE I. ROTHBLUM,[a] JEREMY DUNN,[b] JONATHAN LYTTON,[b] AND JOSEPH Y. CHEUNG[a]

[a]*Weis Center for Research, Geisinger Medical Center, Danville, Pennsylvania 17822, USA*

[b]*Department of Biochemistry and Molecular Biology, University of Calgary Health Sciences Center, Calgary, Alberta, Canada*

KEYWORDS: Na^+/Ca^{2+} exchanger; Ca^{2+} influx; Ca^{2+} efflux; rat myocytes

Controversies exist as to whether the reverse mode of Na^+/Ca^{2+} exchanger (NCX1) operates during excitation–contraction in adult myocytes. Using adenovirus-mediated gene transfer, an approach that is fundamentally different from using transgenic animals, we examined the functional consequences of overexpression and downregulation of NCX1 in adult rat myocytes. To allow time for exogenous genes to be expressed, we first established that adult cardiac myocytes cultured under continued electrical field stimulation conditions retained normal contractile function for at least 72 hours. We next established that infection of myocytes by adenovirus expressing green fluorescent protein (GFP) resulted in greater than 95% infection as ascertained by GFP fluorescence, but contraction at 6, 24, 48, and 72 hours after infection was not affected.

Examined 48 hours after infection, NCX1 protein was threefold higher in myocytes infected with Adv-NCX1 when compared with those infected with Adv-GFP ($P < 0.02$). Under conditions favoring Ca^{2+} efflux (0.6 mM $[Ca^{2+}]_o$), NCX1 myocytes had lower resting $[Ca^{2+}]_i$ ($P < 0.0001$) and sarcoplasmic reticulum (SR) Ca^{2+} contents ($P < 0.001$) and reduced $[Ca^{2+}]_i$ transient ($P < 0.001$) and shortening ($P < 0.0001$) amplitudes. Conversely, under conditions favoring Ca^{2+} influx (5.0 mM $[Ca^{2+}]_o$), NCX1 myocytes exhibited increased $[Ca^{2+}]_i$ transient ($P < 0.0001$) and contraction ($P < 0.001$) amplitudes and elevated SR Ca^{2+} contents ($P < 0.001$). The differences in twitch amplitudes and $[Ca^{2+}]_i$ transient dynamics between GFP and NCX1 myocytes were no longer apparent at intermediate $[Ca^{2+}]_o$ (1.8 mM). Half-time of relaxation from caffeine-induced contracture, an indication of forward $Na^+/$

Address for correspondence: Xue-Oian Zhang, Weis Center for Research, Geisinger Clinic, 100 North Academy Avenue, Danville, PA 17822-2600. Voice: 570-271-5967; fax: 570-271-5886.

xzhang@geisinger.edu

Ca^{2+} exchange activity, was significantly ($P < 0.02$) shorter in NCX1 myocytes. The complex patterns of contractile responses, $[Ca^{2+}]_i$ transients, and SR Ca^{2+} contents suggest that both Ca^{2+} influx and efflux were augmented in NCX1 myocytes.

Infection of myocytes by adenovirus containing the antisense (As) to NCX1 resulted in 32% knockdown of NCX1 by 72 hours ($P < 0.05$) and 66% by 6 days ($P < 0.0005$). When compared with GFP myocytes at 72 hours, AsNCX1 infection resulted in significantly ($P < 0.0001$) increased contraction amplitudes at 0.6 mM $[Ca^{2+}]_o$ and decreased cell shortening at 5 mM $[Ca^{2+}]_o$, a pattern that was exactly opposite of that observed in NCX1-overexpressed myocytes. Whereas $[Ca^{2+}]_i$ transient amplitudes showed a progressive increase as $[Ca^{2+}]_o$ was raised in GFP myocytes, AsNCX1 myocytes exhibited similar $[Ca^{2+}]_i$ transient amplitudes regardless of whether $[Ca^{2+}]_o$ was 0.6, 1.8, or 5 mM. Half-time of relaxation from caffeine-induced contractures was significantly ($P < 0.0002$) longer in AsNCX1 myocytes when compared with GFP myocytes. Collectively, these observations suggest that both Ca^{2+} influx and efflux were depressed in AsNCX1 myocytes.

In summary, overexpression of NCX1 in adult rat myocytes stimulated at high $[Ca^{2+}]_o$ resulted in enhanced Ca^{2+} influx via reverse Na^+/Ca^{2+} exchange, as evidenced by greater SR Ca^{2+} content, larger twitch, and $[Ca^{2+}]_i$ transient amplitudes. Forward Na^+/Ca^{2+} exchanger function also was increased, as indicated by lower resting and diastolic $[Ca^{2+}]_i$ and faster relaxation from caffeine-induced contractures. Downregulation of Na^+/Ca^{2+} exchanger resulted in exactly the opposite patterns of contraction and $[Ca^{2+}]_i$ transients observed in NCX1-overexpressed myocytes, suggesting compromised forward and reverse functions of Na^+/Ca^{2+} exchanger. We conclude that the most consistent explanation for our observations is that the Na^+/Ca^{2+} exchanger was functional in both Ca^{2+} influx and efflux modes in adult rat myocytes.

Volatile Anesthetics and Regulation of Cardiac Na^+/Ca^{2+} Exchange in Neonates versus Adults

Y. S. PRAKASH,[a] LARRY W. HUNTER,[a] INANC SECKIN,[a] AND GARY C. SIECK[a,b]

[a]*Department of Anesthesiology,* [b]*Department of Physiology and Biophysics, Mayo Clinic and Foundation, Rochester, Minnesota 55905, USA*

KEYWORDS: heart; development; halothane; sevoflurane; β-adrenoceptor

INTRODUCTION

Greater sensitivity of the neonatal heart to depression by volatile anesthetics[1,2] may be caused by differences in anesthetic effects on mechanisms that regulate intracellular Ca^{2+} concentration ($[Ca^{2+}]_i$) in neonatal versus adult cardiac myocytes, especially Ca^{2+} influx channels,[3,4] and Na^+/Ca^{2+} exchange (NCX),[4,5] which, depending on the species, may be key to both contraction and relaxation in the neonatal heart.

NCX is normally regulated by several mechanisms including phosphorylation.[5,6] Previous studies in different tissues have shown that cAMP either increases or decreases NCX activity.[5] There is no specific information on relative cAMP effects on influx versus efflux modes, or on postnatal changes in NCX regulation. Volatile anesthetics inhibit the response of adult cardiac myocytes to β-adrenergic stimulation.[7,8] We previously reported that anesthetics inhibit NCX to a greater extent in neonates.[9] Accordingly, we hypothesized that anesthetics produce greater myocardial depression in neonates via greater interference with cAMP regulation of NCX.

METHODS

With approval from the Mayo Institutional Animal Care and Use Committee, hearts were excised under ketamine (60 mg/kg, i.m.) and xylazine (2.5 mg/kg, i.m.) anesthesia from Sprague-Dawley pups (3 days postpartum; neonates; 7 ± 1 g) and adult males (12 weeks; 265 ± 20 g). Ventricular myocytes were enzymatically dissociated as described previously, loaded with 5 µM fluo-3/AM and imaged using a Noran Odyssey XL real-time confocal system.[10] A calibrated online vaporizer was used to add two adult rat minimum alveolar concentration (MAC) halothane (0.50 ± 0.09 mM) to the aerating gas mixture.

Address for correspondence: Y.S. Prakash, M.D., Ph.D., Associate Professor of Anesthesiology, 4-184 W. Jos SMH, Mayo Clinic, Rochester, MN 55905. Voice: 507-255-7560; fax: 507-255-7300.
prakash.ys@mayo.edu

Influx Mode NCX

Cells initially were perfused with normal Tyrode's (mM: 145 NaCl, 4 KCl, 1 $MgCl_2$, 1 $CaCl_2$, 10 glucose, 10 HEPES; pH 7.4) and then "Na-loaded" for 1 minute with 0 Ca^{2+} (5 mM EGTA) and normal Na^+ Tyrode's while inhibiting the sarcoplasmic reticulum (SR) with 5 µM cyclopiazonic acid (CPA) and 10 µM ryanodine.[10] Influx mode NCX then was rapidly activated (<300 milliseconds) by 0 Na^+ and normal Ca^{2+} Tyrode's, and the rate of increase of $[Ca^{2+}]_i$ was measured. After washout with normal Tyrode's, the protocol was repeated with preexposure to 100 nM isoproterenol or 1 mM dibutryl cAMP (dbcAMP) in the presence or absence of 2 MAC halothane.

Efflux Mode NCX

Cells were exposed to 0 Na^+, 0 Ca^{2+} Tyrode's, and 5 µM CPA, allowing $[Ca^{2+}]_i$ levels to increase from continued SR Ca^{2+} "leak." Although plasma membrane Ca^{2+} ATPase was not inhibited, its contribution during this protocol was fairly insignificant. After the $[Ca^{2+}]_i$ levels stabilized, efflux mode NCX was rapidly activated by 0 Ca^{2+}, normal Na^+ Tyrode's, and CPA, and the rapid rate of decrease in $[Ca^{2+}]_i$ levels was recorded. After washout with normal Tyrode's, the protocol was repeated with preexposure to 100 nM isoproterenol or 1 mM dbcAMP in the presence or absence of 2 MAC halothane.

Data were compared using ANOVA with repeated measures and post hoc analyses with age and experimental condition as grouping variables. Statistical significance was established at $P < 0.05$.

RESULTS

In Na^+-loaded myocytes, rapid activation of NCX resulted in a faster rate of increased $[Ca^{2+}]_i$ in neonatal myocytes (76–125 nM s^{-1}) compared with adult myocytes (33–80 nM s^{-1}). Halothane slowed influx to a greater extent in neonates (FIG. 1). Isoproterenol increased NCX-mediated Ca^{2+} influx, but to a greater extent in adults. Subsequent exposure to halothane resulted in less inhibition of NCX in both age groups, but especially in the adult. In contrast, dbcAMP increased Ca^{2+} influx and prevented halothane inhibition of NCX to a greater extent in neonates. The overall effect of isoproterenol versus dbcAMP was not different in adults.

With elevated $[Ca^{2+}]_i$ during inhibited SERCA and NCX, rapid reactivation of NCX resulted in a faster decline in $[Ca^{2+}]_i$ in neonatal myocytes (60–95 nM s^{-1}), compared with adults (19–33 nM s^{-1}). Halothane slowed efflux especially in neonates (FIG. 2). As with influx, isoproterenol increased NCX-mediated influx to a greater extent in adults and blunted the inhibition by halothane. dbcAMP increased Ca^{2+} efflux and prevented halothane inhibition of NCX, but to a greater extent in neonates.

DISCUSSION

Given a smaller SR, NCX-mediated Ca^{2+} influx may be critical to muscle contraction as well as relaxation. NCX is thought to be regulated by cyclic nucleotides.[5] Only a few studies have examined the effects of volatile anesthetics on NCX in ne-

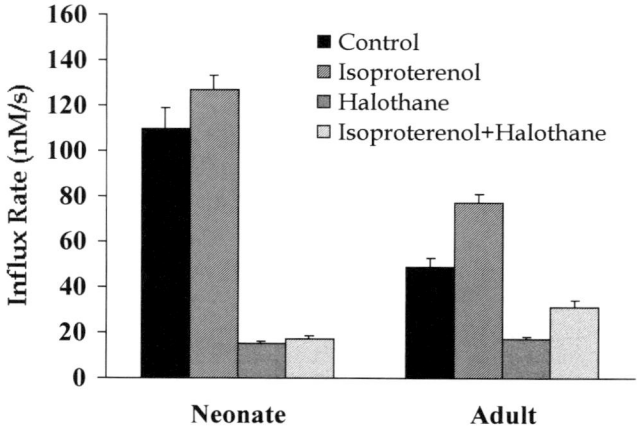

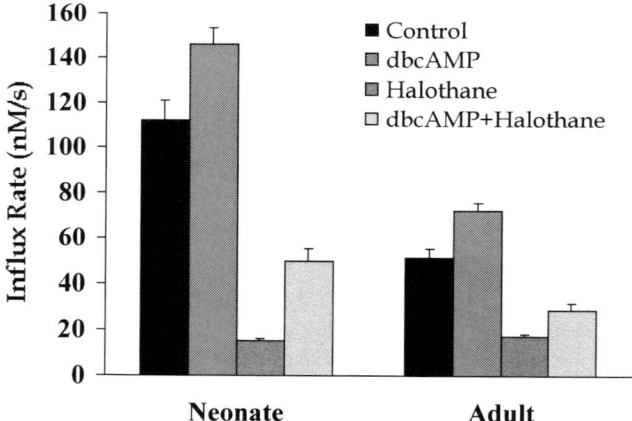

FIGURE 1. Age-related differences in the interaction between halothane and cAMP regulation of influx mode NCX. Halothane decreased influx to a greater extent in neonates. Isoproterenol increased influx, especially in adults, whereas dbcAMP had a greater effect in neonates. Both agents blunted halothane inhibition of influx, but dbcAMP had a greater effect in the neonate. Values ± SE.

onates or adults. Haworth and Goknur[11] found that $^{45}Ca^{2+}$ radioisotope uptake in Na^+-loaded adult rat cardiac myocytes is inhibited by anesthetics. In recent studies, we demonstrated that both halothane and sevoflurane inhibit NCX-mediated Ca^{2+} influx and efflux to a greater extent in neonates[9] compared with adults.[10]

Studies in different cell types indicate that NCX may be either inhibited or accelerated by cAMP. In rat arterial smooth muscle, adenylyl cyclase activation downregulates NCX, whereas in cardiac muscle of some species, β-adrenergic agonists

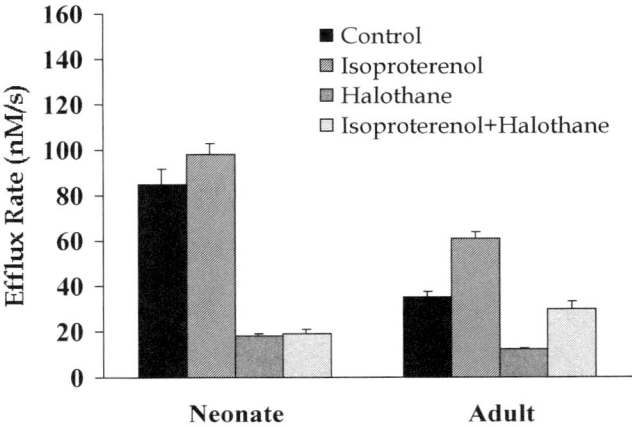
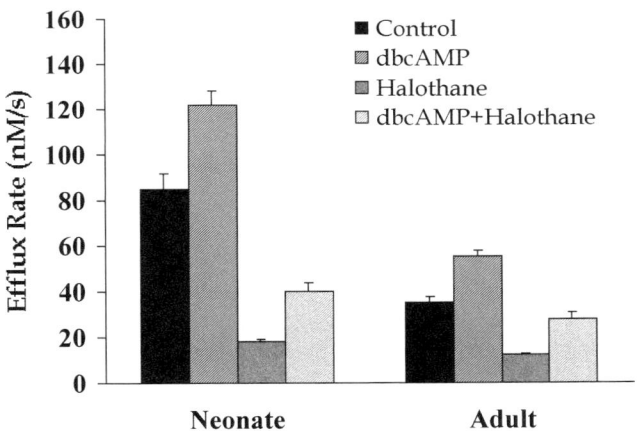

FIGURE 2. Age-related differences in the interaction between halothane and cAMP regulation of efflux mode NCX. As with influx, halothane decreased efflux to a greater extent in neonates. Both isoproterenol and dbcAMP blunted halothane inhibition of efflux, but dbcAMP had a greater effect in the neonate. Values ± SE.

also reduce NCX-mediated Ca^{2+} efflux.[5] However, other studies have reported increased NCX activity with β-adrenergic stimulation.[5] Regardless, because cAMP modulates NCX, and volatile anesthetics interact with the cAMP pathway,[7,8] modulation of NCX by anesthetics is very likely. Our results demonstrate that anesthetic effects on NCX are mediated, at least in part, via the cAMP regulation of NCX. The neonatal heart has been shown to have lower β-adrenergic sensitivity.[12] This may explain our finding that β-adrenoceptor stimulation by isoproterenol produces greater modulation of NCX in adults, rather than neonates. On the other hand, direct anesthetic effects on cAMP-mediated phosphorylation of NCX may be higher in the neonate, and/or the degree of dephosphorylation induced by anesthetics may produce greater changes in NCX activity. Our data showing greater modulation of NCX and prevention of anesthetic-induced inhibition in neonates by dbcAMP is consistent with this scenario.

ACKNOWLEDGMENTS

This work is supported by Grant GM 57816 (Y.S.P.) from the National Institutes of Health, Bethesda, Maryland, and by the Mayo Foundation, Rochester, Minnesota.

REFERENCES

1. KRANE, E.J. & J.Y. SU. 1989. Comparison of the effects of halothane on skinned myocardial fibers from newborn and adult rabbit. II. Effects on sarcoplasmic reticulum. Anesthesiology **71**: 103–109.
2. PALMISANO, B.W., R.W. MEHNER, D.F. STOWE, *et al.* 1994. Direct myocardial effects of halothane and isoflurane. Comparison between adult and infant rabbits. Anesthesiology **81**: 718–729.
3. CHIN, T.K., W.F. FRIEDMAN & T.S. KLITZNER. 1990. Developmental changes in cardiac myocyte calcium regulation. Circ. Res. **67**: 574–579.
4. WETZEL, G.T., F. CHEN & T.S. KLITZNER. 1995. Na^+/Ca^{2+} exchange and cell contraction in isolated neonatal and adult rabbit cardiac myocytes. Am. J. Physiol. **268**: H1195–H1201.
5. BLAUSTEIN, M.P. & W.J. LEDERER. 1999. Sodium/calcium exchange: its physiological implications. Physiol. Rev. **79**: 763–854.
6. PHILLIPSON, K.D. & D.A. NICOLL. 2000. Sodium–calcium exchange: a molecular perspective. Annu. Rev. Physiol. **62**: 111–133.
7. SCHOTTEN, U., C. SCHUMACHER, M. SIGMUND, *et al.* 1998. Halothane, but not isoflurane, impairs the beta-adrenergic responsiveness in rat myocardium. Anesthesiology **88**: 1330–1339.
8. THURSTON, T.A. & S. GLUSMAN. 1993. Halothane myocardial depression: interactions with the adenylyl cyclase system. Anesth. Analg. **76**: 63–68.
9. PRAKASH, Y.S., I. SECKIN, L.W. HUNTER & G.C. SIECK. 2002. Mechanisms underlying greater sensitivity of neonatal cardiac muscle to volatile anesthetics. Anesthesiology **96**: 893–906.
10. SECKIN, I., G.C. SIECK & Y.S. PRAKASH. 2001. Volatile anaesthetic effects on Na^+/Ca^{2+} exchange in rat cardiac myocytes. J. Physiol. **532**: 91–104.
11. HAWORTH, R.A. & A.B. GOKNUR. 1995. Inhibition of sodium/calcium exchange and calcium channels of heart cells by volatile anesthestics. Anesthesiology **82**: 1255–1265.
12. ROBINSON, R.B. 1996. Autonomic receptor-effector coupling during post-natal development. Cardiovasc. Res. **31**: E68–E76.

Modulation of the Na^+/Ca^{2+} Exchanger by Isoprenaline, Adenosine, and Endothelin-1 in Guinea Pig Ventricular Myocytes

YIN HUA ZHANG, ANNABEL K. HINDE, ANDREW F. JAMES, AND JULES C. HANCOX

Department of Physiology and Cardiovascular Research Laboratories, School of Medical Sciences, University of Bristol, Bristol, UK

KEYWORDS: Na^+/Ca^{2+} exchanger; I_{NaCa}; isoprenaline; adenosine; endothelin-1; cardiac myocyte; arrhythmias; protein kinase C; protein kinase A; G protein; inhibitory G protein

BACKGROUND

The Na^+/Ca^{2+} exchanger is an important mechanism involved in controlling Ca homeostasis and ionic current generation in the heart.[1–3] The Na^+/Ca^{2+} exchanger also plays fundamental roles in pathological conditions, such as arrhythmogenesis and heart failure.[4–6] It is well known that neurotransmitters and endogenous peptides modulate the function of the heart through their actions on voltage-gated ion channels and intracellular Ca stores, but until recently there has been relatively little evidence for direct upregulation of ionic current (I_{NaCa}) generated by the exchanger by protein kinase–mediated pathways. This report summarizes recent findings regarding the modulation of Na^+/Ca^{2+} exchanger–mediated current by isoprenaline, adenosine, and endothelin-1 (FIG. 1).

METHODS

Ventricular myocytes were isolated from the hearts of male guinea pigs (400–600 g) using enzymatic and mechanical dispersion and kept in Kraft-Brühe (KB medium) at 4°C until use.[7,8] Solutions and chemicals were as previously described.[7,8] All experiments were performed at 37°C. Recordings were made using the whole-cell mode of the patch-clamp technique. Data were expressed as mean ± SEM.

Address for correspondence: Jules C. Hancox, Department of Physiology, University of Bristol, University Walk, Bristol BS8 1TD, UK. Voice: 44-117-9289028; fax: 44-117-9288923.
jules.hancox@bristol.ac.uk

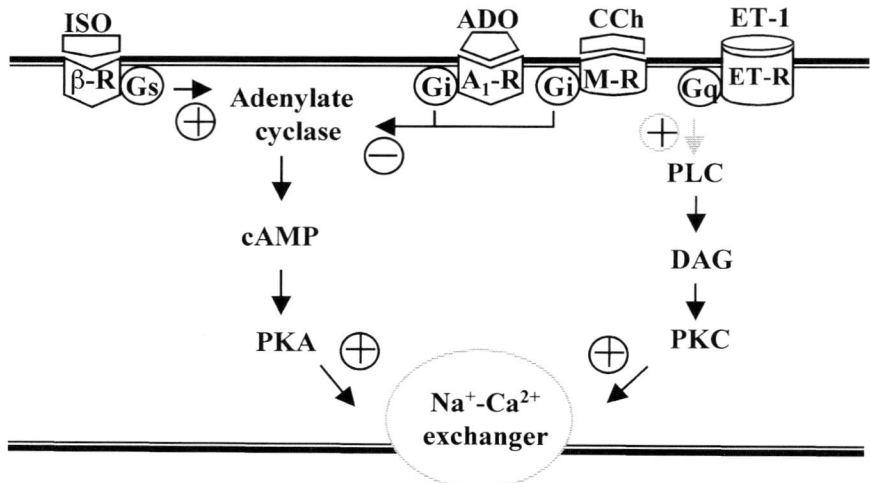

FIGURE 1. Scheme of the effects of ISO, ADO, CCh, and ET-1 and their second-messenger pathways on the modulation of the Na^+/Ca^{2+} exchanger. ISO, isoprenaline; R, receptor; ADO, adenosine; CCh, carbacol; M, muscarinic; ET-1, endothelin-1; G, GTP-dependent protein; PLC, phospholipase C; DAG, diacylglycerol; PKA, protein kinase A; PKC, protein kinase C.

RESULTS

Modulation of the Na^+/Ca^{2+} Exchanger by Pathways Involving Protein Kinase C and Protein Kinase A

For a long time, there has been debate about whether the Na^+/Ca^{2+} exchanger can be modulated by a pathway involved in protein kinase A (PKA). Recently, isoprenaline has been reported to stimulate I_{NaCa} from guinea pig ventricular myocytes *via* a PKA-dependent pathway.[7,9] Similar observations have been made for rat ventricular myocytes and cardiac NCX isoform expressed in *Xenopus* oocytes.[10] In addition, Ruknudin and colleagues[10] have reported that the cardiac isoform of the exchanger (NCX 1.1) is more sensitive to phosphorylation than the renal isoform (NCX 1.3). In contrast with these data, Woo and Morad[11] have reported that activation of PKA by isoprenaline or forskolin attenuates I_{NaCa} from shark ventricular myocytes. Species and experimental conditions may influence results obtained; however, what is clear is that PKA-dependent modulation of I_{NaCa} occurs and therefore may be one consequence of β-adrenergic stimulation of the heart.

Evidence exists that the cloned cardiac isoform of the Na^+/Ca^{2+} exchanger can be phosphorylated by protein kinase C (PKC), and that endothelin-1 (ET-1) alters Ca^{2+} uptake by sarcolemmal vesicles.[12,13] We recently have studied the effect of ET-1 on I_{NaCa} from guinea pig ventricular myocytes. With major interfering currents inhibited, ET-1 significantly increased I_{NaCa}. Application of a PKC activator (phorbol 12-myristate 13-acetate, PMA) mimicked the effect of ET-1. In contrast, the PKC inhibitor chelerythrine abolished the stimulatory effect of ET-1. An inactive phorbol ester,

4-alpha-phorbol-12, 13-didecanoate had no effect on I_{NaCa}. Collectively, these data indicate that ET-1 can activate I_{NaCa} through a PKC-dependent pathway.[8] Similar upregulation of rat ventricular I_{NaCa} by a PKC-dependent pathway has been reported for the α-adrenoceptor agonist phenylephrine.[14] Thus, it appears that mammalian cardiac I_{NaCa} can be increased by both PKA and PKC activation.

Interaction between Isoprenaline and Adenosine or ET-1 in Modulation of the Na^+/Ca^{2+} Exchanger

We have investigated the effects on I_{NaCa} of β-adrenoceptor activation are influenced by coapplication of either adenosine (ADO) or ET-1. When ISO was applied in the presence of adenosine (ADO), ADO abolished the stimulatory effect of ISO, without significantly altering the amplitude of baseline I_{NaCa} when applied on its own.[7] The effect of ADO on the response of I_{NaCa} to ISO was mimicked by the A_1 ADO receptor agonist N_6-cyclopentyladenosine, whereas the effect of ADO on the response of I_{NaCa} to ISO was inhibited by the A_1 ADO receptor antagonist 8-cyclopentyl-1,3-dipropylxanthine. The antiadrenergic effects on I_{NaCa} of ADO were not affected by chelerythrine, nor by the nitric oxide (NO) synthase inhibitor, N (G)-nitro-L-arginine methyl ester. Moreover, in the presence of the PKC activator phorbol 12-myrisate 13-acetate (PMA) or exogenous NO donor sodium nitroprusside, ISO preserved its stimulatory effect on I_{NaCa}. However, prior incubation of myocytes with pertussis toxin prevented the effect of ADO. The antiadrenergic effect of ADO on I_{NaCa} was mimicked by externally applied carbachol (CCh), a muscarinic receptor agonist. These data suggested that ADO antagonized the effect of β-adrenergic stimulation on I_{NaCa} by directly activating inhibitory G-protein (G_i)–linked A_1 receptors in guinea pig ventricular myocytes.[7] In contrast, in further experiments in which the effect of ET-1 and ISO on I_{NaCa} was investigated, ET-1 did not abolish the stimulatory effect of ISO. Both agents stimulated I_{NaCa} when applied alone; however, there was no additive response when concentrations of each agent that produced similar levels of current increase were coapplied.[8] Similar to the effect of ET-1 + ISO, PMA + ISO had no additive effect on I_{NaCa} when coapplied.

CONCLUSION

Growing evidence suggests that β-adrenergic stimulation may increase mammalian cardiac Na^+/Ca^{2+} exchanger activity by a PKA-dependent pathway. This action can be inhibited by activation of inhibitory G protein. Maximally effective concentrations of ET-1 and ISO appear not to be additive in their effects on I_{NaCa} when coapplied, raising the possibility that exchanger activity can be enhanced to only a certain extent, irrespective of whether this enhancement occurs *via* PKA or PKC. We conclude that protein kinase–mediated phosphorylation of the cardiac exchanger may be of physiological and pathophysiological significance.

ACKNOWLEDGMENTS

This work was supported by British Heart Foundation and Wellcome Trust.

REFERENCES

1. EISNER, D.A. & W.J. LEDERER. 1985. Na–Ca exchange: stoichiometry and electrogenicity. Am. J. Physiol. **248:** C189–C202.
2. JANVIER, N.C. & M.R. BOYETT. 1996. The role of Na-Ca exchange current in the cardiac action potential. Cardiovasc. Res. **32:** 69–84.
3. BLAUSTEIN, M.P. & W.J. LEDERER. 1999. Sodium/calcium exchange: its physiological implications. Physiol. Rev. **79:** 763–854.
4. FOZZARD, H.A. 1992. Afterdepolarizations and triggered activity. Basic Res. Cardiol. **87**(Suppl. 2): 105–113.
5. MING, Z., R. ARONSON & C. NORDIN. 1994. Mechanism of current-induced early afterdepolarizations in guinea pig ventricular myocytes. Am. J. Physiol. **267:** H1419–H1428.
6. POGWIZD, S.M., M. QI, W. YUAN, et al. 1999. Upregulation of Na^+/Ca^{2+} exchanger expression and function in an arrhythmogenic rabbit model of heart failure. Circ. Res. **85:** 1009–1019.
7. ZHANG, Y.H., A.K. HINDE & J.C. HANCOX. 2001. Anti-adrenergic effect of adenosine on Na^+-Ca^{2+} exchange current recorded from guinea-pig ventricular myocytes. Cell Calcium **29:** 347–358.
8. ZHANG, Y.H., A.F. JAMES & J.C. HANCOX. 2001. Regulation by endothelin-1 of Na^+-Ca^{2+} exchange current (I_{NaCa}) from guinea-pig isolated ventricular myocytes. Cell Calcium. **30:** 351–360.
9. PERCHENET, L., A.K. HINDE, K.C. PATEL, et al. 2000. Stimulation of Na/Ca exchange by the beta-adrenergic/protein kinase A pathway in guinea-pig ventricular myocytes at 37°C. Pflugers. Arch. **439:** 822–828.
10. RUKNUDIN, A., S. HE, W.J. LEDERER & D.H. SCHULZE. 2000. Functional differences between cardiac and renal isoforms of the rat Na^+-Ca^{2+} exchanger NCX1 expressed in *Xenopus* oocytes. J. Physiol. **529:** 599–610.
11. WOO, S.H. & M. MORAD. 2001. Bimodal regulation of Na^+-Ca^{2+} exchanger by beta-adrenergic signaling pathway in shark ventricular myocytes. Proc. Natl. Acad. Sci. USA **98:** 2023–2028.
12. IWAMOTO, T., Y. PAN, S. WAKABAYASHI, et al. 1996. Phosphorylation-dependent regulation of cardiac Na^+/Ca^{2+} exchanger via protein kinase C. J. Biol. Chem. **271:** 13609–13615.
13. BALLARD, C. & S. SCHAFFER. 1996. Stimulation of the Na^+-Ca^{2+} exchanger by phenylephrine, angiotensin II and endothelin 1. J. Mol. Cell. Cardiol. **28:** 11–17.
14. STENGL, M., K. MUBAGWA, E. CARMELIET, et al. 1998. Phenylephrine-induced stimulation of Na^+/Ca^{2+} exchange in rat ventricular myocytes. Cardiovasc. Res. **38:** 703–710.

Effect of KB-R7943 on Oscillatory Na$^+$/Ca^{2+} Exchange Current in Guinea Pig Ventricular Myocytes

LIBING LI AND JUNKO KIMURA

Department of Pharmacology, Fukushima Medical University, Fukushima, Japan

KEYWORDS: KB-R7943; oscillatory Na$^+$/Ca^{2+} exchange current; ventricular myocytes;

INTRODUCTION

KB-R7943 (2-[2-[4-(4-nitrobenzyloxy)phenyl]ethyl]isothiourea methanesulfonate, KBR) is a potent and relatively selective inhibitor of cardiac Na$^+$/Ca^{2+} exchange current (I_{NCX}).[1,2] It has been reported that KBR inhibits outward I_{NCX} more potently (IC$_{50}$ 0.3 µM) than inward I_{NCX} (IC$_{50}$ 17 µM), when I_{NCX} was unidirectional.[1] However, KBR blocks both directions of I_{NCX} equally (IC$_{50}$ 1 µM), when I_{NCX} is bidirectional.[3] Therefore, KBR is a direction-dependent inhibitor for the unidirectional NCX, but a direction-independent inhibitor for the bidirectional NCX.

When Ca^{2+} overload occurs in cardiac myocytes, I_{NCX} flows as oscillatory currents (I_{OSC}) called a transient inward current.[4] How KBR affects oscillatory I_{NCX} is unknown, however. We investigated the effect of KBR on oscillatory I_{NCX} in guinea pig cardiac myocytes.

MATERIALS AND METHODS

Single ventricular myocytes were dissociated from the guinea pig heart. I_{OSC} was recorded by the whole-cell voltage clamp. The external solution contained (mM) NaCl 140, CaCl$_2$ 1, MgCl$_2$ 1, ouabain 0.02, D600 0.01, HEPES 5 (pH 7.2). The pipette solution contained (mM) NaCl 20 (for inward I_{OSC}) or 10 (for outward I_{OSC}), CsOH 120, aspartic acid 60, MgATP 5, EGTA 0.1, and HEPES 10 (pH 7.2).

RESULTS AND DISCUSSION

Oscillatory I_{NCX} was induced with a high concentration of Na$^+$ (10 or 20 mM) and a low concentration of EGTA 0.1 mM in the pipette solution. Most of the membrane currents other than I_{NCX} were inhibited by appropriate blockers. I_{OSC} was inhibited by 5 µM ryanodine, confirming that it was activated by oscillatory [Ca^{2+}]$_i$ release from sarcoplasmic reticulum. I_{OSC} was insensitive to 1 mM DNDS (4,4′-

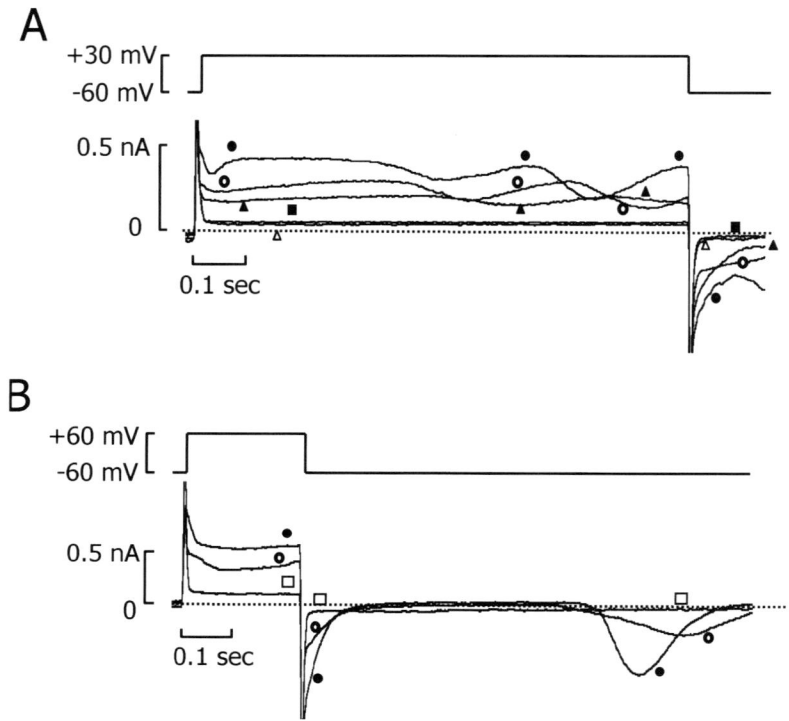

FIGURE 1. Effect of KBR on I_{OSC}. **(A)** Outward I_{OSC}. **(B)** Inward I_{OSC}. Voltage pulses are shown above the current traces. Control (●). KBR at 3 μM (○), 10 μM (▲), 20 μM (□), or 30 μM (△), and 5 mM Ni^{2+} (■) were added cumulatively to the external solution.

dinitrostilbene-2,2'-disulfonic acid), a Cl^- current inhibitor. Outward oscillatory I_{NCX} was induced by depolarizing square pulses from the holding potential of −60 to 30 mV for 1 s (FIG.1). The pulse interval was 30 s, because more frequent depolarizing pulses led to contracture of the cell. Inward I_{OSC} was induced at −60 mV after depolarizing prepulses to 60 mV for 200 ms every 10 s (FIG.1B). KBR up to 30 μM was cumulatively applied in the external solution, and finally 5 mM Ni^{2+} was added to inhibit I_{NCX} completely. KBR suppressed I_{OSC} in a concentration-dependent manner. The current level at 30 μM KBR was the same as that after application of 5 mM Ni^{2+}, suggesting that KBR-sensitive I_{OSC} was I_{NCX}.

FIGURE 2 shows the concentration–response curves of KBR for the outward and inward I_{OSC}. The degree of KBR inhibition was calculated by integrating the current area between I_{OSC} and the basal current at 20 or 30 μM KBR or 5 mM Ni^{2+}. The current was integrated during 800 ms for inward I_{OSC} and 1 s for outward I_{OSC}. The percentage inhibition at a given KBR concentration was calculated against the control. KBR inhibited both outward and inward I_{OSC} in a similar manner with the IC_{50} value of approximately 4 μM.

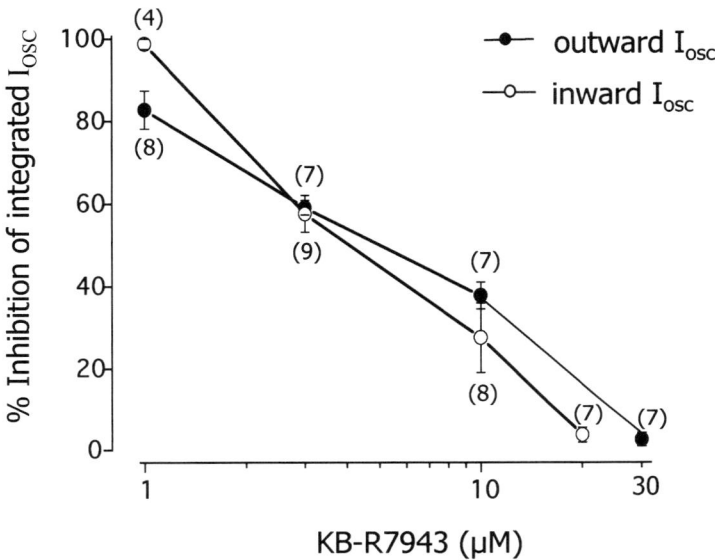

FIGURE 2. Concentration–response curves of KBR for the outward (●) and inward (○) I_{OSC}. Current area was integrated and percentage of inhibition was calculated. IC_{50} values are approximately 4 μM for both I_{OSC}.

We assumed that the "peak" of outward I_{OSC} corresponds to low $[Ca^{2+}]_i$, and the "valley" of outward I_{OSC} corresponds to high $[Ca^{2+}]_i$ (see FIG. 1A), because $[Ca^{2+}]_i$ elevation shifts I-V curve of I_{NCX} in the positive direction and so at a given potential positive to the reversal potential of I_{NCX}, I_{NCX} becomes less outward (or more inward) at higher $[Ca^{2+}]_i$. KB-R7943 inhibited inward and outward oscillatory I_{NCX} direction-independently with IC_{50} of approximately 4 μM. This IC_{50} value is similar to that for inhibiting bidirectional I_{NCX}.[3] In the experiment for unidirectional I_{OSC} (FIG. 1) and bidirectional I_{NCX},[3] both Na^+ and Ca^{2+} were present in the external solution, whereas one or the other ion was absent when the direction-dependent block of KBR was found.[1,2] KBR is a competitive inhibitor for external Ca^{2+} (Watano et al.[1]) and Na^+ (unpublished data). Thus, a direction dependence of the KBR block may be caused by experimental ionic conditions. Outward unidirectional I_{NCX} was induced by raising external Ca^{2+} from 0 to 1 mM, and KBR was applied at 0 Ca^{2+}.[1] In contrast, inward I_{NCX} was induced by replacing external Li^+ with Na^+, and KBR was applied in the absence of Na^+ but presence of Ca^{2+}. KBR applied in the absence of Ca^{2+} could inhibit I_{NCX} more potently than in the presence of Ca^{2+}. Therefore, we propose that the direction-dependent block of KBR is caused by experimental conditions although conformation changes of the exchanger may also be involved.

This proposal is further discussed in the chapter by Kimura et al.[5] in this volume.

REFERENCES

1. WATANO, T., J. KIMURA, T. MORITA & H. NAKANISHI. 1996. A novel antagonist, No. 7943, of the Na^+/Ca^{2+} exchange current in guinea-pig cardiac ventricular cells. Br. J. Pharmacol. **119:** 555–563.
2. IWAMOTO, T., T. WATANO & M. SHIGEKAWA. 1996. A novel isothiourea derivative selectively inhibits the reverse mode of Na^+/Ca^{2+} exchange in cells expressing NCX1. J. Biol. Chem. **37:** 22391–22397.
3. KIMURA, J., T. WATANO, M. KAWAHARA, et al. 1999. Direction-independent block of bi-directional Na^+/Ca^{2+} exchange current by KB-R7943 in guinea-pig cardiac myocytes. Br. J. Pharmacol. **128:** 969–974.
4. LIPP, P. & L. POTT. 1988. Transient inward current in guinea-pig atrial myocytes reflects a change of sodium–calcium exchange current. J. Physiol. **397:** 601–630.
5. KIMURA, J., Y. WATANABE, L. LI & T. WATANO. 2002. Pharmacology of Na/Ca exchanger. Ann. N.Y. Acad. Sci. This volume.

Inhibition of the *Drosophila* Na^+/Ca^{2+} Exchanger, CALX1.1, by KB-R7943

MICHAEL R. ISAAC, CHADWICK L. ELIAS, HOA D. LE,
ALEXANDER OMELCHENKO, MARK HNATOWICH, AND LARRY V. HRYSHKO

*Institute of Cardiovascular Sciences, St. Boniface General Hospital Research Centre,
Department of Physiology, University of Manitoba, Winnipeg, Canada*

KEYWORDS: Na^+/Ca^{2+} exchange; *Drosophila*; CALX1.1; KB-R7943

Since its introduction in 1996, considerable interest has focused on the use of KB-R7943 as a selective inhibitor of the reverse (i.e., Ca^{2+} influx) mode of Na^+/Ca^{2+} exchange. The prospect that KB-R7943 may serve as a lead compound in the development of other, perhaps more potent, pharmacological agents that could be used clinically in the management of, for example, cardiac reperfusion injury, has driven several studies attempting to establish its mechanism of action.

We recently have shown that KB-R7943 preferentially inhibits the reverse mode of Na^+/Ca^{2+} exchange under unidirectional transport conditions.[1] Using the giant, excised patch technique[2,3] and the cloned canine cardiac Na^+/Ca^{2+} exchanger, NCX1.1, we found that inhibition by KB-R7943 was not competitive for Na^+_i and Ca^{2+}_o, and that steady-state current inhibition was insensitive to micromolar levels of regulatory Ca^{2+}_i.[1] After deregulation of NCX1.1 by limited proteolysis with α-chymotrypsin, we found that KB-R7943 was still an effective inhibitor of exchange activity, albeit with reduced potency. These results led us to conclude that KB-R7943 mainly targets the transport machinery of NCX1.1, but that its action is enhanced if the exchanger's large intracellular loop is intact.[1]

In an effort to further explore the actions of KB-R7943 on Na^+/Ca^{2+} exchange activity, in general, and its interaction(s) with the ionic regulatory mechanisms, in particular, we examined the effects of regulatory Ca^{2+} and KB-R7943 on the *Drosophila* Na^+/Ca^{2+} exchanger, CALX1.1. This transporter was specifically chosen for study owing to its unique ionic regulatory properties (i.e., negative regulation by Ca^{2+}_i and inward current inactivation). Complementary RNA encoding *Drosophila* CALX1.1 was prepared and injected into *Xenopus* oocytes as previously described.[1,3] Na^+/Ca^{2+} exchange activity was measured 3–6 days after injection using the giant, excised patch-clamp technique.[2,3]

Address for correspondence: Larry V. Hryshko, St. Boniface Hospital Research Centre, University of Manitoba, Cardiovascular Sciences, 351 Tache Avenue, R3032, Winnipeg, R2H 2A6, Canada. Voice: 204-235-3662; fax: 204-233-6723.
lhryshko@sbrc.ca

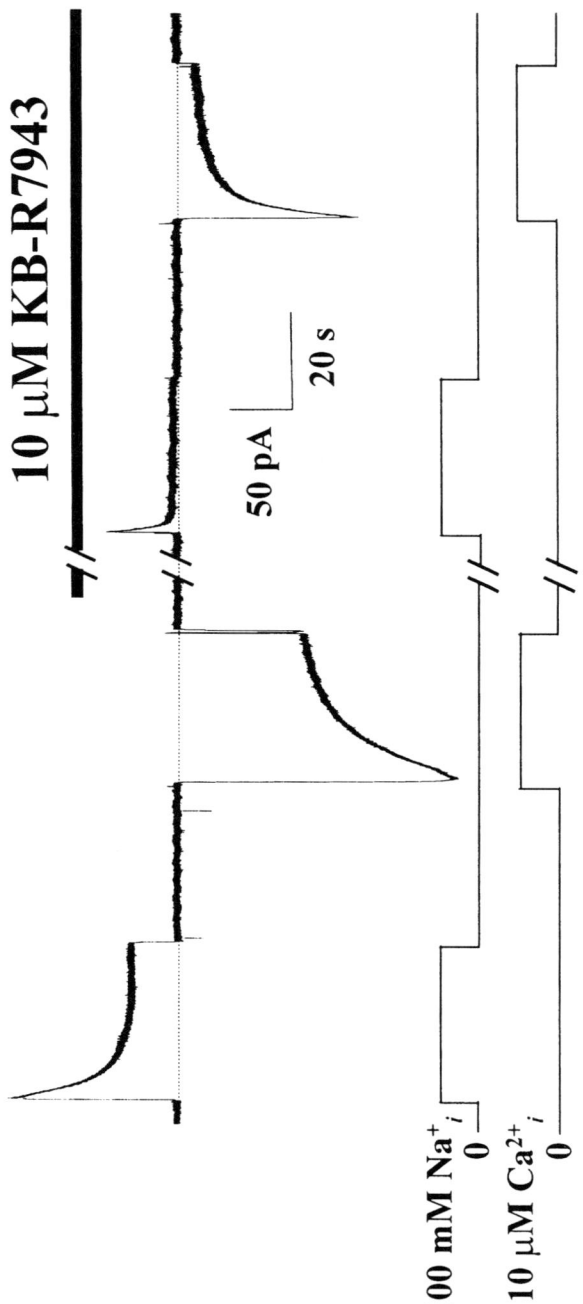

FIGURE 1. Inhibitory properties of 10 μM KB-R7943 on outward and inward Na$^+$/Ca^{2+} currents mediated by CALX1.1.

Pipette solution contained 100 mM Na$^+$ plus 2 mM Ca^{2+} for measuring both inward *and* outward currents from the same patch. Patch pipettes were pulled and polished to an inner diameter of approximately 25–30 µm, solution changes were completed in approximately 200 ms, and all current measurements were obtained at 30°C. (KB-R7943 was generously supplied by Nippon Organon K.K., Osaka, Japan).

We found that KB-R7943 is a weaker inhibitor of CALX1.1- versus NCX1.1-mediated exchange (approximately four- to fivefold), and that the marked mode selectivity observed with NCX1.1 does not extend to the *Drosophila* exchanger (i.e., 90 vs. 23% inhibition of outward vs. inward steady-state currents by 10 µM KB-R7943 for NCX1.1, as compared with 90 vs. 80% inhibition of these steady-state currents for CALX1.1; FIG. 1). Comparison of the effects of regulatory Ca$^{2+}{}_i$ and KB-R7943 suggests that both noncompetitively inhibit CALX1.1-mediated exchange, but that their respective binding is not mutually exclusive. Deregulation of CALX1.1 with α-chymotrypsin resulted in abolition of Ca$^{2+}{}_i$-dependent regulation of exchange activity but did not fully alleviate inhibition by KB-R7943, a result qualitatively similar to that obtained with NCX1.1.[1] Our results strengthen the notion that KB-R7943 primarily targets the transport function(s) of Na$^+$/Ca^{2+} exchangers, and that its effects are augmented by the presence of intact ionic regulatory elements within their large cytoplasmic loops.

ACKNOWLEDGMENTS

This work was supported by a grant from the Heart and Stroke Foundation of Manitoba.

REFERENCES

1. ELIAS, C.L., A. LUKAS, S. SHURRAW, *et al.* 2001. Inhibition of Na$^+$/Ca^{2+} exchange by KB-R7943: transport mode selectivity and antiarrhythmic consequences. Am. J. Physiol. **281:** H1334–H1345.
2. HILGEMANN, D.W. 1990. Regulation and deregulation of cardiac Na$^+$-Ca^{2+} exchange in giant excised sarcolemmal membrane patches. Nature **344:** 242–245.
3. DYCK, C., K. MAXWELL, J. BUCHKO, *et al.* 1998. Structure–function analysis of CALX1.1, a Na$^+$-Ca^{2+} exchanger from *Drosophila*. Mutagenesis of ionic regulatory sites. J. Biol. Chem. **273:** 12981–12987.

Index of Contributors

Ambudkar, S., 176–186
Amende, I., 480–482
Amoroso, S., 408–412
Annunziato, L., 394–404, 408–412
Artigas, P., 31–40
Asteggiano, C., 288–299, 418–420

Bauer, P.J., 325–334
Beaugé, L., 224–236, 288–299
Bell, P.D., 338–341, 342–344
Berberián, G., 288–299, 418–420
Bers, D.M., 157–158, 454–465, 466–471, 478–479, 500–512
Bersohn, M.M., 483–486
Blaustein, M.P., 356–366
Booth, B.J., 117–120
Bortoluzzi, S., 282–284
Boscia, F., 394–404
Bose, R., 345–349, 350–353
Bossuyt, J., 100–102, 197–204
Braden, B.C., 100–102
Brini, M., 376–381
Bullis, B.L., 117–120

Cai, X., 90–93
Caliendo, G., 408–412
Canitano, A., 394–404
Carafoli, E., 282–284, 376–381
Carmack, C.R., 483–486
Castaldo, P., 394–404
Chen, H., 268–281
Cheung, J.Y., 528–529
Choptiany, P., 205–208
Chu, V., 480–482
Church, P.J., 356–366
Colley, N.J., 300–314
Condrescu, M., 214–223
Conway, S.J., 237–247, 268–281
Cooper, C.B., 41–52
Creazzo, T., 268–281
Cross, H.R., 421–430
Czyz, A., 413–417

Danieli, G.A., 282–284

De Groot, S.H.M., 438–445
Despa, S., 157–158
Di Renzo, G.F., 408–412
Diaz-Horta, O., 315–324, 487
DiBattista, E., 117–120
Dibrov, P., 117–120
DiPolo, R., 224–236
Dong, H., 137–141, 159–165, 382–393
Dunn, J., 159–165, 528–529

Elias, C.L., 109–112, 205–208, 543–545
Eriksson, K.S., 405–407
Espinosa-Tanguma, R., 73–76

Fliegel, L., 117–120

Gabellini, N., 282–284
Gadsby, D.C., 31–40
Gaughan, J.P., 466–471
Gille, T., 187–196
Ginsburg, K.S., 157–158
Golovina, V.A., 356–366
Goormaghtigh, E., 97–99
Grinstein, S., 248–258

Haas, H.L., 405–407
Haase, A., 113–116
Haigney, M.C.P., 472–475
Hale, C.C., 100–102, 197–204
Hampton, T.G., 480–482
Hancox, J.C., 535–538
Hanlon, S.U., 472–475
Hantash, B.M., 214–223
Hartung, K., 113–116
Haug-Collet, K., 300–314
Hayashi, H., 248–258
Herchuelz, A., 81–84, 97–99, 315–324, 354–355, 487
Hilgemann, D.W., 142–151
Hill, C.K., 100–102
Hinata, M., 154–156
Hinde, A.K., 535–538

Hnatowich, M., 205–208, 543–545
Houser, S.R., 466–471, 476–477, 478–479
Hryshko, L.V., 109–112, 166–175, 205–208, 543–545
Hunter, L.W., 530–534
Hwang, E.H, 338–341, 342–344

Imaizumi, Y., 154–156
Isaac, M.R., 543–545
Iwamoto, T., 19–30

James, A.F., 535–538
James-Kracke, M., 197–204
Jiang, W., 237–247, 285–287
John, S., 1–10
Jordan, M.C., 259–267
Juhaszova, M., 356–366

Kamagate, A., 81–84, 354–355
Kang, K., 41–52
Kang, T.K., 142–151
Kappl, M., 113–116
Kappler, C., 237–247, 285–287
Kasir, J., 89, 176–186
Kiedrowski, L., 413–417
Kimchi-Sarfaty, C., 176–186
Kimura, J., 154–156, 513–519, 539–542
Kinjo, T.G., 41–52
Kita, S., 19–30
Klimek, A., 488–499
Kneer, P.L., 268–281
Koushik, S.V., 268–281
Kovacs, G., 338–341, 342–344
Kraev, A., 53–59, 382–393
Kruzynska-Frejtag, A., 268–281

Le, H.D., 109–112, 543–545
Leach, S., 94–96
Lederer, J.W., 100–102
Lee, K.L., 488–499
Li, L., 154–156, 513–519, 539–542
Li, X.-F., 64–66, 382–393
Li, Y., 350–353
Litwin, S.E., 446–453
Lytton, J., xiii–xiv, 64–66, 90–93, 94–96, 137–141, 159–165, 382–393, 528–529

MacLennan, D.H., 53–59
Manni, S., 376–381
Margolis, F.L., 67–72
Margolis, J.W., 67–72
Margulies, K.B., 466–471, 476–477, 478–479
Marshall, C., 109–112
Matsuoka, S., 121–132
McCarthy, M.M., 60–63
Mejía-Elizondo, R., 73–76
Menick, D.R., 237–247, 268–281, 285–287
Min, J.-Y., 480–482
Minale, M., 408–412
Mokelke, E.A., 335–337
Moore, E.D.W., 488–499
Morgan, J.P., 480–482
Müller, J.G., 237–247, 285–287
Murphy, E., 421–430
Murtazina, R., 117–120

Nagel, G., 113–116
Nicholas, S.B., 259–267
Nicoll, D.A., 1–10, 11–18, 85–88, 367–375
Noble, D., 133–136, 431–437

Omelchenko A., 109–112, 205–208, 543–545
Opuni, K., 214–223
Ottolia, M., 1–10, 11–18, 85–88

Papa, M., 394–404
Pearson, B., 300–314
Perrot-Sinal, T.S., 60–63
Peti-Peterdi, J., 338–341, 342–344
Pham, C., 418–420
Philipson, K.D., 1–10, 11–18, 85–88, 259–267, 480–482, 483–486
Piacentino, V., III, 466–471, 476–477, 478–479
Pignataro, G., 408–412
Pogwizd, S.M., 454–465
Poljak, R., 100–102
Polumuri, S.K., 60–63, 67–72, 187–196
Porzig, H., 367–375
Prakash, Y.S., 530–534
Price, E.M., 100–102
Prinsen, C., 41–52

INDEX OF CONTRIBUTORS

Pyrski, M., 67–72

Qiu, Z., 1–10
Quednau, B.D., 1–10, 259–267

Rahamimoff, H., 89, 176–186
Reeves, J.P., 214–223
Ren, X., 89, 176–186
Reuter, H., 1–10
Roberts, D.E., 345–349, 350–353
Rogers, R., 268–281
Roos, K.P., 259–267
Ross, R.S., 259–267
Rothblum, L.I., 528–529
Ruknudin, A., 60–63, 67–72, 103–108, 187–196, 209–213
Ruysschaert, J.-M., 97–99

Saavedra-Alanis, V.M., 73–76
Saba, R.I., 97–99
Santagada, V., 408–412
Scatliff, R., 205–208
Schnetkamp, P.P.M., 41–52, 300–314
Schoenmakers, M., 438–445
Schulze, D.H., 60–63, 67–72, 100–102, 103–108, 187–196, 209–213
Schumann, S., 85–88
Scorziello, A., 408–412
Scriven, D.R.L., 488–499
Seckin, I., 530–534
Secondo, A., 408–412
Sellitti, S., 394–404
Sergeeva, O.A., 405–407
Shepherd, N., 237–247
Shigekawa, M., 19–30
Sieck, G.C., 530–534
Singh, D.N., 117–120
Sipido, K.R., 438–445
Siroky, B.J., 338–341, 342–344
Song, J., 528–529
Stanley, E.F., 356–366
Steciuk, M., 142–151
Steenbergen, C., 421–430
Stevens, D.R., 405–407
Stiner, L.M., 77–80
Sturek, M., 335–337
Sullivan, M.F., 480–482
Szászi, K., 248–258
Szerencsei, R.T., 41–52, 300–314

Tadros, G.M., 528–529
Taglialatela, M., 394–404
Taylor, B.E., 197–204
Terracciano, C., 520–527
Thurneysen, T., 367–375
Tibbits, G.F., 109–112
Tortiglione, A., 408–412

Uehara, A., 19–30
Unlap, T., 338–341, 342–344

Van Eylen, F., 81–84, 315–324, 354–355, 487
Verdonck, F., 438–445
Volders, G.A., 438–445
Vos, M.A., 438–445

Wang, J., 268–281
Wang, J.-F., 480–482
Wang, M., 335–337
Watanabe, Y., 154–156, 513–519
Watano, T., 154–156, 513–519
Webel, R., 300–314
Weber, C.R., 157–158, 466–471, 478–479, 500–512
Wheatly, M.G., 77–80
Wiebe, C., 117–120
Williams, I., 338–341, 342–344
Winkfein, R.J., 300–314, 41–52
Withers, P.R., 237–247, 285–287
Wood, P.G., 113–116

Ximenes, H.M., 354–355
Xu, L., 237–247, 285–287
Xue, X.-H., 109–112

Yamamura, H., 154–156
Yoo, S.S., 94–96

Zhang, D., 446–453
Zhang, K., 90–93
Zhang, X.-Q., 528–529
Zhang, Y.H., 535–538
Zhang, Z., 77–80

OHIO UNIVERSITY LIBRARY

Please return this book as soon as you have finished with it. In order to avoid a fine it must be returned by the latest date stamped below. All books are subject to recall after two weeks or immediately if needed for reserve.

JUN 1 3 2005

CF